LONDON MATHEMATICAL SOCIETY LECTURE NOTE SERIES

Managing Editor: Professor M. Reid, Mathematics Institute
University of Warwick, Coventry CV4 7AL, United Kingdom

The titles below are available from booksellers, or from Cambridge University Press at
www.cambridge.org/mathematics

London Mathematical Society Lecture Note Series: 436

Groups, Graphs and Random Walks

Edited by

TULLIO CECCHERINI-SILBERSTEIN
Università degli Studi del Sannio (Benevento), Italy

MAURA SALVATORI
Università degli Studi di Milano, Italy

ECATERINA SAVA-HUSS
Graz University of Technology, Austria

CAMBRIDGE
UNIVERSITY PRESS

University Printing House, Cambridge CB2 8BS, United Kingdom

One Liberty Plaza, 20th Floor, New York, NY 10006, USA

477 Williamstown Road, Port Melbourne, VIC 3207, Australia

314-321, 3rd Floor, Plot 3, Splendor Forum, Jasola District Centre, New Delhi - 110025, India

79 Anson Road, #06-04/06, Singapore 079906

Cambridge University Press is part of the University of Cambridge.

It furthers the University's mission by disseminating knowledge in the pursuit of education, learning and research at the highest international levels of excellence.

www.cambridge.org
Information on this title: www.cambridge.org/9781316604403

© Cambridge University Press 2017

First published 2017

A catalogue record for this publication is available from the British Library

Library of Congress Cataloging in Publication data
Names: Ceccherini-Silberstein, Tullio, editor. | Salvatori, Maura, editor. |
Sava-Huss, Ecaterina, editor. | Woess, Wolfgang, 1954–
Title: Groups, graphs, and random walks / edited by Tullio Ceccherini-Silberstein,
Università degli Studi del Sannio, Italy, Maura Salvatori, Università degli Studi di
Milano, Ecaterina Sava-Huss, Graz University of Technology, Austria.
Description: Cambridge : Cambridge University Press, [2017] |
Series: London mathematical society lecture note series ; 436 | Based on the
workshop "Groups, Graphs and Random Walks," held in Cortona, Italy, on June
2–6, 2014, on the occasion of the 60th birthday of Wolfgang Woess. |
Includes bibliographical references.
Identifiers: LCCN 2016019201 | ISBN 9781316604403 (pbk. : alk. paper)
Subjects: LCSH: Random walks (Mathematics)–Congresses. |
Stochastic processes–Congresses | Arithmetic groups–Congresses.
Classification: LCC QA274.73 .G76 2017 | DDC 519.2/82–dc23
LC record available at https://lccn.loc.gov/2016019201

ISBN 978-1-316-60440-3 Paperback

Contents

Preface

The current volume brings together several contributions from the invited speakers and guests of the workshop 'Groups, Graphs and Random Walks' held in Cortona (Italy) on June 2 to 6, 2014, on the occasion of the sixtieth anniversary of Wolfgang Woess.

Wolfgang was born in Vienna on July 23, 1954, to Friedrich and Elisabeth Woess, both professors at the University of Vienna. His father was also a gifted painter: when visiting Wolfgang, one immediately gets attracted to Friedrich Woess's beautiful watercolor landscapes adorning the walls of his office at the university as well as of his cosy home.

Wolfgang studied mathematics at the Technical University of Vienna, where he obtained his diploma, at the University of Munich, and at the University of Salzburg, where he obtained his PhD under the supervision of Peter Gerl. After a period as an assistant professor at the Montanuniversität Leoben (1984–1989)—including a leave of absence at the University of Rome 'La Sapienza' (1984–1985), where he started a long and fruitful collaboration with the Italian Harmonic Analysis group led by Alessandro Figà-Talamanca—and eleven years as a professor at the University of Milan (1988–1999), he eventually became Professor at the Graz University of Technology in 1999, where he currently serves as the chair of the Institute of Discrete Mathematics.

Wolfgang has been and still is, for many of us, a great teacher, a colleague, and a dear friend. As a teacher, he had thirteen PhD students (essentially from the University of Milan and the Graz University of Technology) and several postdoctoral fellows who have obtained important recognition both at the scientific and the academic levels.

His publications (nearly a hundred) range among various mathematical subjects, including convolution powers of probability measures on groups and asymptotics of random walk transition probabilities (at the very beginning of Wolfgang's research); recurrence, spectral radius and amenability, and spectral computations; boundary theory and harmonic functions; infinite electrical networks; context-free languages and their relations with groups and random walks; infinite graphs and groups; random walks on affine groups, buildings, horocyclic products, and lamplighter groups; finally and more recently, reflected random walks and stochastic dynamical systems; Brownian motion on strip ('quantum') complexes, treebolic spaces and SOL Geometry, and Markov processes on ultra-metric spaces. The long list of collaborators (more than thirty) includes, in order of multiplicity: Massimo Picardello, Laurent Saloff-Coste, Donald Cartwright, Vadim Kaimanovich and his former student Sara Brofferio.

One should also mention his beautiful and masterly written monographs *Random Walks on Infinite Graphs and Groups* (Cambridge University Press, 2000) and *Denumerable Markov Chains—Generating Functions, Boundary Theory, Random Walks on Trees* (European Mathematical Society Publishing House, 2009).

As mentioned, in the present volume we collect some papers contributed by participants to the Cortona conference: the themes are all intimately related to Wolfgang's research interests and scientific production. Here we overview, with a brief description, these contributions.

Growth of Groups and Wreath Products

Laurent Bartholdi (Georg-August University of Göttingen)
The central theme of this survey chapter is the Bartholdi–Erschler construction, via wreath products, of many groups of diverse types of growth: either intermediate, with many different growth functions, or of non-uniform exponential growth. On the way, Bartholdi discusses current hot topics of geometric group theory such as self-similar groups, branch groups, finite-automata groups, rooted trees, complete growth series and the like.

Random Walks on Some Countable Groups

Alexander Bendikov (Wroclaw University) and Laurent Saloff-Coste (Cornell University)
The chapter by Bendikov and Saloff-Coste studies decay of convolution powers of probability measures on non-finitely generated countable groups. Their methods are primarily based on explicit calculations of convolution powers of convex combinations of Haar measures and comparison techniques. It contains, in particular, an interesting collection of precise estimates for random walks on the infinite symmetric group $S^{(\infty)}$.

The Cost of Distinguishing Graphs

Debra Boutin (Hamilton College) and Wilfried Imrich (Leoben University)
Boutin and Imrich study the notion of distinguishing cost of a graph, recently introduced by the first author, which is defined as the smallest size of a vertex set whose set-wise stabilizer in the automorphism group

is trivial, and therefore constitutes a measure of the symmetry of the given graph. Clearly, it exists if and only if the distinguishing number (minimal number of colors needed for a coloring, which is not preserved by any non-trivial automorphism) is at most two. Furthermore, it is always bounded from below by the minimal size of a base (set whose point-wise stabilizer is trivial). Thus, the distinguishing cost could serve as a finer measure of the degree of symmetry for graphs with equal distinguishing number.

A Construction of the Measurable Poisson Boundary
Sara Brofferio (Paris-Sud University)
The chapter by Brofferio addresses an important problem about Poisson boundaries of random walks. Recall that, given a measure μ on a locally compact group G, the Poisson boundary is a measurable G-space (X, ν) with $\mu * \nu = \nu$ such that the Poisson transform $\phi \mapsto f_\phi(g) := \int_X \phi(gx) d\nu(x)$ defines an isometry of $L^\infty(X, \rho * \nu)$ onto the space $\mathcal{H}_\lambda^\infty(G)$ of bounded λ-a.e. μ-harmonic functions on G (here ρ is a probability measure on G equivalent to the Haar measure λ, and a function $f \colon G \to \mathbb{R}$ is termed λ-a.e. μ-harmonic if $f(g) = \int_G f(gh) d\mu(h)$ for λ-a.e. $g \in G$). When μ is supported on a dense countable subgroup Γ of G, there are two notions of Poisson boundary: one (as above) on G and one on Γ endowed with the discrete topology and the counting measure. In this chapter a kind of inductive construction is proposed to obtain the G-Poisson boundary from the Γ-Poisson boundary. Consider the action of Γ on $G \times X$ defined by $\gamma * (g, x) := (g\gamma^{-1}, \gamma x)$. Then Brofferio proves that the quotient space associated with the Γ-invariant sets is a kind of G-Poisson boundary. This is applied to describe the $\mathrm{Aff}(p, \mathbb{R})$ Poisson boundary of the Baumslag–Solitar group $\mathrm{BS}(1, p)$, where $\mathrm{Aff}(p, \mathbb{R})$ is the closure of the usual representation of $\mathrm{BS}(1, p)$ in the group $\mathrm{Aff}(\mathbb{R})$ of affine transformation of the real line.

Structure Trees, Networks and Almost Invariant Sets
Martin J. Dunwoody (University of Southampton)
Stallings' celebrated theorem (1968) about ends of groups states that a finitely generated group G has more than one end if and only if it admits a non-trivial decomposition as an amalgamated free product or an HNN extension over a finite subgroup. In the modern language of Bass–Serre theory, the theorem says that a finitely generated group G has more than one end if and only if it admits a non-trivial (that is, without a global fixed point) action on a simplicial tree with finite edge-stabilizers and without edge-inversions. This fundamental result, together with a question formulated by Wall, was a starting point for Dunwoody's accessibility theory: a finitely generated group G is said to be accessible if the process of iterated nontrivial splittings of G over finite subgroups always terminates in a finite number of steps. Dunwoody (1985) proved

that every finitely presented group is accessible; he later showed that there do exist finitely generated groups that are not accessible. The notion of accessibility was later extended to the graph setting by Thomassen and Woess (1993): a graph is accessible if there is an integer n such that any two ends can be separated by removing at most n edges; in the same chapter, the authors obtain a number of results using structure trees. Dunwoody gives a self-contained account of the theory, initiated with the aforementioned works of Stallings and later developed by Dicks and Dunwoody, of structure trees for edge cuts in networks. Applications include a generalization of the Max-Flow Min-Cut theorem to infinite networks, a short proof of a conjecture of Kropholler, a relative version of Stallings theorem and a generalization of the Almost Stability theorem by Dicks and Dunwoody.

Amenability of Trees

Behrang Forghani (University of Connecticut) and Keivan Mallahi-Karai (Jacobs University of Bremen)

The chapter by Forghani and Mallahi-Karai gives a necessary and sufficient condition for a tree to be amenable. As an application of this result, it is proven that a Galton–Watson tree is, under some specific conditions, almost surely amenable.

Group Walk Random Groups

Agelos Georgakopoulos (University of Warwick)

The chapter by Georgakopoulos discusses a new class of random graphs that combines ideas from random graph theory as well as random walks. Take an infinite graph, and let G_n denote the intrinsic ball of radius n around a root vertex. Then construct the random graph on the boundary $\partial G_n = G_n \setminus G_{n-1}$ by letting random walks start in all the vertices in ∂G_n and connecting $x, y \in \partial G_n$ when the random walk starting in x leaves G_n in y, or vice versa. This is an entirely novel construction of a random graph, somehow interpolating between normal random graphs and random walk interlacements. Georgakopoulos studies various properties of the graph, such as the number of connections between macroscopic parts of ∂G_n as $n \to \infty$, relations to the Poisson boundary of graphs and effective conductances, and the Doob–Naïm's kernel.

Ends of Branching Random Walks on Planar Hyperbolic Cayley Graphs

Lorenz A. Gilch (Graz University of Technology) and Sebastian Müller (Aix-Marseille University)

Properties of branching random walks (BRWs) on a graph G are of interest when the trace of the process (the random subgraph formed of vertices visited by particles of BRW) is a proper subgraph of G. The authors study the trace of transient BRWs when G is a planar hyperbolic

Cayley graph: in this case it is shown that the trace of BRW has, almost surely, infinitely many ends.

Amenability and Ergodic Properties of Topological Groups: From Bogolyubov Onwards

Rostislav Grigorchuk (Texas A&M University) and Pierre de la Harpe (University of Geneva)

The theory of amenable groups emerged from the study of the axiomatic properties of the Lebesgue integral and the discovery of the Hausdorff–Banach–Tarski paradox at the beginning of the last century. The first definition of an amenable group, by the existence of an invariant finitely additive probability measure, is due to von Neumann in his 1929 seminal paper. The term *amenable* was introduced in the 1950s by M.M. Day, who played a central role in the development of the modern theory of amenability by using means and applying techniques from functional analysis. In these references, groups do not have topology; in other words, they are just 'discrete groups'. Amenability is considered explicitly for topological groups in later articles by Dixmier, Fomin and Rickart and (for locally compact groups) by Greenleaf in his influential book. The chapter by Grigorchuk and de la Harpe is a survey on amenability and ergodicity for topological groups (as opposed to locally compact groups), as in Bogolyubov's 1939 paper, emphasizing the characterizations true in general and those true only in the locally compact case. This is of particular importance in view of the recent renewed interest in 'large' groups, for example, topological full groups of Cantor minimal dynamical systems (these groups are sources of infinite simple non-elementary amenable groups).

Schreier Graphs of Grigorchuk's Group and a Subshift Associated to a Non-Primitive Substitution

Rostislav Grigorchuk (Texas A&M University), Daniel Lenz (Friedrich Schiller University Jena) and Tatiana Nagnibeda (University of Geneva)

The authors describe a remarkable connection between a class of Laplace-type operators on Schreier graphs associated with the Grigorchuk group of intermediate growth and a class of Schrödinger operators with potentials defined by sampling over a strictly ergodic aperiodic subshift defined by a non-primitive substitution. This beautifully elucidated connection provides an example of a reduction of a hard (read: not strictly one-dimensional) problem to an easy (read: one-dimensional) problem. From this point of view, it is reminiscent of work on the XY spin chain, which reduces the hard (many-body) problem to an easy (one-particle) effective Hamiltonian via the Jordan–Wigner transformation. This chapter also includes a substantial and clear discussion of the spectral characteristics of aperiodic one-dimensional Schröodinger operators, the definition and properties of the Grigorchuk group, the structure of the substitution τ

and its subshift, and the relationship between this subshift and the graphs associated with the action of Grigorchuk's group on the infinite binary tree.

Thompson's Group F Is Not Liouville

Vadim Kaimanovich (University of Ottawa)

Thompson's group F, introduced by Richard Thompson in 1965, is the (finitely presented, infinite) group consisting of all orientation preserving piecewise-linear dyadic self-homeomorphisms of the closed unit interval $[0,1]$. One of the most important open questions in geometric group theory is whether or not Thompson's group F is amenable. Note that Brin and Squier proved that F does not contain non-abelian free subgroups: as a consequence, if F is not amenable, then it would constitute another counterexample to the so-called 'von Neumann conjecture', which stated that a finitely generated group is amenable if and only if it does not contain non-abelian free subgroups. The first finitely generated (resp. finitely presented) counterexamples to this conjecture were found by Olshanskii in 1980 using his Tarski monsters (resp. by Olshanskii and Sapir in 2002).

The classical Liouville theorem asserts that the only bounded harmonic functions on Euclidean space are the constants. Now, the notion of a harmonic function (based on the mean value property) can in fact be defined for an arbitrary Markov chain and, in particular, for random walks on groups. Given a probability measure μ on a group G, denote by $H^\infty(G,\mu) = \{f \in \ell^\infty(G) : f = f * \mu\}$ the space of all bounded μ-harmonic functions on G. One then says that the random walk (G,μ) is Liouville if $H^\infty(G,\mu)$ consists only of the constant functions. Furstenberg (1973) proved that any group carrying a non-degenerate Liouville random walk is amenable. One calls a group G Liouville if (G,μ) is Liouville for any symmetric and finitely supported probability measure μ on G. In his chapter, Kaimanovich shows that Richard Thompson's group F is not Liouville. More precisely, he proves that the random walk on F driven by any strictly non-degenerate finitely supported probability measure μ has a non-trivial Poisson boundary.

An Alternative Proof of the Subadditive Ergodic Theorem

Anders Karlsson (University of Geneva and Uppsala universitet)

Karlsson's chapter contains a self-contained, clean and short proof of Kingman's celebrated subadditive ergodic theorem. The scheme of the proof follows the classical proof of Birkhoff's ergodic theorem, via the maximal ergodic lemma. There are, however, two important differences: first, the maximal ergodic lemma is replaced by a result by Derriennic (1975), which yields non-positivity of a limit integral $\lim \int_B \frac{a(n,x)}{n} d\mu(x)$ (as opposed to the usual $\int_B a(1,x)d\mu(x)$, which is the same in the case of additive cocycles), and, second, there is a clever argument at the end of the proof to derive everywhere convergence in the subadditive case, from the same statement for additive cocycles.

Boundaries of $\mathbb{Z}^n$-Free Groups

Andrei Malyutin (St. Petersburg Department of V.A. Steklov Mathematical Institute), Tatiana Nagnibeda (University of Geneva) and Denis Serbin (Stevens Institute of Technology)

Given an arbitrary ordered Abelian group Λ (for instance, $\mathbb{Z}$ or $\mathbb{R}$), a Λ-tree is a metric space whose metric takes values in Λ and is subject to certain tree axioms. This notion goes back to the early 1960s, when Lyndon introduced the notion of abstract length functions on groups, and, a few years later, to Chiswell, who related such length functions with group actions on $\mathbb{Z}$- and $\mathbb{R}$-trees – providing a construction of the tree on which the group acts – and to Tits, who explicitly gave the first formal definition of an $\mathbb{R}$-tree. This theory has significantly developed since then, the main problem being addressed is to find the group theoretic information carried by a Λ-tree action, in particular, the structure of Λ-free groups. Malyutin, Nagnibeda, and Serbin study the Poisson boundary of groups acting on $\mathbb{Z}^n$-trees (here $\mathbb{Z}^n$ is equipped with the right lexicographic order). The groups considered in this chapter constitute a natural generalization of free groups; the class of $\mathbb{Z}^n$-free groups includes, in particular, all limit groups, and is closed under taking subgroups, free products and amalgamated free products along maximal cyclic subgroups. The authors provide a construction of the Poisson boundary for these groups directly in terms of the action of the group on the $\mathbb{Z}^n$-tree.

Buildings, Groups of Lie Type, and Random Walks

James Parkinson (University of Sydney)

Parkinson masterly surveys the fascinating theory of random walks on buildings and associated groups of Lie type and Kac–Moody groups. The author does a beautiful job of explaining ideas that are potentially quite technical and provides a comprehensive update of recent results dealing with probability theory on groups of Lie type defined over other p-adic fields, and extensions of these results into the setting of Kac–Moody groups. The unifying feature is the combinatorial-geometrical notion of a building, introduced by Tits in the 1950s in his successful attempt to give a uniform geometric interpretation of semi-simple Lie groups. Here the author focuses on the classes of buildings on which random walks have been studied, including spherical buildings, affine buildings (playing an important role in the study of Lie groups over p-adic fields), and Fuchsian and twin buildings (extensively used in the theory of Kac–Moody groups) and shows how the theory of random walks on buildings leads to limit theorems for random walks on the associated groups.

On Some Random Walks Driven by Spread-out Measures
Laurent Saloff-Coste (Cornell University) and Tianyi Zheng (Stanford University)
The chapter by Saloff-Coste and Zheng concerns heat kernel estimates for random walks on finitely generated groups and focuses on the case when the groups have polynomial volume growth and the initial distribution of the random walk has a suitable (slow) decay. The techniques used allow for the treatment of a variety of examples. The authors also give an application of the method to wreath products.

Topics in Mathematical Cristallography
Toshikazu Sunada (Meiji University)
The volume ends with the survey by Sunada. It is based on his book *Topological Crystallography* and, in connection with tight frames, sheds light, with a new geometric insight, on the relationship of his study on the standard realization of crystal lattices with several different topics such as tight frames in Euclidean space, crystallography, algebraic geometry (rational points in Grassmannians, the Abel–Jacobi map), number theory (quadratic Diophantine equations) and combinatorics (spherical designs).

Acknowledgments
We would like to thank all referees for their most precious collaboration. We also express our deepest gratitude to Sam Harrison, Abigail Walkington, and Clare Dennison at Cambridge University Press, as well as the Copyeditor Alyson Platt and the Project Manager Yassar Arafat, for their most valuable help and constant encouragement at all stages of the editing process.

The present volume is dedicated to Wolfgang Woess who, with his constant encouragement and concern on the one hand and deep and wide knowledge in several areas of mathematics on the other, has been and still is, for many of us, a remarkable reference and source of enlightenment.

Finally, we wish to thank the Istituto Nazionale di Alta Matematica 'Francesco Severi' (INdAM) and the European Science Foundation activity 'Random Geometry of Large Interacting Systems and Statistical Physics' for their financial support.

Rome, Milano and Graz   TCS, MS and ESH

Let me warmly thank all contributors to this volume of research papers with impressive substance in view of my own modest writings. I am particularly grateful to Maura, Ecaterina and Tullio for having undertaken such substantial efforts to organize the wonderful conference in Cortona and to edit these precious proceedings.
Sono davvero commosso.

Graz   WW

Conference Photographs

The conference participants at the Palazzone (photo courtesy of W. Woess).

The conference participants at Piazza Garibaldi in Cortona
(photo courtesy of L. Bartholdi).

Wolfgang Woess with some of his (former) PhD students
(photo courtesy of J. Kloas).

Near the Palazzone (photo courtesy of W. Woess).

Typical math discussions during a coffee break inside the Palazzone
(photo courtesy of A. Georgakopoulos).

At the conference dinner (photo courtesy of L. Bartholdi).

1

GROWTH OF GROUPS AND WREATH PRODUCTS

LAURENT BARTHOLDI

DMA, École Normale Supérieure, 45 rue d'Ulm, 75005 Paris.
Mathematisches Institut, Universität Göttingen, Bunsenstrasse 3-5,
D-37073 Göttingen

To Wolfgang Woess, in fond remembrance of many a visit to Graz

Contents

Introduction

These notes are an expanded version of a mini-course given at Le Louverain, June 24–27, 2014. Its main objective was to gather together useful facts about wreath products, and especially their geometry, in its application to problems and questions about growth of groups. The wreath product is a fundamental construction in group theory, and I hope to help make the reader more familiar with it.

It has proven very useful, in recent years, in better understanding asymptotics of the word growth function on groups, namely the function assigning to $R \in \mathbb{N}$ the number of group elements that may be obtained by multiplying at most R generators. The chapters [4–8] contain many repetitions as well as references to outer literature; by providing a unified treatment of these articles, I may provide the reader with easier access to the results and methods.

I have also attempted to define all notions in their most natural generality, while restricting the statements to the most important or fundamental cases. In this manner, I would like the underlying ideas to

Partially supported by ANR grant ANR-14-ACHN-0018-01 and DFG grant BA4197/6-1.

1

appear more clearly, with fewer details that obscure the line of sight. I avoided as much reference to literature as possible, taking the occasion to reprove some important results along the way.

I have also allowed myself, exceptionally, to cheat. I do so only under three conditions: (1) I clearly mark where there is a cheat; (2) the complete result appears elsewhere for the curious reader; (3) the correct version would be long and uninformative.

I have attempted to make the text suitable for a short course. In doing so, I have included a few exercises, some of which are hopefully stimulating, and a section on open problems. What follows is a brief tour of the highlights of the text.

Wreath Products

The wreath product construction, described in §1, is an essential operation, building a new group W out of a group H and a group G acting on a set X. Assuming[1] that G is a group of permutations of X, the wreath product is the group $W = H \wr_X G$ of H-decorated permutations in G: if elements of G are written in the arrow notation, with elements of X lined in two identical rows above each other and an arrow from each $x \in X$ to its image, then an element of $H \wr_X G$ is an arrow diagram with an element of H attached to each arrow, e.g.

One of the early uses of wreath products is as a classifier for extensions, as discovered by Kaloujnine, see Theorem 1.4: there is a bijective correspondence between group extensions with kernel H and quotient G on the one hand, and appropriate subgroups of the wreath product $H \wr G$, with G seen as a permutation group acting on itself by multiplication. We extend this result to permutational wreath products:

Theorem (Theorem 1.6). *Let G, H be groups, and let G act on the right on a set X. Denote by $\pi \colon H \wr G \to G$ the natural projection. Then the map $E \mapsto E$ defines a bijection between*

$$\left\{ \begin{array}{c} E : E \leq H \text{ and the } G\text{-sets } X \text{ and } H\backslash E \text{ are isomorphic} \\ \text{via a homomorphism } E \to G \end{array} \right\}$$

isomorphism $E \to E'$ of groups intertwining the actions on $H\backslash E$ and $H\backslash E'$

---

[1] There is no need to require the action of G on X to be faithful; this is merely a visual aid. See §1 for the complete definition.

and

$$\left\{ \begin{array}{l} E \leq H \wr G : \pi(E) \ \textit{transitive on } X \ \textit{and} \\ \qquad \ker(\pi) \cap E \xrightarrow{\ \cong\ } H \ \textit{via } f \mapsto f(x) \ \textit{for all } x \in X \end{array} \right\}$$
$$\overline{\qquad\qquad\qquad\qquad \textit{conjugacy of subgroups of } H \wr G \qquad\qquad\qquad\qquad}.$$

The wreath product $H \wr_X G$ is uncountable, if $H \neq 1$ and X is infinite. It contains some important subgroups: $H \wr_X G$, defined as those decorated permutations in which all but finitely many labels are trivial; and $H \wr_X^{\text{f.v.}} G$, defined as those decorated permutations in which the labels take finitely many different values. Clearly $H \wr_X G \subseteq H \wr_X^{\text{f.v.}} G \subseteq H \wr_X G$, and $H \wr_X G$ is countable as soon as H, G, X are countable.

Growth of Groups

Let us summarise here the main notions; for more details, see §2. A choice of generating set S for a group G gives rise to a graph, the *Cayley graph*: its vertex set is G, and there is an edge from g to gs for each $g \in G$, $s \in S$. The path metric on this graph defines a metric d on G called the *word metric*. The Cayley graph is invariant under left translation, and so is the word metric.

One of the most naive invariants of this graph is its *growth*, namely the function $v_{G,S}(R)$ measuring the cardinality of a ball of radius R in the Cayley graph. If the graph exhibits some kind of regularity, then it should translate into some regularity of the function $v_{G,S}$.

For example, Klarner [47, 48] studied the growth of crystals (that expand according to a precise and simple rule) via what turns out to be the growth of an abelian group.

A convenient tool to study various forms of regularity of a function $v_{G,S}$ is the associated generating function $\Gamma_{G,S}(z) = \sum_{R \in \mathbb{N}} (V_{G,S}(R) - V_{G,S}(R-1))z^R$. The regularity of $v_{G,S}$ translates then into a property of $\Gamma_{G,S}$ such as being a rational, algebraic, D-finite, ... function of z.

We may rewrite $\Gamma_{G,S}(z) = \sum_{g \in G} z^{d(1,g)}$; then a richer power series keeps track of more regularity of G:

$$\widehat{\Gamma}_{G,S}(z) = \sum_{g \in G} g z^{d(1,g)}.$$

This is a power series with coefficients in the group ring $\mathbb{Z}G$, and again we may ask whether $\widehat{\Gamma}_{G,S}$ is rational or algebraic[2].

---

[2] The *rational subring* of $\mathbb{Z}G[[t]]$ is the smallest subring of $\mathbb{Z}G[[t]]$ containing $\mathbb{Z}G[t]$ and closed under Kleene's star operation $A^* = 1 + A + A^2 + \cdots$, for all $A(z)$ with $A(0) = 0$.

The *algebraic subring* of $\mathbb{Z}G[[t]]$ is the set of power series that may be expressed as the solution A_1 of a non-trivial system of non-commutative polynomial equations $\{P_1(A_1, \ldots, A_n) = 0, \cdots, P_n(A_1, \ldots, A_n) = 0\}$ with coefficients in $\mathbb{Z}G[t]$. The

If G has an abelian subgroup of finite index [52], or if G is word-hyperbolic [33], then $\widehat{\Gamma}_{G,S}$ is a rational function of z for all choices of S. We give a sufficient condition for $\widehat{\Gamma}_{G,S}$ to be algebraic:

Theorem (Theorem 3.2). *Let* $H = \langle T \rangle$ *be a group such that* $\widehat{\Gamma}_{H,T}$ *is algebraic, and let* F *be a free group. Consider* $G = H \wr F$, *generated by* $S = T \cup \{a\ basis\ of\ F\}$. *Then* $\widehat{\Gamma}_{G,S}$ *is algebraic.*

We then turn to studying the asymptotics of the growth function $v_{G,S}$. Let us write $v \precsim w$ to mean that $v(R) \leq w(CR)$ for some constant $C \in \mathbb{R}_+$ and all $R \geq 0$, and $v \sim w$ to mean $v \precsim w \precsim v$. Then the $\sim$-equivalence class of $v_{G,S}$ is independent of the choice of S, so we may simply talk about v_G.

For 'most' examples of groups, either $v_G(R)$ is bounded by a polynomial in R or $v_G(R)$ is exponential in R. This is, in particular, the case for soluble, linear and word-hyperbolic groups. There exist, however, examples of groups for which $v_G(R)$ admits an intermediate behaviour between polynomial and exponential; they are called groups of *intermediate growth*. The question of their existence was raised by Milnor [58], was answered positively by Grigorchuk [29], and has motivated much group theory in the second half of the twentieth century.

Let $\eta_+ \approx 2.46$ be the positive root of $T^3 - T^2 - 2T - 4$, and set $\alpha = \log 2/\log \eta_+ \approx 0.76$. We shall show that, for *every* sufficiently regular function $f \colon \mathbb{R}_+ \to \mathbb{R}_+$ with $\exp(R^\alpha) \precsim f \precsim \exp(R)$, there exists a group with growth function equivalent to f:

Theorem (Theorem 6.2). *Let* $f \colon \mathbb{R}_+ \to \mathbb{R}_+$ *be a function satisfying*

$$f(2R) \leq f(R)^2 \leq f(\eta_+ R)\ for\ all\ R\ large\ enough.$$

Then there exists a group G *such that* $v_G \sim f$.

Thus, groups of intermediate growth abound, and the space of asymptotic growth functions of groups is as rich as the space of functions. Furthermore, we shall show that there is essentially no restriction on the subgroup structure of groups of intermediate growth. Let us call a group H *locally of subexponential growth* if every finitely generated subgroup of H has growth function $\precsim \exp(R)$. Clearly, if H is a subgroup of a group of intermediate growth then it has locally subexponential growth. We show, conversely:

Theorem (Theorem 7.1). *Let* B *be a countable group locally of subexponential growth. Then there exists a finitely generated group of subexponential growth in which* B *imbeds as a subgroup.*

---

solution is actually rational if furthermore the P_i are of the form $c_{i,0} + \sum_{j=1}^n c_{i,j} A_j$ with $c_{i,j} \in \mathbb{Z}G[t]$.

Finally, it may happen that a group G has exponential growth, namely that the growth rate $\lim_{R\to\infty} v_{G,S}(R)^{1/R}$ is > 1 for all S, but that the infimum of these growth rates, over all S, is 1. Such a group is called of *non-uniform exponential growth*. The question of their existence was raised by Gromov [35, Remarque 5.12]. Again, soluble, linear and word-hyperbolic groups cannot have non-uniform exponential growth; but, again, it turns out that such groups abound. We shall show:

Theorem (Theorem 8.2). *Every countable group may be imbedded in a group of non-uniform exponential growth.*

Furthermore, the group W in which the countable group imbeds may be required to have the following property: there is a constant K such that, for all $R > 0$, there exists a generating set S of W with

$$v_{W,S}(r) \leq \exp(Kr^\alpha) \text{ for all } r \in [0, R].$$

(Self-)Similar Groups and Branched Groups

All the constructions mentioned in the previous subsection take place in the universe of *(self-)similar groups*. Here is a brief description of these groups; see §4 for details.

Just as a self-similar set, in geometry, is a set describable in terms of smaller copies of itself, a *self-similar group* is a group describable in terms of 'smaller' copies of itself. A *self-similar structure* on a group G is a homomorphism $\phi \colon G \to G \wr_X P$ for a permutation group P of X. Thus, elements of G may be recursively written in terms of G-decorated permutations of X. For this description to be useful, of course, the homomorphism ϕ must satisfy some non-degeneracy condition (in particular be injective), and P should be manageable, say finite.

The fact that the copies of G in $G \wr_X P$ are 'smaller' than the original is expressed as follows: there is a norm on G such that, for $g \in G$ and $\phi(g)$ a permutation with labels $(g_x : x \in X)$, the elements g_x are shorter than g, at least as soon as g is long enough. For example, consider G finitely generated, and denote by $\|\cdot\|$ the word norm on G. One requires $\|g_x\| < \|g\|$ for all $x \in X$ and all $\|g\| \gg 1$; this is equivalent to the existence of $\lambda \in (0, 1)$ and $K \geq 0$ such that $\|g_x\| \leq \lambda\|g\| + K$ for all $x \in X, g \in G$.

Furthermore, in cases that interest us, the map ϕ is almost an isomorphism, in that its image $\phi(G)$ has finite index in $G \wr_X P$. Thus, ϕ may be thought of as a *virtual isomorphism* between G and G^X, namely an isomorphism between finite-index subgroups. When one endows G^X with the ℓ^∞ metric $\|(g_x)\| = \max_{x \in X} \|g_x\|$, the aforementioned condition requires that this virtual isomorphism be a contraction. On the other hand, endowing G^X with the ℓ^1 metric $\|(g_x)\| = \sum_{x \in X} \|g_x\|$, the optimal Lipschitz constant of the virtual isomorphism plays a fundamental role in estimating the growth of G.

Similar groups are a natural generalization: one is given a set Ω and a self-map $\sigma: \Omega \circlearrowleft$; for each $\omega \in \Omega$, a group G_ω and a permutation group P_ω of a set X_ω; and homomorphisms $\phi_\omega: G_\omega \to G_{\sigma\omega} \wr_{X_\omega} P_\omega$. Taking for Ω a singleton recovers the notion of a self-similar group. Taking $\Omega = \mathbb{N}$ and $\sigma(n) = n + 1$ defines in full generality a similar group G_0; but it is often more convenient to consider a larger family of groups in which $(G_n)_{n \in \mathbb{N}}$ imbeds. In particular, one obtains a *topological space* of groups, in such a manner that close groups have close properties (for example, their Cayley graphs coincide on a large ball).

Acknowledgments

I am very grateful to Yago Antolin, Laura Ciobanu and Alexey Talambutsa for having organised the workshop in Le Louverain where I presented a preliminary version of this text, and to the participants of the workshop for their perspicacious questions.

A large part of the material is taken from articles written in collaboration with Anna Erschler, and I am greatly indebted to her for generously including me on her projects. It owes much to her energy and enthusiasm that we were able to finish our joint articles.

Hao Chen, Yves de Cornulier and Pierre de la Harpe helped improve these notes by pointing out a number of mistakes and inconsistencies.

Open Problems

This text presents a snapshot of what is known about growth of groups in 2014; a large number of open problems remain. Here are some promising directions for further research.

1. Which groups G are such that, for all generating sets S, the complete growth series $\widehat{\Gamma}_{G,S}$ is a rational function of z?

   This is known to hold for virtually abelian groups, and for word-hyperbolic groups. Conjecturally, this holds for no other group.

   The related question of which groups have a rational (classical) growth function is probably more complicated, see §2.1.

2. Is the analytic continuation $1/\widehat{\Gamma}_{G,S}(1)$ related to the complete Euler characteristic of G, just as $1/\Gamma_{G,S}(1)$ is (under some additional conditions) the Euler characteristic of G? See [74, §1.8] for complete Euler characteristic.

3. Does there exist a group G with two generating sets S_1, S_2 such that $\widehat{\Gamma}_{G,S_1}$ is rational but $\widehat{\Gamma}_{G,S_2}$ is transcendental?

   Such an example could be $G = F_2 \times F_2$. Set $S = \{x, y\}^{\pm 1}$ a free generating set of F_2, and $S_1 = S \times \{1\} \sqcup \{1\} \times S$ and $S_2 = S_1 \sqcup \{(s, s) : s \in S\}$.

   The same properties probably hold for the usual generating series Γ_{G,S_1} and Γ_{G,S_2}. The radius of convergence of Γ_{G,S_1} is $1/3$, but that of Γ_{G,S_2} is unknown.

This problem is strongly related to the 'Matching subsequence problem', which asks for the longest length of a common subsequence among two independently and uniformly chosen words of length n over a k-letter alphabet; see [16]. It is easy to see that, for two uniformly random reduced words of length n in F_2, the longest common subword has length $\approx \gamma n$ for some constant γ, as $n \to \infty$. Thus, a pair $(g, h) \in G$ with $\|g\| = \|h\| = n$ has length $2n$ with respect to S_1, but approximately $(2 - \alpha)n$ with respect to S_2. We might call γ the *Chvátal-Sankoff constant* of F_2.

4. Are there infinite simple groups of subexponential growth?[3]

There is no reason for such groups *not* to exist; but the construction methods described in this text yield groups acting on rooted trees, which therefore are as far as possible from being simple.

There is also no reason for finitely presented groups of subexponential growth *not* to exist; again, the obstacle is probably more our mathematical limitations than fundamental mathematical reasons.

The following question, by de la Harpe [38], is still open at the time of writing: 'Do there exist groups with Kazhdan's property (T) and non-uniform exponential growth?'

Similarly, it is not known whether there are simple finitely generated groups of non-uniform exponential growth, and whether there are finitely presented groups of non-uniform exponential growth.

5. Are there groups whose growth function lies strictly between polynomials and $\exp(R^{1/2})$?

See the discussion in §2.3. There is a superpolynomial function $f(R) \succsim R^{(\log R)^{1/100}}$ such that no group has growth strictly between polynomials and $f(R)$. There is no residually nilpotent group whose growth is strictly between polynomials and $\exp(R^{1/2})$, see Theorem 5.2.

6. What is the asymptotic growth of the first Grigorchuk group? What is its exact growth, for the generating set $\{a, b, c, d\}$? Does the growth series of the Grigorchuk group exhibit some kind of regularity?

Some experiments indicate that this must be the case. For example, consider the quotient G_n of the first Grigorchuk group that acts on $\{0, 1\}^n$. It is a finite group of cardinality $2^{5 \cdot 2^{n-3}+2}$. For $n \leq 7$, the diameter D_n of its Cayley graph (for the natural generating set $\{a, b, c, d\}$) is the sequence $1, 4, 8, 24, 56, 136, 344$ and satisfies the recurrence $D_n = D_{n-1} + 2D_{n-2} + 4D_{n-3}$. If this pattern went on, the growth of the first Grigorchuk group would be asymptotically $\exp(R^{\log 2/ \log \eta_+}) \approx \exp(R^{0.76})$.

---

[3] This question has been answered positively by V. Nekrashevych, see https://arxiv.org/abs/1601.01033.

If a group has subexponential growth, then its growth series is either rational or transcendental, and if the group has intermediate growth, then the growth series must be transcendental; See Theorem 2.1 in §2.1. Thus, Grigorchuk group's growth series is transcendental. Does the series satisfy a functional equation? That would make it akin to the classical partition function $\sum_{n \geq 0} p(n) z^n = \prod_{n \geq 1} (1 - z^n)^{-1}$, which (up to scaling and multiplying by $z^{1/24}$) is a *modular function*.[4] Ghys asked me once: 'Is the growth series $\Gamma(z)$ of Grigorchuk's group modular?'

7. For every $k \in \mathbb{N}$, the *space of marked k-generated groups* $\mathscr{S}_k$ may be defined as the space of normal subgroups of the free group F_k, by identifying $G = \langle s_1, \ldots, s_k \rangle$ with the kernel of the natural map $F_k \to G$ sending generator to generator. It is a compact space. What properties does the set $\mathscr{I}$ of groups of intermediate growth, and the set $\mathscr{N}$ of groups of non-uniform exponential growth, enjoy in this space? For example,

'Is there an uncountable open subset of $\mathscr{S}_k$ in which $\mathscr{N}$, or $\mathscr{I}$, is dense? Is $\mathscr{N}$ dense in the complement of groups of polynomial growth?'

Recall that similar groups are families of groups $(G_\omega)_{\omega \in \Omega}$ indexed by a space Ω. If all G_ω are k-generated, we obtain a map $\Omega \to \mathscr{S}_k$, which under favourable circumstances is continuous. This has been exploited, for example, in [60] to produce groups of non-uniform exponential growth.

It had actually been doubted, before Grigorchuk's discovery [29], whether there were groups of intermediate growth. This text tries to convince the reader that they are abundant. Giving a precise meaning to the above question would quantify, in some manner, the extent to which they are abundant.

Notational Conventions

I try to adhere to standard group-theoretical notation. In particular, the right action of a group element g on a point x is written xg, and a left action is written $^g x$. The stabilizer of x is written G_x. The conjugation action of a group on itself is written $g^h = h^{-1} g h$, and the commutator of two elements is $[g, h] = g^{-1} h^{-1} g h = g^{-1} g^h = h^{-g} h$.

I also introduce a minimal amount of new, 'fancy' notation to represent elements of wreath products or of self-similar groups, and hope that it helps in achieving clarity and conciseness.

---

[4] I.e. a function $A(z) = A(\exp(2\pi i \tau))$ such that the corresponding function $\tau \mapsto A(\exp(2\pi i \tau))$ on the upper half plane is invariant under a finite-index subgroup of $\mathrm{SL}_2(\mathbb{Z})$.

1. Wreath Products

We start with the basic construction. Let H be a group, and let G be a group acting on the right on a set X. We construct two groups:

$$H \wr_X G = \left(\prod_X{}' H \right) \rtimes G \qquad \text{the } \textit{restricted} \text{ wreath product,}$$

$$H \wr\wr_X G = \left(\prod_X H \right) \rtimes G \qquad \text{the } \textit{unrestricted} \text{ wreath product.}$$

Here the unrestricted product $\prod_X H$ may be viewed as the group of functions $X \to H$, with pointwise composition and with left G-action given by pre-composition[5]: $^g f$ is the function given by $(^g f)(x) = f(xg)$. The *restricted* product $\prod'_X H$ is then identified with finitely supported functions $X \to H$. In both cases, this product is a subgroup of the wreath product and is called its *base group*.

In the particular case of $X = G$ with natural right action by multiplication, one calls $H \wr\wr_G G$ the *regular unrestricted wreath product* and writes it simply $H \wr\wr G$; and similarly for the *regular restricted wreath product* $H \wr_G G = H \wr G$.

Assume that the action of G on X is faithful; so that elements of G may be identified with permutations of X. The best way to describe elements of $H \wr\wr_X G$ or its subgroup $H \wr_X G$ is by *decorated permutations*: one writes a permutation of X, decorated by elements of H, such as

$$\tag{1.1}$$

Permutations are multiplied as usual: by stacking them and pulling the arrows tight. Likewise, decorated permutations are multiplied by stacking them and multiplying the labels along the composed arrows. We do not write the labels when they are the identity. Here is a graphical computation of a product:

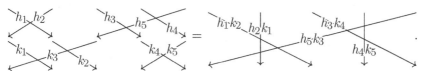

When writing formulæ, we must sometimes depart from the graphical notation, in which permutations are written top-to-bottom or left-to-right and thanks to their arrows there is no ambiguity in knowing in

---

[5] Note that the side of the action changes! It is best to always use the appropriate side, so as to avoid inverses. Recall, however, that every left action can be converted into a right action by setting $f^g := {}^{g^{-1}} f$ and vice versa.

which order to compose the labels. We invariably let permutations act on the right on sets, and thus '$\sigma \tau$' means 'first σ, then τ'.

Exercise 1.1. In the wreath product $W = \{\pm 1\} \wr \mathrm{Sym}(2)$, consider the element

$$h =$$

Show by concatenating the diagram with itself that h has order exactly 4. The group W has order 8; which of the order-8 groups is it?

Let us consider a wreath product $W = H \wr_X G$. In writing elements $w, w' \in W$ algebraically, we may express them in the form $w = fg$ with $f: X \to H$ and $g \in G$, or in the form $w' = gf$. In both cases, the element g and the function f are unique, by definition of the semidirect product. The compositions $(fg)(f'g')$ and $(gf)(g'f')$ are, in all cases, computed using the relation

$$g \cdot f = {}^g f \cdot g,$$

namely

$$(fg)(f'g') = (f \cdot {}^g f')(gg') \text{ and } (gf)(g'f') = (gg')({}^{(g')^{-1}} f \cdot f'). \qquad (1.2)$$

Exercise 1.2. Let R be a ring, viewed as a group under addition, and let RG denote the group ring of G, on which G acts by right multiplication. Show that $R \wr G$ is isomorphic to $RG \rtimes G$. More generally, let X be a G-set; then RX is a G-module. Show that $R \wr_X G$ and $RX \rtimes G$ are isomorphic.

1.1. Actions

Assume now, moreover, that H acts from the right on a set Y. Then there are two natural sets on which $W = H \wr_X G$ acts:

- There is an action on $Y \times X$, given by $(y, x) \cdot fg = (yf(x), xg)$ for $(y, x) \in Y \times X$; it is called the *imprimitive* action;
- There is an action on Y^X, the set of functions $X \to Y$, given by $(\phi \cdot fg)(xg) = \phi(x)f(x)$ for $\phi: X \to Y$; it is called the *primitive* action.

Exercise 1.3. There are natural bijections between the sets $(Z \times Y) \times X = Z \times (Y \times X)$ and $(Z^Y)^X = Z^{Y \times X}$. Assume now that a group G acts on X, a group H acts on Y and a group I acts on Z. Show that the bijections above give isomorphisms between the groups $(I \wr_Y H) \wr_X G$ and $(I \wr_{Y \times X} (H \wr_X G))$ as permutation groups, both of $Z \times Y \times X$ and of $(Z^Y)^X$.

1.2. History

Leo Kaloujnine understood the importance of wreath products in the early 1940s. It is said that he worked, during the Second World War, in a uniform factory and observed a rivet machine; it was made of a rotating ring containing many rotating disks in it. Identifying the movement of the ring with an action of G and each subdisk with an action of H, one sees that motions of the machine are described by wreath product elements. In his dissertation (under Élie Cartan, [44]), he studied the Sylow subgroups of symmetric groups and showed that they were iterated wreath products.

However, this description of maximal p-subgroups of symmetric groups already appears in the classical 1870 treatise by Camille Jordan [43, II.I.41], who implicitly defined wreath products there. This is all the more remarkable as Sylow's theorems were published only two years later [78]!

Kaloujnine returned to the Soviet Union after the war and contributed greatly to the development of mathematics in Ukraine, founding in 1959 the department of algebra and mathematical logic. He is remembered for the following important result classifying group extensions, namely, that the wreath product is a universal object containing all extensions:

Theorem 1.4 (Kaloujnine-Krasner, [45]). *Let G, H be groups. Denote by $\pi \colon H \wr G \to G$ the natural projection. Then the map $E \mapsto E$ defines a bijection between*

$$\frac{\left\{E : 1 \to H \to E \to G \to 1\right\}}{\text{isomorphism of extensions}}$$

and

$$\frac{\left\{E \leq H \wr G : \pi(E) = G \text{ and } \ker(\pi) \cap E \xrightarrow{\cong} H \text{ via } c \in H^G \mapsto c(1)\right\}}{\text{conjugacy of subgroups of } H \wr G}.$$

Exercise 1.5. Prove Theorem 1.4.

Hint: given an extension $1 \to H \to E \xrightarrow{\tau} G \to 1$, choose a set-theoretic section[6] $g \mapsto \tilde{g}$ of τ, and define an imbedding $E \to H \wr G$ by

$$e \mapsto \left(g \mapsto \widetilde{ge(g\tau(e))}^{-1}\right)\tau(e).$$

Theorem 1.4 may be interpreted more abstractly as saying that, if $1 \to H \to E \to G \to 1$ is an exact sequence and $f_0 \colon H \to K$ is a group homomorphism, then f_0 extends naturally to a homomorphism $f \colon E \to K \wr G$. The case $f = \text{id}$ and $H = K$ is exactly the statement of the theorem, and the generalised version is proven in exactly the same

---

[6] Namely, a map satisfying $\tau(\tilde{g}) = g$ for all $g \in G$.

manner. The advantage of this formulation is that one need not require that H be normal in E; and the more general statement is

Theorem 1.6. *Let $H \leq E$ be groups, and let $f_0 \colon H \to K$ be a homomorphism. Then f_0 extends naturally to a homomorphism[7] $f \colon E \to K \wr_{H\backslash E} E / \operatorname{core}(H)$.*

More precisely, let G, H be groups, and let G act on the right on a set X. Denote by $\pi \colon H \wr G \to G$ the natural projection. Then the map $E \mapsto E$ defines a bijection between

$$\left\{ \begin{array}{l} E \colon \quad H \leq E \text{ and the } G\text{-sets } X \text{ and } H\backslash E \text{ are isomorphic} \\ \qquad \text{via a homomorphism } E \to G \end{array} \right\}$$

isomorphism $E \to E'$ of groups intertwining the actions on $H\backslash E$ and $H\backslash E'$

and

$$\left\{ \begin{array}{l} E \leq H \wr G \colon \quad \pi(E) \text{ transitive on } X \text{ and} \\ \qquad \ker(\pi) \cap E \xrightarrow{\cong} H \text{ via } c \in H^X \mapsto c(x) \text{ for all } x \in X \end{array} \right\}$$

conjugacy of subgroups of $H \wr G$

Note that we could have written E instead of $E/\operatorname{core}(H)$ everywhere above. We included it to obtain a smaller group acting on the set X, thus reinforcing the parallel with Theorem 1.4, and expressing E as a subgroup of a presumably easier-to-construct group. In particular, if $[E : H] = d < \infty$, then $E/\operatorname{core}(H)$ is a group of cardinality between d and $d!$, since it acts faithfully and transitively on the set $H\backslash E$ of cardinality d.

Proof. We follow the sketched proof of Theorem 1.4; rather than a section $G \to E$, we choose a right transversal T of H in E; namely, a subset $T \subset E$ such that every $e \in E$ may uniquely be written in the form ht with $h \in H, t \in T$.

For $e \in E$, let $f(e)$ be the following H-decorated permutation of $H\backslash E$: the permutation is given by the natural right-multiplication action of E on $H\backslash E$, and on every edge, say from Ht to Hu if $Ht \cdot e = Hu$, the label is teu^{-1}.

In formulas, this may be written as follows, although all verifications are easier in the 'decorated permutations' form. Denote by π the natural map $E \to E/\operatorname{core}(H)$. Set then $f(e) = c\pi(e)$ with $c \colon H\backslash E \to H$ given, for $t \in T$, by $c(Ht) = teu^{-1}$ for the unique $u \in T$ such that $teu^{-1} \in H$.

Composing the map c by the homomorphism $f_0 \colon H \to K$ gives the first statement.   □

Wreath products have occupied a prominent place in the theory of permutation groups. For illustration, they appear as fundamental

---

[7] Recall that the core of the subgroup H is the intersection of its conjugates. $E/\operatorname{core}(H)$ is the natural permutation group acting faithfully on the left coset space $H\backslash E$.

constructions in the O'Nan–Scott theorem classifying maximal subgroups of the symmetric group. We skip subcases *(3.iii)... (3.viii)*, which are too technical:

Theorem 1.7 (O'Nan-Scott, [70]). *Let G be a maximal subgroup of $\mathrm{Sym}(X)$ for a finite set X. Then either*

    (1) the action of G is not transitive[8]; then $X = Y \sqcup Z$ and $G = \mathrm{Sym}(Y) \times \mathrm{Sym}(Z)$; or

    (2) the action of G is transitive, but not primitive[9]; then $X = Y \times Z$ and $G = \mathrm{Sym}(Y) \wr \mathrm{Sym}(Z)$ in its imprimitive action; or

    (3) the action of G is primitive; and then either

        (3.i) G is an affine group over a finite field; or
        (3.ii) $X = Y^Z$ and $G = \mathrm{Sym}(Y) \wr \mathrm{Sym}(Z)$ in its primitive action; or

    (3.iii) ... (3.viii) ...   □

1.3. Generators for Wreath Products

We now consider restricted wreath products $W = H \wr_X G$ in more detail. We introduce the following notation: if X be a set and H be a group, then for all $x \in X$ and $h \in H$,

we define $h@x\colon X \to H$ by
$$\begin{cases} (h@x)(x) = h, \\ (h@x)(y) = 1 \text{ for all } y \neq x. \end{cases}$$

One has the important formula relating conjugation in W with translation:

$$(h@x)^g = h@xg$$

(check it by drawing the decorated permutations!).

Let S be a generating set for G, and let T be a generating set for H. Let $Y \subseteq X$ be a choice of one representative from each G-orbit on X. Then W is generated by

$$\{t@y : t \in T, y \in Y\} \sqcup S.$$

A good, but also slightly misleading example, is the 'lamplighter group': this is the group $W = \{\pm 1\} \wr \mathbb{Z}$. It is best understood by its primitive action (on the space of functions $\{\pm 1\}^{\mathbb{Z}}$): imagine an infinite street with at each integer position

TRAIN:

...ALMOST...

A MACHINE THAT GRABS THE EARTH BY METAL RAILS AND ROTATES IT UNTIL THE PART YOU WANT IS NEAR YOU

(reproduced from https://xkcd.com)

---

[8] I.e. there are at least two orbits on X.
[9] I.e. there is a non-trivial G-invariant equivalence relation on X.

a lamppost. The lamp there can be 'on' or 'off'. Imagine also that the
street is viewed from the perspective of the 'lamplighter', namely the
person in charge of turning various lamps on and off. The generator s
of $\mathbb{Z}$ means 'move up the street', or equivalently, 'shift all lamps down'
relatively to the lamplighter; and the generator $t@1$ means 'change the
state of the lamp currently in front of the lamplighter'.

This is what is expressed by the cartoon above: relativistically
speaking, it makes no difference to think that a train moves on its tracks,
or that the tracks move under the train. The lamplighter is on the train,
and the lamp configurations are on the track.

The misleading aspect of this example is that one must remember
that, when the lamplighter moves in one direction, the lamps actually
move in the *opposite* direction. In a regular wreath product, this makes
no difference, but in the general case it does.

Definition 1.8. Let $G = \langle S \rangle$ be a finitely generated group acting on
the right on a set X. The associated *Schreier graph* has vertex set X,
and for each $x \in X, s \in S$ an edge from x to xs.   △

In the case $X = G$ of a group G acting on itself by (right or left)
multiplication, one obtains the (right or left) Cayley graph of G. The
group G acts then by graph isometries on its Cayley graph by (left or
right) multiplication.

Exercise 1.9. Consider $G = \mathrm{Sym}(4)$ generated by $\{(1,2),(2,3),(3,4)\}$.
Draw the Schreier graphs of (in order of difficulty):

(1) the action of G on $\{1,2,3,4\}$;
(2) the action on the collection of 2-element subsets of $\{1,2,3,4\}$;
(3) the action of G on itself by conjugation.

It would be a great mistake to think that a path in the permutational
wreath product $W = H \wr_X G$ is a path in the Schreier graph of G
with decorations in H along the path. We will see where exactly the
decorations appear, but their positions are rather at the *inverses* of path
points. To complicate matters, in case G is abelian the map $x \mapsto x^{-1}$
is an automorphism of G, from which the above 'mistaken' description
fortunately gives the correct answer.

Let us fix generating sets $G = \langle S \rangle$ and $H = \langle T \rangle$, and assume for
simplicity that G acts transitively on X. We also choose a base point
$x \in X$. Then W is naturally generated by $T@x \sqcup S$.

We consider first the description of elements of W as fg with
$f \colon X \to H$ and $g \in G$. This is the 'lamplighter moving' version, because
the 'lamp vector' f is not shifted, but the position at which it is changed
varies. By (1.2), the right action on $w = fg$ of

— a generator $s \in S$ gives $ws = f(gs)$;
— a generator $t@x \in T@x$ gives $w(t@x) = (f \cdot t@xg^{-1})g$.

Thus, if $w = (t_0@x)s_1(t_1@x)\cdots s_\ell(t_\ell@x) = fg$, then the support of f is included in $\{x, xs_1^{-1}, \ldots, x(s_1\ldots s_\ell)^{-1}\}$.

We may also describe elements of W as gf; this is the 'earth moving' version, because the 'lamp vector' f is shifted, and the lamplighter always changes the lamp at a fixed position. By (1.2), the right action on $w = gf$ of

— a generator $s \in S$ gives $ws = (gs)^{s^{-1}}f$;
— a generator $t@x \in T@x$ gives $w(t@x) = g(f \cdot t@x)$.

Thus, if $w = (t_0@x)s_1(t_1@x)\cdots s_\ell(t_\ell@x) = gf$, then the support of f is included in $\{xs_1\ldots s_\ell, xs_2\ldots s_\ell, \ldots, xs_\ell, x\}$.

The semidirect product description of $W = H \wr_X G$ gives a presentation of W by generators and relations; see [17]. Generating as above W by $T@Y \sqcup S$, we get

$$W = \langle T@Y \sqcup S \mid \text{relations of } G, \text{relations of } H, \text{ and}$$

$$\forall t, t' \in T, \forall y \neq y' \in Y : [t@y, t'@y'], \text{ and}$$

$$\forall t \in T, \forall y \in Y, \forall g \in G_y : [t@y, g], \text{ and}$$

$$\forall t, t' \in T, \forall y \in Y, \forall g \in (G_y\backslash G)\backslash\{G_y\} : [t@y, (t'@y)^g]\rangle.$$

The exact criterion, assuming $X \neq \emptyset$ and $H \neq 1$, is:

Theorem 1.10 (Cornulier [17, Theorem 1.1]). *The wreath product W is finitely presented if and only if both G, H are finitely presented, G acts on X with finitely generated stabilizers, and G acts diagonally on $X \times X$ with finitely many orbits.* $\square$

Diestel–Leader Graphs

We present a particularly intuitive description of the Cayley graph of 'lamplighter groups' $W = F \wr \mathbb{Z}$, for a finite group F of cardinality q, see [83].

Let $\mathscr{T}$ denote the $(q+1)$-regular tree, choose a basepoint o, and a geodesic ray $\omega\colon \mathbb{N} \to \mathscr{T}$ starting at o. Imagine $\mathscr{T}$ as 'hanging from ω': orient the edges on ω from $\omega(n)$ to $\omega(n+1)$, and orient every other edge from its furthest point to ω to its closest. Define then $h\colon \mathscr{T} \to \mathbb{Z}$ as follows: for each $x \in \mathscr{T}$, there is a unique path in $\mathscr{T}$ from o to x, and set $h(x) = $ (number of edges oriented forward) $-$ (number of edges oriented backward) on this path.[10]

Consider now the following graph $\mathcal{B}(q,q)$. Its vertex set is $\{(x,y) \in \mathscr{T} \times \mathscr{T} : h(x)+h(y) = 0\}$. There is an edge from (x,y) to (x',y') precisely if $\{x, x'\}$ and $\{y, y'\}$ are connected in $\mathscr{T}$. Here is a portion of $\mathcal{B}(2,2)$, with one tree pointing up and one tree pointing down:

---

[10] This is usually called a *Busemann function*.

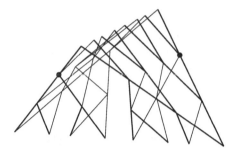

Proposition 1.11. *Consider a wreath product $G = F \wr \mathbb{Z}$ with a finite group F of cardinality q. Denote by s a generator of $\mathbb{Z}$, and consider for G the generating set $S = (F@1)s \sqcup s^{-1}(F@1)$. Then the Cayley graph of (G, S) is $\mathcal{B}(q, q)$.*

Proof. Only in this proof, for a subset A of the integers, let us denote by F^A the set of finitely supported functions $A \to F$. The vertex set of $\mathscr{T}$ may be identified with $\bigsqcup_{n \in \mathbb{Z}} F^{(-\infty, n]} \times \{n\}$, in such a manner that there is an edge from (σ, n) to $(\sigma|_{(-\infty, n-1)}, n-1)$ for all $n \in \mathbb{Z}$ and all $\sigma \in F^{(\infty, n]}$. Vertices on the ray ω are those of the form $(1, n)$ with $n \leq 0$.

Therefore, the vertex set of $\mathcal{B}(q, q)$ may be identified with $\bigsqcup_{n \in \mathbb{Z}} F^{(-\infty, n]} \times F^{(-\infty, -n]} \times \{n\}$. This is easily put in correspondence with $F^{\mathbb{Z}} \times \mathbb{Z}$ via the map $(\sigma, \sigma', n) \mapsto (\tau, n)$ with $\tau \colon \mathbb{Z} \to F$ the finitely supported function given by $\tau(k) = \sigma(k)$ if $k \leq n$ and $\tau(k) = \sigma'(1 - k)$ if $k > n$.

Thus, we put the vertex set of $\mathcal{B}(q, q)$ is bijection with G. It is now routine to check that the generators in S produce the edges of $\mathcal{B}(q, q)$. □

This description of lamplighter groups makes some of their geometric features quite transparent. For example, let us consider a finitely generated group $G = \langle S \rangle$, and its Cayley graph. A vertex $v \in G$ is called a *dead end* if all neighbours of v are at least as close to 1 as v. More generally, let us say v is *on a k-hill* if all paths from v to an element of norm $\|v\| + 1$ has to go through a vertex of norm $\|v\| - k$. In other words, from the top of the hill the only way of going to infinity is to first go down at least k steps.

Consider the two dots on the illustration above. Say one of them is the origin 1. Then the other one is on a 1-hill. More generally, any element in the lamplighter group that is reached from the origin by going down k steps, up $2k$ steps and down again k steps along a reduced path reaches the top of a k-hill.

This is in fact a familiar phenomenon: consider a long street and two remote addresses we want to visit on that street, in whichever order, from our starting point on the street. The shortest way of doing this is to first go to the closest, and then the other one. The *worst* possible place to start is at equal distance from both addresses.

2. Growth of Groups

This section is not a treatise on growth of groups; for that, see rather [53]. We do recall some elementary, basic notions, and provide some motivation for the material to appear later. Let G be a group generated by a finite symmetric set $(S = S^{-1})$. One defines the *word norm* on G by

$$\|g\| = \min\{n : g = s_1 \cdots s_n, \ s_i \in S\},$$

and a distance[11] on G that is invariant under left translation[12]:

$$d(g, h) = \|g^{-1}h\|.$$

Thus, G is viewed as a normed space and as a metric space on which G acts by isometries via left translation.

2.1. Formal Growth

One may be interested in regularity properties of the metric space G; these are best studied via the *growth series*, the formal power series

$$\Gamma_G(z) = \sum_{g \in G} z^{\|g\|} \in \mathbb{Z}[[z]].$$

The natural questions that arise are: What is the domain of convergence of $\Gamma_G(z)$? What can be said of analytic continuations of $\Gamma_G(z)$? What are its singularities? Is the function $\Gamma_G(z)$ rational (i.e. in $\mathbb{Q}(z)$)? or at least algebraic (i.e. there exists a two-variable polynomial $F(y, z) \in \mathbb{Z}[y, z]$ with $F(\Gamma_G(z), z) \equiv 0$)?

The consideration of the power series $\Gamma_G(z)$, and of the above questions, is justified by the answers that have been given:

- $\Gamma_G(z)$ converges in a disk of radius at least $1/\#S$. If S is symmetric, then the convergence radius is in fact at least $1/(\#S - 1)$, with equality if and only G is a free product of $\mathbb{Z}$s and C_2s with its natural generating set, as in Definition 2.9 below.
- If groups G, H are respectively generated by S, T, then the direct product $G \times H$ is naturally generated by $S \sqcup T$. One then has

$$\Gamma_{G \times H}(z) = \Gamma_G(z)\Gamma_H(z),$$

see Proposition 2.4.

---

[11] It is only here that we use the fact that S is symmetric, to obtain $d(g, h) = d(h, g)$; in fact, it suffices in all that follows to assume that S generates G as a monoid.

[12] I.e. $d(tg, th) = d(g, h)$.

- If groups G, H are respectively generated by S, T, then the free product $G * H$ is naturally generated by $S \sqcup T$. One then has

$$\frac{1}{\Gamma_{G*H}(z)} = \frac{1}{\Gamma_G(z)} + \frac{1}{\Gamma_H(z)} - 1,$$

see Proposition 2.5.

- If the group G is virtually abelian [9], or word-hyperbolic [36], or the discrete Heisenberg group [18] $H_3 = \begin{pmatrix} 1 & \mathbb{Z} & \mathbb{Z} \\ & 1 & \mathbb{Z} \\ & & 1 \end{pmatrix}$, then Γ_G is a rational function of z for all choices of the finite generating set S.

- Wreath products give some examples of power series Γ_G that are algebraic functions, as we shall see in Corollary 3.3 below.

- If G is a two-step nilpotent group with cyclic derived subgroup, then there are generating sets for G such that Γ_G is a rational function of z. However, if G is the five-dimensional Heisenberg group $H_5 = \begin{pmatrix} 1 & \mathbb{Z} & \mathbb{Z} & \mathbb{Z} & \mathbb{Z} \\ & 1 & & & \mathbb{Z} \\ & & 1 & & \mathbb{Z} \\ & & & 1 & \mathbb{Z} \\ & & & & 1 \end{pmatrix}$, then there are generating sets for which $\Gamma_G(z)$ is transcendental; see [75].

In respect to this last point, note that, for G nilpotent, the growth of G is polynomial so $\Gamma_G(z)$ converges in the unit disk. It is either rational or transcendental, by the Fatou theorem [21]. More is known:

Theorem 2.1 (Pólya–Carlson [14]). *Let $A(z) = \sum_{n \geq 0} a_n z^n$ be a power series with integer coefficients. If A is not rational, then A does not extend analytically beyond the unit circle.*

Most importantly, the growth series is a convenient object that encodes information on G. A quite satisfactory theory of 'Euler characteristic' has been developed for groups, see [15]. Here is a special case: if G is the fundamental group of a cellular complex $\mathscr{X}$ with contractible universal cover, one declares $\chi(G)$ to be $\chi(\mathscr{X})$. More generally, if G has a finite-index subgroup H, which is the fundamental group of a space $\mathscr{Y}$, one sets $\chi(G) = \chi(\mathscr{Y})/[G : H]$; this makes sense because if G is the fundamental group of $\mathscr{X}$, then H is the fundamental group of a $[G : H]$-sheeted covering of $\mathscr{X}$, whose Euler characteristic is $[G : H]\chi(\mathscr{X})$. In particular, if G is finite, then $\chi(G) = 1/\#G$. This led to the idea that $1/\Gamma_G(z)$ could behave like a Euler characteristic, and that its limit $1/\Gamma_G(1)$ could express $\chi(G)$. This is not always true, but it does hold in some illustrative cases.

Let us compute, for instance, the growth series of a free group F_k generated by a basis.[13] It follows from the formula for free products, or by direct counting if one notes, for all $\ell \geq 1$, that there are $2k(2k-1)^{\ell-1}$ elements of norm ℓ in F_k, that

---

[13] I.e. $S = \{x_1, x_1^{-1}, \ldots, x_k, x_k^{-1}\}$ and F_k may be identified with reduced words over S.

$$\Gamma_{F_k}(z) = \frac{1+z}{1-(2k-1)z}. \tag{1.3}$$

The value $\Gamma_{F_k}(1)$ is uniquely defined by analytic continuation, and one has $1/\Gamma_{F_k}(1) = 1 - k$, in agreement with F_k being the fundamental group of a graph with 1 vertex and k edges. See [23, 32, 51, 73] for more such examples of the '$1/\Gamma_G(1) = \chi(G)$' phenomenon.

2.2. Complete Growth Series

There exists a stronger property than having a rational growth series: a group $G = \langle S \rangle$ has a *rational geodesic combing* if there is a finite directed graph with edge labels in S and a fixed 'initial' vertex, such that the set $L \subseteq S^*$ of words read from the initial vertex along paths in the graph has the following property: L maps bijectively to G by the natural evaluation map of words as elements of G, and the words in L have minimal length among all words in S^* having the same evaluation in G. Kervaire suggested considering the *complete* growth series

$$\widehat{\Gamma}_G(z) = \sum_{g \in G} g z^{\|g\|} \in \mathbb{Z}G[[z]].$$

Note that $\widehat{\Gamma}_G$ depends on the choice of generating set S, even though we do not mention it explicitly. The series $\widehat{\Gamma}_G(z)$ is a power series with coefficients in the group ring, and one may again ask whether it is rational or algebraic. Because $\mathbb{Z}G$ need not be commutative, let us define more precisely these notions; we refer to [68] for details. Let $\Lambda \subseteq \overline{\Lambda}$ be rings. An *algebraic system over* Λ in variables $X_1, \ldots, X_n$ is a non-degenerate[14] n-tuple of polynomials $P_1, \ldots, P_n$ in non-commuting indeterminates $X_1, \ldots, X_n$ and coefficients in Λ. In a *linear system over* Λ, the polynomials are restricted to have degree 1 and contain the indeterminate on the right; i.e. the P_i are sums of monomials all belonging to $\Lambda \cup \bigcup_{1 \le i \le n} \Lambda X_i$. A *solution* is an n-tuple $(f_1, \ldots, f_n) \in \overline{\Lambda}^n$ such that all $P_i(f_1, \ldots, f_n) = f_i$ for all $i = 1, \ldots, n$.

We then say that a power series $F(z) \in \mathbb{Z}G[[z]]$ is *rational*, respectively *algebraic*, if it is the first coordinate of the solution of a linear, respectively algebraic system over the polynomial ring $\mathbb{Z}G[z]$. A more direct definition of the ring of rational functions is that it is the smallest subring of $\mathbb{Z}G[[z]]$ containing $\mathbb{Z}G[z]$ and closed under Kleene's *quasi-inversion*, the operation $F(z)^* = (1 - F(z))^{-1} = 1 + F(z) + F(z)^2 + \cdots$ defined for all $F(z) \in \mathbb{Z}G[[z]]$ with $F(0) = 0$.

Exercise 2.2. If G admits a rational geodesic combing, then its growth series is rational.

---

[14] Let us not detail this too much; suffice it to say that the system must have a unique solution once its initial terms $f_1(0), \ldots, f_n(0)$ have been fixed.

Hint: define one variable X_i for each vertex i of the graph defining the combing, and encode the edges of the graph into polynomials.

Exercise 2.3. If the growth series of G is rational, then G admits a *quasi-geodesic combing*: a language $L \subset S^*$, recognised by a finite graph as above, with the property that, for some constant $C \in \mathbb{N}$, all words $s_1 \ldots s_\ell \in L$ have the property $\ell \leq C\|s_1 \cdots s_\ell\|$.

Hint: consider a polynomial system defining $\widehat{\Gamma}_{G,S}(z)$, make sure that every term in $\mathbb{Z}G$ is accompanied by at least one factor z. Write the terms in $\mathbb{Z}G$ as linear combinations of words over S, yielding a polynomial system over $\mathbb{Z}S^*$. Let C be the maximal length of all these words over S that appear in the polynomial system. A solution to the polynomial system will be a sum of monomials of the form $s_1 \ldots s_\ell z^n$, where $n = \|s_1 \cdots s_\ell\| \geq \ell/C$.

These notions strengthen the ones for the classical growth series: if $\widehat{\Gamma}_G(z)$ is rational or algebraic, then its image under the augmentation map $\mathbb{Z}G \twoheadrightarrow \mathbb{Z}$ is rational or algebraic. On the other hand, statements concerning the complete growth series are usually not much harder to prove than the analogous ones concerning the classical growth series:

Proposition 2.4. *Let the groups G, H and $G \times H$ be respectively generated by S, T and $S \sqcup T$. One then has*

$$\widehat{\Gamma}_{G \times H}(z) = \widehat{\Gamma}_G(z)\widehat{\Gamma}_H(z).$$

Proof. Every element $(g, h) \in G \times H$ satisfies $\|(g, h)\| = \|g\| + \|h\|$; so

$$\widehat{\Gamma}_{G \times H}(z) = \sum_{(g,h) \in G \times H} ghz^{\|g\|+\|h\|} = \sum_{g \in G} gz^{\|g\|} \sum_{h \in H} hz^{\|h\|} = \widehat{\Gamma}_G(z)\widehat{\Gamma}_H(z).$$

$\square$

Proposition 2.5. *Let the groups G, H and $G * H$ be respectively generated by S, T and $S \sqcup T$. One then has*

$$\frac{1}{\widehat{\Gamma}_{G*H}(z)} = \frac{1}{\widehat{\Gamma}_G(z)} + \frac{1}{\widehat{\Gamma}_H(z)} - 1.$$

Proof. Every element of $w \in G * H$ may be uniquely written in the form $w = h_0 g_1 h_1 \cdots g_\ell$ with $h_0 \in H, h_1, \ldots, h_{\ell-1} \in H \setminus \{1\}, g_1, \ldots, g_{\ell-1} \in G \setminus \{1\}, g_\ell \in G$. Thus,

$$\widehat{\Gamma}_{G*H}(z) = \sum_{\ell \geq 0} \widehat{\Gamma}_H(z)((\widehat{\Gamma}_G(z) - 1)(\widehat{\Gamma}_H(z) - 1))^\ell \widehat{\Gamma}_G(z)$$

$$= \widehat{\Gamma}_H(z)\frac{1}{1 - (\widehat{\Gamma}_G(z) - 1)(\widehat{\Gamma}_H(z) - 1)}\widehat{\Gamma}_G(z);$$

so

$$\frac{1}{\widehat{\Gamma}_{G*H}(z)} = \frac{1}{\widehat{\Gamma}_G(z)}(\widehat{\Gamma}_G(z) + \widehat{\Gamma}_H(z) - \widehat{\Gamma}_G(z)\widehat{\Gamma}_H(z))\frac{1}{\widehat{\Gamma}_H(z)}$$
$$= \frac{1}{\widehat{\Gamma}_H(z)} + \frac{1}{\widehat{\Gamma}_G(z)} - 1. \qquad \square$$

These results are generalized to *graph products* in [3]: given a graph Γ with vertex set V and a group G_v for each $v \in V$, the *graph product* of the G_v is

$$G_\Gamma := \underset{v \in V}{*} G_v \Big/ \langle [G_v, G_w] \text{ for each edge } (v, w) \rangle.$$

Recall that a *clique* in a graph is a subset of the vertices, any two of which are connected by an edge. They show:

Proposition 2.6 ([3, Theorem 3.8]). *Let each group G_v have generating set S_v, and consider the generating set $\bigcup_{v \in V} S_v$ of G_Γ. Then*

$$\frac{1}{\widehat{\Gamma}_{G_\Gamma}(z)} = \sum_{\text{clique } W \subseteq V} \prod_{v \in W} \left(\frac{1}{\widehat{\Gamma}_{G_v}} - 1\right).$$

There are few classes of groups in which $\widehat{\Gamma}_G(z)$ is rational for all choices of generating set:

Proposition 2.7 (Liardet [52]). *If G is virtually abelian, then $\widehat{\Gamma}_G(z)$ is rational for all choices of generating set.* $\qquad \square$

Proposition 2.8 (Grigorchuk–Nagnibeda [33]). *If G is word-hyperbolic, then $\widehat{\Gamma}_G(z)$ is rational for all choices of generating set.* $\qquad \square$

Note that there is no need to consider rings such as $\mathbb{Z}G$; the definition is more naturally phrased in terms of a *semiring* such as $\mathbb{N}G$. The polynomials P_i are restricted to be sums of products of monomials, and no subtraction is allowed. The notions of $\mathbb{Z}$-rationality and $\mathbb{N}$-rationality differ subtly, see e.g. [10].

Let us now compute explicitly the complete growth series of a free group. It simplifies a little the notation to answer a slightly more general question. We denote throughout the text the cyclic group of order p by C_p:

Definition 2.9. A *free-like* group is a finite-free product of $\mathbb{Z}$s and C_2s.

Say that a free-like group G has m_1 factors isomorphic to C_2 and m_2 factors isomorphic to $\mathbb{Z}$; then it has a symmetric generating set of size $m_1 + 2m_2$, consisting of one generator for each C_2 and a generator and its inverse for each $\mathbb{Z}$. The group G is characterised by the property that its Cayley graph (see after Definition 1.8) is an $(m_1 + 2m_2)$-regular undirected tree. We call such an S a *natural* generating set for G. $\qquad \triangle$

For instance, the free group $F_{m/2}$ is free-like for m even, and $*^m C_2$ is free-like.

Let G be free-like, and let S denote a natural generating set of G with cardinality m. We shall see that G has rational complete growth series. For ease of notation, we write $\bar{s}$ for s^{-1}. We identify elements of G with *reduced* words over S, i.e. words not containing consecutive $s\bar{s}$.

For all $s \in S$, define $F_s \in \mathbb{Z}G[[z]]$ by

$$F_s = \sum_{\substack{w \in G, \text{ not} \\ \text{starting with } \bar{s}}} wz^{\|w\|};$$

so $\widehat{\Gamma}_G(z) = 1 + \sum_{s \in S} szF_s$. We have the linear system in non-commutative unknowns F_s

$$F_s = 1 + \sum_{t \in S, t \neq \bar{s}} tzF_t,$$

with solution $F_s = (1 - \bar{s}z)\left(1 - \sum_{t \in S} tz + (m-1)z^2\right)^{-1}$, so finally

$$\widehat{\Gamma}_G(z) = \frac{1 - z^2}{1 - \sum_{s \in S} sz + (m-1)z^2}, \tag{1.4}$$

compare with (1.3).

2.3. Asymptotic Growth

We return to a group G with generating set S, and view it as a metric space for the word metric. We consider the volume growth of balls in the metric space (G, d); this is the *growth function* $v_G \colon \mathbb{R}_+ \to \mathbb{N}$ given by

$$v_G(R) = \#\{g \in G : \|g\| \leq R\}.$$

It is naturally related to the formal power series $\Gamma(z)$: indeed $v_G(R)$ is the sum of all coefficients of $\Gamma(z)$ of degree $\leq R$; equivalently, for $R \in \mathbb{N}$ it is the degree-R coefficient of $\Gamma(z)/(1 - z)$. Thus, by Tauberian and Abelian theorems (see e.g. [59]), asymptotics of $v_G(R)$ as $R \to \infty$ may be related to asymptotics of $\Gamma_G(z)$ as $z \to$ the convergence radius. In particular, the function $v_G(R)$ grows as R^d if and only if $\Gamma_G(z)$ converges in the unit disk and has an order-d pole singularity at 1.

The norm $\|\cdot\|$ depends on the choice of generating set S, but only mildly: different choices of generating sets give equivalent norms, and equivalent metrics. If for $v, w \colon \mathbb{R}_+ \to \mathbb{N}$ we write $v \precsim w$ to mean that $v(R) \leq w(CR)$ for a constant $C \in \mathbb{R}_+$ and all $R \gg 0$, and we write $v \sim w$ to mean $v \precsim w \precsim v$, then

Lemma 2.10. *The $\sim$-equivalence class of v_G is independent of the choice of generating set.*

Proof. Let S, S' be two finite generating sets for G, and let us temporarily write $\|g\|_S, \|g\|_{S'}$ and $v_{G,S}, v_{G,S'}$ for the norms and growth functions with respect to S, S'. There exists then a constant $C \in \mathbb{N}$ such that $\|s'\|_S \leq C$ for all $s' \in S'$, and thus $\|g\|_{S'} \leq C\|g\|_S$. This gives $v_{G,S}(R) \leq v_{G,S'}(CR)$. The reverse inequality holds by symmetry.   □

Note, as a consequence, that all exponentially growing functions are equivalent, and that R^d and $C \cdot R^d$ are equivalent as soon as $d > 0$. The exponential growth rate

$$\lambda_{G,S} = \lim v_{G,S}(R)^{1/R} \tag{1.5}$$

is nevertheless worthy of consideration, and will be discussed in §8.

2.4. History

Interest in asymptotic growth of groups dates back at least to the early 1950s, in the works of Krause [50], Efremovich [19] and Švarc [76]; they were seeking coarse invariants of manifolds based on their fundamental group. Milnor noted in [57] that, if G is the fundamental group of a compact riemannian manifold $\mathcal{M}$, then v_G is equivalent to the volume growth of balls in the universal cover of $\mathcal{M}$.

Here is a schematic of the known equivalence classes of growth functions of groups. Note the two dots for the two groups of order 4, respectively the two groups $\mathbb{Z}^4$ and the Heisenberg group H_3 with quartic growth:

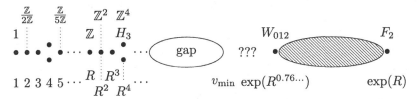

The left of the graph is occupied by finite groups; the growth of a finite group is equivalent to the constant function taking value the order of the group. Abelian groups, and more generally virtually[15] nilpotent groups have polynomial growth of type R^d for an integer d. The converse is a deep result by Gromov:

Theorem 2.11 (Gromov [34]). *A finitely generated group has growth function bounded by a polynomial if and only if it is virtually nilpotent.*   □

It follows also from Gromov's argument that there is a superpolynomial function $v_{\min}(R)$ such that all groups with growth $\precsim v_{\min}$ are

---

[15] A property is said to *virtually* hold if it holds for a finite-index subgroup.

virtually nilpotent; so there are no functions with growth strictly between polynomial and v_{min}. Explicit estimates in [72] imply that one may take $v_{min}(R) = R^{(\log R)^{1/100}}$, although the gap is probably larger. Note also that there need not exist a $\precsim$-largest function v_{min}. If one restricts oneself to virtually residually[16] nilpotent groups, then the gap extends at least to $\exp(R^{1/2})$, see [31]; if one restricts oneself to virtually residually solvable groups, then the gap extends at least to $\exp(R^{1/6})$, see [82].

Milnor asked in 1968, in a famous problem in the '*American Math Monthly*' [58], whether there are groups whose growth function is neither polynomial nor exponential. He also conjectured in that note that groups of polynomial growth are precisely the virtually nilpotent groups. Milnor and Wolf showed in [56, 84] that virtually solvable groups have either polynomial or exponential growth, and the *in*existence of groups with growth between polynomial and exponential became known as 'Wolf's conjecture'. Recall the celebrated 'Tits' alternative' [79]: a finitely generated subgroup of a linear group in characteristic 0 either is virtually solvable or contains a non-abelian free subgroup; from this it follows that linear groups always have polynomial or exponential growth (see furthermore §8).

However, groups of *intermediate growth* exist, and Grigorchuk [29] gave such an example, known as the *first Grigorchuk group* G_{012}; see §4.

The growth of G_{012} is not known, even up to ~-equivalence; conjecturally, it is the same as the growth of the group $W_{012}(C_2)$, which will be introduced in §5. The hatched region above indicates that, in fact, there are many groups of intermediate growth, and that any 'reasonable' function between $\exp(R^{0.76\cdots})$ and $\exp(R)$ is equivalent to the growth function of a group; see Theorem 6.2.

There are at least two arguments for considering asymptotic growth rather than exact growth of groups. Firstly, the asymptotics of the growth function does not depend on the generating set, by Lemma 2.10, so is an invariant of the group itself. Secondly, we expect 'most' growth series to be transcendental power series, so that they are probably difficult to describe, manipulate or expand; this happens, for example, for groups of subexponential growth, whose growth series converges in the unit disk so is either rational or transcendental, by Fatou's theorem (see Theorem 2.1).

3. Growth of Regular Wreath Products

We consider in this section a wreath product $W = H \wr_X G$, and compute its growth series. We assume that generating sets S, T for G, H

---

[16] A property is said to hold *residually* if for every non-trivial element there exists a quotient in which this element remains non-trivial and the property holds.

respectively have been chosen, and that the growth series of G and H are known.

3.1. Wreath Products Over Finite Sets

As a first step, let us suppose that the set X is finite, say $X = \{x_1, \ldots, x_d\}$. Then, as generating set for W, we may take $\{t@x : t \in T, x \in X\} \sqcup S$. For this generating set, we have

$$\Gamma_W(z) = \Gamma_H(z)^{\#X} \Gamma_G(z),$$

and by a small abuse of notation the same relation on complete growth series:

$$\widehat{\Gamma}_W(z) = \left(\prod_{i=1}^{d} \widehat{\Gamma}_H(z)@x_i \right) \widehat{\Gamma}_G(z).$$

Indeed, every element $w \in W$ may uniquely be written in the form $w = (h_1@x_1) \cdots (h_d@x_d)g$ for some $h_1, \ldots, h_d \in H, g \in G$, and the growth series of $\{h_i@x_i : h_i \in H\}$ naturally coincides with that of H. In particular, if $\Gamma_G(z)$ and $\Gamma_H(z)$ are rational, then so is $\Gamma_W(z)$.

Johnson obtained in [42] the same conclusion for more complicated generating sets of W.

3.2. Lamplighter Groups

The next case we consider is $G = X = \mathbb{Z}$, and in particular 'lamplighter groups'. Because the computations will be generalised in the next section, we content ourselves with a brief description of the growth series, and for simplicity assume that H is a finite group. We consider $W = H \wr \mathbb{Z}$, denote a generator of $\mathbb{Z}$ by s, and let W be generated by the set $\{s, s^{-1}\} \sqcup H@1$.

Consider an element $w \in W$. If its image under the natural map $W \to \mathbb{Z}$ is nonnegative, then it may be written minimally in the form $s^{-m}(h_0@1)s(h_1@1) \cdots s(h_p@1)s^{-n}$ with $h_i \in H$, $m, n \geq 0$ and $p \geq m + n$, while if its image in $\mathbb{Z}$ is negative, then it may be written minimally in the form $s^m(h_0@1)s^{-1}(h_1@1) \cdots s^{-1}(h_p@1)s^n$ with $p > m + n$. Furthermore, h_0 must be non-trivial unless $m = 0$, and h_p must be non-trivial unless $n = 0$.

All of these constraints are local and therefore rational, except the long-range relation between m, n, p. However, in terms of computing growth series, the letters in the expression $s^{-m}(h_0@1)s(h_1@1) \cdots s(h_p@1)s^{-n}$ can be permuted at no cost; and the set of expressions of the form

$$(h_0@1)s^{-1}s(h_1@1) \cdots s^{-1}s(h_m@1)s(h_{m+1}@1)$$
$$\cdots s(h_{p-n}@1)ss^{-1}(h_{p-n+1}@1) \cdots ss^{-1}(h_p@1)$$

is indeed a rational language, so that its growth function is rational.

Exercise 3.1. Compute the growth series of $C_2 \wr \mathbb{Z}$ with the standard generators $C_2@1 \cup \{s, s^{-1}\}$. Note that the growth function grows exponentially, at the same rate as Fibonacci numbers. Could you have guessed the appearance of Fibonacci numbers without going through the calculations?

3.3. Regular Wreath Products with Free Groups

We compute in this subsection the complete growth series of a wreath product of the form $W = H \wr G$ for G a free group. In fact, we suppose more generally that G is free-like, see Definition 2.9, so that its Cayley graph for the generating set S is an m-regular tree $\mathscr{T}$. We keep the convention of writing $\bar{s}$ for $s^{-1} \in S$. We suppose as usual that W is generated by $T@1 \sqcup S$.

Consider $w \in W$, written as $w = (h_0@1)g_1(h_1@1) \cdots g_\ell(h_\ell@1)$ with $g_i \in G$ and $h_i \in H$. Following the arguments in §1.3, one may write it as

$$w = \prod_{i=0}^{\ell}(h_i@e_i) \cdot g_1 \cdots g_\ell, \qquad \text{with } e_i = (g_1 \cdots g_i)^{-1}.$$

The *support* of w is the subgraph of $\mathscr{T}$ traced by inverses of prefixes of the word $g_1 \ldots g_\ell$; it is the convex hull of $\{e_0, \ldots, e_\ell\}$ in $\mathscr{T}$. We shall count elements of W by examining their possible supports and summing over them.

For each $s \in S$, let Θ_s denote the set of finite subtrees of $\mathscr{T}$, containing 1 and no element of S except possibly s. Each $\theta \in \Theta_s$ has *outer* vertices, with at most one neighbour in θ, and *inner* vertices, with at least two neighbours in θ. We introduce non-commutative power series $E_s(x, y, z)$ with coefficients in $\mathbb{Z}G[z]$, which count the number of Eulerian cycles[17] in trees Θ_s, weighted by length in z; the variables x, y belong to G and in particular are *not* assumed central. The series $E_s(x, y, z)$ are defined by the algebraic system

$$E_s(x, y, z) = 1 + sy\bar{s}z^2 + sx\left(\prod_{t \in S, t \neq \bar{s}} E_t(x, y, z) - 1\right)\bar{s}z^2. \qquad (1.6)$$

The monomials in E_s are in bijection with (Eulerian cycles tracing) trees in Θ_s; if a tree θ with p edges has inner vertices at $f_1, \ldots, f_n$ and

---

[17] I.e. cycles that traverse each edge once.

non-trivial outer vertices at $f'_1, \ldots, f'_{n'}$, then the monomial corresponding
to it is the product, in some order, of $z^{2p}, x^{f_1}, \ldots, x^{f_n}, y^{f'_1}, \ldots, y^{f'_{n'}}$.
Indeed the equation defining E_s says that a monomial counted by E_s
is either the empty tree (counted as 1) or a single edge from 1 to s
(counted as $sy\bar{s}z^2$), or an edge from 1 to s, followed by $m - 1$ subtrees
counted recursively by E_t for all $t \neq \bar{s}$, which are not all empty. Note
that if θ has p edges, then a minimal closed path that explores all vertices
of θ has length $2p$.

Let $D_s(z)$ denote the sum of $wz^{\|w\|}$ over all elements $w \in W$ which
belong to the base group H^G and whose support is an element of Θ_s.
Such elements may be counted as follows: starting from a support $\theta \in \Theta_s$,
choose a word $g_1 \ldots g_\ell$ in G of minimal length that visits all vertices of
θ; and, each time a vertex is first visited, insert an element of H, which
furthermore must be non-trivial if the vertex is outer. Therefore,

$$D_s(z) = E_s(\widehat{\Gamma}_H(z), \widehat{\Gamma}_H(z) - 1, z). \tag{1.7}$$

Next, let $F_s(z)$ denote the sum of $wz^{\|w\|}$ over all elements $w = fg \in W$
with $f: G \to H$, $g \in G$ not beginning in $\bar{s}$, and whose support does not
contain $\bar{s}$. We have a linear system

$$F_s(z) = \prod_{t \neq \bar{s}} D_t(z) + \sum_{t \neq \bar{s}} \left(\prod_{u \neq t, \bar{s}} D_u(z) \right) tzF_t, \tag{1.8}$$

because in every such element either $g = 1$ and the support explores all
the neighbours t of 1 except $\bar{s}$, or g begins by a generator, say t, and
then its support explores all neighbours of 1 except $\bar{s}, t$, then moves to
t, and continues by an element not starting by $\bar{t}$. Finally,

$$\widehat{\Gamma}_W(z) = \prod_{s \in S} D_s(z) + \sum_{s \in S} \left(\prod_{t \neq s} D_t(z) \right) szF_s(z), \tag{1.9}$$

for the same reasoning as above. Combining Equations (1.6–1.9), we
deduce:

Theorem 3.2. *If H is a finitely generated group whose complete growth
series $\widehat{\Gamma}_H(z)$ is algebraic, and G is a free-like group, then the complete
growth series of W is also algebraic.*

Corollary 3.3 (Parry, [65]). *If H is a finitely generated group whose
growth series $\Gamma_H(z)$ is algebraic, and G is a free-like group, then the
growth series of W is also algebraic.*

*If furthermore $\Gamma_H(z)$ is rational and $m \leq 2$, then $\Gamma_W(z)$ is also
rational.*

*On the other hand, if $m \geq 3$ then $\Gamma_W(z)$ does not belong to the field
generated by z and $\Gamma_H(z)$.*

Proof. Apply the augmentation map $\varpi : g \mapsto 1$ to Equations (1.6–1.9); this gives an algebraic system of degree $\max(m-1, 1)$ expressing $\Gamma_W(z)$ in terms of z and $\Gamma_H(z)$. In particular, for $m = 2$ it is a linear system.

Conversely, assume $m \geq 3$ and let ρ denote the convergence radius of the series of the image of D_s under ϖ. Note that $\lim_{z \to \rho^-} \varpi(D_s)(z)$ is finite: if the limit were infinite, convergence to infinity would be order $(m-1)\times$ itself, a contradiction. Therefore, $\varpi(D_s)$ has a non-pole singularity, so is not in $\mathbb{Q}(z, \Gamma_H(z))$. $\square$

Exercise 3.4. Show that, for the lamplighter group $C_2 \wr \mathbb{Z}$, the complete growth series is not rational.

Hint: use Exercise 2.3.

3.4. Travelling Salesmen

To glimpse at the limit of what can be computed, consider now the case $G = \mathbb{Z}^2$. No property of $\Gamma_W(z)$ is known, and this is due to the fact that there is no good description of words of minimal norm describing group elements.

In fact, the problem can be quite precisely stated as follows. One is given a point p_∞ and a set $\{p_1, \ldots, p_\ell\}$ in $\mathbb{Z}^2$, and is required to find a walk of minimal length on the grid that starts at $(0,0)$, visits all the points $p_1, \ldots, p_\ell$ in some order, and ends at p_∞. This is a classical *travelling salesman* problem and is known to be NP-complete, see [24, 25]. It is a small step to venture that finding a good description of minimal paths is at least as hard as finding those paths' length.

3.5. Asymptotic Growth

Regular wreath products, in non-degenerate cases, all have exponential growth. This is in stark contrast to the case of permutational wreath products, as we will see in §5.

Proposition 3.5. *If $H \neq 1$ and G is infinite, then $W = H \wr G$ has exponential growth.*

Proof. Choose $h \neq 1 \in H$, and without loss of generality assume that h is a generator of H. Since G is infinite, there exists an infinite word $g_1 g_2 \ldots$ that traces a geodesic in the Cayley graph (see after 1.8) of G, with $g_1, g_2, \ldots$ generators of G and also of W. In particular, all $(g_1 \cdots g_i)^{-1}$ are distinct. Consider then, for any $\ell \in \mathbb{N}$, the set of elements

$$\left\{ (h@1)^{\epsilon_0} g_1 (h@1)^{\epsilon_1} \cdots g_\ell (h@1)^{\epsilon_\ell} : \epsilon_0, \ldots, \epsilon_\ell \in \{0, 1\} \right\}.$$

All of these elements have norm at most $2\ell + 1$, and there are $2^{\ell+1}$ such elements. They are all distinct, since when they are rewritten in the form $fg_1 \ldots g_\ell$ one has $f((g_1 \cdots g_i)^{-1}) = h^{\epsilon_i}$ so that the ϵ_i can be recovered from the element. Therefore, $v_W(2\ell + 1) \geq 2^{\ell+1}$. $\square$

4. (Self-)similar Groups

We begin by introducing self-similar groups. They are groups with an additional structure:

Definition 4.1. A group G is *self-similar* if it is endowed with a homomorphism $\phi\colon G \to G \wr_X \mathrm{Sym}(X)$ for some set X. The map ϕ is called the *wreath recursion* of G. △

In this text, we shall always assume that the set X is finite, and shall (unless stated otherwise) also assume that the homomorphism ϕ is injective. A self-similar group is a group G in which elements may be recursively described by a permutation of X, decorated by elements of G itself. In case $X = \{0, 1, \ldots, d-1\}$, we also write elements of $G \wr_X \mathrm{Sym}(X)$ in the form $\langle\!\langle g_0, \ldots, g_{d-1} \rangle\!\rangle \pi$ for group elements $g_0, \ldots, g_{d-1}$ and a permutation $\pi \in \mathrm{Sym}(X)$.

It is essential to understand that being self-similar is an *attribute* of a group, and not a property. Thus, for example, a topological group is a group endowed with a topology; and every group is a topological group, for the discrete and the coarse topology. In the same vein, every group is self-similar, merely for the reason that it is similar to itself. Taking $X = \{0\}$ and $\phi(g) = \langle\!\langle g \rangle\!\rangle$ is uninteresting but not illegal.

4.1. Finite-State Self-Similar Groups

We describe two fundamental constructions of self-similar groups.

For the first, start by a well-understood group F, such as a free group; and choose a (not necessarily injective!) homomorphism $\tilde\phi\colon F \to F \wr_X \mathrm{Sym}(X)$. There exists then a maximal quotient of F on which the map $\tilde\phi$ induces an injective wreath recursion. To wit, one defines an increasing sequence N_i of normal subgroups of F by

$$N_0 = 1, \quad N_{i+1} = \tilde\phi^{-1}(N_i^X), \tag{1.10}$$

and sets $G = F/\bigcup_i N_i$. By construction, the map $\tilde\phi$ induces an injective map $\phi\colon G \to G \wr_X \mathrm{Sym}(X)$.

An important example of a group defined by this method—and which, essentially, cannot be defined differently—is the *first Grigorchuk group*, introduced in [28] and based on [2]. Consider

$$F = \langle a, b, c, d \mid a^2, b^2, c^2, d^2, bcd \rangle \cong C_2 * (C_2 \times C_2), \tag{1.11}$$

and define $\tilde\phi\colon F \to F \wr \mathrm{Sym}(2)$ by

$$\tilde\phi(a) = \diagdown\!\!\!\!\diagup, \qquad \tilde\phi(b) = {\downarrow}a\ {\downarrow}c\,, \qquad \tilde\phi(c) = {\downarrow}a\ {\downarrow}d\,, \qquad \tilde\phi(d) = {\downarrow}\ {\downarrow}b\,.$$

It is straightforward to see that $\tilde\phi$ is a homomorphism—just compute the images of the relators. It is, however, remarkable that one may compute efficiently in G just using this description. We use the same letters a, b, c, d for the corresponding generators of G. As an illustration, let us check that the relation $(ad)^4$ holds in G. Writing the permutation diagrams horizontally, one has

$$\phi((ad)^4) = \left(\times \!\! \longrightarrow \right)^4 = \left(\times \!\! \longrightarrow \times \!\! \longrightarrow \right)^2 = \left(\longrightarrow \right)^2 = \frac{b^2}{b^2} = 1,$$

so $(ad)^4 = 1$ in G because ϕ is injective.

Exercise 4.2. Using similar calculations, compute the exponent of ab and ac in G.

Note that in that example, $G = \langle S \rangle$ for the set $S = \{1, a, b, c, d\}$, which has the property that $\phi(S)$ is contained in $S \times S \times \mathrm{Sym}(2)$. More generally,

Definition 4.3. Let G be a self-similar group. A subset $S \subseteq G$ is *state-closed* if $\phi(S)$ is contained in $S^X \times \mathrm{Sym}(X)$.

An element $g \in G$ is *finite-state* if there exists a state-closed subset of G containing g. A subset of G is *finite-state* if all its elements are finite-state. $\triangle$

Exercise 4.4. Let $\mathbb{Z} = \langle t \rangle$ be endowed with the self-similar structure $\phi(t) = {}^t \!\! \times \!\! t^2$. Show that only t^0 is finite-state.

Lemma 4.5. *The product and inverse of finite-state elements is again finite-state.*

Proof. If g, h are finite-state contained respectively in state-closed sets S, T, then gh^{-1} is finite-state, as it belongs to the finite state-closed set ST^{-1}. $\square$

Therefore, a finitely generated self-similar group G is finite-state if and only if its generators are finite-state, and one may assume that G is generated by a state-closed set.

In that case, the wreath recursion of G may conveniently be represented by an *automaton*, more precisely a *Mealy automaton*. This is a directed graph with vertex set S called its *states*, and with an edge from $s \in S$ to $t \in S$, with label '$x|y$', whenever the decorated permutation $\phi(s)$ maps $x \in X$ to $y \in X$ and has label t on the edge $x \to y$. Thus, in a sense, the graph is the dual of the permutation diagram, with the roles of X and S exchanged. The automaton generating the first Grigorchuk is, with the convention $X = \{\mathbf{0}, \mathbf{1}\}$.

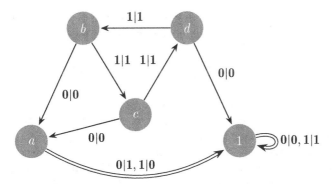

Assume that the self-similar group was obtained as above as a quotient of a self-similar group F with a map $\tilde{\phi} \colon F \to F \wr_X \mathrm{Sym}(X)$. Recall that the *word problem* asks, given a word in the generators of a finitely generated group, to determine whether the group element that it defines is trivial. There are groups, even finitely presented, in which the word problem is unsolvable [62]; however,

Lemma 4.6. *Let F be a finite-state finitely generated self-similar group with wreath recursion $\tilde{\phi}$ and solvable word problem, and G be the maximal quotient of F on which the induced wreath recursion $G \to G \wr_X \mathrm{Sym}(X)$ is injective. Then G also has a solvable word problem.*

Proof. Assume without loss of generality that F is generated by the finite state-closed set S, and denote also by S the corresponding generating set of G. Given a word $w \in S^*$ of length ℓ, it defines a state-closed element of F, belonging to the state-closed set S^ℓ. Consider the corresponding automaton with vertex set S^ℓ.

Let $U \subseteq S^\ell$ denote the set of states that are reachable from $w \in S^\ell$ by arbitrarily long paths. This set is computable: set $U'_0 := \{w\}$, and for $i \geq 0$ set $U'_{i+1} := U'_i \cup \{\text{endpoints of edges starting in } U'_i\}$; then the U'_i form an increasing sequence of subsets of S^ℓ, hence stabilize, say to U''_0. Note that U''_0 is the set of states reachable from w. For all $i \geq 0$, let U''_{i+1} denote those endpoints of edges starting in U''_i; then the U''_i form a decreasing sequence of subsets of S^ℓ, hence stabilize, to U.

The element of G defined by w is trivial in G if and only if both of the edges starting in U''_0 have labels of the form '$x|x$' for some $x \in X$, and all elements of U define trivial elements of F under the evaluation map $S^* \to F$.

More precisely, let $m \in \mathbb{N}$ be minimal such that every element of U may be reached from w by a path of length at most m. Then the conditions above imply that w belongs to the normal subgroup N_m of F, see (1.10). □

The above proof amounts to constructing a Mealy automaton for the action of S^ℓ, and examining it to determine which of its states are trivial in G.

4.2. Linear Groups

Here is another construction of self-similar groups. Consider a group G, a subgroup H, and a homomorphism $\phi_0 \colon H \to G$. By the 'permutational Kaloujnine–Krasner theorem' 1.6, there is a natural extension $\phi \colon G \to G \wr_X \operatorname{Sym}(X)$, with $X = H\backslash G$, in such a manner that $\phi(h) = \langle\!\langle \dots, \phi_0(h), \dots \rangle\!\rangle \dots$ for all $h \in H$, with the '$\phi_0(h)$' in position $H \in H\backslash G$.

Alternatively, this map ϕ may be directly constructed as follows: choose a transversal T of H in G, namely a subset $T \subseteq G$ such that every $g \in G$ may uniquely be written in the form ht with $h \in H, t \in T$. Identify X with T. Let then $\phi(g)$ be the decorated permutation that sends $t \in T$ to $u \in T$ with label $\phi_0(tgu^{-1})$ whenever tgu^{-1} belongs to H.

Here is a fundamental example: choose a prime number p, and consider

$$G = \Gamma_0(p) = \begin{pmatrix} \mathbb{Z} & \mathbb{Z} \\ p\mathbb{Z} & \mathbb{Z} \end{pmatrix} \cap \operatorname{SL}_2(\mathbb{Z}).$$

Consider also the matrix $\Phi = \begin{pmatrix} p & 0 \\ 0 & p^{-1} \end{pmatrix} \in \operatorname{SL}_2(\mathbb{Q})$, and $H = G \cap G^{\Phi^{-1}}$. Set $\phi_0(h) = h^\Phi$. This example generalises naturally to G any matrix group, such as for instance a congruence subgroup of $\operatorname{SL}_n(\mathbb{Z})$ for arbitrary n, or even $\operatorname{SL}_n(\mathbb{Z})$ itself. This shows that the class of linear groups over $\mathbb{Z}$ is contained in the class of self-similar groups.

In fact, the only essential ingredient of the above construction is the element Φ in the *commensurator* of G. Recall that, for a subgroup G of a group L, the commensurator of G is the subgroup of those $x \in L$ such that $G \cap G^x$ has finite index in G and in G^x. If G is an irreducible lattice in a Lie group G, then G may be called *arithmetic* [54] if its commensurator is dense in L; e.g. the commensurator of $\operatorname{SL}_n(\mathbb{Z})$ in $\operatorname{SL}_n(\mathbb{R})$ is $\operatorname{SL}_n(\mathbb{Q})$. Then all arithmetic lattices admit self-similar actions on rooted trees [46].

4.3. Rooted Trees

Let X be a set, and consider the associated *rooted regular tree* $\mathscr{T}$: its vertex set is $X^* = \{x_i \dots x_1 : x_j \in X\} = \bigsqcup_{i \geq 0} X^i$, and it has an edge between $x_{i+1}x_i \dots x_1$ and $x_i \dots x_1$ for all $x_i \in X$. The tree is rooted at the empty word, the unique element of X^0; the set of vertices at distance i from the root is identified with X^i, and the Cartesian product X^∞ is naturally interpreted as the boundary $\partial\mathscr{T}$ of the tree, namely the set of infinite rays emanating from the root. Here for illustration is the top of the binary tree:

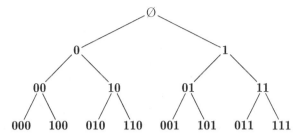

Let W denote the isometry group of $\mathscr{T}$, namely the set of bijections of X^* that fix the root $\emptyset$ and preserve the edge structure of $\mathscr{T}$. Given $g \in W$, let $\sigma \in \mathrm{Sym}(X)$ denote the action of g on $X = X^1$, and for all $x \in X$ define an element $g_x \in W$ by $(x_n \cdots x_1 x)g = (x_n \cdots x_1)g_x\,(x^\sigma)$; namely, g_x describes the action of g on the subtree X^*x as it is carried to X^*x^σ by g.

Lemma 4.7. *The map*

$$\phi: \begin{cases} W & \to\ W \wr \mathrm{Sym}(X) \\ g & \mapsto\ \langle\!\langle g_x : x \in X \rangle\!\rangle \sigma \end{cases}$$

is a group isomorphism.

Proof. Given $g_x \in W$ and $\sigma \in \mathrm{Sym}(X)$, an element $g \in W$ may uniquely be defined by $(x_n \cdots x_1 x)g = (x_n \cdots x_1)g_x\,(x^\sigma)$. This proves that ϕ is bijective.

To see that ϕ is a homomorphism, consider the $\#X$ subtrees below the root. They are permuted according to the permutation part σ of ϕ, and simultaneously acted upon by the decorations g_x. Composition of decorated permutations therefore coincides with composition of tree isometries. □

Let us now start with a self-similar group G. Its wreath recursion $\phi: G \to G \wr_X \mathrm{Sym}(X)$ then defines an action of G on X. Furthermore, the wreath recursion can be 'iterated': one has maps

$$G \xrightarrow{\phi} G \wr_X \mathrm{Sym}(X) \xrightarrow{\phi^X} (G \wr_X \mathrm{Sym}(X)) \wr_X \mathrm{Sym}(X)$$
$$= G \wr_{X^2} (\mathrm{Sym}(X) \wr_X \mathrm{Sym}(X)) \longrightarrow \cdots$$

so that G acts on X^i for all $i \in \mathbb{N}$. Furthermore, these actions are compatible with each other, in the sense that the map $(x_{i+1}, x_i, \ldots, x_1) \mapsto (x_i, \ldots, x_1)$ interlaces the actions on X^{i+1} and X^i. Taking the inverse limit of X^i under these projection maps gives an action on the Cartesian product X^∞. Note that sequences in X^∞ are infinite on the left, namely of the form $(\ldots, x_{i+1}, x_i, \ldots, x_1)$.

The compatibility between the actions on X^i and X^{i+1} precisely means that G acts by tree isometries on the rooted regular tree $\mathscr{T}$ with vertex set X^*.

Note that, even if the wreath recursion ϕ is injective, the action of the G on the tree $\mathcal{T}$ need not be faithful. This is, however, the case for the examples of self-similar action of the first Grigorchuk group (see Proposition 5.9) and of the congruence subgroup $\Gamma_0(p)$.

We introduced the self-similar structure on $\Gamma_0(p)$ and not on $\mathrm{SL}_2(\mathbb{Z})$ because the latter does not act on the *rooted p-regular tree*, in which each vertex has degree $p + 1$ except the root, which has degree p. If we add an edge upwards from the root, and a rooted p-regular tree above it, to the rooted p-regular tree, we obtain a $(p+1)$-regular tree on which the action of $\Gamma_0(p)$ extends to an action of $\mathrm{SL}_2(\mathbb{Z})$. In fact, this action is already well known, see [71, §II.1]: the $(p+1)$-regular tree is the *Bruhat-Tits tree* of $\mathrm{SL}_2(\mathbb{Z}_p)$. Its vertices are homothety classes of lattices $\cong \mathbb{Z}_p^2$ in $\mathbb{Q}_p$, and there is an edge between classes $\mathbb{Q}_p^\times \Lambda$ an $\mathbb{Q}_p^\times \Lambda'$ if they admit representatives $\alpha\Lambda, \alpha'\Lambda'$ with $\alpha\Lambda \subset \alpha'\Lambda'$ and $[\alpha'\Lambda' : \alpha\Lambda] = p$. The group $\mathrm{SL}_2(\mathbb{Q}_p)$ naturally acts on lattices, and $\mathrm{SL}_2(\mathbb{Z}_p)$ acts as the stabilizer of the root $\mathbb{Q}_p^\times \mathbb{Z}_p^2$. The congruence subgroup $\Gamma_0(p)$ fixes an edge adjacent to the root, and the rooted p-regular tree $\mathcal{T}$ is spanned by those lattices of the form $\langle (p^n, 0), (x_0 + x_1 p + \cdots + x_{n-1} p^{n-1}, 1) \rangle$ for $n \in \mathbb{N}$ and $x_0, \ldots, x_{n-1} \in \{0, \ldots, p-1\}$.

Let us remark in passing that obtaining an action on a rooted tree is not spectacular in itself: every countable residually-p group acts on a rooted p-regular tree. Indeed, choose a descending sequence $G = G_0 > G_1 > \cdots$ of subgroups with $[G : G_i] = p^i$ and $\bigcap G_i = 1$. Let the vertices of $\mathcal{T}$ be the set of right cosets of all G_i, with an edge between $G_i g$ and $G_{i+1} g$ for all $i \in \mathbb{N}, g \in G$; and let G act by right multiplication on $\mathcal{T}$.

This action is in general not self-similar, nor is it 'economical', in the sense that the permutation group acting on X^i may have order comparable to $(\#X)^i$ rather than $(\#X!)^{\#X^i}$.

Finally, let us return to the construction of a self-similar group G as a quotient of a self-similar group F so that the wreath recursion becomes injective. Knowing that the action of G is faithful helps in solving the word problem, in case G is not finite-state:

Lemma 4.8. *Let F be a self-similar group with wreath recursion $\tilde{\phi}$ and solvable word problem, and let G be the maximal quotient of F on which the induced wreath recursion is injective. Assume that the action of G on the tree $\mathcal{T}$ is faithful. Then G also has solvable word problem.*

Proof. Let S be a generating set for F, and consider $w \in S^*$. We start two semi-algorithms in parallel; the first one will stop if w is non-trivial in G, and the second one will stop if w is trivial in G.

If w is non-trivial in G, then it will act non-trivially on some vertex of $\mathcal{T}$, and this vertex may be found by enumerating all vertices of $\mathcal{T}$ and computing the action of w on it by applying $\tilde{\phi}$.

If w is trivial, then it belongs to one of the normal subgroups N_i. Going through all $i = 0, 1, \ldots$ in sequence, and iterating i times $\tilde{\phi}$ on

w yields $\#X^i$ elements of F. If all of them are trivial in F, then w is trivial; otherwise, continue with the next i. $\qquad\qquad\qquad\qquad$ □

Here are two fundamental examples of self-similar groups. Let $\mathscr{T} = X^*$ be a rooted regular tree, and let Q be a group acting transitively on X. Consider the iterated wreath products $Q_i = Q \wr_X Q \wr_X \cdots \wr_X Q$, with i factors; these groups act naturally on X^i by the imprimitive action.

On the one hand, there is a natural map $Q_{i+1} \twoheadrightarrow Q_i$, given by deleting the leftmost factor, i.e. naturally mapping $Q_{i+1} = Q \wr_X Q_i$ to $Q_i \cong 1 \wr_X Q_i$. Set then $\overline{G} = \varprojlim Q_i$, the projective limit being taken along these epimorphisms. The self-similarity structure $\phi \colon \overline{G} \to \overline{G} \wr_X Q$ is induced by the identity map $Q_i \xrightarrow{\cong} Q_{i-1} \wr_X Q$. Because it 'peels off' the rightmost factor, it is compatible with the inverse limit. It defines a profinite self-similar group $\overline{G}$.

On the other hand, there is a natural map $Q_i \hookrightarrow Q_{i+1}$, given by inserting a trivial leftmost factor, i.e. naturally mapping $Q_i \cong 1 \wr_X Q_i$ to $Q_{i+1} = Q \wr_X Q_i$. Set then $L = \varinjlim Q_i$, the union ($=$ injective limit) being taken along these monomorphisms. The self-similarity structure $\phi \colon L \to L \wr_X Q$ is induced by the identity map $Q_i \xrightarrow{\cong} Q_{i-1} \wr_X Q$. Because it 'peels off' the rightmost factor, it is compatible with the union. It defines a locally finite group L.

Exercise 4.9. Show that L is a dense subgroup of $\overline{G}$.

A *law* for a group G is a word $w(x_1, x_2, \ldots)$ in variables $x_1, x_2, \ldots$ such that, whenever the elements $x_1, x_2, \ldots$ are replaced by group elements from G, the word evaluates to 1 in G. For example, abelian groups are characterised as those groups satisfying the law $w = [x_1, x_2]$.

Exercise 4.10. Show that L satisfies no non-trivial law.

Hint: it suffices to look at the case $X = \{1, 2\}$. By Theorem 1.4, every finite 2-group imbeds in Q_i for some $i \in \mathbb{N}$, and therefore in G. Finally, the free group is residually 2.

See [1] for a general result about inexistence of group laws, which covers the group L.

4.4. Similar Families of Groups

The notion of a self-similar group may be generalised to a *family* of similar groups.

Definition 4.11. Let Ω be a set, and let $\sigma \colon \Omega \circlearrowleft$ be a map. A *similar family of groups* over Ω is a family $(G_\omega)_{\omega \in \Omega}$ of groups and a family of homomorphisms

$$\phi_\omega \colon G_\omega \to G_{\sigma\omega} \wr_{X_\omega} \mathrm{Sym}(X_\omega),$$

for a family of sets $(X_\omega)_{\omega \in \Omega}$. $\qquad\qquad\qquad\qquad\qquad\qquad\qquad\qquad$ △

Just as before, each group G_ω acts on a tree $\mathscr{T}_\omega$ with vertex set $\bigsqcup_{i \geq 0} X_{\sigma^{i-1}\omega} \times \cdots \times X_\omega$. This rooted tree is now not anymore regular, but it is still *spherically homogeneous*, in that its isometry group is transitive on the set of vertices at a given distance from the root.

As before, there are two fundamental examples of similar families of groups. Let $(X_\omega)_{\omega \in \Omega}$ be a family of sets, and let $(Q_\omega)_{\omega \in \Omega}$ be a family of groups, with Q_ω acting on X_ω, say transitively for simplicity.

Consider the iterated wreath products $Q_{\omega,i} = Q_{\sigma^{i-1}\omega} \wr_{X_{\sigma^{i-2}\omega}} \cdots \wr_{X_\omega} Q_\omega$, with i factors; these groups act naturally on $X_{\sigma^{i-1}\omega} \times \cdots \times X_\omega$ by the imprimitive action.

On the one hand, there is a natural map $Q_{\omega,i+1} \twoheadrightarrow Q_{\omega,i}$, given by deleting the leftmost factor, i.e. naturally mapping $Q_{\omega,i+1} = Q_{\sigma^i\omega} \wr_{X_{\sigma^{i-1}\omega}} Q_{\omega,i}$ to $Q_{\omega,i} \cong 1 \wr_{X_{\sigma^{i-1}\omega}} Q_{\omega,i}$. Set then $\overline{G}_\omega = \varprojlim Q_{\omega,i}$, the projective limit being taken along these epimorphisms. The self-similarity structure $\phi_\omega \colon \overline{G}_\omega \to \overline{G}_{\sigma\omega} \wr_{X_\omega} Q_\omega$ is induced by the identity map $Q_{\sigma,i} \xrightarrow{\cong} Q_{\sigma\omega,i-1} \wr_{X_\omega} Q_\omega$. Because it 'peels off' the rightmost factor, it is compatible with the inverse limit. It defines a profinite self-similar group $\overline{G}_\omega$.

On the other hand, there is a natural map $Q_{\omega,i} \hookrightarrow Q_{\omega,i+1}$, given by inserting a trivial leftmost factor, i.e. naturally mapping $Q_{\omega,i} \cong 1 \wr_{X_{\sigma^{i-1}\omega}} Q_{\omega,i}$ to $Q_{\omega,i+1} = Q_{\sigma^i\omega} \wr_{X_{\sigma^{i-1}\omega}} Q_{\omega,i}$. Set then $L_\omega = \varinjlim Q_{\omega,i}$, the injective limit being taken along these monomorphisms. The self-similarity structure $\phi\omega \colon L_\omega \to L_{\sigma\omega} \wr_{X_\omega} Q_\omega$ is induced by the identity map $Q_{\sigma,i} \xrightarrow{\cong} Q_{\sigma\omega,i-1} \wr_{X_\omega} Q_\omega$. Because it 'peels off' the rightmost factor, it is compatible with the union. It defines a locally finite self-similar group L_ω.

4.5. The Grigorchuk Family G_ω

We shall concentrate particularly on one specific example. Write $\{0, 1, 2\}$ for the three non-trivial homomorphisms $C_2 \times C_2 \to C_2$, identified for definiteness as follows. We view the source $C_2 \times C_2 = \{1, b, c, d\}$ as a subgroup of the group F given in (1.11), and the range $C_2 = \{1, a\}$ in that same group F. The three homomorphisms are then uniquely defined by $\ker(0) = \langle b \rangle$ and $\ker(1) = \langle c \rangle$ and $\ker(2) = \langle d \rangle$. Set then

$$\Omega = \{0, 1, 2\}^\infty, \qquad \sigma(\omega_0 \omega_1 \omega_2 \dots) = \omega_1 \omega_2 \dots .$$

We start by the similar family $(F)_{\omega \in \Omega}$ with maps $\tilde{\phi}_\omega \colon F \to F \wr \mathrm{Sym}(2)$ given by

$$\tilde{\phi}_\omega(a) = \diagup\!\!\!\!\diagdown, \quad \text{and for all } x \in \{b, c, d\}: \quad \tilde{\phi}_\omega(x) = \big\lfloor \omega_0(x) \big\rfloor x ,$$

we define normal subgroups $(N_{\omega,i})_{i \in \mathbb{N}, \omega \in \Omega}$ of F by $N_{\omega,0} = 1$ and $N_{\omega,i+1} = \tilde{\phi}_\omega^{-1}(N_{\omega,i}^2)$, and set $G_\omega = F / \bigcup_{i \in \mathbb{N}} N_{\omega,i}$. This is the same construction as above, and computes G_ω as the maximal quotient of F such that the maps $\tilde{\phi}_\omega \colon F \to F \wr \mathrm{Sym}(2)$ descend to injective maps $\phi_\omega \colon G_\omega \to G_{\sigma\omega} \wr \mathrm{Sym}(2)$.

In particular, letting G denote the Grigorchuk group defined in §4.1, we have isomorphisms

$$G \xlongequal{\sim} G_{(012)^\infty} \xlongequal{\sim} G_{(120)^\infty} \xlongequal{\sim} G_{(201)^\infty}$$

identifying the generating sets as follows:

$$\{a, b, c, d\} \quad \{a, d, c, b\} \quad \{a, b, d, c\} \quad \{a, c, b, d\}.$$

The groups G_ω all act on the binary rooted tree, and it is easy to see that the orbit of the ray $\mathbf{1}^\infty$ is dense. Therefore, the groups G_ω could just as well have been defined by their actions on their respective orbit $\mathbf{1}^\infty G_\omega$. These are naturally graphs, called *Schreier graphs*, with vertex set $\mathbf{1}^\infty G_\omega$, and with an edge from $\mathbf{1}^\infty g$ to $\mathbf{1}^\infty gs$ for each generator $s \in \{a, b, c, d\}$, see Definition 1.8.

Proposition 4.12. *The graph $\mathbf{1}^\infty G_\omega$ is isometric to the half-infinite line $\mathbb{N}$ with multiple edges and loops. Under this identification with $\mathbb{N}$, the action of G_ω is given by*

$$a(2j) = 2j + 1, \qquad a(2j + 1) = 2j,$$

and for all $x \in \{b, c, d\}$, $\quad x(0) = 0, \qquad x(2^i(2j + 1)) = 2^i(2\omega_i(j) + 1).$

Proof. Consider the infinite dihedral group $D = \langle a, x \mid a^2, x^2 \rangle$, and the wreath recursion $\phi \colon D \to D \wr \mathrm{Sym}(2)$ defined by

$$\phi(a) = \times, \qquad \tilde{\phi}(x) = \lfloor a \; \lfloor x \; .$$

It defines a faithful action of D on the binary rooted tree $\mathscr{T}$ with vertex set $\{\mathbf{0}, \mathbf{1}\}^*$, and the action on the ray $\mathbf{1}^\infty$ is isomorphic to the action on the set of cosets $\langle x \rangle \backslash D$, since the stabilizer in D of $\mathbf{1}^\infty$ is $\langle x \rangle$. We abbreviate $\mathbf{1}^\infty =: \bar{1}$. The Schreier graph of the latter is a half-infinite line

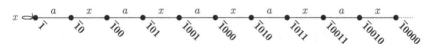

Since the action of generators of D change only a single symbol on sequences in $\{\mathbf{0}, \mathbf{1}\}^\infty$, the identification of the Schreier graph's vertices with $\mathbb{N}$ is explicit: it is the 'Gray code' [27] enumeration starting from the left-infinite word $\bar{1}$. Thus, the sequence $\dots 11x_i \dots x_1$ is identified with the integer $\sum_{j=1}^{i}(1 - x_j)2^{j-1}$, reading the number in base 2 with 0s and 1s switched.

Now, to obtain the Schreier graph of G_ω, one replaces each 'x' edge by a pair of edges labelled by two letters out of $\{b, c, d\}$, and puts loops at the extremities of the edge labelled by the remaining letter. The choice of which letter becomes a loop is determined by the position of the edge on the graph and the sequence ω. $\qquad\square$

For example, here is the Schreier graph of the action of the first Grigorchuk group $G_{012} = \langle a, b, c, d \rangle$ on $\mathbf{1}^\infty G$:

$$(1.12)$$

5. Growth Estimates for Self-similar Groups

One of the purposes of this section is to reprove the following result. Let Ω' denote the subset of Ω consisting of sequences containing infinitely many of each of the symbols $0, 1, 2$.

Theorem 5.1 (Grigorchuk, [30]). *If $\omega \in \Omega'$, and more generally if ω is not ultimately constant, then G_ω has intermediate growth.*

We shall in fact prove much more, in preparation for the construction, in §6, of groups with prescribed growth. We mainly follow [5].

5.1. A Lower Bound via Algebras

We begin by a general lower bound on growth, coming from the theory of Hopf algebras. Recall that the lower central series of a group G is defined by $\gamma_1(G) = G$ and $\gamma_{n+1}(G) = [\gamma_n(G), G]$ for all $n \geq 1$.

Theorem 5.2 (Grigorchuk, [31]). *Let G be a finitely generated group, and assume that there is a subgroup $H < G$ such that $\gamma_n(H) \neq \gamma_{n+1}(H)$ for all $n \in \mathbb{N}$. Then G's growth function satisfies*

$$\gamma_G \succsim \exp(\sqrt{R}).$$

In particular, if G is residually virtually nilpotent, then either G is virtually nilpotent (in which case γ_G is polynomial) or $\gamma_G \succsim \exp(\sqrt{R})$.

Before embarking on the proof, let us set up some algebraic notions. Let $\mathbb{K}$ be a field, and let G be a group. The group ring $\mathscr{A} = \mathbb{K}G$ is the $\mathbb{K}$-vector space with basis G, and multiplication extended linearly. It is a *Hopf algebra*: it admits a *coproduct*, which is an algebra homomorphism $\Delta \colon \mathscr{A} \to \mathscr{A} \otimes \mathscr{A}$ defined on the basis G by $g \mapsto g \otimes g$, a *counit*, which is an algebra homomorphism $\varepsilon \colon \mathscr{A} \twoheadrightarrow \mathbb{K}$ defined on the basis by $g \mapsto 1$; and an *antipode*, which is an antihomomorphism $\sigma \colon \mathscr{A} \to \mathscr{A}$ defined on the basis by $g \mapsto g^{-1}$. Various axioms are satisfied; in particular the coproduct is *coassociative*: $(1 \otimes \Delta) \circ \Delta = (\Delta \otimes 1) \circ \Delta \colon \mathscr{A} \to (\mathscr{A})^{\otimes 3}$, and *cocommutative*: $\Delta = \tau \circ \Delta$, for $\tau \colon \mathscr{A}^{\otimes 2} \to \mathscr{A}^{\otimes 2}$ the map $x \otimes y \mapsto y \otimes x$ flipping both factors. See [77] for details.

Denote by ϖ the kernel of ε, called the *augmentation ideal*. The *associated graded* of $\mathscr{A}$ is the vector space

$$\overline{\mathscr{A}} = \bigoplus_{n \geq 0} \varpi^n / \varpi^{n+1}.$$

Lemma 5.3. *The associated graded $\overline{\mathscr{A}}$ is a graded, cocommutative Hopf algebra.*

Proof. We have $\Delta(\varpi) \leq \mathscr{A} \otimes \varpi + \varpi \otimes \mathscr{A}$; so $\Delta(\varpi^n) \leq \sum_{i=0}^{n} \varpi^i \otimes \varpi^{n-i}$. Now given $\overline{x} \in \varpi^n/\varpi^{n+1}$, choose $x \in \varpi^n$ representing it; write $\Delta(x) = \sum \sum_{i=0}^{n} y_i \otimes z_{n-i}$ with $y_i, z_i \in \varpi^i$; and set $\Delta(\overline{x}) = \sum \sum_{i=0}^{n} \overline{y_i} \otimes \overline{z_{n-i}}$ where $\overline{y_i}, \overline{z_i} \in \varpi^i/\varpi^{i+1}$ are the images of their respective representatives. It is easy to show that this definition does not depend on the choices of x, y_i, z_i; and it defines a coassociative and cocommutative coproduct as $\mathscr{A}$'s coproduct was already coassociative and cocommutative. □

Let H be a Hopf algebra. An element $x \in H$ is called *primitive* if $\Delta(x) = x \otimes 1 + 1 \otimes x$. The set of primitive elements in a Hopf algebra forms a Lie subalgebra of H for the usual bracket $[x, y] = xy - yx$. Conversely, if L is a Lie algebra, then its universal enveloping algebra is a Hopf algebra whose primitive elements are L. Note that in characteristic p, one should consider *restricted* Lie algebras.

Proposition 5.4 ([55, Theorem 6.11]). *Let H be a cocommutative, primitively generated, graded Hopf algebra. Then it is the universal enveloping algebra of its primitive elements.*

Proof. Let P denote the Lie algebra of primitive elements in H. By the universal property of $U(P)$, there is a map $f : U(P) \to H$, which is graded, and surjective because P generates H. We show that f is also injective. Consider a homogeneous element $x \in U(P)$, say of degree n. If $n = 0$, then $x \in \ker(f)$ if and only if $x = 0$. Assume then that f is injective on elements of degree $< n$.

We have $\Delta(x) = 1 \otimes x + x \otimes 1 + y$ for some $y \in U(P)_{<n} \otimes U(P)_{<n}$. If $f(x) = 0$, then $f(y) = 0$; but we had assumed f to be injective on elements of degree $< n$, so $y = 0$ and $x \in P$. By assumption, f is injective on P, so $x = 0$ and therefore f is injective on elements of degree n as well. □

We apply these considerations to $\mathscr{A} = \mathbb{K}G$. First, we identify the primitive elements in $\overline{\mathbb{K}G}$. Let us define the series $\gamma_n^{\mathbb{K}}(G) = \{g \in G \mid g - 1 \in \varpi^n\}$ of normal subgroups of G.

Proposition 5.5. *The space of primitive elements in $\overline{\mathbb{K}G}$ is*

$$\mathscr{L}^{\mathbb{K}}(G) = \bigoplus_{n \geq 1} (\gamma_n^{\mathbb{K}}(G)/\gamma_{n+1}^{\mathbb{K}}(G)) \otimes \mathbb{K}. \tag{1.13}$$

Proof. The natural map $g \mapsto g - 1$ from $\gamma_n^{\mathbb{K}}(G)$ to ϖ^n/ϖ^{n+1} extends to a Lie algebra isomorphism from $\gamma_n^{\mathbb{K}}(G)$ onto primitive elements of $\overline{\mathbb{K}G}$.   □

In fact, $\gamma_n^{\mathbb{K}}(G)$ only depends on the characteristic p of $\mathbb{K}$ and, up to extension of scalars, $\mathscr{L}^{\mathbb{K}}(G)$ depends only on p. Furthermore, it may be identified directly within G, and is a variant of the lower central series [40, 41]. Indeed one has $\gamma_1^p(G) = G$, and if $p = 0$ then

$$\gamma_{n+1}^0(G) = \langle g \in G \mid g^t \in [G, \gamma_n^0(G)] \text{ for some } t > 0\rangle,$$

while if $p > 0$ then

$$\gamma_{n+1}^p(G) = \langle [G, \gamma_n^p(G)]\{x^p \mid x \in \gamma_{\lceil n/p\rceil}(G)\}\rangle.$$

Proposition 5.6. *Let S be such that $S \cup S^{-1}$ generates G, and let $v_{G,S}$ denote the corresponding growth function. Let $w(n) = \dim_{\mathbb{K}} \varpi^n/\varpi^{n+1}$ denote the growth function of $\overline{\mathbb{K}G}$. Then*

$$v_{G,S}(n) \geq w(0) + w(1) + \cdots + w(n) \text{ for all } n \in \mathbb{N}.$$

Proof. Consider first an element $x = 1 - g \in \varpi$, and write $g = s_1^{\epsilon_1} \cdots s_\ell^{\epsilon_\ell}$ as a product of generators and inverses. Using the identities

$$1 - gh = (1 - g) + (1 - h) - (1 - g)(1 - h),$$

$$1 - g^{-1} = -(1 - g) + (1 - g)(1 - g^{-1}),$$

we get $x \equiv \sum_{i=1}^{\ell} \epsilon_i(1 - s_i)$ modulo ϖ^2.

The ideal ϖ^n is generated, qua ideal, by all $x = (1 - g_1) \cdots (1 - g_n)$ with $g_i \in G$. By the above, ϖ^n/ϖ^{n+1} is generated, again qua ideal, by all $x = (1 - s_1) \cdots (1 - s_n)$ with $s_i \in S$. Now ϖ^n/ϖ^{n+1} has trivial multiplication, so the $(1 - s_1) \cdots (1 - s_n)$ also generate ϖ^n/ϖ^{n+1} qua vector space.

This generating set is contained in the linear span of $B_{G,S}(n) \subseteq \mathbb{K}G$; so for $0 \leq i \leq n$ we have $w(i) \leq \dim(\varpi^i \cap \mathbb{K}B_{G,S}(n)) - \dim(\varpi^{i+1} \cap \mathbb{K}B_{G,S}(n))$ and the claim follows.   □

Proof of Theorem 5.2. Consider a subgroup H such that $\gamma_n(H) \neq \gamma_{n+1}(H)$ for all $n \in \mathbb{N}$. Without loss of generality, suppose H is finitely generated. For $n \in \mathbb{N}$, let $\mathcal{P}(n)$ be the set of prime numbers p such that $(\gamma_n(H)/\gamma_{n+1}(H)) \otimes \mathbb{F}_p \neq 0$. Each $\mathcal{P}(n)$ is non-empty because $\gamma_n(H)/\gamma_{n+1}(H)$ is a finitely generated abelian group, and $\mathcal{P}(n+1) \subseteq \mathcal{P}(n)$ because the commutator map $(\gamma_n(H)/\gamma_{n+1}(H)) \times H \to \gamma_{n+1}(H)/\gamma_{n+2}(H)$ is onto. There exists therefore a prime number p such that $\gamma_n(H)/\gamma_{n+1}(H) \otimes \mathbb{F}_p \neq 0$ for all $n \in \mathbb{N}$. In particular, $\gamma_n^p(H) \neq \gamma_{n+1}^p(H)$ for all $n \in \mathbb{N}$.

Let $\mathbb{K}$ be a field of characteristic p. It follows that $\overline{\mathbb{K}H}$ contains a primitive element x_n of degree n for all $n \in \mathbb{N}$, namely $x_n = g_n - 1$ for some $g_n \in \gamma_n^p(H) \backslash \gamma_{n+1}^p(H)$; so $\overline{\mathbb{K}H}$ contains $\pi(n)$ linearly independent elements of degree n, where $\pi(n)$ denotes the partition function. Indeed to every partition $n = i_1 + \cdots + i_k$ with $i_1 \leq \cdots \leq i_k$ associate the element $x_{i_1} \cdots x_{i_k}$; these elements are linearly independent by Proposition 5.4. Therefore, the growth function of $\overline{\mathbb{K}H}$ is at least $\pi(n)$.

It now follows from Proposition 5.6 that $v_{H,S}(R) \geq \pi(R)$ holds for any generating set S of H. Classical results on partitions [37] tell us that $\pi(n) \propto \exp(\sqrt{n})$; so *a fortiori* $v_G(R) \succsim \exp(\sqrt{R})$. $\qquad\square$

Note that, for Grigorchuk's group $G = G_{012}$, the quotients $\gamma_n(G)/\gamma_{n+1}(G)$ have bounded rank, in fact 1 or 2, see [67]; so that no improvement on the lower bound can be obtained using Proposition 5.6.

5.2. Metrics on G_ω

We already saw that the groups G_ω are contracting, namely if $\phi(g) = \langle\!\langle g_0, g_1 \rangle\!\rangle \pi$ then g_0 and g_1 are shorter than g. We shall need a strengthening of this property: we assign norms $\|\cdot\|_\omega$ to the groups G_ω to obtain relations of the form

$$\|g_0\|_{\sigma\omega} + \|g_1\|_{\sigma\omega} \leq \frac{2}{\eta_\omega}\big(\|g\|_\omega + \|a\|_\omega\big) \qquad (1.14)$$

with $\eta_\omega > 2$ as large as possible.

We do this by assigning norms $\in \mathbb{R}_+$ to the generators a, b, c, d of G_ω, and extend $\|\cdot\|_\omega$ to G_ω by the triangular inequality:

$$\|g\|_\omega = \min\{\|s_1\| + \cdots + \|s_n\| : g = s_1 \cdots s_n, \ s_i \in S\}.$$

For this purpose, consider the open 2-simplex

$$\Delta = \left\{ (\beta, \gamma, \delta) \in \mathbb{R}^3 : \max\{\beta, \gamma, \delta\} < \frac{1}{2}, \beta + \gamma + \delta = 1 \right\}.$$

Its extreme points are $\left(\frac{1}{2}, \frac{1}{2}, 0\right)$ and its permutations. A choice of $p_\omega = (\beta, \gamma, \delta) \in \Delta$ defines the following norm on the generators of G_ω:

$$\|a\|_\omega = 1 - 2\max\{\beta, \gamma, \delta\},$$

$$\|b\|_\omega = \beta - \|a\|_\omega, \ \|c\|_\omega = \gamma - \|a\|_\omega, \ \|d\|_\omega = \delta - \|a\|_\omega.$$

In particular, note that the triangular inequality $\|x\|_\omega + \|y\|_\omega \leq \|xy\|_\omega$ holds for $x, y \in \{b, c, d\}$, and is sharp for $x = c, y = d$.

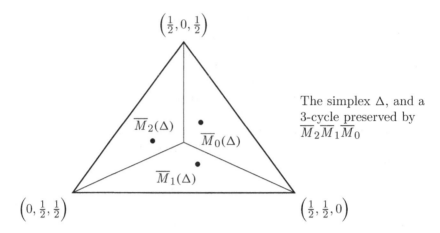

The simplex Δ, and a 3-cycle preserved by $\overline{M}_2\overline{M}_1\overline{M}_0$

We extend the family of groups $(G_\omega)_{\omega\in\{0,1,2\}^{\mathbb{N}}}$ to a family $(G_\omega)_{\omega\in\{0,1,2\}^{\mathbb{Z}}}$; namely, the parameter space is now $\Omega = \{0,1,2\}^{\mathbb{Z}}$ with the two-sided shift map $\sigma\colon \Omega\circlearrowleft$. The group G_ω itself only depends on the restriction of ω to $\mathbb{N}$, but the norm on G_ω depends on the restriction of ω to $-\mathbb{N}$.

Analogously to before, we denote by Ω' the subset of Ω consisting of sequences that contain infinitely many $0,1,2$ in both directions. For $\omega \in \Omega'$, we construct the point $p_\omega \in \Delta$ as follows. Consider the matrices

$$M_0 = \begin{pmatrix} 1 & 1 & 1 \\ 0 & 2 & 0 \\ 0 & 0 & 2 \end{pmatrix}, \qquad M_1 = \begin{pmatrix} 2 & 0 & 0 \\ 1 & 1 & 1 \\ 0 & 0 & 2 \end{pmatrix}, \qquad M_2 = \begin{pmatrix} 2 & 0 & 0 \\ 0 & 2 & 0 \\ 1 & 1 & 1 \end{pmatrix}.$$

Define then

$$\eta\colon \Delta \times \{0,1,2\} \to (2,3], \qquad \overline{M}_\lambda\colon \Delta\circlearrowleft, \qquad \mu\colon \Delta \to (0,\tfrac{1}{3})$$

by setting, for all $\lambda \in \{0,1,2\}$ and all $p \in \Delta$,

$$\eta(p,\lambda) = (1\ 1\ 1)\cdot M_\lambda(p) \quad \text{the } \ell^1\text{-norm of } M_\lambda(p),$$

$$\overline{M}_\lambda(p) = \frac{M_\lambda(p)}{\eta(p,\lambda)} \qquad\qquad \text{the projection of } M_\lambda(p) \text{ to } \Delta,$$

$$\mu(p) = \min\{\beta,\gamma,\delta\} \qquad \text{the minimal distance of } p \text{ to a vertex of } \Delta.$$

Endow Δ with the Hilbert metric $d_\Delta(V_1,V_2) = \log(V_1,V_2;V_-,V_+)$, computed using the cross-ratio of the points V_1, V_2 and the intersections V_-, V_+ of the line containing V_1, V_2 with the boundary of Δ. The transformations $\overline{M}_\lambda$ are projective transformations of Δ, and are therefore contracting:

Lemma 5.7 (Essentially [11]). *Let K be a convex subset of affine space, and let $A\colon K\circlearrowleft$ be a projective map. Then A contracts the Hilbert metric.*

If furthermore, $A(K)$ contains no lines from K (that is, $A(K) \cap \ell \neq K \cap \ell$ for every line ℓ intersecting K), then A is strictly contracting.

Proof. Because A is projective, it preserves the cross-ratio on lines, so we have $d_{A(K)}(A(V_1), A(V_2)) = d_K(V_1, V_2)$ for all $V_1, V_2 \in K$. Furthermore, on the line ℓ through V_1, V_2, the intersection points $\ell \cap \partial A(K)$ are not further from V_1, V_2 than $\{V_+, V_-\} = \ell \cap \partial K$; the Hilbert metric decreases as $V_\pm$ are moved further apart from V_1, V_2, and this gives strict contraction under the condition $A(K) \cap \ell \neq K \cap \ell$. $\qquad\square$

We are ready to define the points $p_\omega \in \Delta$, and therefore the metrics $\| \cdot \|_\omega$. Choose an arbitrary point $p \in \Delta$, and set

$$p_\omega = \lim_{n \to \infty} \overline{M}_{\omega_{-1}} \circ \overline{M}_{\omega_{-2}} \circ \cdots \circ \overline{M}_{\omega_{-n}}(p). \tag{1.15}$$

Note that because the transformations $\overline{M}_\lambda$ are contracting, the limit p_ω is independent of the choice of p. Note also that, by our assumption that the negative part of ω contains infinitely many 0, 1 and 2, the limit p_ω does not belong to the boundary of Δ. From now on, we write

$$\eta_\omega := \eta(p_\omega, \omega_0).$$

Lemma 5.8. *For all $g \in G_\omega$ with $\phi(g) = \langle\!\langle g_0, g_1 \rangle\!\rangle \pi$, we have the inequality (1.14)*

$$\|g_0\|_{\sigma\omega} + \|g_1\|_{\sigma\omega} \leq \frac{2}{\eta_\omega}(\|g\|_\omega + \|a\|_\omega);$$

and furthermore, if $g \notin \{b, c, d\}$ then up to replacing g with a conjugate we have

$$\|g_0\|_{\sigma\omega} + \|g_1\|_{\sigma\omega} \leq \frac{2}{\eta_\omega}\|g\|_\omega.$$

Proof. Without loss of generality, we suppose $\omega_0 = 0$, and write $p_\omega = (\beta, \gamma, \delta)$ and $p_{\sigma\omega} = \overline{M}_{\omega_0} p_\omega = (\beta', \gamma', \delta')$. We have

$$\eta_\omega = 3 - 2\beta, \quad (\beta', \gamma', \delta') = \frac{(\beta + \gamma + \delta, 2\gamma, 2\delta)}{\eta_\omega} = \frac{(1, 2\gamma, 2\delta)}{\eta_\omega}.$$

Thus,

$$\|b\|_{\sigma\omega} + \|\omega_0(a)\|_{\sigma\omega} = \|b\|_{\sigma\omega} = \beta' - \|a\|_{\sigma\omega} = \beta' - (1 - 2\beta') = 3\beta' - 1$$

$$= \frac{3}{\eta_\omega} - 1 = 2\frac{\beta}{\eta_\omega} = \frac{2}{\eta_\omega}\|ab\|_\omega,$$

$$\|c\|_{\sigma\omega} + \|\omega_0(a)\|_{\sigma\omega} = \frac{2\gamma}{\eta_\omega} = \frac{2}{\eta_\omega}\|ac\|_\omega,$$

$$\|d\|_{\sigma\omega} + \|\omega_0(a)\|_{\sigma\omega} = \frac{2\delta}{\eta_\omega} = \frac{2}{\eta_\omega}\|ad\|_\omega.$$

Now given $g \in G_\omega$, write it as a word of minimal norm as $g = a^? x_1 \cdots a x_\ell a^?$, with $x_i \in \{b, c, d\}$, and the $a^?$ mean that the initial and final "a" may be present or absent. Thus, $\|g\|_\omega \geq \|ax_1\|_\omega + \cdots + \|ax_\ell\|_\omega - \|a\|_\omega$. On the other hand, each "a" in the expression of g contributes nothing to g_0 and g_1, while each "x_i" contributes an "x_i" and a "$\omega_0(x_i)$" to g_0 and g_1, in some order. Summing together the inequalities above gives the claimed (1.14).

The second claim follows, because the extra "$\|a\|_\omega$" term occurs only if g both starts and ends with a letter in $\{b, c, d\}$, and this case can be prevented by conjugating g by its last letter. □

Lemma 5.8 can be used to prove statements on G_ω by induction. For example,

Proposition 5.9. *If $\omega \in \Omega'$, then the action of G_ω on the tree $\mathcal{T} = X^*$ is faithful, and it is transitive on each orbit X^i. In particular, G_ω is infinite.*

Proof. We first use induction on $\| \cdot \|_\omega$, simultaneously on all $\omega \in \Omega'$, to show that if $g \in G_\omega$ acts trivially on $\mathcal{T}$, then $g = 1$.

The induction starts by noting that the generators b, c, d act non-trivially by our assumption that ω contains infinitely many $0, 1, 2$. Indeed, without loss of generality consider b; let $k \in \mathbb{N}$ be minimal such that $\omega_k(b) \neq 1$; then b acts non-trivially on X^{k+1}.

Consider then $g \in G_\omega$ acting trivially on $\mathcal{T}$. In particular, g fixes X, so $\phi_\omega(g) = \langle\!\langle g_0, g_1 \rangle\!\rangle$. By Lemma 5.8, both g_0 and g_1 are shorter, so by induction they are trivial; thus $g = 1$ because ϕ_ω is injective.

To check that G_ω acts transitively on X^i, it suffices to show that the stabilizer H of $\mathbf{1}$ acts transitively on $X^{i-1}\mathbf{1}$; because then $X^{i-1}\mathbf{1}G_\omega = X^{i-1}\mathbf{1}\langle a \rangle = X^i$. Now H contains b, c, d, y^a for a letter $y \in \{b, c, d\}$ such that $\omega(y) = a$; and the action of b, c, d, y^a on $w\mathbf{1}$ is $wb\mathbf{1}, wc\mathbf{1}, wd\mathbf{1}, wa\mathbf{1}$ respectively, so that the H-orbit of $w\mathbf{1}$ is $W\mathbf{1}$ for a $G_{\sigma\omega}$-orbit $W \subseteq X^{i-1}$. Again we are done by induction. □

Note that the action of G_ω is still faithful if ω only contains infinitely many of two symbols; it is *not* faithful if ω contains finitely many of two symbols.

5.3. The G_ω Are Infinite Torsion Groups

One of Burnside's questions [13] asks whether there are infinite, finitely generated groups in which every element has finite order. The first such examples were constructed by Golod [26]; here, we show that the groups G_ω are other examples:

Theorem 5.10. *If $\omega \in \Omega'$, then G_ω is an infinite torsion 2-group.*

Proof. The group G_ω is infinite by Proposition 5.9. We prove the claim by induction on $\|g\|$. It is easy to check that the generators a, b, c, d all have order 2. Consider $g \in G_\omega$, with $\phi_\omega(g) = \langle\!\langle g_0, g_1 \rangle\!\rangle \pi$. If $\pi = ()$, then g_0, g_1 are shorter than g by Lemma 5.8 so have finite order, say $2^{n_0}, 2^{n_1}$ respectively. Then g has order $2^{\max\{n_0, n_1\}}$. If $\pi = (0, 1)$, then $g^2 = \langle\!\langle g_0 g_1, g_1 g_0 \rangle\!\rangle$, and $g_0 g_1$ is shorter than g again by Lemma 5.8, so has finite order, say 2^n. Then g has order 2^{n+1}. □

Note that if ω contains finitely many copies of a symbol, then G_ω is not a torsion group anymore. In fact, suppose that ω contains no 0; then $\omega_i(b) = a$ for all i, and the element ab has infinite order, since $\phi_\omega((ab)^{2n}) = \langle\!\langle (ba)^n, (ab)^n \rangle\!\rangle$ for all $n \in \mathbb{Z}$.

We also note that the groups G_ω closely resemble infinitely iterated wreath products; namely the map $\phi_\omega \colon G_\omega \to G_{\sigma\omega} \wr \mathrm{Sym}(2)$ is almost an isomorphism:

Definition 5.11. Let $(G_\omega)_{\omega \in \Omega}$ be a similar sequence of groups. It is called *branched* if every G_ω has a finite-index subgroup K_ω such that

$$K_{\sigma\omega}^{X_\omega} \leq \phi_\omega(K_\omega).$$

If the subgroups K_ω are merely required to be non-trivial, then (G_ω) is called *weakly branched*.

Every group in a (weakly) branched family of groups is also called *(weakly) branched*.   △

Proposition 5.12. *The groups G_ω are branched for all $\omega \in \Omega'$.*

Proof. For each $\omega \in \Omega'$, let $x_\omega \in \{b, c, d\}$ be such that $\omega(x_\omega) = 1$, and set $K_\omega = \langle [x_\omega, a] \rangle^{G_\omega}$. Choose also $y_\omega \in \{b, c, d\} \setminus \{x_\omega\}$. Then $1 \times K_{\sigma\omega}$ is normally generated by $\langle\!\langle 1, [x_{\sigma\omega}, a] \rangle\!\rangle$, and

$$\langle\!\langle 1, [x_{\sigma\omega}, a] \rangle\!\rangle = \phi_\omega([x_\omega, y_\omega^a]) = \phi_\omega([x_\omega, a][x_\omega, a]^{a y_\omega}) \in \phi_\omega(K_\omega);$$

the same computation holds for $K_{\sigma\omega} \times 1$.

We now show that the groups K_ω have a finite index. Consider first the quotient $G_\omega / \langle x_\omega \rangle^{G_\omega}$. This group is generated by two involutions a and y_ω, so is a finite dihedral group, because G_ω is torsion. It follows that $\langle x_\omega \rangle^{G_\omega}$ has finite index in G_ω. Then $\langle x_\omega \rangle^{G_\omega} / K_\omega = \langle x_\omega \rangle K_\omega$ has order 2, so K_ω also has a finite index. □

Proposition 5.13. *If G is a p-torsion weakly branched group, then it contains $L = \bigcup \wr^i C_p$ as a subgroup.*

Sketch of Proof. Let $(G_\omega)_{\omega \in \Omega}$ be a similar family of groups, with $G = G_\omega$, and let $K_\omega \leq G_\omega$ be the subgroups given by the condition that $(G_\omega)_{\omega \in \Omega}$ is branched. For each $\omega \in \Omega$, let $g_\omega \in K_\omega$ be an element of order p, and let $n(\omega) \in \mathbb{N}$ be such that g_ω acts non-trivially on $X_{\sigma^{n(\omega)-1}\omega} \times \cdots \times X_\omega$; let $v_\omega \in X_{\sigma^{n(\omega)-1}\omega} \times \cdots \times X_\omega$ be a point on a non-trivial orbit.

Define then simultaneously and recursively $L_\omega = \langle L_{\sigma^n(\omega)_\omega}@v_\omega, g_\omega\rangle$. It contains an element g_ω permuting p copies of $L_{\sigma^n(\omega)_\omega}$, so is isomorphic to L. $\qquad\square$

If G contains no torsion, or torsion of different primes, analogous (but harder-to-state) results hold. In particular, by Exercise 4.10, branched groups satisfy no law. This recovers a result by Abért [1].

5.4. Lower Growth Estimates for G_ω

The proof of Theorem 5.1 requires upper and lower bounds on the growth function of G_ω. A lower bound is easily provided by Theorem 5.2: the group G_ω is infinite (Proposition 5.9) and torsion (Theorem 5.10), so cannot be virtually nilpotent, since nilpotent groups with finite-order generators are finite. However, in some cases a direct argument also gives the bound $v_{G_\omega}(R) \succsim \exp(\sqrt{R})$.

Define indeed maps $\tilde\theta_\omega \colon F \to F$ by

$$\begin{cases} \tilde\theta_\omega(a) &= aya \text{ if } \omega_0 = \cdots = \omega_{k-1} \neq \omega_k, \text{ and } \omega_0(y) = a, \ \omega_k(y) = 1, \\ \tilde\theta_\omega(x) &= x \text{ for all } x \in \{b, c, d\}. \end{cases}$$

(1.16)

A direct calculation shows that $\tilde\theta_\omega$ induces a map $\theta_\omega \colon G_{\sigma\omega} \to G_\omega$, with

$$\phi_\omega(\theta_\omega(g)) = \bigg\downarrow_* \bigg\downarrow_g$$

for some $* \in \langle a, y\rangle$ using the notation introduced in (1.16). Furthermore, $\langle a, y\rangle$ is a dihedral group of order 2^{k+2}.

Exercise 5.14. Prove that θ_ω is a group homomorphism.

Let us now **cheat**, and assume that the element $*$ is always trivial, rather than an element of a finite group of order 2^{k+2}. If k is bounded, this is unimportant; however, if k is unbounded then an additional argument is really required.

Denote by $B_\omega(R)$ the ball of radius R in G_ω, and abbreviate $v_\omega(R) = v_{G_\omega}(R)$. Consider the map $G_{\sigma\omega}^2 \to G_\omega$ given by $(g_0, g_1) \mapsto \theta_\omega(g_0)^a \cdot \theta_\omega(g_1)$. For R_0, R_1 even, it defines an injective map $B_{\sigma\omega}(R_0) \times B_{\sigma\omega}(R_1) \to B_\omega(2(R_0+R_1))$, hence $v_{\sigma\omega}(R)^2 \leq v_\omega(4R)$ for all R even. We conclude $v_\omega(2 \cdot 4^k) \geq v_{\sigma^k\omega}(2)^{2^k} \geq 5^{2^k}$, so $v_\omega(R) \geq \exp(\frac{1}{2}\log 5\sqrt{R})$.

5.5. Upper Growth Estimates for G_ω

We are ready to give an upper bound on the growth of G_ω. As before, we abbreviate $v_\omega(R) = v_{G_\omega}(R)$ for the growth function of G_ω. We start by a

Lemma 5.15. *For every $A > 0$, there is $B \in \mathbb{N}$ such that, for all $\omega \in \Omega'$, we have $v_\omega(A\mu_\omega) \leq B$.*

Proof. It suffices to bound the number of elements of G_ω whose minimal expression has length $\leq A\mu_\omega$ and has the form $g = ax_1 \cdots ax_\ell$ for some $x_i \in \{b, c, d\}$, since there are at most eight times more elements of norm $\leq A\mu_\omega$, namely all $\{1, b, c, d\}g\{1, a\}$. Now by definition g has norm at least $\ell\mu_\omega$, and there are at most 3^ℓ such gs, so one may take $B = 8 \cdot 3^A$. $\quad\square$

Lemma 5.16. *For all $\omega \in \Omega'$ we have $\eta_\omega \mu_{\sigma\omega} \geq \mu_\omega + \|a\|_\omega$.*

Proof. Write $p_\omega = (\beta, \gamma, \delta)$, and assume without loss of generality $\omega_0 = 0$ so $\mu_{\sigma\omega} = 2/\eta_\omega \min\{\gamma, \delta\}$. Consider the six orderings $\beta < \gamma < \delta$ etc. in turn to check the inequalities

$$\eta_\omega \mu_{\sigma\omega} = 2\min\{\gamma, \delta\} \geq \min\{\beta, \gamma, \delta\} + (1 - \max\{\beta, \gamma, \delta\}) = \mu_\omega + \|a\|_\omega. \quad\square$$

Lemma 5.17. *Let f be a positive sublinear function, namely $f(n)/n \to 0$ as $n \to \infty$. Then f is bounded from above by a concave sublinear function.*

Proof. For every $\theta \in (0, 1)$, let n_θ be such that $f(n) - \theta n$ is maximal. Given $n \in \mathbb{R}$, let $\zeta < \theta$ be such that $n \in [n_\theta, n_\zeta]$ with maximal ζ and minimal θ, and define $\overline{f}(n)$ on $[n_\theta, n_\zeta]$ by linear interpolation between $(n_\theta, f(n_\theta))$ and $(n_\zeta, f(n_\zeta))$. Clearly $\overline{f} \geq f$, and $\overline{f}(n)/n$ is decreasing and coincides infinitely often with $f(n)/n$, so it converges to 0. $\quad\square$

Proposition 5.18. *There is an absolute constant B such that, for all $\omega \in \Omega'$ and all $k \in \mathbb{N}$,*

$$v_\omega(\eta_\omega \cdots \eta_{\sigma^{k-1}\omega}\mu_{\sigma^k\omega}) \leq B^{2^k}.$$

Proof. We *cheat*, in assuming that the functions $v_\omega(R)$ are log-concave, i.e. satisfy $v_\omega(R_0)v_\omega(R_1) \leq v_\omega((R_0 + R_1)/2)^2$. This assumption is in fact harmless, since each function v_ω may be replaced by its log-concave majorand: the smallest log-concave function that is pointwise larger than v_ω, given by Lemma 5.17. For details, see [5, §3.1].

For all $i \in \mathbb{N}$, set $\alpha_i(R) = 18(R + 2)v_{\sigma^i\omega}(R + \mu_{\sigma^i\omega})$.

The proposition will follow once we have proven the inequalities $\alpha_i(R) \leq \alpha_{i+1}(R/\eta_{\sigma^i\omega})^2$ for all i, R; because then

$$v_\omega(\eta_\omega \cdots \eta_{\sigma^{k-1}\omega}\mu_{\sigma^k\omega}) \leq \alpha_0(\eta_\omega \cdots \eta_{\sigma^{k-1}\omega}\mu_{\sigma^k\omega})$$
$$\leq \alpha_1(\eta_{\sigma\omega} \cdots \eta_{\sigma^{k-1}\omega}\mu_{\sigma^k\omega})^2$$
$$\leq \cdots \leq \alpha_k(\mu_k)^{2^k} \leq B^{2^k} \text{ by Lemma 5.15.}$$

To simplify notation, we consider only the case $i = 0$, as all cases are the same. We have

$$\alpha_0(R) = 18(R+2)v_\omega(R+\mu_\omega)$$

$$\leq 18(R+2) \sum_{R_0+R_1 \leq \frac{2}{\eta_\omega}(R+\mu_\omega+\|a\|_\omega)} 2v_{\sigma\omega}(R_0)v_{\sigma\omega}(R_1)$$

$$\leq 36(R+2)^2 \max_{R_0+R_1 \leq \frac{2}{\eta_\omega}(R+\mu_\omega+\|a\|_\omega)} v_{\sigma\omega}(R_0)v_{\sigma\omega}(R_1)$$

$$\leq 6^2(3/\eta_\omega)^2(R+2)^2 v_{\sigma\omega}((R+\mu_\omega+\|a\|_\omega)/\eta_\omega)^2$$

$$\leq \left(18(R+2)/\eta_\omega\, v_{\sigma\omega}(R/\eta_\omega+\mu_{\sigma\omega})\right)^2$$

$$\leq \alpha_1(R/\eta_\omega)^2. \qquad\qquad \square$$

We are now ready to conclude the proof of Theorem 5.1, by showing that the groups G_ω have subexponential growth. There are in fact different methods for this. Let λ_ω be the exponential growth rate of G_ω, as in (1.5); we are to show $\lambda_\omega = 1$.

Let us first assume that ω contains infinitely many $0,1,2$, so there are infinitely many positions $k \in \mathbb{N}$ with $\omega_k = 0$ and with 1 and 2 in ω between these positions. For these k, the point $p_{\sigma^k\omega} \in \Delta$ belongs to the subsimplex $\{\beta < \gamma \wedge \beta < \delta\}$. Thus, $\beta < \frac{1}{3}$ on that subsimplex, $\eta_{\sigma^k\omega} = 3 - 2\beta > 7/3$ is uniformly bounded away from 2, and $\gamma, \delta > \frac{1}{6}$ so $\mu_{\sigma^{k+1}\omega} > \frac{1}{18}$. Thus, by Proposition 5.18

$$\log \lambda_\omega = \lim \frac{\log v_\omega(R)}{R} \leq \liminf \frac{2^k \log B}{\eta_\omega \cdots \eta_{\sigma^{k-1}\omega}\mu_{\sigma^k\omega}} = 0$$

since on a subsequence the $\mu_{\sigma^k\omega}$ are bounded away from 0, all terms $2/\eta_{\sigma^i\omega}$ are bounded by 1, and infinitely many of them are bounded by $6/7$. If ω is not ultimately periodic, then $\mu_{\sigma^k\omega} \to 0$, but the convergence is slow enough (actually, at worst linear) that the same argument applies.

A more "abstract" proof may be obtained by noting that the map $\omega \mapsto \lambda_\omega$ is continuous and bounded by 3, and that the proof of Proposition 5.18 gives $\log \lambda_\omega \leq 2/\eta_\omega \log \lambda_{\sigma\omega}$. Because the action of σ on Ω' is ergodic, we must have $\log \lambda_\omega = 0$ for all $\omega \in \Omega'$.

Let us compute more precisely an upper bound for the growth of the first Grigorchuk group $G_{(012)^\infty}$. Given that the sequence $\omega = (012)^\infty$ is 3-periodic, we can find $p_\omega \in \Delta$ explicitly. The calculation is made even simpler by noting that $p_{\sigma\omega}$ and $p_{\sigma^2\omega}$ are cyclic permutations of p_ω; thus p_ω is the normalised eigenvector of $M_0 \cdot \begin{pmatrix} 0 & 0 & 1 \\ 1 & 0 & 0 \\ 0 & 1 & 0 \end{pmatrix}$, and its spectral radius is $\eta_+ \approx 2.46$, the positive root of the characteristic polynomial $T^3 - T^2 - 2T - 4$. Thus, for the first Grigorchuk group we get $v_\omega(\eta_+^k\mu_\omega) \leq B^{2^k}$ for all k, and therefore

$$v_\omega(R) \precsim \exp(R^{\log 2/\log \eta_+}) \approx \exp(R^{0.76}).$$

6.   Growth of Permutational Wreath Products

The upper and lower bounds on the growth of G_ω are both of inter-mediate type $\exp(R^\alpha)$, but do not match. We consider, in this section, permutational wreath products based on the groups G_ω.

Choose a sequence $\omega \in \Omega'$ and a ray $\xi \in \{0,1\}^\infty$, and consider the ray's orbit $X = \xi\, G_\omega$. Choose a group H. Set then

$$W_\omega(H) := H \wr_X G_\omega.$$

(Even though the notation does not make it clear, the group $W_\omega(H)$ depends on ξ.) We shall show, in this section:

Theorem 6.1.   *If H has subexponential growth, then so does $W_\omega(H)$.*

Theorem 6.2.   *Let $\eta_+ \approx 2.46$ be the positive root of $T^3 - T^2 - 2T - 4$. Let $f\colon \mathbb{R}_+ \to \mathbb{R}_+$ be a function satisfying*

$$f(2R) \le f(R)^2 \le f(\eta_+ R) \text{ for all } R \text{ large enough.} \qquad (1.17)$$

Then there exists $\omega \in \Omega'$ such that $v_{W_\omega(C_2)} \sim f$.

We will give more illustrations of the growth functions that may occur in §6.3. We content ourselves with the following:

Theorem 6.3.   *For any finite group H, the group $W_{012}(H)$ has growth*

$$v_{W_{012}(H)} \sim \exp(R^{\log 2/\log \eta_+}).$$

The proofs of Theorems 6.1 and 6.2 rely on estimates of the *support* $\subset X$ of an element of $W_\omega(H)$ of norm $\le R$. Recall that every element of $W_\omega(H)$ may be written in the form cg with $c\colon X \to H$ and $g \in G_\omega$; its support is $\{x \in X : c(x) \ne 1\}$. To better understand the support of elements of $W_\omega(H)$, let us introduce the following:

Definition 6.4.   Let G be a group acting on the right on a set X with basepoint ξ. For a word $w = w_1 \dots w_\ell \in G^*$, its *inverted orbit* is the set

$$\mathcal{O}(w) = \{\xi\, w_{i+1} \cdots w_\ell : 0 \le i < \ell\}.$$

If, furthermore, G is given with a metric $\|\cdot\|$, then its *inverted orbit growth* is the function $\Delta\colon \mathbb{R}_+ \to \mathbb{N}$ given by

$$\Delta(R) = \max\{\#\mathcal{O}(w) : \|w\| \le R\}. \qquad \triangle$$

We write $\mathcal{O}_\omega(w)$ and $\Delta_\omega(R)$ in the case of $G = G_\omega$ with its metric $\|\cdot\|_\omega$. Thus, for example, taking $\xi = \mathbf{1}^\infty$, the inverted orbit of $acadab$ is

$$\mathcal{O}(acadab) = \{\xi\, acadab, \xi\, cadab, \xi\, adab, \xi\, dab, \xi\, ab, \xi\, b\} = \{\overline{\mathbf{1}010}, \overline{\mathbf{1}}, \overline{\mathbf{1}00}\}$$

see the Schreier graph at the end of §4.

Note that a basepoint ξ is implicit in the definitions; yet,

Exercise 6.5. Assume that G acts transitively on X. Show that $\Delta(R)$ depends only mildly of the choice of ξ, in the following sense: if $\xi, \xi' \in X$ are two choices of a basepoint and $\Delta(R), \Delta'(R)$ are the corresponding inverted orbit growth functions, then there exists a constant $C \in \mathbb{R}$ such that $\Delta(R) \leq \Delta'(R + C)$ and $\Delta'(R) \leq \Delta(R + C)$ for all R.

Proposition 6.6. *There is a universal constant C such that, for all $\omega \in \Omega'$ and all $k \in \mathbb{N}$,*

$$2^k \leq \Delta_\omega(\eta_\omega \cdots \eta_{\sigma^{k-1}\omega}\mu_{\sigma^k\omega}) \leq C\,2^k.$$

Proof. For the upper bound, we note that Lemma 5.8 applies just as well to the group G_ω as to the monoid

$$R := S^*/\{b^2 = c^2 = d^2 = 1, bc = d, cd = b, db = c\},$$

see Equation (1.11). Indeed, in a minimal-length representative of an element of R, the number of "a" is at least the number of b, c, d-letters minus one, and this is the only property required for Lemma 5.8. Now given a word $w \in S^*$, its inverted orbit may be read from the image of w in R. Every element of R has a unique *reduced form*: the reduced form of a word $w \in S^*$ is the word $\overline{w} \in S^*$ obtained by replacing every subword equal to a left-hand side of a relation by the corresponding right-hand side.

Without loss of generality and merely at the cost of increasing the constant C, we may suppose $\xi = 1^\infty$. We claim that the inverted orbit of a word $w \in S^*$ coincides with the inverted orbit of its reduction $\overline{w}$. To see this, consider $w = w_1 \ldots w_\ell \in S^*$, a subword $w_j w_{j+1}$ equal to a left-hand side of a relation, and the word w' obtained by replacing $w_j w_{j+1}$ by the right-hand side of the relation. All terms $\xi w_{i+1} \ldots w_\ell$ with $i \neq j$ clearly appear both in $\mathcal{O}(w)$ and $\mathcal{O}(w')$. For the remaining term in $\mathcal{O}(w)$, we have $\xi w_{i+1} \ldots w_\ell = \xi w_{i+2} \ldots w_\ell$ because w_{i+1} fixes ξ, so this term also belongs to $\mathcal{O}(w')$.

If $w \in F$ satisfies $\phi_\omega(w) = \langle\!\langle w_0, w_1 \rangle\!\rangle\pi$ and $\xi = \xi'0$, then

$$\mathcal{O}_\omega(w) \subseteq \mathcal{O}_{\sigma\omega}(w_0)\mathbf{0} \sqcup \mathcal{O}_{\sigma\omega}(w_1)\mathbf{1},$$

where the inverted orbits $\mathcal{O}_{\sigma\omega}$ are computed with respect to the basepoint ξ'; and similarly if $\xi = \xi'\mathbf{1}$. We therefore get

$$\Delta_\omega(R) \leq \max_{R_0 + R_1 \leq 2/\eta_\omega(R + \|a\|_\omega)} \left(\Delta_{\sigma\omega}(R_0) + \Delta_{\sigma\omega}(R_1)\right).$$

The same argument as in Proposition 5.18 finishes the proof of the upper bound.

For the lower bound, it suffices to exhibit for all $k \in \mathbb{N}$ a word of length at most $\eta_\omega \cdots \eta_{\sigma^{k-1}\omega}\mu_{\sigma^k\omega}$ and inverted orbit of size at least 2^k. For that purpose, define self-substitutions ζ_x of $\{ab, ac, ad\}^*$, for $x \in \{0, 1, 2\}$, by

$$\begin{aligned}
\zeta_0: \quad & ab \mapsto adabac, & ac \mapsto acac, & \quad ad \mapsto adad, \\
\zeta_1: \quad & ab \mapsto abab, & ac \mapsto abacad, & \quad ad \mapsto adad, \\
\zeta_2: \quad & ab \mapsto abab, & ac \mapsto acac, & \quad ad \mapsto acadab,
\end{aligned}$$

and note that for any word $w \in \{ab, ac, ad\}^*$ representing an element of F we have

$$\tilde{\phi}_\omega(\zeta_{\omega 0}(w)) = \begin{cases} \xrightarrow[w]{awa} & \text{if } \zeta_{\omega 0}(w) \text{ contains an even number of ``}a\text{'',} \\[2mm] \begin{smallmatrix} a & a \\ \times \\ w & w \end{smallmatrix} & \text{if } \zeta_{\omega 0}(w) \text{ contains an odd number of ``}a\text{''.} \end{cases}$$

In particular, $\zeta_{\omega 0}$ induces a homomorphism $G_{\sigma\omega} \to G_\omega$.

By induction, we see that for any non-trivial $w \in \{ab, ac, ad\}^*$ (representing an element of $G_{\sigma^k \omega}$) we have

$$\Delta_\omega(\zeta_{\omega 0} \cdots \zeta_{\omega_{k-1}}(w)) \geq 2^k.$$

Note then that, if $Z \in \mathbb{N}^3$ count the numbers of ab, ac, ad respectively in w, then $M_x^t Z$ counts the numbers of ab, ac, ad respectively in $\zeta_x(w)$. Indeed without loss of generality consider $x = 0$; then every ab in w contributes one each of ab, ac, ad to $\zeta_x(w)$, while every ac and ad in w contributes two copies of itself to $\zeta_x(w)$.

Let $as \in \{ab, ac, ad\}$ be such that $\|as\|_{\sigma^k \omega}$ is minimal – recall the notation p_ω from (1.15): if $p_{\sigma^k \omega} = (\beta, \gamma, \delta)$ and $\beta \leq \gamma, \delta$ then $s = b$, etc. Let W be the basis vector in $\mathbb{R}^3$ with a "1" at the position that as has in $\{ab, ac, ad\}$: if $\beta \leq \gamma, \delta$ then $W = (1, 0, 0)^t$, etc. Set $w = \zeta_{\omega 0} \cdots \zeta_{\omega_{k-1}}(as)$. We have $\Delta_\omega(w) \geq 2^k$, and

$$\begin{aligned}
\|w\|_\omega &= Z^t p_\omega = W^t M_{\omega_{k-1}} \cdots M_{\omega 0} p_\omega \\
&= \eta_\omega \cdots \eta_{\sigma^{k-1}\omega} W^t p_{\sigma^k \omega} = \eta_\omega \cdots \eta_{\sigma^{k-1}\omega} \mu_{\sigma^k \omega}. \qquad \square
\end{aligned}$$

Finally, let us introduce the "choice of inverted orbits growth" function, first generally for a group G, with given metric $\|\cdot\|$, acting on a set X with basepoint ξ:

$$\Sigma(R) := \#\{\mathcal{O}(w) : \|w\| \leq R\}.$$

This function counts the number of subsets that may occur as an inverted orbit of a word of length at most R. Because $\mathcal{O}(w)$ is a subset of cardinality at most $R + 1$ of the Schreier graph X, and furthermore lies in the ball of radius R about ξ in X, we get the crude estimate $\Sigma(R) \leq \binom{v_{X, \xi}(R) + R}{R}$ based on the growth function $v_{X, \xi}$ of balls centered at ξ in the graph X. However, in the particular case of the groups G_ω, we can do better:

Proposition 6.7. *There is an absolute constant D such that for all $\omega \in \Omega'$ and all $k \in \mathbb{N}$ we have*

$$\Sigma_\omega(\eta_\omega \cdots \eta_{\sigma^{k-1}\omega} \mu_{\sigma^k \omega}) \leq D^{2^k}.$$

Proof. Consider $w \in G_\omega^*$ with $\tilde\phi_\omega(w) = \langle\!\langle w_0, w_1 \rangle\!\rangle$. The inverted orbit of w is determined by the inverted orbits of w_0 and w_1, two words of total $\sigma\omega$-length at most $2/\eta_\omega(\|w\|_\omega + \|a\|_\omega)$ by Lemma 5.8. Therefore,

$$\Sigma_\omega(\eta_\omega R) \leq \sum_{R_0 + R_1 \leq \frac{2}{\eta_\omega}(R + \|a\|_\omega)} \Sigma_{\sigma\omega}(R_0)\Sigma_{\sigma\omega}(R_1),$$

and the same argument as in Proposition 5.18 applies. $\qquad\square$

6.1. The Growth of $W_\omega(H)$

We start by general estimates on the growth of a permutational wreath product:

Proposition 6.8. *Let H be a group with growth function v_H, and suppose that v_H is log-concave.*

Let G be a group acting transitively on a set X with basepoint ξ, and let v_G denote the growth function of G. Denote the inverted orbit growth of G on (X, ξ) by Δ, and denote its inverted orbit choice growth by Σ.

Consider the wreath product $W = H \wr_X G$, generated by $S \cup T@\xi$ for the generating sets S, T of G, H respectively. Then

$$v_G(R)v_H(R/\Delta(R))^{\Delta(R)} \leq v_W(3R),$$

$$v_W(R) \leq v_G(R)v_H(R/\Delta(R))^{\Delta(R)}(2R)^{\Delta(R)}\Sigma(R),$$

$$v_W(R) \leq v_G(R)(\#H)^{\Delta(R)}\Sigma(R) \text{ if } H \text{ is finite.}$$

Note that the assumption that v_H be log-concave is mild, owing to Lemma 5.17.

Proof. We begin by the lower bound. For every $R \in \mathbb{N}$, consider a word $w \in G^*$ of norm $\leq R$ realizing the maximum $\Delta(R)$; write $\mathcal{O}(w) = \{x_1, \ldots, x_k\}$ for $k = \Delta(R)$. Choose then k elements $a_1, \ldots, a_k$ of norm $\leq R/k$ in A. Define $f \in \sum_X H$ by $f(x_i) = a_i$, all unspecified values being 1. Then $wf \in W$ may be expressed as a word of norm $R + |a_1| + \cdots + |a_k| \leq 2R$ in the standard generators of W, by inserting $a_1@\xi, \ldots, a_k@\xi$ appropriately into the word w.

Furthermore, different choices of a_i yield different elements of W. Finally multiplying wf with an arbitrary $g \in G$ of length at most R, we obtain $v_G(R)v_H(R/k)^k$ elements in the ball of radius $3R$ in W.

For the upper bound, consider a word w of norm R in W, and let $f \in \sum_X H$ denote its value in the base of the wreath product. The support of f has cardinality at most $\Delta(R)$, and may take at most $\Sigma(R)$ values.

Write then $\sup(f) = \{x_1, \ldots, x_k\}$ for some $k \leq \Delta(R)$, and let $a_1, \ldots, a_k \in H$ be the values of w at its support; write $\ell_i = \|a_i\|$.

Since $\sum \ell_i \leq R$, the norms of the different elements on the support of f define a composition of a number not greater than R into at most k

summands; such a composition is determined by k "marked positions" among $R+k$, so there are at most $\binom{R+k-1}{k}$ possibilities, which we bound crudely by $(2R)^k$. Each of the a_i is then chosen among $v_H(\ell_i)$ elements, and (by the assumption that v_H is log-concave) there are $\prod v_H(\ell_i) \leq v_H(R/k)^k$ total choices for the elements in H.

We have now decomposed w into data that specify it uniquely, and we multiply the different possibilities for each of the pieces of data. Counting the possibilities for the value of w in G, the possibilities for its support in X, and the possibilities for the elements in H on its support, we get

$$v_W(R) \precsim v_G(R)v_A(R/k)^k(2R)^k\Sigma(R),$$

which is maximised by $k = \Delta(R)$.

Finally, if H is finite, then we may more simply bound the possible values of f by $(\#H)^k$. □

Corollary 6.9. *Let G, H be groups of subexponential growth. Let G act transitively on a set X with basepoint ξ, with sublinear inverted orbit growth and subexponential inverted orbit choice growth. Then the wreath product $W = H \wr_X G$ has subexponential growth.* □

Proof of Theorem 6.1. Assume that H has subexponential growth; then by Lemma 5.17 there exists a log-concave subexponentially growing function v_H bounding the growth of H from above.

By Propositions 6.6 and 6.6, the function Δ is sublinear and the function Σ is subexponential. By Proposition 5.18, the growth of G_ω is subexponential. Corollary 6.9 then shows that $W_\omega(H)$ has subexponential growth. □

In the special case of H finite and $G = G_\omega$, Proposition 6.8 gives the

Corollary 6.10. *Let H be a non-trivial finite group. There are then two absolute constants $F, E > 1$ such that the growth function v of $W_\omega(H) = H \wr_X G_\omega$ satisfies*

$$E^{2^k} \leq v(\eta_\omega \cdots \eta_{\sigma^{k-1}\omega}\mu_{\sigma^k\omega}) \leq F^{2^k}.$$

Proof. Take together the upper bound on the growth of G_ω from Proposition 5.18, the bounds on the inverted orbit growth from Proposition 6.6, and the choices for the inverted orbits from Proposition 6.7. The conclusion follows from Proposition 6.8. □

Proof of Theorem 6.3. This follows directly from Corollary 6.10, using the fact that $\eta_{\sigma^i\omega} = \eta_+$ for all $i \in \mathbb{N}$. □

6.2. Proof of Theorem 6.2

Our approach will be to construct, out of the function f satisfying (1.17), a sequence $\omega \in \{2, 012\}^\infty$ with long stretches of 2 when f grows fast, and long stretches of 012 when f grows slowly.

We start by introducing some shorthand notation. For a finite sequence $\omega = \omega_0 \ldots \omega_{n-1} \in \{0,1,2\}^n$ and $p \in \Delta$, we write by extension

$$\overline{M}_\omega = \overline{M}_{\omega_0} \cdots \overline{M}_{\omega_{n-1}} : \Delta \circlearrowleft$$

and

$$\eta(p, \omega_0 \ldots \omega_{n-1}) = \eta(p, \omega_0)\eta(\overline{M}_{\omega_0}p, \omega_1) \cdots \eta(\overline{M}_{\omega_0 \ldots \omega_{n-2}}p, \omega_{n-1}).$$

For $\omega \in \{0,1,2\}^{\mathbb{Z}}$, recall the construction of $p_\omega \in \Delta$ from (1.15), and $\eta_\omega = \eta(p_\omega, \omega_0)$ and $\mu_\omega = \mu(p_\omega)$. Assume that a sequence $\omega \in \{0,1,2\}^{\mathbb{Z}}$ is under construction, and that there exists $k \in \mathbb{N}$ such that ω_i has been determined for all $i \leq k$. Then p_ω, η_ω and μ_ω are determined, and so are $p_{\sigma^i \omega}, \eta_{\sigma^i \omega}, \mu_{\sigma^i \omega}$ for all $i \leq k$. We abbreviate

$$p_i = p_{\sigma^i \omega}, \qquad \eta_i = \eta_{\sigma^i \omega}, \qquad \mu_{\sigma^i \omega}.$$

Lemma 6.11. *If the restriction to $\mathbb{N}$ of the sequence ω has the form*

$$\omega = (012)^{i_1} 2^{j_1} (012)^{i_2} 2^{j_2} (012)^{i_3} 2^{j_3} \ldots, \tag{1.18}$$

with $i_1, j_1, i_2, j_2, \cdots \geq 1$, then the μ_k are all bounded away from 0.

Proof. In fact, the image of $\overline{M}_{012}$ is the open triangle spanned by $(\frac{1}{3}, \frac{1}{3}, \frac{1}{3})$, $(\frac{2}{7}, \frac{2}{7}, \frac{3}{7})$ and $(\frac{4}{17}, \frac{6}{17}, \frac{7}{17})$, so after each 012 the μ_k belongs to $(\frac{4}{17}, \frac{1}{3})$.

The image of that triangle under $\overline{M}_{2^i}$ is contained in the convex quadrilateral spanned by $(\frac{1}{3}, \frac{1}{3}, \frac{1}{3})$, $(\frac{4}{17}, \frac{6}{17}, \frac{7}{17})$, $(\frac{1}{4}, \frac{1}{4}, \frac{1}{2})$ and $(\frac{1}{5}, \frac{3}{10}, \frac{1}{2})$, so $\mu_k \in (\frac{1}{5}, \frac{1}{3})$ for all k. $\qquad\square$

The heart of the argument is the following lemma, which shows that η approaches very quickly its limiting values 2 and η_+ as the sequence ω contains long segments of 2 or of 012:

Lemma 6.12. *There exist constants $A' \leq 1$, $B' \geq 1$ such that,*

1. *For all $p \in \Delta$ and all $n \in \mathbb{N}$,*

$$\eta(p, (012)^n) \geq \eta_+^{3n} A';$$

2. *For all $p \in \Delta$ and all $n \in \mathbb{N}$,*

$$\eta(p, 2^n) \leq 2^n B'.$$

Proof. Let $\mathcal{U}$ denote the image of $\overline{M}_{012}$, and let $p_+ \in \mathcal{U}$ denote the fixed point of $\overline{M}_{012}$. Note first that $\eta(-, 012)$ is differentiable at p_+, and that $\overline{M}_{012}$ is uniformly contracting on $\mathcal{U}$; let $\rho < 1$ be such that $\overline{M}_{012}$ is ρ-Lipschitz on $\mathcal{U}$, and let D be an upper bound for the derivative of $\log \eta(-, 012)$ on $\mathcal{U}$. Recall that $\eta_+^3 = \eta(p_+, 012)$.

For all $k \in \mathbb{N}$, write $p_k = \overline{M}_{(012)^k}(p)$. For $k \geq 1$ we have $\|p_k - p_+\| \leq \rho^{k-1}$, so $|\log \eta(p_k, 012) - 3 \log \eta_+| < D\rho^{k-1}$, while for $k = 0$ we write $|\log \eta(p, 012) - 3 \log \eta_+| < 3 \log 3$. Therefore,

$$|\log \eta(p, (012)^n) - 3n \log \eta_+| \leq 3 \log 3 + \sum_{k=1}^{n-1} |\log \eta(p_k, 012) - 3 \log \eta_+|$$
$$\leq 3 \log 3 + D/(1 - \rho)$$

is bounded over all n and p. The estimate (6.12) follows, with $A' = 3^3 \exp(D/(1 - \rho))$.

For the second part, consider $p_k = \overline{M}_{2^k}(p)$, and note that p_k converges at exponential (Euclidean) speed to a point p_∞ on the side $\{\delta = \frac{1}{2}\}$ of $\partial\Delta$, since $(\partial\overline{M}_2/\partial\delta)(*, *, \frac{1}{2}) = \frac{1}{2}$; so we have $\|p_k - p_\infty\| < \rho^{k-1}$ for some $\rho < 1$. As above, $\eta(-, 2)$ is differentiable in a neighbourhood $\mathcal{U}$ of $\{\delta = \frac{1}{2}\}$, and the derivative of $\log \eta(-, 2)$ is bounded on $\mathcal{U}$, say by D. Recall that $\eta(p, 2) = 2$ for all $p \in \partial\Delta$. The same computation as above yields (6.12), with $B' = 3 \exp(D/(1 - \rho))$. $\square$

We now reformulate the statement of Theorem 6.2 as follows. Set $g(R) = \log f(R)$, so that we have

$$g(2R) \leq 2g(R) \leq g(\eta_+ R) \tag{1.19}$$

for all R large enough. For simplicity (since growth is only an asymptotic property) we assume that (1.19) holds for all R. Without loss of generality (since we are allowed to replace g by an equivalent function), we also assume that g is increasing and satisfies $g(1) = 1$.

We are ready to construct ω. Fix arbitrarily the value of ω on its negative part, say $\omega_{|-\mathbb{N}} = (012)^{-\infty}$. This determines an initial metric $p_0 \in \Delta$. Out of the function g, we will construct a sequence ω such that, for constants A, B, we have

$$A \leq \frac{g(\eta(p_0, \omega_0 \ldots \omega_{k-1}))}{2^k} \leq B \text{ for all } k; \tag{1.20}$$

in fact, it will suffice to obtain this inequality for a set of values $k_0, k_1, \ldots$ of k such that $\sup_i(k_{i+1} - k_i) < \infty$. Indeed, the orbit p_i of p_0 in Δ' will remain bounded, so we will have $\mu(p_i) \in [C, 1]$ for some $C > 0$. By Corollary 6.10,

$$E^{2^k} \leq v(\eta(p_0, \omega_0 \ldots \omega_{k-1})\mu_k) \leq v(Bg^{-1}(2^k)),$$
$$v(ACg^{-1}(2^k)) \leq v(\eta(p_0, \omega_0 \ldots \omega_{k-1})\mu_k) \leq F^{2^k}$$

and therefore $v(R) \sim \exp(g(R)) = f(R)$.

Proof of (1.20). We will extend the sequence ω to be, on the positive integers, of the form (1.18),

$$\omega = (012)^{i_1} 2^{j_1} (012)^{i_2} 2^{j_2} (012)^{i_3} 2^{j_3} \ldots,$$

with $i_1, j_1, i_2, j_2, \cdots \geq 1$. The μ_k are bounded away from 0 by Lemma 6.11.

Assuming by induction that $\omega' = \omega_0 \ldots \omega_{k-1}$ has been constructed, we repeat the following:

- while $g(\eta(p_0, \omega')) < 2^k$, we append 012 to ω';
- while $g(\eta(p_0, \omega')) > 2^k$, we append 2 to ω'.

For our induction hypothesis, we assume that the stronger condition

$$\frac{1}{2}2^k \leq g(\eta(p_0, \omega')) \leq 2^k$$

holds for each k of the form $i_1 + j_1 + \cdots + i_m + j_m$, and that

$$2^k \leq g(\eta(p_0, \omega')) \leq 3^3 2^k$$

holds for each k of the form $i_1 + j_1 + \cdots + i_m$; these conditions apply whenever ω is a product of "syllables" $(012)^{i_t}$ and 2^{j_t}.

Consider first the case $\frac{1}{2}2^k \leq g(\eta(p_0, \omega')) \leq 2^k$; and let n be minimal such that $g(\eta(p_0, \omega'(012)^n)) > 2^{k+3n}$. Then, for all $i \in \{1, \ldots, n\}$, Lemma 6.12(6.12) gives $\eta(p_0, \omega'(012)^i) \geq \eta(p_0, \omega')\eta_+^{3i} A'$. Let $u \in \mathbb{N}$ be minimal such that $A' \geq \eta_+^{-u}$; this, combined with $g(\eta_+ R) \geq 2g(R)$, gives

$$g(\eta(p_0, \omega'(012)^i)) \geq g(\eta(p_0, \omega')\eta_+^{3i-u}) \geq 2^{-1-u}2^{k+3i}.$$

By minimality of n, we have $g(\eta(p_0, \omega'(012)^{n-1})) \leq 2^{k+3(n-1)}$; since g is sublinear and $\eta \leq 3$, we get

$$g(\eta(p_0, \omega'(012)^n)) \leq 3^3 2^{k+3n}.$$

Consider then the case $2^k \leq g(\eta(p_0, \omega')) \leq 3^3 2^k$, which is similar, and let n be minimal such that $g(\eta(p_0, \omega'2^n)) < 2^{k+n}$. Then, for all $i \in \{1, \ldots, n\}$, Lemma 6.12(6.12) gives $\eta(p_0, \omega'2^i) \leq \eta(p_0, \omega')2^i B'$; this, combined with $g(2R) \leq 2g(R)$, gives

$$g(\eta(p_0, \omega'2^i)) \leq 3^3 2^{k+i} B'.$$

By minimality of n, we have $g(\eta(p_0, \omega'2^{n-1})) \geq 2^{k+n-1}$; because g is increasing, we get

$$g(\eta(p_0, \omega'2^n)) \geq \frac{1}{2}2^{k+n}.$$

We have proved the claim (1.20), with $A = 2^{-1-u}$ and $B = 3^3 B'$.   □

Remark 6.13. The construction of ω from f is algorithmic, in the following sense. The initial point p_0 may be computed to arbitrary precision by an algorithm. If there exists an algorithm that computes values of f, then there is an algorithm that, with $k \in \mathbb{N}$ as input, computes $g(\eta(p_0, \omega_0 \ldots \omega_{k-1}))$ to arbitrary precision; so there is an algorithm that computes the digits of ω.

It then follows via Theorem 1.10 that the groups G_ω and $W_\omega(H) = H \wr_X G_\omega$ are recursively presented, for recursively presented H.

6.3. Illustrations

We now consider illustrations of Theorem 6.2, and examples of growth functions that may occur for groups $W_\omega(C_2)$. In fact, the examples can be constructed in both directions: either choose a "nice" function f that satisfies (1.17), or choose a "nice" sequence ω and estimate the corresponding growth using Corollary 6.10, Lemma 6.11 and Lemma 6.12. We follow both approaches.

- For every $\alpha \in [\log 2/\log \eta_+, 1]$, there is a group of growth $\sim \exp(R^\alpha)$. Furthermore, for a dense set of α in that interval, there exists a periodic sequence ω such that $W_\omega(C_2)$ has growth $\sim \exp(R^\alpha)$.

  For every $\alpha \le \beta \in [\log 2/\log \eta_+, 1)$, one may construct a function f satisfying (1.17) that coincides, on arbitrarily large intervals, sometimes with the function $\exp(R^\alpha)$ and sometimes with the function $\exp(R^\beta)$. Therefore, there is a group whose growth function accumulates both at $\exp(R^\alpha)$ and at $\exp(R^\beta)$. This recovers a result by Brieussel, see [12].

- There exist groups of growth $\sim \exp(R/\log R)$, of growth $\sim \exp(R/\log\log R)$, of growth $\sim \exp(R/\log \cdots \log R)$.

- Consider conversely the sequence $\omega = (012)2^1(012)2^2(012)2^3(012)2^4 \ldots$ Among the first k entries, approximately $\sqrt{k}$ instances of 012 will have been seen; therefore $\eta(p_0, \omega_0 \ldots \omega_{k-1}) \approx 2^{k+\mathcal{O}(1)\sqrt{k}}$. This gives a growth function of the order of

$$\exp\left(R/\exp(\mathcal{O}(1)\sqrt{\log R})\right).$$

  Consider next the sequence $\omega = (012)2^1(012)2^2(012)2^4(012)2^8 \ldots$ Among the first k entries, approximately $\log k$ instances of 012 will have been seen; therefore, $\eta(p_0, \omega_0 \ldots \omega_{k-1}) \approx 2^{k+\mathcal{O}(1)\log k}$. This gives a growth function of the order of

$$\exp\left(R/(\log R)^{\mathcal{O}(1)}\right).$$

  Consider further the sequence $\omega = (012)2^{2^1}(012)2^{2^2}(012)2^{2^4}(012)2^{2^8} \ldots$ Among the first k entries, approximately $\log\log k$ instances of 012 will have been seen; therefore, $\eta(p_0, \omega_0 \ldots \omega_{k-1}) \approx 2^{k+\mathcal{O}(1)\log\log k}$. This gives a growth function of the order of

$$\exp\left(R/(\log\log R)^{\mathcal{O}(1)}\right).$$

  These constructions generalise easily to give "nice" sequences ω such that $W_\omega(C_2)$ has growth of the order of $\exp(R/(\log \cdots \log R)^{\mathcal{O}(1)})$.

- Consider the Ackermann function

$$A(m,n) = \begin{cases} n+1 & \text{if } m = 0, \\ A(m-1,1) & \text{if } m > 0 \text{ and } n = 0, \\ A(m-1, A(m, n-1)) & \text{if } m > 0 \text{ and } n > 0, \end{cases}$$

and set $B(n) = A(n,n)$. Then there exists a group whose growth function is $\sim \exp(R/B(R)^{-1})$; this last growth function is faster than any subexponential primitive-recursive function.

Considering $\omega = (012)2^{A(0,0)}(012)2^{A(1,1)}(012)2^{A(2,2)} \ldots$ gives a group W_ω whose growth function is subexponential, but at least as fast as $\exp(R/B(R)^{-1})$;

7. Imbeddings and Subgroups

A classical result by Higman, Neumann and Neumann [39] states that every countable group imbeds in a finitely generated group. It was then shown that many properties of the group can be inherited by the imbedding: in particular, solvability (Neumann-Neumann [61]), torsion (Phillips [66]), residual finiteness (Wilson [80]) and amenability (Olshansky-Osin [63]).

Seen the other way round, these results show that there is little restriction, apart from being countable, on the subgroups of a finitely generated group; even if that group is furthermore assumed to be residually finite, amenable or solvable.

On the other hand, very strong restrictions exist on the subgroups of a virtually nilpotent group: they are all finitely generated, for example; see Exercise 7.6

Since by Gromov's Theorem 2.11 the finitely generated virtually nilpotent groups are precisely the groups of polynomial growth, we naturally ask what conditions are imposed on the subgroups of a group of subexponential word growth. As we shall see, there are essentially none.

The certainly do not share the above property of virtually nilpotent groups: there are torsion branched groups of subexponential growth such as the group G_ω. They contain iterated wreath products, and as per Proposition 5.13, also contain infinitely generated subgroups.

Let us say that a group has *locally subexponential growth* if all of its finitely generated subgroups have subexponential growth. Clearly, if G has subexponential growth, then all its subgroups have locally subexponential growth. This is the only restriction, and the objective of this section is to prove the following result:

Theorem 7.1 ([5]). *Let B be a group. Then there exists a finitely generated group of subexponential growth in which B imbeds as a subgroup if and only if B is countable and locally of subexponential growth.*

For example, this implies that there is a group of subexponential growth containing $\mathbb{Q}$ as a subgroup.

7.1. Neumann's Proof

By way of motivation, we start with the classical result by Higman and the Neumanns:

Theorem 7.2 (Higman, B.H. Neumann and H. Neumann [39]). *Every countable group imbeds in a finitely generated group.*

We shall not follow the original proof (which proceeds by a sequence of "HNN extensions"), but rather that by the two Neumanns [61], which uses wreath products. It follows immediately from combining the following two propositions:

Proposition 7.3. *Every countable group B imbeds in the commutator subgroup $[G, G]$ of a countable group G.*

Proposition 7.4. *For every countable group G, there exists a 2-generated group W such that $[G, G]$ imbeds in $[W, W]$.*

Proof of Proposition 7.3. Consider the following subgroup G of the unrestricted wreath product $B \wr \mathbb{Z}$. The group G is generated by $\mathbb{Z}$ and, for all $b \in B$, the function $f_b \colon \mathbb{Z} \to B$ defined by $f_b(m) = b^m$. Denoting by t the generator of $\mathbb{Z}$, we see that $[t, f_b]$ is the constant function b; so B is in fact imbedded in $[t, G]$.   □

Exercise 7.5. Could we also have defined $f_b(m) = b$ for $m \geq 0$ and $f_b(m) = 1$ for $m < 0$? What would be the advantages and disadvantages of this alternative construction?

Proof of Proposition 7.4. Consider the following subgroup W of the unrestricted wreath product $G \wr \mathbb{Z}$. Denote by u a generator of $\mathbb{Z}$, and by $\{b_1, b_2, \dots\}$ a generating set of G. Choose a sparse-enough sequence of elements $x_1, x_2, \dots$ of $\mathbb{Z}$, and define $f \colon \mathbb{Z} \to G$ by $f(x_i) = b_i$, all other values being trivial. The group W is then $W = \langle f, u \rangle$.

Let us spell out below what it means to be "sparse enough". We write $\mathbb{Z}$ additively. Because as per Proposition 7.3 we only need to imbed $[t, G]$ in W, we may set $f(0) = t$ and a sufficient condition on the $x_i \in \mathbb{Z}$ is that $x_i \neq 0$ for all i; all x_i are distinct; and $x_i + x_j \notin \{0, x_k\}$ for all $i, j, k \in \mathbb{N}$. One then sees that $[f, f^{u^{-x_i}}]$ is a function supported only at 0, with value $[t, b_i]$ there. This defines the imbedding of $[t, G]$ in W.   □

Note that the group W contains the standard wreath product $B \wr \mathbb{Z}$, so always has exponential growth.

The above construction shows that every countable solvable group imbeds in a 2-generated solvable group.

Exercise 7.6. Show first that *not* every countable nilpotent group imbeds in a 2-generated nilpotent group.

Show then that every finitely generated nilpotent group imbeds in a 2-generated nilpotent group.

Here is a useful, small improvement on Theorem 7.2:

Theorem 7.7.   *Let $p \geq 5$ be an integer. Then every countable group imbeds in a 2-generated group both of whose generators have order p.*

Proof.   Piggybacking on Propositions 7.3 and 7.4, it suffices to consider a 2-generated group $G = \langle x, y \rangle$, and to imbed $[G, G]$ into a 2-generated group W with generators of order p. Write $C_p = \langle t | t^p \rangle$, and define $f : C_p \to G$ by

$$f(1) = x, \quad f(t^{-1}) = y, \quad f(t^2) = y^{-1}x^{-1}, \quad f(g) = 1 \text{ for all other } g \in C_p.$$

Consider then the group $W = \langle t, ft \rangle$. It is clearly generated by two elements of order p, since in $(ft)^p$ all the coordinates contain some cyclic permutation of the product $x \cdot y \cdot y^{-1}x^{-1}$. Now W contains f, so it also contains $[f, f^t]$, which is the function $C_p \to G$ taking value $[x, y]$ at 1 and 1 elsewhere. Furthermore, conjugating $[f, f^t]$ by an arbitrary word in f and f^t, one obtains the function taking value an arbitrary conjugate of $[x, y]$ at 1; so W contains $[G, G]@1$.   □

Exercise 7.8.   Where have we used the assumption that $p \geq 5$? Can you improve the above result to arbitrary $p \geq 3$? Can you imbed any countable group into a group generated by an involution and an element of order p?

7.2. Finite-Valued Permutational Wreath Products

Our goal is, starting from a countable group B locally of subexponential growth, to construct a finitely generated group W of subexponential growth containing B. We take inspiration from Neumann's proof given above, with two modifications: first, we consider permutational wreath products rather than standard ones; secondly, we consider *finite-valued permutational wreath products*:

Definition 7.9.   Let H be a group acting on a set X, and let H be a group. Their *finite-valued permutational wreath product* is the group $H \wr_X^{\text{f.v.}} G$, defined as the extension of functions $X \to H$ with finite image by G:

$$H \wr_X^{\text{f.v.}} G = \{(\phi, g) \in H^X \times G : \#\phi(X) < \infty\}. \qquad \triangle$$

Note that it is a subgroup, because if $(\phi, g)^{-1}(\phi', g') = (\phi'', g^{-1}g')$ then $\phi''(X) \subseteq \phi(X)^{-1}\phi'(X)$ is finite. Clearly, we have

$$H \wr_X G \leq H \wr_X^{\text{f.v.}} G \leq H \wr G.$$

We also introduce a condition on imbeddings that guarantees control on growth:

Definition 7.10. Let B be a group. A group G is called *hyper-B* if it is a directed union of finite extensions of finite powers of B.   △

Clearly, if a group B is locally of subexponential growth and a group G is hyper-B, then G is also locally of subexponential growth. Indeed, for every finite subset S of G there is a finite extension of a finite power of B that contains S.

Lemma 7.11. *Let G be a hyper-B group, and let H be a hyper-G group. Then H is hyper-B.*

Proof. Consider $h \in H$; then h belongs to a finite extension of a finite power of G, which may be assumed of the form $G \wr F$ for a finite group F. Let us write $h = \phi f$ with $\phi \colon F \to G$ and $f \in F$; then $\phi(f)$ belongs for all $f \in F$ to a finite extension of a finite power of B, which can be assumed to be the same for all f. This extension may be assumed to be of the form $B \wr E$ for a finite group E. It follows that h belongs to $B \wr_{E \times F} (E \wr F)$, a finite extension of a finite power of B; so H is hyper-B.   □

Lemma 7.12. *If H is a hyper-B group and U is locally finite, then $H \wr^{\text{f.v.}} U$ is a hyper-B group.*

Proof. We first show that $H \wr^{\text{f.v.}} U$ is hyper-H. By hypothesis, U is a directed union of finite subgroups E. The partitions $\mathscr{P}_0$ of U into finitely many parts also form a directed poset; and for every such partition $\mathscr{P}_0$ and every finite subgroup $E \leq U$ there exists a finite partition $\mathscr{P}$ of U that is invariant under E and refines $\mathscr{P}_0$, namely the wedge (= least upper bound) of all E-images of $\mathscr{P}_0$.

Consider now the directed poset of pairs $(E, \mathscr{P})$ consisting of finite subgroups $E \leq U$ and E-invariant partitions of U. Consider the corresponding subgroups $H^{\mathscr{P}} \rtimes E$ of $H \wr^{\text{f.v.}} U$. If $(E, \mathscr{P}) \leq (E', \mathscr{P}')$ then $H^{\mathscr{P}} \rtimes E$ is naturally contained in $H^{\mathscr{P}'} \rtimes E'$, so these subgroups of $H \wr^{\text{f.v.}} U$ form a directed poset, which exhausts $H \wr^{\text{f.v.}} U$.

It follows that $H \wr^{\text{f.v.}} U$ is a hyper-H group, and we are done by Lemma 7.11.   □

7.3. Imbedding in the Derived Subgroup

Our main goal, in this section, is to prove the following proposition, which replaces Proposition 7.3.

Proposition 7.13. *Let B be a group. Then there exists a hyper-B group G such that $[G, G]$ contains B as a subgroup.*

If B is infinite, then G may further be supposed to have the same cardinality as B.

Lemma 7.14. *Let B be a group. Then there exists a subgroup C of B, containing $[B, B]$, such that B/C is torsion and $C/[B, B]$ is free abelian.*

Proof. $B/[B, B] \otimes_{\mathbb{Z}} \mathbb{Q}$ is a $\mathbb{Q}$-vector space, hence has a basis, call it X. It generates a free abelian group $\mathbb{Z}X$ within $B/[B, B]$, whose full preimage in B we call C. Then $B/C \otimes_{\mathbb{Z}} \mathbb{Q} = 0$ so B/C is torsion. □

We set up the following notation for the proof of Proposition 7.13. We choose a subgroup $C \leq B$ as in Lemma 7.14 and write $T := B/C$. We choose a basis X of $C/[B, B]$, for every $x \in X$ we choose an element $b_x \in C$ representing it, and we define a homomorphism $\theta_x \colon C \to \langle b_x \rangle \subseteq B$, trivial on $[B, B]$, by $\theta_x(b_x) = b_x^{-1}$ and $\theta_x(b_y) = 1$ for all $y \neq x \in X$. In particular, we have for all $b \in C$

$$b \cdot \prod_{x \in X} \theta_x(b) \in [B, B]$$

and the product is finite.

We write $\pi \colon B \to T$ the natural projection, and define a section σ of π with the following property, which we single out as a lemma:

Lemma 7.15. *There exists a set-theoretic section $\sigma \colon T \to B$ such that, for every $t \in T$, the subset $\{\sigma(tu)\sigma(u)^{-1} : u \in T\}[B, B]$ of $B/[B, B]$ is finite.*

Proof. Because every abelian torsion group is the direct sum of its p-subgroups, we may first define the section on each of T's p-components, and then extend it to T multiplicatively (in any order).

We therefore suppose that T is a p-group. Recall the notation $\Omega_n(T) = \{t \in T : t^{p^n} = 1\}$. Each quotient $V_n := \Omega_n(T)/\Omega_{n-1}(T)$ is a vector space over $\mathbb{F}_p$, and the homomorphism $t \mapsto t^p$ induces an injective linear map $V_n \to V_{n-1}$. Choose inductively subsets $X_1, X_2, \ldots$ of T such that X_n maps to a basis of V_n and such that $t \mapsto t^p$ induces an injective map $X_n \to X_{n-1}$. Set $X = \bigcup_n X_n$, and give an arbitrary total order to X. Choose for each $x \in X$ a π-preimage $\sigma(x) \in B$.

Given that T is torsion, every element $t \in T$ belongs to $\Omega_n(T)$ for some $n \in \mathbb{N}$, so can be uniquely written as a product $t = x_1^{\alpha_1} \cdots x_n^{\alpha_n}$ with ordered $x_i \in X$ and $0 < \alpha_i < p$ for all i. Extend then σ by $\sigma(t) = \sigma(x_1)^{\alpha_1} \cdots \sigma(x_n)^{\alpha_n}$.

Consider now $t \in T$, and write it in the form $t = x_1^{\alpha_1} \cdots x_n^{\alpha_n}$ as above. Extend $\{x_1, \ldots, x_n\}$ to a finite subset $Y = \{x_1, \ldots, x_n, \ldots, x_s\}$ of X, by adding all p^jth powers of all x_i to it. The set

$$T' := \{x_1^{\gamma_1} \cdots x_s^{\gamma_s} : 0 \leq \gamma_i < p\}$$

is then a finite subgroup of T. Consider next $u \in T$, and note that it may be written uniquely in the form $u = y_1^{\beta_1} \cdots y_m^{\beta_m} z$ with $y_i \in X \setminus Y$ and $z \in T'$. Then $tu = y_1^{\beta_1} \cdots y_m^{\beta_m}(zt)$ and this representation is unique; so $\sigma(tu)\sigma(u)^{-1}[B, B]$ belongs to the finite set $\{\sigma(t')[B, B] : t' \in T'\}$. □

Exercise 7.16. Rephrase Lemma 7.15 in terms of the cohomology of T with coefficients in $\mathbb{Z}X$.

Let F be a locally finite group of cardinality $> \#X$, and fix an imbedding of X in $F \setminus \{1\}$. As a first step, we consider the group $G_0 = B \wr^{\text{f.v.}} (T \times F)$, and define a map $\Phi_0 \colon B \to G_0$ as follows:

$$\Phi_0(b) = (\phi, \pi(b), 1) \text{ with } \phi(t,f) = \begin{cases} b & \text{if } f = 1, \\ \theta_f(\sigma(t)b\sigma(t\pi(b))^{-1}) & \text{if } f \in X, \\ 1 & \text{otherwise.} \end{cases}$$
(1.21)

Lemma 7.17. *The map Φ_0 is well defined and is an injective homomorphism.*

Proof. To see that Φ_0 is well defined, note that the argument $\sigma(t)b\sigma(t\pi(b))^{-1}$ belongs to $\ker(\pi) = C$, so that θ_f may be applied to it. Note then that, by Lemma 7.15, the expression $\sigma(t)b\sigma(t\pi(b))^{-1}$ takes finitely many values in $C/[B, B]$, so that $\phi(t,f)$ takes finitely many values for varying t and fixed f. Finally, $\theta_f(\sigma(t)b\sigma(t\pi(b))^{-1}) = 1$ except for finitely many values of $f \in X$. In summary, the function $\phi \in B^{T \times F}$ is such that $\phi(t,f)$ takes only finitely many values.

It is clear that Φ_0 is injective: if $b \neq 1$ and $\Phi_0(b) = (\phi, \pi(b), 1)$, then $\phi(1,1) = b \neq 1$. It is a homomorphism because all θ_f are homomorphisms. $\square$

Lemma 7.18. *We have $\Phi_0(C) \leq [G_0, G_0]$.*

Proof. If $b \in [B, B]$, then clearly $\Phi_0(b) \in [G_0, G_0]$. Because C is generated by $[B, B] \cup \{b_x\}_{x \in X}$, it suffices to consider $b = b_x$.

We define $g \in G_0$ by

$$g = (\psi, 1, 1) \text{ with } \psi(t,f) = \begin{cases} b_x & \text{if } f = 1, \\ 1 & \text{otherwise.} \end{cases}$$

Then $\Phi_0(b_x) = (\phi, 1, 1)$ with $\phi(t,1) = b_x$ and $\phi(t,x) = b_x^{-1}$, all other values being trivial, according to (1.21); so, as was to be shown,

$$\Phi_0(b_x) = (\phi, 1, 1) = (^x\psi^{-1} \cdot \psi, 1, 1) = [(1, 1, x^{-1}), g] \in [G_0, G_0]. \quad \square$$

We finally define

$$G = G_0 \wr^{\text{f.v.}} (\mathbb{Q}/\mathbb{Z})$$

and a map $\Phi \colon B \to G$ by

$$\Phi(b) = (\phi, 0) \text{ with } \phi(r) = \Phi_0(b) \text{ for all } r \in \mathbb{Q}/\mathbb{Z}.$$

Lemma 7.19. *The map Φ is an injective homomorphism, and $\Phi(B) \leq [G, G]$.*

Proof. Clearly Φ is an injective homomorphism, as Φ_0 is an injective homomorphism by Lemma 7.17.

We identify $\mathbb{Q}/\mathbb{Z}$ with $\mathbb{Q} \cap [0,1)$. For every $n \in \mathbb{N}$, consider the map $\Psi_n \colon B \to G$ defined by

$$\Psi_n(b) = (\phi, 0) \text{ with } \phi(r) = \begin{cases} \Phi_0(b) & \text{if } r \in [0, 1/n), \\ 1 & \text{otherwise;} \end{cases}$$

so $\Phi = \Psi_1$. We know from Lemma 7.18 that $\Psi_n(C)$ is contained in $[G, G]$.

Consider now $b \in B$. Since B/C is torsion, there exists $n \in \mathbb{N}$ such that $b^n \in C$. We define $g \in G$ by

$$g = (\psi, 0) \text{ with } \psi(r) = \Phi_0(b)^{\lfloor rn \rfloor} \text{ for } r \in [0, 1) \cap \mathbb{Q}.$$

Let us write $h = \Phi_0(b)$, and consider the element $[(1, 1/n), g] \cdot \Psi_n(b^n) = (\phi, 0)$. If $r \in [0, 1/n)$ then $\phi(r) = \psi(r - 1/n)^{-1} \psi(r) h^n = h$, while if $r \in [1/n, 1)$ then $\phi(r) = \psi(r - 1/n)^{-1} \psi(r) = h$; therefore

$$\Phi(b) = [(1, 1/n), g] \cdot \Psi_n(b^n) \in [G, G]. \qquad \square$$

Proof of Proposition 7.13. The first assertion is simply Lemma 7.19. For the last one: if B is infinite, we wish to find a subgroup H of G with the same cardinality as B, such that Φ maps into $[H, H]$. For each $b \in B$, choose a finite subset S_b of G such that $\Phi(b) \in [\langle S_b \rangle, \langle S_b \rangle]$, and a subgroup G_b, containing S_b, that is virtually a finite power of B. Consider the group H generated by the union of all the G_b. As soon as B is infinite, all G_b have the same cardinality as B, and so does H. $\quad \square$

7.4. Spreading, Stabilizing, Rectifiable Sequences

We also extend Proposition 7.4 by replacing the wreath product $G \wr \mathbb{Z}$ by permutational wreath products of the form $G \wr_X P$ for a group P acting on a set X, and considering subgroups of the form $W = \langle f, P \rangle$ for a function $f \colon X \to G$.

We already encountered in Proposition 6.8 sufficient conditions for such a group W to be of subexponential growth, in case f is finitely generated. As we shall see, the group W may also have subexponential growth if f is *infinitely* supported, but its support is sufficiently sparse, in a sense that we describe now.

In this section, we assume a finitely generated group $P = \langle S \rangle$ acting on the right on a set X has been fixed. We use the same notation for X as a set and as a *Schreier graph*, namely as the graph with vertex set X, and for all $x \in X, s \in S$ an edge labelled s from x to xs, see Definition 1.8. We denote by d the path metric on this graph.

Definition 7.20. A sequence $(x_0, x_1, \dots)$ in X is *spreading* if for all R there exists N such that if $i, j \geq N$ and $i \neq j$ then $d(x_i, x_j) \geq R$. $\quad \triangle$

Example 7.21. If all x_i lie in order on a geodesic ray starting from x_0 (for example, if X itself is a ray starting from x_0) and for all i we have $d(x_0, x_{i+1}) \geq 2d(x_0, x_i)$, then (x_i) is spreading.

Exercise 7.22. Show that a sequence $(x_0, x_1, \ldots)$ in X is spreading if and only if for all R there is N such that if $i \neq j$ and $i \geq N$ then $d(x_i, x_j) \geq R$.

Definition 7.23. A sequence (x_i) in X *locally stabilises* if for all R there is N such that if $i, j \geq N$ then the S-labelled radius-R balls centered at x_i and x_j in X are isomorphic as labelled graphs. $\triangle$

Definition 7.24. A sequence of points (x_i) in X is *rectifiable* if for all i, j there is $g \in P$ with $x_i g = x_j$ and $x_k g \neq x_\ell$ for all $k \notin \{i, \ell\}$. $\triangle$

For example, if $X = \mathbb{Z}$ and $P = \mathbb{Z}$ acting by translations, then $\Sigma = \{2^i : i \in \mathbb{N}\}$ is rectifiable, as $2^j - 2^i = 2^\ell - 2^k$ only has trivial solutions $i = k, j = \ell$ and $i = j, k = \ell$. It is also spreading and locally stabilizing.

Exercise 7.25. Show that the sequence $\Sigma = (x_i) \subseteq X$ is rectifiable if and only if for all i, j there exists $g \in P$ with $x_i g = x_j$ and $\Sigma \cap \Sigma g \subseteq \{x_j\} \cup \text{fixed.points}(g)$.

Definition 7.26. Fix a point $z \in X$. A sequence (g_i) in P is *parallelogram-free* at z if, for all i, j, k, ℓ with $i \neq j$ and $j \neq k$ and $k \neq \ell$ and $\ell \neq i$ one has $z g_i^{-1} g_j g_k^{-1} g_\ell \neq z$. $\triangle$

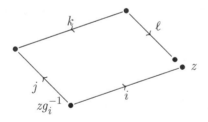

Lemma 7.27. *If $z \in X$ and (g_i) is parallelogram-free at z, then $(z g_i^{-1})$ is a rectifiable sequence in X.*

Proof. Set $x_i = z g_i^{-1}$ for all $i \in \mathbb{N}$. Given $i, j \in \mathbb{N}$, consider $g = g_i g_j^{-1}$, so $x_i g = x_j$. If furthermore we have $x_k g = x_\ell$, then we have $z g_k^{-1} g_i g_j^{-1} g_\ell = z$, so either $k = i$, or $i = j$, which implies $k = \ell$, or $j = \ell$, which implies $k = i$, or $\ell = k$. In all cases $k \in \{i, \ell\}$ as was to be shown. $\square$

It is clear that, if P is finitely generated and X is infinite, then it admits spreading and locally stabilizing sequences. Indeed every sequence contains a spreading subsequence, and every sequence contains a stabilizing subsequence, and a subsequence of a spreading or locally stabilizing sequence is again spreading, respectively locally stabilizing.

We will content ourselves with the following rectifiable sequences, based on the Grigorchuk groups $G_\omega = \langle a, b, c, d \rangle$, with $\omega \in \Omega'$; for example, the first Grigorchuk group G_{012}. Recall the description of G_ω from §4.5, and, in particular, the Schreier graph of its action on 1^∞ in (1.12). We construct explicitly a spreading, locally stabilizing, rectifiable sequence for the action of G_ω on X: for all $i \in \mathbb{N}$, let us define

$$x_i = 1^\infty 0^i,$$

the point at distance 2^i from the origin on the Schreier graph.

Lemma 7.28. *For all $\omega \in \Omega$ containing infinitely many $0, 1, 2s$, and for all $i, j \in \mathbb{N}$,*

1. *the marked balls of radius $2^{\min(i,j)}$ in X around x_i and x_j coincide;*
2. *the distance $d(x_i, x_j)$ is $|2^i - 2^j|$;*
3. *there exists $g_{i,j} \in G_\omega$ of length $|2^i - 2^j|$ with $x_i g_{i,j} = x_j$ and $x_k g_{i,j} \neq x_\ell$ for all $(k, \ell) \neq (i, j)$.*

Proof. (1), (2) Consider the map θ_ω from Equation (1.16). It maps the stabilizer of 1^∞ in $G_{\sigma\omega}$ to the stabilizer of 1^∞ in G_ω, and therefore defines a self-map of X by sending $1^\infty g$ to $1^\infty \theta_\omega(g)$. A direct calculation shows that it sends $x \in X$ to $x0$.

Given that θ_ω is 2-Lipschitz on words of even length in $\{a, b, c, d\}$, it maps the ball of radius n around x to the ball of radius $2n$ around $x0$. Its image is in fact a net in the ball of radius $2n$: two points at distance 1 in the ball of radius n around x will be mapped to points at distance 1 or 3 in the image, connected either by a segment $\underline{\quad a \quad}$ or by a path $\underline{\quad}^x_y\underline{\quad}$ for some $\{x, y\} \subset \{b, c, d\}$. In particular, the 2^n-neighbourhoods of the balls about the x_m coincide for all $m \geq n$.

(3) Note, first, that there exists $g_{i,j}$ with $x_i g_{i,j} = x_j$, because the rays ending in 1^∞ form a single orbit. Note, also, that we have $x_k g_{i,j} = x_\ell$ for either finitely many $(k, \ell) \neq (i, j)$ or for all but finitely many (k, ℓ), because there is a level N at which the decomposition of $g_{i,j}$ consists entirely of generators; if the entry at 0^N of $g_{i,j}$ is trivial or "d", then all but finitely many of the x_k are fixed; while otherwise (up to increasing N by at most one) we may assume it is an "a"; then $0^{N+1} g_{i,j} = 0^N 1$, so $x_k \neq x_\ell$ for all $k > N + 1$.

We use the following property of the Grigorchuk groups G_ω: for every finite sequence $u \in \{0, 1\}^*$ there exists an element $h_u \in G_\omega$ whose fixed points are precisely those sequences in $\{0, 1\}^\infty$ that do not start with u. One may take for h_u the element $[a, b]$, $[a, c]$ or $[a, d]$ inside the copy of the branching subgroup of G_ω that acts on $\{0, 1\}^\infty u$, see Proposition 5.12.

If the entry at 0^N of $g_{i,j}$ is trivial, then we multiply $g_{i,j}$ with h_{0^M} for some $M > \max(N, i)$, so as to fall back to the second case.

Then, for each pair $(k, \ell) \neq (i, j)$ with $x_k g_{i,j} = x_\ell$, we multiply $g_{i,j}$ with h_{10^ℓ}, to destroy the relation $x_k g_{i,j} = x_\ell$.

The resulting element $g_{i,j}$ satisfies the required conditions.   □

7.5. Subexponential Growth of Wreath Products

The next step in the proof is an argument controlling the growth of a subgroup of the form $W = \langle P, f \rangle \leq G \wr_X P$, for a function $f \colon X \to G$ with sparse-enough (but infinite!) support.

We select a finitely generated group P of subexponential growth acting on a set X with sublinear inverted orbit growth (see Definition 6.4; recall, from Exercise 6.5, that the property of having sublinear orbit growth is independent of a choice of basepoint) and subexponential inverted orbit choice growth. These are the hypotheses for Corollary 6.9, which guarantee that $G \wr_X P$ has subexponential growth as soon as G has subexponential growth. The main example we have in mind is a Grigorchuk group $P = G_\omega$ acting on $X = \mathbf{1}^\infty P$, for $\omega \in \{0, 1, 2\}^\infty$ containing infinitely many times each symbol, see Lemma 7.28.

We also assume that there are rectifiable sequences in X, and (using the results of the previous section) we fix a rectifiable, spreading, locally stabilizing sequence $(x_0, x_1, \dots)$ of elements of X.

Finally, we fix a countable group G, and a finite or infinite sequence of elements $(b_1, b_2, \dots)$ generating G.

Definition 7.29.   For an increasing finite or infinite sequence $0 \leq n(1) < n(2) < \dots$ of integers, define $f \colon X \to G$ by

$$f(x_{n(1)}) = b_1, \qquad f(x_{n(2)}) = b_2, \qquad \dots, \qquad f(x) = 1 \text{ for other } x.$$

The group $W_{(n)} = W_{n(1), n(2), \dots}$ is then defined as the subgroup $\langle P, f \rangle$ of the unrestricted wreath product $G \wr_X P$.   △

If all but finitely many b_i are trivial, then $W_{(n)}$ has subexponential growth as then G is finitely generated and $W_{(n)}$ is a subgroup of $G \wr_X P$. However, if $(n(i))$ is sparse enough, then $W_{(n)}$ may have subexponential growth even if f has infinite support:

Proposition 7.30.   *If G has locally subexponential growth, then there exists an infinite sequence $(n(i))$ such that the group $W_{(n)}$ has subexponential growth.*

The proof of Proposition 7.30 follows from a stronger, and independently interesting, statement: arbitrarily large balls in $W_{(n)}$ are approximable by groups of the form $W_{(n(1), \dots, n(i))}$, which have subexponential growth by the remark above:

Proposition 7.31.   *Assume that the sequence (x_i) in X is spreading and locally stabilizing, and that all elements b_i have the same order.*

Then for every increasing sequence $(m(i))$ there exists an increasing sequence $(n(i))$ with the following property: the ball of radius $m(i)$ in $W_{(n)}$ coincides with the ball of radius $m(i)$ in $W_{n(1),n(2),\ldots,n(i)}$, via the natural identification $P \leftrightarrow P, f \leftrightarrow f$ between $W_{(n)}$ and $W_{n(1),\ldots,n(i)}$.

Furthermore, the term $n(i)$ depends only on the previous terms $n(1),\ldots,n(i-1)$, on the initial terms $m(1),\ldots,m(i)$, and on the ball of radius $m(i)$ in the subgroup $\langle b_1,\ldots,b_{i-1}\rangle$ of G.

Proof. Choose $n(i)$ such that $d(x_j, x_k) \geq m(i)$ for all $j \neq k$ with $k \geq n(i)$, and such that the balls of radius $m(i)$ around $x_{n(i)}$ and x_j coincide for all $j > n(i)$.

Consider then an element $h \in W_{(n)}$ in the ball of radius $m(i)$, and write it in the form $h = (c, g)$ with $c \colon X \to G$ and $g \in P$. The function c is a product of conjugates of f by words of length $< m(i)$. Its support is therefore contained in the union of balls of radius $m(i) - 1$ around the x_j, with j either $\geq n(i)$ or of the form $n(k)$ for $k < i$. In particular, the entries of c are in $\langle b_1,\ldots,b_{i-1}\rangle \cup \bigcup_{j \geq i}\langle b_j\rangle$. For $j > n(i)$, the restriction of c to the ball around x_j is determined by the restriction of c to the ball around $x_{n(i)}$, via the identification $b_i \mapsto b_j$, because the neighbourhoods in X coincide and all cyclic groups $\langle b_j\rangle$ are isomorphic.

It follows that the element $h \in W$ is uniquely determined by the corresponding element in $W_{n(1),\ldots,n(i)}$. $\square$

Proof of Proposition 7.30. Let $Z = \langle z\rangle$ be a cyclic group whose order (possibly ∞) is divisible by the order of all the b_is. We replace G by $G \times Z$ and each b_i by $b_i z$, so as to guarantee that all generators in G have the same order.

Let ϵ_i be a decreasing sequence tending to 1. We now construct a sequence $m(i)$ inductively, and obtain the sequence $n(i)$ by Proposition 7.31, making always sure that $m(i)$ depends only on $m(j), n(j)$ with $j < i$.

Denote by v_i the growth function of the group $W_{n(1),\ldots,n(i)}$. Because the group $W_{n(1),\ldots,n(i)}$ is contained in $G \wr_X P$, it has subexponential growth. Therefore, there exists $m(i)$ such that

$$v_i(m(i)) \leq \epsilon_i^{m(i)}.$$

By Proposition 7.31, the terms $n(i+1), n(i+2), \ldots$ can be chosen in such a manner that the balls of radius $m(i)$ coincide in $W_{(n)}$ and $W_{n(1),\ldots,n(i)}$.

Denote now by w the growth function of $W_{(n)}$. We then have $w(m(i)) \leq \epsilon_i^{m(i)}$ for all $i \in \mathbb{N}$. Therefore,

$$w(R) \leq \epsilon_i^{R+m(i)} \text{ for all } R > m(i),$$

so $\limsup \sqrt[R]{w(R)} \leq \epsilon_i$ for all $i \in \mathbb{N}$. Thus, the growth of $W_{(n)}$ is subexponential. $\square$

Finally, the rectifiability of the sequence (x_i) guarantees that functions with singleton support and arbitrary values in $[G, G]$ belong to $W_{(n)}$ for all sequences n:

Lemma 7.32. *If the sequence (x_i) is rectifiable, then $[W_{(n)}, W_{(n)}]$ contains $[G, G]$ as a subgroup for all choices of $n = n(1) < n(2) < \cdots$.*

Proof. We denote by $\iota \colon G \to G^X \rtimes P$ the imbedding of G mapping the element $b \in G$ to the function $X \to G$ with value b at x_0 and 1 elsewhere. We abbreviate $W = W_{(n)}$. We shall show that $[W, W]$ contains $\iota([G, G])$. For this, denote by H the subgroup $\iota([G, G]) \cap [W, W]$.

We first consider an elementary commutator $g = [b_i, b_j]$. Let $g_i, g_j \in P$ respectively map x_i, x_j to x_0, and be such that $g_i g_j^{-1}$ maps no x_k to x_ℓ with $k \neq \ell$, except for $x_i g_i g_j^{-1} = x_j$. Consider $[f^{g_i}, f^{g_j}] \in [W, W]$; it belongs to G^X, and has value $[b_i, b_j]$ at x_0 and is trivial elsewhere, so equals $\iota(g)$ and therefore $\iota(g) \in H$.

We next show that H is normal in G^X. For this, consider $h \in H$. It suffices to show that $h^{\iota(b_i)}$ belongs to H for all i. Now $h^{\iota(b_i)} = h^{f^{g_i}}$ belongs to H, and we are done. $\square$

We are now ready to complete the proof of Theorem 7.1. By Proposition 7.13, the countable, locally subexponentially growing group B imbeds in $[G, G]$ for a countable, locally subexponentially growing group G. Let $(b_1, b_2, \dots)$ be a generating set for G. By Proposition 7.30, there exists an increasing sequence $(n(i))$ such that the group $W = W_{(n)}$ has subexponential growth. By Lemma 7.32, $[G, G]$ imbeds in $[W, W]$, so B imbeds in $[W, W]$ and we are done.

Exercise 7.33. Give examples of sequences (x_i) that are only spreading, or only stabilizing, and such that $W_{(n)}$ has exponential growth, even when the sequence $(n(i))$ grows arbitrarily fast.

8. Groups of Non-Uniform Exponential Growth

Recall that $v_{G,S}(R)$ denotes the growth function of a group G generated by a finite set S. The *volume entropy* of (G, S) is

$$\lambda_{G,S} := \lim_{R \to \infty} \frac{\log v_{G,S}(R)}{R}.$$

The limit exists because the function $v_{G,S}$ is submultiplicative (namely, $v_{G,S}(R_1 + R_2) \leq v_{G,S}(R_1) v_{G,S}(R_2)$.) Indeed, apply the following lemma to the sequence $(\log v_{G,S}(n))_{n \geq 1}$:

Lemma 8.1 (Fekete [22]). *Let (a_n) be a subadditive sequence: $a_{n+m} \leq a_n + a_m$ for all $m, n \geq 1$. Then $\lim a_n/n$ exists and equals $\inf a_n/n$.*

Proof. Set $A = \inf a_n/n$ and consider any $B > A$. Choose $k \geq 1$ such that $a_k/k < B$. Every $n \in \mathbb{N}$ may be written in the form $n = rk + s$ with $r \in \mathbb{N}$ and $s \in \{0, \ldots, k-1\}$. Thus,

$$\frac{a_n}{n} = \frac{a_{rk+s}}{n} \leq \frac{ra_k + a_s}{n} = \frac{a_k}{k}\frac{n-s}{n} + \frac{a_s}{n},$$

so $\limsup_{n\to\infty} a_n/n \leq a_k/k \leq B$. Since $B > A$ was arbitrary, $\limsup a_n/n \leq L$ and we are done. $\square$

Furthermore, the following are equivalent: G has exponential word growth; $\lambda_{G,S} > 0$ for some generating set S; and $\lambda_{G,S} > 0$ for all generating sets S.

Let G be a group of exponential growth. Note that, even though $\lambda_{G,S} > 0$ for all S, one might have $\lambda_{G,S_i} \to 0$ along a sequence of generating sets S_i. It is easy to see that this cannot happen for G a free group of rank $k \geq 2$; indeed then each S_i contains a subset of cardinality k generating a free subgroup, so $\lambda_{G,S_i} \geq \log(2k-1) > 0$ for all i. Let us say that a finitely generated group G has *uniform exponential growth* if $\inf_S \lambda_{G,S} > 0$, and *non-uniform exponential growth* if $\lambda_{G,S} > 0$ for all S yet $\inf_S \lambda_{G,S} = 0$.

The existence of groups of non-uniform exponential growth is asked by Gromov in [35, Remarque 5.12]; see [38] for a survey. There have been quite a few positive results: Osin showed in [64] that virtually solvable groups have uniform exponential growth unless they are virtually nilpotent; Eskin, Mozes and Oh obtained the same result in [20] for finitely generated linear groups in characteristic 0; Koubi showed in [49] that word-hyperbolic groups have exponential growth unless they are virtually cyclic.

A *Golod-Shafarevich group* is a residually-p group such that the associated Hopf algebra $\bigoplus_{n\geq 0} \varpi^n/\varpi^{n+1}$ from §5.1 has exponential growth; equivalently, the Lie algebra $\bigoplus_{n\geq 1} \gamma_n(G)/\gamma_{n+1}(G)$ from (1.13) has exponential growth. Since by Proposition 5.6 the growth of $\overline{\mathbb{k}G}$ is always a lower bound for the growth of G, such groups have uniformly exponential growth.

Among groups G of uniform exponential growth, one may ask whether the infimal entropy $\inf_{\langle S\rangle = G} \lambda_{G,S}$ is realised. Recall that a group G is *Hopfian* if it is not isomorphic to a proper quotient of itself. Sambusetti proved in [69] that if G is a free product, $G = G_1 * G_2$ with G_1 non-Hopfian and G_2 non-trivial, then $\inf_{\langle S\rangle = G} \lambda_{G,S}$ is not attained.

It was widely suspected, since the appearance of groups of intermediate growth, that examples of groups of non-uniform exponential growth should exist. The first examples of groups of non-uniform exponential growth were exhibited by Wilson, see [81]. We now give a simple construction showing that such groups abound:

Theorem 8.2 ([7, Theorem E]).   *Every countable group may be imbedded in a group of non-uniform exponential growth.*

Furthermore, let $\eta_+ \approx 2.46$ denote the positive root of the polynomial $T^3 - T^2 - 2T - 4$. Then the group W in which the countable group imbeds may be required to have the following property: there is a constant K such that, for all $R > 0$, there exists a generating set S of W with

$$v_{W,S}(r) \leq \exp(Kr^{\log 2/\log \eta_+}) \text{ for all } r \in [0, R].$$

Proof.   Let B be a countable group. By Theorem 7.7, one may imbed B into a group G generated by two elements s, t of order 5. Without loss of generality, assume that G has exponential growth (if needed, replace first B by $B \times F_2$).

The group W in which G imbeds is the wreath product $G \wr_X G_{012}$ of G with the first Grigorchuk group. We also consider $A = C_5 \times C_5 = \langle s', t' \rangle$, and the wreath product $W' = A \wr_X G_{012}$.

We consider the points $x_0 = \mathbf{0}^\infty$ and $x_i = \mathbf{0}^\infty \mathbf{10}^i$ for all $i \geq 1$ in the Schreier graph X, and the generating sets $S_i = \{a, b, c, d, s@x_0, t@x_i\}$ of W and $S' = \{a, b, c, d, s'@x_0, t'@x_0\}$ of W'.

Note that the sequence (x_i) is spreading and locally stabilizing (see Definitions 7.20 and 7.23); better, for all R the radius-R balls around x_0 and x_i in X are isomorphic as labelled graphs for all i large enough, because the action on a sequence $\in \{\mathbf{0}, \mathbf{1}\}^\infty$ of an element of G_{012} of length R depends only on the last $\lceil \log_2(R) \rceil$ symbols of the sequence.

We now claim that, for all $R \in \mathbb{N}$, there exists i such that the balls of radius R in the Cayley graphs of (W, S_i) and (W', S') coincide. By Theorem 6.3, there exists a constant K such that, for all $R \in \mathbb{N}$, the ball of radius R in the Cayley graph of (W', S') has cardinality $v_{W',S'}(R) \leq \exp(KR^{\log 2/\log \eta_+})$. Assuming the claim, we get $v_{W,S_i}(R) \leq \exp(KR^{\log 2/\log \eta_+})$, from which the second claim of the theorem follows.

For every $\epsilon > 0$ there exists R such that $KR^{\log 2/\log \eta_+} < \epsilon R$, so $v_{W,S_i}(R) \leq \exp(\epsilon R)$ and $\lambda_{W,S_i} \leq \epsilon$ for some i. It follows then that W has non-uniform exponential growth.

It remains to prove the claim. Given $R \in \mathbb{N}$, let i be large enough so that the distance between x_0 and x_i in X is at least $2R$. Consider a word w in S_i of length $\leq R$, and let w' be the corresponding word in S' obtained by replacing $s@x_0, t@x_i$ respectively by $s'@x_0, t'@x_0$. We show that w represents the identity in W if and only if w' represents the identity in W'.

Write $w = (c, g)$ in W, with $c \colon X \to G$ and $g \in G_{012}$. Similarly, write $w' = (c', g)$ in W', with $c' \colon X \to A$ and the same $g \in G_{012}$. Note that the support of c is contained in the union of the balls of radius R around x_0 and x_i, and these balls are isomorphic and disjoint. Therefore, c can be written in the form $c = c_1 c_2$ with $c_1 \colon X \to \langle s \rangle$ and $c_2 \colon X \to \langle t \rangle$, and c_1, c_2 have disjoint support so they commute. The function c' may correspondingly be written as $c' = c_1' c_2'$ with $c_1' \colon X \to \langle s' \rangle$ obtained by composing c_1 with the isomorphism $\langle s \rangle \to \langle s' \rangle$, and $c_2' \colon X \to \langle t' \rangle$ obtained by composing the isomorphism from the radius-R ball around x_0 to the radius-R ball around x_i, the map c_2, and the isomorphism $\langle t \rangle \to \langle t' \rangle$. Therefore, $c' = 1$ if and only if $c = 1$, so the balls in W and W' are isomorphic. $\qquad\square$

References

[1] Miklós Abért, *Group laws and free subgroups in topological groups*, Bull. London Math. Soc. **37**(2005), no. 4, 525–34, available at arXiv:math.GR/0306364. MR2143732.

[2] Stanislav V. Alešin, *Finite automata and the Burnside problem for periodic groups*, Mat. Zametki **11** (1972), 319–28. MR46#265.

[3] Daniel Allen, Megan Cream, Kate Finlay, John Meier, and Ranjan Rohatgi, *Complete growth series and products of groups*, New York J. Math. **17** (2011), 321–9. MR2811067 (2012h:20091)

[4] Laurent Bartholdi Anna G. Erschler, *Growth of permutational extensions*, Invent. Math. **189** (2012), no. 2, 431–55, DOI 10.1007/s00222-011-0368-x, available at arXiv:math/1011.5266. MR2947548

[5] ———, *Imbeddings into groups of intermediate growth*, Groups Geom. Dyn. **8** (2014), no. 3, 605–20, DOI 10.4171/GGD/241, available at arXiv:math/1403.5584. MR3267517

[6] ———, *Groups of given intermediate word growth*, Ann. Inst. Fourier **64** (2014), no. 5, 2003–36, available at arXiv:math/1110.3650.

[7] ———, *Ordering the space of finitely generated groups*, Ann. Inst. Fourier **65** (2015), no. 5, 2091–144, DOI 10.5802/aif.2984, available at arXiv:math/1301.4669.

[8] ———, *Distortion of imbeddings of groups of intermediate growth into metric spaces*. Proc. Amer. Math. Soc. **145** (2017), no. 5, 1943–1952.

[9] Michael Benson, *Growth series of finite extensions of $\mathbb{Z}^n$ are rational*, Invent. Math. **73** (1983), 251–69.

[10] Jean Berstel, *Sur les pôles et le quotient de Hadamard de séries N-rationnelles*, C. R. Acad. Sci. Paris Sér. A-B **272** (1971), A1079–A1081 (French). MR0285521 (44 #2739).

[11] Garrett Birkhoff, *Extensions of Jentzsch's theorem*, Trans. Amer. Math. Soc. **85** (1957), 219–27. MR0087058 (19,296a)

[12] Jérémie Brieussel, *Growth behaviors in the range e^{r^α}* Afr. Mat. **25** (2014), no. 4, 1143–1163.

[13] William S. Burnside, *On an unsettled question in the theory of discontinuous groups*, Quart. J. Pure Appl. Math. **33** (1902), 230–8.

[14] Fritz Carlson, *Über Potenzreihen mit ganzzahligen Koeffizienten*, Math. Z. **9** (1921), 1–13, DOI 10.1007/BF01378331 (German).

[15] Ian M. Chiswell, *Euler characteristics of groups*, Math. Z. **147** (1976), no. 1, 1–11. MR0396785 (53 #645)

[16] Václáv Chvatal David Sankoff, *Longest common subsequences of two random sequences*, J. Appl. Probability **12** (1975), 306–15. MR0405531 (53 #9324)

[17] Yves de Cornulier, *Finitely presented wreath products and double coset decompositions*, Geom. Dedicata **122** (2006), 89–108, DOI 10.1007/s10711-006-9061-4. MR2295543 (2008e:20040)

[18] Moon Duchin and Michael Shapiro, *Rational growth in the Heisenberg group* (2014), preprint, available at arXiv:1411.4201.

[19] Vadim A. Efremovich, *The proximity geometry of Riemannian manifolds*, Uspekhi Mat. Nauk **8** (1953), 189.

[20] Alex Eskin, Shahar Mozes, and Hee Oh, *Uniform exponential growth for linear groups*, Int. Math. Res. Not. **31** (2002), 1675–83. MR1916 428

[21] Pierre Fatou, *Sur les séeries entiéeres à coefficients entiers*, C. R. Acad. Sci., Paris **138** (1904), 342–4 (French).

[22] Mihály Fekete, *Über die Verteilung der Wurzeln bei gewissen algebraischen Gleichungen mit ganzzahligen Koeffizienten*, Math. Z. **17** (1923), no. 1, 228–49 (German). MR1544613

[23] William J. Floyd and Steven P. Plotnick, *Growth functions on Fuchsian groups and the Euler characteristic*, Invent. Math. **88** (1987), no. 1, 1–29.

[24] Michael R. Garey, David S. Johnson, and Larry J. Stockmeyer, *Some simplified NP-complete graph problems*, Theoret. Comput. Sci. **1** (1976), no. 3, 237–67. MR0411240 (53 #14978)

[25] Michael R. Garey and David S. Johnson, *The rectilinear Steiner tree problem is NP-complete*, SIAM J. Appl. Math. **32** (1977), no. 4, 826–34. MR0443426 (56 #1796)

[26] Evguenĭ S. Golod, *On nil-algebras and finitely approximable p-groups*, Izv. Akad. Nauk SSSR Ser. Mat. **28** (1964), 273–6.

[27] Frank Gray, *Pulse code communication*, 1953. US Patent 2,632,058.

[28] Rostislav I. Grigorchuk, *On Burnside's problem on periodic groups*, Функционал. Анал. и Приложен. **14** (1980), no. 1, 53–54. English translation: Functional Anal. Appl. **14** (1980), 41–3. MR81m:20045

[29] ———, *On the Milnor problem of group growth*, Dokl. Akad. Nauk SSSR **271** (1983), no. 1, 30–3. MR85g:20042

[30] ———, *Degrees of growth of finitely generated groups and the theory of invariant means*, Изв. Акад. Наук **48** (1984), no. 5, 939–985. English translation: Math. USSR-Izv. **25** (1985), no. 2, 259–300. MR86h:20041

[31] ———, *On the Hilbert-Poincaré series of graded algebras that are associated with groups*, Мат. Сб. **180** (1989), no. 2, 207–25, 304. English translation: Math. USSR-Sb. **66** (1990), no. 1, 211–29. MR90j:20063

[32] ———, *Growth functions, rewriting systems and Euler characteristic*, Mat. Zametki **58** (1995), no. 5, 653–68, 798. MR97d:20031a

[33] Rostislav I. Grigorchuk and Tatiana Nagnibeda, *Complete growth functions of hyperbolic groups*, Invent. Math. **130** (1997), no. 1, 159–88. MR98i:20038

[34] Mikhael L. Gromov, *Groups of polynomial growth and expanding maps*, Inst. Hautes Études Sci. Publ. Math. **53** (1981), 53–73.

[35] ———, *Structures métriques pour les variétés riemanniennes*, CEDIC, Paris, 1981. Edited by J. Lafontaine and P. Pansu.

[36] ———, *Hyperbolic groups*, Essays in group theory, Math. Sci. Res. Inst. Publ., vol. 8, Springer, New York, 1987, pp. 75–263, DOI 10.1007/978-1-4613-9586-7_3. MR919829 (89e:20070)

[37] Geoffrey H. Hardy and Srinivasa Ramanujan, *Asymptotic formulae in combinatory analysis*, Proc. London Math. Soc. **S2-17**, no. 1, 75, DOI 10.1112/plms/s2-17.1.75. MR1575586

[38] Pierre de la Harpe, *Uniform growth in groups of exponential growth*, Geom. Dedicata **95** (2002), 1–17.

[39] Graham Higman, Bernard H. Neumann, and Hanna Neumann, *Embedding theorems for groups*, J. London Math. Soc. **24** (1949), 247–54. MR0032641 (11,322d)

[40] Stephen A. Jennings, *The structure of the group ring of a p-group over a modular field*, Trans. Amer. Math. Soc. **50** (1941), 175–85.

[41] ———, *The group ring of a class of infinite nilpotent groups*, Canad. J. Math. **7** (1955), 169–87.

[42] David L. Johnson, *Rational growth of wreath products*, Groups—St. Andrews 1989, Vol. 2, London Math. Soc. Lecture Note Ser., vol. 160, Cambridge University Press, 1991, pp. 309–15, DOI 10.1017/CBO9780511 661846.005. MR1123986 (92g:20045)

[43] Camille Jordan, *Traité des substitutions et des équations algébriques*, Les Grands Classiques Gauthier-Villars. [Gauthier-Villars Great Classics], Éditions Jacques Gabay, Sceaux, 1989 (French). Reprint of the 1870 original. MR1188877 (94c:01039)

[44] Lev A. Kaloujnine, *La structure des p-groupes de Sylow des groupes symétriques finis*, Ann. École Norm. Sup. (3) **65** (1948), 239–76.

[45] Lev A. Kaloujnine and Marc Krasner, *Le produit complet des groupes de permutations et le problème d'extension des groupes*, C. R. Acad. Sci. Paris **227** (1948), 806–8 (French). MR0027758 (10,351e)

[46] Michael E. Kapovich, *Arithmetic aspects of self-similar groups*, Groups Geom. Dyn. **6** (2012), no. 4, 737–54, DOI 10.4171/GGD/172. MR2996409

[47] David A. Klarner, *Mathematical crystal growth. I*, Discrete Appl. Math. **3** (1981), no. 1, 47–52. MR604265 (82e:05016)

[48] ———, *Mathematical crystal growth. II*, Discrete Appl. Math. **3** (1981), no. 2, 113–17. MR607910 (83a:05018)

[49] Malik Koubi, *Croissance uniforme dans les groupes hyperboliques*, Ann. Inst. Fourier (Grenoble) **48** (1998), no. 5, 1441–53. MR99m:20080

[50] Hans Ulrich Krause, *Gruppenstruktur und Gruppenbild*, Thesis, Eidgenössische Technische Hochschule, Zürich, 1953. MR15,99b

[51] Jacques Lewin, *The growth function of a graph group*, Comm. Algebra **17** (1989), no. 5, 1187–91, DOI 10.1080/00927878908823782. MR993397 (90c:20040)

[52] Fabrice Liardet, *Croissance des groupes virtuellement abéliens*, PhD Thesis, 1996.

[53] Avinoam Mann, *How groups grow*. London Mathematical Society Lecture Note Series, 395. Cambridge University Press, Cambridge, 2012. x+199 pp. ISBN: 978-1-107-65750-2

[54] Grigori A. Margulis, *Discrete subgroups of semisimple Lie groups*, Ergebnisse der Mathematik und ihrer Grenzgebiete (3) [Results in Mathematics and Related Areas (3)], vol. 17, Springer-Verlag, Berlin, 1991. MR1090825 (92h:22021)

[55] John W. Milnor, *Growth of finitely generated solvable groups*, J. Differential Geom. **2** (1968), 447–9.

[56] ——, *A note on curvature and fundamental group*, J. Differential Geom. **2** (1968), 1–7.

[57] ——, *Problem 5603*, Amer. Math. Monthly **75** (1968), 685–6.

[58] John W. Milnor and John C. Moore, *On the structure of Hopf algebras*, Ann. of Math. (2) **81** (1965), 211–64. MR0174052 (30 #4259)

[59] Melvyn B. Nathanson, *Asymptotic density and the asymptotics of partition functions*, Acta Math. Hungar. **87** (2000), no. 3, 179–95. MR2001b:11091

[60] Volodymyr V. Nekrashevych, *A minimal Cantor set in the space of 3-generated groups*, Geom. Dedicata **124** (2007), 153–90. MR2318543 (2008d:20075)

[61] Bernard H. Neumann and Hanna Neumann, *Embedding theorems for groups*, J. London Math. Soc. **34** (1959), 465–79. MR0163968 (29 #1267)

[62] Pet'r S. Novikov, *On algorithmic unsolvability of the problem of identity*, Doklady Akad. Nauk SSSR (N.S.) **85** (1952), 709–12 (Russian). MR0052436 (14,618h)

[63] Alexander Yu. Ol'shanskiĭ and Denis V. Osin, A *quasi-isometric embedding theorem for groups*, Duke Math. J. **162** (2013), no. 9, 1621–48, DOI 10.1215/00127094-2266251. MR3079257

[64] Denis V. Osin, *The entropy of solvable groups*, Ergodic Theory Dynam. Systems **23** (2003), no. 3, 907–18. MR1992 670

[65] Walter R. Parry, *Growth series of some wreath products*, Trans. Amer. Math. Soc. **331** (June 1992), no. 2, 751–9.

[66] Richard E. Phillips, *Embedding methods for periodic groups*, Proc. London Math. Soc. (3) **35** (1977), no. 2, 238–56. MR0498874 (58 #16896)

[67] Alexander V. Rozhkov, *Lower central series of a group of tree automorphisms*, Mat. Zametki **60** (1996), no. 2, 225–37, 319.

[68] Arto Salomaa and Matti Soittola, *Automata-theoretic aspects of formal power series*, Springer-Verlag, 1978.

[69] Andrea Sambusetti, *Minimal growth of non-Hopfian free products*, C. R. Acad. Sci. Paris Séer. I Math. **329** (1999), no. 11, 943–6. MR2000j:20056

[70] Leonard L. Scott, *Representations in characteristic p*, The Santa Cruz Conference on Finite Groups (Univ. California, Santa Cruz, Calif., 1979), Proc. Sympos. Pure Math., vol. 37, Amer. Math. Soc., Providence, R.I., 1980, pp. 319–31. MR604599 (82e:20052)

[71] Jean-Pierre Serre, *Trees*, Springer-Verlag, Berlin, 1980. Translated from the French by John Stillwell. MR82c:20083

[72] Yehuda Shalom and Terence Tao, *A finitary version of Gromov's polynomial growth theorem*, Geom. Funct. Anal. **20** (2010), no. 6, 1502–47, DOI 10.1007/s00039-010-0096-1. MR2739001 (2011m:20100)

[73] Neville F. Smythe, *Growth functions and Euler series*, Invent. Math. **77** (1984), 517–31.

[74] John R. Stallings, *Centerless groups—an algebraic formulation of Gottlieb's theorem*, Topology **4** (1965), 129–134. MR0202807 (34 #2666)

[75] Michael Stoll, *Rational and transcendental growth series for the higher Heisenberg groups*, Invent. Math. **126** (1996), no. 1, 85–109. MR98d:20033

[76] Albert S. Švarc, *A volume invariant of coverings*, Dokl. Akad. Nauk SSSR **105** (1955), 32–4 (Russian).

[77] Moss E. Sweedler, *Hopf algebras*, Mathematics Lecture Note Series, W. A. Benjamin, New York, 1969. MR0252485 (40 #5705)

[78] Ludwig Sylow, *Théorèmes sur les groupes de substitutions*, Math. Ann. **5** (1872), 584–94, DOI 10.1007/BF01442913.

[79] Jacques Tits, *Free subgroups in linear groups*, J. Algebra **20** (1972), 250–70.

[80] John S. Wilson, *Embedding theorems for residually finite groups*, Math. Z. **174** (1980), no. 2, 149–57, DOI 10.1007/BF01293535. MR592912 (81m:20041)

[81] ———, *On exponential and uniformly exponential growth for groups*, Invent. Math. **155** (2004), no. 2, 287–303.

[82] ———, *The gap in the growth of residually soluble groups*, Bull. Lond. Math. Soc. **43** (2011), no. 3, 576–82, DOI 10.1112/blms/bdq124. MR2820146 (2012f:20105)

[83] Wolfgang Woess, *Lamplighters, Diestel–Leader graphs, random walks, and harmonic functions*, Combin. Probab. Comput. **14** (2005), no. 3, 415–33, DOI 10.1017/S0963548304006443. MR2138121 (2006d:60021)

[84] Joseph A. Wolf, *Growth of finitely generated solvable groups and curvature of Riemanniann manifolds*, J. Differential Geom. **2** (1968), 421–46.

2

RANDOM WALKS ON SOME COUNTABLE GROUPS

ALEXANDER BENDIKOV[1,*] AND LAURENT SALOFF-COSTE[2,†]

[1]Institute of Mathematics, University of Wrocław, Wrocław, Poland
[2]Department of Mathematics, Cornell University, Ithaca, New York, USA

To our friend Wolfgang Woess, on his sixtieth birthday

Abstract
We study the decay of convolution powers of probability measures on some discrete groups that are not finitely generated such as the infinite symmetric group $G = \mathbb{S}^{(\infty)}$ and algebraic direct sums of finitely generated groups $G = \sum \Gamma_i$, e.g., with $\Gamma_i = \mathbb{Z}/p_i\mathbb{Z}$ for some sequence of integer p_i or $\Gamma_i = \mathbb{Z}^d$ for all i.

Contents

1. Introduction

Let G be a countable group equipped with the counting measure. For any finite subset B of G, we let $|B|$ be the number of group elements contained B. Let ϕ be a probability density on G. We say that ϕ is symmetric if $\phi(x) = \phi(x^{-1})$, $x \in G$. The measure ϕ induces a random walk on G that starts at the neutral element e and whose law at time n is the nth convolution power $\phi^{(n)}$. Understanding the large time behavior of the functions $n \to \phi^{(n)}(e)$ and $(n, x) \mapsto \phi^{(n)}(x)$ is of interest from many different viewpoints. For instance, the random walk driven by ϕ is recurrent if and only if $\sum \phi^{(n)}(e) = \infty$, the group G is nonamenable if and only if there exists a symmetric ϕ such that $\limsup_{n \to \infty} (\phi^{(n)}(e))^{1/n} < 1$. In many cases, the behavior of the

---

[*]Research partially supported by Polish Government Scientific Research Fund, Grant 2012/05/B/ST 1/00613.
[†]Research partially supported by NSF grant DMS- 1404435.

probability of return, $\phi^{(n)}(e)$, captures important information about the algebraic properties of the group. When ϕ is symmetric, $n \mapsto \phi^{(2n)}(e)$ is nonincreasing and tends to zero unless the support of ϕ generates a finite group. Roughly speaking, the larger the group and the more spread-out the measure, the faster $\phi^{(n)}(e)$ decays.

On a finitely generated infinite group, symmetric random walks driven by a probability density with finite generating support form a natural class of random walks. For results describing the behavior of $n \mapsto \phi^{(n)}(e)$ in this context, we refer the reader to [15–18, 21, 22]. The fundamental result of [16] asserts that the behavior of the probability of return of a random walk in this natural class is, in some sense, an invariant of the finitely generated group that supports the walk. Some results regarding spread-out measures on finitely generated groups are given in [2, 5, 16].

The present work focuses on random walks on countable groups that are not finitely generated. A fundamental difference with the case of finitely generated groups is that, on a general countable group (e.g., on $\mathbb{Q}$), there exists no obvious "canonical" class of random walks. Instead, depending on the particular structure of the group, there might be large natural families of random walks that present a continuum of behaviors. One of the most basic examples of countable nonfinitely generated group is the group R_2 of rational numbers of the form $m2^{-k}$, $m \in \mathbb{Z}$, $k = \mathbb{Z}$. A natural symmetric generating set is $S = \{\pm 2^{-k}, k = 1, 2, \dots\}$. For any probability distribution $\mathfrak{p} = (p_k)$ on $\{0, 1, 2, \dots\}$, the symmetric probability density $\phi_{\mathfrak{p}} = p_0 \delta_{\{0\}} + \sum_{k=1}^{\infty} \frac{p_k}{2}(\delta_{\{2^{-k}\}} + \delta_{\{-2^{-k}\}})$ defines a natural random walk on R_2. The transience and recurrence of such random walks is studied in [9]. Unfortunately, we will have very little to say about this example except to point out that little is known about the behavior of $\phi_{\mathfrak{p}}^{(n)}(e)$ and how it depends on $\mathfrak{p}$. The fundamental structure of R_2 emphasized above is that R_2 is the increasing union $R_2 = \cup_{i=0}^{\infty} 2^{-i}\mathbb{Z}$ of copies of the finitely generated group $\mathbb{Z}$. Any countable group is (in many different ways!) the increasing countable union of a family of finitely generated groups, but this type of structure is not always very natural.

When a countable group G is the increasing countable union of a collection of finitely generated groups Γ_i, $G = \cup_{i \in I} \Gamma_i$, indexed by a countable ordered set I, it is natural to pick a symmetric probability density ϕ_i on each Γ_i (say, with finite generating support) and consider the walks driven by the probability densities $\phi_{\mathfrak{p}} = \sum_{i \in I}^{\infty} p_i \phi_i$, where $\mathfrak{p} = (p_i)_1^{\infty}$ is a probability distribution on I. Unless some additional structural hypotheses are made, very little is known about how to approach the study of the behavior of $n \mapsto \phi_{\mathfrak{p}}^{(n)}(e)$ in this context.

Earlier works concerning examples of random walks on countable groups include [7, 9, 13, 14]. These works contain additional references. The works of Kesten and Spitzer [13] and Fereig and Molchanov [9] are based on the use of Fourier transform and deal with abelian examples.

The works of Lawler and of Bofferio and Woess deal mostly with the infinite symmetric group $\mathbb{S}^{(\infty)}$, that is, the group of permutations of the integers that leave all but finitely many integers fixed. All these works focus on the dichotomy between transient and recurrent behaviors.

The present work is closer in spirit to [3] and [5, 16, 21]. Our aim is to obtain results describing the decay behavior of $n \mapsto \phi^{(n)}(e)$ as precisely as possible in a variety of situations. The technique we will use is based on two simple ideas: (1) often, there are very specific examples that can be studied using rather elementary but ad hoc techniques; (2) once a good example (or a collection of good examples) is understood, one can use comparison techniques to understand the behavior of many more examples including examples that are quite different from the ones studied in the first step.

1.1. Union of Finite Groups

One interesting special case that we will consider at some length is the case when $G = \cup_{i \in I} \Gamma_i$ and the Γ_i's are finite groups. The following three examples are of this type.

Example 1.1. The infinite symmetric group $\mathbb{S}^{(\infty)}$ is the group of those bijections of $\{1, 2, \dots\}$ that move only finitely many integers. Obviously, this is the increasing union of all finite symmetric groups, $\mathbb{S}^{(\infty)} = \cup_{i=1}^{\infty} \mathbb{S}_i$. There are subtleties in studying models of random walks on $\mathbb{S}^{(\infty)}$. For instance, we are able to determine quite precisely which $\mathfrak{p} = (p_i)$ make $\sum_1^{\infty} p_i \phi_i$ recurrent or transient when ϕ_i is the "random transposition" measure on $\mathbb{S}_i$, but we are not able to do so when ϕ_i is the "adjacent transposition" measure. This last case points toward interesting open questions.

Example 1.2. The infinite special linear group with coefficients in the finite field $\mathbb{Z}/p\mathbb{Z}$ can be defined as the increasing union of $\mathrm{SL}_i(p)$. Thinking in terms of matrices, to go from i to $i + 1$, add a column to the right and a row to the bottom. Identify $\mathrm{SL}_i(p)$ in $\mathrm{SL}_{i+1}(p)$ as those matrices with 0 entries in the last row and column except for the bottom right entry, which is 1.

Example 1.3. Let $B(3, r)$ denotes the Burnside group with r generators and all element of order 3. This is a finite group of order $3^{\binom{r}{3} + \binom{r}{2} + \binom{r}{1}}$. Ordering the generators of each $B(3, i)$, we obviously have consistent natural injective homomorphisms from $B(3, i)$ to $B(3, i+1)$, $i \geq 1$. Then, the group $B(3, \infty) = \cup_1^{\infty} B(3, i)$ is the increasing union of the groups $B(3, i)$. This group comes with a natural infinite sequence of generators $\{g_1, g_2, \cdots\}$ such that $\{g_1, \dots, g_i\}$ are the generators of $B(3, i)$.

In each of these examples, there are a variety of interesting ways to pick symmetric probability densities ϕ_i supported on Γ_i. Setting

$\phi_{\mathfrak{p}} = \sum_I p_i \phi_i$ as above, one would like to study how the decay of $n \mapsto \phi_{\mathfrak{p}}^{(n)}(e)$ depends on $\mathfrak{p}$ and on the choice of the ϕ_i's.

1.2. Direct Sums

Another special case of interest is when $G = \sum_I G_i$ is the direct sum of finitely generated groups G_i, $i \in I$, that is,

$$G = \{(g_i)_{i \in I} : g_i \in G_i, g_i = e_i \text{ for all but finitely many } i\}.$$

In this case, assuming that $I = \cup_{j \in J} I_j$ with I_j, $j \in J$, being an increasing family of finite sets, we also have $G = \cup_{j \in J} \left(\sum_{i \in I_j} G_i \right)$.

However, the most natural families of probability densities on $G = \sum_I G_i$ associated with the direct sum structure are defined as follows. On each G_ℓ, let ψ_ℓ be a probability density and set

$$\phi_\ell : G \to [0,1], \quad g = (g_i) \mapsto \phi_\ell(g) = \begin{cases} \psi_\ell(g_\ell) & \text{if } g_i = e_i \text{ for all } i \neq \ell \\ 0 & \text{otherwise.} \end{cases}$$
(2.1)

Then, for any choice of a probability density $\mathfrak{p} = (p_i)_I$, set

$$\phi_{\mathfrak{p}} = \sum_I p_i \phi_i. \tag{2.2}$$

Here are two simple examples of this situation.

Example 1.4. Given a finitely generated group Γ, let $G = \Gamma^{(\infty)}$ be the sum of countably many copies Γ_i of Γ. In this case, it is very natural to pick a probability density ψ on Γ, set $\psi_i = \psi$ on each Γ_i and consider the probability densities ϕ_i and $\phi_{\mathfrak{p}} = \sum p_i \phi_i$ on G defined by (2.1)–(2.2).

Example 1.5. Given a sequence of integers m_i, $i = 1, 2, \ldots$, set $\Gamma_i = \mathbb{Z}/m_i\mathbb{Z}$ and $\psi_i(\pm 1) = 1/2$, $\psi_i(x) = 0$ if $x \neq \pm 1$. In this case, the group $G = \sum_{i=1}^{\infty} \Gamma_i$ is both a direct sum and an increasing union of finite groups.

1.3. Other Examples

There are, of course, many other types of examples. One of the simplest is the group $U(\infty)$ of all (infinite) upper-triangular matrices with 1 along the diagonal and finitely many nonzero integer entries in the upper half. This group can obviously be viewed as the increasing union of the upper-triangular nilpotent groups $U(n)$. Let $g_i^{\pm 1}$ have all nondiagonal entries equal to zero except the $(i, i+1)$-entry, which equals ± 1. The elements $(g_i^{\pm 1})_1^{\infty}$ yield a natural symmetric generating set. Set $\phi_i = \frac{1}{3}(\delta_e + \delta_{g_i} + \delta_{g_i^{-1}})$. On $U(\infty)$, one can consider the family of probability measures $\phi_{\mathfrak{p}} = \sum_1^{\infty} p_i \phi_i$ with $\mathfrak{p} = (p_i)_1^{\infty}$ as before. We know little about these measures except for the obvious fact that $\phi_{\mathfrak{p}}^{(n)}(e) \leq \overline{\phi}_{\mathfrak{p}}^{(n)}(0)$ where

$\overline{\phi}_{\mathfrak{p}}$ is the measure naturally associated to $\mathfrak{p}$ on $\mathbb{Z}^{(\infty)}$, the direct sum of countably many copies of of $\mathbb{Z}$. We know a lot about $\overline{\phi}_{\mathfrak{p}}^{(n)}(0)$. See Section 6.

Understanding whether or not the upper-bound $\phi_{\mathfrak{p}}^{(n)}(e) \leq \overline{\phi}_{\mathfrak{p}}^{(n)}(0)$ is sharp for some interesting $\mathfrak{p}$ is an open question.

Two groups that are much easier to understand than $U(\infty)$ are $H_1(\infty)$ and $\mathbf{H}$ described below.

Example 1.6. The group $H_1(\infty)$ is a nilpotent group of class 2 generated by the elements $x_i^{\pm 1}, y_i^{\pm 1}$, $i = 1, \ldots$, and such that all pairs $(x_i, x_j), (x_i, y_j)$ and (y_i, y_j) commute except x_i, y_i, which satisfies $[x_i, y_i] = z$ for all $i \geq 1$ with z central. The subscript 1 in $H_1(\infty)$ refers to the dimension of the center. The group $H_1(\infty)$ can be viewed as the increasing union of the usual Heisenberg groups $H_1(2n + 1)$. It can also be identified as the set $\mathbb{Z}^{(\infty)} \times \mathbb{Z}^{(\infty)} \times \mathbb{Z}$ with group law $(x, y, z) \cdot (x', y', z') = (x + x', y + y', z + z' + x \cdot y')$ where $x \cdot y' = \sum_1^\infty x_i y_i$. This makes good sense because the sequences x, y' are eventually 0.

Let $[H_1(3)]^{(\infty)}$ be the direct sum of countably many copies of the Heisenberg group $H_1(3) = U(3)$. Let ϕ_i be the probability measure defined by $\phi_i(x_i^{\pm}) = \phi_i(y_i^{\pm 1}) = \phi_i(e) = 1/5$ in $H_1(\infty)$, and set $\phi_{\mathfrak{p}} = \sum_1^\infty p_i \phi_i$ with $\mathfrak{p} = (p_i)_1^\infty$ as before.

Similarly, let $\widetilde{\phi}_{\mathfrak{p}}$ (resp. $\overline{\phi}_{\mathfrak{p}}$) be the measure on $[H_1(3)]^{(\infty)}$ (resp. the measure on $[\mathbb{Z}^2]^{(\infty)}$, the direct sum of countably many copies of $\mathbb{Z}^2$) naturally associated with $\mathfrak{p}$. Because we have natural surjective group homomorphisms

$$[H_1(3)]^{(\infty)} \to H_1(\infty) \to [\mathbb{Z}^2]^{(\infty)},$$

it immediately follows that

$$\widetilde{\phi}_{\mathfrak{p}}^{(n)}(e) \leq \phi_{\mathfrak{p}}^{(n)}(e) \leq \overline{\phi}_{\mathfrak{p}}^{(n)}(0).$$

This yields good bounds on $\phi_{\mathfrak{p}}^{(n)}(e)$ based on the bounds for direct sums obtained in Section 6.

Example 1.7. The group $\mathbf{H}$ is also a nilpotent group of class 2. It is generated by $x_i^{\pm 1}$, $i \geq 1$, where all pairs (x_i, x_j) commute except (x_i, x_{i+1}), which satisfies $[x_i, x_{i+1}] = z_i$ with z_i central, $i \geq 1$. This is a subgroup of $[H_1(3)]^{(\infty)}$. Namely, if we write $g \in [H_1(3)]^{(\infty)}$ as a sequence of 3×3 upper-triangular matrices g_i with upper-entries X_i, Y_i, and Z_i, then $\mathbf{H}$ is the subgroup of those sequences such that $Y_i = X_{i+1}$, $i \geq 1$. In addition, there is again an obvious surjective homomorphism onto $\mathbb{Z}^{(\infty)}$. Let ϕ_i be the probability measure $\phi_i = \frac{1}{3}(\delta_e + \delta_{x_i} + \delta_{x_i^{-1}})$ on $\mathbf{H}$. Let $\mathfrak{p} = (p_i)_1^\infty$ be a probability distribution such that there are constants $0 < a \leq 1 \leq b < \infty$ such that $ap_i \leq p_{i+1} \leq bp_i$. With these notations and hypotheses, the comparison techniques of [16] easily yield that, for fixed a, b, there are constants $c, C \in (0, \infty)$ such that for all $\mathfrak{p}$ satisfying $a p_i \leq p_{i+1} \leq b p_i$ and all n, we have

$$c \widetilde{\phi}_{\mathfrak{p}}^{(n)}(e) \leq \phi_{\mathfrak{p}}^{(n)}(e) \leq \overline{\phi}_{\mathfrak{p}}^{(n)}(0).$$

Here, $\widetilde{\phi}_{\mathfrak{p}}$ is the probability measure on $\mathbb{Z}^{(\infty)}$ naturally associated to $\mathfrak{p}$ and $\frac{1}{3}(\delta_0 + \delta_1 + \delta_{-1})$, and $\overline{\phi}_{\mathfrak{p}}$ is the measure on $[H_1(3)]^{(\infty)}$ associated with the probability distribution $\mathfrak{p}$ and the measure $\frac{1}{5}(\delta_e + \delta_x + \delta_{x^{-1}} + \delta_y + \delta_{y^{-1}})$ where x (resp. y) is the matrix in $H_1(3)$ with trivial entries except the $(1,2)$-entry (resp. the $(2,3)$-entry) which is 1. Again, this gives good bounds on $\phi_{\mathfrak{p}}^{(n)}(e)$ based on the bounds for direct sums obtained in Section 6.

2. Selected Technical Tools

Let G be a countable group and ϕ be a probability density on G. Set

$$h_t^{\phi}(x) = e^{-t} \sum_0^{\infty} \frac{t^n}{n!} \phi^{(n)}. \tag{2.3}$$

h_t^{ϕ} is the probability density of the continuous time random walk driven by ϕ at time $t > 0$. The following well-known proposition says that $n \mapsto \phi^{(2n)}(e)$ and $t \mapsto h_t^{\phi}(e)$ have essentially the same behavior when ϕ is symmetric. We will make frequent use of this result.

Proposition 2.1 ([16, Prop. 3.2]). *Assume that ϕ is symmetric. Then, for $n = 1, 2, \ldots$*

$$\phi^{(2n+2)}(e) \leq 2h_{2n}^{\phi}(e) \quad and \quad h_{4n}^{\phi}(e) \leq e^{-2n} + \phi^{(2n)}(e).$$

The Dirichlet form associated to a symmetric probability density ϕ is

$$\mathcal{E}_{\phi}(f, g) = \frac{1}{2} \sum_{x,y \in G} |f(xy) - f(x)|^2 \phi(y). \tag{2.4}$$

The following Proposition indicates that if two symmetric probability densities ϕ_i, $i = 1, 2$ have comparable Dirichlet forms $f \mapsto \mathcal{E}_{\phi_i}(f, f)$, $i = 1, 2$, then the behaviors of $n \mapsto \phi_i^{(2n)}(e)$, $i = 1, 2$ are essentially the same.

Proposition 2.2 ([16, Lem. 3.1 & Prop. 3.2]). *Assume that ϕ_i, $i = 1, 2$ are symmetric and that there is a constant A such that, for all functions f with finite support on G, $\mathcal{E}_{\phi_1}(f, f) \leq A \mathcal{E}_{\phi_2}(f, f)$. Then we have*

- $\forall t > 0, \quad h_t^{\phi_2}(e) \leq h_{t/A}^{\phi_1}(e)$.
- *For all $n = 1, 2, \ldots$, $\phi_2^{(4n+2)}(e) \leq 2e^{-2n} + 2\phi_1^{(2\lfloor n/A \rfloor)}(e)$.*

Given two nonnegative functions f, g of $t > 0$ or $n = 1, 2, \ldots$ in a neighborhood of 0 or infinity, we say that $f \asymp g$ if there are constants $c_i \in (0, \infty)$, $i = 1, 2$ such that $c_1 f(t) \leq g(t) \leq c_2 f(t)$ in a neighborhood of 0 or infinity. We say that $f \simeq g$ if there are constants $c_i \in (0, \infty)$, $i = 1, \ldots, 4$ such that $c_1 f(c_2 t) \leq g(t) \leq c_3 f(c_4 t)$ in a neighborhood of 0 or infinity. For instance, $f(t) = \exp(-t^{1/3})$ and $g(t) = \exp(-5t^{1/3} + t^{1/4} + 2)$ satisfy $f \simeq g$ near infinity. We will only use this notation when at least one of the functions f or g is monotone. Using this notation, we can state the following two-sided version of the previous proposition.

Proposition 2.3 ([16, Lem. 3.1 & Prop. 3.2]). *Assume that ϕ_i, $i = 1, 2$ are symmetric and that there are constants $a, A \in (0, \infty)$ such that, for all functions f with finite support on G, $a\mathcal{E}_{\phi_2}(f, f) \leq \mathcal{E}_{\phi_1}(f, f) \leq A\mathcal{E}_{\phi_2}(f, f)$. Then we have*

$$h_t^{\phi_2}(e) \simeq h_t^{\phi_1}(e) \text{ at infinity}$$

and

$$\phi_2^{(2n)}(e) \simeq \phi_1^{(2n)}(e) \text{ at infinity.}$$

Given a probability density ϕ on a countable group G, let R_ϕ be the operator of convolution on the right by ϕ, $R_\phi : f \mapsto R_\phi f = f * \phi$ acting on finitely supported function and, by extension on the spaces $L^p(G)$, $1 \leq p \leq \infty$. With this notation, h_t^ϕ is the convolution density of the semigroup of operators $e^{-t(I - R_\phi)}$, that is, $e^{-t(I - R_\phi)} f = f * h_t^\phi$.

3.   Random Walks on Increasing Union of Finite Groups

This section focuses on group G that can be viewed as the increasing union of a countable family of finite groups G_i. Namely, we assume that there is a countable infinite family of finite subgroups $G_i \subset G$, $i = 1, 2 \ldots$, such that $G_i \subset G_j$ for any $i \leq j$ and $G = \cup_1^\infty G_i$. Note that we assume that the sequence of subgroups G_i is strictly increasing so that G is infinite. See Examples 1.1, 1.2, 1.3 and 1.5 described in the introduction. This case is also the case considered in [14] and in [7].

3.1.   On Diagonal Upper Bounds: Convex Combinations of Commuting Probability Densities

For each i, let v_i be a probability density on G_i (with respect to the unnormalized counting measure). By extension, we view all v_i as probability densities on G as well.

In this section, we make the following rather restrictive assumptions.

(a) for all $i, j \in I$, $v_i * v_j = v_j * v_i$, i.e., the v_i's define commuting convolution operators on G.

(b) For each i there is an integer T_i such that $\sup_{x \in G_i}\{h_{T_i}^{v_i}(x)\} \leq C/|G_i|$. The time T_i provides a rough notion of "time to equilibrium" for the random walk driven by v_i on the finite group G_i.

We note that hypothesis (b) does not imply convergence of $v_i^{(k)}$ to the uniform probability u_i on G_i, nor even the convergence of $h_t^{v_i}$ to u_i.

For any probability measure $\mathfrak{p} = (p_i)$ on $\mathbb{N}$, we consider the probability measure $v_{\mathfrak{p}}$ on G defined by

$$v_{\mathfrak{p}} = \sum_1^\infty p_i v_i. \tag{2.5}$$

Because the operators R_{v_i} commute, we have

$$e^{-t(I-R_{v_{\mathfrak{p}}})} = \prod_I e^{-tp_i(I-R_{v_i})}$$

and

$$h_t^{v_{\mathfrak{p}}}(e) = \left[\prod_I^* h_{tp_i}^{v_i}\right](e) \quad \text{(convolution product evaluated at } e\text{)}.$$

This leads immediately to the following estimate.

Proposition 3.1. *Under the assumptions* (a)–(b) *introduced above, we have*

$$\forall t > 1, \quad h_t^{v_{\mathfrak{p}}}(e) \leq C \inf\left\{|G_i|^{-1} : i \in I, tp_i \geq T_i\right\}.$$

Example 3.2. On $\mathbb{S}^{(\infty)} = \cup_1^\infty \mathbb{S}_i$, let v_i be the random transposition measure on $\mathbb{S}_i$, that is, $v_i(x) = 2/n(n-1)$ if $x = (k,\ell)$ with $1 \leq k < \ell \leq i$ a transposition and 0 otherwise. In this case (see, e.g., [19]),

$$h_t^{v_{\mathfrak{p}}}(e) \leq C\frac{1}{i!} \text{ for } t \geq T_i/p_i, \quad T_i = 2i \log i.$$

Suppose that $p_i \simeq i^{-a}$, $a > 1$. Then we have $h_t^{v_{\mathfrak{p}}}(e) \leq \frac{C}{i!}$ for $t = 2i^{a+1}\log i$. This implies that (with different C, c)

$$\forall t > 0, \quad h_t^{v_{\mathfrak{p}}}(e) \leq C\exp(-ct^{1/(1+a)}(\log t)^{a/(1+a)}).$$

As we shall see later, this result is a little bit off and we do not know what the behavior of $h_t^{v_{\mathfrak{p}}}(e)$ is exactly in this case.

Remark 3.3. The anonymous referee points out that, when the measures v_i are symmetric, we have $\mathcal{E}_{v_{\mathfrak{p}}} \geq p_i \mathcal{E}_{v_i}$ for any i and assumption (a) can be omitted in the proposition stated above. By the comparison techniques of [8, 16] (i.e., Proposition 2.2), the assumption (b) alone implies the stated conclusion.

The next subsection proposes a more general and more powerful technique.

3.2. On Diagonal Bounds: Convex Combinations of Haar Measures and Comparison

We now explain how to apply the two basic ideas described in the introduction in the context of countable increasing union of finite groups. Let $G = \cup_1^\infty G_i$ as before and let u_i denote the uniform probability density on the finite group G_i. For any probability $\mathfrak{p} = (p_i)$ on $\{1, 2, \ldots\}$, set

$$\sigma_k = \sigma_k(\mathfrak{p}) = \sum_{i \geq k} p_i, \ k = 1, \ldots,$$

and consider the probability density $u_\mathfrak{p}$ on G defined by

$$u_\mathfrak{p} = \sum_1^\infty p_i u_i. \tag{2.6}$$

The behavior of these measures is studied in [3] using the observation that

$$u_\mathfrak{p}^{(n)} = \sum_{i_1,\ldots,i_n} p_{i_1} \cdots p_{i_n} u_{\max\{i_1,\ldots,i_n\}}$$

$$= \sum_k \left(\left(\sum_{i \leq k} p_i \right)^n - \left(\sum_{i < k} p_i \right)^n \right) u_k$$

$$= \sum_k \left((1 - \sigma_{k+1})^n - (1 - \sigma_k)^n \right) u_k.$$

Hence

$$u_\mathfrak{p}^{(n)}(e) = \sum_2^\infty (1 - \sigma_k)^n \left(\frac{1}{|G_{k-1}|} - \frac{1}{|G_k|} \right). \tag{2.7}$$

We write this in the form of a Laplace transform, namely,

$$u_\mathfrak{p}^{(n)}(e) = \int_0^\infty e^{-ns} d\mathcal{F}(s) = n \int_0^\infty e^{-ns} \mathcal{F}(s) ds \tag{2.8}$$

where

$$\mathcal{F}(s) = \frac{1}{|G_{k-1}|} \text{ for } \log \frac{1}{1 - \sigma_k} \leq s < \log \frac{1}{1 - \sigma_{k-1}}, \ k = 2, 3 \ldots.$$

The following less-precise version is often easier to handle.

Proposition 3.4. *We have*

$$u_\mathfrak{p}^{(n)}(e) \simeq \int_0^\infty e^{-ns} dF(s) = n \int_0^\infty e^{-ns} F(s) ds$$

where

$$F(s) = \frac{1}{|G_{k-1}|} \ for \ \sigma_k \leq s < \sigma_{k-1}, \ k = 1, 2, \ldots.$$

Many explicit examples illustrating this proposition are discussed in [4], and we will discuss some subtle use (2.7) below. What make this computation very valuable is that, by using comparison techniques, they lead to interesting estimate for other random walk models. See Proposition 3.7 below along with its applications.

Example 3.5. Because $G_k \subset G_{k+1}$, we must have $|G_{k+1}| \geq 2|G_k|$. Hence, if $g(k) = |G_k|$, the slowest possible growth rate of g is exponential. Let us assume that $\sup_k \frac{g(k+1)}{g(k)} \leq A$ so that $g(s) \simeq e^s$. Assume further that $\sigma_k = \xi(k)$ where ξ is a continuous decreasing function such that $\sup_k \frac{\xi(k)}{\xi(k+1)} \leq A$. Under these assumptions, Proposition 3.4 is very easy to use given that

$$F \simeq \frac{1}{g \circ \xi^{-1}}.$$

For instance, if $\xi(s) = c_\alpha s^{-\alpha}$ (hence $p_k \simeq k^{-\alpha-1}$) then

$$u_{\mathfrak{p}}^{(n)}(e) \simeq e^{-n^{1/\alpha}}.$$

As a second example, consider the case when $g(k) \asymp ce^{\beta k}$ and $\xi(s) = c_\alpha e^{-\alpha s}$. Then

$$u_{\mathfrak{p}}^{(n)}(e) \simeq n^{-\beta/\alpha}.$$

Next, for each i, we fix a symmetric probability density ϕ_i on G_i and we let λ_i be the second smallest eigenvalue of $I - R_{\phi_i}$ on $L^2(G_i)$. By definition λ_i is given by

$$\lambda_i = \inf\left\{\frac{\mathcal{E}_{\phi_i}(f,f)}{\|f - u_i(f)\|^2} : f \in \ell^2(G_i), \ f \text{ not a constant function}\right\}.$$

Given a sequence $(\phi_i)_1^\infty$ as above and a probability law $\mathfrak{p} = (p_i)$ on $\{1, 2, \ldots\}$, we set

$$\phi_{\mathfrak{p}} = \sum_1^\infty p_i \phi_i. \tag{2.9}$$

Lemma 3.6. *Referring to the above setup and notation and assuming that at least one λ_i is positive, define $\mathfrak{q} = (q_i)$ by setting*

$$\lambda(\mathfrak{p}) = \sum \lambda_i p_i, \quad q_i = \lambda(\mathfrak{p})^{-1}\lambda_i p_i.$$

Then we have

$$\lambda(\mathfrak{p})\mathcal{E}_{u_{\mathfrak{q}}} \leq \mathcal{E}_{\phi_{\mathfrak{p}}} \leq 2\mathcal{E}_{u_{\mathfrak{p}}}.$$

Proof. Simply observe that $\mathcal{E}_{\phi_{\mathfrak{p}}} = \sum p_i \mathcal{E}_{\phi_i}$ and that $\lambda_i \mathcal{E}_{u_i} \leq \mathcal{E}_{\phi_i} \leq 2\mathcal{E}_{u_i}$. The last inequality expresses the fact that the nonzero eigenvalues of $I - R_{\phi_i}$ are in the interval $[\lambda_i, 2]$. $\qquad\square$

Proposition 3.7. *Referring to the above setup and notation, there are constants $c, C \in (0, \infty)$ and an integer k such that*

$$\phi_{\mathfrak{p}}^{(2kn)}(e) \leq C u_{\mathfrak{q}}^{(2n)}(e)$$

and

$$u_{\mathfrak{p}}^{(2kn)}(e) \leq C \phi_{\mathfrak{p}}^{(2n)}(e).$$

Proof. Applies Lemma 3.6 and Proposition 2.3. $\qquad\square$

Let us point out that Proposition 3.7 provides excellent control of $\phi_{\mathfrak{p}}$ when the family $(G_i, \phi_i)_{i \in I}$ has the property that $\lambda_i \asymp 1$. We give two explicit examples.

Example 3.8. Consider the infinite dimensional special linear group with coefficients in the finite field $\mathbb{Z}/p\mathbb{Z}$ defined as the increasing union of $\mathrm{SL}_i(p)$. Thinking in terms of matrices, to go from i to $i+1$, add a column to the right and a row to the bottom. Identify $\mathrm{SL}_i(p)$ in $\mathrm{SL}_{i+1}(p)$ as those matrices with 0 entries in the last row and column except for the bottom right entry, which is 1. Let ϕ_i be the "random transvection" measure on $\mathrm{SL}_i(p)$ (see [10, 19] for a detailed description). Then $\lambda_i \asymp 1$ and $\phi_{\mathfrak{p}}^{(n)}(e) \simeq u_{\mathfrak{p}}^{(n)}(e)$ for every choice of $\mathfrak{p}$.

Example 3.9. M. Kassabov [11] proved that the symmetric groups $\mathbb{S}_i$ can be made a family of expanders by an appropriate choice of symmetric generating sets $B_i \subset \mathbb{S}_i$ of uniformly bounded cardinal. Assuming as we may that each B_i contains the identity element, let ϕ_i be the uniform probability measure on B_i. Then the associated spectral gap sequence λ_i satisfies $\lambda_i \asymp 1$ and we have $\phi_{\mathfrak{p}}^{(n)}(e) \simeq u_{\mathfrak{p}}^{(n)}(e)$ for every choice of $\mathfrak{p}$. See also [12], which provides further examples of this type involving finite simple groups.

4. Examples on $\mathbb{S}^{(\infty)}$

In this section, we write $\mathbb{S}^{(\infty)} = \cup_1^\infty \mathbb{S}_n$ where $\mathbb{S}_n$ is the symmetric group on $\{1, \ldots, n\}$ objects.

4.1. Behavior of Convex Combinations of Uniforms

Fix $\mathfrak{p} = (p_i)_n$ and $\sigma_k = \sum_{i \geq k} p_i$. The computation of the previous section yields

$$u_{\mathfrak{p}}^{(n)}(e) \simeq \int_0^\infty e^{-ns} dF(s) \text{ where } F(s) = 1/k! \text{ if } s \in [\sigma_k, \sigma_{k-1}). \quad (2.10)$$

For the case when $p_k = c_\alpha k^{-\alpha}$ and $p_k = 2^{-k+1}$, the computations are straightforward and given in [3]. One obtains

$$- \log u_{\mathfrak{p}}^{(n)}(e) \simeq \begin{cases} (\log n)(\log \log n) & \text{if } p_k = 2^{-k+1} \\ n^{1/\alpha}(\log n)^{1-1/\alpha} & \text{if } p_k = c_\alpha k^{-\alpha}, \ \alpha > 1. \end{cases}$$

Next, we treat the more difficult case when (recall that $\Gamma(n+1) = n!$)

$$p_k \asymp \sigma_k \asymp \frac{k^a}{k!} \asymp \frac{1}{\Gamma(k - a - 1)}.$$

As we shall see, this case allow us to obtain both recurrent walks and transient walks and to capture the transition from one case to the other as a function of the parameter a. In this particular context, for s small enough, we have

$$c_1 s \left(\frac{\log \log 1/s}{\log 1/s} \right)^a \leq F(s) \leq C_1 s \left(\frac{\log \log 1/s}{\log 1/s} \right)^{a-1}.$$

By [6, Theorem 1.7.1'], for $n \geq 2$, we have

$$c_2 n^{-1} \left(\frac{\log \log n}{\log n} \right)^a \leq u_{\mathfrak{p}}^{(n)}(e) \leq C_2 n^{-1} \left(\frac{\log \log n}{\log n} \right)^{a-1}.$$

The following result is more precise.

Theorem 4.1. *On* $\mathbb{S}^{(\infty)}$ *with* $p_k \asymp \frac{k^a}{k!}$, $a \in \mathbb{R}$ *fixed, we have*

$$\forall n, \ u_{\mathfrak{p}}^{(n)}(e) \leq C_1 n^{-1} \left(\frac{\log \log n}{\log n} \right)^{a-1}.$$

Further, for any fixed constants $0 < c < C < \infty$, *there exists* $c_1 > 0$ *such that for* $n \in \bigcup_k (c \frac{k!}{k^a}, C \frac{k!}{k^a})$, *we have*

$$u_{\mathfrak{p}}^{(n)}(e) \geq c_1 n^{-1} \left(\frac{\log \log n}{\log n} \right)^{a-1}.$$

Finally,

$$\forall n, \ u_{\mathfrak{p}}^{(n)}(e) \geq c_2 n^{-1} \frac{(\log \log n)^{a+1}}{(\log n)^a}$$

and there is a constant $c_3 > 1$ *such that this lower bound is sharp up to a multiplicative constant for integers* n *in* $\bigcup_k (c_3 \frac{k!}{k^a \log k}, c_3^2 \frac{k!}{k^a \log k})$.

Proof. The upper bound has already been proved above. For the lower bounds, we go back to the formula

$$u_{\mathfrak{p}}^{(n)}(e) = \sum_1^\infty (1 - \sigma_k)^n \left(\frac{1}{(k-1)!} - \frac{1}{k!} \right) \simeq \sum_1^\infty \frac{e^{-n\sigma_k}}{(k-1)!}.$$

Set $A_k = \frac{e^{-n\sigma_k}}{(k-1)!}$ and consider

$$\frac{A_k}{A_{k+1}} = e^{-n(\sigma_k - \sigma_{k+1})}\frac{k!}{(k-1)!} = k e^{-n(\sigma_k - \sigma_{k+1})}.$$

For a fixed n, let $\ell = \ell_n$ be the integer such that

$$\frac{\ell!}{\ell^a} \leq n < \frac{(\ell+1)!}{(\ell+1)^a}.$$

For $k \in \{m_a, \ldots, \ell-1\}$ with m_a a sufficient large integer, we have

$$2n(\sigma_k - \sigma_{k+1}) \geq n\sigma_k \geq c\sigma_\ell^{-1}\sigma_{\ell-1} \asymp (1 + 1/\ell)^{-a}\ell \geq \ell/2$$

for some $c > 0$. Hence $A_k/A_{k+1} \leq \ell e^{-\ell/4}$. For n large enough (hence ℓ large enough as well), $A_k/A_{k+1} \leq 1/2$ for all $k \in \{m_a, \ldots, \ell-1\}$. It follows that

$$\sum_1^{\ell-1} A_k \leq CA_\ell.$$

For $k \geq \ell+1$, $n(\sigma_k - \sigma_{k+1}) \leq C$. Hence $A_k/A_{k+1} \geq e^{-C}k \geq 2$ for n large enough and

$$\sum_{\ell+2}^\infty A_k \leq C'A_{\ell+1}.$$

This shows that

$$u_{\mathrm{p}}^{(n)}(e) \simeq \frac{e^{-n\sigma_{\ell_n}}}{(\ell_n-1)!} + \frac{e^{-n\sigma_{\ell_n+1}}}{\ell_n!}.$$

In particular, for any fixed constant $c \in (0, 1)$, and any n large enough such that $n \in (c\sigma_k^{-1}, (c\sigma_k)^{-1})$ for some k, we must have $k = \ell_n$ or $k = \ell_n + 1$. In both cases, it follows that

$$u_{\mathrm{p}}^{(n)}(e) \asymp \frac{1}{(k-1)!} \asymp n^{-1}\left(\frac{\log\log n}{\log n}\right)^{a-1}.$$

However, the union of the intervals $(c\sigma_k^{-1}, (c\sigma_k)^{-1})$, $k = 1, 2, \ldots$, does not cover all large integers.

To understand the behavior of $u_{\mathrm{p}}^{(n)}(e)$ when n is away from the special values σ_k^{-1}, for $n \in (\sigma_\ell^{-1}, \sigma_{\ell+1}^{-1})$, $\ell = \ell_n$, we define $t = t_n$ by

$$n = \sigma_{\ell_n}^{-1}t_n = \sigma_\ell^{-1}t.$$

Note that

$$t \in (1, \sigma_\ell\sigma_{\ell+1}^{-1}) \text{ and } \sigma_\ell\sigma_{\ell+1}^{-1} \asymp \ell.$$

We also set

$$\Phi(s) = \frac{1}{s}\left(\frac{\log\log s}{\log s}\right)^{a-1}$$

and observe that

$$\Phi(n) \asymp t^{-1}\Phi(\sigma_\ell^{-1}) \asymp t^{-1}\frac{1}{(\ell-1)!}.$$

Because we have

$$\frac{c_1}{(\ell-1)!}\left(e^{-c_2 t} + \frac{1}{\ell}e^{-c_2 t(\sigma_{\ell+1}/\sigma_\ell)}\right) \leq u_{\mathfrak{p}}^{(n)}(e)$$

$$\leq \frac{c_3}{(\ell-1)!}\left(e^{-c_4 t} + \frac{1}{\ell}e^{-c_4 t(\sigma_{\ell+1}/\sigma_\ell)}\right)$$

and that $t\sigma_{\ell+1}/\sigma_\ell \leq c_5$ and $\frac{1}{(\ell-1)!} \asymp \Phi(\sigma_\ell^{-1})$, we obtain that

$$ct\left(e^{-Ct} + \frac{1}{\ell}\right)\Phi(n) \leq u_{\mathfrak{p}}^{(n)}(e) \leq Ct\left(e^{-ct} + \frac{1}{\ell}\right)\Phi(n).$$

For any fixed $b > 0$, the function $g_\ell(s) = se^{-bs} + s/\ell$ satisfies $g(s) \simeq 1$ for s near 1 and s near ℓ and it attained a minimum of order $\ell^{-1}\log\ell$ near $s \asymp \log\ell$. Near 1, $g_\ell(s) \simeq se^{-s}$ and on the interval $(\log\ell, \ell)$, $g_\ell(s) \simeq s/\ell$. Because for any n, $\ell = \ell_n \simeq \frac{\log n}{\log\log n}$, we obtain that

$$u_{\mathfrak{p}}^{(n)}(e) \geq c\Phi(n)\frac{(\log\log n)^2}{\log n} \simeq \frac{1}{n}\left(\frac{\log\log n}{\log n}\right)^a \log\log n.$$

Further, this lower bound is achieved (up to a multiplicative constant) at values of n of the form $n \asymp \sigma_\ell^{-1}\log\ell$.   □

4.2. Walks Based on Small Generating Sets of $\mathbb{S}_n$

In this section, we study random walks on $\mathbb{S}^{(\infty)}$ that are of the form

$$\phi_{\mathfrak{p}} = \sum_1^\infty p_i\phi_i$$

where the sequence ϕ_i is a natural sequence of probability measures ϕ_i where ϕ is supported on the symmetric group $\mathbb{S}_i$. One of the most natural examples is when

$$\phi_i(\tau) = \begin{cases} 1/i \text{ if } \tau = e \\ 2/i^2 \text{ if } \tau = (p,q), 1 \leq p < q \leq i \\ 0 \text{ otherwise.} \end{cases}$$

This probability measure on $\mathbb{S}_i$ is known as "random transposition."

It is important to note that the measure $\phi_{\mathfrak{p}} = \sum_i p_i \phi_i$ can be written in many different ways. In particular, we can collect all occurrences of the transposition (p, q), $p \leq q$, as follows.

$$\phi_{\mathfrak{p}} = \left(\sum_1^\infty \frac{p_i}{i^2}\right)\delta_e + \sum_{q=2}^\infty \left((q-1)\sum_{i \geq q}\frac{2p_i}{i^2}\right)\frac{1}{q-1}\sum_{1 \leq p < q}\delta_{(p,q)}.$$

In other words,

$$\phi_{\mathfrak{p}} = \omega_{\tilde{\mathfrak{p}}} = \sum \tilde{p}_j \omega_j$$

where

$$\tilde{p}_1 = \sum_1^\infty i^{-2}p_i, \ \ \tilde{p}_j = 2(j-1)\sum_{i \geq j}^\infty i^{-2}p_i$$

and

$$\omega_1 = \delta_e, \omega_j = (j-1)^{-1}\sum_{1 \leq k < j}\delta_{(k,j)}.$$

The measures $\omega_{\tilde{\mathfrak{p}}}$ are considered in [7] (model μ_1).

In addition to the map $\mathfrak{p} \mapsto \tilde{\mathfrak{p}}$, consider the maps

$$\mathfrak{p} \mapsto \mathfrak{p}_\pm, \ \ (p_\pm)_i = c_\pm(\mathfrak{p})^{-1}(1+i)^{\pm 1}p_i$$

where $c_\pm(\mathfrak{p}) = \sum_i(1+i)^{\pm 1}p_i$ (assuming the sum is finite).

It is well known (e.g., see [19]) that the spectral gap of the measure ϕ_i and ω_i are both of order $1/i$. Hence, Lemma 3.6 yields

$$cc_-(\mathfrak{p})\mathcal{E}_{u_{\mathfrak{p}_-}} \leq \mathcal{E}_{\phi_{\mathfrak{p}}} \leq 2\mathcal{E}_{u_{\mathfrak{p}}}$$

and

$$cc_-(\mathfrak{p})\mathcal{E}_{u_{\mathfrak{p}_-}} \leq \mathcal{E}_{\omega_{\mathfrak{p}}} \leq 2\mathcal{E}_{u_{\mathfrak{p}}}.$$

Obviously, we also have $\mathcal{E}_{\phi_{\mathfrak{p}}} = \mathcal{E}_{\omega_{\tilde{\mathfrak{p}}}}$. This means that $\phi_{\mathfrak{p}}^{(n)}(e) \simeq \omega_{\mathfrak{p}}^{(n)}(e)$ if and only if $\tilde{p}_i \asymp p_i$. This will happen whenever p_i decays as a (summable) power function.

Proposition 4.2 (Random transposition). *Referring to the above setup and notation on $\mathbb{S}^{(\infty)}$, there is a constant c such that for any probability $\mathfrak{p} = (p_i)$*

$$cc_-(\mathfrak{p})\mathcal{E}_{u_{\mathfrak{p}_-}} \leq \mathcal{E}_{\phi_{\mathfrak{p}}} \leq 2\mathcal{E}_{u_{\tilde{\mathfrak{p}}}}.$$

In general, for each $\mathfrak{p}$, there are constants C, k such that

$$\phi_{\mathfrak{p}}^{(kn)}(e) \leq Cu_{\mathfrak{p}_-}^{(n)}(e)$$

and

$$u_{\tilde{\mathfrak{p}}}^{(kn)}(e) \leq C\phi_{\mathfrak{p}}^{(n)}(e).$$

In particular, when $\mathfrak{p}$ *is such that* $\tilde{p}_i \asymp (1+i)^{-1} p_i$, *we have*

$$\omega_{\mathfrak{p}_-}^{(n)}(e) \simeq \phi_{\mathfrak{p}}^{(n)}(e) \simeq u_{\mathfrak{p}_-}^{(n)}(e).$$

Corollary 4.3. *Referring to the above setup and notation on* $\mathbb{S}^{(\infty)}$, *assume that* $\mathfrak{p}$ *is such that* $p_i \asymp \frac{i^a}{i!}$.

1. *The walk associated with* $\phi_{\mathfrak{p}}$ *is recurrent if and only if* $a \le 3$. *Further, for* $n \in \bigcup_k (ck!\, k^{-a}, Ck!\, k^{-a})$,

$$\phi_{\mathfrak{p}}^{(n)}(e) \simeq n^{-1} \left(\frac{\log\log n}{\log n} \right)^{a-2}.$$

2. *The walk associated with* $\omega_{\mathfrak{p}}$ *is recurrent if and only if* $a \le 2$. *Further, for* $n \in \bigcup_k (ck!\, k^{-a}, Ck!\, k^{-a})$,

$$\omega_{\mathfrak{p}}^{(n)}(e) \simeq n^{-1} \left(\frac{\log\log n}{\log n} \right)^{a-1}.$$

Remark 4.4. A typical case when $\tilde{p}_i \not\asymp (1+i)^{-1} p_i$ is $p_i \asymp (1+i)^{-a}$, $a > 1$. Indeed, in this case, $\tilde{p}_i \asymp p_i$. As $-\log(u_{\mathfrak{p}}^{(n)}(e)) \simeq n^{1/a}(\log n)^{1-1/a}$, Proposition 4.2 gives

$$cn^{1/(1+a)}(\log n)^{1-1/(1+a)} \le -\log(\phi_{\mathfrak{p}}^{(n)}(e)) \le Cn^{1/a}(\log n)^{1-1/a}.$$

Note that in this case, we also have $-\log(\phi_{\mathfrak{p}}^{(n)}(e)) \simeq -\log(\omega_{\mathfrak{p}}^{(n)}(e))$.

Remark 4.5. In [7], four different models are considered. As mentioned above the model μ_1 in [7] is (essentially) what we call $\omega_{\mathfrak{p}}$ here. For the other models μ_i, $i = 2, 3, 4$ of [7], we have $\mathcal{E}_{\mu_1} \le C\mathcal{E}_{\mu_i}$ as long as the coefficients p_i in these models satisfy

$$\limsup_i p_i/p_{i+1} < 1. \tag{2.11}$$

Under condition (2.11), we also have $\tilde{p}_i \asymp (1+i)^{-1} p_i$ and

$$\forall\, n, \quad \mu_i^{(n)}(e) \simeq u_{\mathfrak{p}_+}^{(n)}(e), \quad i = 1, 2, 3, 4.$$

Example 4.6. Another example of interest (there are many on $\mathbb{S}^{(\infty)}$), is built on adjacent transpositions. On $\mathbb{S}_i$, set $\nu_i = (i-1)^{-1} \sum_{1 \le j < i} \delta_{(j,j+1)}$. Set

$$\nu_{\mathfrak{p}} = \sum p_i \nu_i = \sum_{j=1}^{\infty} \check{p}_j \delta_{(j,j+1)}, \quad \check{p}_j = \left(\sum_{i \ge j+1} ((i-1)^{-1} p_i \right).$$

The spectral gap of ν_i is known to be of order $1/i^3$ (see [19] and the references therein). This implies that there are k, C such that

$$u_{\mathfrak{p}}^{(kn)}(e) \le Cv_{\mathfrak{p}}^{(n)}(e), \quad v_{\mathfrak{p}}^{(kn)}(e) \le Cu_{\mathfrak{p}_-3}^{(n)}(e)$$

where $\mathfrak{p}_{-x}$ denotes the probability proportional to $(1+i)^{-x} p_i$. Even in the best-case scenario where $\check{p}_i \asymp (1+i)^{-1} p_i$, we only get the bounds

$$u_{\mathfrak{p}_{-1}}^{(kn)}(e) \leq C v_{\mathfrak{p}}^{(n)}(e), \quad v_{\mathfrak{p}}^{(kn)}(e) \leq C u_{\mathfrak{p}_{-3}}^{(n)}(e).$$

However, on each $\mathbb{S}^i$, we also have $c i^{-2} \mathcal{E}_{\omega_i} \leq \mathcal{E}_{v_i} \leq C \mathcal{E}_{\omega_i}$. This gives

$$\omega_{\mathfrak{p}}^{(kn)}(e) \leq C v_{\mathfrak{p}}^{(n)}(e), \quad v_{\mathfrak{p}}^{(kn)}(e) \leq C \omega_{\mathfrak{p}_{-2}}^{(n)}(e).$$

Now, if we assume that $\tilde{p}_i \asymp (1+i)^{-1} p_i$, we have

$$\omega_{\mathfrak{p}_{-2}}^{(n)}(e) \asymp u_{\mathfrak{p}_{-2}}^{(n)}(e).$$

In particular, if $\check{p}_i \asymp \tilde{p}_i \asymp (1+i)^{-1} p_i$, we have

$$u_{\mathfrak{p}_{-1}}^{(kn)}(e) \leq C v_{\mathfrak{p}}^{(n)}(e), \quad v_{\mathfrak{p}}^{(kn)}(e) \leq C u_{\mathfrak{p}_{-2}}^{(n)}(e).$$

Still, we are not able to describe precisely the behavior of $v_{\mathfrak{p}}^{(n)}(e)$. In fact, we are not able to capture the transition between recurrence and transience for $v_{\mathfrak{p}}$ with $p_i \asymp \frac{i^a}{i!}$ in terms of the parameter a.

5. Around the Recurrence/Transience Dichotomy

This section abstracts and extends the results obtained in Theorem 4.1. We consider $G = \bigcup_k G_k$ (increasing union of finite groups and study $u_{\mathfrak{p}}^{(n)}(e)$ under the hypothesis that

$$\sigma_k(\mathfrak{p}) = \sigma_k = \frac{1}{\xi(|G_k|)}$$

where ξ is a regularly varying function of positive index. In addition, we assume that the sequence $|G_{k+1}|/|G_k|$ is weakly increasing in the sense that

$$\forall k \leq m, \quad \frac{|G_{k+1}|}{|G_k|} \leq C \frac{|G_{m+1}|}{|G_m|}. \tag{2.12}$$

We already know that we always have (see (2.7))

$$u_{\mathfrak{p}}^{(n)}(e) \simeq \sum_1^\infty \frac{e^{-n\sigma_k}}{|G_{k-1}|}.$$

However, under the present hypotheses, we claim that

$$u_{\mathfrak{p}}^{(n)}(e) \simeq \frac{e^{-n\sigma_\ell}}{|G_{\ell-1}|} + \frac{e^{-n\sigma_{\ell+1}}}{|G_\ell|} \quad \text{for} \quad n \in [\xi(|G_\ell|), \xi(|G_{\ell+1}|)]. \tag{2.13}$$

To prove this claim, set $A_k = e^{-n\sigma_k}/|G_{k-1}|$ and observe that

$$\frac{A_k}{A_{k+1}} = e^{-n(\sigma_k - \sigma_{k+1})} \frac{|G_k|}{|G_{k-1}|}.$$

For n, ℓ related as in (2.13), we have

$$\frac{A_k}{A_{k+1}} \leq C \frac{|G_\ell|}{|G_{\ell-1}|} \exp\left(-\frac{\xi(|G_\ell|)}{\xi(|G_{\ell-1}|)}\right) \text{ for } k \leq \ell - 1$$

and

$$\frac{A_k}{A_{k+1}} \geq e^{-1} \frac{|G_\ell|}{|G_{\ell-1}|} \text{ for } k \geq \ell + 1.$$

It follows that for n large enough (this forces ℓ to large as well), we have

$$\frac{A_k}{A_{k+1}} \begin{cases} \leq 1/2 \text{ if } k < \ell \\ \geq 2 \text{ if } k > \ell. \end{cases}$$

This proves (2.13).

Theorem 5.1. *Assume that (2.12) is satisfied and that $\mathfrak{p}$ is such that $\sigma_k(\mathfrak{p}) = 1/\xi(|G_k|)$ where ξ is regularly varying of positive index α. Assume further that $|G_k|/|G_{k-1}| = \gamma(|G_k|)$ where γ is a slowly varying function. Then*

$$\forall n, \ u_{\mathfrak{p}}^{(n)}(e) \leq \frac{C\gamma \circ \xi^{-1}(n)}{\xi^{-1}(n)}.$$

Furthermore, for any fixed constants $0 < c < C < \infty$, we have

$$\forall n \in \bigcup_k (c\xi(|G_k|), C\xi(|G_k|)), \ u_{\mathfrak{p}}^{(n)}(e) \asymp \frac{\gamma \circ \xi^{-1}(n)}{\xi^{-1}(n)}.$$

In particular, the walk associated with $u_{\mathfrak{p}}$ is recurrent if $\alpha > 1$ and transient if $\alpha < 1$.

Proof. For the upper bound, it is best to use (2.8) and to observe that, in the present case, the function $\mathcal{F}(s)$ in (2.8) is bounded above by

$$\mathcal{F}(s) \leq \frac{\gamma \circ \xi^{-1}(1/s)}{\xi^{-1}(1/s)}.$$

It follows from (2.8) and [6, Theorem 1.7.1] that

$$u_{\mathfrak{p}}^{(n)}(e) \leq C \frac{\gamma \circ \xi^{-1}(n)}{\xi^{-1}(n)}.$$

For $n \asymp \xi(|G_\ell|)$, the estimate (2.13) gives

$$u_{\mathfrak{p}}^{(n)}(e) \asymp \frac{1}{|G_\ell|} \frac{|G_\ell|}{|G_{\ell-1}|} \asymp \frac{C\gamma \circ \xi^{-1}(n)}{\xi^{-1}(n)}.$$

$\square$

Example 5.2. The Burnside group $B(3, i)$ has order $3^{\binom{i}{3}+\binom{i}{2}+\binom{i}{1}} \simeq e^{r^3}$. This gives the following behaviors of $u_{\mathfrak{p}}^{(n)}(e)$. If $p_k \simeq k^{-\alpha-1}$, $\alpha > 0$, then

$$u_{\mathfrak{p}}^{(n)}(e) \simeq \exp\left(-n^{3/(3+\alpha)}\right).$$

If $p_k \simeq e^{-k}$ then

$$\log(u_{\mathfrak{p}}^{(n)}(e)) \simeq -(\log n)^3.$$

In order to apply Theorem 5.1, we note that

$$\frac{|B(3, r)|}{|B(3, r-1)|} = \gamma(|B(3, r)|)$$

where γ is slowly varying and $\log \gamma(t) \asymp (\log t)^{2/3}$. If $p_k = |B(3, k)|^{-\beta}$, $\beta > 0$, then Theorem 5.1 yields

$$c_1 n^{-1/\beta} \exp(-C_1 (\log n)^{2/3}) \le u_{\mathfrak{p}}^{(n)}(e) \le C_2 n^{-1/\beta} \exp(-c_2 (\log n)^{2/3})$$

as long as $n \asymp |B(3, k)|^\beta$ for some k.

Next, set $\phi_r = (1+r)^{-1} \sum_0^r \delta_{w_i}$ where $w_0 = e$ and w_i, $1 \le i \le r$, are the canonical generators of $B(3, r)$. We are interested in the behavior of

$$\phi_{\mathfrak{p}} = \sum p_i \phi_i, \quad \mathfrak{p} = (p_i)_1^\infty$$

on $B(3, \infty)$. We note that

$$\phi_{\mathfrak{p}} = \sum p_i \phi_i = \sum_{r=0}^{\infty} \left(\sum_{j \ge r} (1+j)^{-1} p_j \right) \delta_{w_r}.$$

Further, it is known that the spectral gap of ϕ_i on $B(3, i)$ is of order $1/i$. See [19, 20]. Hence, setting $\bar{p}_i = \sum_{j \ge i} (1+i)^{-1} p_i$

$$u_{\bar{\mathfrak{p}}}^{(kn)}(e) \le C \phi_{\mathfrak{p}}^{(n)}(e), \quad \phi_{\mathfrak{p}}^{(kn)}(e) \le C u_{\mathfrak{p}-1}^{(n)}(e).$$

In particular, if $\bar{p}_i \asymp (1+i)^{-1} p_i$ then

$$\phi_{\mathfrak{p}}^{(n)}(e) \simeq u_{\mathfrak{p}-1}^{(n)}(e).$$

This happens in particular if $p_i = A^i$, $A > 1$.

6. Direct Sums of Finitely Generated Groups

In this section, we consider the case when $G = \sum_I G_i$ is a (countable) direct sum. Recall that the elements of G are simply the sequences $g = (g_i)_{i \in I}$ such that $g_i = e_i$ for all but finitely many i in I.

In this situation, for each i, we fix a probability measure ψ_i on G_i and we fix a probability law $\mathfrak{p} = (p_i)$ on I. For each i, define a probability density ϕ_i on G by setting

$$\phi_i(y) = \begin{cases} \psi_i(y_i) & \text{if } y = (y_j)_{j \in I} \text{ with } y_j = e_j \text{ for all } i \neq j \\ 0 & \text{otherwise.} \end{cases}$$

We set $\phi_{\mathfrak{p}} = \sum p_i \phi_i$. Note that we use the same notation as in the previous section with a somewhat different meaning. The semigroup $e^{-t(I-R_{\phi_{\mathfrak{p}}})}$ is given by convolution with measures that we call $h_t^{\mathfrak{p}}$. Obviously,

$$h_t^{\mathfrak{p}} = e^{-t} \sum_0^\infty \frac{t^n}{n!} \phi_{\mathfrak{p}}^{(n)}.$$

By Proposition 2.1, we know that

$$\phi_{\mathfrak{p}}^{(n)}(e) \simeq h_n^{\mathfrak{p}}(e). \tag{2.14}$$

In the present setting, this is a very useful fact as we will show that the direct sum structure makes the behavior of $h_t^{\mathfrak{p}}(e)$ easily accessible.

On G_i, we have $e^{-t(I-R_{\psi_i})} f = f * h_{i,t}$ with $h_{i,t} = e^{-t} \sum_0^\infty (t^n/n!) \psi_i^{(n)}$. This can be used to express $e^{-t(i-R_{\phi_i})} f$ as follows. For each i, $x \in G$ and $y \in G_i$ set $x_y^i = (z_j)_{j \in I}$ where $z_j = x_j$ if $j \neq i$ and $x_i = y$. For any finitely supported function f on G, defines $f_x^i : G_i \mapsto \mathbb{R}$ by setting $f_x^i(y) = f(x_y^i)$. Then,

$$e^{-t(I-R_{\phi_i})} f(x) = e^{-t(I-R_{\psi_i})} f_x^i(x_i).$$

Further, viewing all R_{ϕ_i}'s as commuting convolution operators on G, we have

$$I - R_{\phi_{\mathfrak{p}}} = \sum_I p_i (I - R_{\phi_i})$$

and thus

$$h_t^{\mathfrak{p}}(x) = \prod_I h_{i, tp_i}(x_i), \quad x = (x_i).$$

We need the following simple lemma.

Lemma 6.1. *On a finite or countable group Z, let ψ be a probability measure such that $\psi^{(n)}(e) \leq \eta < 1$ for all n. Then $h_t = e^{-t} \sum_0^\infty (t^n/n!) \psi^{(n)}$ satisfies*

$$e^{-t} \leq h_t(e) \leq e^{-(1-c_\eta)t}, \quad t \in (0,1)$$

where $c_\eta \in (0,1)$ is the unique solution to $se^{s-1} = \eta$.

Proof. The lower bound is obvious. For the upper bound, we write $h_t(e) \leq e^{-t}(1 + \eta(e^t - 1))$. Set $v(t) = 1 + \eta(e^t - 1)$. Then $v(0) = 1$ $v'(t) = \eta e^t \leq c_\eta e^{c_\eta t}$ on $(0, 1)$. Hence $v(t) \leq e^{c_\eta t}$. This gives $h_t(e) \leq e^{-(1-c_\eta)t}$ as desired. □

Proposition 6.2. *Let* $G = \sum_I G_i$. *Fix* $\eta \in (0, 1)$. *Let* ψ_i *be a probability density on* G_i *such that for all* $i \in I$ *and* n, $\psi_i^{(n)}(e_i) \leq \eta$. *Assume further that* $\log h_{i,t}(e_i) \simeq -H(t)$ *for* $t \geq 1$. *Then*

$$\log h_t^{\mathrm{p}}(e) \simeq -t \sum_{k:tp_k \leq 1} p_k - \sum_{k:p_k t \geq 1} H(tp_k).$$

Proof. Write

$$\log h_t^{\mathrm{p}}(e) = \sum_I \log h_{i,tp_i}(e_i)$$

$$= \sum_{tp_i \leq 1} \log h_{i,tp_i}(e_i) + \sum_{i:tp_i>1} \log h_{i,tp_i}(e_i).$$

To obtain the desired result, use Lemma 6.1 in the first sum and the assumption in the second sum. □

6.1. Examples on Direct Sums

Example 6.3. Consider the case when $G = \sum_I G_i$ with $G_i = \mathbb{Z}^d$, $\psi_i = \psi_0$ with ψ_0 symmetric with finite generating support on $\mathbb{Z}^d$ so that $h_{i,t}(0) \simeq t^{-d/2}$ for $t \geq 1$. Hence we can take $H(t) = \log t$, $t > 1$.

Case (a): Set $p_i = 2^{-i}$. We compute

$$\sum_{k:p_k t \leq 1} p_k = \sum_{k \geq \log_2 t} 2^{-k} \simeq 1/t.$$

To compute $\sum_{p_k t \geq 1} \log(tp_k)$, we fix an integer ℓ and consider all the integers k such that

$$\ell \leq \log(tp_k) < \ell + 1,$$

that is,

$$\ell - \log t \leq -k \log 2 < \ell + 1 - \log t.$$

Hence,

$$\sum_{p_k t \geq 1} \log(tp_k) \simeq \sum_{\ell \leq \log t} \ell \simeq (\log t)^2.$$

Applying Proposition 6.2, we obtain

$$\log h_{\mathrm{p},t}(e) \simeq -(\log t)^2.$$

Case (b): Set $p_i = c_\sigma i^{-\sigma}$ with $\sigma > 1$. Then

$$\sum_{k:tp_k<1} p_k = c_\sigma \sum_{k>(c_\sigma t)^{1/\sigma}} k^{-\sigma} \sim \frac{c_\sigma}{\sigma-1}(c_\sigma t)^{-1+1/\sigma} = \frac{c_\sigma^{1/\sigma}}{\sigma-1} t^{-1+1/\sigma}.$$

To compute $\sum_{k:p_k t\geq 1} \log(tp_k)$, fix an integer ℓ and consider the values of k such that $\log tp_k \in [\ell, \ell+1)$. We must have

$$\ell - \log c_\sigma t \leq -\sigma \log k < \ell - \log c_\sigma t + 1,$$

that is,

$$(c_\sigma t)^{1/\sigma} e^{-(\ell+1)/\sigma} < k \leq (c_\sigma t)^{1/\sigma} e^{-\ell/\sigma}.$$

Hence, by Proposition 6.2,

$$\sum_{k:p_k t\geq 1} \log(tp_k) \simeq \sum_{\ell\leq \log t} t^{1/\sigma} e^{-\ell/\sigma} \ell \simeq t^{1/\sigma}.$$

This gives

$$\log h_{p,t}(e) \simeq -t^{1/\sigma}.$$

We note that these results apply in fact not only when $G_0 = \mathbb{Z}^d$ but also when G is a group with polynomial volume growth. This includes all finitely generated nilpotent groups.

Example 6.4. Consider the case when $G_i = G_0$ with G_0 being a polycyclic group with exponential volume growth. Let $\psi_i = \psi_0$ with ψ_0 being a symmetric probability density with finite generating support on G_0. In this case, we have $H(s) \simeq s^{1/3}$. See, e.g., [21, Sect. VII.5] and [1]).
   Case (a): Set $p_i = 2^{-i}$. Then $\sum_{k:tp_k<1} p_k \simeq 1/t$ and

$$\sum_{p_k t\geq 1} H(tp_k) \simeq t^{1/3}\left(\sum_{k\leq \log_2 t} p_k^{1/3}\right) \simeq t^{1/3}.$$

Hence

$$\log h_{p,t}(e) \simeq -t^{1/3}.$$

This result generalizes immediately to the case where G_0 and ψ_0 are such that $H(s) \simeq s^a$ for some fixed $a \in (0,1]$ (see, e.g., [15] for such examples). In this case, $\log h_{p,t}(e) \simeq -t^a$.
   Case (b): Set $p_i = c_\sigma i^{-\sigma}$ with $\sigma > 1$. Then $\sum_{k:tp_k<1} p_k \simeq t^{-1+1/\sigma}$ as in Example 6.3(b). Further

$$\sum_{p_k t\geq 1} H(tp_k) \simeq t^{1/3}\left(\sum_{k\leq t^{1/\sigma}} k^{-\sigma/3}\right) \simeq \begin{cases} t^{1/3} & \text{if } \sigma > 3 \\ t^{1/3}\log t & \text{if } \sigma = 3 \\ t^{1/\sigma} & \text{if } \sigma < 3 \end{cases}$$

By Proposition 6.2, it follows that

$$\log h_{\mathrm{p},t}(e) \simeq - \begin{cases} t^{1/3} & \text{if } \sigma > 3 \\ t^{1/3} \log t & \text{if } \sigma = 3 \\ t^{1/\sigma} & \text{if } \sigma < 3. \end{cases}$$

This result generalizes immediately to the case where G_0 and ψ_0 are such that $H(s) \simeq s^a$ for some fixed $a \in (0,1]$ (see, e.g., [15] for such examples). In this case,

$$\log h_{\mathrm{p},t}(e) \simeq - \begin{cases} t^a & \text{if } \sigma > 1/a \\ t^a \log t & \text{if } \sigma = 1/a \\ t^{1/\sigma} & \text{if } \sigma < 1/a. \end{cases}$$

Our next two examples treat examples where $G = \sum_I G_i$ and the G_i's are finite groups.

Example 6.5. Consider the case when $G = \sum_I G_i$ with $G_i = G_0$, a finite group, and $\psi_i = \psi_0$ with ψ_0 symmetric with finite generating support on G_0. Then we can take $H(t) = 1$, $t > 1$.

Case (a): Set $p_i = 2^{-i}$. We have

$$\sum_{k:p_k t \leq 1} p_k \simeq 1/t \quad \text{and} \quad \sum_{p_k t \geq 1} 1 \simeq \log(t).$$

Applying Proposition 6.2, we obtain

$$\log h_{\mathrm{p},t}(e) \simeq - \log t.$$

In this case, a more precise estimate is desirable.

Case (b): Set $p_i = c_\sigma i^{-\sigma}$ with $\sigma > 1$. Then $\sum_{k:tp_k<1} p_k \sim \frac{c_\sigma^{1/\sigma}}{\sigma-1} t^{-1+1/\sigma}$. Further $\sum_{k:p_k t \geq 1} 1 \simeq t^{1/\sigma}$ as well. Hence, by Proposition 6.2,

$$\log h_t^{\mathrm{p}}(e) \simeq -t^{1/\sigma}.$$

In case (b) the behavior obtained when G_0 is finite is the same as when G_0 has polynomial volume growth.

Example 6.6. Consider the case when $G_i = \mathbb{Z}/m_i\mathbb{Z}$ is a cyclic finite group. On each of these finite group we let ψ_i be the probability density that is uniform on $\{-1, +1\}$. In this case, we observe that

$$\log h_{i,t}(0) \simeq - \begin{cases} t & \text{if } t \in (0, 1] \\ \log(1 + t) & \text{if } t \in (1, m_i^2) \\ \log m_i & \text{if } t \geq m_i^2. \end{cases}$$

See, e.g., [19].

Set $p_i = 2^{-i}$. We will consider three cases depending on the behavior of the sequence m_i. We assume that $m_i \simeq m(i)$ where m is an increasing function and $\log \circ m$ is regularly varying at infinity, of index $\alpha \in [0, \infty)$. In all cases,

$$\sum_{k:p_k t \leq 1} p_k \simeq 1/t.$$

Case (a): Assume that $\alpha \in [0,1]$ and that $\limsup_{t \to \infty} t^{-1} \log \circ m(t) < \infty$. Then, for any $t > 1$

$$\sum_{k:p_k t \geq m_k^2} \log m_k \simeq \log(t) \log[m(\log t)].$$

Further,

$$\sum_{k:p_k t \in [1, m_k^2)} \log(tp_k) \leq C \sum_{k \in [1, \log t)} \log m(k) \leq C \log(t) \log[m(\log t)].$$

Hence, we obtain

$$\log h_t^{\mathrm{p}}(e) \simeq - \log(t) \log[m(\log t)].$$

Case (b): Assume that $\alpha \in [1, \infty)$ and that $\liminf_{t \to \infty} t^{-1} \log \circ m(t) > 0$. Then

$$\sum_{k:p_k t \geq m_k^2} \log m_k \leq C \log(t)^2.$$

Further, using the computation form Example 6.3(a), we have

$$\sum_{k:p_k t \in [1, m_k^2)} \log(tp_k) \simeq \log(t)^2.$$

Hence, we obtain

$$\log h_t^{\mathrm{p}}(e) \simeq - \log(t)^2.$$

6.2. Comparison Inequalities

In this last subsection, we give a simple result that allows us to extend the estimates described above to measures that are not exactly adapted to the direct sum structure of the underlying group.

We continue to assume that $G = \sum_I G_i$ as earlier in this section and consider

$$\phi_{\mathrm{p}} = \sum_i p_i \phi_i,$$

where each ϕ_i is a symmetric probability measure on G_i with associated Dirichlet form $\mathcal{E}_{\phi_i}$ and set $\mathcal{E}_{\phi_i} = \mathcal{E}_i$.

Assume further that each ϕ_i satisfies the pseudo-Poincaré inequality

$$\forall \, y \in G_i, \quad \sum_{x \in G_i} |f(xy) - f(x)|^2 \leq P_i \rho_i(y)^2 \mathcal{E}_i(f,f). \qquad (2.15)$$

Example 6.7. If for each i there is a finite symmetric generating set S_i of G_i such that $\phi_i \geq \epsilon_i > 0$ on S_i then (2.15) holds true with $\rho_i(y)$ being the word-length distance between e_i and y in G_i based on the generating set S_i and $P_i = \epsilon_i^{-1}$.

Given $y \in G_i$, let $y^{\downarrow i} \in G = \sum_I G_i$ be the element $y^{\downarrow i} = (y_j)$ with $y_j = e_j$ if $j \neq i$ and $y_i = y$. Given a function $f \in \ell^2(G)$, $G = \sum_I G_i$, set

$$\mathcal{E}_i(f,f) = (1/2) \sum_{x \in G, y \in G_i} |f(xy^{\downarrow i}) - f(x)|^2 \phi_i(y).$$

Lemma 6.8. *Assume* (2.15). *For any* $y = (y_i) \in G$, *set*

$$\rho_{\mathrm{p}}(y) = \left(\sum_i \frac{P_i}{p_i} \rho_i(y_i)^2 \right)^{1/2}.$$

Then we have

$$\forall y \in G, \quad \sum_{x \in G} |f(xy) - f(x)|^2 \leq \rho_{\mathrm{p}}(y)^2 \mathcal{E}_{\phi_{\mathrm{p}}}(f,f). \qquad (2.16)$$

Note that $\rho_{\mathrm{p}}(y)$ is finite for any $y = (y_i)$ because $y_i = e$ for all but finitely many i.

Proof. To simplify notation, we write $\rho = \rho_{\mathrm{p}}$. Given $y = (y_i)$, let y^{*i} be the element in G with coordinates y_j^{*i} given by $y_j^{*i} = y_j$ if $j \leq i$ and $y_j^{*i} = e_j$ if $j > i$. Then

$$f(xy) - f(x) = \sum_i f(xy^{*i}) - f(xy^{*(i-1)})$$

(note that only finitely many terms are nonzero in this sum) and, by Minkowski inequality,

$$\|f(\cdot y) - f(\cdot)\|_2 = \| \sum_i f(\cdot y^{*i}) - f(\cdot y^{*(i-1)}) \|_2 \leq \sum_i \|f(\cdot y^{\downarrow i}) - f(\cdot)\|_2.$$

Next, using (2.15), we have

$$\sum_i \|f(\cdot y^{\downarrow i}) - f(\cdot)\|_2 \leq \sum_i P_i^{1/2} \rho_i(y_i) \mathcal{E}_i(f,f)^{1/2}$$

$$= \sum_i (P_i/p_i)^{1/2} \rho_i(y_i)(p_i \mathcal{E}_i(f,f))^{1/2}$$

$$\leq \rho(y)^{1/2} \left(\sum_i p_i \mathcal{E}_i(f,f) \right)^{1/2}$$

$$= \rho(y)^{1/2} \mathcal{E}_{\phi_{\mathrm{p}}}(f,f)^{1/2}.$$

$\square$

Proposition 2.2 and 6.3 together yield the following statement.

Proposition 6.9. *Referring to the setting and notation introduced above, let v be a symmetric probability measure on $G = \sum_I G_i$ such that $v(\rho_p^2) < \infty$. Then there are C, k such that*

$$\phi_p^{(kn)}(e) \leq C \left(v^{(n)}(e) + e^{-n/C} \right).$$

References

[1] G. Alexopoulos. A lower estimate for central probabilities on polycyclic groups. *Canad. J. Math.*, 44(5):897–910, 1992.

[2] A. Bendikov and L. Saloff-Coste. Random walks on groups and discrete subordination. *Math. Nachr.*, 285(5-6):580–605, 2012.

[3] Alexander Bendikov and Barbara Bobikau. Long time behavior of random walks on abelian groups. *Colloq. Math.*, 118(2):445–64, 2010.

[4] Alexander Bendikov, Barbara Bobikau, and Christophe Pittet. Spectral properties of a class of random walks on locally finite groups. *Groups Geom. Dyn.*, 7(4):791–820, 2013.

[5] Alexander Bendikov and Laurent Saloff-Coste. Random walks driven by low moment measures. *Ann. Probab.*, 40(6):2539–88, 2012.

[6] N. H. Bingham, C. M. Goldie, and J. L. Teugels. *Regular Variation*, Volume 27 of *Encyclopedia of Mathematics and its Applications*. Cambridge University Press, 1989.

[7] Sara Brofferio and Wolfgang Woess. On transience of card shuffling. *Proc. Amer. Math. Soc.*, 129(5):1513–19, 2001.

[8] Persi Diaconis and Laurent Saloff-Coste. Comparison techniques for random walk on finite groups. *Ann. Probab.*, 21(4):2131–56, 1993.

[9] Nabil′ Fereǐg and S. A. Molčanov. Random walks on abelian groups with an infinite number of generators. *Vestnik Moskov. Univ. Ser. I Mat. Mekh.*, (5):22–9, 1978.

[10] Martin Hildebrand. Generating random elements in $SL_n(\mathbf{F}_q)$ by random transvections. *J. Algebraic Combin.*, 1(2):133–50, 1992.

[11] Martin Kassabov. Symmetric groups and expander graphs. *Invent. Math.*, 170(2):327–54, 2007.

[12] Martin Kassabov, Alexander Lubotzky, and Nikolay Nikolov. Finite simple groups as expanders. *Proc. Natl. Acad. Sci. USA*, 103(16): 6116–119 (electronic), 2006.

[13] H. Kesten and F. Spitzer. Random walk on countably infinite Abelian groups. *Acta Math.*, 114:237–65, 1965.

[14] Gregory F. Lawler. Recurrence and transience for a card shuffling model. *Combin. Probab. Comput.*, 4(2):133–42, 1995.

[15] C. Pittet and L. Saloff-Coste. On random walks on wreath products. *Ann. Probab.*, 30(2):948–77, 2002.

[16] Ch. Pittet and L. Saloff-Coste. On the stability of the behavior of random walks on groups. *J. Geom. Anal.*, 10(4):713–37, 2000.

[17] Laurent Saloff-Coste. Probability on groups: random walks and invariant diffusions. *Notices Amer. Math. Soc.*, 48(9):968–77, 2001.

[18] Laurent Saloff-Coste. Analysis on Riemannian co-compact covers. In *Surveys in differential geometry. Vol. IX*, Surv. Differ. Geom., IX, pp. 351–84. Int. Press, Somerville, MA, 2004.

[19] Laurent Saloff-Coste. Random walks on finite groups. In *Probability on Discrete Structures*, Volume 110 of *Encyclopaedia Math. Sci.*, pp. 263–346. Springer, Berlin, 2004.

[20] Richard Stong. Eigenvalues of the natural random walk on the Burnside group $B(3, n)$. *Ann. Probab.*, 23(4):1950–60, 1995.

[21] N. Th. Varopoulos, L. Saloff-Coste, and T. Coulhon. *Analysis and Geometry on Groups*, Volume 100 of *Cambridge Tracts in Mathematics*. Cambridge University Press, 1992.

[22] Wolfgang Woess. *Random Walks on Infinite Graphs and Groups*, Volume 138 of *Cambridge Tracts in Mathematics*. Cambridge University Press, 2000.

3

THE COST OF DISTINGUISHING GRAPHS

DEBRA BOUTIN[1] AND WILFRIED IMRICH[2]

[1]Hamilton College, Clinton, NY 13323, USA
[2]Montanuniversität Leoben, 8700 Leoben, Austria

Dedicated to Wolfgang Woess on the occasion of his sixtieth birthday

Abstract

In a graph, a set of vertices that is stabilized setwise by only the trivial automorphism is called a distinguishing class. Not every graph has such a set, but if it does, we call its minimum size the distinguishing cost. Many families of graphs have such sets, and for some families the distinguishing costs are surprisingly small.

The chapter begins with a survey of results about the distinguishing cost for finite and infinite graphs. Then it focuses on infinite graphs with finite cost and presents a new upper bound on the distinguishing cost of graphs with linear growth, two ends, and infinite automorphism group. It ends with the problem of the existence of one-ended, connected, locally finite infinite graphs of linear growth and countably infinite automorphism group.

Key words: Distinguishing number, automorphism, determining set, determining number, infinite graph

AMS subject classification (2000): 05C15, 05C25.

Contents

1. Introduction

A significant number of recent papers investigate vertex colorings that are only preserved by the identity automorphism. Such colorings are called *distinguishing*. If two colors suffice, the coloring is said to be *two-distinguishing*, and the graph is said to be *two-distinguishable*. Given a two-distinguishing coloring, the set of vertices of either of the two colors is stabilized setwise only by the trivial automorphism. Such a set is called a *distinguishing class*, and the minimum size of a distinguishing class of a graph G is called the *distinguishing cost*; we denote it by $\rho(G)$.

The concept of a distinguishing class is different from that of a determining set. A *determining set* is a set of vertices whose pointwise stabilizer is the identity. Such a set is also called a *base for the action of the automorphism group*. A determining set need not be a distinguishing class, but every distinguishing class is a determining set. The minimum size of a determining set for G is called the *determining number*, $\mathrm{Det}(G)$. Clearly $\mathrm{Det}(G) \leq \rho(G)$.

By [10], an infinite, connected, locally finite graph G with infinite automorphism group has finite distinguishing cost (and thus finite determining number) if and only if the automorphism group is countable. In such a case, $\rho(G)$ cannot be larger than $3 \cdot \mathrm{Det}(G)$. However, for finite graphs $\rho(G)$ and $\mathrm{Det}(G)$ can be very far apart. We show this by constructing a family of finite graphs in which the distinguishing cost is a large multiple of the determining number.

Such constructions can easily be extended to finite graphs of any size and infinite graphs. Because, given a nontrivial graph G and a path P of arbitrary length or a one-sided infinite path R, then the graph H, obtained from G and P, respectively R, by joining one end vertex of P, respectively the end vertex of R, to all vertices of G, has the same cost and determining number as G. This is so because every automorphism of H fixes all vertices not in G, and every automorphism of G extends to one of H.

Our main result is a new bound on the distinguishing cost of infinite, connected, 2-ended, locally finite graphs with infinite automorphism group and linear growth.

For infinite graphs the proofs rely on facts about structure, ends, and automorphisms of infinite, locally finite graphs given by Halin [14], and more recently in [16]. For finite graphs, Cartesian products are the main tool.

To gain perspective, we also present a short survey of results about the distinguishing cost of finite and infinite graphs, the relationship between $\rho(G)$ and $\mathrm{Det}(G)$, and a partial list of classes of graphs that are two-distinguishable.

2. Preliminaries

Distinguishing two-colorings are a special case of distinguishing d-colorings. By a coloring of the vertices of a graph G with d colors we mean a labeling with the integers $1, \ldots, d$. If no nontrivial automorphism preserves the colors we call this a *d-distinguishing coloring*. A graph with such a coloring is called *d-distinguishable*. This concept was introduced by Albertson and Collins in [3] and has spawned a wealth of results for both finite and infinite graphs.

In many families of graphs, all but a small finite number of members are two-distinguishable. For example hypercubes Q_n are two-distinguishable for $n \geq 4$, Cartesian powers K_3^n for $n \geq 3$, and G^n for connected graphs $G \neq K_2, K_3$, where $n \geq 2$ [1, 9, 18, 19]. Other examples are Kneser graphs $KG_{n,k}$ for $n \geq 6, k \geq 2$ [2], and three-connected planar graphs (with seven small exceptions) [17].

Families of infinite two-distinguishable graphs include homogeneous trees of arbitrary degree [4], the hypercube of arbitrary transfinite dimension [15], denumerable, vertex-transitive graphs of connectivity one [21], the denumerable random graph [15], and connected, locally finite graphs of polynomial growth with the property that every automorphisms moves infinitely many vertices [20].

As noted previously, the color classes in a two-distinguishable coloring of a graph G are called *distinguishing classes*, and the minimum size of a distinguishing class is the *cost of two-distinguishing G*, or the *distinguishing cost of G*, and is denoted by $\rho(G)$. This concept was introduced and studied by Boutin in [5], motivated by the observation that the size of the smaller color class in a two-distinguishing coloring of a graph can be extremely small. For example, she showed that the hypercube Q_{2^k} of dimension 2^k can be two-distinguished by coloring $k+2$ vertices black[†] and the others white [7]. Moreover, there are infinite graphs with infinite automorphism group and finite distinguishing cost. In particular, if G is a connected, locally finite graph with infinite automorphism group, then $\rho(G)$ is finite precisely when $\mathrm{Aut}(G)$ is countable [10].

In both the finite and the infinite case, there is a connection between the minimum size of a determining set and that of a distinguishing class. Recall that a *determining set* for a graph G is a set whose point stabilizer in $\mathrm{Aut}(G)$ is trivial, that the *determining number* $\mathrm{Det}(G)$ is the minimum size of a determining set of G, and that $\mathrm{Det}(G) \leq \rho(G)$. It is possible that $\mathrm{Det}(G) = \rho(G)$, but we show that in finite graphs $\rho(G)$ can be an arbitrarily large multiple of $\mathrm{Det}(G)$.

We then provide a new upper bound for $\rho(G)$ when G is a connected, locally finite graph with linear growth, infinite automorphism group and two ends. This is our main result. Of course, it also implies an upper bound on $\mathrm{Det}(G)$.

3. Finite Graphs

Most of the results in this section pertain to Cartesian products and Cartesian powers of graphs. Many proofs invoke the structure of the automorphism group of Cartesian products. The kth Cartesian power of

---

[†] In a two-coloring we assign black to the smaller class and white to the other, unless both classes have the same size.

a graph G, denoted G^k, is the Cartesian product of G with itself k times. A good reference for both products and powers is [13]. In particular, automorphisms of Cartesian products are discussed beginning on page 128. Recall that a nontrivial graph G is *prime* with respect to the Cartesian product if it cannot be written as the Cartesian product of two smaller graphs. In particular, if G and H are relatively prime, that is, if they have no common nontrivial factor, then the automorphism group of $G \square H$ is the direct product of the automorphism groups of G and H. To be more precise, in this case, $\varphi \in \text{Aut}(G \square H)$ if and only if there is $\alpha \in \text{Aut}(G)$ and $\beta \in \text{Aut}(H)$ so that for all $(g, h) \in V(G \square H)$, $\varphi(g, h) = (\alpha(g), \beta(h))$. When useful, we write $\varphi = \gamma_{\alpha, \beta}$.

We begin with the hypercube Q_k, which is the kth Cartesian power of the complete graph K_2. In 2004, Bogstad and Cowan [9] showed that for $k \geq 4$, Q_k is two-distinguishable. The two-distinguishing coloring produced by their method gave a distinguishing class of size $k+2$. Imrich asked how small a distinguishing class could be for hypercubes. In the first paper to address the question of distinguishing cost, Boutin found upper and lower bounds within a factor of 2 of each other. The theorem is stated below.

Theorem 3.1. [5] *For $k \geq 5$, $\lceil \log_2 k \rceil + 1 \leq \rho(Q_k) \leq 2\lceil \log_2 k \rceil - 1$.*

Later, precise values for the distinguishing cost of certain hypercubes and certain Kneser graphs were obtained.

Theorem 3.2. [7] *For $m \geq 3$, $\rho(Q_{2^m}) = \rho(Q_{2^m-1}) = \rho(Q_{2^m-2}) = m + 2$.*

Theorem 3.3. [7] *$\rho(KG_{2^m-1,2^{m-1}-1}) = m + 1$.*

Before further examining graphs with a small distinguishing cost, it is natural to consider how large the distinguishing cost might be. Clearly, in a two-distinguishing coloring, one of the color classes must be at most half the size of the vertex set. This bound can be achieved as the following theorem shows.

Theorem 3.4. [12] *Let G be a graph with trivial automorphism group and H the lexicographic product of G and K_2. Then $\rho(H) = \frac{1}{2}|V(H)|$.*

Proof. First observe that H is obtained from G by replacing every vertex v by an edge $v_1 v_2$, and by replacing every edge uv by the edges $u_i v_j$, where $i, j \in \{1, 2\}$. Clearly the transposition of any pair $\{v_1, v_2\}$, where all other vertices are fixed, is an automorphism. To break it we have to assign different colors to v_1 and v_2, say black and white. Doing this for all pairs we see that H consists of a copy of G that has only black vertices and a copy that has only white vertices. Clearly the two copies of G cannot be interchanged. The observation that G is asymmetric completes the proof. $\square$

We wish to remark that for each two-distinguishing coloring of H above, though there is no nontrivial automorphism that preserves the color classes, there is one that switches them.

At the end of this section we will consider the graphs $H = K_{2^m} \,\square\, G$, where G is an asymmetric graph of order m. For graphs in this family we also have $\rho(H) = \frac{1}{2}|V(H)|$.

The fact that Q_k is K_2^k was instrumental in proving Theorem 3.1. For the cost of K_3^k a detour via the determining number of Cartesian powers of graphs was taken. For such powers we have the following theorem regarding determining numbers.

Theorem 3.5. [6] *Let G be a connected graph of order n that is prime with respect to the Cartesian product. Then*

$$\max\left\{\mathrm{Det}(G), \left\lceil (\log_n k + \log_n |\mathrm{Aut}(G)|) \right\rceil\right\} \le \mathrm{Det}(G^k)$$
$$\le \mathrm{Det}(G) + \left\lceil \log_n k \right\rceil.$$

Note that since G is fixed, $\mathrm{Det}(G)$, $|V(G)|$, and $|\mathrm{Aut}(G)|$ are constant. Thus, both upper and lower bounds on $\mathrm{Det}(G^k)$ grow logarithmically with k.

For graphs on at least three vertices meeting other mild hypotheses, the connection with distinguishing cost is provided by the next result. This will imply that both upper and lower bounds on $\rho(G^k)$ also grow logarithmically with k.

Theorem 3.6. [8] *If G^k is a two-distinguishable Cartesian power of a prime, connected graph G on at least three vertices with $\mathrm{Det}(G) \le k$ and $\max\{2, \mathrm{Det}(G)\} < \mathrm{Det}(G^k)$, then $\mathrm{Det}(G^k) \le \rho(G^k) \le \mathrm{Det}(G^k) + 1$.*

Using the known value of $\mathrm{Det}(K_3^k)$ from [6], the following theorem was an immediate consequence.

Theorem 3.7. [8] *For* $k \ge 3$, $\left\lceil \log_3(2k+1) \right\rceil + 1 \le \rho(K_3^k) \le \left\lceil \log_3(2k+1) \right\rceil + 2$.

We continue with a new result about finite graphs with large distinguishing cost.

Theorem 3.8. *If H is an asymmetric graph on m vertices, then $\rho(K_{2^m} \,\square\, H) = m2^{m-1}$.*

Proof. Let $V(H) = \{h_1, \ldots, h_m\}$. Recall that all complete graphs are prime with respect to the Cartesian product. Because K_{2^m} has more vertices than H, it cannot be a factor of it. Hence K_{2^m} and H are relatively prime. Using the notation from the beginning of this section, given that H is asymmetric, all automorphisms of $K_{2^m} \,\square\, H$ are of the form $\gamma_{\alpha,\mathrm{id}} = \gamma_\alpha$. That is, $\gamma_\alpha \in \mathrm{Aut}(K_{2^m} \,\square\, H)$ permutes the sets $H^z = \{(z, h_i) \mid h_i \in H\}$, $z \in V(K_{2^m})$, taking $H^x = \{(x, h_i) \mid h_i \in H\}$ to

$H^{\alpha(x)} = \{(\alpha(x), h_i) \mid h_i \in H\}$. The sets $H^z = \{(z, h_i) \mid h_i \in H\}$ are called H-fibers. Notice that γ_α, when applied to a vertex, only changes the first coordinate.

There are 2^m distinct two-colorings of H. Thus, we may assign distinct two-colorings to each of the 2^m H-fibers of $K_{2^m} \,\square\, H$. Any nontrivial automorphism of $K_{2^m} \,\square\, H$ (which is necessarily induced by a nontrivial automorphism of K_{2^m}) must nontrivially permute H-fibers. Because each H-fiber has a distinct coloring, no such nontrivial automorphism preserves the coloring. Thus, the coloring is distinguishing.

Note that the coloring produces two equal-size color classes of size $m2^{m-1}$, because $\sum_{i=0}^{m} \binom{m}{i} i = m2^{m-1}$. Thus, $\rho(K_{2^m} \,\square\, H) \le m2^{m-1}$.

Now, consider a two-coloring c of $K_{2^m} \,\square\, H$ with fewer than $m2^{m-1}$ black vertices. Because having distinct two-colorings of H-fibers implies $m2^{m-1}$ black vertices, not all two-colorings of H-fibers under c can be distinct. Thus, there is some $x \ne y \in V(K_{2^m})$ so that H^x and H^y have the same two-coloring. Let $\alpha = (x\ y)$ be the transposition of x and y in K_{2^m}. Then γ_α simply switches H^x and H^y in $K_{2^m} \,\square\, H$, fixing all other vertices. This tells us that γ_α is a nontrivial automorphism that preserves the coloring and thus the coloring is not distinguishing. We conclude that $\rho(K_{2^m} \,\square\, H) = m2^{m-1}$. $\qquad\square$

Clearly, $\rho(K_{2^m} \,\square\, H) = \frac{1}{2}|V(K_{2^m} \,\square\, H)|$. Furthermore, there is again an automorphism that interchanges the color classes.

It is easy to see that $\mathrm{Det}(K_{2^m}) = 2^m - 1$. Hence $\rho(K_{2^m}) = m2^{m-1} = \frac{m}{2}(2^m - 1) + \frac{m}{2} > \frac{m}{2} \cdot \mathrm{Det}(K_{2^m})$. We thus have the following corollary.

Corollary 3.9. *To any given integer n there are infinitely many graphs G such that $\rho(G) > n \cdot \mathrm{Det}(G)$.*

Proof. The assertion of the corollary is true for any graph $K_{2^m} \,\square\, H$, where H is an asymmetric graph on m vertices and $\frac{m}{2} > n$. $\qquad\square$

4. Infinite Graphs

We already mentioned that there are many families of infinite graphs that are two-distinguishable. In this section we survey the results previously known about the distinguishing cost of some of these graph families. In particular, we focus on the class of infinite, connected, locally finite graphs, denoted Γ.

An infinite graph is called *locally finite* if all its degrees are finite. For such graphs we have the following fundamental result.

Theorem 4.1. [10] *Let $G \in \Gamma$ with infinite automorphism group. Then $\rho(G)$ is finite if and only if $\mathrm{Aut}(G)$ is countable.*

Note that the above theorem implies that, within Γ, a countably infinite automorphism group also implies a finite determining number.

Along the way to proving Theorem 4.1, the authors prove the following.

Theorem 4.2. [10] *Let* $G \in \Gamma$ *have countably infinite automorphism group. Let* $\mathrm{Det}(G) = n$. *Then*

$$\rho(G) \le \left\lceil \frac{5n}{2} \right\rceil - b(n) - 1,$$

where $b(n)$ *denotes the number of 1s in the base-2 representation of* n.

Another result pertains to graphs of linear growth. A graph has linear growth if the number of vertices of distance at most n from a given vertex grows linearly with n. To be more precise, recall that the set of vertices $u \in V(G)$ for which $d(u, v) \le n$ is called *ball of radius* n *centered at* v, and denoted $B_v(n)$. For later reference we also define the *sphere* $S_v(n)$ *of radius* n *centered at* v as the set of vertices $u \in V(G)$ for which $d(u, v) = n$.

A connected, locally finite, infinite graph G is said to have *linear growth* if there exists a vertex v and a constant, real number c, such that $|B_v(n)| \le cn$, for all $n \in \mathbb{N}$. The definition is independent of the choice of v, but c may have to be replaced by a different constant if v is changed. We say that G has linear growth with *growth constant* c. Notice that $c > 1$, because $B_v(n)$ has at least $n + 1$ vertices.

In [10] the authors studied the distinguishing cost for graphs in Γ with linear growth, with the following result:

Theorem 4.3. [10] *Let* $G \in \Gamma$ *with countably infinite automorphism group and linear growth with growth constant* c. *Then* $\rho(G) \le \max(3, c + 1)$.

The bound is sharp for $c = 2$. To see this, consider the two-sided infinite path $P_{\aleph_0}$. It has growth constant 2, and we know that it can be distinguished by just three black vertices, but not by two. Hence $\rho(P_{\aleph_0}) = 3$, which is the bound given by the theorem. However, in the next section we improve the bound for $c \ge 3$.

4.1. A New Bound for Linear Growth

One of the main ingredients of the original proof of Theorem 4.3 is the fact, see [16, Corollary 3.8], that every graph $G \in \Gamma$ with countably infinite automorphism group has a finite determining set, and that every nontrivial automorphism of G moves infinitely many vertices. A subgroup of $\mathrm{Aut}(G)$ is said to have *infinite motion* if each of its nontrivial automorphisms moves infinitely many vertices of G. The graph G is said to have infinite motion if $\mathrm{Aut}(G)$ does.

We use the notation $\mathrm{Aut}(G)_v$ for the stabilizer of vertex v, that is, for the subgroup of $\mathrm{Aut}(G)$ that consists of all automorphisms that fix v.

It will be useful to know that by [10, Lemma 2.3], in a graph in Γ with a finite determining set, vertex stabilizers are finite and have infinite motion.

The proof of Theorem 4.3 constructs, for a given, fixed v, a two-coloring that is not preserved by any nontrivial automorphism in $\mathrm{Aut}(G)_v$. To improve on the conclusion of Theorem 4.3, we use a similar procedure, along with the conclusion of Proposition 4.4 below. Later we will make use of this proposition in our investigation of graphs with two ends.

Proposition 4.4. *Let v be a vertex of a graph $G \in \Gamma$ such that $\mathrm{Aut}(G)_v$ is a finite group with infinite motion. If G has linear growth with growth constant c, then*

$$\rho(G) \leq \max(3, c).$$

Further, the distinguishing coloring with cost $\rho(G)$ can be chosen such that v is colored black.

Proof. For $c < 3$ the arguments in the proof of Theorem 4.3 from [10] yield the bound $\rho(G) \leq 3$.

Hence, we can assume that $c \geq 3$. Fix a vertex v so that $|B_v(n)| \leq cn$. We need the following facts from the proof of Theorem 4.3 in [10]. As the proofs are relatively short, we include them for the benefit of the reader and the sake of completeness.

Fact 1 Infinitely many spheres with center v have size at most c.

For, if only finitely many have size at most c, then there is an index n_0 such that $|S_v(n)| > c$ for $n > n_0$. If $|B_v(n_0)| \geq cn_0$, then $|B_v(n)| > cn$ for $n > n_0$. Hence $cn_0 + 1 - |B_v(n_0)|$ is a positive integer, we denote it by d. Clearly

$$|B_v(n_0+d)| \geq |B_v(n_0)| + d(c+1)| = (cn_0 + 1 - d + (c+1)d) > c(n_0 + d),$$

which violates the fact that the growth constant is c.

Fact 2 There is a natural number b, $b \leq c$, such that infinitely many spheres with center v have size b.

This follows immediately from Fact 1. Notice that $b \neq 1$; otherwise G could not have infinite motion.

Fact 3 There is a sequence of spheres $R_i = S_v(n_i)$, for $n_1 < n_2 < \cdots$ of size b, such that, for $j < i$, the distance between arbitrary vertices in R_i and R_j is larger than the maximal distance between any two vertices in $\cup_{j<i} R_j$, for each $i > 1$.

Fact 4 We can assume without loss of generality that every nontrivial element of $\mathrm{Aut}(G)_v$ acts nontrivially on each of the R_i.

$\mathrm{Aut}(G)_v$ has infinite motion by [10, Lemma 7]. Hence, if $\alpha \in \mathrm{Aut}(G)_v$ fixes $S_v(n)$, then it fixes $B_v(n)$. Thus, again by infinite motion, such an α acts nontrivially on all but finitely many of the R_i. Without loss of generality, we may assume that $\mathrm{Aut}(G)_v - \{\mathrm{id}\}$ acts nontrivially on all R_i.

Fact 5 There is a bijection $\phi_i : R_1 \rightarrow R_i$ so that for each $\alpha \in \text{Aut}(G)_v$, $\alpha | R_1 = \phi_i^{-1}(\alpha | R_i)\phi_i$.

For each $i > 1$, fix a bijection ϕ_i from R_1 to R_i. Let $\phi_1 = \text{id}$ on R_1 and denote the group $\phi_i^{-1}(\text{Aut}(G)_v | R_i)\phi_i$ by A_i. The A_i are subgroups of $\text{Sym}(R_1)$. Furthermore, we chose an $\alpha \in \text{Aut}(G)_v$ and consider the infinite set of permutations $\{\beta_i = \phi_i^{-1}(\alpha | R_i)\phi_i\} \subseteq \text{Sym}(R_1)$. As $\text{Sym}(R_1)$ has only finitely many elements, infinitely many of the β_i must identical. Let J be the set of indices for which this is the case. Notice, this set need not include 1. Let j_0 be the smallest of these indices. Clearly, for $j \in J$ all

$$\gamma_j = \phi_{j_0}\phi_j^{-1}(\alpha | R_j)\phi_j\phi_{j_0}^{-1}$$

are identical on R_{j_0}, and

$$\gamma_{j_0} = \phi_{j_0}\phi_{j_0}^{-1}(\alpha | R_{j_0})\phi_{j_0}\phi_{j_0}^{-1} = \alpha | R_{j_0} .$$

Notice that the $\phi_j\phi_{j_0}^{-1}$ are bijections from R_{j_0} to R_j. Hence, we can assume without loss of generality that J is $\mathbb{N}$; in other words, that all β_i are identical and that $\beta_1 = \alpha | R_1$.

Because $\text{Aut}(G)_v$ is finite, proceeding successively, we can assume that the assertion of Fact 5 holds.

Having established these facts, we now suppose $\alpha \in \text{Aut}(G)_v$ fixes a vertex $\phi_i(u)$ in R_i. Then $\alpha(u) = \phi_i^{-1}\alpha\phi_i(u) = \phi_i^{-1}(\alpha(\phi_i(u))) = \phi_i^{-1}\phi_i(u) = u$. That is, if $\alpha \in \text{Aut}(G)_v$ fixes $\phi_i(u)$ in R_i, then α fixes u in R_1.

Let $v_1, v_2, \ldots, v_b$ be the vertices of R_1, and $X = \{v, \phi_1(v_1), \phi_2(v_2), \ldots, \phi_{b-1}(v_{b-1})\}$. Color the vertices of X black, and all other vertices in the graph white. Suppose $\alpha \in \text{Aut}(G)$ preserves color classes.

Suppose first that $b > 2$. By our choice of distances between the spheres $R_1, \ldots, R_{b-1}$, the vertex v is uniquely identified by its distances to the remaining (at least two) vertices of X. Because α preserves X, it must fix v. And because α must also preserve distance to v, α cannot permute the remaining vertices of X. Thus, α fixes each of $\phi_1(v_1), \ldots, \phi_{b-1}(v_{b-1})$. By our earlier argument, α therefore fixes each of $v_1, \ldots, v_{b-1}$. Given that $\alpha \in \text{Aut}(G)_v$ must preserve R_1, a sphere centered at v with b vertices, α also fixes the remaining vertex v_b of R_1. But the spheres were chosen so that every nontrivial automorphism in $\text{Aut}(G)_v$ acts nontrivially on each sphere, and in particular on R_1. Thus, α itself is trivial. And X is a distinguishing class of size $b \leq c$ for G.

If $b = 2$, then we have colored only two vertices black, namely v and v_1. Hence, to make sure that they cannot be interchanged we choose a third vertex x such that the distances between any two vertices from the set $\{v, v_1, x\}$ are different, and color it black.   □

An immediate consequence is the following theorem, which improves Theorem 4.3 for $c \geq 3$.

Theorem 4.5. *Let $G \in \Gamma$ with countably infinite automorphism group and linear growth with growth constant c, then*

$$\rho(G) \leq \max(3, c).$$

Proof. Because $G \in \Gamma$ has countably infinite automorphism group, it has a finite determining set [14], and thus finite vertex stabilizers [10, Lemma 2.3]. Furthermore, because G also has linear growth, $\mathrm{Aut}(G)$ and $\mathrm{Aut}(G)_v$ have infinite motion [10, Lemma 4.2]. Application of Proposition 4.4 completes the proof. □

4.2. Graphs with Two Ends

In this subsection we prove auxiliary results that we will use, together with Proposition 4.4, for the proof of our main theorem.

Good references for the ends of infinite graphs are [11, 14]. However, the following basic definitions and properties should be sufficient. To define ends of a graph we first observe that every connected, infinite, locally finite graph must have a one-sided infinite path. Such paths are called *rays*. A subgraph of a ray that is itself a ray is called a *tail*. Two rays are *equivalent* if, for every finite set $F \subseteq V(G)$, both have a tail in the same component of $G - F$. This relation is indeed an equivalence relation, and the resulting equivalence classes are called *ends*. For example, a finite graph has no ends, the two-sided infinite path $P_{\aleph_0}$ has two ends, $P_{\aleph_0} \square P_{\aleph_0}$ has one, and the infinite homogeneous tree T_d of degree $d > 2$ has infinitely many ends.

To show how ends relate to subgraphs of G, let us consider an end e and a finite $F \subseteq V(G)$. Take a ray $R \in e$. Because F is finite, R can meet F only finitely often. Hence, all but finitely many vertices of R must be in the same component, say C, of $G - F$. In other words, a tail of R is in the infinite component C. It is easy to see that any other ray in e also has a tail in C.

Note that each infinite connected component of $G - F$ contains a ray and therefore an end. Thus, if G has only two different ends, say e and f, they correspond to the same infinite component C of $G - F$ only when C is the sole infinite component of $G - F$. Otherwise, that is, when $G - F$ has exactly two infinite components, say C_1 and C_2, one of the ends corresponds to C_1 and the other to C_2.

Lemma 4.6. *Let $G \in \Gamma$ and $|\mathrm{Aut}(G)| = \aleph_0$. Then the orbits of all vertices of G are infinite.*

Proof. By [14], a connected, locally finite graph, with countably infinite automorphism group has a finite determining set, say D. Suppose no vertex of D has infinite orbit. Then the orbit $\mathcal{O}(D)$ of D is finite. Because $\mathrm{Aut}(G)$ is infinite, and $\mathcal{O}(D)$ is an orbit that is finite, there must be two distinct automorphisms, say α and β, that agree in their action on $\mathcal{O}(D)$.

Thus, their actions also agree on the determining set D. By definition of determining set, this means $\alpha = \beta$, contrary to assumption.

So there is at least one vertex of D, say v, with an infinite orbit. Because G is locally finite, every ball of finite radius around v contains only finitely many vertices. Thus, since the orbit of v is infinite, to every positive integer d there is an automorphism α such that $d(v, \alpha(v)) > d$.

Let w be any vertex of G distinct from v, and let $m = d(v, w)$. To prove that w has an infinite orbit, we show that there are automorphisms $\alpha_i, i \in \mathbb{N}$, such that the vertices $\alpha_i(w), i \in \mathbb{N}$ are pairwise different. We define the α_i by induction, beginning with $\alpha_1 = \mathrm{id}$. Suppose $\alpha_1, \alpha_2, \ldots, \alpha_k$ are already defined and the vertices in $W_k = \{\alpha_i(w) : 1 \le i \le k\}$ are distinct. We know there is an automorphism α such that

$$d(v, \alpha(v)) > \max\{d(v, \alpha_i(w)) : 1 \le i \le k\} + m.$$

Then $\alpha(w) \notin W_k$. Hence, setting $\alpha_{k+1} = \alpha$, the elements in W_{k+1} are clearly pairwise distinct. $\qquad\square$

The following immediate result generalizes the preceding work to any infinite subset of $\mathrm{Aut}(G)$.

Corollary 4.7.   *Let $G \in \Gamma$ have the property that $|\mathrm{Aut}(G)| = \aleph_0$. If A is any infinite set of automorphisms of G, then all vertices of G have infinite A orbits.*

Lemma 4.8.   *Let $G \in \Gamma$ be 2-ended with countably infinite automorphism group and linear growth. Then, for every finite set of vertices F, every nontrivial automorphism α of G moves infinitely many vertices in each infinite component of $G - F$.*

Proof.   In the following, we show that, if there exists a nontrivial automorphism of G that moves only finitely many vertices of some end of G, then to any finite set of vertices F there exists a nontrivial $\alpha \in \mathrm{Aut}(G)$ such that $\alpha(F) = F$. By [14, Theorem 6], this would mean $\mathrm{Aut}(G)$ is uncountable, a contradiction to our hypothesis.

Because G has two ends, there is a finite separating set S of G such that $G - S$ has precisely two infinite components C_1 and C_2. We wish to show first that there are infinitely many automorphisms that do not switch ends. If only finitely many automorphisms preserve ends, then there are infinitely many automorphisms $\gamma, \gamma_1, \gamma_2, \ldots$ that do switch ends. Consequently, the infinitely many automorphisms $\gamma\,\gamma_1, \gamma\,\gamma_2, \ldots$ do not switch ends. Further, using the methods of the proof of Lemma 4.6, we conclude that there are infinitely many automorphisms that do not switch ends and also move any finite set arbitrarily far.

Suppose F is a finite set of vertices and infinitely many automorphisms move F into $C_2 - B_v(m)$ for an arbitrary, but fixed $B_v(m)$. Then their inverses (again, infinitely many) move F into $C_1 - B_v(m)$.

By [16], given that $G \in \Gamma$ has infinite automorphism group, it has infinite motion. That is, every nontrivial automorphism of G moves an infinite number of vertices. Thus, each must move an infinite number of vertices in at least one of C_1 and C_2.

Assume now that a nontrivial $\alpha \in \mathrm{Aut}(G)$ moves only finitely many vertices of C_1. Let v be a vertex in the finite set S that separates C_1 from C_2, and let m be the maximum distance from v of any of the finitely many of vertices in C_1 that are moved by α. Then α fixes every vertex in $C_1 - B_v(m)$. Now, let F be any finite set of vertices of G. By previous remarks, since all vertices have infinite orbit, and there are automorphisms that do not switch ends but also move a finite set arbitrarily far, there is $\beta \in \mathrm{Aut}(G)$ such that $\beta(F) \subset C_1 - B_v(m)$. Then $\beta^{-1}\alpha\beta$ is a nontrivial automorphism that fixes F pointwise. $\square$

4.3. The Main Result

We now have all prerequisites for the proof of the main result.

Theorem 4.9.   *Let $G \in \Gamma$ be a 2-ended graph with countably infinite automorphism group and linear growth with growth constant c. Then*

$$\rho(G) \leq \max\left(3, \frac{c}{2}\right).$$

Proof.   Because G has linear growth with growth rate c, there exists $v \in V(G)$ so that $|B_v(n)| \leq cn$ for all $n \in \mathbb{N}$. Because G has two ends, there is an $m \in \mathbb{N}$ so that for all $m \geq N$, $G - B_v(m)$ has exactly two infinite components. Let C_1 and C_2 be the infinite components of $G - B_v(m)$, and R_1 and R_2 the boundary points of $B_v(m+1)$ that are in C_1, respectively in C_2. By choice of m, R_1 and R_2 are disjoint and their union is $S_v(m+1)$. Note that the growth rate of one of C_1, C_2 is at most $\frac{c}{2}$; otherwise, the growth rate of G would be larger than c. We choose the notation such that C_1 has growth constant at most $\frac{c}{2}$.

We now observe that C_1, together with v and all shortest paths between v and C_1, is a graph with growth constant at most $\frac{c}{2}$. We denote it $\bar{C}$. ($\bar{C}$ is the convex hull of $C_1 \cup \{v\}$.) Clearly $\mathrm{Aut}(\bar{C})_v$ contains the restriction of $\mathrm{Aut}(G)_v$ to $\bar{C}$. Given that we are only concerned with automorphisms arising from $\mathrm{Aut}(G)$, we will restrict our consideration to these automorphisms; denote this subgroup of $\mathrm{Aut}(\bar{C})_v$ by H. Because we may consider $H \leq \mathrm{Aut}(G)_v$, H has infinite motion, and by Proposition 4.4, we infer that $\bar{C}$ has a distinguishing coloring of cost $\max(3, \frac{c}{2})$ under the action of H and that v can be chosen as a member of the minimum size distinguishing class.

By Proposition 4.4 we can choose a vertex set S such that $v \in S$; all other vertices of S are in C_1, and S is a distinguishing class for the action of H on $\bar{C}$. Color the vertices of S black and all other vertices of G white. Using the methods of the proof of Proposition 4.4 we may assume that

v is distinguished by its distances to the other vertices of S. Thus, any automorphism α that preserves S must fix v. Also, α cannot interchange C_1 and C_2, because all vertices in C_2 are white. Thus, we may consider $\alpha \in H$. Because S is a distinguishing class for the action of H on $\bar{C}$, α acts trivially on $\bar{C}$ and thus on C_1. But the only automorphism that moves at most a finite number of vertices in C_1 is the trivial automorphism. Thus, S is a distinguishing class for G. □

Note that Proposition 4.4 can be slightly improved for graphs with two ends. One simply applies it twice to the two ends of G and uses the fact that the vertex v for which $|B_v(n)| \leq c$ is the same in both ends:

Proposition 4.10. *Let v be a vertex of a 2-ended graph $G \in \Gamma$ such that $\mathrm{Aut}(G)_v$ is a finite group with infinite motion. If G has linear growth with growth constant c, then $\rho(G) \leq \max(5, c-1)$.*

5. Outlook

Superficially it looks like we have the bound $\max\{3, c\}$ for one-ended graphs, and $\max\{3, \frac{c}{2}\}$ for two-ended graphs. But there is a catch. We have not been able to find a graph in Γ that has linear growth, only one end, and countably infinite automorphism group.

We show in the following that all graphs in Γ with linear growth and countably infinite automorphism group have at most two ends. Hence, the nonexistence of one-ended graphs of linear growth and countably infinite automorphism group would mean that we always have the bound $\max\{3, \frac{c}{2}\}$.

For vertex transitive graphs this is indeed the case. It is well known that vertex transitive graphs with infinite automorphism group have 1,2 or infinitely many ends, and that, when there are infinitely many ends, their growth is exponential. Hence, in the case of linear growth, we have at most two ends. In fact, it is not hard to show that vertex transitive graphs of linear growth must have two ends, so for them the bound is $\max\{3, \frac{c}{2}\}$.

However, without vertex transitivity we can only show that graphs in Γ with linear growth have at most two ends. We do this with two lemmas.

Lemma 5.1. *Every graph $G \in \Gamma$ with infinite automorphism group has 1, 2, or infinitely many ends.*

Proof. We have to show that a graph $G \in \Gamma$ with infinite $\mathrm{Aut}(G)$ has infinitely many ends if it has more than two ends. If G has more than two ends, there is a finite set F such that $G - F$ has at least three infinite components, say r, $r > 2$.

If G has only finitely many ends there is a finite set S such that $G - S$ has the maximum number of infinite components, say $s > 2$. We also admit $S = F$. We show that there is an automorphism α such that $\alpha(F)$ is contained in an infinite component of $G - S$ and that S is completely contained in an infinite component of $G - \alpha(F)$.

To see this, let F' be the union of F with all finite components $G - F$ and S' be the union of S with all finite components of $G - S$. We now observe that the orbit of every vertex is infinite by Lemma 4.6. Hence there exists an $\alpha \in \text{Aut}(G)$ such that the distance from $\alpha(F')$ to S' is larger than the diameter of F. This means that $\alpha(F)$ is completely contained in an infinite component of $G - S$, say $C_{\alpha(F)}$, and S in an infinite component of $G - \alpha(F)$.

Consider $S \cup \alpha(F)$ and the components of $G - \{S \cup \alpha(F)\}$. Any infinite component of $G - S$ different from $C_{\alpha(F)}$ and any infinite component of $G - \alpha(F)$ different from $C_{\alpha(F)}$ is a component of $G - \{S \cup \alpha(F)\}$, because every path between any two of them must go through F, $\alpha(F)$, or both F and $\alpha(F)$.

Hence, $G - \{S \cup \alpha(F)\}$ has at least $(s-1) + (r-1) = s + r - 2 \geq s + 1$ infinite components, contrary to the maximality of s.

Lemma 5.2. *A graph with linear growth and an infinite orbit cannot have more than two ends.*

Proof. Suppose G is a graph of linear growth with growth constant c, an infinite orbit, and more than two ends. By Lemma 5.1 it has infinitely many ends, and thus a set of $c + 1$ nonequivalent rays $\{U_1, \ldots, U_{c+1}\}$. Because they are nonequivalent, each pair of rays is disjoint after a finite number of vertices. Thus, there is an integer M_{ij} so that U_i and U_j are disjoint after their M_{ij}^{th} vertex. Let $M = \max\{M_{i,j}\}_{i,j \in [c+1]}$. We can conclude that $U_1, \ldots, U_{c+1}$ are pairwise disjoint after their M^{th} vertex. By disjointness, $|S_v(n)| \geq c + 1$ for all $n > M$. Further, $|S_v(n)| \geq 1$ for $n \leq M$. Thus, we have $|B_v(n)| \geq M + (c+1)(n - M)$, which can be seen, for n large enough, to violate the fact that the growth constant is c. Thus, given G has linear growth, it cannot have more than two ends. $\square$

Theorem 5.3. *All graphs in Γ that have linear growth and countably infinite automorphism group have one or two ends.*

Proof. By Lemma 4.6 we have an infinite orbit and we can apply Lemma 5.2. $\square$

This provides the background for the following problem.

Problem *Does there exist a one-ended graph in Γ that has linear growth and countably infinite automorphism group?*

As mentioned earlier, G cannot be vertex transitive. In fact, G must have infinitely many orbits. Furthermore, all vertex stabilizers must be finite (see, for example [10, Lemma 2.3]), all orbits must be infinite by

Lemma 4.6, and G must have infinite motion by [10, Lemma 4.2]. Finally, we add without proof that all automorphisms must have finite order.

Acknowledgment

We wish to thank the referee for numerous helpful and insightful comments.

References

[1] M. O. Albertson, Distinguishing Cartesian powers of graphs, *Electron. J. Combin.* 12 (2005), Note 17, 5 pp.

[2] M. O. Albertson and D. L. Boutin, Using determining sets to distinguish Kneser graphs, *Electron. J. Combin.* 14 (2007), no. 1, Research Paper 20, 9 pp.

[3] M. O. Albertson and K. L. Collins, Symmetry breaking in graphs, *Electron. J. Combin.* 3 (1996), no. 1, Research Paper 18, 17 pp.

[4] L. Babai, Asymmetric trees with two prescribed degrees, *Acta Math. Acad. Sci. Hungar.* 29 (1977), 193–200.

[5] D. L. Boutin, Small label classes in 2-distinguishing labelings, *Ars Math. Contemp.* 1 (2008), no. 2, 154–64.

[6] D. L. Boutin, The determining number of a Cartesian product, *J. Graph Theory* 61 (2009), no. 2, 77–87.

[7] D. L. Boutin, The cost of 2-distinguishing selected Kneser graphs and hypercubes, *J. Combin. Math. Combin. Comput.* 85 (2013), 161–71.

[8] D. L. Boutin, The cost of 2-distinguishing Cartesian powers, *Electron. J. Combin.* 20 (2013), no. 1, Paper 74, 13 pp.

[9] B. Bogstad and L. J. Cowen, The distinguishing number of the hypercube, *Discrete Math.* 283 (2004), 29–35.

[10] D. L. Boutin and W. Imrich, Infinite graphs with finite 2-distinguishing cost, *Electron. J. Combin.*, forthcoming.

[11] R. Diestel, *Graph Theory.* Electronic edition. Springer, 2005.

[12] M. Goff, Personal communication.

[13] R. Hammack, W. Imrich, and S. Klavžar, *Handbook of product graphs,* second edition. *Discrete Mathematics and its Applications* (Boca Raton). CRC Press, Boca Raton, FL, 2011. xviii+518 pp.

[14] R. Halin, Automorphisms and endomorphisms of infinite locally finite graphs, *Abh. Math. Sem. Univ. Hamburg,* 39 (1973), 251–83.

[15] W. Imrich, S. Klavar, and V. Trofimov, Distinguishing infinite graphs, *Electron. J. Combin.* 14 (2007), no. 1, Research Paper 36, 12 pp.

[16] W. Imrich, S. M. Smith, T. W. Tucker, and M. E. Watkins, Infinite motion and the distinguishing number of graphs and groups, *J. Algebraic Comb.,* DOI 10.1007/s10801-014-0529-2.

[17] T. Fukuda, S. Negami, and T. W. Tucker, 3-connected planar graphs are 2-distinguishable with few exceptions, *Yokohama Math. J.* 54 (2008), no. 2, 143–53.

[18] W. Imrich and S. Klavžar, Distinguishing Cartesian powers of graphs, *J. Graph Theory* 53 (2006), no. 3, 250–60.

[19] S. Klavžar and X. Zhu, Cartesian powers of graphs can be distinguished by two labels, *European J. Combin.* 28 (2007), no. 1, 303–10.

[20] F. Lehner, Distinguishing graphs with intermediate growth, *Combinatorica*, 2015, 36 (2016), no. 3, 333–47.

[21] S. M. Smith, T. W. Tucker, and M. E. Watkins, Distinguishability of infinite groups and graphs, *Electron. J. Combin.* 19 (2012), no. 2, Paper 27, 10 pp.

Added in Proof

Problem 2 was solved by J. Carmesin, F. Lehner, R.G. Möller, who proved that every graph in Γ with linear growth and countable automorphism group has two ends. This means that the bound of Theorem 4.9 always holds. The paper will be entitled "Tree decompositions and automorphism groups of 1-ended graphs", and is still in preparation.

4

A CONSTRUCTION OF THE MEASURABLE POISSON BOUNDARY: FROM DISCRETE TO CONTINUOUS GROUPS

SARA BROFFERIO

Université Paris-Sud, Laboratoire de Mathématiques et IUT de Sceaux, 91405 Orsay Cedex, France

To Wolfgang Woess, on his sixtieth birthday, a special thanks for his continuous and kind support during all my mathematical career

Abstract

Let Γ be a dense countable subgroup of a locally compact continuous group G, and μ a probability measure on Γ. Two spaces of harmonic functions are naturally associated with μ: the space of μ-harmonic functions on the countable group Γ and the space of μ-harmonic functions seen as functions on G defined a.s. with respect to its Haar measure λ. Correspondingly we have two natural Poisson boundaries: the Γ-Poisson boundary and the G-Poisson boundary. Since boundaries on the countable group are quite well understood, a natural question is to ask how the G-boundary is related to the Γ-boundary.

In this chapter we introduce a general technique that allows us to build the G-Poisson boundary from the Γ-boundary. As an application, we determine the Poisson boundary of the closure of the Baumslag–Solitar group in the group of real matrices. In particular we show that, under suitable moment conditions and assuming that the action on $\mathbb{R}$ is not contracting, this boundary is the p-solenoid.

Contents

An important topic in the study of random walks on groups is the study of harmonic functions relative to a measure μ on a group G, i.e. of the functions f on the group such that

$$f(g) = \int_G f(g\gamma)\,d\mu(\gamma). \tag{4.1}$$

The Poisson boundary is, in this setting, the measurable space that gives the integral representation of all bounded harmonic functions. This space encodes the asymptotic information contained in all random walk paths of law μ. A natural problem is to determine when this space is trivial and, if it is not, to exhibit a geometric model.

After the works of Blakwell, Choquet and Deny on abelian groups and the seminal papers of Furstenberg in the 1960s, much progress has been made on these questions. In particular when the harmonic functions live on a *countable discrete group* Γ, a complete theory has been developed by Derriennic [5], Kaimanovich and Vershik [15], allowing us to construct the Poisson boundary (or at least decide whether it is trivial) for large classes of groups.

In the more general case where the measure μ is supported on a locally compact group G, the situation is more complex and one has to decide on which space harmonic functions live. A natural choice is to consider harmonic functions as a subspace of the space $L^\infty(G, \lambda)$ of essentially bounded functions with respect to the Haar measure λ of the group. If the measure μ is spread out (and thus well adapted to the continuous structure) satisfactory general results have been obtained for Lie groups. The case where the measure μ is not necessarily smooth, is far from being completely understood. Some results have been obtained for particular classes of groups (e.g. Nilpotent groups [10, 2], NA groups [17], etc.). Abstract constructions have also been proposed, but they do not allow us, in general, to construct geometric models for the boundary, nor to check whether it is trivial. I refer to the survey of M. Babillot [1] for a precise and complete overview of the subject and a more detailed bibliography.

In antithesis to the case of a smooth measure, we may consider a purely atomic measure μ, supported on a countable subgroup Γ that we can suppose is dense in the continuous group G. In such a situation, harmonic functions can be seen both as functions on the discrete group Γ and as measurable functions on the continuous group G.

When the Poisson boundary of the discrete group Γ is known (so that we can describe Γ-harmonic functions), several natural questions concerning G-measurable harmonic functions arise:

- Which Γ-harmonic functions can be extended to a G-harmonic function?
- How are the Γ-Poisson boundary and the G-Poisson boundary related?
- If we know how G acts on the Γ-Poisson boundary, can we give conditions that imply there are no nontrivial G-harmonic functions?

The goal of this manuscript is to investigate these questions. We are particularly interested in groups of matrices with rational entries, embedded as subgroups of groups of real matrices. In this case the Poisson boundaries of the countable subgroups are well understood [4], while there are still many open questions concerning the Poisson boundaries of the corresponding real groups (see section 1 for more detailed examples).

In section 2, we give a general construction of the G-Poisson boundary as a space of Γ-ergodic components in the product of G and the Γ-boundary (Proposition 2.3). We use this construction to determine the real boundary in the case of the Baumslag Solitar group $BS(1,p)$, embedded as a dense subgroup of

$$\left\{ \begin{bmatrix} p^m & b \\ 0 & 1 \end{bmatrix} \mid m \in \mathbb{Z}, b \in \mathbb{R} \right\} = \mathbb{R} \rtimes \mathbb{Z}.$$

In particular, if μ is dilating on $\mathbb{R}$ it is known that the $BS(1,p)$- Poisson boundary is the p-adic field $\mathbb{Q}_p$ (thus there is no "real" component in the boundary); however the real Poisson boundary is not trivial and is given by the p-solenoid

$$[0,1) \times \mathbb{Z}_p = (\mathbb{R} \times \mathbb{Q}_p) \big/ \mathbb{Z}(1/p)$$

where the action of $\mathbb{Z}(1/p)$ on $\mathbb{R} \times \mathbb{Q}_p$ is the diagonal action (Corollary 3.1).

Acknowledgment

I would like to thank Vadim Kaimanovich, Jean-François Quint and Bertrand Deroin for enlightening discussions. A special thanks also to Wolfgang Woess for his continuous and kind support during all my mathematical career.

1. G-Harmonic Functions and G-Poisson Boundary

This section is a brief introduction to measurable Poisson boundary, following Babillot [1] and Kaimanovich [13].

G-Harmonic Functions

Let G be a locally compact second countable (thus metrizable and complete) group. Let $\mathfrak{G}$ be the Borel σ-algebra of G and λ the right Haar measure.

Let μ be a probability on G such that the closed semigroup generated by the support of μ is the whole group G.

We say that a function $f \in L^\infty(G,\lambda)$ is μ-*harmonic on* (G,λ) (or G-*harmonic*) if

$$f(g) = \int_G f(g\gamma)d\mu(\gamma) \qquad \text{for } \lambda\text{-almost all } g \in G.$$

We denote by $H^\infty_\lambda(G)$ the subspace of G-harmonic functions in $L^\infty(G,\lambda)$.

It can be shown, using left convolution by identity approximations of G, that any $f \in H^\infty_\lambda(G)$ is λ-a.e. limit of harmonic functions that are left uniformly continuous on G. In this sense, the space of G-harmonic functions is determined by the behavior of continuous ones. In particular, if all continuous harmonic functions are constant, then $H_\lambda^\infty(G)$ is trivial. We denote by $H_{\mathrm{luc}}^\infty(\Gamma)$ the space of left uniformly continuous G-harmonic functions.

Random Walks and Invariant Map

Harmonic functions can be seen as asymptotic values of random walks in the following way. Let $(\Omega, \mathbb{P}) = (G, \mu)^{\mathbb{N}}$ be the space of random steps and consider the right random walk

$$r_n(\omega) = \omega_1 \cdots \omega_n.$$

Let f be a bounded G-harmonic function. Notice that because the function f is defined only λ-almost surely, the process $f(gr_n(\omega))$ is well defined only for λ-almost all g. For this reason the starting point g has to be chosen according to ρ, a probability law on G with bounded density with respect to λ. Then the random process $f(gr_n(\omega))$ is well defined on the space $(G \times \Omega, \rho \times \mathbb{P})$, and since f is harmonic, it is a bounded martingale. Thus the limit

$$\lim_{n \to \infty} f(gr_n(\omega)) =: Z_f(g, \omega) \text{ exists } \rho(dg)\mathbb{P}(d\omega)\text{-almost surely.} \qquad (4.2)$$

Let T be the shift on Ω; then is easily checked that

$$Z_f(g, \omega) = Z_f(g\omega_1, T\omega) \qquad \rho(dg)\mathbb{P}(d\omega)\text{-almost surely}$$

that is, Z_f is a bounded measurable *invariant map* on $G \times \Omega$. In fact (4.2) defines an isometry of $H_\lambda^\infty(G)$ onto the subspace of measurable invariant maps of $L^\infty(G \times \omega, \rho \times \mathbb{P})$. The reverse map is given by

$$f_Z(g) := \mathbb{E}(Z(g, \omega)) \qquad \rho(dg)\text{-almost surely.}$$

Poisson Transform and G-Poisson boundary

Take a measurable space $(X, \mathfrak{X}, \nu)$ endowed with a measurable G-action and a μ-stationary probability measure ν. The *Poisson transform*

$$\mathcal{P}_\nu : \phi \mapsto f_\phi(g) := \int \phi(g \cdot x) d\nu(x)$$

maps any bounded function ϕ in $L^\infty(X, \rho * \nu)$ to a μ-harmonic function f_ϕ of $H_\lambda^\infty(\Gamma)$.

Notice that the Poisson transform is not well defined as a map on $L^\infty(X, \nu)$. In fact, since ν is not in general G-quasi invariant (i.e. $g * \nu$ is not in general absolutely continuous with respect to ν), two functions that coincide ν-a.s. can have different images.

If the Poisson transform is an **isometry** of $L^\infty(X, \rho * \nu)$ onto $H_\lambda^\infty(G)$ then we say that (X, ν) is the (G, μ)-*Poisson boundary*. It can be shown that the Poisson boundary is unique as a G-measurable space and that $(X, \rho * \nu)$ is a Lebesgue space (cf. [1] propositions 2.26 and 2.28).

If X is the G-Poisson boundary then there exists a measurable *boundary map* **bnd** : $\Omega \to X$ such that for every harmonic function $f \in H_\lambda^\infty(G)$ there exists $\phi_f \subset L^\infty(X, \rho * \nu)$ with

$$\phi_f(g \cdot \mathbf{bnd}(\omega)) = \lim_{n \to \infty} f(gr_n(\omega)) \qquad \rho(dg)\mathbb{P}(d\omega) - \text{a.s.}$$

(cf. [1] proposition 2.26. See also the proof of Lemma 2.1). Thus,

$$f(g) = \int \phi_f(g \cdot x)d\nu(x) \qquad \rho(dg) - \text{a.s.} \qquad \text{and}$$

$$Z_f(g, \omega) = \phi_f(g \cdot \mathbf{bnd}(\omega)) \qquad \rho(dg)\mathbb{P}(d\omega) - \text{a.s.}$$

The μ-invariant measure ν on X is then the image of $\mathbb{P}$ under **bnd**. The boundary map is G-equivariant in the sense that $\mathbf{bnd}(\omega) = \omega_1 \cdot \mathbf{bnd}(T\omega)$.

Countable Group Γ

Suppose now that the group $G = \Gamma$ is countable. The Haar measure λ is then the counting measure and one can choose ρ with nonzero mass in all elements $g \in \Gamma$. This means that all the equalities above hold for all $g \in \Gamma$.

In this particular case (and under the hypothesis that the support of μ generates Γ as a semigroup), the stationary measure ν on X is Γ-quasi invariant and $\mathcal{P}_\nu$ is well defined on $L^\infty(X, \nu)$ itself.

The fact that μ is absolutely continuous with respect to λ_Γ is also fundamental for the study of Poisson boundary based on entropy [5, 15]. This complete theory has permitted us to determine a geometric model of the Poisson boundary for large classes of countable groups.

Countable Subgroup Γ of a Continuous G

In this note we are interested in the case when the measure μ is supported on a countable subgroup Γ of a continuous group G and in particular when Γ is dense in G. Then a *continuous* harmonic function f on G is uniquely determined by the values $f(\gamma)$ for $\gamma \in \Gamma$. Thus, f can also be seen as a Γ-harmonic function. In other words, the restriction to Γ is an isometric embedding of $H_{\text{luc}}^\infty(G)$ into $H_\lambda^\infty(\Gamma)$. In particular, if (X, ν) is the Γ-Poisson boundary then there exists ϕ in $L^\infty(X, \nu)$ such that

$$f(\gamma) = \int_X \phi(\gamma \cdot x) \qquad \forall \gamma \in \Gamma.$$

However, in general there is no such integral representation for $f(g)$ when g is not in Γ, as X is not a priori a G-space.

In conclusion, the Γ-Poisson boundary contains in principle all the information about the G-Poisson boundary. But in order to extract this information one needs to answer two related questions:

- Determine the G-action on (an extension of) X adapted to the action of G on $H_\lambda^\infty(G)$.
- Determine which are the functions in $L^\infty(X, \nu)$ whose Poisson transform can be extended to G.

Examples: Linear Groups with Rational Coefficients

We are in particular interested in the case where the Γ-Poisson boundary is known, but G-harmonic functions are not completely understood. Here some examples.

Affine Groups

The real affine group $\mathrm{Aff}(\mathbb{R})$ is the group of real maps $(b, a) : x \mapsto ax + b$ with $a \in \mathbb{R}_+^*$ and $b \in \mathbb{R}$; that is the group of matrices

$$\mathrm{Aff}(\mathbb{R}) = \left\{ \begin{bmatrix} a & b \\ 0 & 1 \end{bmatrix} \mid a \in \mathbb{R}_+^*, b \in \mathbb{R} \right\} = \mathbb{R} \rtimes \mathbb{R}_+^*.$$

Harmonic functions on $\mathrm{Aff}(\mathbb{R})$ have been widely studied and some results are known also without continuity assumptions on the measure μ. In particular, under the log-moment assumptions:

$$\mathbb{E}(|\log a|) < \infty \text{ and } \mathbb{E}(\log^+ b) < \infty,$$

it is known that:

- If $\mathbb{E}(\log a) = 0$ the $\mathrm{Aff}(\mathbb{R})$-Poisson boundary is trivial ([18], see also [1, sect.4.5]).
- If $\mathbb{E}(\log a) < 0$ the $\mathrm{Aff}(\mathbb{R})$-Poisson boundary is $\mathbb{R}$ with the μ-invariant measure ν given by the law of

$$Z_\infty = \sum_{n=1}^\infty a_1 \cdots a_{n-1} b_n \tag{4.3}$$

where (b_n, a_n) are i.i.d. with law μ ([17], see also [1, thm 5.7]).

If $\mathbb{E}(\log a) > 0$ and the measure is spread out then the $\mathrm{Aff}(\mathbb{R})$-Poisson boundary is trivial. But it is still not known what happens if $\mathbb{E}(\log a) > 0$ and the measure μ is supported on a countable subgroup Γ.

On the other hand, using entropic criteria, the Γ-Poisson boundaries are well understood. If $\Gamma = \mathrm{Aff}(\mathbb{Q})$, the group of affine maps with rational coefficients, and under suitable moment conditions, the $\mathrm{Aff}(\mathbb{Q})$-Poisson boundary is given by the product of the p-adic fields $\mathbb{Q}_p$ where the sum (4.3) converges a.s., that is

$$\prod_{p : \mathbb{E}(\log |a|_p) < 0} \mathbb{Q}_p,$$

where we use the convention that $\mathbb{Q}_\infty = \mathbb{R}$ (see [3]).

This property was first proved by V. Kaimanovich [13] in the case of the Baumslag-Solitar group

$$BS(1,p) = \left\langle \begin{bmatrix} p^{\pm 1} & \pm 1 \\ 0 & 1 \end{bmatrix} \right\rangle$$

$$= \left\{ \begin{bmatrix} p^m & qp^n \\ 0 & 1 \end{bmatrix} \mid m, n \text{ et } q \in \mathbb{Z} \right\} = \mathbb{Z}\left(\frac{1}{p}\right) \rtimes \mathbb{Z}$$

for some prime p. In this particular case, the $BS(1,p)$-Poisson boundary is $\mathbb{R}$ if $\mathbb{E}(\log a) = -\mathbb{E}(\log |a|_p) < 0$ and $\mathbb{Q}_p$ if $\mathbb{E}(\log |a|_p) = -\mathbb{E}(\log a) < 0$.

It is then natural to ask which harmonic functions can be extended to (continuous) harmonic functions of the closure of $BS(1,p)$ in Aff($\mathbb{R}$), that is to

$$\text{Aff}(p,\mathbb{R}) = \left\{ \begin{bmatrix} p^m & b \\ 0 & 1 \end{bmatrix} \mid m \in \mathbb{Z}, b \in \mathbb{R} \right\} = \mathbb{R} \rtimes \mathbb{Z}.$$

It turns out that, even if the $BS(1,p)$-Poisson boundary is $\mathbb{Q}_p$, the real Poisson boundary is not trivial. In Corollary 3.1 we will construct the Aff($p,\mathbb{R}$)-Poisson boundary as a p-solenoid.

The unpublished manuscript [16] of J.-F. Quint presents a similar example of a dynamic system acting in noncontacting way on the torus and constructs harmonic functions on the unstable variety.

As we will see in Corollary 2.4 this kind of construction is possible as the action of $BS(1,p)$ on Aff($p,\mathbb{R}$) × $\mathbb{Q}_p$ has a discrete orbit. It is still not clear to me what happens when the action of Γ on the product of G and the Γ-Poisson boundary is dense.

Question. For instance, let Aff($1/2, 1/3$) be the countable subgroup generated by the affinities

$$\left\langle \begin{bmatrix} 3^{\pm 1} & \pm 1 \\ 0 & 1 \end{bmatrix}, \begin{bmatrix} 2^{\pm 1} & \pm 1 \\ 0 & 1 \end{bmatrix} \right\rangle = \left\{ \begin{bmatrix} 2^{m_2}3^{m_3} & q2^{n_2}3^{n_3} \\ 0 & 1 \end{bmatrix} \mid m_i, n_i \text{ et } q \in \mathbb{Z} \right\}.$$

Suppose $\mathbb{E}(\log |a|_\infty) > 0$; thus the Γ-Poisson boundary is equal to $\mathbb{Q}_2$, $\mathbb{Q}_3$ or $\mathbb{Q}_2 \times \mathbb{Q}_3$ (according to the sign of $\mathbb{E}(\log |a|_2)$ and $\mathbb{E}(\log |a|_3)$) and has no real component. Is then the Aff($\mathbb{R}$)-Poisson boundary trivial?

Semi-Simple Groups

Similar questions arise for semi-simple groups. Take, for instance, a measure μ supported on $SL_2(\mathbb{Q})$. Then the $SL_2(\mathbb{Q})$-boundary is the product of the $\mathbb{Q}_p$-projective lines for all primes p such that the support of μ is not contained in a compact subgroup of $SL_2(\mathbb{Q}_p)$ (see [4]). In particular for

$$\Gamma = SL_2(\mathbb{Z}(1/2))$$

$$= \left\{ \begin{bmatrix} a & b \\ c & d \end{bmatrix} \mid ad - cd = 1, a, b, c \text{ et } d \in \mathbb{Z}/2^m \text{ for some } m \in \mathbb{Z} \right\}$$

the Γ-Poisson boundary is $\mathbb{P}^1(\mathbb{R}) \times \mathbb{P}^1(\mathbb{Q}_2)$. It is natural to expect that the $SL_2(\mathbb{R})$-Poisson boundary should be $\mathbb{P}^1(\mathbb{R})$, however I am not aware of any proof of this fact. See also [1] section 1.7.4, for a similar example.

2. From Γ-Boundaries to G-Boundaries

Construction of a G-Action on a Γ-Space

Let $(X, \mathfrak{X}, \nu)$ be a Γ-measurable Lebesgue space equipped with a measure ν that is Γ-quasi invariant. Suppose that Γ is contained in a locally compact group G. We want to construct a sort of minimal class of functions on X, on which G acts in such a way that the restriction to Γ of this action coincides with the Γ-action.

Consider the product space $(G \times X, \mathfrak{G} \times \mathfrak{X}, \rho \times \nu)$ and define the Γ-action on $G \times X$

$$\gamma \star (g, x) := (g\gamma^{-1}, \gamma \cdot x). \qquad (4.4)$$

Let $\mathfrak{I}$ be the σ-algebra of $(\Gamma, \star)$-invariant functions of $G \times X$ that is the class of the functions ϕ such that $\rho(dg) \times \nu(dx)$-almost surely

$$\phi(g, x) = \phi(g\gamma^{-1}, \gamma x) \qquad \forall \gamma \in \Gamma. \qquad (4.5)$$

The σ-algebra $\mathfrak{I}$ is complete because $\rho \times \nu$ is $(\Gamma, \star)$-quasi invariant and Γ is countable. Rokhlin's correspondence associates with the σ-algebra $\mathfrak{I}$ a partition of $G \times X$, such that the functions in $\mathfrak{I}$ are constant on the elements of the partition. Because ρ is in the class of the Haar measure, we can choose the partition η to be G-equivariant, for the G-action on $G \times X$ given by left multiplication on the G component. In fact we have the following:

Lemma 2.1.

1. *There exists a countable family* $\{\phi_n\}_{n \in \mathbb{N}}$ *of bounded functions dense in* $L^1(G \times X, \mathfrak{I}, \rho \times \nu)$ *such that for any* $x \in X$ *the function* $\phi_n(\cdot, x)$ *is continuous on* G *and such that (4.5) hold for all* $(g, x) \in G \times X$.
2. *Let* η *be the partition defined by the equivalence relation*

$$(g_1, x_1) \sim (g_2, x_2) \Leftrightarrow \phi_n(\gamma g_1, x_1) = \phi_n(\gamma g_2, x_2) \,\, \forall n \in \mathbb{N} \text{ and } \gamma \in \Gamma. \qquad (4.6)$$

   Then η *is a measurable partition (i.e. countably generated) and the associated complete* σ-*algebra coincides with* $\mathfrak{I}$. *In particular, for all* $\phi \in \mathfrak{I}$ *there exists* $\widetilde{\phi}$ *defined on* $\widetilde{X} = G \times X/\eta$ *such that* $\phi = \widetilde{\phi} \circ \eta$ $\rho \times \nu$-*a.s.*
3. *If* G *acts on* $G \times X$ *by the left multiplication on the* G *component, then such a partition is* G-*equivariant, i.e.*

$$g_0 \cdot \eta(g, x) = \eta(g_0 g, x) \qquad \forall g_0, g \in G \text{ and } x \in X.$$

Proof.   **1.** Because $(G \times X, \mathfrak{I}, \rho \times \nu)$ is a Lebesgue space, there exists a countable family $\{\varphi_i\}$ of bounded functions dense in the L^1-norm. Set $\varphi_i(g, x) \equiv 0$ on the set of the (g, x) on which (4.5) does not hold, in order to obtain a family of functions $\star$-invariant everywhere.

Take an approximation of the identity on G, i.e. a sequence of nonnegative continuous functions α_n whose supports shrink to the identity e and such that $\|\alpha_n\|_1^\lambda = 1$. Let

$$\varphi_i^n(g, x) = \int_G \alpha_n(h)\varphi_i(hg, x)d\lambda(h).$$

It is easy to check that the φ_i^n are still $\star$-invariant. By classical results, for any $x \in X$, the functions $\varphi_i^n(\cdot, x)$ are continuous and converge to $\varphi_i(\cdot, x)$ in $L^1(G, \rho)$ when n goes to ∞. Then

$$\lim_{n \to \infty} \left\| \varphi_i^n - \varphi_i \right\|_1^{\rho \times \nu} = \lim_{n \to \infty} \int_X \left\| \varphi_i^n(\cdot, x) - \varphi_i(\cdot, x) \right\|_1^\rho d\nu(x)$$

$$= \int_X \lim_{n \to \infty} \left\| \varphi_i^n(\cdot, x) - \varphi_i(\cdot, x) \right\|_1^\rho d\nu(x) = 0$$

since $\left\| \varphi_i^n(\cdot, x) - \varphi_i(\cdot, x) \right\|_1^\rho$ is bounded by $2\|\varphi_i\|_\infty^{\rho \times \nu}$.

Thus $\{\varphi_i^n\}_{i,n}$ is a countable family of G-continuous functions dense in $L^1(G \times X, \mathfrak{I}, \rho \times \nu)$.

2. Let $\{I_i\}$ be a countable family of intervals of $\mathbb{R}$ that separates the points and let

$$B(n, i) := \phi_n^{-1}(I_i) \subseteq G \times X.$$

Then the partition η defined in (4.6) is generated by $\{\gamma B(n, i) | \gamma \in \Gamma, n, i \in \mathbb{N}\}$; in fact

$$\eta(x, g) = \bigcap_{(g,x) \in \gamma B(n,i)} \gamma B(n, i) \bigcap_{(g,x) \notin \gamma B(n,i)} \gamma B(n, i)^c.$$

Given that by step **1** the complete σ-algebra generated by the sets $B(n, i)$ is $\mathfrak{I}$, by Rohlin's correspondence, we can conclude that any function of $\mathfrak{I}$ is almost surely constant on the elements of the partition.

3. Finally to prove G-equivariance of η, we need to verify that for every $g_0, g \in G$ and $x \in X$, we have that $(g', x') \in \eta(g, x)$, i.e.

$$\phi_n(\gamma g', x') = \phi_n(\gamma g, x) \quad \forall n \in \mathbb{N} \text{ and } \gamma \in \Gamma$$

if and only if $g_0 \cdot (g', x') = (g_0 g', x') \in \eta(g_0 g, x)$, i.e.

$$\phi_n(\gamma_0 g_0 g', x') = \phi_n(\gamma g_0 g, x) \quad \forall n \in \mathbb{N} \text{ and } \gamma_0 \in \Gamma.$$

This follows from the fact that the functions ϕ_n are G-continuous and that Γ is dense in G, letting $\gamma \to \gamma_0 g_0$ (resp. $\gamma_0 \to \gamma g_0^{-1}$). $\qquad \square$

If η is defined as in (4.6), let

$$\widetilde{X} = G \times X / \eta$$

be the space of the η components. The projection $\eta \colon G \times X \to \widetilde{X}$ defines a natural σ-algebra on $\widetilde{X}$

$$\widetilde{\mathfrak{I}} = \left\{ A \subset G \times X \,|\, \eta^{-1}(A) \in \mathfrak{G} \times \mathfrak{X} \right\}.$$

By the previous lemma the completion of $\eta^{-1}(\widetilde{\mathfrak{I}})$ is $\mathfrak{I}$.

We have just proved that $\widetilde{X}$ has a natural structure of a G-space:

$$g_0 \cdot \eta(g, x) = \eta(g_0 g, x). \tag{4.7}$$

Because the functions ϕ_n are $\star$-invariant everywhere, η allows us to "transfer" the Γ-action from G to X, in the sense that

$$\eta(g\gamma, x) = \eta(g, \gamma \cdot x) \quad \forall (g, x) \in G \times X.$$

In particular, the action of Γ on X and on its projection on $\widetilde{X}$ are related by

$$\gamma \cdot \eta(e, x) = \eta(\gamma, x) = \eta(e, \gamma \cdot x).$$

Let $\widetilde{\nu} = \eta_*(\delta_e \times \nu)$ be the image on $\widetilde{X}$ of the measure $\delta_e \times \nu$ by η; that is,

$$\widetilde{\nu}(\widetilde{\phi}) := \int_X \widetilde{\phi}(\eta(e, x)) d\nu(x). \tag{4.8}$$

By (4.7) we have then that the image by η of $\rho \times \nu$ is $\rho * \widetilde{\nu}$:

$$\eta_*(\rho \times \nu)(\widetilde{\phi}) = \int_{G \times X} \widetilde{\phi}(\eta(g, x)) d\rho(g) d\nu(x)$$

$$= \int_{G \times X} \widetilde{\phi}(g \cdot \eta(e, x)) d\rho(g) d\nu(x)$$

$$= \rho * \widetilde{\nu}(\widetilde{\phi}).$$

In conclusion,

Corollary 2.2. *The projection η induces an isometry between $L^\infty(G \times X, \mathfrak{I}, \rho \times \nu)$ and $L^\infty(\widetilde{X}, \widetilde{\mathfrak{I}}, \rho * \widetilde{\nu})$.*

In the next section, using this measure theoretical construction, we will build the G-boundary on the Γ-boundary and prove that, if the $\star$-action has a fundamental domain, this fundamental domain is the G-boundary. However, it is not clear to me how to construct a geometric model of this measure space when the Γ-action is "dense."

An interesting case is, for instance, when G acts on X and this action coincides with the Γ-action. Then $L^\infty(X, \rho * \nu)$ embeds isometrically in $L^\infty(G \times X, \mathfrak{I}, \rho \times \nu)$. In fact, if $\psi \in L^\infty(X, \rho * \nu)$, then

$$\phi_\psi(g, x) := \psi(g \cdot x)$$

is clearly $\star$-invariant and this embedding is an isometry since

$$\|\phi_\psi\|_\infty = \lim_{p\to\infty} \left(\int \phi_\psi(g,x)^p \rho(dg)\nu(dx) \right)^{1/p}$$
$$= \lim_{p\to\infty} \left(\int \psi(y)^p \rho * \nu(dy) \right)^{1/p} = \|\psi\|_\infty.$$

Question. However, it is not clear under which conditions this map is surjective, that is when $\tilde{X}$ coincide with X.

For instance, as a toy model, take $G = (\mathbb{R}, +)$, $X = \mathbb{R}$ and $\Gamma = \mathbb{Q}$. For which measure ν does $\tilde{X} = \mathbb{R}$? This is true by if ν is a.c. with respect to the Lebesgue measure, but what happens for other measures?

What happens if $G = SL_2(\mathbb{R})$ $X = \mathbb{P}^1(\mathbb{R})$ and $\Gamma = SL_2(\mathbb{Z}(1/2))$ (or $\Gamma = SL_2(\mathbb{Q})$)?

From Γ-Boundaries to G-Boundaries

Suppose that the measure ν on X is μ-stationary. For every bounded function ϕ in $L^\infty(G \times X, \rho \times \nu)$, define the Poisson transform:

$$\mathcal{P}_\nu : \phi \mapsto f_\phi(g) = \int \phi(g,x)d\nu(x) \text{for } \lambda(dg)\text{-almost all } g.$$

If $\phi \in \mathfrak{I}$ then f_ϕ is a bounded μ-harmonic function on $L^\infty(G,\lambda)$, as required. Indeed

$$f_\phi(g) = \int \phi(g,x)\nu(dx) = \int \phi(g, \gamma \cdot x)\nu(dx)\mu(d\gamma) =$$
$$= \int \phi(g\gamma, x)\nu(dx)\mu(d\gamma) = \int f_\phi(g\gamma)\mu(d\gamma).$$

The following proposition shows that all G-harmonic functions can be written in such a way.

Proposition 2.3. *If (X,ν) is the Poisson boundary of (Γ, μ) then for every μ-harmonic function f in G, there exists a bounded function $\phi \in \mathfrak{I}$ such that $f = f_\phi$ in $L^\infty(G,\lambda)$.*

In this case $\mathcal{P}_\nu$ is an isometry from $L^\infty(G \times X, \mathfrak{I}, \rho \times \nu)$ onto $H_\lambda^\infty(G)$. In other words, $(\tilde{X}, \tilde{\nu})$ is the G-Poisson boundary.

Proof. Let $\omega \in (\Omega, \mathbb{P}) = (\Gamma^{\mathbb{N}}, \mu^{\otimes \mathbb{N}})$ and $r_k = r_k(\omega) = \omega_1 \cdots \omega_k$ be the right random walk on Γ of law μ. The process $f(gr_k(\omega))$ is a bounded martingale on the space $(G \times \Omega, \rho \times \mathbb{P})$; thus it converges almost surely. If $\mathbf{bnd}\,\Omega \to X$ is the boundary map

$$\lim_{n\to\infty} f(gr_k(\omega)) = \phi(g, \mathbf{bnd}(\omega)) \tag{4.9}$$

$\rho \times \mathbb{P}$-almost surely. Thus $\phi(g, \mathbf{bnd}(\omega))$ is $G \times \Omega$ measurable and, since $\nu = \mathbf{bnd}^{-1}\mathbb{P}$, the function $\phi(g,x)$ is $G \times X$-measurable. Furthermore, given that Γ is countable, for $\rho \times \mathbb{P}$-almost all (g, ω)

$$\lim_{n \to \infty} f(g\gamma r_k(\omega)) = \phi(g\gamma, \mathbf{bnd}(\omega)) \quad \text{for all } \gamma \in \Gamma.$$

Since X is a μ-boundary, notice that

$$\omega_1 \mathbf{bnd}(T\omega) = \mathbf{bnd}(\omega)$$

where T is the shift on Ω. Take γ_1 in the support of μ, then the event $\gamma_1 = \omega_1$ has a positive measure and is conditioned to this event

$$\phi(g\gamma_1^{-1}, \gamma_1 \mathbf{bnd}(T\omega)) = \phi(g\gamma_1^{-1}, \mathbf{bnd}(\omega))$$
$$= \lim_{n \to \infty} f(g\gamma_1^{-1}\gamma_1 r_n(T\omega)) = \phi(g, \mathbf{bnd}(T\omega)).$$

Because $T\omega$ is independent of ω_1 and of same law as ω and that the support of μ generates Γ, we can conclude that $\phi \in \mathfrak{I}$.

Lastly, let us check that the Poisson transform is an isometry. In fact.

$$\|f_\phi\|_\infty^\rho = \lim_{p \to \infty} \left(\int |f_\phi(g)|^p \, d\rho(g) \right)^{1/p}$$
$$\leq \lim_{p \to \infty} \left(\int \int |\phi(g, x)|^p \, d\nu(x) d\rho(g) \right)^{1/p} = \|\phi\|_\infty^{\rho \times \nu}.$$

On the other hand by the bounded convergence theorem

$$\|\phi\|_p^{\rho \times \nu} = \left(\int \int |\phi(g, x)|^p \, d\nu(x) d\rho(g) \right)^{1/p}$$
$$= \left(\int \int |\lim_{n \to \infty} f(g r_n(\omega))|^p \, d\mathbb{P}(\omega) d\rho(g) \right)^{1/p}$$
$$= \lim_{n \to \infty} \left(\int \int |f(g r_n(\omega))|^p \, d\rho(g) d\mathbb{P}(\omega) \right)^{1/p}$$
$$\leq \left(\int (\|f\|_\infty^\rho)^p \, d\mathbb{P}(\omega) \right)^{1/p} = \|f\|_\infty^\rho$$

since ρ is G-quasi invariant.   □

G-Poisson Boundary as Γ-Ergodic Diagonal Components

Another way to express the result of Proposition 2.3 is to say that the G-Poisson boundary coincides with the space of ergodic components of Γ on $(G \times X)$ with respect to the action $\star$ defined in (4.4).

Observe that the action $\star$ is, in reality, the standard left diagonal action of Γ on $G \times X$:

$$\gamma \overset{\mathrm{d}}{\cdot} (g, x) = (\gamma g, \gamma \cdot x).$$

In fact the two actions are conjugated by the map $\pi : (g, x) \mapsto (g^{-1}, x)$, that is an isomorphism of the measure space of $(G \times X, \rho \times \nu)$ that preserves the class of measure. Thus the space $L^\infty(G \times X, \mathfrak{I}, \rho \times \nu)$ coincides (via π) with the space of bounded functions of $(G \times X, \rho \times \nu)$ that project on $\Gamma\backslash(G \times X)$. In particular, the G-Poisson boundary is trivial if and only if the (diagonal) action of Γ on $(G \times X, \rho \times \nu)$ is ergodic.

Conversely if the action of Γ on $G \times X$ is "measurably discrete," that is, there exists a fundamental domain Δ, then it is possible to identify the G-Poisson boundary with this geometric model:

Corollary 2.4. *Suppose there exists a measurable fundamental domain $\Delta \in \mathfrak{G} \times \mathfrak{X}$ for the action $\star$ of Γ on $G \times X$ (or equivalently for the diagonal action) that is*

- $\rho \times \nu(\Gamma \star \Delta) = 1$
- $\rho \times \nu(\Delta \cap \bigcup_{\gamma \in \Gamma - \{e\}} \gamma \star \Delta) = 0.$

Let $\mathfrak{D}$ be the restriction of the σ-algebra $\mathfrak{G} \times \mathfrak{X}$ to Δ. Then $L^\infty(\Delta, \mathfrak{D}, \rho \times \nu)$ is isometric to $L^\infty(G \times X, \mathfrak{I}, \rho \times \nu)$. The measurable space $(\Delta, \mathfrak{D})$ with the induced G-action

$$g_0 * \phi(g, x) := \sum_{\gamma \in \Gamma} \phi(g_0 g \gamma^{-1}, \gamma \cdot x) 1_\Delta(g_0 g \gamma^{-1}, \gamma \cdot x)$$

for all $\phi \in L^\infty(\Delta, \mathfrak{D}, \rho \times \nu)$ and the μ-invariant measure defined by

$$\tilde{\nu}(\phi) := \sum_{\gamma \in \Gamma} \int \phi(\gamma^{-1}, \gamma \cdot x) 1_\Delta(\gamma^{-1}, \gamma \cdot x) \nu(dx)$$

is the G-Poisson boundary.

Proof. The map

$$A \mapsto \Gamma \star A$$

induces an isometry of $L^\infty(\Delta, \mathfrak{D}, \rho \times \nu)$ onto $L^\infty(G \times X, \mathfrak{I}, \rho \times \nu)$.

In fact, *if A is a nontrivial set of Δ then $\Gamma \star A$ is a non trivial set of $\mathfrak{I}$.* Clearly $\Gamma \star A \in \mathfrak{I}$ and it has nonzero measure. Let $B \subset \Delta$ be a nontrivial set such that $\rho \times \nu(A \cap B) = 0$. We claim that $\rho \times \nu(\Gamma \star A \cap \Gamma \star B) = 0$; in fact, the measure $\rho \times \nu$ being quasi-invariant $\rho \times \nu(\gamma \star A \cap \gamma \star B) = 0$ and if $\gamma_1 \neq \gamma_2$

$$\rho \times \nu(\gamma_1 \star A \cap \gamma_2 \star B) \leq \rho \times \nu \left(\gamma_1 \star \left(\Delta \cap \bigcup_{\gamma \in \Gamma - \{e\}} \gamma \star \Delta \right) \right) = 0.$$

The isometry is surjective. Let $I \in \mathfrak{I}$, then we claim that $I = \Gamma \star (I \cap \Delta)$. Indeed

$$\Gamma \star (I \cap \Delta) = \bigcup_\gamma \gamma \star I \cap \gamma \star \Delta = \bigcup_\gamma (I \cap \gamma \star \Delta) = I \cap \Gamma \star \Delta.$$

Observe that if $A \subseteq \Delta$ then

$$1_{\Gamma \star A}(g, x) = \sum_{\gamma \in \Gamma} 1_A(g\gamma^{-1}, \gamma \cdot x)$$

and the sum has only one term for $\rho \times \nu$-almost all (g, x). It can easily be seen that the projection of ν on $\mathfrak{D}$ is

$$\tilde{\nu}(A) = \sum_{\gamma \in \Gamma} \int 1_A(\gamma^{-1}, \gamma \cdot x)\nu(dx) = \nu(\Gamma \star A).$$

$\square$

3.   *G*-Poisson Boundary of Baumslag–Solitar Group

Corollary 3.1.   *Let p be a prime number and consider the Baumslag–Solitar group*

$$BS(1, p) = \left\langle \begin{bmatrix} p^{\pm 1} & \pm 1 \\ 0 & 1 \end{bmatrix} \right\rangle.$$

Let μ be an irreducible measure on $BS(1, p)$ with first logarithmic moment on $\mathbb{R}$ and $\mathbb{Q}_p$. Suppose that

$$\phi_p = \int_\Gamma \log|a(\gamma)|_p \, d\mu(\gamma) < 0$$

where $\gamma = \begin{bmatrix} a(\gamma) & b(\gamma) \\ 0 & 1 \end{bmatrix}$; that is the $BS(1, p)$-Poisson boundary is $X = \mathbb{Q}_p$. Let

$$\mathrm{Aff}(p, \mathbb{R}) = \left\{ \begin{bmatrix} p^m & b \\ 0 & 1 \end{bmatrix} \Big| m \in \mathbb{Z}, b \in \mathbb{R} \right\} = \mathbb{R} \rtimes \mathbb{Z}$$

be the closure of $BS(1, p)$ in $\mathrm{Aff}(\mathbb{R})$. Then the $\mathrm{Aff}(p, \mathbb{R})$-Poisson boundary is the p-solenoid:

$$\Delta = \{(g, x) \in \mathrm{Aff}(\mathbb{R}) \times \mathbb{Q}_p | a(g) = 1; 0 \le b(g) < 1; |x|_p \le 1\} = [0, 1) \times \mathbb{Z}_p,$$

equipped with the $\mathrm{Aff}(p, \mathbb{R})$-action on $\phi \in L^\infty(\Delta, \rho \times \nu)$:

$$(b, p^m) \cdot \phi(x_\infty, x_p) = \sum_{\beta \in \mathbb{Z}(1/p)} 1_\Delta \cdot \phi(p^m x_\infty + b - \beta, p^m x_p + \beta),$$

and the invariant measure

$$\tilde{\nu}(\phi) := \sum_{\beta \in \mathbb{Z}(1/p) \cap [0, 1)} \int \phi(\beta, x - \beta) 1_{\mathbb{Z}_p + \beta}(x)\nu(dx).$$

Proof. We just need to prove that Δ is a fundamental domain. In fact, for any $x \in \mathbb{Q}_p$, let $\alpha(x) \in \mathbb{Z}(1/p)$ such that $|x - \alpha(x)|_p \leq 1$. The choice of α is unique up to the sum with an integer. It is easily checked that, for every $(b, x) \in \mathbb{R} \times \mathbb{Q}_p$, the unique $k \in \mathbb{Z}(1/p)$ such that $|x + k|_p \leq 1$ and $b - k \in [0, 1)$ is $k = [b + \alpha(x)] - \alpha(x)$. Thus

$$\gamma \star ((b, p^m), x) \in \Delta \Leftrightarrow \gamma = ([b + \alpha(p^m x)] - \alpha(p^m x), p^m).$$

$\square$

To illustrate how the previous corollary can be used to study the behavior of harmonic functions on $BS(1, p)$, consider, for example,

$$\phi(g, x) = 1_{[0,1) \times \{1\}}(g) 1_{p\mathbb{Z}_p}(x)$$

and the associated harmonic function:

$$f(b, p^m) = \int \sum_{\beta \in \mathbb{Z}(1/p)} 1_{[0,1)}(b - \beta) 1_{p\mathbb{Z}_p}(p^m x + \beta) \nu(dx).$$

Then we have

- f is periodic of period p on the b coordinate

$$f(pk + b, p^m) = \int \sum_{\beta \in \mathbb{Z}(1/p)} 1_{[0,1)}(b - \beta) 1_{p\mathbb{Z}_p}(p^m x + \beta + pk) \nu(dx)$$
$$= f(b, p^m)$$

- $\lim_{m \to +\infty} f(b, p^m) = 1$ if $b \in [0, 1) + p\mathbb{Z}$. In fact $\|f\|_\infty = 1$ and $b \in [0, 1)$

$$f(b, p^m) \geq \int 1_{[0,1)}(b) 1_{p\mathbb{Z}_p}(p^m x) \nu(dx) = \nu(p^{1-m}\mathbb{Z}_p) \to 1$$

when $m \to +\infty$
- $\lim_{m \to +\infty} f(b, p^m) = 0$ if $b \notin [0, 1) + p\mathbb{Z}$ in fact

$$f(b, p^m) \leq \int \sum_{\beta \in \mathbb{Z}(1/p)} 1_{[0,1)}(b - \beta) 1_{p\mathbb{Z}_p}(p^m x + \beta) 1_{p^{1-m}\mathbb{Z}_p}(x) \nu(dx) +$$
$$+ (1 - \nu(p^{1-m}\mathbb{Z}_p))$$
$$\leq \sum_{\beta \in \mathbb{Z}(1/p)} 1_{[0,1)}(b - \beta) 1_{p\mathbb{Z}_p}(\beta) + (1 - \nu(p^{1-m}\mathbb{Z}_p))$$
$$= \sum_{k \in \mathbb{Z}} 1_{[0,1)}(b - pk) + (1 - \nu(p^{1-m}\mathbb{Z}_p))$$
$$= 1_{[0,1)+p\mathbb{Z}}(b) + (1 - \nu(p^{1-m}\mathbb{Z}_p)).$$

References

[1] **Babillot, Martine:** *An introduction to Poisson boundaries of Lie groups.* Probability measures on groups: recent directions and trends, 1–90, Tata Inst. Fund. Res., Mumbai, 2006.

[2] **Breuillard, Emmanuel:** *Equidistribution of random walks on nilpotent Lie groups and homogeneous spaces.* Thesis (PhD)- Yale University. ProQuest LLC, Ann Arbor, MI, 2004. 162 pp.

[3] **Brofferio, Sara:** *The Poisson boundary of random rational affinities,* Ann. Inst. Fourier 56, (2006), 499–515.

[4] **Brofferio, Sara, and Schapira, Bruno:** *Poisson boundary of $GL_d(\mathbb{Q})$,* Israel J. Math. 185, (2011), 125–40.

[5] **Derriennic, Yves:** *Entropie, théorèmes limite et marches aléatoires,* in Probability measures on groups VIII (Oberwolfach, 1985), LNM 1210, pp. 241–84, Springer, Berlin (1986).

[6] **Elie, Laure:** *Noyaux potentiels associés aux marches aléatoires sur les espaces homogènes. Quelques exemples clefs dont le groupe affine,* in Théorie du potentiel (Orsay, 1983), volume 1096 of Lectures Notes in Math., 223–60, Springer, Berlin, 1984.

[7] **Furman, Alex:** *Random walks on groups and random transformations,* Handbook of dynamical systems, vol. 1A, pp. 931–1014, Amsterdam: North-Holland (2002).

[8] **Furstenberg, H.:** *A Poisson formula for semi-simple Lie groups,* Ann. of Math. 77 (1963), 335–86.

[9] **Furstenberg, H.:** *Boundary theory and stochastic processes on homogeneous spaces,* in Harmonic analysis on homogeneous spaces (Proc. Sympos. Pure Math., Vol. XXVI, Williams Coll., Williamstown, Mass., 1972), pp. 193–229, Amer. Math. Soc., Providence, R.I. (1973).

[10] **Guivarc'h, Yves:** *Extension d'un théorème de Choquet-Deny à une classe de groupes non abéliens* Séminaire KGB sur les Marches Aléatoires (Rennes, 1971–1972) 41–59. Astérisque, 4, Soc. Math. France, Paris, 1973.

[11] **Guivarc'h, Yves:** *Quelques proprits asymptotiques des produits de matrices alatoires.* (French) Eighth Saint Flour Probability Summer School–1978 (Saint Flour, 1978), pp. 177–250, Lecture Notes in Math., 774, Springer, Berlin, 1980.

[12] **Guivarc'h, Y., and Raugi, A.:** *Frontière de Furstenberg, propétés de contraction et thórèmes de convergence.*

[13] **Kaimanovich, V. A.:** *The Poisson formula for groups with hyperbolic properties,* Ann. of Math. (2) 152, (2000), 659–92.

[14] **Kaimanovich, V. A.:** *Lyapunov exponents, symmetric spaces and a multiplicative ergodic theorem for semisimple Lie groups,* Zap. Nauchn. Sem. Leningrad. Otdel. Mat. Inst. Steklov. (LOMI) 164 (1987), Differentsialnaya Geom. Gruppy Li i Mekh. IX, 29–46, 196–97; translation in J. Soviet Math. 47 (1989), no. 2, 2387–98.

[15] **Kaimanovich, V. A. and Vershik, A. M.:** *Random walks on discrete groups: boundary and entropy,* Ann. Probab. 11, (1983), 457–90.

[16] **Quint, J-F:** *Choquet-Deny theorem for critical measures on the group ax + b*, Unpublished

[17] **Raugi, Albert:** *Fonctions harmoniques sur les groupes localement compacts base dnombrable.* (French) Bull. Soc. Math. France Mm. No. 54 (1977), 5–118.

[18] **Raugi, Albert:** *Périodes des fonctions harmoniques bornées*, Seminar on Probability, Rennes 1978 (French), Exp. No. 10, 16, Univ. Rennes, Rennes (1978).

5

STRUCTURE TREES, NETWORKS AND ALMOST INVARIANT SETS

MARTIN J. DUNWOODY

University of Southampton, Southampton, United Kingdom

For Wolfgang Woess, in gratitude

Abstract

A self-contained account of the theory of structure trees for edge cuts in networks is given. Applications include a generalisation of the Max-Flow Min-Cut theorem to infinite networks and a short proof of a conjecture of Kropholler. This gives a relative version of Stallings theorem on the structure of groups with more than one end. A generalisation of the Almost Stability theorem is also obtained, which provides information about the structure of the Sageev cubing.

Contents

1. Introduction

Let X be a connected graph. A subset A of the vertex set VX is defined to be a *cut* if δA is finite. Here δA is the set of edges with one vertex in A and one vertex in $A^* = VX - A$. A ray R in X is an infinite sequence $x_1, x_2, \ldots$ of distinct vertices such that x_i, x_{i+1} are adjacent for every i. If A is an edge cut and R is a ray, then there exists an integer N such that for $n > N$ either $x_n \in A$ or $x_n \in A^*$. We say that A separates rays $R = (x_n), R' = (x'_n)$ if for n large enough either $x_n \in A, x'_n \in A^*$ or $x_n \in A^*, x'_n \in A$. We define $R \sim R'$ if they are not separated by any edge cut. It is easy to show that $\sim$ is an equivalence relation on the set ΦX of rays in X. The set $\Omega X = \Phi X / \sim$ is the set of edge ends of X. An edge cut A separates ends ω, ω' if it separates rays representing ω, ω'. A cut A separates an end ω and a vertex $v \in VX$ if for any ray representing ω, R is eventually in A and $v \in A^*$ or vice versa.

1991 *Mathematics Subject Classification.* 20F65 (20E08).

Key words and phrases. Structure trees, tree decompositions, group splittings.

The number $e(G)$ of a finitely generated group is the number of ends of a Cayley graph of X with respect to a finite generating set S. It turns out that $e(G)$ does not depend on which generating set S is chosen, and that it is always one of $0, 1, 2$ or the cardinal number c. If a finitely generated group G has more than one end, then there is a cut $A \subset G$ (the vertex set of any Cayley graph), which separates two rays. Thus both A and A^* are infinite. The fact that δA is finite is equivalent to the fact that the symmetric difference $A + As$ is finite for each $s \in S$, and it is not hard to see that this is equivalent to requiring that $A + Ag$ is finite for every $g \in G$. A set A with these properties is called a *proper almost invariant set*. Thus a subset A of G is said to be *almost invariant* if the symmetric difference $A + Ag$ is finite for every $g \in G$. In addition A is said to be *proper* if both A and $A^* = G - A$ are infinite. Clearly the finitely generated group G has more than one end if and only if it has a proper almost invariant subset. This provides a way of extending our definition to arbitrary groups. We say that a group G has more than one end if it has a proper almost invariant set.

Theorem 1.1. *A group G contains a proper almost invariant subset (i.e. it has more than one end) if and only if it has a non-trivial action on a tree with finite edge stabilizers.*

This result was proved by Stallings [31] for finitely generated groups and was generalized to all groups by Dicks and Dunwoody [3]. The action of a group G on a tree is *trivial* if there is a vertex that is fixed by all of G. Every group has a trivial action on a tree.

Let T be a tree with directed edge set ET. If e is a directed edge, then let $\bar{e}$ denote e with the reverse orientation. If e, f are distinct directed edges, then write $e > f$ if the smallest subtree of T containing e and f is as below.

Suppose the group G acts on T. We say that g *shifts* e if either $e > ge$ or $ge > e$. If for some $e \in ET$ and some $g \in G$, g shifts e, then G acts non-trivially on a tree T_e obtained by contracting all edges of T not in the orbit of e or $\bar{e}$. In this action there is just one orbit of edge pairs. Bass-Serre theory tells us that either $G = G_u *_{G_e} G_v$ where u, v are the vertices of e and they are in different orbits in the contracted tree T_e, or G is the HNN-group $G = G_u *_{G_e}$ if u, v are in the same G-orbit. If either case occurs we say that G *splits over* G_e.

If there is no edge e that is shifted by any $g \in G$, (and G acts without involutions, i.e. there is no $g \in G$ such that $ge = \bar{e}$), then G must fix a vertex or an end of T. If the action is non-trivial, it fixes an end of T, i.e.

G is a union of an ascending sequence of vertex stabilizers, $G = \bigcup G_{v_n}$, where $v_1, v_2, \ldots$ is a sequence of adjacent vertices and $G_{v_1} \leq G_{v_2} \leq \ldots$ and $G \neq G_{v_n}$ for any n.

Thus Theorem 1.1 could be restated as

Theorem 1.2 ([31], [3]). *A group G contains a proper almost invariant subset (i.e. it has more than one end) if and only if it splits over a finite subgroup or it is countably infinite and locally finite.*

If a group splits over a finite subgroup, then it is possible to choose a generating set S so that the Cayley graph has more than one end. However, for a countably infinite locally finite group there is no Cayley graph with more than one end.

The if part of the theorem is fairly easy to prove. We now prove a stronger version of the if part, following [2].

Let H be a subgroup of G. A subset A is H-*finite* if A is contained in finitely many right H-cosets, i.e. for some finite set F, $A \subseteq HF$. A subgroup K is H-finite if and only if $H \cap K$ has finite index in K. Let T be a G-tree and suppose there is an edge e and vertex v.

We say that e *points at* v if there is a subtree of T as below. We write $e \twoheadrightarrow v$.

Let $G[e, v] = \{g \in G \mid e \twoheadrightarrow gv\}$.

If $h \in G$, then $G[e, v]h = G[e, h^{-1}v]$, since if $e \twoheadrightarrow gv$, $e \twoheadrightarrow gh(h^{-1}v)$.

It follows from this that If $K = G_v$, then $G[e, v]K = G[e, v]$. Also if $H = G_e$, then $HG[e, v] = G[e, v]$.

If $v = \iota e$, then $G_e = H \leq K = G_v$ and if $A = G[e, \iota e]$, then $A = HAK$.

Consider the set $Ax, x \in G$. If $g \in A, gx \notin A$, then $e \twoheadrightarrow gv, \bar{e} \twoheadrightarrow gxv$. This means that e is on the directed path joining gxv and gv. This happens if and only if $g^{-1}e$ is on the path joining xv and v. There are only finitely many directed edges in the G-orbit of e in this path. Hence $g^{-1} \in FH$, where F is finite, and $H = G_e$, and $g \in HF^{-1}$. Thus $A - Ax^{-1} = HF^{-1}$, i.e. $A - Ax^{-1}$ is H-finite. It follows that both $Ax - A$ and $A - Ax$ are H-finite and so $A + Ax$ is H-finite for every $x \in G$, i.e. A is an H-*almost invariant set*.

If the action on T is non-trivial, then neither A nor A^* is H-finite. We say that A is *proper*.

Peter Kropholler has conjectured that the following generalization of Theorem 1.1 is true for finitely generated groups.

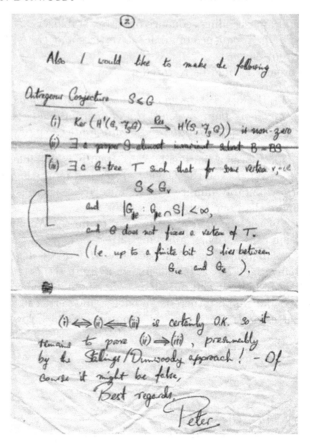

Conjecture 1.3. *Let G be a group and let H be a subgroup. If there is a proper H-almost invariant subset A such that $A = AH$, then G has a non-trivial action on a tree in which H fixes a vertex v and every edge incident with v has an H-finite stabilizer.*

We have seen that the conjecture is true if H has one element. The conjecture has been proved for H and G satisfying extra conditions by Kropholler [22], Dunwoody and Roller [15], Niblo [26] and Kar and Niblo [20].

If G is the triangle group $G = \langle a, b \,|\, a^2 = b^3 = (ab)^7 = 1 \rangle$, then G has an infinite cyclic subgroup H for which there is a proper H-almost invariant set. Note that in this case G has no non-trivial action on a tree, so the condition $A = AH$ is necessary in Conjecture 1.3.

A discussion of the Kropholler conjecture is given in [28]. I first learned of this conjecture in a letter Peter wrote to me in January 1988, a page of which is shown here.

We give a proof of the conjecture when G is finitely generated over H, i.e. it is generated by H together with a finite subset.

I am very grateful to Peter Kropholler for enjoyable discussions and a very helpful email correpondence about his conjecture.

The main tool in proving Conjecture 1.3 is the theory of structure trees in connected graphs, which was initiated in [8] and [3]. In the next section a fairly self-contained account of this theory is given. In fact the theory is extended to apply to networks, and it is shown that the sequence of structure trees obtained for a network is uniquely determined. It is this property that is crucial in proving the Kropholler conjecture.

Another very interesting aspect of extending the theory to networks is that one can obtain non-trivial result for a finite network. Results for finite networks such as the Max-Flow Min-Cut Theorem (MFMC) and the existence of a Gomory-Hu tree are shown to be special cases of our results for more general networks. It is also the case that Stallings theorem on the structure of groups with more than one end also follows from the theory developed here. It is very pleasing (to me at least) that there is a theory that includes both the Stallings theorem and the MFMC.

If A is an almost invariant set, and $M = \{B | B =_a A\}$ so that for $B, C \in M, B + C = F$ where F is finite, then the Almost Stability theorem of [3] shows that M is the vertex set of a G-tree T.

We can define a metric on M. For $B, C \in M$ define $d(B, C) = |B+C|$, and this is a geodesic metric on T. In the final section a generalisation of this result is proved. If H is a subgroup of G, and $A = HAH$ is an H-almost invariant subset, we now put $M = \{B | A+B = HF\}$ where F is finite, then M is a right G-set, i.e. it admits an action of G by right multiplication. If for $B, C \in M$ we put $d(B, C)$ to be the number of cosets Hx in $B+C$, then we have a metric on M as before. It is again shown that this is a metric on a tree, thus giving the H-almost stability theorem.

At the risk of appearing self-indulgent, I record the history of the theory of structure trees.

As noted above, in his breakthrough work [31] on groups with more than one end, in the late 1960s, Stallings showed that a finitely generated group has a Cayley graph (corresponding to a finite generating set) with more than one end if and only if it has a certain structure. At about that time Bass and Serre (see [3] or [30]) developed their theory of groups acting on trees and it was clear that the structure of a group with more than one end, as in Stallings theorem, was associated with an action on a tree. In [7] I gave a proof of Stallings' result by constructing a tree on which the relevant group acted. This involved showing that if the finitely generated group G had more than one end, then there is a subset $B \subset G$ such that both B and B^* are infinite, δB is finite and the set $\mathcal{E} = \{gB | g \in G\}$ is almost nested. Here we define a cut in a graph X to be a subset A of VX such that δA if finite, where δA is the set of edges with one vertex in A and one vertex in $A^* = VX - A$. The set of all cuts is denoted $\mathcal{B}X$.

A set $\mathcal{E}$ of cuts is almost nested if for every $A, B \in \mathcal{E}$ at least one corner of A and B is finite. A corner of A, B is one of the four sets $A \cap B, A^* \cap B, A \cap B^*, A^* \cap B^*$.

In [8] I gave a stronger result by showing that if a group G acts on a graph X with more than one end, then there exists a subset $B \in \mathcal{B}X$ such that B and B^* are both infinite and for any $g \in G$ the sets B and gB are nested, i.e. at least one of the four corners is empty. The set of all such gB can be shown to be the edge set of a tree, called a structure tree.

This result was further extended by Warren Dicks and myself [3]. In Chapter 2 of that book it is shown that for any graph X the Boolean ring $\mathcal{B}X$ has a particular nested set of generators invariant under the automorphism group of G. At the time I thought that the result when applied to finite graphs was of little interest. This was partly because an action of a group on a finite tree is always trivial, i.e. there is always a vertex of the tree fixed by the whole group. This is not the case for groups acting on infinite trees: the theory of such actions is the subject matter of Bass-Serre theory. Also for a finite graph X, there is always a nested set of generators for $\mathcal{B}X$ consisting of single elements subsets. The belated realisation that the theory developed in [3] might be of some significance for finite networks occurred only recently.

In 2007 Bernhard Krön asked me if one could develop a theory of structure trees for graphs with more than one vertex end rather than more than one edge end. These are connected graphs that have more than one infinite component after removing finitely many vertices. We were able to develop such a theory in [14]. In the course of our work on this, we realised that we could develop a theory of structure trees for finite graphs that generalised the theory of Tutte [33], who obtained a structure tree result for two-connected finite graphs that are not three-connected. The theory for vertex cuts is more complicated than that for edge cuts. In 2008 I learned about the cactus theorem for min-cuts from Panos Papasoglu. This theory, due to Dinits, Karsanov and Lomonosov [5] (see also [16]) is for finite networks. It is possible, with a bit more work, to deduce the cactus theorem from the proof of Theorem 2.2. Evangelidou and Papasoglu [19] have obtained a cactus theorem for edge cuts in infinite graphs, giving a new proof of Stallings theorem. In [4] Diekert and Weiss gave a definition for thin cuts, which is equivalent to the one given in [3], but which made more apparent the connection with the Max-Flow Min-Cut theorem. I also had a very helpful email exchange with Armin Weiss. Weiss told me about Gomory-Hu trees that are structure trees in finite networks.

Thinking about these matters finally led me to think about structure trees for edge cuts in finite graphs and networks and the realisation that the theory developed in [3] might be of some interest when applied to finite networks.

In Section 2 the theory for finite networks is recalled. The theory is then generalised to arbitrary networks. For any network N we obtain a canonically determined sequence of trees T_n that provide complete information about the separation of a pair s, t where each of s and t is either a vertex or an end of X. It is possible to obtain all such information from a single tree T_n if X is *accessible*. A graph is accessible if there is an integer n such that any two ends can be separated by removing at most n edges. This definition is due to Thomassen and Woess [32]. Other ways of defining accessibility of graphs are discussed. There are locally finite vertex transitive graphs that are inaccessible. Such graphs are constructed in [9] or [10].

The situation for edge cuts contrasts with the situation for vertex cuts. Thus there is a canonically determined sequence of trees that separates a pair s, t in the set of vertices or ends of the graph X. For vertex cuts, one can only find a canonically defined structure tree that separates a pair κ-inseparable sets or a pair of vertex ends, where κ is the smallest integer for which it is possible to separate such a pair.

Structure tree theory has been used by several authors to classify infinite graphs that have more than one end and which satisfy different transitivity condition. For example, Macpherson [24] used a structure tree to classify infinite locally finite distance transitive graphs, and Möller [25] used these methods to classify infinite ended locally finite graphs for which the automorphism group acted transitively on the ends. In [32] Thomassen and Woess obtain a number of results using structure trees. They show, for example, that if r is prime, then a connected, r-regular, 1-transitive graph with more than one end, is a tree.

This chapter incorporates two papers [12] and [13] that have appeared on arXiv. I am very grateful to Peter Kropholler and Armando Martino who made a careful study of the earlier papers. I have included their suggestions and corrections in this version. I thank Alex Margolis for his careful comments and suggestions.

2. Networks and Structure Trees

2.1. Finite Networks

In this subsection we define our terminology, but restrict attention to networks based on finite graphs. We recall the Max-Flow Min-Cut theorem and state the result that our more general theory gives for finite networks. We illustrate the theory with examples.

We define a network N to be a finite simple, connected graph X and a map $c : EX \rightarrow \{1, 2, \dots\}$.

Let $s, t \in VX$. An (s, t)-*flow* in N is a map $f : EX \rightarrow \{0, 1, 2, \dots\}$ together with an assignment of a direction to each edge e so that its vertices are ιe and τe and the following holds.

(i) For each $e \in EX$, $f(e) \le c(e)$.

(ii) If we put $f^+(v) = \Sigma(f(e)|\iota e = v)$ and $f^-(v) = \Sigma(f(e)|\tau e = v)$, then for every $v \in VX$, $v \ne s$, $v \ne t$, we have $f^+(v) = f^-(v)$. That is, at every vertex except s or t, the flow into that vertex is the same as the flow out.

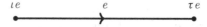

It is easy to show that in an (s, t)-flow, $f^+(s) - f^-(s) = -(f^+(t) - f^-(t))$. The *value* of the flow is defined to be $|f| = |f^+(s) - f^-(s)|$. We define a *cut* in X to be a subset A of VX, $A \ne \emptyset$, $A \ne VX$. If A is a cut, then so is its complement A^*. If N is a network and $A \subset VX$ is a cut, then the capacity $c(A)$ of A is the sum $c(A) = \Sigma\{c(e)|e = (u, v), u \in A . v \in A^*\}$. We define δA to be the set of edges with one vertex in A and one in A^*, so that $c(A)$ is the sum of the values $c(e)$ as e ranges over the edges of δA. We could replace each edge e of X with $c(e)$ edges joining the same two vertices and then have a theory in which the capacity of a cut is the number of edges in δA.

In Figure 5.1 a network is shown, together with a max-flow (which has value 7), together with a corresponding min-cut.

Theorem 2.1 (The Max-Flow Min-Cut theorem [17]).   *The maximum value of an (s, t)-flow is the minimal capacity of a cut separating s and t.*

In the proof of this result it is shown that one obtains a min-cut from a max-flow as the set of vertices that are connected to s by a path in which each edge has some unused capacity. Thus in Figure 5.1, bottom diagram, the min-cut vertices are the ones to the left of the dotted line and are shown in bold.

In this chapter it is shown that for any finite network there is a uniquely determined network based on a *structure tree* that provides a convenient way of encoding the minimal flow between any pair of vertices. Specifically we will prove the following theorem.

Theorem 2.2.   *Let $N(X)$ be a finite network. There is a uniquely determined network $N(T)$ based on a tree T and an injective map $v : VX \to VT$, such that the maximum value of an (s, t)-flow in X is the maximum value of a (vs, vt)-flow in $N(T)$. Also, for any edge $e' \in ET$, there are vertices $s, t \in VX$ such that e' is on the geodesic joining vs and vt and $c(e')$ is the capacity of a minimal (s, t)-cut.*

An example of a network and its structure tree are shown in Figure 5.2. Thus in this network the max-flow between u and p is 12. One can read off a corresponding min-cut by removing the corresponding edge from the structure tree. Thus a min-cut separating u and p is $\{q, r, s, t, u, v, w\}$. The map v need not be surjective. In our example there is a single vertex

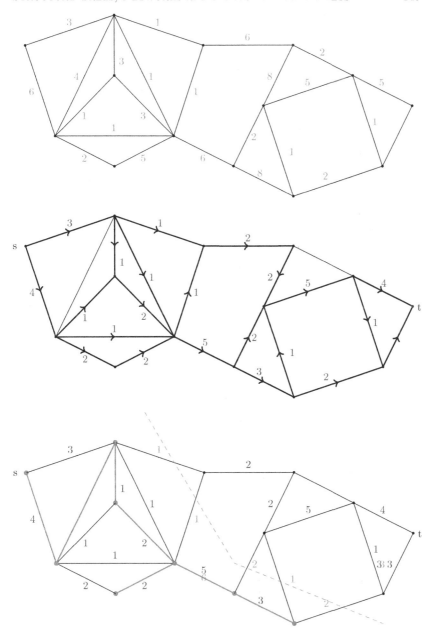

Figure 5.1 Max-Flow Min-Cut theorem

z that is not in the image of ν shown in bold. One can get a structure tree for which ν is bijective by contracting one of the four edges incident with this vertex. The tree then obtained is a Gomory-Hu tree [18]. The structure tree constructed in the proof of Theorem 2.2 is uniquely determined and is therefore invariant under the automorphism group of

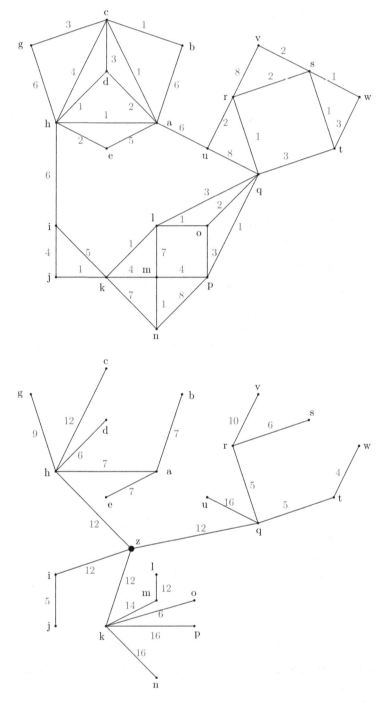

Figure 5.2 Network and structure tree

the network. The tree obtained by contracting one of the four edges is no longer uniquely determined as one gets a different tree for each of the four choices. In some cases this would mean that the structure tree did not admit the automorphism group of the network. Thus, for example, if the automorphism group of X is transitive on VX and $c(e) = 1$ for every edge, then the structure tree would have n vertices of degree one, where $n = |VX|$ and one vertex of degree n. Clearly this structure tree will admit the automorphism group of X, but if one edge is contracted to get a tree with n vertices, then the new tree will not admit the automorphism group.

Not every min-cut separating a pair of vertices can be obtained from the structure tree. The min-cuts obtained are the ones that are optimally nested with the cuts of equal or smaller capacity. In our example there are four cuts of capacity 12 corresponding to edges in the structure tree incident with z. However, there are other cuts of capacity 12. Thus there are two min-cuts in the structure tree separating k and h, but there are in fact four min-cuts separating k and h. In [5] it is shown that the min-cuts separating two vertices correspond to the edge cuts in a cactus, which is a connected graph in which each edge belongs to at most one cycle. The cactus of min-cuts separating k and h is a 4-cycle.

2.2.   The Algebra of Cuts

Let N be a network based on the graph X. We now allow X to be infinite. Thus N is a simple connected graph with a map $c : EX \rightarrow \{1, 2, \ldots\}$. If A is a cut, i.e. a subset of VX for which δA is finite, then let $c(A) = \Sigma\{c(e)|e \in \delta A\}$. Note that we do not assume that X is locally finite. It is convenient from here on to allow $\emptyset$ and VX to be cuts. Thus the set of cuts is a Boolean ring $\mathcal{B}X$.

A ray R in X is an infinite sequence $x_1, x_2, \ldots$ of distinct vertices such that x_i, x_{i+1} are adjacent for every i. If A is an edge cut, and R is a ray, then there exists an integer N such that for $n > N$ either $x_n \in A$ or $x_n \in A^*$. We say that A separates rays $R = (x_n), R' = (x'_n)$ if for n large enough either $x_n \in A, x'_n \in A^*$ or $x_n \in A^*, x'_n \in A$. We define $R \sim R'$ if they are not separated by any edge cut. It is easy to show that $\sim$ is an equivalence relation on the set ΦX of rays in X. The set $\Omega X = \Phi X / \sim$ is the set of edge ends of X. An edge cut A separates ends ω, ω' if it separates rays representing ω, ω'. A cut A separates an end ω and a vertex $v \in VX$ if for any ray representing ω, R is eventually in A and $v \in A^*$ or vice versa.

A cut A is defined to be *thin with respect to* $u, v \in VX \cup \Omega X$ if it separates some $u, v \in VX \cup \Omega X$ and $c(A)$ is minimal among all the cuts that separate u and v. A cut is defined to be *thin* if it is thin with respect to u, v for some $u, v \in VX \cup \Omega X$.

A cut A is defined to be *tight* if both A and A^* are connected, i.e if $x, y \in A$ then there is a path joining x, y whose vertices are all in A, and similarly for A^*.

Proposition 2.3. *A thin cut is tight.*

Proof. Let A be thin with respect to u, v. It is easy to see that if A separates u, v then some component C of A or A^* must separate u, v. If C is a component of A, then $\delta C \subset \delta A$ and if C, D are distinct components then δC and δD are disjoint. Thus if A is thin then $C = A$. The result follows.   □

It is shown in [8] that there are only finitely many tight cuts C with a fixed capacity such that δC contains a particular edge. The proof of this in [32] is neater and it is reproduced here for completeness. By replacing each edge with capacity $c(e)$, by $c(e)$ edges joining the same pair of vertices, we can assume that every edge has capacity one.

Proposition 2.4. *For any $e \in EX$, there are only finitely many tight cuts A with $|\delta A| = c(A) = k$ such that $e \in \delta A$.*

Proof. The proof is by induction on k. For $k = 1$ there is nothing to prove. So assume $k > 1$. We can assume that $e = xy$ is in some tight k-cut, i.e. a cut A such that δA has k edges. Hence $X - e$ has a path P from x to y. Now every tight k cut that contains e also contains an edge of P. By the induction hypothesis there are only finitely many tight $(k-1)$-cuts in $X - e$ containing an edge of P, and we are done.   □

If A, B are cuts, then the sets $A \cap B, A^* \cap B, A^* \cap B, A^* \cap B, A \cap B^*$ are also cuts. These sets are called the *corners* of A, B. This term is suggested by Figure 5.3. Two corners are called opposite or adjacent as suggested in this figure. We say two cuts A, B are *nested* if one $A \cap B$, $A^* \cap B, A^* \cap B, A^* \cap B, A \cap B^*$ is empty. A set $\mathcal{E}$ of cuts is said to be nested if any two elements of $\mathcal{E}$ are nested.

Two cuts that are not nested are said to *cross*.

Lemma 2.5. *Let B be a cut. There are only finitely many tight cuts A with capacity n that cross B.*

Proof. Let F be a finite connected subgraph of X that contains δB. If A crosses B, then both A and A^* contain a vertex of δB. Hence F contains an edge of δA. The lemma follows from Proposition 2.4.   □

We consider sets $\mathcal{E}$ satisfying the following conditions:

  (i) If $A \in \mathcal{E}$, then $A^* \in \mathcal{E}$.
  (ii) The set $\mathcal{E}$ is nested.
  (iii) If $A, B \in \mathcal{E}$ and $A \subset B$, then there are only finitely many $C \in \mathcal{E}$ such that $A \subset C \subset B$.

The following result was first obtained explicitly in [7].

Theorem 2.6. *If $\mathcal{E}$ is a set satisfying conditions (i), (ii) and (iii), then there is a tree $T = T(\mathcal{E})$ such that the directed edge set is $\mathcal{E}$.*

Proof. Consider the set of maps $\alpha : \mathcal{E} \to \mathbb{Z}_2$ satisfying the following:

(a) If $\alpha(A) = 1$, then $\alpha(A^*) = 0$. If $\alpha(A) = 0$, then $\alpha(A^*) = 1$.
(b) If $\alpha(A) = 1$ and $A \subset B$, then $\alpha(B) = 1$.

The vertex set of T will be a subset of the set of all maps satisfying (a) and (b).

Put $ET = \mathcal{E}$ and for $A \in \mathcal{E}$, put $\iota A = \alpha$ where $\alpha(B) = 1$ if $A \subseteq B$ or if $A^* \subset B$. Put $\tau A = \iota A^*$ Thus $\tau A = \beta$ where $\beta(B) = 1$ if $A^* \subseteq B$ or if $A \subset B$. Then ιA and τA take the same value on every B except if $B = A$ or $B = A^*$. We define VT to be the set of all functions which are ιA for some $A \in \mathcal{E}$. It is fairly easy to check that ιA satisfies conditions (a) and (b). If $u = \iota A$ and $v = \tau B$, then the directed edges in a path joining u and v consist of the set $\{C \in \mathcal{E} | A \subseteq C \subseteq B\}$. Using (iii) this set is totally ordered by inclusion and if it is finite by (ii) so is the unique geodesic joining u and v. Thus T is a tree. □

If $\mathcal{E}$ is a nested set of cuts, satisfying (i) (ii) and (iii), in a graph X and there is a bound on the capacity of cuts in $\mathcal{E}$, then the set of all maps $\alpha : \mathcal{E} \to \mathbb{Z}_2$ satisfying (a) and (b) can be identified with $VT \cup \Omega T$. Thus a ray in T corresponds to a strictly $E_1 \supset E_2 \supset \ldots$ of cuts in $\mathcal{E}$ and if we put $\alpha(E) = 1$ if for i sufficiently large $E \supset E_i$. It follows from the finite interval condition (iii) that the intersection of all the E_i's is the empty set and that for any cut A for i sufficiently large either $E_i \subset A$ or $E_i \subset A^*$. If $\alpha : \mathcal{E} \to \mathbb{Z}_2$ satisfies (a) and (b) and there is a unique minimal $B \in \mathcal{E}$ for which $\alpha(B) = 1$, then $\alpha = \iota B$. If there is no such B then we can find a strictly $E_1 \supset E_2 \supset \ldots$ of cuts in $\mathcal{E}$ such that $\alpha E_i = 1$ for every i. Thus α corresponds to a ray in T.

We can identify a vertex v of X with a map $v : \mathcal{B}X \to \mathbb{Z}_2$. Thus $v(A) = 1$ if $v \in A$ and $v(A) = 0$ if $v \notin A$. Restricting to $\mathcal{E}$ will give a vertex of T. Thus there is a map $v : VX \to VT$ such that $v(\alpha), v(\beta)$ differ only on the cuts separating α and β.

Note that there may be vertices of T which are not in the image of v. Each directed edge e of X will give a finite directed path in T consisting of those cuts $A \in \mathcal{E}$ for which $\iota e \in A$ and $e \in \delta A$. A ray in X will correspond to a path in T by concatenating the paths for each edge. It may be the case that this path may backtrack. It will determine a ray in T unless it visits a particular vertex of T infinitely many times, in this case. v maps the end containing the ray to the vertex visited infinitely often. Note that because of Proposition 2.4 it cannot visit two distinct vertices infinitely many times. Thus we can extend v so that it is a map $v : VX \cap \Omega X \to VT \cap \Omega T$.

If $\mathcal{E} \subset \mathcal{B}X$ satisfies the above conditions, then there is a tree $T(\mathcal{E})$. If G is the automorphism group of X and $\mathcal{E}$ is a G-set, then T which is a

G-tree, is called a *structure tree* for X. If $T = T(\mathcal{E})$ is a structure tree for X, then the map $\nu : VX \to VT$ defined above is a G-map.

We now show that a structure tree determines a decomposition of the graph X, in the same way that a group G acting on a tree determines a decomposition of the group G.

Let $T = T(\mathcal{E})$ be a structure tree for X corresponding to a nested set of tight cuts $\mathcal{E}$ satisfying the finite interval condition (iii). Let $v \in VT$.

Let $\nu : VX \to VT$ be the map defined earlier. We define a graph X_v as follows. We take $EX_v \subset EX$ to be the set of edges which lie in δC for some edge C of T incident with v, together with those edges e such that ν maps both vertices of e to v. We take VX_v to be the set of vertices of these edges, but we identify vertices x, y if they both lie in C^* when C has initial vertex v.

Each vertex x of X for which $\nu x = v$ is a vertex of VX_v. Such a vertex is called a ν-vertex, but there may be no such vertices. There is another vertex of X_v for each cut C with initial vertex v, and this vertex is obtained by identifying all the vertices of δC that are in C^*. Such a vertex is called a ρ-vertex. A ρ-vertex has degree $|\delta C|$.

It is fairly easy to see that X_v is connected. Thus any two vertices of X are joined by a path in X. If x, y are two vertices of X that become vertices of X_v after carrying out the identifications just described, then the path in X will become a path p in X_v if we delete any edges that are not in X_v. Here we use the fact that C^* is connected, and when p enters C^* at vertex w it must leave C^* at a vertex w' that is identified with w in X_v.

In a similar way a ray in X_v corresponds to a ray in X. If the ray passes through a vertex corresponding to the cut C, then the two incident edges will both lie in δC. There will be a path in C^* joining the corresponding vertices before they are identified. For each such vertex that is visited by the ray we can add in this path to obtain a ray in X. This ray will belong to an end $\omega \in \Omega X$ such that $\nu \omega = v$.

We regard the graphs X_v for each $v \in VT$ as the factors in the decomposition for X. We now describe how they fit together to give X. For each edge $e \in EX$ there are only finitely many $E \in \mathcal{E}$ such that $e \in \delta E$. These edges form the edges of the geodesic in T joining νu and νv where u, v are the vertices of e. Suppose there are $k(e)$ such edges. Now form a new graph X' in which each edge e is subdivided into $k(e)$ edges. We can extend $\nu : VX \to VT$ to a map also denoted $\nu : VX' \to VT$ which can be extended to a graph morphism. It will now be the case that ν is surjective. For each cut $A \in \mathcal{E}$, there is a cut $A' \in \mathcal{B}X'$ in which $\delta A'$ consists of those edges of X' that are mapped to the edge A of ET under the extended morphism $\nu : X' \to T$. We have that $A' \cap VX = A$ and $|\delta A'| = |\delta A|$.

Clearly we have a nested set of cuts $\mathcal{E}'$ that will be the edge set of a structure tree T which is isomorphic to T, and can be identified with T

in a natural way. For $v \in VT$ the graph X'_v will be a subdivision of the graph X_v. Each edge of X_v that joins two ρ-vertices is subdivided into two edges in which the centre vertex is a v-vertex in X'.

It is easier to use X' rather than X to understand the structure of the graphs X_v. This is because every edge of X' lies in at most one δA for $A \in \mathcal{E}'$ and if u, v are the vertices of $A \in ET = ET'$, then X_u and X_v are the only factors in the decomposition of X that contain the edge e. If G is a group acting on X, then it will also act on T. For $v \in VT$ the stabiliser G_v will act on X_v. Two directed edges of X'_v will lie in the same G-orbit if and only if they lie in the same G_v-orbit. If u, v are distinct vertices of T, then edge sets of X'_v and X'_u are disjoint unless $u.v$ are adjacent in T in which case the intersection consists of the edges that map to the edge with vertices vu and vv in T. It follows that if X has finitely many G-orbits, then each X_v has finitely many G_v-orbits. We have a decomposition of X in which the factors are X_v.

Lemma 2.7. *If A, B are crossing thin cuts, with $c(A) = m$, $c(B) = n$, then after relabelling A as A^* and B as B^* if necessary, both $A \cap B^*$, $A^* \cap B$ are thin cuts with capacities m, n respectively.*

Proof. We refer to Figure 5.3. Suppose $m \leq n$. Suppose A is thin with respect to x, y with $x \in A$ and B is thin with respect to x', y' with $x' \in B$. After possible relabelling we can assume $a \leq b$, $c \leq d$. If $a < c$, then $c(A \cap B) < n$ and $c(A^* \cap B) < n$ and so B is not thin since one of these two corners separates x' and y'. If $c < a$, then A is not thin. Hence $a = c$. If $a < b$, then $c(A \cap B) = 2a + f < a + e + f + b = m$, and so $x \in A \cap B^*$ and $c(A \cap B^*) = a + e + c = m$ and $f = 0$. Also $c(A^* \cap B) = a + e + d = n$, and $x' \in A^* \cap B$ and so it is thin, and we are done. If $a = b$, then $m = 2a + e + f \leq a + e + f + d = n$ and so $b \leq d$. Thus $a = b = c \leq d$. If e is not 0, then $c(A \cap B) < m, c(A^* \cap B^*) < n$ and the lemma follows easily. If $e = 0$ and $f \neq 0$, then $c(A \cap B^*) < m, c(A^* \cap B) < n$ and the lemma follows if we relabel A as A^*. But if $e = f = 0$, then $c(A \cap B) = c(A \cap B^*) = m$ and $c(A^* \cap B) = c(A^* \cap B^*) = n$. In this situation it is not possible that two adjacent corners of A, B are not thin. Thus for one pair of opposite corners we have that both corners are thin. By relabelling we can assume these corners are $A \cap B^*$ and $A^* \cap B$ and the lemma is proved. ☐

Lemma 2.8. *Let A, B, C be cuts.*

(i) *Let A, B be not nested and let C be nested with both A and B, then C is nested with every corner of A, B.*

(ii) *If C is nested with A, then C is nested with two adjacent corners of A and B.*

Proof. For (i) by possibly relabelling A as A^* and/or B as B^* and/or C as C^* we can assume either

(a) $C \subset A$ and $C \subset B$ or

(b) $C \subset A$ and $C^* \subset B$.

If (a) then $C \subset A \cap B$ and C is contained in the complement of each of the other corners. If (b), then $B^* \subset C \subset A$, and so A, B are nested, which contradicts our hypothesis.

For (ii) if $A \subset C$, then $A \cap B \subset C$ and $A \cap B^* \subset C$.   □

Let $\mathcal{C}$ be a set of cuts. Let A be a cut and let $M(A, \mathcal{C})$ be the set of cuts in $\mathcal{C}$ which are not nested with A. Set $\mu(A, \mathcal{C}) = |M(A, \mathcal{C})|$.

Lemma 2.9. *Let $\mathcal{C}$ be a nested set of tight cuts. Let A be a tight cut which is not nested with some $B \in \mathcal{C}$. Let $\mu(A) = \mu(A, \mathcal{C})$ be the number of cuts in $\mathcal{C}$ that are not nested with A. then*

$$\mu(A \cap B, \mathcal{C}) + \mu(A \cap B^*, \mathcal{C}) < \mu(A, \mathcal{C}).$$

Proof. If $C \in \mathcal{C}$ is nested with A, then it is nested with both A and B and so it is nested with $A \cap B$ and $A \cap B^*$ by Lemma 2.8. If C is not nested with A, then it must be nested with one of $A \cap B$ and $A \cap B^*$. For if, say, $C \subset B$ then $B^* \subset C^*$ and so $A \cap B^* \subset C^*$. Thus C is not nested with at most one of $A \cap B$ and $A \cap B^*$ and the lemma follows, since B is counted on the right but not on the left.   □

Let $\mathcal{C}_n$ be the set of thin cuts with capacity n.

Theorem 2.10. *There is a uniquely defined nested set of thin cuts $\mathcal{E}$ in which $\mathcal{E}_n = \{E \in \mathcal{E} | c(E) \leq n\}$ constructed inductively as follows:-*
$\mathcal{E}_1 = \mathcal{C}_1$.
If $\mathcal{D}_n = \{A \in \mathcal{C}_n | \mu(A, \mathcal{E}_{n-1}) = 0\}$, then $\mathcal{E}_n = \mathcal{E}_{n-1} \cup \mathcal{D}'_n$, where $\mathcal{D}'_n$ consists of all those cuts $D \in \mathcal{D}_n$ satisfying

(*) *D is thin with respect to some $u, v \in VX \cup \Omega X$ and $\mu(D, \mathcal{D}_n)$ takes the minimal value among all $D \in \mathcal{D}_n$ that are thin with respect to u, v.*

A cut in $\mathcal{D}_n$ satisfying (*) is said to be *optimally nested with respect to u, v*.

Proof. This is an argument from [4]. Put $\mu(A) = \mu(A, \mathcal{D}_n)$. Let $A, B \in \mathcal{D}_n$ be not nested. Each corner of A, B is nested with every $e \in \mathcal{E}_{n-1}$ by Lemma 2.8. Suppose A is optimally nested with respect to x, y and B is optimally nested with respect to x', y'. Here x, y, x', y' are elements of $VX \cup \Omega X$. Each of x, y, x', y' determines a corner of A, B. There are two possibilities.

(i) The sets x, y determine opposite corners, and x', y' determine the other two corners.

(ii) There is a pair of opposite corners such that one corner is determined by one of x, y and the opposite corner is determined by one of x', y'.

In case (i) A and B separate both pairs x, y and x', y'. Because A, B are optimally nested with respect to x, y and x', y', we have $\mu(A) = \mu(B)$. But now both $A \cap B$ and $A^* \cap B^*$ separate x, y say and $c(A \cap B) = c(A^* \cap B^*) = n$ by Lemma 2.7 so that both the corners are in $\mathcal{D}_n$, and $\mu(A \cap B) + \mu(A^* \cap B^*) < \mu(A) + \mu(B) = 2\mu(A)$, by Lemma 2.8, since if an element of $\mathcal{C}_n$ is not nested with both $A \cap B$ and $A^* \cap B^*$ it is not nested with both A and B and if it is not nested with one of $A \cap B, A^* \cap B^*$ then it is not nested with one of A and B. The strict equality follows because $A \in \mathcal{D}_n$ separates x, y and is not nested with B but both $A \cap B, A^* \cap B^*$ are nested with A and B. Since both $A \cap B, A^* \cap B^*$ separate x and y we have a contradiction.

In case (ii) suppose these corners are $A \cap B$ and $A^* \cap B^*$, and that $x \in A \cap B, y' \in A^* \cap B^*$. But then $A \cap B$ separates x and y and $A^* \cap B^*$ separates x' and y'. Given that A is optimally nested with respect to x and y we have $\mu(A \cap B) \geq \mu(A)$ and because B is optimally nested with respect to x' and y' we have $\mu(A^* \cap B^*) \geq \mu(B)$. But it follows, as in the previous paragraph, that $\mu(A \cap B) + \mu(A^* \cap B^*) < \mu(A) + \mu(B)$ and so we have a contradiction. Thus $\mathcal{D}'_n$ is a nested set and the proof is complete.

Note that $\mathcal{E}$ is uniquely defined, as no choices are made in its construction. This is very important in applications. It means that $\mathcal{E}$ is invariant under the automorphism group of the graph.   □

Recall that $\mathcal{B}X$ is the Boolean ring of all cuts. Let $\mathcal{B}_n X$ be the ring generated by all cuts A such that $c(A) \leq n$.

Theorem 2.11.   *For every $u, v \in VX \cup \Omega X$ that can be separated by a cut, $\mathcal{E}$ contains a cut A that is thin with respect to u, v. The set $\mathcal{E}_n$ generates $\mathcal{B}_n X$ and is the directed edge set of a tree.*

Proof.   To prove the first statement we need to show that for every u, v there is a cut A that is thin with respect to u, v that is nested with every cut $E \in \mathcal{E}_{n-1}$, where $n = c(A)$, i.e. we need to show that there is a cut in $\mathcal{D}_n$ that is thin with respect to u, v. We know that there is a cut B that is thin with respect to u, v. Let $k = \mu(B, \mathcal{E}_{n-1})$. If $k = 0$, then we take $A = B$. If $k \geq 1$, then let $C \in \mathcal{E}_{n-1}$ cross B. We know from Lemma 2.7 that one of the corners (say $B \cap C$) of B, C is thin with respect to u, v. By Lemma 2.9 we have $\mu(B \cap C) + \mu(B \cap C^*) < k$. Thus if $\mu(B) = k > 0$ we can find a cut $B \cap C$ which is thin with respect to u, v for which $\mu(B \cap C) < k$. Thus there must be a cut A which is thin with respect to u, v for which $\mu(A) = 0$.

Let $\mathcal{B}_n$ be the subring of $\mathcal{B}$ generated by $\mathcal{E}_n$. Clearly $\mathcal{B}_n \subseteq \mathcal{B}_n$ is the subring generated by all cuts with capacity at most n. We want to show that $\mathcal{B}_n = \mathcal{B}_n$. Let A be a cut with $c(A) = n$. We show that $A \in \mathcal{B}_n$ by induction on $\mu(A) = \mu(A, \mathcal{E}_n)$. Suppose that $\mu(A) = 0$ so that A is nested with every cut in $\mathcal{E}_n$. In this case if $A \notin \mathcal{E}_n$ then A determines a vertex v of $T = T(\mathcal{E}_n)$. Thus we define $v : \mathcal{E}_n \to \mathbb{Z}_2, vE = 0,$ if $E \subset A,$ or $E \subset A^*, vE = 1,$ if $E^* \subset A,$ or $E^* \subset A^*.$

We have said that $\tau E = v$ if $E \subset A$ and E is a maximal element of $\mathcal{E}_n$ with this property.

Let X_v be the graph defined earlier.

The cut A will become a cut A_v in X_v. Thus A_v will consist of all v-vertices $x \in A$ such that $vx = v$ and also and also all ρ-vertices y corresponding to cuts $C \in \mathcal{E}_n$ such that $C^* \subset A$. It will then be the case that $\delta A = \delta A_v$. Now $A \in B_n$ if and only if either A_v or A_v^* consists of finitely many ρ-vertices. If this is not the case, then both A_v and A_v^* either contain infinitely many ρ-vertices or at least one v-vertex. However if say A_v consists only of infinitely many ρ-vertices, then A_v is a connected subset of X_v consisting of infinitely many vertices of bounded degree. Such a set must contain the vertices of a ray by König's Lemma. It follows that if A is not in B_n then it must separate two elements of $VX \cup \Omega X$. But these two elements must be separated by an element of $\mathcal{E}_n$ which is not the case, and so we have a contradiction. Thus A_v or A_v^* consists of finitely many ρ-vertices and so $A \in B_n$.

If $\mu(A) > 0$ and $E \in \mathcal{E}_n$ is not nested with A, then $A = A \cap E + A \cap E^*,$ and both $A \cap E$ and $A \cap E^*$ are in B_n by induction on $k = \mu(A)$. The theorem is proved. $\square$

If every pair $x, y \in VX \cup \Omega X$ that can be separated by a cut in $\mathcal{E}_n$, then $\mathcal{B}X = \mathcal{B}_n X$. If $\mathcal{B}X = \mathcal{B}_n X$ and $vx = vy = v$, then x, y cannot be separated in X or X_v. This means that either both x, y and every other vertex of X_v have infinite degree and X_v has one end or $x = y$ and X_v contains at most one v-vertex. If v is not in the image of v, and $\mathcal{B}X = \mathcal{B}_n X$, then X_v will be a one-ended graph in which each vertex has degree at most n.

Thomassen and Woess [32] define a graph to be *accessible* if there is an integer n such that any two ends can be separated by removing at most n edges. Alternative ways of defining accessibility are suggested by Theorem 2.11.

Definition 2.12.   A graph X is said to be $\mathcal{B}$-accessible if $\mathcal{B}X = \mathcal{B}_n X$ for some n. A graph X is said to be $\mathcal{E}$-accessible if $\mathcal{E}$ satisfies the finite interval condition (iii) of Theorem 2.6.

A graph X is $\mathcal{B}$-accessible if and only every pair $x, y \in VX \cup \Omega X$ that can be separated by a cut can be separated by a cut in $\mathcal{E}_n$. There is then a structure tree T_n and a map $v : VX \cup \Omega X \to VT_n \cup \Omega T_n$ such that $vx \neq vy$ if and only if x, y are not separated by any cut.

A graph X is $\mathcal{E}$-accessible if and only if $\mathcal{E}$ is the directed edge set of a structure tree T for which there is a map $v : VX \cup \Omega X \to VT \cup \Omega T$ in which $vx = vy$ if and only if x, y are not separated by any cut A. If

X is $\mathcal{E}$-accessible, then for every $v \in VT$ the graph X_v has the following structure:

> X_v has at most one end. Every vertex of infinite degree is a v vertex. The vertices of finite degree consist of ρ-vertices and at most one ν-vertex.

Clearly it follows from Theorem 2.14 that X is accessible if $\mathcal{B}X = \mathcal{B}_n X$ for some n. If every vertex of X has bounded degree, then $\mathcal{B}X = \mathcal{B}_n X$ for some n if and only if X is $\mathcal{E}$-accessible, and all three definitions of accessible are equivalent. Wall [34] defined a finitely generated group to be accessible if any process of successively splitting the group over finite subgroups eventually terminates with factors that are finite or one-ended. By Bass-Serre theory this is equivalent to saying that the group has an action on a tree with every edge group finite and every vertex group is finite or one-ended.

As proved in [32] a finitely generated group is accessible if and only if its Cayley graph (with respect to a finite generating set) is accessible. I proved in [8] that finitely presented groups are accessible and in [9] I gave an example of a finitely generated group that is not accessible. Thus there are vertex transitive locally finite graphs that are not accessible.

It is fairly easy to construct graphs that are $\mathcal{E}$-accessible but not $\mathcal{B}$-accessible or accessible. A graph that is $\mathcal{B}$-accessible is both accessible and $\mathcal{E}$-accessible.

Theorem 2.13. *A locally finite accessible graph is $\mathcal{E}$-accessible.*

Proof. Suppose that any two ends of X are separated by an n-cut, so that X is accessible. Let $T = T_n$ be the structure tree with edge set $\mathcal{E}_n$. For each $v \in VT$, X_v has at most one end. We have to show that any edge of X lies in finitely many δA for $A \in \mathcal{E}$. We know that any edge of X' lies in at most two graphs X'_v and so any edge of X lies in finitely many X_v. It therefore suffices to show that any edge of X_v lies in finitely δA for $A \in \mathcal{E}$. Such an A can be regarded as a cut in X_v. Thus it suffices to prove the theorem when X has at most one end. Clearly the result is true if X is finite. By König's lemma, if X is infinite then it has a ray and so in our case it has one end. There will be an infinite sequence of elements $A_i \in \mathcal{E}$ such that $A_{i+1} \subset A_i$ whose intersection is empty. Because X has one end A_i^* is finite. If $A_i \in \mathcal{E}_{n(i)}$, then the edges in $T_{n(i)}$ that separate vertices of A_i^* form a finite subtree and this will be a subtree of every tree T_m for $m > n(i)$. It follows that every edge of X lies in finitely many δA for $A \in \mathcal{E}$.   □

I think it ought to be possible to drop the locally finite condition in the last theorem.

In [11] I showed that a vertex transitive locally finite planar graph is accessible. A locally finite graph X has a Freudenthal compactification

$\mathcal{F}(X)$ in which distinct ends correspond to distinct points. Richter and Thomassen [27] show that if X is connected with a locally finite planar graph then $\mathcal{F}(X)$ can be embedded in S^2, and this embedding has a uniqueness property if X is three-connected. If A is a tight cut in such a graph X, then there is a simple closed curve in S^2 that intersects $\mathcal{F}(X)$ in a finite set of points consisting of a single point in each edge of δA. The set $\mathcal{E}$ will correspond to a set of non-intersecting simple closed curves. The graph X will be $\mathcal{E}$-accessible if and only if every edge of X intersects finitely many of the simple closed curves. The structure tree corresponding to $\mathcal{E}$ will then be the dual graph to the set of curves.

It would be interesting to know if every planar graph is $\mathcal{E}$-accessible. If this was the case, then a planar graph of bounded vertex degree would be $\mathcal{B}$-accessible. A locally finite $\mathcal{E}$-accessible graph has a unique decomposition in which the factors are one ended or finite.

Combining the two previous theorems we have the following:

Theorem 2.14. *Let $N(X)$ be a network in which X is an arbitrary connected graph. For each $n > 0$, there is a network $N(T_n)$ based on a tree $T = T_n$ and a map $\nu : VX \cup \Omega X \to VT \cup \Omega T$, such that $\nu(VX) \subset VT$ and $\nu x = \nu y$ for any $x, y \in VX \cup \Omega X$ if and only if x, y are not separated by a cut A with $c(A) \leq n$.*

The network $N(T_n)$ is canonically determined and is invariant under the automorphism group of $N(X)$.

For a finite network Theorem 2.14 reduces to Theorem 2.2. For a finite network the structure tree of Theorem 2.2 will become a Gomory-Hu tree by contracting certain edges. Thus a Gomory-Hu tree has the same properties as our tree except that the map $\nu : VX \to VT$ is a bijection. One obtains a Gomory-Hu tree from our structure tree by choosing for each vertex $v \in VT$ that is not in the image of ν an incident edge of maximal capacity and then contracting all those edges. If we choose the edges to contract in the way just described one will preserve the property that for any $s, t \in VX$, there is a minimal cut separating s, t in the geodesic joining $\nu s, \nu t$ in T. Note that while the tree of Theorem 2.2 is uniquely determined there may be more than one Gomory-Hu tree. This has already been noted in the tree of Figure 5.2.

Let n be the smallest capacity of a cut in $N(X)$. It can be seen fairly easily from Figure 5.3 that if A, B are cuts with $c(A) = c(B) = n$ and A is not nested with B, then $n = 2m$ and δA partitions $\delta A = (\delta(A \cap B) \cap \delta A) \cup \delta(A \cap B^*) \cap \delta A)$, where each of $\delta(A \cap B) \cap \delta A$ and $\delta(A \cap B^*) \cap \delta A$, contain m edges. If C is also a cut with capacity n that is not nested with A, then one can show that the partition of δA given by C is the same as that corresponding to B. This result is crucial in the cactus representation of mincuts by Dinits, Karzanov and Lomonosov [5]. A cactus is a simple graph in which every edge lies in at most one cycle. In the cactus representation there is a cactus K and a mincut in $N(X)$

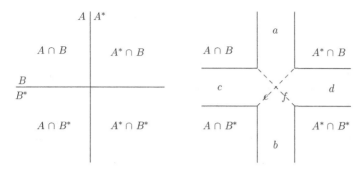

Figure 5.3 Crossing cuts

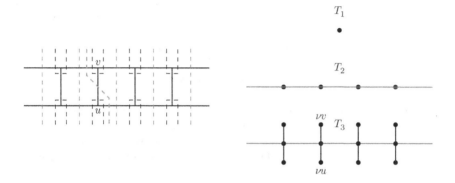

Figure 5.4 Cutting up a ladder

corresponds to a tight cut in K with capacity at most 2. In our notation the mincuts in $N(X)$ are the elements of $C_n = \mathcal{D}_n$ and the elements of $\mathcal{D}'_n = \mathcal{E}_n$ correspond to the tight cuts E in K in which δE has one edge or it consists of adjacent edges of a cycle.

Evangelidou and Papasoglu [19] use a similar cactus argument but for minimal cuts separating ends of a graph to give a proof of Stallings theorem.

We illustrate our results with some simple examples.

In the example of Figure 5.4, the graph X is an infinite ladder, and every edge has capacity one. The two ends of X and some vertices are separated in T_2 and all ends and vertices in T_3. The vertices of T_3 on the horizontal central line are not in the image of ν. Note the cuts A, A^* given by the bolder dashed line are not thin with respect to the two ends of the graph since $c(A) = 3$ and the two ends are separated by a cut with capacity 2. However, A is thin with respect to the two vertices u, v of one rung of the ladder. However, A is not optimally nested with respect to u, v since it is not nested with the cut αA where alpha is the automorphism swapping the two sides of the ladder. The cuts given by

$$T_1 = T_2 \qquad \bullet$$

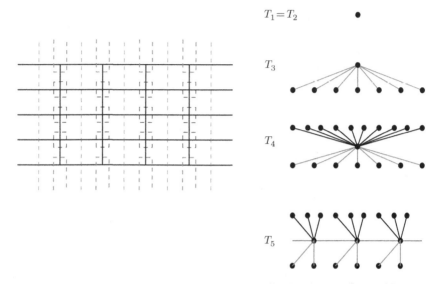

$$T_3$$

$$T_4$$

$$T_5$$

Figure 5.5 Cutting up another ladder

the blue black dashed lines are nested with every thin cut and so are optimally nested with respect to any pair of vertices that they separate.

In the example of Figure 5.5, all the vertices are separated in T_4 and all ends and vertices in T_5. There is a vertex of infinite degree in each of T_3 and T_4.

A finitely generated group G is said to have more than one end if a Cayley graph $X = X(G, S)$ of G corresponding to a finite generating set S has more than one end.

Theorem 2.15 (Stallings theorem). *If G is a finitely generated group with more than one end, then G has a non-trivial action on a tree T with finite edge stabilizers.*

Proof. Let n be the smallest integer for which there is a cut A such that $|\delta A| = n$ which separates two ends s, t. The structure tree $T = T_n$ will have the required property. Here we use the fact that the construction of T is canonical and so is invariant under the action of automorphisms. Thus the action of G on X gives an action on T. Each edge of T is a cut C with $|\delta C| \le n$. The stabilizer of C will permute the edges of δC and will therefore be finite. We also have to show that action is non-trivial. We know there is a cut D in ET that separates a pair of ends. For such a cut both D and D^* are infinite. The action of G on X is vertex transitive. There exists $g \in G$ such that the vertices of $g\delta D$ are contained in D and an element $h \in G$ such that the vertices of $h\delta D$ are contained in D^*. It then follows that for $x = g, x = h$ or $x = gh$ we have xD is a proper subset of D or D is a proper subset of xD. It follows from elementary Bass-Serre theory that x cannot fix a vertex of T. $\quad\square$

This proof is essentially that of [21].

In any tree T if p is a vertex and Q is a set of unoriented edges, then there is a unique set of vertices P such that $v \in P$ then the geodesic $[v, p]$ contains an odd number of edges from Q. We then have $\delta P = Q$. Note that $\mathcal{B}T = \mathcal{B}_1 T$ and every element of $\mathcal{B}T$ is uniquely determined by the set Q together with the information for a fixed $p \in VT$ whether $p \in A$ or $p \in A^*$. The vertex p induces an orientation $\mathcal{O}_p$ on the set of pairs $\{e, \bar{e}\}$ of oriented edges by requiring that $e \in \mathcal{O}$ if e points at p. For $A \in \mathcal{B}T$, A is uniquely determined by δA together with the orientation $\mathcal{O}_p \cap \delta A$ of the edges of δA.

In X it is the case that a cut A is uniquely determined by δA together with the information for a fixed $p \in VX$ whether $p \in A$ or $p \in A^*$.

Since $\mathcal{B}_n X$ is generated by $\mathcal{E}_n = ET_n$, the cut A can be expressed in terms of a finite set of oriented edges of T_n. This set is not usually uniquely determined. Thus if ν is not surjective, and v is not in the image of ν, and the set of edges incident with v is finite, then VX is the union of these elements in $\mathcal{B}X$. The empty set is the intersection of the complements of these sets. Orienting the edges incident with v towards v gives the empty set and orienting them away from v gives all of VX. However there is a canonical way of expressing an element of $\mathcal{B}_n X$ in terms of the generating set $\mathcal{E}_n$. To see this let $A \in \mathcal{B}_n X - \mathcal{B}_{n-1} X$. There are only finitely many $C \in \mathcal{E}_n$ with which C is not nested. This number is $\mu(A, \mathcal{E}_n) = \mu(A)$. We use induction on $\mu(A)$. Our induction hypothesis is that there is a canonically defined way of expressing A in terms of the $\mathcal{E}_n$. Any two ways of expressing A in terms of $\mathcal{E}_n$ differ by an expression which gives the empty set in terms of $\mathcal{E}_n$. Such an expression will correspond to a finite set of vertices, each of which has finite degree in T_n and none of which is in the image of ν. The canonical expression is obtained if there is a unique way of saying whether or not each such vertex is in the expression for A. Thus the canonical expression for A is determined by a set of vertices of VT which consists of the vertices of $\nu(A)$ together with a recipe for deciding for each vertex which is not in the image of ν whether it is in the expression for A.

Suppose $\mu(A) = 0$, so that A is nested with every $C \in \mathcal{E}_n$, and neither A nor A^* is empty. If $A \in \mathcal{E}_n$, then this gives an obvious way of expressing A in terms of the $\mathcal{E}_n$. If A is not in $\mathcal{E}_n$, then it corresponds to a unique vertex $z \in VT_n$. Thus because $\mu(A) = 0$, A induces an orientation of the edges of $\mathcal{E}_n$. To see this, let $C \in \mathcal{E}_n$, then just one of $C \subset A$, $C^* \subset A$, $C \subset A^*$, $C^* \subset A^*$ holds. From each pair C, C^* we can choose C if $C \subset A$ or $C \subset A^*$ and we choose C^* if $C^* \subset A$ or $C^* \subset A^*$. Let $\mathcal{O}$ be this subset of $\mathcal{E}$. Then if $C \in \mathcal{O}$ and $D \in \mathcal{E}$ and $D \subset C$, then $D \in \mathcal{O}$. This means that the orientation $\mathcal{O}$ determines a vertex z in VT_n. Intuitively the edges of $\mathcal{O}$ point at the vertex z. It can be seen that A or A^* will be the union of finitely many edges E of $\mathcal{E}_n = ET_n$, all of which have $\tau E = z$. If A is such a union, then we use this to express $A = C_1 \cup C_2, \cdots \cup C_k$. If A is not such a union, but $A^* = C_1 \cup C_2, \cdots \cup C_k$, then we write

$A = (C_1 \cup C_2, \cdots \cup C_k)^* = C_1^* \cap C_2^* \cap \cdots \cap C_k^*$. Note that this gives a unique way of expressing cuts corresponding to a vertex z of finite degree not in the image of ν. The vertex z is included in the expression for A^* if and only if only finitely many cuts in $\mathcal{E}_n$ incident with z and pointing at z are subsets of A. Suppose then that the hypothesis is true for elements $B \in \mathcal{B}_n X$ for which $\mu(B) < \mu(A)$. Let $C \in \mathcal{E}_n$ be not nested with A. Then $\mu(A \cap C) + \mu(A \cap C^*) \leq \mu(A)$. Thus each of $A \cap C$ and $A \cap C^*$ can be expressed in a unique way in terms of the $\mathcal{E}_n$. If at most one of these expressions involves C, then we take the expression for A to be the union of the two expressions for $A \cap C$ and $A \cap C^*$. If both of the expressions involve C, then we take the expression for A to be the union of the two expression with C deleted. The expression obtained for A is independent of the choice of C. In fact the decomposition will involve precisely those C for which C occurs in just one of the decompositions for $A \cap C$ and $A \cap C^*$. We therefore have a canonical decomposition for A. To further clarify this proof observe the following. The edges C, which are not nested with A, form the edge set of a finite subtree F of T_n. If $EF \neq \emptyset$ we can choose C so that it is a twig of F, i.e. so that one vertex z of F is only incident with a single edge C of F. By relabelling C as C^* if necessary we can assume that $\mu(A \cap C) = 0$. The vertex determined by $A \cap C$ as above is z, and we have spelled out the recipe for if this vertex is to be included in the expression for A. The induction hypothesis gives us a canonical expression for $A \cap C^*$, which together with the expression for $A \cap C$ gives the expression for A.

2.3. Flows in Networks

In this subsection we give a version of the Max-Flow Min-Cut theorem for arbitrary networks that reduces to the usual theorem for a finite network. Let X, N be as before. For $s, t \in VX \cup \Omega X$ an (s, t)-*flow* in N is a map $f : EX \to \{0, 1, 2, \dots\}$ together with an assignment of a direction to each edge e for which $f(e) \neq 0$ so that its vertices are ιe and τe and the following holds.

(i) For each $v \in VX$ there are only finitely many incident edges e for which $f(e) \neq 0$.

(ii) If $f^+(v) = \Sigma(f(e)|\iota e = v)$ and $f^-(v) = \Sigma(f(e)|\tau e = v)$, then $f^+(v) = f^-(v)$ for every $v \neq s, t$.

(iii) For every cut A that does not separate s, t if we put $f^+(A) = \Sigma(f(e)|e \in \delta A, \iota e \in A)$ and $f^-(A) = \Sigma(f(e)|e \in \delta A, \iota e \in A^*)$, then we have $f^+(A) = f^-(A)$. That is, for every cut that does not separate s, t, the flow into the cut is the same as the flow out.

Proposition 2.16.   *For any (s, t)-flow and any cut A such that $s \in A$, $t \in A^*$, the value $f^+(A) - f^-(A)$ does not depend on A. This value is denoted $|f|$.*

Proof. Let A, B be cuts separating s, t. Because $A \cap B$ also separates s, t, it suffices to prove that $f^+(A) - f^-(A) = f^+(B) - f^-(B)$ when $A \subset B$. Let $e \in \delta A$. Either $e \in \delta B$ or $e \in \delta(B \cap A^*)$. If $e' \in \delta B$ is not in δA then $e' \in \delta(B \cap A^*$ and $\delta(B \cap A^*)$ partitions into those edges with both vertices in A and those with both vertices not in A. Because $A^* \cap B$ does not separate s, t, $f^+(A^* \cap B) = f^-(A^* \cap B)$ and the value of $f^+ - f^-$ on the edges of $\delta(A^* \cap B)$ that are in A is minus the value on the edges not in A. The symmetric difference of δA and δB consists of the edges in $\delta(A^* \cap B)$ and it follows that $f^+(A) - f^-(A) = f^+(B) - f^-(B)$. □

Let $T = T_n$ be the structure tree for $N(X)$. We have a network $N(T)$ based on T in which each edge has capacity at most n. If $u, v \in VT$ there is a unique maximal (u, v)-flow $f(u, v)$ in which $|f(u, v)|$ is the minimal capacity of an edge in the geodesic path $[u, v]$ joining u, v. An (s, t)-flow f in N such that $|f| \leq n$, induces a (vs, vt)-flow $\bar{f}$ in T such that $|f| = |\bar{f}|$. In T an end corresponds to a set of rays. For any vertex of T there is a unique ray starting at that vertex and belonging to the end. If $u \in VT$ and $v \in \Omega T$, then there is a unique ray starting at u and representing v. While if $u, v \in \Omega T$, then there is a unique two-ended path in T whose ends represent u and v. In each case we get a (u, v)-flow in T by assigning a constant value on the directed edges of the path, provided this constant is less than or equal to the capacity of each edge in the path. Every (u, v)-flow is of this type. A maximal (u, v)-flow $f(u, v)$ is obtained by taking this constant to be the minimal capacity of an edge in the path.

Theorem 2.17 (MFMC). *Let N be a network based on a graph X. Let $s, t \in VX \cup \Omega X$. The maximum value of an (s, t)-flow is the minimal capacity of a cut separating s and t.*

Proof. Let n be the minimal capacity of a cut in X separating s, t. In the structure tree $T = T_n$ there is a flow from vs to vt with the property that the value of the flow is n. We have to show that each such flow corresponds to a flow in X.

If $s, t \in VX$, then this follows from the usual proof of the theorem, which we repeat here. Suppose we have an (s, t)-flow f in N. Let $e_1, e_2, \ldots e_k$ be a path p joining s and t with the following property. Each edge e_i is given an orientation in the flow f. This orientation will not usually be the same as that of going from s to t. We say that p is an f-augmenting path if for each e_i for which ιe is s or a vertex of e_{i-1} we have $f(e_i) < c(e_i)$, and for each edge e_i for which $\iota e = t$ or ιe is a vertex of e_{i+1} we have $f(e_i) \neq 0$. For any flow augmenting path p we get a new flow f^* as follows.

(i) If $e \in EX$ is not in the path p, then $f^*(e) = f(e)$.
(ii) If e is in the path p and $f(e) = 0$, then orient e so that ιe is s or a vertex of e_{i-1}, and put $f^*(e) = 1$. Recall that we are assuming that $c(e) \neq 0$ for every $e \in EX$ and so we have $f^*(e) \leq c(e)$.

(iii) If e is in the path p and ιe is t or a vertex of e_{i+1} (so that $\bar{e} = e_{i+1}$) and $f(e) \neq 0$, then $f^*(e) = f(e) - 1$.

(iv) If e is in the path p and ιe is s or a vertex of e_{i-1} (so that $e = e_i$), then $f^*(e) = f(e) + 1$.

The effect of changing f to f^* is to increase the flow along the path p. We have $|f^*| = |f| + 1$.

Let $S_f \subset VX$ be the set of vertices that can be joined to s by a flow augmenting path. If $t \in S_f$, then we can use the flow augmenting path joining s and t to get a new flow f^*. We keep repeating this process until we obtain a flow f for which S_f does not contain t. But now S_f is a cut separating s and t. Also if $e \in \delta S_f$, then we have $\iota e \in S_f$ and $f(e) = c(e)$, since otherwise we can extend the f-augmenting path from s to ιe to an f-augmenting path to τe. Thus $|f| = c(S_f)$. But n is the minimal capacity of a cut separating s and t and so $|f| \geq n$. But also $|f|$ must be less than the capacity of any cut separating s and t and so $|f| = n$, and S_f is a minimal cut separating s and t.

If $s \in VX$ and $t \in \Omega X$, then we can build up a flow from s to t in the following way. Let D be a cut in ET separating s and t, so that $s \in D$ and $c(D) \geq n$. Let X_D be the graph defined as follows. The edge set EX_D consists of all edges e of X which have at least one vertex in D, so that either $e \in \delta D$ or e has both vertices in D. The vertex set VX_D consists of the vertices of EX_D, except that we identify all such vertices that are in D^*. Let this vertex be denoted d^*. Thus in X_D the edges incident with d^* are the edges of δD. Since $c(D) \geq n$, then as in the previous case there is a flow f_D from s to d^* such that $|f_D| = n$. Let X_{D^*} be the graph defined as for X_D, using D^* instead of D. We now have a vertex $d \in VX_{C^*}$ whose incident edges are the edges of δD. Now choose another edge $E \neq D$, such that $E^* \subset D^*$, separating s, t. Thus $s \in E$. Now form a graph $X(D^*, E)$ whose edge set consists of those edges that have at least one vertex in $D^* \cap E$ and whose vertex set is the set of vertices of the set of edges except that we identify the vertices that are in D and also identify the vertices that are in E^*. Thus in $X(D^*, E)$ there is a vertex d whose incident edges are those of δD and a vertex e^* whose incident edges are those of δE. The flow f already constructed takes certain values on the edges of δD. We can find a (d, e^*) flow which takes these same values on δD. This flow together with the original flow will give an (s, e^*)-flow also denoted f such that $|f| = n$. We can keep on repeating this process and obtain the required (s, t)-flow.

If s, t are both in ΩX, choose a minimal cut M separating s and t, so that $s \in M, t \in M^*$. Let X_s be the graph defined as follows. The edge set EX_s consists of all edges e of X which have at least one vertex in M, so that either $e \in \delta M$ or e has both vertices in M. The vertex set VX_s consists of the vertices of EX_s, except that we identify all such vertices that are in M^*. Let this vertex be denoted m_t. Thus in X_s the edges incident with m_t are the edges of δM. If $c(M) = n$, then by the previous case there is a flow f_s from s to m_t such that $|f_s| = n$. If we carry out a similar construction for M^* we obtain a flow f_t from m_s to t with $|f_t| = n$. We can then piece these flows together to obtain a flow in X from s to t with $|f| = n$.   □

The following interesting fact emerges from the above proof in the case when $s, t \in VX$. If $s, t \in VX$, the cuts in $C \in ET_n$ such that $s \in C, t \in C^*$ form a finite totally ordered set. It is the geodesic in ET_n joining vs and vt. Let D be the smallest minimal cut with this property. Then $S_f \subseteq D$, since for any vertex $u \in D^*$ there can be no f-augmenting path joining s and u. But this must mean that $S_f = D$, since $S_f \in \mathcal{B}_n X$ which is generated by ET_n. Although the maximal flow between s, t is not usually unique, the smallest minimal cut separating s, t is unique. The way of obtaining D by successively increasing the flow between s and t is obviously not a canonical process, as we choose flow augmenting paths to increase the flow.

One might think that one could use a structure tree approach to reduce any question about cuts and flows to one about cuts and flows in a structure tree. However, this is not always possible. Thus earlier in this section it is shown that a cut A in $\mathcal{B}_n X$ has a canonical representation in terms of $\mathcal{E}_n$ and therefore corresponds to a cut A' in $\mathcal{B}_n T$. However the capacity of A is not usually the same as the capacity of A'. Thus in Figure 5.4 the cut A corresponding to the dashed line has capacity 3. This cut is the union of a cut (with capacity 2) and a cut (with capacity 3) and in T_3 corresponds to a cut with capacity 5.

3.   Almost Invariant Sets

3.1.   Relative Structure Trees

We prove Conjecture 1.3 in the case when G is finitely generated over H, i.e. G is generated by $H \cup S$ where S is finite.

First, we explain the strategy of the proof. Suppose that we have a non-trivial G-tree T in which every edge orbit contains an edge which has an H-finite stabiliser, and suppose there is a vertex $\bar{o}$ fixed by H. Let T_H be an H-subtree of T containing $\bar{o}$ and every edge with H-finite stabiliser. The action of H on T_H is a trivial action, since it has a vertex fixed by H, and so the orbit space $H \backslash T_H$ is a tree, which might well be finite, but must have at least one edge. Our strategy is to show that if G is finitely generated over H and there is an H-almost invariant set A satisfying $AH = A$, then we can find a G-tree T with the required properties by first deciding what $H \backslash T_H$ must be and then lifting to get T_H and then T.

We show that if G is finitely generated over H, then there is a G-graph X of which there is a vertex with stabiliser H and in which a proper H-almost invariant set A satisfying $AH = A$ corresponds to a proper set of vertices with H-finite coboundary. It then follows from Theorem 2.14, that there is a sequence of structure trees for $H\backslash X$. We choose one of these to be $H\backslash T_H$, and show that we can lift this to obtain T_H and then T itself.

For example, if $G = H*_K L$, then there is a G-tree Y with one orbit of edges and a vertex $\bar{o}$ fixed by H, and every edge incident with $\bar{o}$ has H-finite stabiliser. Suppose that K, L are such that these are the only edges with H-finite stabilisers. Then $H\backslash T_H$ has two vertices and one edge. When we lift to T_H we obtain an H-tree of diameter two in which the middle vertex $\bar{o}$ has stabiliser H. The tree T is covered by the translates of T_H.

On the other hand, if $G = L *_K H$ where K is finite, and T is as above, then every edge of T is H-finite and so T_H is T regarded as an H-tree. The fact that our construction gives a canonical construction for $H\backslash T_H$ means that when we lift to T_H and T we will get the unique tree that admits the action of G.

We proceed with our proof.

Lemma 3.1. *The group G is finitely generated over H if and only if there is a connected G-graph X with one orbit of vertices, and finitely many orbits of edges, and there is a vertex o with stabiliser H.*

Proof. Suppose G is generated by $H\cup S$, where S is finite. Let X be the graph with $VX = \{gH|g \in G\}$ and in which EX is the set of unordered pairs $\{\{gH, gsH\}, g \in G, s \in S\}$. We then have that X is vertex transitive, there is a vertex $o = H$ with stabilizer H and $G\backslash X$ is finite. We have to show that X is connected. Let C be the component of X containing o. Let G' be the set of those $g \in G$ for which $gH \in C$. Clearly $G'H = G'$ and $G's = G'$ for every $s \in S$. Hence $G' = G$ and $C = X$. Thus X is connected.

Conversely let X be a connected G-graph and $VX = Go$ where $G_o = H$. Suppose EX has finitely many G-orbits, $Ge_1, Ge_2, \ldots, Ge_r$ where e_i has vertices o and $g_i o$. It is not hard to show that G is generated by $H \cup \{g_1, g_2, \ldots, g_r\}$. □

Let $A \subset G$ be a proper H-almost invariant set satisfying $AH = A$. Let G be finitely generated over H, and let X be a G-graph as in the last lemma. There is a subset of VX corresponding to A, which is also denoted A. For any $x \in G$, $A + Ax$ is H-finite. In particular this is true if $s \in S$. This means that δA is H-finite. Note that neither A nor $A^* = VX - A$ is H-finite. Thus a proper H-almost invariant set corresponds to a proper subset of VX such that δA is H-finite.

From the previous section (Theorem 2.10) we know that $\mathcal{B}(H\backslash X)$ has a uniquely determined nested set of generators $\mathcal{E} = \mathcal{E}(H\backslash X)$. For $E \in \mathcal{E}$, let $\bar{E} \subset VX$ be the set of all $v \in VX$ such that $Hv \in E$. Let C be a component of $\bar{E}$.

Lemma 3.2. *For* $h \in H$, $hC = C$ *or* $hC \cap C = \emptyset$. *Also* $HC = \bar{E}$, $h\delta C \cap \delta C = \delta C$ *or* $h\delta C \cap \delta C = \emptyset$ *and* $H \backslash \delta C = \delta E$.

Proof. Let $h \in H$. Then hC is also a component of $\bar{E}$, since $HC \subseteq E$. Thus either $hC = C$ or $hC \cap C = \emptyset$. Let K be the stabiliser of C in H. If $v \in C$, then $hv \in C$ if and only if $h \in K$. Thus $K \backslash C$ injects into $H \backslash C = E$ and $K \backslash \delta C$ injects into δE. But E is connected, and so the image HC is E. It follows that there is a single H-orbit of components.   □

Let $\bar{\mathcal{E}}(H, X)$ be the set of all such C, and let $\bar{\mathcal{E}}_n(H, X)$ be the subset of $\bar{\mathcal{E}}(H, X)$ corresponding to those C for which δC lies in at most n H-orbits.

Lemma 3.3. *The set* $\bar{\mathcal{E}}(H, X)$ *is a nested set. The set* $\bar{\mathcal{E}}_n(H, X)$ *is the edge set of an H-tree.*

Proof. Let $C, D \in \bar{\mathcal{E}}_n(H, X)$. Then HC, HD are in the nested set $\mathcal{E}$. Suppose $HC \subset HD$, then $C \subset D$ or $C \cap D = \emptyset$. It follows easily that $\bar{\mathcal{E}}(H, X)$ is nested. It was shown in [7] that a nested set $\mathcal{E}$ is the directed edge set of a tree if and only if it satisfies the finite interval condition, i.e. if $C, D \in \mathcal{E}$ and $C \subset D$, then there are only finitely many $E \in \mathcal{E}$ such that $C \subset E \subset D$. Thus we have to show that $\bar{\mathcal{E}}_n(H, X)$ satisfies the finite interval condition. If $C \subset D$ and $C \subseteq E \subseteq D$ where $C, E, D \in \bar{\mathcal{E}}_n(H, X)$, then $HC \subseteq HE \subseteq HD$. But $\mathcal{E}_n(H, X)$ does satisfy the finite interval condition and $HC = HE$ implies $C = E$. Now let $C \cap D = \emptyset$ and suppose that $o = H \in C^* \cap D^*$. There are only finitely many $E \in \bar{\mathcal{E}}_n$ such that $C \subset E$ and $o \in E^*$ or such that $D \subset E^*$ and $o \in E$. Each $E \in \bar{\mathcal{E}}_n$ such that $C \subset E \subset D^*$ has one of these two properties.   □

Let $\bar{T} = \bar{T}(H)$ be the tree constructed in the last lemma. Let $T = H \backslash \bar{T}$. Note that in the above $\bar{T}(H)$ is the Bass-Serre H-tree associated with the quotient graph $T(H) = H \backslash \bar{T}(H)$ and the graph of groups obtained by associating appropriate labels to the edges and vertices of this quotient graph (which is a tree). Clearly the action of H on $T(H)$ is a trivial action in that H fixes the vertex $\bar{o} = \nu o$. The stabilisers of edges or vertices on a path or ray beginning at $\bar{o}$ will form a non-increasing sequence of subgroups of H.

We now adapt the argument of the previous section to show that if $A \subset VX$ is such that δA lies in at most n H-orbits, then there is a canonical way of expressing A in terms of the set $\bar{\mathcal{E}}(H, X)$. In this case we have to allow unions of infinitely many elements of the generating set. Our induction hypothesis is that if δA lies in at most n H-orbits, then A is canonically expressed in terms of $\bar{\mathcal{E}}_n(H, X)$. First note that there are only finitely many H-orbits of elements of $\bar{\mathcal{E}}_n = \bar{\mathcal{E}}_n(H, X)$ with which A is not nested. This is because if $C \in \bar{\mathcal{E}}_n$ is not nested with A

and F is a finite connected subgraph of $H\backslash X$ containing Ho and all the edges of $H\delta A$, then $H\delta C$ must contain an edge of F and there are only finitely many elements of $\mathcal{E}_n$ with this property. We now let $\mu(A)$ be the number of H-orbits of elements of $\bar{\mathcal{E}}_n$ with which A is not nested. If $\mu(A) = 0$, then A is nested with every $C \in \bar{\mathcal{E}}_n$. This then means that if neither A nor A^* is empty and it is not already in $\bar{\mathcal{E}}_n$, then A determines a vertex z of $\bar{T}_n$ and either A or A^* is the union (possibly infinite) of edges of T_n that lie in finitely many H-orbits. If A is such a union, then we use this union for our canonical expression for A. If A is not such a union, then A^* is; we have $A^* = \bigcup\{C_\lambda|\lambda \in \Lambda\}$, where each C_λ has $\tau C_\lambda = z$ and the edges lie in finitely many H-orbits. We write $A = (\bigcup\{C_\lambda|\lambda \in \Lambda\})^* = \bigcap\{C_\lambda^*|\lambda \in \Lambda\}$. Note that this gives a canonical way of expressing cuts corresponding to a vertex that is not in the image of ν and whose incident edges lie in finitely many H-orbits. Suppose then that the hypothesis is true for elements B for which $\mu(B) < \mu(A)$. Let $C \in \bar{\mathcal{E}}_n$ be not nested with A. Then $\mu(A \cap HC) + \mu(A \cap HC^*) \leq \mu(A)$. Thus each of $A \cap HC$ and $A \cap HC^*$ can be expressed in a unique way in terms of the $\mathcal{E}_n$. We take the expression for A to be the union of the two expressions $A \cap HC$ and $A \cap HC^*$ except that we include hC for $h \in H$, only if just one of the two expressions involves hC.

If $g \in G$, then $g\bar{T}(H)$ is a (gHg^{-1})-tree. It is the tree $\bar{T}(gHg^{-1})$ obtained from the G-graph X by using the vertex go instead of o. We now show that there is a G-tree T which contains all of the trees $g\bar{T}(H)$.

We know that the action of the group G on X is vertex transitive and that X has a vertex o fixed by H. Also G is generated by $H \cup S$ where S is finite.

Clearly there is an isomorphism $\alpha_g : \bar{T}(H) \to \bar{T}(gHg^{-1})$ in which $D \mapsto gD$.

Suppose now that $\nu o \neq \nu(go)$. Let A, B be H-almost invariant sets satisfying $AH = A, BH = B$ and let $g \in G$. We regard A, B as subsets of VX, so that δA and δB are H-finite.

Suppose that $o \in gB^*$ and $go \in A^*$. The following lemma is due to Kropholler [22], [23]. We put $K = gHg^{-1}$.

Lemma 3.4.   *In this situation $\delta(A \cap gB)$ is $(H \cap K)$-finite.*

Proof.   Let $x \in G$. We show that the symmetric difference $(A \cap gB)x + (A \cap gB)$ is $(H \cap K)$-finite. Since A, B are H-almost invariant, there are finite sets E, F such that $A + Ax \subseteq HE$ and $B + Bx \subseteq HF$. We then have

$$(A \cap gB)x + (A \cap gB) = Ax \cap (gBx + gB) + (Ax + A) \cap gB$$

$$= Ax \cap gHF + g(g^{-1}HE \cap B).$$

Now $Ax \cap gHF$ is K-finite, but it is also H-finite because gH is contained in A^*, given that $go \in A^*$. A set, which is both H-finite and

K-finite is $H \cap K$-finite. Thus $Ax \cap gHF$ is $(H \cap K)$-finite. Similarly using the fact that $g^{-1}o \in B^*$, it follows that $g^{-1}HE \cap B$ is $H \cap (g^{-1}Hg)$-finite, and so $g(g^{-1}HE \cap B)$ is $(H \cap K)$-finite. Thus $A \cap gB$ is $(H \cap K)$-almost invariant. But this means that $\delta(A \cap gB)$ is $(H \cap K)$-finite.   □

What this lemma says is that if A, gB are not nested then there is a special corner—sometimes called the *Kropholler corner*—which is $(H \cap K)$-almost invariant.

Notice that in the above situation all of $\delta A, \delta(A \cap gB^*)$ and $\delta(A \cap gB)$ are H-finite. If we take the canonical decomposition for A, then it can be obtained from the canonical decompositions for $A \cap gB$ and $A \cap gB^*$ by taking their union and deleting any edge that lies in both. Also $\delta(gB)$ is K-finite and the decomposition for gB can be obtained from those for $gB \cap A$ and $gB \cap A^*$. But the edges in the decomposition for $A \cap gB$ which is $(H \cap K)$-almost invariant are the same in both decompositions.

We will now show that it follows from Lemma 3.4 that the set $G\bar{\mathcal{E}}_n$ is a nested G-set which satisfies the final interval condition, and so it is the edge set of a G-tree. We have seen that $\bar{\mathcal{E}}_n$ is a nested H-set where $\mathcal{E}_n = H\backslash\bar{\mathcal{E}}_n$ is the uniquely determined nested subset of $\mathcal{B}_n(H\backslash X)$ that generates $\mathcal{B}_n(H\backslash X)$ as an abelian group. It is the edge set of a tree $T_n(H\backslash X)$.

If $A, B \in \bar{\mathcal{E}}_n$ and A, gB are not nested for some $g \in G$, then by Lemma 3.4 there is a corner—the Kropholler corner—which we take to be $A \cap gB$, for which $\delta(A \cap gB)$ is $(H \cap K)$-finite. We then have canonical decompositions for $A \cap gB$ and $A \cap gB^*$ as above. This is illustrated in Figure 5.6. The labels a, b, c, d, e, f are for sets of edges joining the indicated corners. In this case the letters do not represent edges of X but elements of $\bar{\mathcal{E}}_n$. Although each $E \in \bar{\mathcal{E}}_n$ comes with a natural direction, in the diagram we only count the unoriented edges, i.e. we count the number of edge pairs (E, E^*). In the diagram, $A \cap gB$ is always taken

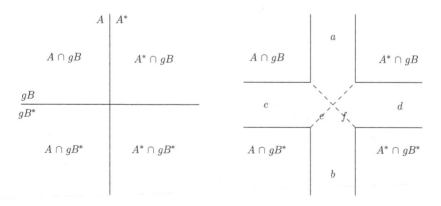

Figure 5.6 Crossing cuts

to be the Kropholler corner. Thus we have that any pair contributing to a, f or e must be $(H \cap K)$-finite. Any pair contributing to e or b must be H-finite and any pair contributing to e or d must be K-finite.

We have that $a + e + f + b = 1$ and $c + e + f + d = 1$. Suppose that the Kropholler corner $A \cap B$ is not empty. It is the case that each of o and go lies in one of the other three corners. We know that $o \in gB^*$, $go \in A^*$. If $o \in A \cap gB^*$ and $go \in A^* \cap gB$, then $a = c = 1$ and $e = f = b = d = 0$ and $A^* \cap gB^* = \emptyset$. If $o \in A^* \cap gB$ and $go \in A^* \cap gB^*$, then $a = d = 1$ and $A \cap gB^* = \emptyset$, while if both o and go are in $A^* \cap gB^*$, then either $a = d = 1$ and $A \cap gB^* = \emptyset$ or $a = c = 1$ and $A^* \cap gB = \emptyset$ or $f = 1$ and both $A \cap gB^*$ and $A^* \cap gB$ are empty, so that $A = gB$. In all cases A, gB are nested.

We need also to show that $G\bar{\mathcal{E}}_n$ satisfies the finite interval condition. Let $g \in G$ and let $K = gHg^{-1}$. Consider the union $\bar{\mathcal{E}} \cup g\bar{\mathcal{E}}$. This will be a nested set. In fact it will be the edge set of a tree that is the union of the trees $T(H)$ and $T(K)$. In the diagram the very thick edges are the edges that are just in $T(H)$. The thick edges are the ones that are in $T(K)$. The other edges are in both $T(H)$ and $T(K)$. An edge is in the geodesic joining o and go if and only if it has stabiliser containing $H \cap K$, it will also lie in both $T(H)$ and $T(K)$ if and only if its stabiliser contains $H \cap K$ as a subgroup of finite index. It may be the case that $T(H)$ and $T(K)$ have no edges in common. An edge lies in both trees if and only if it has a stabiliser that is $(H \cap K)$-finite. If there are such edges then they will be the edge set of a subtree of both trees. They will correspond to the edge set $\bar{\mathcal{E}}(H \cap K)$.

It follows that $T(H)$ is always a subtree of a tree constructed from a subset of $G\bar{\mathcal{E}}_n$ that contains $\bar{\mathcal{E}}_n$. If $T(H)$ and $T(K)$ do have an edge in common, then $T(H) \cup T(K)$ will be a subtree of the tree we are constructing. If $e \in EX$ has vertices go and ko and there is some $C \in G\bar{\mathcal{E}}_n$ that has $e \in \delta C$, then $C \in gET(g^{-1}Hg) \cap kET(k^{-1}Hk)$. If there is no such C, i.e. there is no cut $C \in G\bar{\mathcal{E}}_n$ that separates o and $k^{-1}go$ then $T(H) = k^{-1}gT(H)$. As there is a finite path connecting any two vertices u, v in X, it can be seen that there are only finitely many edges in $G\bar{\mathcal{E}}_n$ separating u and v since any such edge must separate the vertices of one of the edges in the path. Thus $G\bar{\mathcal{E}}_n$ is the edge set of a tree.

We say that a G-tree T is reduced if for every $e \in ET$, with vertices ιe and τe we have that either ιe and τe are in the same orbit, or G_e is a proper subgroup of both $G_{\iota e}$ and $G_{\tau e}$.

Theorem 3.5. *Let G be a group that is finitely generated over a subgroup H. The following are equivalent:*

(i) *There is a proper H-almost invariant set $A = HAK$ with left stabiliser H and right stabiliser K, such that A and gA are nested for every $g \in G$.*

(ii) *There is a reduced G-tree T with vertex v and incident edge e such that $G_v = K$ and $G_e = H$.*

Proof. It is shown that (ii) implies (i) in the Introduction.

Suppose then that we have (i). We will show that there is a G-tree— in which G acts on the right—which contains the set $V = \{Ax | x \in G\}$ as a subset of the vertex set. Let $x \in G$, then $A + Ax$ is a union of finitely many cosets $Hg_1, Hg_2, \ldots, Hg_k$. Then $\{g_1^{-1}A, g_2^{-1}A, \ldots, g_k^{-1}A\}$ is the edge set of a finite tree F. We know that the set $\{gA | g \in G\}$ is the edge set of a G-tree T provided we can show that it satisfies the finite interval condition. But this must be the case as the edges separating vertices A and Ax will be the edges of F. □

Theorem 3.6. *Let G be a group and let H be a subgroup, and suppose G is finitely generated over H. There is a proper H-almost invariant subset A such that $A = AH$, if and only if there is a non-trivial reduced G-tree T in which H fixes a vertex and every edge orbit contains an edge with an H-finite edge stabilizer.*

Proof. The only if part of the theorem is proved in Theorem 3.5. In fact it is shown there that if G has an action on a tree with the specified properties, then there is a proper H almost invariant set A for which $HAH = A$.

Suppose then that G has an H-almost invariant set A such that $AH = A$. Since G is finitely generated over H, we can construct the G-graph X as above, in which A can be regarded as a set of vertices for which δA lies in finitely many H-orbits. Let this number of orbits be n. Then we have seen that there is a G-tree $\bar{T}_n$ for which H fixes a vertex $\bar{o}$ and every edge is in the same G-orbit as an edge in $\bar{T}(H)$. The edges in this tree are H-finite. The set A has an expression in terms of the edges of $\bar{T}(H)$. Finally we need to show that the action on $\bar{T}_n$ is non-trivial. If G fixes $\bar{o}$, then $v(A)$ consists of the single vertex o and so A is not proper. In fact the fact that A is proper ensures that no vertex of $\bar{T}_n$ is fixed by G.

It can be seen from the above that $\bar{T}(H) \cap \bar{T}(g^{-1}Hg) = \bar{T}(H \cap gHg^{-1})$ so that if $e \in ET(H)$, and $g \in G_e$, then $e \in \bar{T}(gHg^{-1})$ and so G_e is H-finite. □

The Kropholler conjecture follows immediately from the last theorem.

4. H-Almost Stability

Let G be a group with subgroup H, and let T be a G-tree.

Let $\bar{A} \subset VT$ be such that $\delta\bar{A} \subset ET$ consists of finitely many H-orbits of edges e such that G_e is H-finite. Also let H fix a vertex of T. Note that $\delta\bar{A}$ consists of whole H-orbits, so that $e \in \delta\bar{A}$ implies $he \in \delta A$ for every $h \in H$. The fact that G_e is H-finite for $e \in \delta\bar{A}$ follows from the fact that $\delta\bar{A}$ is H-finite. If H_e is the stabiliser of $e \in \delta\bar{A}$, then $[G_e : H_e]$ is finite.

Let $v \in VT$, and let $A = A(v) = \{g \in G | gv \in \bar{A}\}$. Note that $A(xv) = A(v)x^{-1}$, so that the left action on T becomes a right action on the sets $A(v)$. If $x \in G$ and $[v, xv]$ is the geodesic from v to xv, then $g \in A + Ax$ if and only if the geodesic $[gv, gxv]$ contains an odd number of edges in $\delta\bar{A}$. If $[v, xv]$ consists of the edges $e_1, e_2, \ldots, e_r$, then $ge_i \in \delta\bar{A}$ if and only if $Hge_i \in \delta\bar{A}$. It follows that $H(A + Ax) = A + Ax$. It is also clear that for each e_i there are only finitely many cosets Hg such that $Hge_i \in \delta\bar{A}$. Thus A is H-almost invariant. We also have $A(v)H = A(v)$ if H fixes v.

For each $e \in ET$, let $d(e)$ be the number of cosets Hg such that $Hge \in \delta\bar{A}$. We see that $d(e) = d(xe)$ for every $x \in G$ and so we have a metric on VT, that is invariant under the action of G. We will show that if G has an H-almost invariant set such that $HAH = A$ then there is a G-tree with a metric corresponding to this set.

From now on we are interested in the action of G on the set of H-almost invariant sets. But note that we are interested in the action by right multiplication. The Almost Stability theorem [3], also used the action by right multiplication. Let $A \subset G$ be H-almost invariant and let $HA = A$. For the moment we do not assume that $AH = A$.

Let $M = \{B | B =_a A\}$ so that for $B, C \in M, B + C = HF$ where F is finite.

Note that for $H = \{1\}$ it follows from the Almost Stability theorem that M is the vertex set of a G-tree.

We define a metric on M. For $B, C \in M$ define $d(B, C)$ to be the number of H-cosets in $B + C$.

This is a metric on M, given that $(B + C) + (C + D) = (B + D)$, and so an element which is in $B + D$ is in just one of $B + C$ or $C + D$. Thus $d(B, D) \le d(B, C) + d(C, D)$.

Also G acts on M by right multiplication and this action is by isometries, since $(B + C)z = Bz + Cz$. Let Γ be the graph with $V\Gamma = M$ and two vertices are joined by an edge if they are distance one apart. Every edge in Γ corresponds to a particular H-coset. There are exactly $n!$ geodesics joining B and C if $d(B, C) = n$, since a geodesic will correspond to a permutation of the cosets in $B + C$. The vertices of Γ on such a geodesic form the vertices of an n-cube.

The edges corresponding to a particular coset Hb disconnect Γ, because removing this set of edges gives two sets of vertices, B and B^*, where B is the set of those $C \in M$ such that $Hb \subset C$.

It was pointed out to me by Graham Niblo that Γ is related to the 1-skeleton of the Sageev cubing introduced in [29]. For completeness we describe this connection.

Let G be a group with subgroup H and let $A = HA$ be an H-almost invariant subset. Let

$$\Sigma = \{gA | g \in G\} \cup \{gA^* | g \in G\}.$$

We define a graph Γ'. A vertex V of Γ' is a subset of Σ satisfying the following conditions:

(1) For all $B \in \Gamma'$, exactly one of B, B^* is in V.
(2) If $B \in V, C \in \Sigma$ and $B \subseteq C$, then $C \in V$.

Two vertices are joined by an edge in Γ' if they differ by one element of Σ. For $g \in G$, there is a vertex V_g consisting of all the elements of Σ that contain g. Then Sageev shows that there is a component Γ^1 of Γ' that contains all the V_g.

By (1) for each $V \in \Sigma$ either $A \in V$ or $A^* \in V$ but not both. Let Σ_A be the subset of Σ consisting of those $V \in \Sigma$ for which $A \subset V$. The edges joining Σ_A and Σ_A^* in Γ^1 form a hyperplane. Each edge in the hyperplane joins a pair of vertices that differ only on the set A. For each xA there is a hyperplane joining vertices that differ only on the set xA. Clearly G acts transitively on the set of hyperplanes.

With V as above, consider the subset A_V of G

$$A_V = \{x \in G | x^{-1} A \in V\}.$$

Then $HA_V = A_V$ and $A_{V_1} = A$. Also $A_V + A$ is the union of those cosets Hx for which V and V_1 differ on $x^{-1}A$, which is finite. Thus $A_V \in V\Gamma$.

Thus there is a map $V\Gamma^1 \to V\Gamma$ in which $V \mapsto A_V$. This map is an injective G-map.

If the set A is such that A and gA are nested for every $g \in G$, then Γ^1 is a G-tree. Thus $V\Gamma$ contains a G-subset that is the vertex set of a G-tree.

In Γ a hyperplane consists of edges joining those vertices that differ only by a particular coset Hx. Every edge of Γ belongs to just one hyperplane. The group G acts transitively on hyperplanes. The hyperplane corresponding to Hx has stabilizer $x^{-1}Hx$.

Suppose now that A is H-almost invariant with $HAK = A$. Here H is the left stabiliser and K is the right stabiliser of A, and we assume that $H \leq K$, so that in particular $HAH = A$. Note that it follows from the fact that A is H-almost invariant that it is also K almost invariant. Suppose that G is finitely generated over K. We saw in the previous section that there is a G-tree T in which A uniquely determines a set $\bar{A}$ of vertices with H-finite coboundary $\delta\bar{A}$. Here $T = T_n$ for n is sufficiently large that in the graph X—as defined in the previous section—the set $\delta\bar{A}$ is contained in at most n H-orbits of edges. Note that if e is an edge of

$\bar{T}(H) = \bar{\mathcal{E}}(H, X)$, then δe is H_e-finite, and will consist of finitely many H_e-orbits. It is then the case that $[G_e : H_e]$ is finite, as δe will consist of finitely many G_e-orbits, each of which is a union of $[G_e : H_e]$ H_e-orbits of edges.

We also know that K fixes a vertex $\bar{o}$ of T, and that $H\delta\bar{A} = \delta\bar{A}$. Thus $\delta\bar{A}$ consists of finitely many H-orbits of edges. We can contract any edge whose G-orbit does not intersect $\delta\bar{A}$. We will then have a tree that has the properties indicated at the beginning of this section. Thus $\bar{A} \subset VT$ is such that $\delta\bar{A} \subset ET$ consists of finitely many H-orbits of edges e such that G_e is H-finite. We see that the metric d on M is the same as the metric defined on VT. Explicitly we have proved the following theorem in the case when G is finitely generated over K.

Theorem 4.1. *Let G be a group with subgroup H and let $A = HAK$ where $H \leq K$ and A is H-almost invariant. Let M be the G-metric space defined above. Then there is a G-tree T such that VT is a G-subset of M and the metric on M restricts to a geodesic metric on VT. If $e \in ET$, then some edge in the G-orbit of e has H-finite stabiliser.*

This is illustrated in Figures 5.7 and 5.8.

Proof. It remains to show that the theorem for arbitrary G follows from the case when G is finitely generated over K. Thus if F is a finite subset of G, then there is a finite convex subgraph C of Γ containing AF. We can use the graph X of the previous section for the subgroup L of G generated by $H \cup F$ to construct an L-tree which has a subtree $S(F)$ with vertex set contained in VC. These subtrees have the nice property that if $F_1 \subset F_2$ then $S(F_1)$ is a subtree of $S(F_2)$. They therefore fit together nicely to give the required G-tree. We give a more detailed argument for why this is the case. We follow the approach of [1].

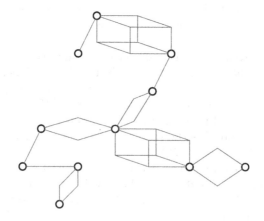

Figure 5.7 Finding a tree in M

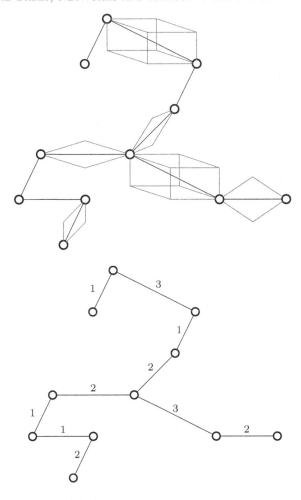

Figure 5.8 Finding a tree in M

Let M' be the subspace of M consisting of the single G-orbit AG. Define an inner product on M' by $(B.C)_A = \frac{1}{2}(d(A,B) + d(A,C) - d(B,C))$.

This turns M' into a 0-hyperbolic space, i.e. it satisfies the inequality

$$(B.C)_A \geq min\{(B.D)_A, (C.D)_A\}$$

for every $B, C, D \in M'$. This is because we know that if $L \leq G$ is finitely generated over H, then there is an L-tree which is a subspace of M. But A, B, C, D are vertices of such a subtree which is 0-hyperbolic. It now follows from [1], Chapter 2, Theorem 4.4 that there is a unique $\mathbb{Z}$-tree VT (up to isometry) containing M'. The subset of VT consisting of vertices of degree larger than 2 will be the vertices of a G-tree and can be regarded as a G-subset of M containing M'.   □

References

[1] Ian Chiswell, *Introduction to Λ-trees*, World Scientific, 2001.

[2] Warren Dicks, *Group, trees and projective modules*, Springer Lecture Notes **790** (1980),

[3] Warren Dicks and M.J. Dunwoody, *Groups acting on graphs*, Cambridge University Press, 1989. Errata http://mat.uab.es/~dicks/

[4] V. Diekert and A. Weiss, *Context free groups and their structure trees*, arXiv:1202.3276.

[5] E.A. Dinits, A.V. Karzanov and M.V. Lomonosov, *On the structure of a family of minimal weighted cuts in a graph*. Studies in Discrete Optimization [Russian], [290–306, Nanka, Moscow (1976).

[6] M.J. Dunwoody, *Cutting up graphs*, Combinatorica **2** (1982) 15–23.

[7] M.J. Dunwoody, *Accessibility and groups of cohomological dimension one*, Proc. London Math. Soc. **38** (1979) 193–215.

[8] M.J. Dunwoody, *The accessibility of finitely presented group*, Invent. Math. **81** (1985) 449–57. 193–215.

[9] M.J. Dunwoody, *An inaccessible group*, London Math. Soc. Lecture Note Series **181** (1991) 173–8.

[10] M.J. Dunwoody, *Inaccessible groups and protrees*, J. Pure Applied Al. **88** (1993) 63–78.

[11] M.J. Dunwoody, *Planar graphs and covers*, arXiv 193–215.

[12] M.J. Dunwoody, *Structure trees and networks*, arXiv:1311.3929.

[13] M.J. Dunwoody, *Almost invariant sets*, arXiv 1409–6782.

[14] M.J. Dunwoody and B. Krön, *Vertex Cuts*, J. Graph Theory (published on line 2014).

[15] M.J. Dunwoody and M. Roller, *Splitting groups over polycyclic-by-finite subgroups*, Bull. London Math. Soc. **23** (1989) 29–36.

[16] T. Fleiner and A. Frank, *Aquick proof for the cactus representation of mincuts*, EGRES Quick-Proof No. 2009–03.

[17] L.R. Ford, Jr. and D.R. Fulkerson, *Maximal flow through a network*, Canadian Journal of Mathematics **8** (1956) 399–404.

[18] R.E. Gomory and T. C. Hu *Multi-terminal network flows*. Journal of the Society for Industrial and Applied Mathematics, **9**, (1961) 551–70.

[19] A. Evangelidou and P. Papasoglu, *A cactus theorem for edge cuts*, International J. of Algebra and Computation **24** (2014) 95–107 .

[20] A. Kar and G.A. Niblo, *Relative ends ℓ^2-invariants and property T*, arXiv:1003.2370.

[21] B. Krön, *Cutting up graphs revisited*, Groups Complex. Cryptol. **2** (2010) 213–21.

[22] P.H. Kropholler, *An analogue of the torus decomposition theorem for certain Poincaré groups*, Proc. London Math. Soc. (3) **60** (1990) 503–29.

[23] P.H. Kropholler, *A group theoretic proof of the torus theorem*, London Math. Soc. Lecture Note Series **181** (1991) 138–58.

[24] H.D. Macpherson, *Infinite distance transitive graphs of finite valency* Combinatorica **2** (1982) 62–9.

[25] R.G. Möller, *Ends of graphs*, Math. Proc. Camb. Phil. Soc **111** (1992) 255–66.

[26] G.A. Niblo, *A geometric proof of Stallings theorem on groups with more than one end,* Geometriae Dedicata **105**, 61–76 (2004).

[27] R.B. Richter and C. Thamassen, 3-connected planar spaces uniquely embed in the sphere, Trans. Amer. Math. Soc. **354** (2002) 4585–95.

[28] G. Niblo and M. Sageev, *The Kropholler conjecture.* In *Guido's Book of Conjectures,* Monographies de L'Enseignement Mathématique, 40. L'Enseignement Mathématique, Geneva, 2008.

[29] M. Sageev, *Ends of group pairs and non-positively curved cube complexes,* Proc. London Math. Soc. (3) **71** (1995) 585–617.

[30] J.-P. Serre, *Trees.* Translated from the French by John Stillwell. Springer-Verlag, Berlin-New York, 1980.

[31] J.R. Stallings, *Group theory and three-dimensional manifolds.* Yale Mathematical Monographs, **4.** Yale University Press, New Haven, Conn.-London, 1971.

[32] C. Thomassen and W. Woess, *Vertex-transitive graphs and accessibility,* J. Combin. Theory Ser. B **58** (1993) 248–68.

[33] W.T. Tutte, *Graph Theory.* Cambridge University Press, 1984.

[34] C.T.C. Wall. *Pairs of relative cohomological dimension one,* J. Pure Appl. Algebra **1** (1971) 141–54.

6

AMENABILITY OF TREES

BEHRANG FORGHANI[1] AND KEIVAN MALLAHI-KARAI[2]

[1]Department of Mathematics, University of Connecticut, Connecticut, USA
[2]Mathematics Department, Jacobs University of Bremen, Germany

Dedicated to Wolfgang Woess on the occasion of his sixtieth birthday

Abstract

We will give a criterion for the amenability of arbitrary locally finite trees. The criterion is based on the trimming operator, which is defined on the space of trees. As an application, we obtain a necessary and sufficient condition for that amenability of Galton–Watson trees.

Contents

Introduction

The notion of amenability for groups emerged out of von Neumann's effort [vN29] in 1929 to find the underlying reason for Hausdorff's paradox. It is his observation that once the dimension of a Euclidean space E exceeds two, the group of isometries of E will contain a copy of the free group on two generators and hence fails to be amenable. This copy of the free group can then be used to carry out various paradoxical decompositions, which are analogous to Hausdorff's original result.

Since its inception, the theory of amenable groups has been explored and the notion of amenability has been extended to a broad class of algebraic objects. Zimmer [Zim78] found that a non-amenable group may have actions that share many properties of the actions of amenable groups and thus initiated the theory of amenable group actions.

Amenability of a discrete group can also be formulated in terms of its Cayley graph. Before giving this definition, let us recall the notion of the Cayley graph for a (finitely generated) group. Recall that if G is a group generated by a (symmetric) set Σ of generators, then Cayley graph $\Gamma = \Gamma(G, \Sigma)$ is an undirected graph with the underlying set of

G as its vertex set in which vertices $g_1, g_2 \in G$ form an edge when $g_1^{-1} g_2 \in \Sigma$. For instance, one can readily see that the Cayley graph of a free group on k generators with respect to the standard generating set is isomorphic to the (unique) $2k$-regular tree. For a subset A of vertices of Γ, the boundary ∂A, by definition, consists of those vertices in A that have a neighbor in $V(\Gamma) \backslash A$. A family $A_n \subseteq G$ is then called a Følner family, if $|\partial A_n|/|A_n| \to 0$, as $n \to \infty$. It is a classical result that a finitely generated group G is amenable iff such a Følner family exists, see [Føl55]. This definition lends itself to using many other methods to establish the amenability of a group. For instance, in various works by Bartholdi, Kaimanovich, Nekrashevych, Amir, Angel, and Virag ([Kai05], [BKN10], [AAV13]) random walks have been used to prove the amenability of several self-similar groups. Many definitions and statements about amenable groups carry over almost verbatim to amenable graphs. For instance, Gerl [Ger88] showed that a connected graph is amenable if and only if the spectral radius of simple random walk on the graph is strictly less than 1. This result is a generalization of Kesten's result for amenable groups [Kes59].

Note that the Cayley graph of a group G is homogeneous, i.e., its automorphism group (containing G as the subgroup of "internal symmetries") acts transitively on the vertex set of G. In particular, the only trees that can be realized as Cayley graphs are those of free products of cyclic group, which, except in trivial cases, are non-amenable. In this note, we will take up the question of characterizing amenability for arbitrary locally finite trees. Our point of departure is a group of results proved by Gerl and Woess who investigated this property for trees that do not have degree-one vertices. Recall that a branch is a vertex with degree at least three. It is clear that a graph which contains arbitrarily long paths without branches is amenable. Gerl [Ger86] proved a tree without any leaves with uniformly bounded degrees is amenable if and only if there are arbitrarily long paths without any branch as induced subgraphs. Later, Woess [Woe00, p. 114] improved this result by dropping the uniform finiteness condition. (see Theorem 2.7). One can easily see that the assumption that T has not vertices of degree one cannot be dropped (see Example 2.8).

Our first theorem extends this characterization to arbitrary trees and, *en passant*, also supplies a rather elementary and "probability-free" proof of that result too. In order to state the theorem, we will need to introduce some new terminology. Our central new concept is the *trimming* operator Θ defined on the space of countable trees. Intuitively, trimming a tree amounts to removing all vertices of degree 1. Hence, the set of fixed points of this operator are precisely the trees without leaves (but see Example 3.3). Using the trimming operator, we will define *inessential subtrees* of a tree which are, roughly speaking, finite subtrees hanging from a vertex of the main tree. Next to long paths, inessential subtrees can provide another source of Følner sets for infinite trees. Our theorem roughly says that these are the only underlying reasons for the amenability of a tree:

Theorem 1. *Let T be an infinite tree. Then T is amenable if and only if T contains arbitrarily large inessential trees or for some $k \geq 1$, $\Theta^k(T)$ has arbitrarily long paths. Moreover, the former is always the case if T can be trimmed indefinitely.*

This theorem can then be applied in a probabilistic setting. There are many models for random trees. The oldest and perhaps the most well-known one is the family of Galton–Watson trees. In this context, we have the following theorem:

Theorem 2. *The Galton–Watson tree $\mathcal{T}$ associated to the probability distribution $(p_i)_{i \geq 0}$ is almost surely amenable, if and only if $\mathbb{P}[X_I \leq 1] = p_0 + p_1 > 0$.*

This chapter is organized as follows. In Section 1, we will define the graph-theoretical terminology that is freely used throughout the chapter. Section 2 is devoted to stating general facts about amenability. A simple proof of Theorem 1 in the special case of trees without leaves is also given in this section. In Section 3, the trimming operator and inessential trees are introduced and studied. Finally, in Section 4, we show an application of Theorem 1 in the context of Galton–Watson trees.

Acknowledgment

Authors would like to thank V. Kaimanovich for his useful comments on the first draft of this chapter. We would like to thank W. Woess for his comments about the history of amenability of trees and pointing out reference [Woe00].

1. Preliminaries

In this section, we will define the basic graph-theoretic terminology and set the notations used in this chapter. A (undirected) graph G consists of a non-empty set $V(G)$ called the vertices and a family of 2-element subsets of $V(G)$, called the edges of G, and denoted by $E(G)$. For brevity, the edge $\{u, v\}$ with vertices (also called endpoints) u and v will be denoted by uv, hence $uv = vu$. The set of neighbors of a vertex v, denoted by $N(v)$, consists of the vertices $u \in V(G)$ with $uv \in E(G)$. Correspondingly, for a subset $A \subseteq V(G)$, we set $N(A) = \bigcup_{v \in A} N(a)$. The degree of a vertex is given by $\deg v = |N(v)|$, where $|X|$ denotes the cardinality of set X. All the graphs considered in this chapter are assumed to be locally finite, that is, $\deg v < \infty$ for all $v \in V(G)$. A leaf is a vertex of degree 1. The set of leaves of a graph G is denoted by $L(G)$. For two leaves $u, v \in G$, we write $u \sim v$ if $N(u) = N(v)$. A branch is a vertex with degree strictly more than 2. For a non-empty subset $A \subseteq V(G)$, the induced subgraph $G[A]$ is the graph with $V(G[A]) = A$ and $E(G[A]) = \{uv \in E(G) : u, v \in A\}$. Similarly, for a non-empty subset R of edges of G, the edge-induced subgraph $G(R)$ has R as the set of edges and the set of endpoints of R as the vertices. A path of length n between two vertices u and v is a finite sequence of vertices $x_0 = u, x_1, \ldots, x_n = v$ such that x_i and x_{i+1} are neighbors for $i = 0, 1, \cdots n - 1$. A tree T is a graph such that for any two vertices $u, v \in V(T)$, there exists a unique path joining u to v. This path (viewed as in induced subgraph of T) will be denoted by $[uv]$. We say that a graph G contains arbitrarily long paths without branches if for any n, there exist vertices $v_1, \ldots, v_n \in V(G)$ such that v_i is connected to v_{i+1} for $1 \le i \le n$ and for $1 \le i \le n$, the degree of v_i in G is exactly 2.

2. Amenability of Graphs

In this section, we recall the definition of amenability for graphs and groups. Moreover, we will give a complete characterization of amenable trees without leaves.

Definition 2.1. Let A be a subgraph of graph G. The *boundary* of A consists of those vertices of A which are connected to at least one vertex outside A. We will denoted the boundary by ∂A. Hence,

$$\partial A = \{v \in V(A) : vu \in E(G), \text{ for some } u \in V(A^c)\},$$

where $|A| = |V(A)|$.

We will also need the following definition.

Definition 2.2. Define the *isoperimetric number* or *Cheeger constant* of a graph G as follows

$$i(G) = \inf\left\{ \frac{|\partial A|}{|A|} : A \text{ is a non-empty and finite subgraph of } G \right\}.$$

The graph G is called *amenable* if $i(G) = 0$.

Equivalently, the graph G is amenable if and only if there is a sequence of finite subgraphs $(A_n)_{n \ge 1}$ of G such that

$$\lim_{n \to \infty} \frac{|\partial A_n|}{|A_n|} = 0.$$

Such a sequence $(A_n)_{n \ge 1}$ witnessing the amenability is called a the Følner set. Let us make two remarks about the Følner set: first, one can always exchange a Følner set with one that consists of finite connected graphs. This follows from the following easy lemma:

Lemma 2.3. *If A is a finite subgraph of G, such that $|\partial A| \leq \epsilon |A|$, then there is connected subgraph B of A, such that $|\partial B| \leq \epsilon |B|$.*

Proof. Let $A_1, A_2, \cdots, A_n$ are finite connected components of A. Because $A_i s$ are pairwise disjoint, we have $|\partial A| = |\partial A_1| + \cdot + |\partial A_n|$. Therefore, $|\partial A_1| + \cdots + |\partial A_n| = |\partial A| \leq \epsilon |A| = \epsilon |A|_1 + \cdots + \epsilon |A_n|$. Hence, there exists $1 \leq i \leq n$ such that $|\partial A_i| \leq \epsilon |A_i|$. $\qquad\square$

Second, the Følner set can be chosen to exhaust $V(G)$:

Proposition 2.4. *Let $(A_n)_{n \geq 1}$ be a Følner sequence for the graph G. Then there exists Følner sequence $(A'_n)_{n \geq 1}$ such that $\bigcup_{n \geq 0} A'_n = G$.*

Proof. Let B_n be a sequence of finite subgraphs such that their union is the whole graph G. By induction, define $A'_n = B_n \cup A_{k_n}$, where $\sqrt{|A_{k_n}|} \geq |B_n|$. Then

$$\frac{|\partial A'_n|}{|A'_n|} \leq \frac{|B_n| + |\partial A_{k_n}|}{|A_{k_n}|} \to 0$$

as n goes to infinity. $\qquad\square$

We can now give the definition of amenability for countable groups.

Definition 2.5. A countable group G is called amenable whenever its Cayley graph admits a Følner set, i.e., there exists a sequence of finite subsets $(A_n)_{n \geq 1}$ such that

$$\lim_n \frac{|g A_n \triangle A_n|}{|A_n|} = 0$$

for every $g \in G$, where $g A_n = \{ga : a \in A_n\}$ and $A_n \triangle g A_n$ is the symmetric difference of two sets A_n and $g A_n$.

Let G be a finitely generated group with a symmetric finite set Σ. The Cayley graph of the group G is a graph whose vertices are elements of G and two vertices of g_1 and g_2 are connected if there is an element $s \in \Sigma$ such that $g_1 s = g_2$. A finite generated group G is thus amenable if and only if its Cayley graph is amenable.

Proposition 2.6. *Let T be a tree which does not have any vertex with degree less than 3. Then for every finite subtree A, we have $|A| \leq 2|\partial A|$. Consequently, T is not amenable.*

Proof. Let A be a finite subtree of T. Define $K_i = \{v \in A : \deg v = i\}$. Hence, the boundary of A at least includes the vertices with degree 1 and 2, hence $|\partial A| \geq |K_1| + |K_2|$. It is well known that

$$2|E(A)| = \sum_{v \in V(G)} deg(v) = \sum_{i \geq 1} i|K_i|. \qquad (6.1)$$

On the other hand

$$|V(A)| = \sum_{i \geq 1} |K_i|. \tag{6.2}$$

Since A is a tree, $|E(G)| = |V(A)| - 1$. Equalities (6.1) and (6.2) now imply

$$|K_1| = \sum_{i \geq 3} (i - 2)|K_i| + 2.$$

We now have $2|\partial A| \geq |K_1| + |K_2| + \sum_{i \geq 3} (i - 2)|K_i| \geq \sum_{j \geq 1} |K_j|.$ □

As mentioned, Woess [Woe00, p. 114] classified amenable trees without any leaves. Here we provide an alternative proof for this result.

Theorem 2.7. [Woe00, p. 114] *Let T be an infinite tree with no leaves. Then T is amenable if and only if T contains arbitrarily long paths without any branch.*

Proof. Let T contain arbitrary long paths without branches. For each n, let A_n be a finite subtree of T with exactly n vertices with degree 2. Hence, $|A_n| = n + 1$ and $|\partial A_n| \leq 2$. Consequently, (A_n) is a Følner set.

Assume T does not contain arbitrarily long paths without branches. Let T' be a tree obtained after removing all vertices of T whose degrees are 2. Because T does not have any leaves, by Proposition 2.6, T' is not amenable. If A is a finite subtree of T, then corresponding subtree A' in T' has the same boundary as A and clearly $|A'| \leq |A|$. In addition, each edge of A' is obtained by removing at most d vertices, where d is the longest path without any branch. In other words, $|A'| \leq d|A|$, and

$$\frac{|\partial A'|}{|A'|} \leq \frac{|\partial A|}{|A|} \leq d \frac{|\partial A'|}{|A'|}. \tag{6.3}$$

Combining the preceding inequalities and the fact T' is not amenable imply non-amenability of T.

The preceding theorem is not true if the trees are allowed to have infinitely many leaves, as the following example shows.

Example 2.8. Let T be a tree which can be obtained by attaching one vertex and one edge to the Cayley graph of $\mathbb{Z}$ with respect to the generating set $\{-1, 1\}$ (see, Figure 6.1). Then T is amenable, but T does not contain arbitrarily long paths without any branch.

Remark 2.9. One has to note that some properties of amenability for graphs may diverge from amenability of groups. For instance, it is known that every subgroup of a (discrete) amenable group is amenable. The analogous property does not hold for trees. This can be easily seen as follows: let T be a 3-regular tree and T' be the tree obtained by adding an infinite ray Z (the graph with vertices $1, 2, \ldots$ where i and j are

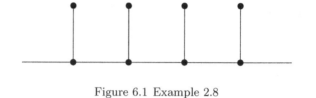

Figure 6.1 Example 2.8

Figure 6.2 Example 2.8 after trimming

adjacent when $|i - j| = 1$) to one of the vertices of T. In other words, let T' be the graph obtained by taking the disjoint union of T and Z and identifying vertex 1 of Z with one of the vertices of T. Clearly T' contains arbitrary long paths without any branch and is hence amenable, but it contains T as a subtree which is not amenable. One can modify this example to construct an example of a tree T' and a subtree T with the same set of ends as T such that T' is amenable, but T is not.

3. Trimming and Inessential Subtrees

In this section, we will define certain operators on the space of trees. These definitions will later be used to give a criterion for amenability of trees.

Definition 3.1. Let T be an infinite tree. The trimming operator $\Theta(T)$ is defined by

$$\Theta(T) = T[V(T)\backslash L(T)].$$

In other words, $\Theta(T)$ is the tree obtained by removing all the leaves of T together with their incident edges.

Remark 3.2. We will always view $\Theta(T)$ as an induced subtree of T. For instance, T has no leaves iff $\Theta(T) = T$. Note that

$$T \supset \Theta(T) \supset \Theta^2(T) \supset \cdots$$

is a decreasing sequence of subtrees of T. Also define, $\Theta^0(T) = T$ and $\Theta^{k+1}(T) = \Theta(\Theta^k(T))$, for $k \geq 0$.

We say that the trimming stops in finite time if this sequence stabilizes at some point, i.e., if there exists $k \geq 0$ such that $\Theta^k(T)$ does not have any leaves, which is equivalent to $\Theta^l(T) = \Theta^k(T)$ for all $l \geq k$. Otherwise, we say that T can be trimmed indefinitely.

Example 3.3. Although Θ is defined on the space of trees, it also induces a map on the space $\mathcal{T}$ of isomorphism classes of trees. Let us

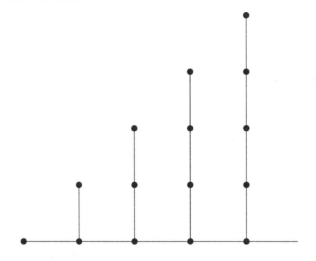

Figure 6.3 Example 3.3

denote the isomorphism class of a tree T by $[T]$. We can now study the dynamics of Θ on $\mathcal{T}$ and pose various questions about it. For instance, finite trees $[T]$ are exactly those trees whose orbit contains the empty tree (the tree with one vertex and no edge). For a more interesting example, let T be the tree with $V(T) = \{(i, j) \in \mathbb{Z}^2 : 0 \le j \le i\}$, and the edges between $(i, 0), (i+1, 0)$ for $i \ge 0$ and (i, j) and $(i, j+1)$ for every $i \ge 1$ and $0 \le j \le i - 1$, see Figure 6.3. It is easy to see that T can be trimmed indefinitely, while $\Theta(T)$ is isomorphic to T. In other words, $[T]$ is a fixed point for Θ. Clearly, if $L(T) = \emptyset$, then $\Theta(T) = T$. It is an interesting question to characterize those trees with $L(T) \ne \emptyset$, for which $\Theta([T]) = [T]$. One can analogously define for any $n \ge 1$ the tree T_n by

$$V(T_n) = \{(i, j) \in \mathbb{Z}^2 : 0 \le j \le ni\},$$

with the edges similar to those of T above. In this case, one can see that $[T_n]$ is a periodic point of Θ with the smallest period n.

We will now define the notion of inessential subtree and use it to prove some of the properties of finitely trimmable trees.

Definition 3.4.   Let T_0 be a finite (connected) subtree of an infinite tree T with at least one edge. We say that T_0 is inessential if the edge-induced subgraph $G(E(T) \backslash E(T_0))$ is connected, and hence is a tree.

   For an inessential subtree T_0 of T, set $\overline{T_0} = G(E(T) \backslash E(T_0))$, and note that $E(G)$ is the disjoint union of the edge set of the trees T_0 and $\overline{T_0}$. This implies that T_0 and $\overline{T_0}$ have exactly one vertex in common. We call this vertex the root of T_0 and denote it by $\mathsf{root}(T_0)$. Note that since

root(T_0) is adjacent to a vertex in T_0, as well as one in $\overline{T_0}$, its degree is
at least 2. Let us derive some immediate consequences of this definition.

Proposition 3.5. *If T_1 and T_2 are inessential subtrees of an infinite
tree T with root(T_1) = root(T_2), then $T_1 \cup T_2$ is also an inessential
subtree of T.*

Proof. Because T_1 and T_2 have a common vertex, $T_1 \cup T_2$ is connected,
and hence a subtree of T. Also $\overline{T_1 \cup T_2} = \overline{T_1} \cap \overline{T_2}$ is the intersection of
two trees that have a common vertex root(T_1) = root(T_2), and is hence
connected.   □

Proposition 3.6. *An infinite tree has an inessential subtree if and
only if it has a leaf.*

Proof. Let v be a leaf of T and w be the unique vertex of T adjacent
to v. Then the single edge vw is an inessential subtree of T. Conversely,
if T' is an inessential subtree of a tree T, then T' has at least two
vertices of degree 1. Call them u' and v'. We claim that at least one of
u' and v' has degree 1 in T. If, on the contrary, u' and v' are adjacent
to vertices u and v in $V(T) \backslash V(T')$, respectively, then the unique path
from u to v will contain both $u', v' \in V(T')$. Now consider the path
joining u and v in $G(E(T) \backslash E(T'))$. The unique path between u' and v'
in T must contain this path, and will hence depart T'. This contradicts
the assumption that T' is connected.   □

Lemma 3.7. *If $\Theta(T)$ contains an inessential subtree with k vertices,
then T contains an inessential tree with at least $k+1$ vertices.*

Proof. Let T_0 be an inessential subtree of $\Theta(T)$ and v be one of its
leaves. Because v is not a leaf of T, it must be connected to at least one
vertex of $V(T) \backslash V(\Theta(T))$, which is automatically a leaf of T. The tree
added by adding all such vertices w and the corresponding edges vw to
T_0 is connected and hence an inessential subtree of T with at least $k+1$
vertices.   □

Proposition 3.8. *Let T be an infinite tree that can be trimmed
indefinitely. Then T is amenable.*

Proof. We show that T contains an inessential subtree with arbitrarily
large number of vertices. Since $\Theta^k(T)$ contains a leaf, hence it contains
an inessential subtree by Proposition 3.6. Now, a repeated application
of Lemma 3.7 shows that T contains an inessential tree T_0 with at least
k vertices. Let S be the set of all vertices of T_0 except for the root. It is
clear that $|S| = k - 1$ and $\partial S = \{$root(T_0)$\}$. This shows that

$$\frac{|\partial S|}{|S|} = \frac{1}{k-1}.$$

Hence, by letting $k \to \infty$, we obtain a sequence of Følner sets in the tree. □

Lemma 3.9. *Let T be an infinite tree such that $\Theta(T)$ is amenable. Then T is amenable.*

Proof. Let S be a connected subgraph of $\Theta(T)$ with $\dfrac{|\partial S|}{|S|} < \epsilon$. Set $\overline{S}$ to be the connected subgraph of T obtained by adding all of the leaves of T that are connected to a vertex of S. Note that since the only vertices of $T \setminus \Theta(T)$ are leaves of T, the boundary of $\overline{S}$ in T is equal to the boundary of S in $\Theta(T)$. Since $|\overline{S}| \geq |S|$, we have $\dfrac{|\partial \overline{S}|}{|\overline{S}|} \leq \epsilon$. □

We can now prove the main result of this section which is the generalization of Theorem 2.7.

Theorem 3.10. *Let T be an infinite tree. Then T is amenable if and only if T contains arbitrarily large inessential trees or for some $k \geq 1$, $\Theta^k(T)$ has arbitrarily long paths without branch. Moreover, the former is always the case if T can be trimmed indefinitely.*

Proof. The same argument as in Proposition 3.8 shows that if T contains arbitrarily large inessential trees, then it is amenable. Hence, without loss of generality, we can assume that there exists $k \geq 1$ such that $\Theta^k(T)$ does not have any leaves but contains arbitrarily long paths. The proof now follows by induction on k. For $k = 0$, we can take these long paths without branch as Følner sets. Since $\Theta^k(T) = \Theta^{k-1}(\Theta(T))$, by induction hypothesis, we obtain that $\Theta(T)$ is amenable. Hence, using Lemma 3.9, T is amenable.

Let us now prove the converse. Assume that there exists $k \geq 0$ such that $\Theta^k(T)$ contains no leaves and does not contain arbitrarily long paths, and that the largest inessential tree of T has cardinality R. We will show that T is nonamenable.

We will assume that $\Theta^k(T) = \Theta^{k+1}(T) = T'$, i.e., T can only be trimmed k times and T' does not have arbitrarily long paths. This implies that T is obtained from T' by adding a (possibly infinite) number of inessential trees, each attached at a distinct vertex of T.

Let us now assume that A is a connected subgraph of T. Let $A' = A \cap V(T')$. We claim that

$$|A'| \geq \frac{1}{R}|A|.$$

Also the cardinality of the boundary of A' in T' is the same as the cardinality of the boundary of A in T. This implies that

$$\frac{|\partial A'|}{|A'|} \leq R\frac{|\partial A|}{|A|}.$$

We have now reduced the problem to the case that the tree does not have any vertices of degree 1.   □

4. Application: Amenability of Random Trees

In this section, we will consider the question of amenability for Galton–Watson trees. First we will give some basic definitions regarding the Galton–Watson process. For details, the reader is referred to [AN72].

Let $\pi = (p_n)_{n \geq 0}$ be a distribution on the set of non-negative integers. We will define the Galton–Watson process associated to π as a distribution on the set of rooted labeled trees. We will start by setting up the notations that will be used in this section.

A Galton–Watson tree is always designated with a distinguished vertex refereed to as the root and denoted by $\emptyset$. Set

$$\mathcal{I} = \{\emptyset\} \cup \bigcup_{j=1}^{\infty} \mathbb{N}^j$$

where $\mathbb{N}$ is the set of positive integers. For each $I = (i_1, \ldots, i_k) \in \mathcal{I}$, we define its length (or generation) by $|I| = k$. We will also set $|\emptyset| = 0$. A set $\mathcal{J} \subseteq \mathcal{I}$ is called inductive if it satisfies the following properties:

(1) $\emptyset \in \mathcal{J}$.
(2) For each $k \geq 1$ and $I = (i_1, \ldots, i_k) \in \mathcal{J}$, we have $\hat{I} := (i_1, \ldots, i_{k-1}) \in \mathcal{J}$.
(3) For each $k \geq 1$ and $I = (i_1, \ldots, i_k) \in \mathcal{J}$, if $i_k \geq 2$, then $(i_1, \ldots, i_{k-1}, i_k - 1) \in \mathcal{J}$.

If $\hat{J} = I$, we say that I is an ancestor of J, and that J is a descendant of I. Intuitively, an inductive set is a set that is closed with respect to the ancestor operation and moreover, the set of offsprings of any vertex are always labeled from 1 to k for some $k \geq 0$. Let $X_I, I \in I$ be a sequence of independent identically distributed random variables with distribution π. A random rooted tree is constructed as follows: the root $v_\emptyset$ has $X_\emptyset$ direct offsprings (also called children) denoted by $v_{(i_1)}$, for $1 \leq i_1 \leq X_\emptyset$. These vertices are called the first generation. From here, the construction continues inductively. Assume that the vertices of generation ℓ have been constructed. Each vertex in generation ℓ is of the form v_I for some $I = (i_1, \ldots, i_\ell)$, hence $|I| = \ell$. The vertex v_I has X_I children, namely $v_{(i_1, \ldots, i_\ell, i_{\ell+1})}$, where $i_{\ell+1}$ ranges from 1 to X_I. It is easy to see that the set of vertices of the tree $\mathcal{T}$ thus constructed is of the form $V(\mathcal{T}) = \{v_J : J \in \mathcal{J}\}$, where $\mathcal{J}$ is an inductive set. We will denote the set of vertices in generation ℓ by $V(\mathcal{T})_\ell$ and set $W_\ell = |V(\mathcal{T})_\ell|$. The rooted subtree of $\mathcal{T}$ consisting of all the offsprings of vertex v_I rooted at v_I will be denoted by $\mathcal{T}^I$. The (finite) rooted subtree of $\mathcal{T}$ consisting of all vertices of the first k generations will be denoted by $\mathcal{T}_k$.

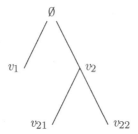

Remark 4.1. Before we proceed to the proof, a few remarks are in order. There is a vast literature on the Galton–Watson process. Let $m = \mathbb{E}[X_I] = \sum_{j=0}^{\infty} jp_j$ be the expected number of children of any vertex. It is a classical theorem that if $p_1 \neq 1$, then $\mathcal{T}$ is almost surely an infinite tree iff $m > 1$. If $p_1 = 1$, then T will be isomorphic to an infinite path and hence amenable. From now on we will always exclude this case. The cases $m = 1$ and $m < 1$ are usually referred to as the critical and subcritical case. Since every finite tree is by definition amenable, we can condition on the non-extinction of $\mathcal{T}$.

Proof of Theorem 2. First note that if $p_0 = p_1 = 0$, then $\mathcal{T}$ is almost surely infinite and the degree of every vertex, except possibly for the root, is at least 3. Such a tree is clearly non-amenable. We will now show that if $p_0 > 0$ or $p_1 > 0$, then the tree is amenable. Let $m = \sum_{k=1}^{\infty} kp_k$. By the above remark, we can assume that $m > 1$. First assume that $m < \infty$. We know that $\mathbb{E}[W_n] = m^n$. We will distinguish two cases:

Case 1: Assume that $p_0 = 0$ and $0 < p_1 < 1$. We will start by two observations. First, it is easy to see that in this case W_n is a non-decreasing sequence. Moreover, $W_{n+1} = W_n$ if each vertex in generation n has exactly one offspring. This implies that if $k \geq 1$,

$$\mathbb{P}[W_{n+1} > W_n | W_n = k] = 1 - p_1^k \geq 1 - p_1 > 0.$$

Hence $\mathbb{P}[W_{n+1} > W_n] \geq 1 - p_1$, and an application of Borel–Cantelli shows that $\mathbb{P}[W_n \to \infty] = 1$. Second, with probability $q = p_1^{d+1} > 0$ each vertex in the the first d generations has exactly one child, i.e., the tree $\mathcal{T}_{d+1}$ is isomorphic to P_{d+1}.

For any vertex v_I in the nth generation, consider the rooted subtree $\mathcal{T}^I$ at v_I. Note that of $\mathcal{T}^I$ are i.i.d. random trees with the same distribution as $\mathcal{T}$. By the second observation above, each $\mathcal{T}^I$ has probability q of being isomorphic to a path of length $d + 1$. Let A_n be the event that at least one of these subtrees is isomorphic to a path of length $d + 1$. For a fixed r, we have

$$\mathbb{P}[A_n] \geq \mathbb{P}[A_n | W_n > r]\mathbb{P}[W_n > r]$$
$$= (1 - (1 - q)^r)\mathbb{P}[W_n > r] \to 1 - (1 - q)^r,$$

as $n \to \infty$. Given that r is arbitrary, we have $\mathbb{P}[A_n] \to 1$, as $n \to \infty$. This means that with probability 1, $\mathcal{T}$ contains a path of length $d + 1$ for every $d \geq 1$, which proves the almost sure amenability of $\mathcal{T}$.

Case 2: Let us now consider the case that $p_0 > 0$. Note that in this case there is no guarantee that $W_n \to \infty$ as $n \to \infty$. Fix $d \geq 1$. We will show that, with probability 1, the isoperimetric constant of $\mathcal{T}$ is at most $1/d$. The large Følner sets in this case arise from the following dichotomy: for an appropriate value of $n \gg 1$: (a) either there are "many" vertices in generation n, in which case, with high probability, the subtree of $\mathcal{T}$ starting from one of them must terminate exactly after d generations, i.e., the first d generations starting from one of these vertices must be a finite tree with all vertices of degree 1 in generation d, or, (b) there are "few" vertices in generation n, which implies that the first $n - 1$ generation of the graph forms a large set with a small boundary. Let us make this idea precise. Fix $d \geq 1$, choose $s > 0$ such that $p_s > 0$. Let $T_{s,d}$ denote the (deterministic) finite rooted s-ary tree of depth d, that is, a rooted tree, where starting from root up to generation $d - 1$, each vertex has exactly s children, but the vertices in generation d have no any children. Let q be the probability that $\mathcal{T}_{d+1}$ is isomorphic $T_{s,d}$. Since $p_0 > 0, p_s > 0$, we have $q > 0$.

Let E denote event that $\mathcal{T}$ is finite, or equivalently, that $W_n = 0$ for $n \gg 1$. It suffices to show that $\mathcal{T}$ is amenable conditioned on E^c. Let A_d denote the event that $\mathcal{T}$ contains a subtree with isoperimetric constant at most $1/d$. For any $r \geq 1$, choose $n \geq rd$, and note that given that $\mathcal{T}$ is infinite and $W_n > r$, A_d will take place if at least one of the subtrees starting from one of the vertices in generation d is isomorphic to $T_{s,d}$. Because there are at least r vertices in generation n, we have

$$\mathbb{P}[A_d | E^c \cap \{W_n > r\}] \geq 1 - (1 - q)^r.$$

On the other hand, if there are at most r vertices in generation n, then the subtree $\mathcal{T}_{n-1}$, which has at least n vertices, has a boundary of size at most r, implying that its isoperimetric constant is at most $r/n \leq 1/d$. Hence,

$$\mathbb{P}[A_d | E^c \cap \{W_n \leq r\}] = 1.$$

Combining the two cases, we have that for given $r, d \geq 1$,

$$\mathbb{P}[A_d | E^c] \geq 1 - (1 - q)^r.$$

Because r is arbitrary and $q > 0$, we deduce that for any $d \geq 1$, $\mathbb{P}[A_d | E^c] = 1$, implying that $\mathcal{T}$ almost surely has a set with isoperimetric constant at most $1/d$, for every d, proving the almost sure amenability of $\mathcal{T}$ in this case. $\square$

References

[AAV13] Gideon Amir, Omer Angel, and Bálint Virág. Amenability of linear-activity automaton groups. *J. Eur. Math. Soc. (JEMS)*, 15(3):705–30, 2013.

[AN72] Krishna B. Athreya and Peter E. Ney. *Branching processes*. Springer-Verlag, New York-Heidelberg, 1972. Die Grundlehren der mathematischen Wissenschaften, Band 196.

[BKN10] Laurent Bartholdi, Vadim A. Kaimanovich, and Volodymyr V. Nekrashevych. On amenability of automata groups. *Duke Math. J.*, 154(3):575–98, 2010.

[Føl55] Erling Følner. On groups with full Banach mean value. *Math. Scand.*, 3:243–54, 1955.

[Ger86] Peter Gerl. Eine isoperimetrische Eigenschaft von Bäumen. *Österreich. Akad. Wiss. Math.-Natur. Kl. Sitzungsber. II*, 195(1-3): 49–52, 1986.

[Ger88] Peter Gerl. Amenable groups and amenable graphs. In *Harmonic analysis (Luxembourg, 1987)*, volume 1359 of *Lecture Notes in Math.*, pp. 181–90. Springer, Berlin, 1988.

[Kai05] Vadim A. Kaimanovich. "Münchhausen trick" and amenability of self-similar groups. *Internat. J. Algebra Comput.*, 15(5-6):907–37, 2005.

[Kes59] Harry Kesten. Full Banach mean values on countable groups. *Math. Scand.*, 7:146–156, 1959.

[vN29] John von Neumann. Zur allgemeinen theorie des maßes. *Fundamenta Mathematica*, 13:73–116, 1929.

[Woe00] Wolfgang Woess. *Random walks on infinite graphs and groups*, vol. 138 of *Cambridge Tracts in Mathematics*. Cambridge University Press, 2000.

[Zim78] Robert J. Zimmer. Amenable ergodic group actions and an application to Poisson boundaries of random walks. *J. Functional Analysis*, 27(3):350–72, 1978.

7

GROUP-WALK RANDOM GRAPHS

AGELOS GEORGAKOPOULOS*

Mathematics Institute, University of Warwick, CV4 7AL, United Kingdom

Dedicated to Wolfgang Woess, on the occasion of his sixtieth birthday

Abstract

We introduce a construction that gives rise to a variety of 'geometric' finite random graphs, and describe connections to the Poisson boundary, Naim's kernel, and Sznitman's random interlacements.

Contents

1. Introduction

The purpose of this chapter is to introduce a new construction of 'geometric' finite random graphs, called group-walk random graphs (GWRGs from now on), and describe the rich connections to other objects, including the Poisson boundary, Naim's kernel, and Sznitman's random interlacements. GWRGs do not only yield new interesting examples of random graphs, but, as we will argue, they can be thought of as a tool for studying groups.

---

*Supported by EPSRC grant EP/L002787/1. This project has received funding from the European Research Council (ERC) under the European Unions Horizon 2020 research and innovation programme (grant agreement No 639046). The author would like to thank the Isaac Newton Institute for Mathematical Sciences, Cambridge, for support and hospitality during the programme 'Random Geometry' where work on this chapter was undertaken.

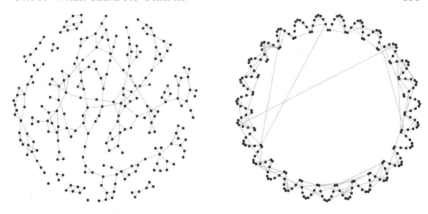

Figure 7.1 A sample GWRG produced by computer simulation by A. Janse van Rensburg. The host graph is a ternary tree, and $n = 5$. Both drawings depict the same graph: the right-hand figure respects the natural cyclic ordering of the leaves of the host tree, while the left-hand one makes the component structure clearer.

We start by introducing the simplest special case of GWRGs. Let G be an infinite homogeneous tree, rooted at a vertex o, called the *host* graph; we will later allow G to be an arbitrary locally finite Cayley graph, or even a more general graph. Let $G_n := G[\{v \in V(G) \mid d(v, o) \leq n\}]$ be the ball of radius n centered at o, and define the boundary ∂G_n to be the set $\{v \in V(G) \mid d(v, o) = n\}$ of vertices at distance exactly n from o.

We construct a random graph R_n as follows. The vertex set of R_n is the deterministic set ∂G_n. The edge set of R_n is constructed by the following process. We start an independent simple random walk in G_n from each vertex $v \in \partial G_n$, and stop it upon its first return to ∂G_n, letting $v^\dagger$ denote the vertex in ∂G_n where this random walk was stopped. We then put an edge in R_n joining v to $v^\dagger$ for each $v \in \partial G_n$.

We could stop the construction here and declare R_n to be our GWRG, but it is more interesting to consider the following evolution: let $R_n^1 := R_n$, and for $i = 2, 3, \dots$ let R_n^i be the union of R_n^{i-1} with an independent sample of R_n^1; or in other words, R_n^i is the random graph obtained as above when we start i independent particles at each vertex in ∂G_n.

An important observation from [15] is that these random graphs R_n^i have the following scale-invariance property. Let C, D be two *branches* of our tree G, where a branch is a component of $G \backslash e$ for some edge e. Then, for any fixed i, we have the following observation.

Observation 1.1 ([15]). *The expected number of edges of R_n^i from branch C to branch D converges as $n \to \infty$. The limit is always > 0, and it is finite if and only if $C \cap D = \emptyset$.*

This might at first sight look surprising, as the number of vertices of R_n^i inside each of C, D grows exponentially with n, yet no rescaling is involved in Observation 1.1.

The same construction can be repeated when instead of the binary tree the host graph is any infinite graph. Then Observation 1.1 has a generalisation, but in order to formulate it we need the Poisson boundary $\mathcal{P}$: instead of the 'branches' of Observation 1.1 we have to talk about subgraphs 'converging' to a measurable subset of $\mathcal{P}$. This is explained in greater detail in Section 2, where we also elaborate more on the general construction of GWRGs and their variants.

The construction of GWRGs was motivated by a measure space introduced in [15], called the effective conductance measure, which is closely related to the Poisson boundary. It is a generalisation of effective conductance for electrical networks, and it is important for the study of Dirichlet harmonic functions on infinite graphs. More details about this measure are given in Section 3.

I expect that GWRGs can unify many existing models of geometric random graphs, while introducing new ones, and offer new tools for analysing them including the Poisson boundary, as well as the notion of graphons [5]. Conversely, GWRGs provide an additional tool for indirectly studying groups, just like random walks; see Section 8.3 for more.

The study of random graphs is currently one of the most active branches of graph theory. By far the most studied random graph model is that of Erdős & Rényi (ER) [10], in which every pair of vertices is joined with an edge with the same probability, and independently of each other pair.

In recent years, many models of *geometric* random graphs have been emerging [16, 22]. The idea now is to embed the set of vertices (possibly randomly) into a geometric space—usually the euclidean or hyperbolic plane, and their higher-dimensional analogs—and then to independently join each pair of vertices with a probability that decays as the distance between the vertices in the underlying space grows.

One advantage of these geometric random graphs compared to the ER model is that they can approximate real-life networks much more realistically, but they are also of great theoretical interest given the impact of the ER model. A disadvantage is that there is an infinity of such models, obtained by varying the underlying geometry, the way the points are embedded, and the connection probability as a function of distance, and no canonical choice is available.

I like thinking of GWRGs as geometric random graphs, where the underlying geometry is a Cayley graph G of an arbitrary finitely generated group.[1] Although there is a huge variety for the underlying

---

[1] The Cayley graph of a group Γ with respect to some symmentric generating set S is the graph with vertex set Γ in which xy is an edge whenever $y = xs$ holds for some $s \in S$.

'geometry', there are tools for their analysis that apply to all cases and will be discussed here.

This chapter is written in survey style although the material reviewed is quite new and partly under development, the main aim being to make the open problems of the project accessible to other researchers willing to get involved. A lot of the material is drawn from the paper [15] which is still in progress. New here is the definition of GWRGs and some observations about them.

2. The General Construction of GWRGs

In the Introduction we chose the host graph G to be a tree, the reason being that Observation 1.1 is easier to state in that case. Let us now consider the general case where the host graph G is arbitrary, and see how Observation 1.1 generalises, which will lead us to the definition of the effective conductance measure.

The construction of R_n^i can be repeated verbatim, except that the number of particles we start at a vertex $v \in \partial G_n$ in round i is equal to the vertex degree $d_{G_n}(v)$ of v in the ball G_n; the reason will become apparent later. However, we could instead of starting exactly $d_{G_n}(v)$ particles at v in round i, start a random number of particles following some distribution with expectation $d_{G_n}(v)$, the most natural candidate being the Poisson distribution; the following discussion remains valid for this variant of GWRGs.

Observation 1.1, which is easy to prove in the case of trees, now becomes a substantial theorem, but in order to formulate it we need to involve the concept of the Poisson boundary of G to extend the above notion of branch in the correct way.

The Poisson boundary of an infinite (transient) graph G is a measurable, Lebesgue–Rohlin, space $\mathcal{P} = \mathcal{P}(G)$, endowed with a family of probability measures $\{\mu_v \mid v \in V(G)\}$, such that every bounded harmonic function $h : V(G) \to \mathbb{R}$ can be represented by integration on $\mathcal{P}$: we have $h(v) = \int_{\mathcal{P}} \hat{h} d\mu_v(\eta)$ for a suitable boundary function $\hat{h} : \mathcal{P} \to \mathbb{R}$. This can be thought of as a discrete version of Poisson's integral representation formula:

$$h(z) = \int_0^1 \frac{1 - |z|^2}{|e^{2\pi i\theta} - z|^2} \hat{h}(\theta) d\theta = \int_0^1 \hat{h}(\theta) d\nu_z(\theta),$$

recovering every continuous harmonic function $h : \mathbb{D} \to \mathbb{R}$ in terms of its boundary values $\hat{h} : \mathbb{S}^1 \to \mathbb{R}$, except that we replaced $\mathbb{D}$ with a transient graph.

Triggered by the work of Furstenberg [12] who introduced the concept, the study of the Poisson boundary of Cayley graphs has grown into a very active research field; see [11] for a survey including many references. Although there is a straightforward abstract construction of the Poisson boundary $\mathcal{P}(G)$ of any Cayley graph, given a concrete G it is desirable to identify $\mathcal{P}(G)$ with a geometric boundary. This pursuit can, however, be very hard, although some general criteria are available [17, 18].

As an example, we remark that the Poisson boundary of a regular tree can be identified with its set of ends, and the Poisson boundary of a regular tessellation of the hyperbolic plane can be identified with its circle at infinity. More generally, the Poisson boundary of any non-amenable, bounded-degree, Gromov-hyperbolic graph coincides with its hyperbolic boundary [1, 2]. The Poisson boundary of a 1-ended bounded-degree planar graph can be identified with a circle [13].

Now back to Observation 1.1, letting G be an arbitrary transient graph, we consider measurable subsets X, Y of $\mathcal{P}(G)$. One can associate with these sets sequences of vertex sets $(X_n)_{n\in\mathbb{N}}, (Y_n)_{n\in\mathbb{N}}$, where $X_n, Y_n \subseteq \partial G_n$ in the above notation, such that for random walk on G from any starting vertex, the events of converging to X and visiting infinitely many of the X_n coincide up to a set of measure zero, and similarly for Y and the Y_n. The first statement of Observation 1.1 generalises as shown in the following theorem.

Theorem 2.1 ([15]). *The expected number of edges of R_n^i from X_n to Y_n converges as $n \to \infty$.*

For the second statement of Observation 1.1 we remark that the limit is infinite when $X \cap Y$ has positive measure, and there are, rather rare, interesting cases where the limit is infinite independently of the choice of X, Y as long as they have positive measure: any lamplighter graph over a transient graph has this property [15].

In order to understand why Theorem 2.1 (or Observation 1.1) is true, it is helpful to consider the well-known relationship between random walks and electrical networks as introduce by Doyle and Snell [9]. Think of G_n as an electrical network with boundary nodes ∂G_n, at which we impose a constant potential $v(b) = 1$ for every $b \in \partial G_n$. This Dirichlet problem has the unique trivial solution $v(x) = 1$ for every $x \in V(G_n)$. Now, the aforementioned relationship tells us that the solution to any such Dirichlet problem can be obtained as follows: we start $d_{G_n}(b)v(b)$ random walk particles at each boundary node b, and stop them upon their first re-visit to ∂G_n. Then letting $v(x)$ be the expected number of visits to x by all those particles divided by the degree $d(x)$ solves the Dirichlet problem [14]. But as our Dirichlet problem has constant boundary values, we then expect $d(x)$ visits to each interior vertex x. This implies that if we start $d_{G_n}(b)$ random walkers at each vertex $b \in \partial G_n$, stop them upon their first re-visit to ∂G_n, and observe the parts of their trajectories inside G_m for $m < n$, then the situation we observe will be similar as if we had performed the same process on G_m instead of G_n. This is the central observation for the proof of Theorem 2.1.

Theorem 2.1 will be crucial in the next section.

3. The Effective Conductance Measure $\mathcal{C}$

Theorem 2.1 relates our GWRGs to the Poisson boundary, but in fact the connection is more intricate. Before elaborating on this, we recall Douglas' [8] formula

$$E(h) = \int_0^{2\pi} \int_0^{2\pi} (\hat{h}(\eta) - \hat{h}(\zeta))^2 \Theta(\eta, \zeta) d\eta d\zeta,$$

expressing the (Dirichlet) energy of a harmonic function h on the unit disc $\mathbb{D}$ in the complex plain from its boundary values $\hat{h}$ on the circle $\mathbb{S}^1 = \partial\mathbb{D}$, where $\Theta(\eta, \zeta) := 16\pi \sin^{-2} \frac{\eta - \zeta}{2}$. The physical intuition here is that $E(h)$ is the power dissipated by a circular metal plate, on the boundary circle of which some potential $\hat{h} : \mathbb{S}^1 \to \mathbb{R}$ is imposed by an external source or field.

A discrete variant of this formula is the following, expressing the energy dissipated by a finite electrical network with a set B of 'terminal nodes', i.e. vertices that are connected to external power sources, in terms of the voltages at B and certain 'effective conductances' relative to B (see [15] for details).

$$E(h) = \sum_{a,b \in B} (h(a) - h(b))^2 C_{ab}. \tag{7.1}$$

In [15] we prove the following statement, providing a general formula for the Dirichlet energy of harmonic functions on a graph, which is similar to Douglas' formula.

Theorem 3.1 ([15]). *For every transient graph G, there is a measure $\mathcal{C}$ on $\mathcal{P}(G)^2$ such that for every harmonic function $h : V(G) \to \mathbb{R}$ with boundary function $\hat{h}$, the energy $E(h)$ equals*

$$D(\hat{h}) := \int_{\mathcal{P}(G)^2} \left(\hat{h}(\eta) - \hat{h}(\zeta) \right)^2 \mathcal{C}(\eta, \zeta).$$

This measure $\mathcal{C}$, which we call the *effective conductance measure*, can be thought of as a continuous analogue of effective conductance in a finite electrical network due to the similarity of the above formula with (7.1); in the next section we will elaborate more on this.

The construction of $\mathcal{C}$ is based on Theorem 2.1: we set the value $\mathcal{C}(X, Y)$ to be the limit value returned by that theorem, and apply Caratheodory's extension theorem to show that this defines a measure on $\mathcal{P}(G)^2$. This explains the relationship between GWRGs and $\mathcal{C}$, and also my motivation in introducing the former.

4. Doob's Formula and Naim's Kernel

Theorem 3.1 was motivated by a similar result of Doob [7], which generalises Douglas' formula to arbitrary Green spaces. I will refrain from repeating the definition of Green spaces here, which can be found in [7], and suffice it to say that they generalise Riemannian manifolds. In a sense, our Theorem 3.1 is a discrete version of Doob's result. But rather than working with the Poisson boundary —which first appeared the year after Doob's paper [12]—Doob worked with the Martin boundary, which is closely related to the Poisson boundary (the latter can be defined as the support of the former, but the former is also endowed with a topology). Moreover, Doob's approach to the effective conductance measure was different from ours: rather than working with the measure, Doob was working with its density, namely the Naim kernel. Let me explain this further, as it will be of interest later.

Naim [20] introduced the formula

$$\Theta(x, y) := \frac{G(x, y)}{G(x, o)\, G(o, y)},$$

where x, y are points of a Green space X, and $G(\cdot, \cdot)$ denotes the corresponding Green's function (see [6] for definitions). The same formula, however, makes perfect sense when x, y are points of a transient graph G, in which case $G(x, y)$ denotes the discrete *Green's function*, i.e. the expected number of visits to y by random walk from x.

Naim proved that $\Theta(x, y)$ can be extended to pairs of points $\{\eta, \zeta\}$ in the Martin boundary of X, by taking a limit

$$\Theta(\eta, \zeta) := \lim_{x \to \eta, y \to \zeta} \Theta(x, y). \tag{7.2}$$

The convergence of this limit is only clear when one of x, y is fixed and the other converges to a point η in the boundary, as it reduces to the well-known convergence of the Martin kernel $K(x, y) := \frac{G(x,y)}{G(o,y)}$ [26]. The two-sided convergence is a puzzling fact, and in fact is not true everywhere but 'almost everywhere': the exact statement proved by Naim [20] is too technical to state here precisely. It involves Cartan's fine topology. As far as I know, this convergence has not been proved for graphs; we will return to this issue below.

Our intuitive interpretation of $\Theta(\eta, \zeta)$ is that it denotes the 'effective conductance density' between the boundary points η, ζ. To support this intuition, we remark in [15] that if x, y are boundary vertices of a finite electrical network, then $\Theta(x, y)$ does indeed coincide with the effective conductance between x and y if the Green function G is defined with respect to random walk killed at the boundary.

Doob's formula for the Dirichlet energy of a harmonic function h on X, with boundary extension $\hat{h}$ on the Martin boundary $M(X)$ of X (see e.g. [26] for the definition) reads

$$D(\hat{h}) := \int_{M(X)^2} \left(\hat{h}(\eta) - \hat{h}(\zeta)\right)^2 \Theta(\eta, \zeta)\, d\mu_o(\eta)\, d\mu_o(\zeta).$$

Notice the similarity to the formula of Theorem 3.1 and Douglas' (7.1).

As we show in [15], our effective conductance measure $\mathcal{C}$ is absolutely continuous with respect to the square of harmonic measure μ_o on $\mathcal{P}$, and so we can define a kernel Θ' on $\mathcal{P}^2$ by the Radon–Nykodym derivative $\frac{\partial \mathcal{C}}{\partial \mu_o^2}(\eta, \zeta)$. The above discussion suggests that Θ' should coincide with Θ if Θ can be defined by a limit similar to (7.2). Now instead of trying to imitate Naim's proof of the convergence of (7.2), which is rather technical, we propose the following problem.

Problem 4.1. *Let G be a transient graph and $o \in V(G)$. Let $(x_n)_{n\in\mathbb{N}}$ and $(y_n)_{n\in\mathbb{N}}$ be independent simple random walks from o. Then $\lim_{n,m\to\infty} \Theta(x_n, y_m)$ exists almost surely.*

This is seemingly easier than trying to prove the convergence of (7.2), since, for example, it is easier to prove the Martingale convergence theorems than Fatou's theorem. If it is true, then one could further ask the following.

Problem 4.2. *Let G be a transient graph and $o \in V(G)$. Let $(x_n)_{n\in\mathbb{N}}$ and $(y_n)_{n\in\mathbb{N}}$ be independent simple random walks from o, and X, Y measurable subsets of $\mathcal{P}(G)$. Then $\mathcal{C}(X, Y) = \mathbb{E}\lim_{n,m\to\infty} \mu_{x_n}(X)\mu_{y_n}(Y) \Theta(x_n, y_m)$, where $\mathcal{C}$ denotes the effective conductance measure.*

The factors $\mu_{x_n}(X), \mu_{y_n}(Y)$ in the above limit essentially 'condition' the random walks to converge to X, Y respectively. This last problem is motivated by our belief that $\Theta' = \Theta$.

5. Random Interlacements

Recall that we have defined $\mathcal{C}(X, Y)$ as the limit of the expected number of edges of our GWRG R_n^1 from X_n to Y_n. In doing so, we were only interested in the starting and finishing vertices of our random walks, ignoring the exact trajectories inside the host graph G. In this section we remark that the distributions of these trajectories converge, in a certain sense, to the intensity measure of the random interlacement model as introduced by Sznitman [23] for $G = \mathbb{Z}^d$ and generalised to arbitrary transient G by Teixera [24].

Given a transient graph G, the random interlacement on G is a Poisson point process on the space of two-way infinite trajectories in G modulo the time shift. It is governed by a σ-finite measure ν on $(W^*, \mathcal{W}^*)$, where W^* denotes the set of equivalence classes of two-way infinite walks in G with respect to the time shift, and $\mathcal{W}^*$ is the canonical sigma-algebra on W^*. We remark that this measure ν can also be obtained from the process we used to construct our GWRG R_n^1 as follows (the original definition of Sznitman will be given below).

Let C be a cylinder set of W^* defined by a finite walk Z, i.e. C is the set of two-way infinite walks containing Z as a subsequence. Let $\mathcal{P}_n$

be the random process from the definition of R_n^1, that is, a collection of particles performing random walk starting at ∂G_n and stopped upon the first revisit to ∂G_n, with $d_{\partial G_n}(x)$ particles started at each $v \in \partial G_n$. We set

$$\mu(C) := \lim_n \mathbb{E}[\#\{\text{trajectories in } \mathcal{P}_n \text{ containing } Z \text{ as a subwalk}\}],$$
(7.3)

and extend μ to a measure on $(W^*, \mathcal{W}^*)$ using e.g. Caratheodory's extension theorem. It turns out [15] that

$$\mu = \nu.$$

An interesting consequence of this is the following relation between the effective conductance measure $\mathcal{C}$ and the random interlacement intensity measure, ν:

Corollary 5.1. *For every two measurable subsets X, Y of $\mathcal{P}(G)$, we have*

$$\mathcal{C}(X, Y) = \nu(W_{XY}^*),$$

where W_{XY}^* denotes the set of elements of W^* all initial subwalks of which meet infinitely many X_n (as defined in Section 2) and all final subwalks of which meet infinitely many Y_n.

In order to explain why this is true we need to recall the definition of the intensity measure ν from [24].

To begin with, given a finite subset K of $V(G)$, we define the *equilibrium measure* e_K on K by

$$e_K(x) = \mathbb{1}_{x \in K} c_x \mathbb{P}_x[A \mid \text{random walk never returns to} K],$$

where c_x is the vertex degree of x.

Let π^* be the canonical projection from the set of two-way infinite walks W to W^*. Then ν is defined as the unique measure on $(W^*, \mathcal{W}^*)$ satisfying, for every finite subset K of $V(G)$,

$$\mathbb{1}_{W_K^*} \cdot \nu = \pi^* \circ Q_K.$$
(7.4)

(Another way to state this formula is $\nu(\mathbb{1}_{W_K^*} \cdot A) = Q_K(\pi^{*-1}(A))$, where $\pi^{*-1}(A)$ returns those walks in the π^*-preimage of A that enter K at time 0 for the first time.)

Here, Q_K is a finite measure on the space W_K of two-way infinite walks meeting K, given by the formula

$$Q_K[(X_{-n})_{n \geq 0} \in A, X_0 = x, (X_n)_{n \geq 0} \in B]$$
$$= \mathbb{P}_x[A \mid \text{no return to } K]e_K(x)\mathbb{P}_x[B],$$

where A, B are measurable subsets of the space W_+ of one-way infinite walks. Let me explain why ν coincides with μ as claimed above. The main idea is to think of the equilibrium measure $e_K(x)$, which is proportional to the probability of escaping K, as the probability for a random walker coming from 'infinity' to enter K at x; this intuition is justified by the reversibility of our walks, and coming from infinity can be made precise using the process used to construct GWRGs (and μ).

To make this more precise, let G_n denote the ball of radius n around o, and suppose that $K \subset V(G_n)$ for some $K \subset V(G)$. Let G_n^* be the graph obtained from G by contracting the complement of G_n into a single vertex $*_n$. Then the reversibility of our random walk implies that, for every $x \in K$, letting $\mathbb{P}_y$ denote the law of a random walk $X_0, X_1, \ldots X_\tau$ from y stopped the first time $\tau > 0$ when $X_\tau \in K \cup \{*_n\}$, we have

$$c_x \mathbb{P}_x[X_\tau = *_n] = c_{*_n} \mathbb{P}_{*_n}[X_\tau = x].$$

Now note that, by the transience of G, the limit as n goes to infinity of the left-hand side converges to $e_K(x)$, while the limit of the right-hand side is closely related to our definition of μ.

Random interlacements have been studied extensively, and researchers have found interesting applications in the study of the vacant set for random walk on discrete tori [19]. This connection to the Poisson boundary has apparently not been observed before.

6.   Graphons

Graphons were recently introduced [5] as a notion of limit for sequences of dense finite graphs, and they have already had a seminal impact on combinatorics. Formally, a graphon is a symmetric, measurable function $w : [0,1]^2 \to [0,1]$.

Every graphon naturally gives rise to a family of random graphs G_n on n vertices for every $n \in \mathbb{N}$: we sample n independent, uniformly distributed points from $[0,1]$ to be the vertices of G_n, and join vertices x, y with probability $w(x, y)$.

Graphs sampled this way are dense, i.e. have average degree of order n. But a variant of the above sampling method introduced in [4] produces graphs of average degree $o(n)$.

Now note that formally, our effective conductance measure $\mathcal{C}$ is similar to a graphon, as it is a measure on a square. In particular, we can sample a random graph from it. We expect these graphs to be closely related to our GWRGs.

Problem 6.1. *Show that for every (transient) Cayley graph G, the sequence of corresponding GWRGs converges to a sparse graphon in the sense of [4].*

7. Simulation Data

Computer simulations on GWRG for certain concrete host graphs G, performed by Chris Midgley [19], suggest that certain properties are heavily influenced by G, while convergence with n is fast enough that simulation data can help make explicit predictions.

In most simulations the host graph was the 2- or 3-dimensional grid $\mathbb{Z}^2$ or $\mathbb{Z}^3$, the infinite binary tree T_2, a hyperbolic planar graph T° obtained from the binary tree by adding a cycle C_n joining the vertices at distance n from the root for every n, and the lamplighter graph L over $\mathbb{Z}$.

The outcomes of these simulations, with 10,000 random graphs R_n^1 generated in each case for n up to 8 (where the exponential growth of T° and L become computationally demanding), suggest the following:

- The number of isolated vertices of R_n^1 is asymptotically proportional to the number of vertices $|\partial G_n|$ of R_n^1. The same is true for the number of components of R_n^1. The corresponding leading coefficients are very similar (possibly converging to the same number) when G is $\mathbb{Z}^2$ or T°, slightly different when $G = L$, and very different when $G = \mathbb{Z}^3$.
- The expected diameter of the largest connected component of R_n^1 seems to be proportional to $\log(|\partial G_n|)$.

In a further experiment of [19] on the infinite binary tree T_2, random graphs R_n^i were generated for $2 \leq n \leq 8$ for all values of i until the first time that R_n^i becomes connected. The data suggest (quite clearly) that the average time i till connectedness is roughly $0.26n$, i.e. linear in n. (The exact outcome was $i = (0.26 \pm 0.003)k + (0.90 \pm 0.02)$ with an adjusted R^2 value of 0.9993.)

8. Further Problems

8.1. Dirichlet Harmonic Functions

Let G be a graph on which all harmonic functions with finite Dirichlet energy are constant; the class of such graphs is denoted by $\mathcal{O}_{HD}$. By Theorem 3.1, we can introduce the following trichotomy for such graphs:

(i) $\mathcal{P}(G)$ is trivial (in other words, G has the Liouville property), or
(ii) $\mathcal{P}(G)$ is not trivial, and $\mathcal{C}(X, Y) = \infty$ for every two measurable $X, Y \subseteq \mathcal{P}(G)$, or
(iii) none of the above holds, but the integral of Theorem 3.1 is infinite for every boundary function $\hat{h}$.

Recall that the property of being in $\mathcal{O}_{HD}$ is quasi-isometry invariant for graphs of bounded degree [25], and therefore, given a group Γ, it is independent of the choice of the Cayley graph of Γ. A well-known open problem asks whether the Liouville property too is independent of the choice of the Cayley graph of Γ. The above trichotomy suggests the following refinement of that problem for groups in $\mathcal{O}_{HD}$.

Problem 8.1. *Let G be a Cayley graph in $\mathcal{O}_{HD}$. Do all finitely generated (in other words, locally finite) Cayley graphs of the group of G have the same type with respect to the above trichotomy?*

We remark that Cayley graphs of all three types exist: $\mathbb{Z}^3$ is of type (i), lamplighter graphs over any transient Cayley graph are of type (ii) [15], while all tessellations of hyperbolic 3-space $\mathbb{H}^3$ are of type (iii).

8.2. GWRGs

None of the results suggested by the simulations of Section 7 have been proved rigorously, and it would be interesting to do so, especially if the methods involved can be applied to large families of host graphs. Results on the expected number of components or isolated vertices of R_n^1 could be within reach.

The interesting meta-problem is to find natural properties of the GWRGs that are universal in the sense that they are true independently of the host graph G (under some restriction, e.g. G being vertex-transitive), as well as properties that do depend on G and the dependence can be explained. For example, what can be said about R_n if the host G is hyperbolic?

Erdős Rényi random graphs $G(n, p)$ are known to display a sharp threshold for their connectedness at the value $p = \ln n/n$ for the presence of each edge. This threshold coincides with the threshold for having an isolated vertex [10]. This motivates the question of whether similar behaviour is observed for GWRGs. To make this more precise, define the random variable $\tau_n := \min\{i \mid G_n^i \text{ is connected}\}$. Our first question is how concentrated τ_n is (we interpret high concentration as a sharp phase transition). The second question is whether the threshold for connectedness coincides with that for absence of isolated vertices: letting $\tau_n^* := \min\{i \mid G_n^i \text{ has no isolated vertices}\}$,

Problem 8.2. *Is $\lim_n \frac{\mathbb{E}\tau_n}{\mathbb{E}\tau_n^*} = 1$?*[2]

This might depend on the host graph G, and the answer might be positive for every 'nice' host, e.g. any Cayley graph.

For any host graph G, and $n \in \mathbb{N}$, consider the ball G_n as a random rooted graph G_n^* by rooting it at a uniformly chosen vertex in the boundary ∂G_n. Then, as suggested by Gourab Ray (private communication), if G_n^* converges in the local weak sense, then our

---

[2] This question was answered in the negative when the host graph is the binary tree by J. Haslegrave (private communication) after this chapter was accepted.

GWRG R_n^i should converge for every i in the Benjamini–Schramm sense
[3]. The relationship between this limit and the host graph could be
interesting. (But even if G_n^* does not converge, it will have sub-sequential
limits and the situation is not less interesting.)

A particular case of interest is where G is the grid $\mathbb{Z}^d$, and we start
particles at ∂G_n according to the Poisson distribution in the construction
of R_n^i. The aforementioned limit coincides then with the long-range
percolation model as defined e.g. in [21] (this connection was again
noticed by Gourab Ray).

It would be interesting to compare our GWRGs induced by certain
simple Cayley graphs, like tessellations of the hyperbolic plane, to other
geometric random graphs from the literature:

Problem 8.3. *Show that other geometric random graph constructions
(e.g. those appearing in [22]) can be obtained as special cases of (or
approximated by) GWRGs by choosing an appropriate underlying graph.*

8.3. Groups

Our main objective is to understand the interplay between typical
properties of GWRGs and their host groups. In particular, we have:

Problem 8.4. *Which properties of the random graphs are determined
by the group of the host graph and do not depend on the choice of a
generating set?*

A further problem in this vein is the following.

Problem 8.5. *Are the properties of transience, Liouvilleness, exis-
tence of harmonic Dirichlet functions, and amenability on the host
Cayley graph detectable by the asymptotic behaviour of the corresponding
GWRGs?*

Answers to these problems would make GWRG a tool for studying
group-theoretical questions, on a par with the study of random walks on
groups.

9. Conclusions

We introduced a construction of 'geometric' random graphs (GWRGs),
and established strong links to the Poisson boundary of their host graphs,
the effective conductance measure and Naim's kernel, and to random
interlacements. This project will be successful if we can exploit these
connections in order to make conclusions about one of these objects by
studying the other. Another major aim is to relate GWRGs to existing
models of geometric random graphs. Last but not least, we would like
to understand the effect of various properties of the host group on the
typical graph-theoretic properties of GWRGs.

Acknowledgement

I am grateful to Christophe Garban for suggesting the connection to Random Interlacements, which gave rise to Section 5, and to A. Janse van Rensburg and C. Midgley for their simulations. I would like to thank Gourab Ray for the aforementioned remarks, and Remco van der Hofstad for triggering Problem 8.2.

References

[1] A. Ancona. Negatively curved manifolds, elliptic operators, and the Martin boundary. *The Annals of Mathematics*, 125(3):495, 1987.

[2] A. Ancona. Positive harmonic functions and hyperbolicity. In *Potential Theory Surveys and Problems*, vol. 1344 of *Lecture Notes in Mathematics*, pp. 1–23. 1988.

[3] I. Benjamini and O. Schramm. Recurrence of distributional limits of finite planar graphs. *Electronic Journal of Probability*, 6, 2001.

[4] C. Borgs, J.T. Chayes, H. Cohn, and Y. Zhao. An Lp theory of sparse graph convergence I: limits, sparse random graph models, and power law distributions. Preprint 2014.

[5] C. Borgs, J.T. Chayes, L. Lovász, V.T. Sós, and K. Vesztergombi. Convergent sequences of dense graphs I: Subgraph frequencies, metric properties and testing. *Advances in Mathematics*, 219(6):1801–51, 2008.

[6] M. Brelot and G. Choquet. Espaces et lignes de Green. *Ann. Inst. Fourier*, 3:119–263, 1951.

[7] J. L. Doob. Boundary properties of functions with finite Dirichlet integrals. *Ann. Inst. Fourier*, 12:573–621, 1962.

[8] Jesse Douglas. Solution of the problem of Plateau. *Trans. Am. Math. Soc.*, 33(1):263–321, 1931.

[9] P.G. Doyle and J.L. Snell. *Random Walks and Electrical Networks*. Carus Mathematical Monographs 22, Mathematical Association of America, 1984.

[10] P. Erdös and A. Rényi. On random graphs I. *Publ. Math. Debrecen*, 6:290–7, 1959.

[11] A. Erschler. Poisson–Furstenberg boundaries, large-scale geometry and growth of groups. In *Proceedings of the ICM*, pp. 681–704, 2010.

[12] H. Furstenberg. A Poisson formula for semi-simple Lie groups. *The Annals of Mathematics*, 77(2):335–86, 1963.

[13] A. Georgakopoulos. The boundary of a square tiling of a graph coincides with the poisson boundary. To appear in *Invent. Math.*, DOI 10.1007/s00222-015-0601-0.

[14] A. Georgakopoulos. Electrical networks from the mathematical viewpoint. Lecture notes. In preparation.

[15] A. Georgakopoulos and V. Kaimanovich. In preparation.

[16] R. Van Der Hofstad. Random graphs and complex networks. Lecture Notes, 2013.

[17] V. Kaimanovich. The poisson formula for groups with hyperbolic properties. *The Annals of Mathematics*, 152(3): 659–92, 2000.

[18] V.A. Kaimanovich and A.M. Vershik. Random walks on discrete groups: boundary and entropy. *The Annals of Probability*, 11(3):457–90, 1983.

[19] C. Midgley. Random graphs from groups, 2014. Undergraduate research project. University of Warwick.

[20] L. Naïm. Sur le rôle de la frontière de R.S. Martin dans la théorie du potentiel. *Annales Inst. Fourier*, 7:183–281, 1957.

[21] C.M. Newman and L.S. Schulman. One-dimensional $1/|ji|s$ percolation models: the existence of a transition for $s \leq 2$. *Commun. Math. Phys.*, 104:547–71, 1986.

[22] M. Penrose. *Random Geometric Graphs*. Oxford University Press, 2003.

[23] A.-S. Sznitman. Vacant set of random interlacements and percolation. *Annals of Mathematics*, 171(3):2039–2087, 2010.

[24] A. Teixeira. Interlacement percolation on transient weighted graphs. 14:1604–27, 2009.

[25] C. Thomassen. Resistances and currents in infinite electrical networks. *J. Combin. Theory (Series B)*, 49:87–102, 1990.

[26] Wolfgang Woess. *Denumerable Markov Chains. Generating Functions, Boundary Theory, Random Walks on Trees*. EMS Textbooks in Mathematics. European Mathematical Society (EMS), Zürich, 2009.

8

ENDS OF BRANCHING RANDOM WALKS ON PLANAR HYPERBOLIC CAYLEY GRAPHS

LORENZ A. GILCH[1] AND SEBASTIAN MÜLLER[2]

[1] Know Center GmbH, Inffeldgasse 13, A-8010 Graz, Austria
[2] Aix Marseille University, CNRS, Centrale Marseille, I2M, Marseille, France

To Professor Woess on the occasion of his sixtieth birthday

Abstract
We prove that the trace of a transient branching random walk on a planar hyperbolic Cayley graph has a.s. continuum many ends and no isolated end.

Contents

1. Introduction

A *branching random walk* (BRW) is a growing cloud of particles on some graph G. In this note we consider BRWs in discrete time, which are defined as follows. The process starts with one particle in the root o of the graph G. At each time step each particle splits into offspring particles, which then move one step according to a random walk on G. Particles branch and move independently of the other particles and the history of the process. A BRW is therefore driven by two classical stochastic processes: Galton–Watson processes and random walks. Under the assumption that the underlying Galton–Watson process survives the number of particles grows exponentially. If the return probabilities of the underlying random walk decay subexponentially the effect of the growing particles overshadows the transience of the spatial dynamic, and the underlying graph will eventually be full of particles. However, if the return probabilities of the random walk decay exponentially as well, then there is a critical growth rate of the Galton–Watson process where the

2000 *Mathematics Subject Classification*. Primary: 60J80; Secondary: 60J10, 05C80.
Key words and phrases. branching random walk, hyperbolic groups.

205

two exponential effects cancel out. Above the critical value the BRW is again *recurrent*, i.e. every finite set is visited infinitely many times with positive probability; below and at the critical value every finite set is eventually free of particles and the BRW is called *transient*. In the transient case, the set of visited vertices and traversed edges defines a proper random subgraph of G and its properties become of interest. This subgraph is called the *trace* Tr of the BRW.

Let $\mu = (\mu_k)_{k \geq 0}$ be the probability distribution that describes the branching, i.e. each particle produces k offspring with probability μ_k. The expected number of offspring is denoted by $m = \sum_{k \geq 0} k\mu_k$. In this note we assume the underlying graph $G = G(\Gamma, S)$ to be the Cayley graph of a finitely generated group Γ with respect to a finite symmetric generating set S. The movement of the particles is governed by a driving measure q on $S \cup \{e\}$, where e is the group identity. The driving measure q defines a transition kernel $P = (p(x, y))_{x,y \in \Gamma}$ by $p(x, y) = q(x^{-1}y)$ for all $x, y \in \Gamma$. The corresponding n-step transition probabilities are denoted by $p^{(n)}(x, y)$.

We make the following standing assumptions.

Assumption 1.1.

- The underlying Galton–Watson process is supercritical, that is, $m > 1$. Furthermore, we assume that $\mu_0 = 0$ and $\mu_1 > 0$.
- The driving measure q of the random walk on G is symmetric, i.e. $q(s) = q(s^{-1})$ for all $s \in S$, and satisfies $supp(q) = S \cup \{e\}$.

These assumptions are to some extent chosen to improve the presentation. The assumptions that are really necessary are that $m > 1$ and that the driving measure q is symmetric.

The spectral radius, $\rho = \rho(P) = \limsup_{n \to \infty} (p^{(n)}(e, e))^{1/n}$, is a crucial quantity in the study of BRWs: a BRW on a Cayley graph is transient if and only if $m\rho \leq 1$. This is a consequence of the classification of recurrent groups and Kesten's amenability criterion, see also [8] for an alternative proof. We speak of a critical BRW if $m\rho = 1$.

It was shown in [3] that the trace of a transient BRW on a Cayley graph G is a.s. transient for simple random walk but recurrent for BRW. Therefore, the trace of a transient BRW is a.s. a proper subgraph of G. It is believed that the trace shares many properties with infinite percolation clusters in the non-unicity phase. In the case of BRWs on free groups (or regular trees) it even turns out that the law of the trace of a BRW is the law of an infinite cluster of some invariant percolation, see [2, 3]. However, the situation is not as clear for other Cayley graphs, especially one-ended Cayley graphs.

This note is devoted to the following property of invariant percolation and its analogue for the trace of BRWs. A bond percolation is a probability measure on subsets (also called configurations) ω of the edge set E of the graph G. If the distribution of ω is invariant under the left action of Γ, we call the percolation $(\Gamma\text{-})$invariant. A percolation is called insertion-tolerant if $A \cup \{\mathfrak{e}\}$ has positive probability for all $\mathfrak{e} \in E$ and all edge sets A having positive probability. For every invariant insertion-tolerant percolation process on a non-amenable Cayley graph G that has a.s. infinitely many clusters, we have that a.s. every infinite cluster has continuum many ends, no isolated end, and is transient for simple random walk, see Theorem 8.32 in [12] where this is stated in more generality.

On groups with infinitely many ends, there are various ways to see that the trace has infinitely many ends. In some cases even the Hausdorff dimension of the set of ends can be calculated; see [11] for free groups and [6] for free products of groups.

As pointed out in [7] there is an elegant argument using symmetry that the trace of a subcritical BRW, i.e. $m < 1/\rho$, has infinitely many ends. This argument extends to the critical case if $\sum_{n \geq 1} n m^n p^{(n)}(e, e) < \infty$. However, this condition does not apply to the most interesting graphs like Cayley graphs of hyperbolic groups and the following conjecture remains open.

Conjecture 1.2 (I. Benjamini). *Let G be any non-amenable vertex transitive graph and assume the transition kernel P of the underlying random walk to be symmetric. Then the trace of a transient BRW has infinitely many ends.*

Remark 1.3. The assumption of symmetry is crucial, as there are non-symmetric driving measures that induce one-ended traces, see [7].

We answer this conjecture affirmative for BRWs on planar hyperbolic Cayley graphs. We briefly recall the definition of a Gromov hyperbolic group. A geodesic triangle consists of three elements $u, v, w \in \Gamma$ together with three geodesic paths connecting each pair of these vertices. Let $\delta > 0$. A geodesic triangle is called δ-thin if each side is contained in a δ-neigbourhood of the two other sides. A graph is called hyperbolic if it satisfies the thin triangle condition for some δ. A hyperbolic group is a finitely generated group whose Cayley graphs are hyperbolic.

Theorem 1.4. *Under Assumption 1.1 the trace of a transient BRW on a planar hyperbolic Cayley graph has a.s. continuum many ends and no isolated end.*

The proof uses the concept of unimodular random graphs. This class of graphs can be considered as stochastic generalizations of Cayley graphs and are also known as stochastic homogeneous graphs. We will use the fact that the trace of a BRW is a unimodular random graph and therefore gives rise to the application of the generalized Mass-Transport principle. This property is used to prove that a.s. every trace that has at least 3 ends has no isolated end, see Proposition 3.1. The main issue will be to prove that the trace has a.s. no isolated end, see Proposition 3.2, and hence infinitely many ends. The proof of this proposition uses crucially the planarity of the Cayley graph, a recent result in [9] that Ancona's Inequality still holds true for the Green function at the radius of convergence, and Proposition 3.1.

In Section 2 we give some background on unimodular random graphs (URG) and random walks on hyperbolic groups. We believe that the information given there is sufficient to follow the proof of Theorem 1.4 in Section 3. However, for readers who are not familiar with the concept of URGs and with random walks on hyperbolic groups, it might be useful to consult some of the references given along Section 2.

2. Preparations

2.1. Definition and Preliminaries

We use the standard notation for a locally finite graph $G = (V, E)$: V is the set of vertices, E is the set of edges, and we write $x \sim y$ if $(x, y) \in E$. The distance between two vertices is the length of a shortest path between these vertices and will be denoted by $d(\cdot, \cdot)$. We write (G, o) for a rooted graph with root o.

Let Γ be a finitely generated group with group identity e; group operations are written multiplicatively. The group Γ together with some finite symmetric generating set S induces a Cayley graph $G = G(\Gamma, S)$ whose vertex set equals Γ and $x \sim y$ if and only if $x^{-1}y \in S$. The group identity e of Γ will be identified with the root o of the Cayley graph G.

Let q be a probability measure on the generating set $S \cup \{e\}$. The corresponding random walk $(S_n)_{n \geq 0}$ on G is a Markov chain with transition probabilities $p(x, y) = q(x^{-1}y)$ for $x, y \in \Gamma$. Equivalently, the random walk (starting in x) can be described as $S_n = xX_1 \cdots X_n$, $n \geq 1$, where the X_i's are i.i.d. random variables with distribution q.

Besides the definition of BRWs given in Section 1 there is another powerful description of BRWs. This definition is based on the concept of tree-indexed random walks introduced in [4]. Let $(\mathbb{T}, \mathbf{r})$ be a rooted infinite tree. The tree-indexed random walk can be described as a marking (or labelling) of the rooted tree $(\mathbb{T}, \mathbf{r})$. For any vertex $v \neq \mathbf{r}$ denote by v^- the neighbour of v closest to $\mathbf{r}$. Label the edges of $\mathbb{T}$ with i.i.d. random variables X_v's with distribution q; the random variable X_v is the label of the edge (v^-, v). These labels correspond to the steps of the tree-indexed walk and the positions of 'particles' are given by $S_v = x \cdot \prod_{i=1}^n X_{v_i}$ where $\langle v_0 = \mathbf{r}, v_1, \ldots, v_n = v \rangle$ is the unique geodesic from $\mathbf{r}$ to v at level n.

A tree-indexed random walk becomes a BRW if the underlying tree $\mathbb{T}$ is a realization of a Galton–Watson process. We call $\mathbb{T}$ the family tree of the BRW.

2.2. Unimodular Random Graphs

In this note we only give the essentials needed for our proofs. We invite the reader to consult [3, 13] for more details on the connection between BRWs and unimodular random graphs, and [1] for a more general introduction to the concept of URGs.

A rooted isomorphism between two rooted graphs (G, o) and (G', o') is an isomorphism of G onto G' which maps o to o'. We denote by $\mathcal{G}_*$ the space of isomorphism classes of rooted graphs and write $[G, o]$ for the equivalence class that contains (G, o). In the same way one defines the space $\mathcal{G}_{**}$ of isomorphism classes of graphs with an ordered pair of distinguished vertices. That is, (G_1, o_1, o_2) and (G_2, o'_1, o'_2) are isomorphic if and only if there is an isomorphism from G_1 onto G_2, which maps o_1 to o'_1 and o_2 to o'_2. The spaces $\mathcal{G}_*$ and $\mathcal{G}_{**}$ can be equipped with metrics that turn them into separable and complete metric spaces.

A Borel probability measure ν on $\mathcal{G}_*$ is called *unimodular* if it obeys the mass-transport principle (MTP): for every Borel function of the form $f : \mathcal{G}_{**} \to [0, \infty]$, we have

$$\int \sum_{x \in V} f(G, o, x) d\nu([G, o]) = \int \sum_{x \in V} f(G, x, o) d\nu([G, o]). \qquad (8.1)$$

Realizations of unimodular measures are called *unimodular random graphs*.

An important class of unimodular measures arises from Galton–Watson processes. The Galton–Watson tree is defined inductively: start with one vertex, the root $\mathbf{r}$ of the tree. Then the number of offspring of each particle (vertex) is distributed according to μ. Edges are between vertices and their offspring. We denote by **GW** the corresponding measure on the space of rooted trees. In this construction the root clearly plays a special role and **GW** is not unimodular. However, if we bias the distribution such that the probability that the root has degree $k + 1$ is proportional to $\frac{\mu_k}{k+1}$ we obtain a unimodular measure **UGW**. When we use the **UGW** measure instead of the standard **GW** measure to define the family tree of the BRW, we denote the BRW by UBRW.

Due to the description of the BRW as a tree-indexed random walk the unimodularity of **UGW** carries over to the trace: the trace of a UBRW on a Cayley graph is a unimodular random graph; see Theorem 3.7 in [3]. This property makes the UBRW more natural to consider than the original BRW and we will prove Theorem 1.4 for UBRWs. However, it is not difficult to see that it then also holds true for BRWs.

2.3. Ends of Graphs

Consider a locally finite graph $G = (V, E)$. A *ray* is a sequence $\pi = \langle x_0, x_1, \ldots \rangle$ of distinct vertices such that $x_i \sim x_{i+1}$ for all $i \geq 0$. For any finite set F of vertices we consider its complement $G \backslash F$, which is the graph induced by the vertex set $V \backslash F$. This graph consists of finitely many connected components. Every ray π must have all but finitely many points in exactly one component; we say that π *ends up* in that component. Two ends are *equivalent* if they end up in the same connected component for all choices of F. *Ends* are equivalence classes of rays and we denote by ϑG the set of ends. Let F be a finite vertex set and C be some component of $G \backslash F$. We write ϑC for the set of ends whose rays end up in C. The space of ends ϑG can be equipped with a discrete topology in the following way. For any finite set F and any end w there is precisely one component of $G \backslash F$ whose completion contains w. Varying F yields a neighbourhood base for w. An *isolated end* is an end that is isolated in this topology.

2.4. Hyperbolic Groups and Random Walks

From now on we assume that the underlying group Γ is hyperbolic, and the generating set S induces a planar Cayley graph $G = G(\Gamma, S)$. Both assumptions, hyperbolicity and planarity, are crucial.

Let us first collect several classic facts about hyperbolic groups and random walks; we refer to the survey [5] for an excellent introduction. An elementary hyperbolic group is either finite or has two ends. We will focus on the case of non-elementary hyperbolic groups since random walks on them are transient. Define the *Green functions*

$$G_r(x, y) = \sum_{n=0}^{\infty} p^{(n)}(x, y) r^n, \quad x, y \in \Gamma,$$

for all $r \leq R := 1/\rho$. It is proved in [9] that for finite-range random walks on hyperbolic groups Ancona's Inequalities hold up to the radius of convergence: there exists some $C > 0$ such that for any $x, z \in \Gamma$ and for any y on a geodesic segment from x to z we have

$$G_r(x, z) \leq C G_r(x, y) G_r(y, z) \quad \forall r \in [1, R].$$

Symmetry of the random walk implies, see Lemma 2.1 in [10], that

$$\lim_{d(e,x) \to \infty} G_R(e, x) = 0.$$

Eventually we obtain, see Lemma 2.2 in [10], that the Green functions of the random walk decay exponentially, that is, there exist some constants C_1 and $\varrho < 1$ such that for all $x, y \in \Gamma$

$$G_R(x, y) \leq C_1 \varrho^{d(x,y)}. \tag{8.2}$$

In the case of planar Cayley graphs, the Cayley 2-complex is the 2-complex such that the one-skeleton is given by the Cayley graph G and the 2-cells are bounded by loops in G. The 2-complex is homeomorphic to the hyperbolic disc and it can be endowed with an orientation. This orientation is used implicitly at several points in the proof of Proposition 3.2. Moreover, the Gromov hyperbolic boundary ∂G can be identified with the unit circle.

3. Proof of Theorem 1.4

The following result holds true for traces of transient BRWs on Cayley graphs (or even URGs). Its proof is an adaptation of the one for invariant percolation, see Proposition 8.33 in [12]. As the arguments are short we can present the details. We stick as close as possible to the notations in [12].

Proposition 3.1. *Consider the trace of a transient symmetric UBRW on a Cayley graph. Almost surely every trace that has at least three ends has no isolated end.*

Proof. For each $n \in \mathbb{N}$ let A_n be the union of all vertex sets $A \subset \mathtt{Tr}$ such that the diameter $diam(A) \leq n$ (in the metric in $\mathtt{Tr}$) and $\mathtt{Tr} \backslash A$ has at least three infinite components. If $\mathtt{Tr}$ has at least three ends then $A_n \neq \emptyset$ for all but finitely many n. We assume from now on that $\mathtt{Tr}$ has at least three ends.

Let be $n \in \mathbb{N}$. For any vertex x in $\mathtt{Tr}$ let

$$C(x) = \{y \in A_n : d(x, y) = \min_{z \in A_n} d(x, z)\}$$

be the set of vertices in A_n that are closest to x in the metric in $\mathtt{Tr}$. We define the Borel function $F : \mathcal{G}_{**} \to [0, \infty]$:

$$F(\mathtt{Tr}, x, y) = \begin{cases} \frac{1}{|C(x)|} \mathbb{1}_{\{y \in C(x)\}}, & \text{if } A_n \neq \emptyset, \\ 0, & \text{otherwise.} \end{cases}$$

The function F is well defined since $F(\mathtt{Tr}, o, x)$ is invariant under isomorphisms. Given that the law ν of the rooted trace $[\mathtt{Tr}, o]$ is unimodular we can apply the MTP to obtain that the expected mass received by o is at most 1:

$$\int \sum_{x \in \mathtt{Tr}} F(\mathtt{Tr}, x, o) d\nu([\mathtt{Tr}, o]) = \int \sum_{x \in \mathtt{Tr}} F(\mathtt{Tr}, o, x) d\nu([\mathtt{Tr}, o]) \leq 1.$$

Assume that ξ is an isolated end of $\mathtt{Tr}$ and let us show that this leads to a contradiction. We describe ξ by a geodesic ray $\gamma = \langle o, x_1, x_2, \ldots \rangle$ starting at o. There exists some finite set $F_0 = \{y \in \mathtt{Tr} \mid d(o, y) \leq R_0\}$ for some $R_0 \in \mathbb{N}$ such that there is a connected component U_0 of $G \backslash F_0$ whose completion contains only ξ on the boundary. Choose now $n \in \mathbb{N}$

large enough such that there exists some $x \in \gamma$ with $x \in U_0 \cap A_n$. If we choose $R \in \mathbb{N}$ large enough and set $F = \{y \in \mathtt{Tr} \mid d(o, y) \le R\}$ then the connected component U of $G \backslash F$ whose completion contains only ξ on the boundary satisfies $U \cap A_n = \emptyset$. Now we set $B = F \cap U_0$ which is finite. This provides that all paths from U to all other connected components of $G \backslash F$ must pass through B. As for all $y \in U \cap \gamma$ we have that $d(y, x) \le d(y, F_0)$ we see that a finite subset of vertices of B gets all the mass from all the vertices in $U \cap \gamma$. As $U \cap \gamma$ contains infinitely many vertices the set B receives infinite mass. Using the property that 'everything shows at the root', see [1, Lemma 2.3], we obtain that $\int \sum_{x \in \mathtt{Tr}} F(\mathtt{Tr}, x, o) d\nu([\mathtt{Tr}, o]) = \infty$, a contradiction. Hence, almost surely every trace with $A_n \ne \emptyset$ does not have isolated ends. Because this holds for all but finitely many $n \in \mathbb{N}$, we obtain that almost surely a trace with isolated ends can have at most two ends.   □

Proposition 3.2.   *Under Assumption 1.1 the trace of a transient UBRW on a planar hyperbolic Cayley graph has a.s. no isolated end.*

Proof.   We start with some preparations. Since the Gromov boundary ∂G can be identified with a circle there exist infinite geodesics γ_1, γ_2 and γ_3 starting from o with distinct boundary points ξ_1, ξ_2 and ξ_3. Denote by $\gamma = \bigcup_{i=1}^{3} \{\gamma_i\}$, where $\{\gamma_i\}$ is the set of vertices in γ_i. We define ∂G_{ij} to be the part of the Gromov boundary that is between ξ_i and ξ_j and S_{ij} to be the set of vertices between γ_i and γ_j. Let K be some large positive constant to be chosen later. Geodesics in hyperbolic groups either converge to the same boundary point or diverge exponentially. Hence, for K sufficiently large there exist $x_1 \in S_{12}, x_2 \in S_{23}$, and $x_3 \in S_{31}$ such that $d(x_i, \gamma) = K$ for all $i \in \{1, 2, 3\}$.

Let $i \in \{1, 2, 3\}$ and consider the sphere $S(x_i, n)$ of radius n around x_i in the Cayley graph metric. Since the γ_i's are geodesics and triangles are thin we know that there exists some constant C_2 such that $|S(x_i, n) \cap \gamma| \le C_2 n$. Now, start a BRW in x_i and denote by $\mathtt{Tr}_i$ the trace of the BRW started in x_i. Using Markov's Inequality and Inequality (8.2) we obtain that

$$\mathbb{P}[|\mathtt{Tr}_i \cap \gamma| \ge 1] \le \mathbb{E}[|\mathtt{Tr}_i \cap \gamma|] \le \sum_{y \in \gamma} G_m(x_i, y) \le \sum_{n \ge K} C_1 C_2 n \varrho^n, \quad (8.3)$$

which tends to 0 as K tends to infinity. We note that $G_m(x_i, y)$ is the expected number of particles visiting y when starting a BRW with one particle at x_i.

We assume now that $\mathbb{P}[\mathtt{Tr}$ has an isolated end$] = c > 0$ and show that this yields a contradiction. Inequality (8.3) allows us to choose K sufficiently large such that for $i \in \{1, 2, 3\}$,

$$\mathbb{P}[\mathtt{Tr}_i \text{ has an isolated end}, |\mathtt{Tr}_i \cap \gamma| = 0] > 0.$$

We start the UBRW in o and condition the UBRW on the event that at time $N = \max\{d(o, x_i) : i \in \{1, 2, 3\}\}$ in each of the vertices x_1, x_2 and x_3 there is exactly one particle and no particle elsewhere. This is possible since we assume that $\mu_1 > 0$ and $q(e) > 0$. So what happens at times $n \geq N$ has the same distribution as we start three independent BRWs in x_1, x_2 and x_3. Eventually, using the planarity of G we get that Tr has with positive probability at least three distinct ends including at least one isolated end, which yields a contradiction to Proposition 3.1.   □

It remains to show that the trace has a.s. continuum many ends.

Corollary 3.3. *Under Assumption 1.1 the trace of a transient UBRW on a planar hyperbolic Cayley graph has a.s. continuum many ends.*

Proof. Due to Proposition 3.2 the trace Tr must have infinitely many ends because otherwise each end would be isolated. Moreover, each infinite connected component of $\text{Tr}\backslash B(o, n)$ must contain at least two ends; otherwise, such a component would contain an isolated end. Thus, the number of ends is at least of order $|2^{\mathbb{N}}|$, which proves the claim since Tr is of bounded degree.   □

Acknowledgement

The authors are grateful to Elisabetta Candellero and Matthew Roberts for comments on a first version of this chapter. The research was supported by the exchange programme Amadeus-Amadée 31473TF.

References

[1] D. Aldous and R. Lyons. Processes on unimodular random networks. *Electron. J. Probab.*, 12:no. 54, 1454–508, 2007.

[2] I. Benjamini, R. Lyons, and O. Schramm. Unimodular random trees. *Preprint, http://arxiv.org/abs/1207.1752*, 2012.

[3] I. Benjamini and S. Müller. On the trace of branching random walks. *Groups Geom. Dyn.*, 6(2):231–47, 2012.

[4] I. Benjamini and Y. Peres. Markov chains indexed by trees. *Ann. Probab.*, 22(1):219–43, 1994.

[5] D. Calegari. The ergodic theory of hyperbolic groups. *Contemp. Math*, pages 15–52, 2013.

[6] E. Candellero, L. A. Gilch, and S. Müller. Branching random walks on free products of groups. *Proc. Lond. Math. Soc. (3)*, 104(6):1085–120, 2012.

[7] E. Candellero and M. I. Roberts. The number of ends of critical branching random walks. *http://arxiv.org/abs/1401.0429*, 2014.

[8] N. Gantert and S. Müller. The critical branching Markov chain is transient. *Markov Process. and Rel. Fields.*, 12:805–14, 2007.

[9] S. Gouëzel. Local limit theorem for symmetric random walks in Gromov-hyperbolic groups. *J. Amer. Math. Soc.*, 27(3):893–928, 2014.

[10] S. Gouëzel and S. P. Lalley. Random walks on co-compact Fuchsian groups. *Ann. Sci. Éc. Norm. Supér. (4)*, 46(1):129–73, 2013.

[11] I. Hueter and S. P. Lalley. Anisotropic branching random walks on homogeneous trees. *Probab. Th. Rel. Fields*, 116(1):57–88, 2000.

[12] R. Lyons, with Y. Peres. *Probability on Trees and Networks*. Cambridge University press, In preparation. Current version available at http://mypage.iu.edu/~rdlyons/.

[13] S. Müller. Interacting growth processes and invariant percolation. *Ann. Appl. Prob.*, 25(1):268–86, 2015.

9

AMENABILITY AND ERGODIC PROPERTIES OF TOPOLOGICAL GROUPS: FROM BOGOLYUBOV ONWARDS

ROSTISLAV GRIGORCHUK[1] AND PIERRE DE LA HARPE[2]

[1]Department of Mathematics, Mailstop 3368, Texas A&M University, College Station, TX 77843-3368, USA

[2]Section de Mathématiques, University of Geneva, 2-4, Rue du Lièvre, Case Postale 64, 1211 Genève 4, Suisse

For Wolfgang Woess

Abstract

The purpose of this expository and historical article is to revisit the notions of amenability and ergodicity, and to point out that they appear for topological groups that are not necessarily locally compact in articles by Bogolyubov (1939), Fomin (1950), Dixmier (1950), and Rickert (1967).

Contents

2000 *Mathematics Subject Classification.* 43A07

Key words and phrases. Amenable topological groups, B-amenable groups, extremely amenable groups, Bogolyubov, Fomin, Kolmogorov

The visit of the first author to Geneva, where much of this work was done, was supported by the Swiss National Science Foundation grant 200020_141329. The first author acknowledges support of NSF grant DMS-1207699. We are grateful to Claire Anantharaman, Ivan Babenko, Dana Bartosova, Bachir Bekka, Yves Cornulier, Julien Melleray, Vladimir Pestov, John Roe, and Klaus Schmidt for useful conversations and mail exchanges, as well as to the referee for his constructive criticism of a first version of this chapter.

1. Introduction

The notion of amenability for groups has several historical origins. Most authors cite the article [vNeu–29] of von Neumann, sometimes complemented by a note [Tars–29] of Tarski, developed in [Tars–38]. Von Neumann and Tarski were motivated by the paradox of Banach and Tarski [BaTa–24], and its origin in a decisive observation of Hausdorff [Haus–14, page 469]. There is a good and comprehensive exposition of this subject in [Wago–85]. In these references, groups do not have topology; in other words they are just 'discrete groups' (even if topological groups appear in Wagon's book, they appear only marginally). Amenability is considered explicitly for topological groups in later articles [Dixm–50, Fomi–50, Rick–67]. But attention has often been restricted to two particular classes of topological groups: *discrete* groups (and semi-groups), as in [HeRo–63, § 18], and *locally compact groups*, as in the influential book of Greenleaf [Gree–69].

The first goal of this chapter is to indicate that the notion of amenability has at least one other independent origin. Indeed, we were surprised to discover that topological groups G, such that there exist left-invariant means on $\mathcal{C}^b(G)$, are already the main subject of the 1939 article of Bogolyubov [Bogo–39], where appears the class of *topological groups* that we call 'B-amenable' in Definition 3.8 below; the 'B' refers to Bogolyubov. Here, $\mathcal{C}^b(G)$ stands for the Banach space of bounded continuous real-valued functions on G.

The second goal of this chapter is to expose and survey various aspects of amenability as they occur for topological groups that need not be locally compact. Results stated (and often proved) below are not original.

1.1. Plan of the Chapter

We add some historical comments in Section 2. In Section 3, we review the basic notions: amenability and the related fixed point property; and also B-amenability; for locally compact groups, the three notions are equivalent. Section 4 is about their hereditary properties. Section 5 contains three examples of amenable topological groups that are not locally compact: the unitary group of a separable infinite dimensional Hilbert space, the symmetric group Sym(**N**) of the positive integers, and the general linear group GL(V) of a vector space of countable infinite dimension over a finite field (with their Polish topologies). Section 6 provides more examples which have indeed a stronger property: they are extremely amenable. In Section 7, we discuss several definitions of ergodicity for actions on compact spaces; they agree for second countable locally compact groups, but not in general, as shown in Section 8 by an example of Kolomogorov involving Sym(**N**). The final Section 9 describes a characterization, due to Fomin, of ergodicity in terms of unitary representations.

It is convenient to agree that **all topological spaces and groups appearing in this chapter are assumed to be Hausdorff.**

2. Some Historical Comments

Let G be a topological group and X a non-empty compact space on which G acts continuously by homeomorphisms. Denote by $\mathcal{P}(X)$ the space of probability measures on X, by $\mathcal{P}^G(X)$ its subspace of G-invariant probability measures, and by $\mathcal{E}^G(X)$ the subspace of indecomposable G-invariant probability measures on X (the definitions of indecomposable measure, and of related ergodic notions, are recalled in Section 7). In the classical case of G the additive group $\mathbf{R}$ of real numbers and X a metrizable compact space, Bogolyubov and Krylov established in [KrBo–37] three fundamental results, that we quote essentially as Fomin does in [Fomi–50, § 2]:

(1) $\mathcal{P}^{\mathbf{R}}(X)$ is a non-empty convex compact subspace in the appropriate topological vector space.

(2) $\mathcal{P}^{\mathbf{R}}(X)$ is the convex closure of the space $\mathcal{E}^{\mathbf{R}}(X)$ of ergodic invariant probability measures; in particular $\mathcal{E}^{\mathbf{R}}(X)$ is non-empty. ('Ergodic' is used today, but 'transitive' is used in [KrBo–37, Bogo–39, Fomi–50].)

(3) For every $\mu \in \mathcal{E}^{\mathbf{R}}(X)$, there exists an $\mathbf{R}$-invariant Borel subset $E_\mu \subset X$ with the following properties: $\mu(E_\mu) = 1$, and $\mu'(E_\mu) = 0$ for all $\mu' \in \mathcal{E}^{\mathbf{R}}(X)$, $\mu' \neq \mu$.

See also [Oxto–52] for $G = \mathbf{Z}$ (instead of $\mathbf{R}$).

The generalization from $\mathbf{R}$ to other topological groups raises several problems and justifies new notions, in particular that of amenable groups, for which (1) holds (Proposition 3.6 below). When (1) holds, (2) holds without further restriction on G, as it is indicated in [Bogo–39]; this follows alternatively from the Krein–Milman theorem [KrMi–40]. Result (3) holds more generally when G is a second countable locally compact group, as exposed in [Vara–63] and [GrSc–00, Theorem 1.1(3)], see Section 7; but (3) does not hold for all groups, as shown by the example of Kolmogorov discussed in Section 8.

It is likely that Bogolyubov, and perhaps later also Fomin, were not aware of von Neumann's and Tarski's 1929 articles when they wrote [Bogo–39] and [Fomi–50]. Moreover, Bogolyubov's 1939 article had very little impact at the time (see [Anos–94], in particular the bottom of Page 10). Indeed, neither [Bogo–39] nor [Fomi–50] appears in the lists of references of any of [Dixm–50], [HeRo–63], [Rick–67], [Gree–69], or [Wago–85].

Nevertheless, in his 1939 article, Bogolyubov shows that (1) above holds for every topological group which is B-amenable, and the *same proof* shows that this extends to amenable topological groups. In § 2 of [Fomi–50], Fomin shows that the following classes of groups are B-amenable: compact groups (his Theorem 3), groups containing a cocompact closed B-amenable subgroup (Theorem 4), abelian groups (Theorem 5), and solvable groups (Corollary 2). For Fomin's work in general during the 1950s, see [Ale–76].

A possible other origin of amenability could be an article [Ahlf–35] on Nevanlinna theory, where Ahlfors defines 'regularly exhaustible' open Riemann surfaces, i.e. surfaces S with a nested sequence $\Omega_1 \subset \Omega_2 \subset \cdots$ of domains with compact closures and smooth boundaries, such that $S = \bigcup_{n=1}^{\infty} \Omega_n$, and such that $\lim_{n \to \infty} \frac{\text{length}_g(\partial \Omega_n)}{\text{area}_g(\Omega_N)} = 0$, where the subscript g refers to some metric in the conformal class defined by the complex structure on S. Ahlfors has used such sequences to define averaging processes, as Følner did later with 'Følner sequences' in groups. The notion of regular exhaustion has natural formulations in Riemannian geometry; from several possible references on this subject, we will only quote [Roe–88]. The connection between Ahlfors' regular exhaustions and amenability seems rather obvious now, but we are not aware of any discussion of it in the literature before the 1980s. Apparently, the connection could only be observed after amenability was recognized as a metric property of both groups and spaces.

Though we will not discuss it here, we note that the notion of amenability has been extended to many other objects than groups, including semi-groups, associative algebras, Banach algebras, operator algebras (nuclearity, exactness, injectivity), metric spaces, equivalence relations, group actions, foliations, and groupoids.

3. Amenability, Fixed Point Property, and B-amenability

Let X be a topological space. Let $\mathcal{C}^b(X)$ denote the Banach space of **bounded continuous real-valued functions** on X, with the sup norm defined by $\|f\| = \sup_{x \in X} |f(x)|$ for $f \in \mathcal{C}^b(X)$. When X is compact, every continuous function on X is bounded, and we rather write $\mathcal{C}(X)$ for $\mathcal{C}^b(X)$.

Let G be a topological group. For $a \in G$ and $f \in \mathcal{C}^b(G)$, define $_af$ and f_a by $_af(g) = f(a^{-1}g)$ and $f_a(g) = f(ga^{-1})$ for all $g \in G$. Observe that, for a given $f \in \mathcal{C}^b(G)$, the map $G \longrightarrow \mathcal{C}^b(G)$, $a \longmapsto {}_af$ need not be continuous (example with $G = \mathbf{R}$ and $f(t) = \sin(\pi t^2)$ for all $t \in \mathbf{R}$). Let $\mathcal{C}^b_{\mathrm{ru}}(G)$ denote the subspace of $\mathcal{C}^b(G)$ of those functions f for which $_af$ depends continuously of a, i.e. the space of[1] **bounded right-uniformly**

---

[1] Here is a justification of the word 'right'. The **right-uniform structure** on G is rightfully the uniform structure with entourages of the form $U_V = \{(g, h) \in G \times G \mid hg^{-1} \in V\}$, for some neighbourhood V of 1 in G; this structure is invariant by right

continuous functions on G. This is a Banach space, indeed a closed subspace of the Banach space $\mathcal{C}^{\mathrm{b}}(G)$. [We use a^{-1} in the definition of $_af$, so that $(a, f) \longmapsto {}_af$ defines a left-action of G on $\mathcal{C}^{\mathrm{b}}(G)$ and $\mathcal{C}^{\mathrm{b}}_{\mathrm{ru}}(G)$; some authors use a rather than a^{-1}; this does not change the definition of $\mathcal{C}^{\mathrm{b}}_{\mathrm{ru}}(G)$.] Similarly, the space $\mathcal{C}^{\mathrm{b}}_{\ell u}(G)$ of **bounded left-uniformly continuous functions** on G, i.e. the space of those $f \in \mathcal{C}^{\mathrm{b}}(G)$ for which f_a depends continuously on a, is a closed subspace of $\mathcal{C}^{\mathrm{b}}(G)$. We denote by $\mathcal{C}^{\mathrm{b}}_u(G)$ the intersection $\mathcal{C}^{\mathrm{b}}_{\mathrm{ru}}(G) \cap \mathcal{C}^{\mathrm{b}}_{\ell u}(G)$.

Let E be a linear subspace of $\mathcal{C}^{\mathrm{b}}(G)$ containing the constant functions. A **mean** on E is a linear form M on E that is **positive**, i.e. $M(f) \geq 0$ whenever $f(g) \geq 0$ for all $g \in G$, and **normalized**, i.e. $M(1) = 1$, where 1 denotes both the number $1 \in \mathbf{R}$ and the corresponding constant function on G. Equivalently, a mean is a linear form on E such that

$$\inf_{g \in G} f(g) \leq M(f) \leq \sup_{g \in G} f(g) \quad \forall \ f \in E.$$

Observe that a mean M on E is bounded of norm 1. Assume moreover that E is such that $_af$ is in E for all $f \in E$ and $a \in G$. A mean M on E is **left-invariant** if $M(_af) = M(f)$ for all $f \in E$ and $a \in G$. **Right-invariant** means are defined similarly. Assume that E is such that both $_af$ and f_a are in E for all $f \in E$ and $a \in G$; a mean on E is **bi-invariant** if it is both left-invariant and right-invariant.

Proposition 3.1. *For a topological group G, the following properties are equivalent:*

(i) there exists a left-invariant mean on $\mathcal{C}^{\mathrm{b}}_{\mathrm{ru}}(G)$,
(ii) there exists a right-invariant mean on $\mathcal{C}^{\mathrm{b}}_{\ell u}(G)$,

and they imply

(iii) there exists a bi-invariant mean on $\mathcal{C}^{\mathrm{b}}_u(G)$.

Proof. For a function f defined on G, denote by $f^\vee$ the function $g \longmapsto f(g^{-1})$. Observe that $(f^\vee)^\vee = f$, and that, for $f \in \mathcal{C}^{\mathrm{b}}(G)$, we have $f \in \mathcal{C}^{\mathrm{b}}_{\mathrm{ru}}(G)$ if and only if $f^\vee \in \mathcal{C}^{\mathrm{b}}_{\ell u}(G)$.

---

multiplications. Hence a function $f : G \longrightarrow \mathbf{R}$ is **right-uniformly continuous** if, for all $\varepsilon > 0$, there exists a neighbourhood V of 1 in G such that

$$(g, h) \in U_V \quad \Rightarrow \quad |f(h) - f(g)| < \varepsilon,$$
$$\textit{i.e.} \qquad a \in V \quad \Rightarrow \quad |f(ag) - f(g)| < \varepsilon \ \forall \ g \in G,$$
$$\textit{i.e.} \qquad a \in V \quad \Rightarrow \quad \|_af - f\| < \varepsilon.$$

This is why we use 'right' here, as, for example, in [Rick–67], even though, in the last inequality, a appears on the *left* of f; other authors use 'left' at the same place [Zimm–84, page 136].

If there exists a left-invariant mean M on $C^{\mathrm{b}}_{\mathrm{ru}}(G)$, then $f \longmapsto M(f^{\vee})$ is a right-invariant mean on $C^{\mathrm{b}}_{\ell\mathrm{u}}(G)$. Hence (i) implies (ii). Similarly, (ii) implies (i).

Assume that there exist a left-invariant mean M_{ℓ} on $C^{\mathrm{b}}_{\mathrm{ru}}(G)$ and a right-invariant mean M_{r} on $C^{\mathrm{b}}_{\ell\mathrm{u}}(G)$. For $f \in C^{\mathrm{b}}_{u}(G)$, define first $F_f :$ $G \longrightarrow \mathbf{R}$ by $F_f(a) = M_{\ell}(f_a)$; then $F_f \in C^{\mathrm{b}}_{u}(G)$; define now a linear form M on $C^{\mathrm{b}}_{u}(G)$ by $M(f) = M_{\mathrm{r}}(F_f)$. Then M is a bi-invariant mean on $C^{\mathrm{b}}_{u}(G)$. This shows that (i) or/and (ii) implies (iii). $\qquad\square$

We do not know whether (iii) implies (i) and (ii). It does for locally compact groups: see 'References for the proof' after Proposition 3.10.

Definition 3.2. A topological group G is **amenable** if it has Properties (i) and (ii) of the previous proposition.

We write LCTVS as a shorthand for 'locally convex topological vector space', here on the field of real numbers. Let C be a convex subspace of an LCTVS; a transformation g of C is **affine** if

$$g(cx + (1 - c)y) = cg(x) + (1 - c)g(y)$$

for all $x, y \in C$ and $c \in \mathbf{R}$ with $0 \le c \le 1$. An action of a topological group G on C is **continuous** if the corresponding map $G \times C \longrightarrow C$ is continuous, and **affine** if $x \longmapsto g(x)$ is affine for all $g \in G$. Group actions below are **actions from the left**, unless explicitly written otherwise.

Definition 3.3. A topological group G has the **fixed point property** **(FP)** if every continuous affine action of G on a non-empty compact convex set in an LCTVS has a fixed point.

Definition 3.3 plays a fundamental role in a work of Furstenberg [Furs–63].

It is straightforward to check that compact groups have Property (FP). For abelian groups, the following fixed-point theorem appeared first in [Mark–36] and [Kaku–38]; a convenient reference is [Edwa–65, Theorem 3.2.1].

Theorem 3.4 (Markov–Kakutani theorem). *Every abelian group has Property (FP).*

The result carries over from abelian groups to solvable groups, by Proposition 4.1(3) below, and from solvable groups to topological solvable groups, obviously. Therefore, we have also:

Corollary 3.5. *Every solvable topological group has Property (FP).*

Proposition 3.6. *For a topological group G, the following three properties are equivalent:*

(i) G has Property (FP);

(ii) G is amenable;
(iii) for every non-empty compact space X and every continuous action
   of G on X, there exists a G-invariant probability measure on X.

Before the proof, we recall two standard facts from functional analysis.

Unit Balls in Duals of Banach Spaces. For a real Banach space E, we denote by E^* its Banach space dual, and E_1^* the unit ball in E^*. On E^* and E_1^*, we consider also the weak-$*$-topology, for which E^* is an LCTVS and E_1^* a compact space (this is the Banach–Alaoglu theorem).

Barycentres. Let C be a non-empty compact convex subspace of an LCTVS E, and let μ be a probability measure on C. Then there exists a unique point $b_\mu \in E$ such that $f(b_\mu) = \int_C f(x)d\mu(x)$ for every continuous linear form f on E; moreover $b_\mu \in C$. The point b_μ is called the **barycentre** of μ, or the *resultant* of μ, and μ *represents* b_μ. Recall the following formulation of the Krein–Milman theorem: every point of C is the barycentre of a probability measure on C, which is supported by the closure of the set of extreme points of C; see [Phel–66, Section 1].

Proof. (i) $\Rightarrow$ (ii) The set Mean(G) of all means on $\mathcal{C}_{ru}^b(G)$ is a subspace of the unit ball $\mathcal{C}_{ru}^b(G)_1^*$, and is closed for the weak-$*$-topology, so that Mean(G) is a compact convex set. Moreover the natural action of G on Mean(G) is affine and continuous (in the sense recalled just before Definition 3.3).

If G has Property (i), there exists a G-invariant probability measure μ on Mean(G). Such a measure has a barycentre $M \in$ Mean(G). Since M is G-invariant (by uniqueness of the barycentre), G has Property (ii).

[It is even more straightforward to check that (i) implies (iii), because if G acts on X, then G acts on the space $\mathcal{P}(X)$ of probability measures on X, that is naturally a compact convex set.]

(ii) $\Rightarrow$ (iii) We reformulate the argument of [Bogo–39], written there for $\mathcal{C}^b(G)$; the *same argument* applies equally well to $\mathcal{C}_{ru}^b(G)$, as can be read for example in [Rick–67].

Let ν be a probability measure on X. For $f \in \mathcal{C}(X)$, define a function

$$F_f : G \longrightarrow \mathbf{R}, \quad g \longmapsto \int_X f(gx)d\nu(x).$$

Then F_f is bounded and right-uniformly continuous on G. Let moreover $a \in G$. Then

$$F_{(af)}(g) = \int_X f(a^{-1}gx)d\nu(x) = F_f(a^{-1}g) = \big(a(F_f)\big)(g)$$

for all $g \in G$.

Let M be left-invariant mean on $\mathcal{C}_{ru}^b(G)$. Define a linear form

$$\mu : \mathcal{C}(X) \longrightarrow \mathbf{R}, \quad f \longmapsto M(F_f).$$

Then μ is a normalized positive linear form on $\mathcal{C}(X)$, i.e. μ can be seen as a probability measure on X. Since M is left-invariant, we have

$$\mu(_af) = M\big(F_{(_af)}\big) = M\big(_a(F_f)\big) = M(F_f) = \mu(f)$$

for all $f \in \mathcal{C}(X)$ and $a \in G$, i.e. the measure μ is G-invariant.

(iii) $\Rightarrow$ (i) Let C be a compact convex set on which G acts continuously, by affine transformations. If G has Property (iii), there exists a G-invariant probability measure μ on C. The barycentre $b_\mu \in C$ of μ is fixed by G. $\square$

Remark 3.7.   (1) The equivalence of (i) and (ii) appears in many places, for example in [Eyma–75], who quotes [Day–61] (as much as we know the first article where the arguments are found, but applied to abstract groups only) and Rickert (as much as we know the first author who applies the arguments to topological groups); see more precisely [Rick–67, Theorem 4.2].

(2) For a direct proof of (ii) $\Rightarrow$ (i), see also [BeHV–08, Theorem G.1.7].

(3) There is a natural embedding $G \longrightarrow \mathcal{C}^{\mathrm{b}}_{\mathrm{ru}}(G)^*_1$, $g \longmapsto (F \mapsto F(g))$, which is continuous for the weak-$*$-topology on the range (recall that the subscript 1 stands for 'unit ball'). By definition, the **universal equivariant compactification** $\gamma_u G$ of G is the closure of the image of this embedding, and the natural action of G on $\gamma_u G$ is continuous. The properties of Proposition 3.6 are moreover equivalent to

(iv) the natural action of G on its universal equivariant compactification $\gamma_u G$ has an invariant probability measure,

as it is shown in [BaBo–11].

(4) Finally, Properties (i) to (iv) are equivalent to

(v) every continuous action of G on the Hilbert cube has an invariant probability measure,

as has been established in [AlMP–11], and previously in [BoFe–07] for countable groups. For a countable group Γ, amenability is also equivalent to the following property [GiHa–97]:

(vi) every action by homeomorphisms of Γ on the Cantor middle-third space has an invariant probability measure.

The end of this section is devoted to a notion stronger than amenability. It was introduced by Bogolyubov [Bogo–39], without a name, and appears in [Rick–67, Definition 2.1], unfortunately with a name used most often later for the notion of our Definition 3.2; we call it B-amenability (Definition 3.8). Historically, B-amenability came before amenability for topological groups. Amenability, equivalent to Property (FP) of Definition 3.3, has been so far more often useful than B-amenability, and deserves the shorter name. (See also Remark 3.15.)

Definition 3.8. A topological group G is **B-amenable** if there exists a left-invariant mean on $\mathcal{C}^{\mathrm{b}}(G)$.

Proposition 3.9. *A B-amenable topological group is amenable.*

Proof. If G is a topological group, the space $\mathcal{C}^{\mathrm{b}}_{\mathrm{ru}}(G)$ is a G-invariant subspace of $\mathcal{C}^{\mathrm{b}}(G)$, and the restriction to this subspace of a G-invariant mean on $\mathcal{C}^{\mathrm{b}}(G)$ is a G-invariant mean on $\mathcal{C}^{\mathrm{b}}_{\mathrm{ru}}(G)$. $\square$

Propositions 3.10 and 5.2 establish that the converse holds for locally compact groups, but not in general.

Proposition 3.10. *Let G be a locally compact group. Then G is B-amenable if and only if G is amenable.*

Reference for the Proof. See [Gree–69, Theorem 2.2.1 page 26, and page 29]. Greenleaf shows there that the following properties of a locally compact group G are equivalent: existence of a left-invariant mean on each of

  (i) the space $L^{\infty}(G)$ of essentially bounded Borel measurable functions (modulo equality in complements of locally null sets),
 (ii) the space $\mathcal{C}^{\mathrm{b}}(G)$,
(iii) the space $\mathcal{C}^{\mathrm{b}}_{\mathrm{ru}}(G)$,
 (iv) the space $\mathcal{C}^{\mathrm{b}}_{u}(G)$.

Moreover, Properties (i) to (iv) are equivalent to

  (*) there exists a bi-invariant mean on E,

for E any one of the spaces in (i) to (iv) above.

Indeed, a substantial part of the early theory of amenability on locally compact groups is to show that infinitely many definitions are equivalent with each other. $\square$

Example 3.11. Compact groups are B-amenable, since the normalized Haar measure on a compact group G provides a mean on $\mathcal{C}^{\mathrm{b}}(G)$ that is both left- and right-invariant.

Proposition 3.12. *For a topological group G, the following two properties are equivalent:*

  (i) *G is B-amenable,*
 (ii) *for all $n \geq 1$, $f^{(1)}, \ldots, f^{(n)} \in \mathcal{C}^{\mathrm{b}}(G)$, $a_1, \ldots, a_n \in G$, and $t \in \mathbf{R}$, such that $f^{(1)} - a_1 f^{(1)} + \cdots + f^{(n)} - a_n f^{(n)} \geq t$, we have $t \leq 0$.*

On the Proof. The non-trivial implication is (ii) $\Rightarrow$ (i). It is a consequence of the Hahn–Banach theorem; see [Dixm–50, Théorème 1]. $\square$

This has the following consequence, for which we refer again to Dixmier [Dixm–50, Théorème 2(α)]:

Proposition 3.13. *Abelian topological groups are B-amenable.*

By Proposition 4.3, we have also the following corollary (compare with 3.5):

Corollary 3.14. *Solvable topological groups are B-amenable.*

Remark 3.15 (on terminology).   In a first version of the present article, we used 'strongly amenable' for 'B-amenable'. But this was unfortunate, because the terminology is already used for groups of which the universal minimal proximal flow is trivial, and this is a different notion, as there is an example of Furstenberg which shows that a solvable group need not be strongly amenable [Glas–76, page 28]. Strong amenability plays its role in more recent work [MeNT–16, Section 4].

The terminology 'u-amenable' and 'amenable' is used in [Harp–73], instead of 'amenable' and 'B-amenable' here.

4. Hereditary Properties

In most of this section, we address topological groups in general. However, groups are assumed to be locally compact in Proposition 4.4, and metrizable in Corollary 4.6.

A topological group G is the **directed union** of a family $(H_\alpha)_{\alpha \in A}$ of closed subgroups of G if the following conditions hold: (1) $G = \bigcup_{\alpha \in A} H_\alpha$; (2) for every $\alpha, \beta \in A$, there exists $\gamma \in A$ such that $H_\alpha \cup H_\beta \subset H_\gamma$; (3) G has the topology of the inductive limit of the H_α's. [Note that the index set A is a directed set for the preorder defined by $\alpha \leq \beta$ if $H_\alpha \subset H_\beta$.]

Proposition 4.1 (on amenability).   *Let G be a topological group.*

(1) If G is amenable, then every open subgroup of G is amenable.

(2) If G is a directed union of a family $(H_\alpha)_{\alpha \in A}$ of closed subgroups, and if each H_α is amenable, then G is amenable.

(3) If G has an amenable closed normal subgroup N such that the quotient G/N is amenable, then G is amenable.

(4) Let H be a topological group such that there exists a continuous homomorphism $H \longrightarrow G$ with dense image; if H is amenable, then so is G.

(5) If H is a dense subgroup of G, then G is amenable if and only if H, endowed with the induced topology, is amenable.

On the Proof.   For each claim, there is at least one property of Proposition 3.6 that makes the proof rather straightforward. As a sample, we check the last claim, and refer to [Rick–67, Section 4] for (1) to (4).

(5) Because right-uniformly continuous functions can be extended from H to G, the restriction of functions provides an isomorphism of Banach spaces $\mathcal{C}^b_{ru}(G) \longrightarrow \mathcal{C}^b_{ru}(H)$, by which these spaces can be identified; let us denote it (them) by $\mathcal{C}^b_{ru}$.

On the one hand, a left-G-invariant mean on $\mathcal{C}_{\mathrm{ru}}^{\mathrm{b}}$ is obviously a left-H-invariant mean on $\mathcal{C}_{\mathrm{ru}}^{\mathrm{b}}$. On the other hand, since the action of G on $\mathcal{C}_{\mathrm{ru}}^{\mathrm{b}}$ is continuous, a left-H-invariant mean on this space is also left-G-invariant. Claim (5) follows. $\qquad\square$

Remark 4.2. (1) In the first claim of Proposition 4.1, 'open' cannot be replaced by 'closed' (see Corollary 5.4). See, however, Propositions 4.4 and 4.5.

(2) Let G be a group with two Hausdorff topologies $\mathcal{T}_s, \mathcal{T}_w$ such that $G_s := (G, \mathcal{T}_s)$ and $G_w := (G, \mathcal{T}_w)$ are topological groups, and $\mathcal{T}_s$ stronger than $\mathcal{T}_w$. If G_s is amenable, then G_w is amenable, as it follows from Claim (4) applied to the continuous identity homomorphism id $: G_s \longrightarrow G_w$.

Suppose, for example, that $\mathcal{T}_s$ is the discrete topology and that G_s is amenable; this is the case if G is abelian, or more generally solvable, by Theorem 3.4. Then G_w is amenable for every topology $\mathcal{T}_w$ making G a topological group.

(3) The proof of Claim (5) *cannot* be adapted to Proposition 4.3. Indeed the analogue of Claim (5) *does not* hold for B-amenability (Remark 5.3(1)).

(4) The following result of Calvin C. Moore (1979) is reminiscent of Claim (5), but the proof uses completely different notions. Let G be the group of real points of an **R**-algebraic group, H an amenable topological group, and $\varphi : H \longrightarrow G$ a continuous homomorphism; if H is amenable and $\varphi(H)$ Zariski-dense in G, then G is amenable. We refer to [Zimm–84, Theorem 4.1.15].

Proposition 4.3 (on B-amenability). *Let G be a topological group.*

(1) If G is B-amenable, then every open subgroup of G is B-amenable.

(2) If G is a directed union of a family $(H_\alpha)_{\alpha \in A}$ of closed subgroups, and if each H_α is B-amenable, then G is B-amenable.

(3) If G has a B-amenable closed normal subgroup N such that the quotient G/N is B-amenable, then G is B-amenable.

(4) Let H be a topological group such that there exists a continuous epimorphism $H \twoheadrightarrow G$; if H is B-amenable, then so is G.

Reference for the Proof. As for Proposition 4.1, proofs are straightforward. In [Rick–67], see respectively Theorems 3.2, 2.4, 2.6, and 2.2. There are related results in the older article by Dixmier [Dixm–50]. $\qquad\square$

Amenability of closed subgroups of amenable groups is more subtle. Let us first recall the following result in the classical setting of locally compact groups.

Proposition 4.4. *Let G be a locally compact group and H a closed subgroup. If G is amenable, then so is H.*

On the Proof. We mention here two proofs of this statement.

The proof of [Zimm–84, Proposition 4.2.20] uses induction from H to G for actions on compact convex sets. Since Haar measure is an essential ingredient of induction, this cannot be used for groups that are not locally compact.

The proof of [Gree–69, Theorem 2.3.2] has three steps. We denote by $H\backslash G$ the space of H-cosets of the form Hg; we could use G/H instead, but this would impose on us right-invariant means on $\mathcal{C}^{\mathrm{b}}(\cdot)$, rather than left-invariant means.

For the first step, G is assumed to be second countable. Then there exists a Borel transversal T for $H\backslash G$, so that the multiplication map $H \times T \longrightarrow G$, $(h,t) \longmapsto ht$, is a Borel isomorphism. This can be used first to extend functions on H to functions on G, and then to show that amenability is inherited from G to H (we refer to Greenleaf's book for details). For the second step, G is assumed to be σ-compact. Then G has a compact normal subgroup N such that G/N is second countable (Kakutani–Kodaira theorem [KaKo–44]). The first step and Proposition 4.1(3) imply again that amenability is inherited from G to H. The final step makes use of the fact that every locally compact group contains an open σ-compact subgroup.

In Rickert's article, the proposition is proved with additional hypothesis only, essentially that the quotient of G by its connected component is a compact group [Rick–67, Section 7]. □

Corollary 5.4 below shows that Proposition 4.4 *does not* carry over to arbitrary topological groups. In particular this proposition is unlikely to have a completely straightforward proof.

However, with appropriate extra hypothesis, B-amenability is inherited by closed subgroups. This is shown by the following proposition, which is Theorem 3.4 in [Rick–67].

Proposition 4.5. *Let G be a topological group, and H a closed subgroup. Assume that $H\backslash G$ is paracompact and the fibration $\pi : G \longrightarrow H\backslash G$ is locally trivial.*

If G is B-amenable, then so is H.

Proof. Because $G \longrightarrow H\backslash G$ is locally trivial, there exist an open cover $(U_\iota)_{\iota \in I}$ of $H\backslash G$ and a family of continuous sections $(\sigma_\iota : U_\iota \longrightarrow G)_{\iota \in I}$ such that $\pi\sigma_\iota(x) = x$ for all $\iota \in I$ and $x \in U_\iota$. Given that $H\backslash G$ is paracompact, there is a partition of unity $(\varphi_\iota)_{\iota \in I}$ subordinate to $(U_\iota)_{\iota \in I}$. For $f \in \mathcal{C}^{\mathrm{b}}(H)$, define $F_f : G \longrightarrow \mathbf{R}$ by

$$F_f(g) = \sum_{\iota \in I} \varphi_\iota(\pi(g)) \, f\left(g(\sigma\pi(g))^{-1}\right).$$

The function F_f is well-defined, continuous because the family of the supports of the φ_ι 's is locally finite, and obviously bounded. The assignment $f \longmapsto F_f$ is a linear map from $\mathcal{C}^{\mathrm{b}}(H)$ to $\mathcal{C}^{\mathrm{b}}(G)$, that respects positivity and constant functions. Moreover, for $f \in \mathcal{C}^{\mathrm{b}}(H)$ and $a \in H$, we have

$$_a(F_f) = F_{(_a f)}.$$

Indeed

$$\left(_a(F_f)\right)(g) = \sum_{\iota \in I} \varphi_\iota(\pi(a^{-1}g)) \, f\left(a^{-1}g(\sigma\pi(a^{-1}g))^{-1}\right)$$

$$= \sum_{\iota \in I} \varphi_\iota(\pi(g)) \, f\left(a^{-1}g(\sigma\pi(a^{-1}g))^{-1}\right) = \left(F_{_a f}\right)(g)$$

for all $g \in G$.

Suppose that there exists a left-G-invariant mean M on $\mathcal{C}^{\mathrm{b}}(G)$. The assignment $m : f \longmapsto M(F_f)$ is obviously a mean on $\mathcal{C}^{\mathrm{b}}(H)$. Since M is invariant, we have

$$m(_a f) = M(F_{(_a f)}) = M(_a(F_f)) = M(F_f) = m(f)$$

for all $f \in \mathcal{C}^{\mathrm{b}}(H)$ and $a \in H$, i.e. m is a left-H-invariant mean on $\mathcal{C}^{\mathrm{b}}(H)$.   □

Corollary 4.6. *Let G be a metrizable topological group. If G contains a non-amenable discrete subgroup Γ, then G is not B-amenable.*

Proof. Since G is metrizable, so is $\Gamma \backslash G$, hence $\Gamma \backslash G$ is paracompact.

Let $p : G \longrightarrow \Gamma \backslash G$ denote the canonical projection. Let V be a neighbourhood of 1 in G such that $V^{-1} = V$ and $\Gamma \cap V^2 = \{1\}$. For every $g \in G$, the open subsets γVg, for $\gamma \in \Gamma$, are pairwise disjoint; it follows that $p^{-1}(p(Vg))$ is homeomorphic to the product $\Gamma \times Vg$, and therefore that the fibration p is locally trivial.

Hence G is not B-amenable by Proposition 4.5.   □

Concerning the hypothesis of Corollary 4.6, let us recall a theorem due to Birkhoff and Kakutani; for a topological group G, the three following conditions are equivalent:

(i) G is metrizable as a topological space;
(ii) the topology of G can be defined by a left-invariant metric;
(iii) G has a countable basis of neighbourhoods of 1.

See, for example, [BTG5-10, § 3, no 1, pages IX.23-24], or [CoHa–16].

Example 4.7. In Proposition 4.5, the property of local triviality is not automatic, as the following example, from [Karu–58], shows.

Consider the circle group $T = \mathbf{R}/\mathbf{Z}$, its subgroup C of order 2, the compact group $G = T^{\mathbf{N}}$, and the closed subgroup $H = C^{\mathbf{N}}$ of G. Then G is locally connected, as a product of (locally) connected groups, while H and $H \times (H \backslash G)$ are not locally connected (indeed H is totally disconnected). It follows that G and $H \times (H \backslash G)$ are not locally homeomorphic, and in particular that the projection $G \longrightarrow H \backslash G$ is not locally trivial.

5. Examples

Propositions 5.2, 5.5, and 5.6 provide examples of topological groups that are amenable and are not B-amenable. By necessity, these groups are not locally compact (Proposition 3.10).

Let $\mathcal{H}$ be an infinite dimensional complex Hilbert space. We denote by $\mathrm{U}(\mathcal{H})_{\mathrm{str}}$ its *unitary group*, endowed with the strong topology (equivalently: with the weak topology); as is well known, this is a topological group, and it is not locally compact. For the strong and weak topologies on sets of operators, and their properties, see e.g. [Dixm–57]. If $\mathcal{H}$ is separable, $\mathrm{U}(\mathcal{H})_{\mathrm{str}}$ is a Polish group, i.e. it is separable and its topology can be defined by a complete metric. As a curiosity, we note that, for a separable Hilbert space $\mathcal{H}$, the strong topology is the unique topology for which $\mathrm{U}(\mathcal{H})$ is a Polish group [AtKa–12].

Lemma 5.1. *Let $\mathcal{H}$ be an infinite dimensional separable complex Hilbert space, and $\mathrm{U}(\mathcal{H})_{str}$ its unitary group.*

Every countable group Γ is isomorphic to a discrete subgroup of $\mathrm{U}(\mathcal{H})_{str}$.

Proof. Suppose first that Γ is an infinite group. We can identify $\mathcal{H}$ with the Hilbert space $\ell^2(\Gamma)$ of complex-valued functions ξ on Γ such that $\sum_{\gamma \in \Gamma} |\xi(\gamma)|^2 < \infty$. Let

$$\lambda : \Gamma \longrightarrow \mathrm{U}(\ell^2(\Gamma))_{\mathrm{str}}$$

be the left-regular representation of Γ, defined by $(\lambda(\gamma)\xi)(\gamma') = \xi(\gamma^{-1}\gamma')$ for all $\gamma, \gamma' \in \Gamma$ and $\xi \in \ell^2(\Gamma)$.

For $\gamma \in \Gamma$, let $\delta_\gamma \in \ell^2(\Gamma)$ denote the unit vector defined by $\delta_\gamma(\gamma) = 1$ and $\delta_\gamma(\gamma') = 0$ if $\gamma' \neq \gamma$; observe that $\lambda(\gamma)\delta_1 = \delta_\gamma$. Set

$$U_\gamma = \left\{ g \in \mathrm{U}(\ell^2(\Gamma))_{\mathrm{str}} \ \mid \ \|(g - \lambda(\gamma))\delta_1\| < \sqrt{2}/2 \right\}.$$

On the one hand, U_γ is open in $\mathrm{U}(\ell^2(\Gamma))_{\mathrm{str}}$ and $\lambda(\gamma) \in U_\gamma$. On the other hand, since $\|\delta_\gamma - \delta_{\gamma'}\| = \sqrt{2}$ for $\gamma, \gamma' \in \Gamma$, $\gamma \neq \gamma'$, the U_γ 's are pairwise disjoint. Hence $\lambda(\Gamma)$ is a discrete subgroup of $\mathrm{U}(\ell^2(\Gamma))_{\mathrm{str}}$.

If Γ is a finite group, we can apply the previous argument to the direct product $\Gamma \times \mathbf{Z}$, and use the embedding $\Gamma \simeq \Gamma \times \{0\} \subset \mathbf{Z}$. □

Note. More generally, for every locally compact group G, the left-regular representation $G \longrightarrow \mathcal{U}(L^2(G))_{\mathrm{str}}$ is both a continuous homomorphism and a homeomorphism of G onto a closed subgroup of $\mathcal{U}(L^2(G))_{\mathrm{str}}$ [BeHV–08, Exercise G.6.4].

In general, $L^2(G)$ need not be separable. It is when G is separable.

Proposition 5.2 ([Harp–73]). *Let $\mathcal{H}$ be an infinite dimensional separable complex Hilbert space. The topological group $\mathrm{U}(\mathcal{H})_{str}$ is amenable and is not B-amenable.*

Remark. Proposition 6.3 establishes that $U(\mathcal{H})_{\mathrm{str}}$ has a property much stronger than amenability.

Proof. Let $(e_n)_{n\geq 1}$ be an orthonormal basis of $\mathcal{H}$. For each $n \geq 1$, we identify the compact Lie group $U(n)$ to the subgroup of $U(\mathcal{H})_{\mathrm{str}}$ of those unitary operators leaving invariant the linear span V_n of $\{e_1, \ldots, e_n\}$ and coinciding with the identity on the orthogonal complement $V_n^{\perp}$. Let $U(\infty)$ denote the union of the compact groups in the nested sequence

$$U(1) \subset \cdots \subset U(n) \subset U(n+1) \subset \cdots,$$

with the inductive limit topology. The group $U(\infty)$ is dense in $U(\mathcal{H})_{\mathrm{str}}$.

The later claim follows from Kaplansky's density theorem, essentially the version of [Pede-79, Theorem 2.3.3]. Pedersen formulates his Theorem 2.3.3 in terms of a C*-subalgebra A of $L(\mathcal{H})$. More generally, his proof applies without change to an involutive subalgebra A of $L(\mathcal{H})$ such that $\exp(x) \in A$ for all $x \in A$. In our case, A is the algebra spanned by the identity operator and $\bigcup_{n\geq 1} M_n$, where M_n stands for the the finite-dimensional algebra $M_n = \{x \in L(\mathcal{H}) \mid x(V_n) \subset V_n \text{ and } x(V_n^{\perp}) = \{0\}\}$.

For the inductive limit topology, the group $U(\infty)$ is a topological group. It is not locally compact; indeed, it is not a Baire group, because $U(n)$ has empty interior in $U(\infty)$ for all n. (For other topological properties of $U(\infty)$, see e.g. [Hans-71, Theorem 4.8]). Since the compact groups $U(n)$ are amenable, so is $U(\infty)$ by Proposition 4.1(2). It follows from Proposition 4.1(5) that $U(\mathcal{H})_{\mathrm{str}}$ is amenable.

By Lemma 5.1, the group $U(\mathcal{H})_{\mathrm{str}}$ contains non-amenable discrete subgroups. It follows from Corollary 4.6 that $U(\mathcal{H})_{\mathrm{str}}$ is not B-amenable.

The proof of the proposition is complete. Let us, however, reproduce the argument of [Harp-73], that is a different proof that $U(\mathcal{H})_{\mathrm{str}}$ is not B-amenable.

Suppose ab absurdo that there exists a left-invariant mean M on the space $C^b(U(\mathcal{H})_{\mathrm{str}})$. Let ξ be a unit vector in $\mathcal{H}$. For every bounded operator S on $\mathcal{H}$, the function

$$f_{S,\xi} : \begin{cases} U(\mathcal{H})_{\mathrm{str}} & \longrightarrow & \mathbf{R} \\ g & \longmapsto & \mathrm{Re}\left(\langle g^{-1}Sg\xi \mid \xi\rangle\right) \end{cases}$$

is bounded and continuous. Let $L(\mathcal{H})$ denote the algebra of all bounded operators on $\mathcal{H}$. Observe that, for $h \in U(\mathcal{H})_{\mathrm{str}}$, we have

$$_h f_{S,\xi}(g) = \mathrm{Re}\left(\langle g^{-1}hSh^{-1}g\xi \mid \xi\rangle\right)$$

for all $g \in U(\mathcal{H})_{\mathrm{str}}$, i.e. $_h f_{S,\xi} = f_{hSh^{-1},\xi}$.

Consider the linear form

$$\tau_\xi : \begin{cases} L(\mathcal{H}) & \longrightarrow & \mathbf{R} \\ S & \longmapsto & M(f_{S,\xi}). \end{cases}$$

Because M is left-invariant, we have, for all $S \in L(\mathcal{H})$ and $h \in U(\mathcal{H})_{str}$,

$$\tau_\xi(hSh^{-1}) = M(f_{hSh^{-1},\xi}) = M(_hf_{S,\xi}) = M(f_{S,\xi}) = \tau_\xi(S),$$

and therefore also $\tau_\xi(Sh) = \tau_\xi(hS)$.

Every operator in $L(\mathcal{H})_{str}$ is a linear combination of unitaries.[2] Hence $\tau_\xi(ST) = \tau_\xi(TS)$ for all $S, T \in L(\mathcal{H})$. Because the identity operator is a sum of two commutators, i.e. since $id_\mathcal{H}$ is of the form $S_1 T_1 - T_1 S_1 + S_2 T_2 - T_2 S_2$ (see e.g. Problem 186 in [Halm–67]), we have $\tau_\xi(id_\mathcal{H}) = 0$. But this is preposterous, because $\tau_\xi(id_\mathcal{H}) = M(1) = 1$. Hence $U(\mathcal{H})_{str}$ is not B-amenable. $\square$

Remark 5.3. (1) Let the notation be as in the proof above. Because the group $U(\infty)$ is B-amenable (by Claim (2) of Proposition 4.3) and dense in $U(\mathcal{H})_{str}$, Proposition 5.2 justifies Remark 4.2(3).

(2) Proposition 5.2 has the following offspring. Let $\mathcal{M}$ be a von Neumann algebra, realized as a weakly closed $*$-subalgebra of $L(\mathcal{H})$, for some separable Hilbert space $\mathcal{H}$. Let $U(\mathcal{M})_{str}$ be its unitary group, with the strong topology. Then $U(\mathcal{M})_{str}$ is amenable if and only if $\mathcal{M}$ is injective [Harp–79]. For a C*-algebra A, there is a similar characterization of nuclearity of A in terms of amenability of the unitary group $U(A)$ of A, with the norm topology [Pate–92].

(3) Consider the Banach algebra $L(\mathcal{H})$ of all bounded operators on $\mathcal{H}$, with the usual operator norm. Let $GL(\mathcal{H})_{norm}$ be its general linear group, with its topology as an open subset of $L(\mathcal{H})$. We denote by $U(\mathcal{H})_{norm}$ the unitary group of $\mathcal{H}$, with the topology induced by the operator norm; it is a closed subgroup of $GL(\mathcal{H})_{norm}$. The groups $U(\mathcal{H})_{norm}$ and $GL(\mathcal{H})_{norm}$ are not amenable. Also, the group $GL(\ell^p)_{norm}$ is not amenable, for p with $1 \leq p < \infty$. For this (and more), see [Harp–73] and [Pest–06, Examples 3.6.13-15].

From the proof of Proposition 5.2, let us extract the following observation. It appeared in [Harp–82, page 489].

Corollary 5.4. *A closed subgroup of an amenable topological group need not be amenable.*

Let $Sym(\mathbf{N})$ denote the full *symmetric group* of the positive integers, with its standard Polish topology (any infinite countable set would do instead of $\mathbf{N}$). There is a curiosity similar to that noted just before Lemma 5.1: the group $Sym(\mathbf{N})$ has a unique topology making it a Polish group [Gaug–67, Kall–79], indeed a unique topology making it

---

[2] Let A be a C*-algebra with unit. Every $x \in A$ is a linear combination of four unitaries. Indeed, since $x = \frac{1}{2}(x + x^*) + \frac{1}{2i}(ix - ix^*)$, it is enough to check that a self-adjoint element of norm at most 1 is a linear combination of two unitaries. If $x^* = x$ and $\|x\| \leq 1$, then $u = x + i\sqrt{1 - x^2}$ and $u^* = x - i\sqrt{1 - x^2}$ are unitary, and $x = \frac{1}{2}(u + u^*)$.

a separable Hausdorff group [KeRo-07, Theorem 1.11]. We denote by $\mathrm{Sym}_f(\mathbf{N})$ the subgroup of $\mathrm{Sym}(\mathbf{N})$ of permutations with finite support; it is a locally finite group, dense in $\mathrm{Sym}(\mathbf{N})$.

As in Proposition 5.2, we have:

Proposition 5.5. *The topological group* $\mathrm{Sym}(\mathbf{N})$ *is amenable and is not B-amenable.*

Proof. The proof is analogous to that of Proposition 5.2. On the one hand, since $\mathrm{Sym}(\mathbf{N})$ contains a dense subgroup that is locally finite and therefore amenable, $\mathrm{Sym}(\mathbf{N})$ is amenable by Proposition 4.1. On the other hand, $\mathbf{N}$ can be identified (as a set) with a non-amenable countable group Γ, and the metrizable group $\mathrm{Sym}(\Gamma)$ contains Γ as a discrete closed subgroup, so that $\mathrm{Sym}(\mathbf{N}) \simeq \mathrm{Sym}(\Gamma)$ is not B-amenable, by Corollary 4.6. $\square$

Let V be a vector space over a finite field $\mathrm{GF}(q)$. Assume that the dimension of V is infinite and countable. Observe that V is a countable infinite set, and that the *general linear group* $\mathrm{GL}(V)$ is a subgroup of $\mathrm{Sym}(V) \simeq \mathrm{Sym}(\mathbf{N})$.

Moreover, each of the equations

$$g(0) = 0, \quad \text{with } 0 \text{ the origin of } V,$$
$$g(\lambda v) = \lambda g(v), \quad \text{with } \lambda \in \mathrm{GF}(q)^\times \text{ and } v \in V,$$
$$g(v + w) = g(v) + g(w), \quad \text{with } v, w \in V,$$

defines a closed subset of $\mathrm{Sym}(V)$. Hence $\mathrm{GL}(V)$ is a closed subgroup of $\mathrm{Sym}(V)$; in particular $\mathrm{GL}(V)$ itself is a Polish group.

Proposition 5.6. *The topological group* $\mathrm{GL}(V)$ *is amenable and is not B-amenable.*

Proof. This is one more variation on the same proof, as for Propositions 5.2 and 5.5. Let $(e_n)_{n \in \mathbf{N}}$ be a basis of V. For every $n \in \mathbf{N}$, denote by V_n the linear span of $\{e_1, e_2, \ldots, e_n\}$, and by GL_n the subgroup of $\mathrm{GL}(V)$ consisting of those elements g such that $g(V_n) = V_n$ and $g(e_k) = e_k$ for all $k > n$. We have a nested sequence

$$\mathrm{GL}_1 \subset \cdots \subset \mathrm{GL}_n \subset \mathrm{GL}_n \subset \cdots \subset \mathrm{GL}_f := \bigcup_{n=1}^{\infty} \mathrm{GL}_n$$

of which the union GL_f is locally finite, in particular amenable, and dense in $\mathrm{GL}(V)$. Hence $\mathrm{GL}(V)$ is amenable.

The space V has a basis $(e_\gamma)_{\gamma \in F}$ indexed by a non-abelian free group F. Each $\gamma_0 \in F$ can be viewed as an element of $\mathrm{GL}(V)$ mapping e_γ to $e_{\gamma_0 \gamma}$ for all $\gamma \in F$. This shows that $\mathrm{GL}(V)$ contains a discrete closed subgroup isomorphic to F, and in particular that $\mathrm{GL}(V)$ is not B-amenable. $\square$

Example 5.7. Let **k** be a commutative ring, with unit. Denote by $\mathcal{J}(\mathbf{k})$ the *substitution group of formal power series over* **k**, with elements of the form $f(x) = x + \sum_{i \geq 2} a_i x^i$, where $a_i \in \mathbf{k}$, and with substitution for the group law. For what follows, and much more, about groups of this kind, see [Babe–13].

For each $n \geq 1$, denote by $\mathcal{J}^{n+1}(\mathbf{k})$ the normal subgroup of $\mathcal{J}(\mathbf{k})$ defined by the equations $a_2 = a_3 = \cdots = a_{n+1} = 0$ and by $\mathcal{J}_n(\mathbf{k})$ the quotient $\mathcal{J}(\mathbf{k})/\mathcal{J}^{n+1}(\mathbf{k})$. There is a natural bijection between $\mathcal{J}_n(\mathbf{k})$ and $\mathbf{k}^n$. When **k** is a topological ring, we use this bijection to define a topology on $\mathcal{J}_n(\mathbf{k})$; it is a group topology. Then $\mathcal{J}(\mathbf{k})$ is also a topological group, with the topology of the inverse limit $\varprojlim_n \mathcal{J}_n(\mathbf{k})$. This topology is interesting even if the ring **k** is discrete; note that the group $\mathcal{J}(\mathbf{k})$ is profinite when the ring **k** is finite.

It is known that, for every topological commutative ring **k**, the group $\mathcal{J}(\mathbf{k})$ is amenable [BaBo–11]. According to [Babe–13, page 61], it is not known whether the group $\mathcal{J}(\mathbf{Z})$ is B-amenable.

6. Extreme Amenability

Several of the examples we know of topological groups that are amenable and not locally compact have a property stronger than amenability: that of extreme amenability. The notion appeared in the mid 1960s. The subject became more important with later articles, such as those of Gromov–Milman [GrMi–83], written in the late 1970s, and Kechris–Pestov–Todorcevic [KePT–05]. For indications on the development of the subject, see the introduction of [Pest–06].

Definition 6.1. A topological group G is **extremely amenable** if every continuous action of G on a compact space has a fixed point.

In the next proposition, Items (1) to (4) are reminiscent of similar items in Propositions 4.1 and 4.3.

Proposition 6.2 (hereditary properties of extreme amenability). *Let* G *be a topological group.*

(0) If G is extremely amenable, then G is amenable.

(1) If G is extremely amenable, every open subgroup of G is extremely amenable.

(2) If G is a directed union of a family $(H_\alpha)_{\alpha \in A}$ of closed subgroups, and if each H_α is extremely amenable, then G is extremely amenable.

(3) If G has an extremely amenable closed normal subgroup N such that the quotient G/N is extremely amenable, then G is extremely amenable.

(4) Let H be a topological group such that there exists a continuous homomorphism H $\longrightarrow$ G with dense image; if H is extremely amenable, then so is G.

(5) A closed subgroup of an extremely amenable group NEED NOT be amenable.

Proof. Claims (0) and (4) are straightforward. Claim (1) appears in [BoPT–11, Lemma 13].

(2) Let X be a non-empty compact G-space. For each $\alpha \in A$, the set X^α of H_α-fixed points is closed and therefore compact in X, and non-empty by hypothesis on H_α. The intersection $\bigcap_{\alpha \in A} X^\alpha$ is non-empty, by compactness of X. Because this intersection coincides with the set of G-fixed points in X, the group G is extremely amenable.

For (3), see [Pest–06, Corollary 6.2.10].

A smart example confirming (5) is the group Aut(**Q**) of order-preserving permutations of the set **Q** of rational numbers (see next proposition), which contains a non-abelian free discrete subgroup [Pest–98, Theorem 8.1]. Note that a non-trivial locally compact closed subgroup of an extremely amenable group *is not* extremely amenable, see Proposition 6.5 below.   □

Proposition 6.3.   *The following groups are extremely amenable.*

(1) The unitary group U($\mathcal{H}$)$_{str}$ *of an infinite dimensional separable complex Hilbert space $\mathcal{H}$, with the strong topology.*

(2) The group Aut(**Q**) *of order-preserving permutations of* **Q**, *with the topology of pointwise convergence on the discrete space* **Q**.

(3) The group $\mathcal{H}^+([0,1])$ of orientation-preserving homeomorphisms of the closed unit interval, with the compact-open topology.

(4) The isometry group of the Urysohn space.

(5) The group Aut(X, μ) *of all measure-preserving automorphisms of a standard non-atomic finite or infinite and sigma-finite measure space, with the weak topology.*

(6) The group $L^0(X, \mathcal{B}, \mu; G)$ of all measurable maps from a Lebesgue space with a non-atomic probability measure $(X, \mathcal{B}, \mu)$ to a second-countable compact group G, up to equality μ-almost everywhere, with the topology of convergence in measure.

References for the Proof.   The case of U($\mathcal{H}$)$_{str}$ is shown in [GrMi–83]; for the translation of their result in terms of Definition 6.1; see, for example, [Pest–06, Section 2.2]. For Aut(**Q**) and $\mathcal{H}^+([0,1])$, see [Pest–98].

For the Urysohn space and its isometry group, see [Pest–06], in particular Theorem 5.3.10 (the isometry group of the Urysohn space, with its standard Polish topology, is a Levy group) and Theorem 4.1.3 (every Levy group is extremely amenable).

For Aut(X, μ), see [GiPe–07, Theorem 4.2]. The case of $L^0(X, \mathcal{B}, \mu; G)$ is due to Eli Glasner [Glas–98] and Furstenberg–Weiss [unpublished]; see [Pest–06, Section 4.2].

For some of these groups, extreme amenability can also be shown by the arguments of [Mel's–13]. □

Remark 6.4. Propositions 5.2 and 6.3(1) show that an extremely amenable group need not be B-amenable.

Concerning (1), recall on the one hand that the group U$(\mathcal{H})_{\mathrm{str}}$ is known to have Kazhdan's Property (T) [Bekk–03]. On the other hand, an amenable *locally compact* group which has Property (T) is compact.

Concerning (2), note that not only Aut$(\mathbf{Q})$ with the indicated topology is extremely amenable, but moreover every action by homeomorphisms of the group Aut$(\mathbf{Q})$ with the *discrete topology* on a compact *metrizable space* has a fixed point [RoSo–07, Corollary 7].

Concerning (3), recall that Thompson's group F is a dense subgroup of $\mathcal{H}^+([0, 1])$. The only consequence of (3) we can state is that is does not exclude that F is amenable. (This is a repetition of Remark 12 of [Pest–02].)

For the next proposition, we use the following notation.

Sym$(\mathbf{N})$ is the symmetric group of $\mathbf{N}$, with its standard Polish topology, as in Proposition 5.5.

Let p be a positive number with $1 \le p < \infty$ and $p \ne 2$. Let ℓ^p be the Banach space of sequence $(z_n)_{n \ge 1}$ of complex numbers such that $\|z\| := \left(\sum_{n \ge 1} |z_n|^p\right)^{1/p} < \infty$. We denote by U$(\ell^p)$ the group of linear isometries of ℓ^p, with the strong topology.

Proposition 6.5. *The following groups are NOT extremely amenable.*

(1) *Any locally compact group $G \ne \{1\}$.*
(2) *The symmetric group* Sym$(\mathbf{N})$.
(3) *The group $\mathcal{H}(C)$ of homeomorphisms of the Cantor space, with the compact-open topology.*
(4) *The unitary group* U(ℓ^p), *for p with $1 \le p < \infty$ and $p \ne 2$.*
(5) *The group* GL(V), *for $V \simeq$ GF$(q)^{(\mathbf{N})}$ as in Proposition 5.6.*

References for the Proof. Every locally compact group admits a free action on a suitable compact space [Veec–77]; and (1) follows.

For (2) and Sym$(\mathbf{N})$, see [Pest–98], or [Pest–06, Section 2.4]. As Glasner and Weiss have observed, the natural action of Sym$(\mathbf{N})$ on the compact space LO $\subset \mathbf{N}^2$ of all linear orders on $\mathbf{N}$ has no fixed point; details in [Pest–06, Example 2.4.6, page 47].

For (3), observe that the action of $\mathcal{H}(C)$ on C has no fixed point. For more on actions of this group on compact spaces, see [GlWe–03].

For (4), let us reproduce the argument of [Pest–06, Example 3.6.15]. Let $(e_n)_{n \in \mathbf{N}}$ be the canonical basis of ℓ^p. For every sequence $(t_n)_{n \in \mathbf{N}}$

in the compact group $\prod_{n\in\mathbf{N}} T_n$, where each T_n is a copy of $T = \{z \in \mathbf{C} \mid |z| = 1\}$, and every $\sigma \in \mathrm{Sym}(\mathbf{N})$, there is an isometry of ℓ^p mapping e_n to $t_n e_{\sigma(n)}$ for all $n \in \mathbf{N}$. Moreover, every isometry of ℓ^p is of this form; the proof is like that of Banach, for the ℓ^p-space of real sequence and for $t_n \in \{-1, 1\}$ for all n [Bana–32, Chap. XI, § 5, page 178]. In other terms, the unitary group $\mathrm{U}(\ell^p)$ of ℓ^p is a semi-direct product

$$\mathrm{U}(\ell^p) = \left(\prod_{n\in\mathbf{N}} T_n\right) \rtimes \mathrm{Sym}(\mathbf{N}).$$

Now Claim (4) follows from (2) and Proposition 6.2(4).

For (5), see [Pest–06, Example 6.7.17].   □

To conclude this section, we quote a result that extends Proposition 6.3(1). Compare with the way Remark 5.3(2) extends part of Proposition 5.2.

Proposition 6.6. *A countably decomposable von Neumann algebra M is injective if and only if its unitary group $\mathrm{U}(M)$, with the weak topology, is the direct product of a compact group and an extremely amenable group.*

In particular, an infinite dimensional factor M is injective if and only if $\mathrm{U}(M)$ is extremely amenable, and the same holds for M a properly infinite von Neumann algebra.

References for the proof:. [GiPe–07, Theorem 3.3] and [GiNg–13].   □

7. On the Definition of Ergodicity

Let us agree on the following terminology. Consider a Borel action of a topological group G on a Borel space $(X, \mathcal{B})$. A Borel subset $A \subset X$ is **invariant** by G if $gA = A$ for all $g \in G$.

Assume that we have moreover a G-invariant probability measure μ on $(X, \mathcal{B})$. A Borel subset $A \subset X$ is μ-**essentially invariant** by G if $\mu(gA \,\Delta\, A) = 0$ for all $g \in G$ (where Δ indicates a symmetric difference). For 'μ-essentially invariant', Varadarajan uses 'μ-invariant' [Vara–63, page 196], Maitra 'μ-almost invariant' [Mait–77], and Phelps 'invariant (mod μ)' [Phel–66].

Definition 7.1. Let G be a topological group acting as above on a Borel space $(X, \mathcal{B})$ with a G-invariant probability measure μ.

The action is **w-ergodic** if, for every invariant set $A \in \mathcal{B}$, either $\mu(A) = 0$ or $\mu(X \smallsetminus A) = 0$; in this situation, we say also that the invariant measure μ is w-ergodic.

The action is **s-ergodic** if, for every μ-essentially invariant set $A \in \mathcal{B}$, either $\mu(A) = 0$ or $\mu(X \smallsetminus A) = 0$; in this situation, we say also that the invariant measure μ is s-ergodic.

Some authors use 'ergodic' for our 'w-ergodic' (see, for example, [Zimm–84, beginning of Chapter 2]), others use 'ergodic' for our 's-ergodic' (see, for example, [Phel–66, Chapter 10]). Proposition 7.7 shows that, in a standard setting, the two notions coincide.

Remark 7.2. Consider a Borel action of a topological group G on a Borel space $(X, \mathcal{B})$, and a G-invariant probability measure μ on X. Assume that μ is s-ergodic. Then every Borel subset A in X such that $\mu(A) = 1$ is μ-essentially invariant by G.

Indeed, let $g \in G$. The subset $g(A) \smallsetminus (A \cap g(A))$ is contained in $X \smallsetminus A$, hence is negligible for μ. Similarly, $A \smallsetminus (A \cap g(A)) = g(g^{-1}(A) \smallsetminus (A \cap g^{-1}(A)))$ is negligible for μ, because μ is G-invariant. Hence $\mu(A \triangle g(A)) = 0$ for all $g \in G$.

Suppose moreover that X is a compact space, and that $\mathcal{B}$ is the σ-algebra of Borel subsets of X, i.e. the σ-algebra of subsets of X generated by the open subsets of X. The space $\mathcal{P}(X)$ of probability measures on X is a compact convex subspace of the dual space of $\mathcal{C}(X)$, with the weak-$*$-topology, and the space $\mathcal{P}^G(X)$ of G-invariant probability measures is a compact convex subspace of $\mathcal{P}(X)$.

Definition 7.3. With the notation above, a G-invariant measure $\mu \in \mathcal{P}^G(X)$ is **indecomposable** if it is in the subset of extreme points $\mathcal{E}^G(X)$ of $\mathcal{P}^G(X)$.

In other words, μ is **decomposable** if there exist two distinct measures $\mu_1, \mu_2 \in \mathcal{P}^G(X)$ and a constant c with $0 < c < 1$ such that $\mu = c\mu_1 + (1 - c)\mu_2$.

The following proposition appears in [Bogo–39].

Proposition 7.4. *Let G be a topological group acting continuously by homeomorphisms on a compact space X, and let μ be a G-invariant probability measure on X. The following two properties are equivalent:*

(i) μ is indecomposable,
(ii) μ is s-ergodic;

and imply the following third property

(iii) μ is w-ergodic.

Note. We will add another property equivalent to (i) and (ii) in Proposition 9.3.

It is recalled below that (iii) does imply (ii) when G is second countable locally compact (Proposition 7.7), but not in general (Proposition 8.1(3)).

Proof. For (i) $\Leftrightarrow$ (ii), we reproduce the proof of [Bogo–39]. This proof has remained the standard one: see, for example, [Walt–82, Theorem 610]. Walters writes his proof for one continuous map $T : X \longrightarrow X$, but it works without change in the present situation. For (ii) $\Rightarrow$ (i), see also [Phel–66, Proposition 10.4] (Proposition 12.4 of the Second Edition).

(i) $\Rightarrow$ (ii) We prove the contraposition. Assume that there exists a μ-essentially G-invariant subset A of X with $0 < \mu(A) < 1$. Then $\mu = \mu(A)\mu_1 + (1 - \mu(A))\mu_2$, with

$$\mu_1(B) = \frac{1}{\mu(A)}\mu(B \cap A), \quad \mu_2(B) = \frac{1}{1 - \mu(A)}\mu(B \cap (X \smallsetminus A)),$$

for all Borel subsets B of X.

(ii) $\Rightarrow$ (i) Assume (again by contraposition) that there exist two distinct G-invariant probability measures μ_1, μ_2 of which $\mu = c_1\mu_1 + c_2\mu_2$ is a convex combination. For every Borel subset B of X with $\mu(B) = 0$, we have $\mu_1(B) = 0 = \mu_2(B)$, i.e. μ_1, μ_2 are absolutely continuous with respect to μ. By the Radon–Nikodym theorem (one of many convenient references is [Rudi–66, Theorem 6.9]), there exist well-defined and unique functions $f_1, f_2 \in L^1(X, \mu)$, with non-negative values, such that $\mu_1 = f_1\mu$ and $\mu_2 = f_2\mu$. By uniqueness, f_1 and f_2 are G-invariant. Set

$$A = \{x \in X \mid f_1(x) > f_2(x)\} \text{ and } B = \{x \in X \mid f_1(x) \le f_2(x)\}.$$

Then A, B are μ-essentially G-invariant Borel subsets of X of positive measure, and constitute a partition of X.

The implication (ii) $\Rightarrow$ (iii) is trivial. $\square$

Definition 7.5. Let $(\Omega, \mathcal{B})$ be a Borel space; assume that it is a standard Borel space, i.e. that there exist an isomorphism of $(\Omega, \mathcal{B})$ with a Borel subset of a complete separable metric space. Let G be a topological group acting in a Borel way on $(\Omega, \mathcal{B})$; assume that there exists a probability measure μ on $(\Omega, \mathcal{B})$ which is a G-**quasi-invariant**, i.e. such that, for all $A \subset \mathcal{B}$ and $g \in G$, we have $\mu(A) = 0$ if and only if $\mu(gA) = 0$.

Let $(Y, \mathcal{C})$ be another standard Borel space (the most important case here is $Y = \mathbf{R}$, with the usual Borel σ-algebra). A Borel function $f : \Omega \longrightarrow Y$ is μ-**essentially** G-**invariant** if, for each $g \in G$, we have $f(g\omega) = f(\omega)$ for μ-almost all $\omega \in \Omega$; it is G-**invariant** if, for each $g \in G$, we have $f(g\omega) = f(\omega)$ for all $\omega \in \Omega$.

Lemma 7.6 (Lemma 3.3 in [Vara–63], or Lemma 2.2.16 in [Zimm–84]). *Let G be a second countable locally compact group acting in a Borel way on a standard Borel space Ω. Assume that Ω has a G-quasi-invariant probability measure μ.*

Let Y be a standard Borel space and $f : \Omega \longrightarrow Y$ be a Borel function. Assume that f is essentially G-invariant. Then there exists a Borel function $\tilde{f} : \Omega \longrightarrow Y$ which is G-invariant, and $\tilde{f}(\omega) = f(\omega)$ for μ-almost all $\omega \in \Omega$.

In particular, if $A \subset \Omega$ is a μ-essentially G-invariant Borel subset, there exists a G-invariant Borel set $\tilde{A} \subset \Omega$ such that $\mu(\tilde{A}\triangle A) = 0$.

On the Proof. Note that the particular case follows from the result on functions, with $Y = \mathbf{R}$ and f the characteristic function of A.

It is crucial here that G is locally compact (hence G has a Haar measure) and 'not too large' (more precisely 'second countable' in [Zimm–84] and [Vara–63]), because the proof relies very strongly on Fubini's theorem, and Fubini's theorem holds for Haar measure with appropriate conditions only.   □

Proposition 7.7. *Let G be a second countable locally compact group acting in a Borel way on a standard Borel space Ω. Assume that Ω has a G-quasi-invariant probability measure μ.*

Then the action is s-ergodic if and only if it is w-ergodic.

Proof. The proposition follows from the lemma and the definitions.   □

We quote now the following decomposition theorem. It can be seen as an elaboration of results going back to von Neumann, in the early 1930s. Two convenient references are [Vara–63], for invariant measures, and [GrSc–00], for quasi-invariant measures.

Theorem 7.8 (Ergodic Decomposition Theorem). *Let G be a second countable locally compact group acting in a Borel way on a standard Borel space $(X, \mathcal{B})$; assume that $\mathcal{P}(X)^G$ is non-empty (equivalently that $\mathcal{E}(X)^G$ is non-empty). Denote by $(Y, \mathcal{C})$ the standard Borel space with $Y = \mathcal{E}(X)^G$ and $\mathcal{C}$ its Borel σ-algebra.*

Then there exists a family $(p_y)_{y \in Y}$ of probability measures on Y, with the following properties:

(1) for every Borel subset B in X, the map $y \longmapsto p_y(B)$ from Y to $[0, 1]$ is Borel;

(2) for every $y \in Y$, the measure p_y is G-invariant and ergodic;

(3) for $y, y' \in Y$ with $y \neq y'$, the measures p_y and $p_{y'}$ are mutually singular;

(4) for every $\mu \in \mathcal{P}(X)^G$, there exists a probability measure ν on $(Y, \mathcal{C})$ such that $\mu(B) = \int_Y p_y(B) d\nu(y)$ for every $B \in \mathcal{B}$.

Note: A fortiori, the theorem holds for a continuous action of G on a metrizable compact space.

Corollary 7.9. *Let G be a second-countable locally compact group acting continuously by homeomorphisms on a metrizable compact space X.*

For every $\mu \in \mathcal{E}^G(X)$, there exists a Borel subset A_μ of X such that

$$\mu(A_\mu) = 1 \quad \text{and} \quad \mu'(A_\mu) = 0 \text{ for all } \mu' \in \mathcal{E}^G(X) \text{ with } \mu' \neq \mu.$$

On the Proof. This follows from a version of the Ergodic Decomposition Theorem more comprehensive than that of Theorem 7.8, see [GrSc–00].

Theorem 1.4 of [GrSc–00] is such a theorem for a *countable* discrete group, say Γ. It is shown there that

(a) the sub-σ-algebra $\mathcal{T}$ of $\mathcal{B}$ consisting of Γ-invariant sets is countably generated, say generated by a countable sub-algebra $\mathcal{T}'$;

(b) for every $x \in X$, the set $[x]_\mathcal{T} = \bigcap C$, where the intersection is over all $C \in \mathcal{T}'$ with $x \in C$, is a Borel subset of X (and it coincides with the intersection $\bigcap C$ over all $C \in \mathcal{T}$ with $x \in C$);

(c) $[x]_\mathcal{T} \in \mathcal{T}$, i.e. $[g(x)]_\mathcal{T} = [x]_\mathcal{T}$ for all $x \in X$ and $g \in \Gamma$;

(d) there exist a Γ-invariant Borel subset $X_0 \subset X$, with $\eta(X \setminus X_0) = 0$ for all $\eta \in \mathcal{P}^\Gamma(X)$, and a surjective map $p : X_0 \longrightarrow Y_\Gamma$, $x \longmapsto p_x$, measurable with respect to $\mathcal{T}$ and $\mathcal{C}$, such that

$$p_x\left([x]_\mathcal{T}\right) = 1,$$
$$p_{x'}\left([x]_\mathcal{T}\right) = 0,$$

for all $x, x' \in X_0$ with $p(x') \neq p(x)$.

Then, for $\mu \in \mathcal{E}^\Gamma(X)$, it suffices to set:

(e) $A_\mu = [x]_\mathcal{T}$, with $x \in X_0$ such that $p(x) = \mu$.

[Note that, in [GrSc–00], the set $[x]_\mathcal{T}$ is defined as an intersection over $C \in \mathcal{T}$ with $x \in C$; but $[x]_\mathcal{T}$ is Borel because this is also an intersection over C in the countable subalgebra $\mathcal{T}'$. We are grateful to K. Schmidt for an e-mail clarifying this point for us.]

In the more general case of a second-countable locally compact group G, we refer to [GrSc–00, Theorem 5.2]. Choose a countable dense subgroup Γ of G. Define X_0 as above, in terms of Γ and, for $x \in X_0$, define also p_x and $[x]_\mathcal{T}$ in terms of Γ. Then we have:

(f) $[g(x)]_\mathcal{T} = [x]_\mathcal{T}$ for all $g \in G$ and $x \in X$ (compare with (c));

(g) $\mathcal{E}^G(X) = \mathcal{E}^\Gamma(X)$;

(h) $p_x \in Y = \mathcal{E}^G(X)$ (compare with (d));

(i) $p_x\left([x]_\mathcal{T}\right) = 1$ and $p_{x'}\left([x]_\mathcal{T}\right) = 0$ for all $x, x' \in X_0$ with $p(x') \neq p(x)$ (as in (d)).

[The sets $[x]_\mathcal{T}$ may depend on the choice of Γ, but we will not discuss this further here.]

The article [GrSc–00] is written for the case of quasi-invariant measures. For the particular case of invariant measures, as discussed in this article, we could equally have quoted an earlier article by Varadarajan. Specifically, we refer to [Vara–63, Theorem 4.2] for the fact that our Borel space $(Y, \mathcal{C})$ is standard. Our sets A_μ, with $\mu \in \mathcal{E}^G(X)$, correspond to the sets X_e, with e in the space $\mathcal{J}$ of ergodic G-invariant measures, in [Vara–63]. □

We end this section with a simpler version of the previous corollary, for comparison with Propositions 8.2 and 8.3 below.

Corollary 7.10. *Let G be a second-countable locally compact group acting continuously by homeomorphisms on a metrizable compact space X. Assume that there exist two distinct ergodic probability measures $\mu_1, \mu_2 \in \mathcal{E}^G(X)$.*

Then there exists two G-invariant Borel subsets $A_1, A_2 \subset X$ such that

$$\begin{aligned} \mu_1(A_1) &= 1, & \mu_1(A_2) &= 0 \\ \mu_2(A_1) &= 0, & \mu_2(A_2) &= 1. \end{aligned} \tag{9.1}$$

Proof. This is a consequence of Theorem 7.8, since, on the one hand, two distinct measures in $\mathcal{E}^G(X)$ are mutually singular, and, on the other hand, two measures μ_1 and μ_2 are mutually singular precisely when there exist two Borel subsets A_1, A_2 in X for which the equalities of (9.1) hold. $\qquad\square$

8. Kolmogorov's Example

The example of this section is due to Kolmogorov. It appears in [Fomi–50]; it has been revisited in [Vara–63], with reference to Kolmogorov, and in [Mait–77], without. Note that Fomin was a student of Kolmogorov [Ale–76].

Consider a number p with $0 \le p \le 1$ and the measure λ_p on $Y := \{0,1\}$ defined by $\lambda_p(0) = p$ and $\lambda_p(1) = 1 - p$. For $i \in \mathbf{Z}$, let $(Y_i, \lambda_{p,i})$ be a copy of (Y, λ_p). Let $X = \prod_{i \in \mathbf{Z}} X_i = \{0,1\}^{\mathbf{Z}}$ be the product of the Y_i's, and $\mathcal{B}$ the usual σ-algebra; $(X, \mathcal{B})$ is a standard Borel space. Let μ_p be the product probability measure $\prod_{i \in \mathbf{Z}} \lambda_{p,i}$, which is called a Bernoulli measure. We denote by $S : X \longrightarrow X$ the corresponding **Bernoulli shift**, defined by $(Sx)_i = x_{i+1}$ for all $i \in \mathbf{Z}$; observe that the measure μ_p is preserved by S. Recall that the transformation S of (X, μ_p) is s-ergodic, indeed strongly mixing; see e.g. [Walt–82, Theorems 1.12 and 1.30].

Denote as in Proposition 5.5 by $\mathrm{Sym}(\mathbf{Z})$ the full symmetric group of $\mathbf{Z}$, with its standard Polish topology, and by $\mathrm{Sym}_f(\mathbf{Z})$ the subgroup of $\mathrm{Sym}(\mathbf{Z})$ of permutations with finite support. Recall that $\mathrm{Sym}_f(\mathbf{Z})$ is countable, locally finite, and dense in $\mathrm{Sym}(\mathbf{Z})$. The natural action of $\mathrm{Sym}(\mathbf{Z})$ on X is continuous, and preserves μ_p.

Denote as in Definition 7.3 by $\mathcal{P}^{\mathrm{Sym}_f(\mathbf{Z})}(X)$ the compact convex set of $\mathrm{Sym}_f(\mathbf{Z})$-invariant probability measures on X; it is a compact convex set in the dual space of $\mathcal{C}(X)$, with the weak-$*$-topology, and $\mathcal{E}^{\mathrm{Sym}_f(\mathbf{Z})}(X)$ is the set of its extreme points. Similarly for $\mathcal{E}^{\mathrm{Sym}(\mathbf{Z})}(X)$. Since $\mathrm{Sym}_f(\mathbf{Z})$ is a subgroup of $\mathrm{Sym}(\mathbf{Z})$, the space $\mathcal{P}^{\mathrm{Sym}(\mathbf{Z})}(X)$ is contained in $\mathcal{P}^{\mathrm{Sym}_f(\mathbf{Z})}(X)$. Since $\mathrm{Sym}_f(\mathbf{Z})$ is dense in $\mathrm{Sym}(\mathbf{Z})$, and the latter acts continuously on $\mathcal{P}(X)$, we have indeed

$$\mathcal{P}^{\mathrm{Sym}(\mathbf{Z})}(X) = \mathcal{P}^{\mathrm{Sym}_f(\mathbf{Z})}(X).$$

In terms of extreme points of compact convex sets, we have consequently

$$\mathcal{E}^{\text{Sym}(\mathbf{Z})}(X) = \mathcal{E}^{\text{Sym}_f(\mathbf{Z})}(X). \tag{9.2}$$

Proposition 8.1. *Consider as above the natural actions of the Polish group* $\text{Sym}(\mathbf{Z})$ *and of its subgroup* $\text{Sym}_f(\mathbf{Z})$ *on the compact space* $X = \{0, 1\}^{\mathbf{Z}}$.

(1) *The set* $\mathcal{E}^{\text{Sym}_f(\mathbf{Z})}(X)$ *coincides with the set* $\{\mu_p\}_{0 \le p \le 1}$ *of Bernoulli measures.*

(2) *Similarly,* $\mathcal{E}^{\text{Sym}(\mathbf{Z})}(X) = \{\mu_p\}_{0 \le p \le 1}$; *in particular, for every* $p \in [0, 1]$, *the Bernoulli measure* μ_p *is invariant and s-ergodic for the action of* $\text{Sym}(\mathbf{Z})$.

(3) *Consider an integer* $k \ge 1$, *a finite sequence* $(p_j)_{j=1,\dots,k}$ *with* $0 < p_1 < \cdots < p_k < 1$, *positive constants* $c_1, \dots, c_k$ *with* $c_1 + \cdots + c_k = 1$, *and the* $\text{Sym}(\mathbf{Z})$*-invariant probability measure* $\mu = c_1 \mu_{p_1} + \cdots + c_k \mu_{p_k} \in \mathcal{P}^{\text{Sym}(\mathbf{Z})}(X)$. *Then* μ *is w-ergodic. If* $k \ge 2$, μ *is not s-ergodic.*

Proof. Claim (1) is known as a result of de Finetti. The original article seems to be [dFin–37], but we rather refer to [Fell–71, Section VII.4, pages 228–9].

Claim (2) follows by Equality (9.2). It is the way de Finetti's result is quoted in [Fomi–50, Section 2.4].

For (3), observe that the $\text{Sym}(\mathbf{Z})$-orbits on X are easily described:

(a) two one-point orbits, one with 0 's only, the other with 1 's only,
(b) for each $k \ge 1$ two countable infinite orbits $\{(x_i)_{i \in \mathbf{Z}} \in X \mid \sum_{i \in \mathbf{Z}} x_i = k\}$ and $\{(x_i)_{i \in \mathbf{Z}} \in X \mid \sum_{i \in \mathbf{Z}} (1 - x_i) = k\}$,
(c) and the uncountable orbit, that we denote by X', of sequences that have infinitely many 0's and infinitely many 1 's.

In particular, the complement N of X' in X is countable, the partition $X = X' \sqcup N$ is $\text{Sym}(\mathbf{Z})$-invariant, and $\text{Sym}(\mathbf{Z})$ is transitive on X'. It follows that every $\text{Sym}(\mathbf{Z})$-invariant subset of X is either inside N or contains X'.

Because N is countable and the measure μ of (3) without atoms, $\mu(N) = 0$. Hence the action of $\text{Sym}(\mathbf{Z})$ on (X, μ) is w-ergodic.

If $k \ge 2$, the measure μ is by definition decomposable, i.e. not s-ergodic (Proposition 7.4). □

Proposition 8.2. *Corollary 7.10 does not extend to the situation of the group* $\text{Sym}(\mathbf{Z})$ *acting on the compact space* X.

More precisely, there exists a $\text{Sym}(\mathbf{Z})$*-invariant Borel subset* X' *of* X *such that, for every* $p \in \,]0, 1[$, *we have* $\mu_p \in \mathcal{E}^{\text{Sym}(\mathbf{Z})}(X)$ *and* $\mu_p(X') = 1$.

Proof. Let $A \subset X$ be a $\text{Sym}(\mathbf{Z})$-invariant Borel subset. Suppose that there exists some $p \in \,]0, 1[$ such that $\mu_p(A) \ne 0$. In the previous proof, we have checked that $X \smallsetminus A$ is countable. Hence, for every $p \in]0,1[$, we have $\mu_p(A) = 1$. $\square$

Despite the failure of Corollary 7.10 in situations such as that of the previous proposition, we have the following result, that is Theorem 6 of [Fomi–50]:

Proposition 8.3 (Fomin). *Let G be a topological group acting continuously by homeomorphisms on a compact space Ω. Assume that there exist two distinct s-ergodic G-invariant probability measures on Ω, say ν_1 and ν_2.*

Then there exists a Borel partition of Ω in two subsets A_1, A_2 that are ν_1-essentially invariant and ν_2-essentially invariant by G, and such that

(1) $\nu_1(A_1) = 1$ and $\nu_2(A_1) = 0$,
(2) $\nu_1(A_2) = 0$ and $\nu_2(A_2) = 1$.

When G is a second countable locally compact group, recall we gave a much stronger conclusion in Corollary 7.10; in particular, the conclusion of Proposition 8.3 holds with A_1, A_2 actually G-invariant, a condition stronger than the above conditions of essential invariance.

Before giving an illustration of Proposition 8.3 with Kolmogorov's example, we recall the following well-known fact. For $p \in]0,1[$, let E_p denote the Borel subset of X consisting of sequences $(x_i)_{i \in \mathbf{Z}}$ such that $\lim_{k \to \infty} \frac{1}{2k+1} \sum_{i=-k}^{k} x_i = p$, i.e. of sequences in which the 1 's have density p. Note that E_p is invariant by $\mathrm{Sym}_f(\mathbf{Z})$, but not by $\mathrm{Sym}(\mathbf{Z})$.

Proposition 8.4. *Let $p, q \in]0,1[$ with $p \neq q$, and let $E_p, E_q \subset X$ be as above. Then $\mu_p(E_p) = 1$ and $\mu_p(E_q) = 0$.*

Proof. Let $\varphi_0 : X \longrightarrow \mathbf{R}$ be defined by $\varphi_0(x) = x_0$. Then

$$\int_X \varphi_0(x) d\mu_p(x) = p,$$

by definition of μ_p. By Birkhoff's ergodicity theorem, the limit

$$\varphi_0^*(x) := \lim_{k \to \infty} \frac{1}{2k+1} \sum_{i=-k}^{k} \varphi_0(S^i x)$$

exist for μ_p-almost all $x \in X$, and defines a μ_p-almost everywhere constant function φ_0^* of essential value p (because the shift S is s-ergodic on (X, μ_p)). Observe that, for $x \in X$, we have $\varphi_0^*(x) = p$ if and only if $x \in E_p$. Hence $E_p = X$, up to μ_p-negligible sets; otherwise said: $\mu_p(E_p) = 1$.

Since $E_p \cap E_q = \emptyset$, we have $\mu_p(E_q) = 0$. $\square$

If we particularize the pair (G, Ω) to the pair $(G = \mathrm{Sym}(\mathbf{Z}), X = \{0, 1\}^{\mathbf{Z}})$ of Kolmogorov's example, and ν_1, ν_2 to the Bernoulli measures $\mu_{1/3}, \mu_{2/3}$, then the conclusion of Proposition 8.3 holds for the subsets $E_{1/3}, E_{2/3}$ of Proposition 8.4; for the essential invariance of these two subsets, see Remark 7.2.

Let us finally mention a generalization of Equality (9.2), from just before Proposition 8.1, and of Part (1) of the same proposition. Consider a measure space $(Y, \mathcal{C})$ and the product space $(X, \mathcal{B})$, with $X = Y^{\mathbf{Z}}$, with the natural action of the groups $\mathrm{Sym}(\mathbf{Z})$ and $\mathrm{Sym}_f(\mathbf{Z})$. For every probability measure λ on $(Y, \mathcal{C})$, let $\tilde{\lambda}$ be the probability measure on $(X, \mathcal{B})$ that is the product of copies of λ indexed by $\mathbf{Z}$; we denote by $\mathrm{Bern}(X)$ the set of measures of the form $\tilde{\lambda}$; observe that $\mathrm{Bern}(X) \subset \mathcal{P}^{\mathrm{Sym}(\mathbf{Z})}(X)$. Then:

(1) $\mathcal{P}^{\mathrm{Sym}(\mathbf{Z})}(X) = \mathcal{P}^{\mathrm{Sym}_f(\mathbf{Z})}(X)$ [HeSa–63, Theorem 3.2], and therefore $\mathcal{E}^{\mathrm{Sym}(\mathbf{Z})}(X) = \mathcal{E}^{\mathrm{Sym}_f(\mathbf{Z})}(X)$.

(2) $\mathcal{E}^{\mathrm{Sym}_f(\mathbf{Z})}(X) = \mathrm{Bern}(X)$ [HeSa–63, Theorem 5.3].

9. Fomin's Representations

In [Fomi–50, § 1], Fomin proves the equivalence (i) ⇔ (ii) of Proposition 7.4 by adding one more equivalent condition, of independent interest, in terms of unitary representations. The object of this section is to describe this condition. Interaction between ergodic theory and unitary representations were pointed out earlier by Koopman [Koop–31].

Let G be a topological group acting continuously by homeomorphisms on a compact space X. Let $\mathcal{C}(X, \mathbf{T})$ denote the group of all continuous functions from X to the compact group $\mathbf{T}$ of complex numbers of modulus one. We consider the natural action of G on $\mathcal{C}(X, \mathbf{T})$, defined by $(g(\varphi))(x) = \varphi(g^{-1}(x))$ for all $g \in G$, $\varphi \in \mathcal{C}(X, \mathcal{T})$, and $x \in X$. In reference to Fomin, we denote by $\mathcal{F}$ the corresponding semi-direct product $\mathcal{C}(X, \mathbf{T}) \rtimes G$ (Fomin's notation is P). We do not furnish the group $\mathcal{F}$ with any topology.

Consider a G-invariant probability measure μ on X, and the complex Hilbert space $L^2_{\mathbf{C}}(X, \mu)$. For $(\varphi, g) \in \mathcal{F}$, define a linear operator $\rho_\mu(\varphi, g)$ on $L^2_{\mathbf{C}}(X, \mu)$ by

$$\left(\rho_\mu(\varphi, g)\xi\right)(x) = \varphi(x)\xi(g^{-1}(x))$$

for all $\xi \in L^2_{\mathbf{C}}(X, \mu)$ and $x \in X$. The following proposition, which is now straightforward to check, is Theorem 1 in [Fomi–50].

Proposition 9.1. *Let $G, X, \mathcal{F}$, and μ be as above. Then $(\varphi, g) \longmapsto \rho_\mu(\varphi, g)$ defines a unitary representation of the group $\mathcal{F}$ on the Hilbert space $L^2_{\mathbf{C}}(X, \mu)$.*

For every essentially bounded function $\psi \in L^\infty_{\mathbf{C}}(X, \mu)$, we denote by M_ψ the multiplication operator $\xi \longmapsto \varphi\xi$ on $L^2_{\mathbf{C}}(X, \mu)$.

Lemma 9.2. *With the notation above, let S be a continuous linear operator on $L^2_{\mathbf{C}}(X, \mu)$ that commutes with $\rho_\mu(\varphi, g)$ for all $(\varphi, g) \in \mathcal{F}$.*

Then $S = M_\psi$ for some $\psi \in L^\infty_{\mathbf{C}}(X, \mu)$; moreover ψ is μ-essentially G-invariant (in the sense of Definition 7.5).

Proof. By hypothesis, S commutes with $\rho_\mu(\varphi, 1) = M_\varphi$ for every $\varphi \in \mathcal{C}(X, \mathbf{T})$, and therefore with sums of products of such multiplication operators. By the Stone–Weierstrass theorem, S commutes also with M_φ for every continuous function $\varphi : X \longrightarrow \mathbf{C}$. By a standard argument, it follows that there exists $\psi \in L^\infty_{\mathbf{C}}(X, \mu)$ such that $S = M_\psi$; see [BourTS, Chap. II, § 3, no 3, Lemma 3].

Given that $S = M_\psi$ commutes with $\rho_\mu(1, g)$ for all $g \in G$, the function ψ is μ-essentially G-invariant. $\qquad\square$

The following proposition is Theorem 2 in [Fomi–50]. Note that Property (ii) below coincides with Property (ii) of Proposition 7.4.

Proposition 9.3. *Let $G, X, \mathcal{F}$, and μ be as above. The following properties are equivalent:*

(i) the representation ρ_μ is irreducible,
(ii) the measure μ is s-ergodic.

Proof. For (i) $\Rightarrow$ (ii), we prove the contraposition. If μ is not s-ergodic, there exists a μ-essentially invariant Borel subset $A \subset X$ with $0 < \mu(A) < 1$. The subspace of $L^2_{\mathbf{C}}(X, \mu)$ of functions which vanish outside A is invariant by ρ_μ, and therefore the representation is reducible.

The converse implication (ii) $\Rightarrow$ (i) follows from Lemma 9.2 and Schur's lemma (for which we refer to [BeHV–08, Theorem A.2.2]). $\qquad\square$

References

[Ahlf–35] Lars Ahlfors, *Zur Theorie der Überlagerungsflächen*, Acta Math. **65** (1935), 157–94.

[Ale–76] Pavel Sergeyevich Aleksandrov and others, *In memory of Sergei Vasil'evic Fomin*, Russian Math. Surveys **31**:3 (1976), 205–20.

[AlMP–11] Yousef Al-Gadid, Brice R. Mbombo, and Vladimir Pestov, *Sur les espaces test pour la moyennabilité*, C.R. Math. Acad. Sci. Soc. R. Can. **33**:3 (2011), 65–77.

[Anos–94] Dmitri Anosov, *On the contribution of N.N. Bogolyubov to the theory of dynamical systems*, Russian Math. Surveys **49**:5 (1994), 1–18.

[AtKa–12] Alexandru G. Atim and Robert R. Kallman, *The infinite unitary and related groups are algebraically determined Polish groups*, Topology and its Appl. **159** (2012), 2831–40.

[Babe–13] Ivan K. Babenko, *Algebra, geometry and topology of the substi-tution group of formal power series*, Russian Math. Surveys **68** (2013), 1–68.

[BaBo–11] Ivan K. Babenko and Semen A. Bogatyi, *The amenability of the substitution of formal power series*, Izv. Math. **75** (2011), 239–52. [See also Pestov's review of this article, MR2830241 (2012e:43002).]

[Bana–32] Stefan Banach, *Théorie des opérations linéaires*, Hafner, 1932.

[BaTa–24] Stefan Banach and Alfred Tarski, *Sur la décomposition des ensembles de points en parties respectivement congruentes*, Fund. Math. **6** (1924), 244–77 [Banch, Oeuvres, Vol. I, 118–48 and 125–7].

[Bekk–03] Bachir Bekka, *Kazhdan's property (T) for the unitary group of a separable Hilbert space*, Geom. Funct. Anal. **13** (2003), 509–20.

[BeHV–08] Bachir Bekka, Pierre de la Harpe, and Alain Valette, *Kazhdan's property (T)*, Cambridge University Press, 2008.

[BoPT–11] Manuel Bodirsky, Michael Pinsker, and Todor Tsankov, *Decid-ability of definability*, 26th Annual IEEE Symposium on Logic in Computer Science – LICS 2011, IEEE Computer Soc., Los Alamitos, CA, 2011, pp. 321-8.

[BoFe–07] Semeon A. Bogatyi and Vitaly V. Fedorchuk, *Schauder's fixed point and amenability of a group*, Topol. Methods Nonlinear Anal. **29**:2 (2007), 383–401.

[Bogo–39] Nikolay N. Bogolyubov, *On some ergodic properties of continuous groups of transformations*, 1939. Published first in Ukrainian (in Scientific Notes of Kiev State University of T.G. Shevchenko, Physics - Mathematics zbirnyk, **4**, N. 5 (1939), 45–52), and then in Russian (In 'Selected works', Vol. 1, Kiev 1969, pp. 561–9). [English translation in [BogoSW], Part III, pages 455–62. Strangely enough, Bogolyubov's article appears there without any precise reference to the Russian version or the Ukrainian original. For these, see [Anos–94].]

[BogoSW] Nikolay N. Bogolyubov, *Selected works*, Volume 2 of the series 'Classics of soviet mathematics'. Volume 2 has four parts (each one a volume by itself) which are:
I. Dynamical theory.
II. Quantum and classical statistical mechanices.
III. Nonlinear mechanics and pure mathematics. Edited by V.S. Vladimirov, translated from the Russian by V.M. Million-shchikov, Gordon and Breach, 1995.
IV. Quantum field theory.

[BTG5-10] Nicolas Bourbaki, *Topologie générale, chapitres 5 à 10*. Hermann, 1974.

[BourTS] Nicolas Bourbaki, *Théories spectrales*, Hermann, 1967.

[CoHa–16] Yves Cornulier and Pierre de la Harpe, *Metric geometry of locally compact groups*, Tracts in Mathematics 25, European Mathematical Society, 2016.

[Day–61] Mahlon Marsh Day, *Fixed-point theorems for compact sets*, Illinois J. Math. **5** (1961), 585–90. Correction, Illinois J. Math. **8** (1964), 713.

[Dixm–50] Jacques Dixmier, *Les moyennes invariantes dans les semi-groupes et leurs applications*, Acta Sci. Math. Szeged **12** (1950), 213–27.

[Dixm–57] Jacques Dixmier, *Les algèbres d'opérateurs dans l'espace hilbertien (algèbres de von Neumann)*, Gauthier-Villars, 1957. [2nd edn, 1969.]

[Edwa–65] Robert Edmund Edwards, *Functional analysis, theory and applications*, Holt, Rinehart and Wilson, 1965.

[Eyma–75] Pierre Eymard, *Initiation à la théorie des groupes moyennables*, in 'Analyse harmonique sur les groupes de Lie (Sém. Nancy-Strasbourg, 1973-1975)', Lecture Notes in Math. **497** (Springer, 1975), 89–107.

[Fell–71] William Feller, *An introduction to probability theory and its applications, Volume II*, 2nd edn, J. Wiley & Sons, 1971.

[dFin–37] Bruno de Finetti, *La prévision : ses lois logiques, ses sources subjectives*, Annales de l'I.H.P. **7**(1) (1937), 1–68.

[Fomi–50] Sergei V. Fomin, *On measures invariant under certain groups of transformations*, Izvestiya Akak. Nauk. SSSR. Ser. Mat. **14** 1950), 261–74. [English translation: Amer. Math. Soc. Translations, Series 2, Vol. **51** (1966), 317–32.]

[Furs–63] Harry Furstenberg, *A Poisson formula for semi-simple Lie groups*, Ann. Math. **22** (1963), 335–86.

[Gaug–67] Edward D. Gaughan, *Topological group structures of infinite symmetric groups*, Proc. Nat. Acad. Sci. U.S.A. **58** (1967), 907–10.

[GiHa–97] Thierry Giordano and Pierre de la Harpe, *Moyennabilité des groupes dénombrables et actions sur les espaces de Cantor*, C.R. Acad. Sci. Paris Sér. I Math. **324** (1997), 1255–8.

[GiPe–07] Thierry Giordano and Vladimir Pestov, *Some extremely amenable groups related to operator algebras and ergodic theory*, J. Inst. Math. Jussieu **6** (2007), 279–315.

[GiNg–13] Thierry Giordano and Ping W. Ng, *Some consequences of von Neumann algebra uniqueness*, J. Functional Analysis **264** (2013), 1112–24.

[Glas–76] Shmuel Glasner, *Proximal flows*, Lecture Notes in Math. **517**, Springer, 1976.

[Glas–98] Eli Glasner, *On minimal actions of Polish groups*, Topology and its Appl. **85** (1998), 119–25.

[GlWe–03] Eli Glasner and Benjamin Weiss, *The universal minimal system for the group of homeomorphisms of the Cantor set*, Fund. Math. **176** (2003), 277–89.

[Gree–69] Frederick P. Greenleaf, *Invariant means on topological groups*, van Nostrand, 1969.

[GrSc–00] Gernot Greschonig and Klaus Schmidt, *Ergodic decomposition of quasi-invariant probability measures*, Colloq. Math. **84/85** (2000), 495–514.

[GrMi–83] Mikhael Gromov and Vitali Milman, *A topological application of the isoperimetric inequality*, Amer. J. Math **105** (1983) 843–54.

[Halm–67] Paul R. Halmos, *A Hilbert space problem book*, van Nostrand, 1967. [2nd edn, Springer, 1982.]

[Hans–71] Vagn Lundsgaard Hansen, *Some theorems on direct limits of expanding sequences of manifolds*, Math. Scand. **29** (1971), 5–36.

[Harp–73] Pierre de la Harpe, *Moyennabilité de quelques groupes topologiques de dimension infinie*, C.R. Acad. Sci. Paris Sér. A-B **277** (1973), A1037–Q1040.

[Harp–79] Pierre de la Harpe, *Moyennabilité du groupe unitaire et propriété P de Schwartz des algèbres de von Neumann*, in 'Algèbres d'opérateurs (Sém., Les Plans-sur-bex, 1978)', Lecture Notes in Math. **725** (Springer, 1979), 220–7.

[Harp–82] Pierre de la Harpe, *Classical groups and classical Lie algebras of operators*, in 'Operator algebras and applications', Part I (Kingston, Ont., 1980), Proc. Sympos. Pure Math. **38** (Amer. Math. Soc. 1982), 477–513.

[Haus–14] Felix Hausdorff, *Grundzüge der Mengenlehre*. Veit, 1914.

[HeRo–63] Edwin Hewitt and Kenneth A. Ross, *Abstract harmonic analysis, Volume I*. Die Grundlehren der mathematischen Wissenschaften **115**, Springer, 1963.

[HeSa–63] Edwin Hewitt and Leonard J. Savage, *Symmetric measures on Cartesian products*, Trans. Amer. Math. Soc. **80** (1955), 470–501.

[Kaku–38] Shizuo Kakutani, *Two fixed-point theorems concerning bicompact convex sets*, Proc. Imp. Acad. **14** (1938), 242–5.

[KaKo–44] Shizuo Kakutani and Kunihiko Kodaira, *Über das Haarsche Mass in der lokal bikompakten Gruppe*, Proc. Imp. Acad. Tokyo **20** (1944), 444–50. [*Shizuo Kakutani: Selected Papers, Vol. 1*, Robert R. Kallman Editor, Birkhäuser (1986), 68–74.]

[Kall–79] Robert R. Kallman, *A uniqueness result for the infinite symmetric group*, in 'Studies in analysis' (Academic Press, 1979), 321–2.

[Karu–58] Takashi Karube, *On the local cross-sections in locally compact groups*, J. Math. Soc. Japan **10**⁴ (1958), 343–7.

[KePT–05] Alexander S. Kechris, Vladimir Pestov, and Stevo Todorcevic, *Fraïssé limits, Ramsey theory, and topological dynamics of automorphism groups*, Geom. Funct. Anal. **15** (2005), 106–89.

[KeRo–07] Alexander S. Kechris and Christian Rosendal, *Turbulence, amalgamation, and genereic automorphisms of homogeneous structures*, Proc. London Math. Soc. (3) **94** (2007), 302–50.

[Koop–31] Bernard Osgood Koopman, *Hamiltonian systems and transformations in Hilbert space*, Proc. N.A.S. **17** (1931), no. 5. 315–8.

[KrMi–40] Mark Grigor'evich Krein and David Milman, *On extreme points of regular convex sets*, Studia Math. **9** (1940), 133–8.

[KrBo–37] Nicolas Kryloff and Nicolas Bogoliouboff, *La théorie générale de la mesure dans son application à l'étude des systèmes dynamiques de la mécanique non linéaire*, Annals of Math. **38** (1937), 65–113.

[Mait–77] Ashok Maitra, *Integral representations of invariant measures*, Trans. Amer. Math. Soc. **229** (1977), 209–25.

[Mark–36] Andreï Markov, *Quelques théorèmes sur les ensembles abéliens*, Dokl. Akad. Nauk. SSSR (N.S:) **10** (1936), 311–14. [This Andreï Markov (1903–1979) is the son of Andreï Markov (1856–1922) known for 'Markov chains'.]

[MeTs–13] Julien Melleray and Todor Tsankov, *Generic representations of abelian groups and extreme amenability*, Israel J. Math. **198** (2013), 129–67.

[MeNT–16] Julien Melleray, Lionel Nguyen van Thé, and Todor Tsankov, *Polish groups with metrizable universal minimal flows*, Int. Math. Res. Not. IMRN 2016, no. 5, 1285–307.

[vNeu–29] John von Neumann, *Zur allgemeinen Theorie des Masses*, Fund. Math. **13** (1929) 73–116, 333 [Collected works, Vol. I, 599–643].

[Oxto–52] John C. Oxtoby, *Ergodic sets*, Bull. Amer. Math. Soc. **58** (1952), 116–36.

[Pate–92] Alan L.T. Paterson, *Nuclear C*-algebras have amenable unitary groups*, Proc. Amer. Math. Soc. **114** (1992), 719–21.

[Pede–79] Gert K. Pedersen, *C*-algebras and their automorphism groups*, Academic Press, 1979.

[Pest–98] Vladimir Pestov, *On free actions minimal flows, and a problem by Ellis*, Trans. Amer. Math. Soc. **250** (1998), 4149–65.

[Pest–02] Vladimir Pestov, *mm-spaces and groups actions*, L'Ens. Math. **48** (2002), 209–36.

[Pest–06] Vladimir Pestov, University Lecture Series **40**, Amer. Math. Soc., 2006.

[Phel–66] Robert R. Phelps, *Lectures on Choquet's theorem*, van Nostrand, 1966. [2nd edn, Lecture Notes in Math. **1757**, Springer, 2001.]

[Rick–67] Neil W. Rickert, *Amenable groups and groups with the fixed point property*, Trans. Amer. Math. Soc. **127** (1967), 221–32.

[Roe–88] John Roe, *An index theorem on open manifolds. I, II*, J. Differential Geom. **27**, no. 1 (1988), 87–113, 115–36.

[RoSo–07] Christian Rosendal and Slawomir Solecki, *Automatic continuity of homomorphisms and fixed points on metric compacta*, Israel J. Math. **162** (2007), 349–71.

[Rudi–66] Walter Rudin, *Real and complex analysis*, McGraw-Hill, 1966.

[Tars–29] Alfred Tarski, *Sur les fonctions additives dans les classes abstraites et leur application au problème de la mesure*, C.R. Soc. Sc. Varsovie **22** (1929), 114–17. Collected Papers, Vol. 1, pp. 245–8.

[Tars–38] Alfred Tarski, *Algebraische Fassung des Massproblems*, Fund. Math. **31** (1938), 47–66. Collected Papers, Vol. 2, pp. 453–72.

[Vara–63] Veeravalli S. Varadarajan *Groups of automorphisms of Borel spaces*, Trans. Amer. Math. Soc. **109** (1963), 191–220.

[Veec–77] William A. Veech, *Topological dynamics*, Bull. Amer. Math. Soc. **83** (1977), 775–830.

[Wago–85] Stan Wagon, *The Banach-Tarski paradox*, Cambridge University Press, 1985. Second Edition co-authored with Grzegorz Tomkowicz, October 2016.

[Walt–82] Peter Walters, *An introduction to ergodic theory*, Springer, 1982.

[Zimm–84] Robert J. Zimmer, *Ergodic theory and semisimple groups*, Birkhäuser, 1984.

Added in Proof

We are grateful to Maxime Gheysens and Nicolas Monod for having brought to our attention the following facts and references. Compare the definition below to that of the fixed point property, 3.3. We use "compact convex set" for "non-empty compact convex subset of a locally convex topological real vector space".

Definition 1. Let G be a topological group. An action by homeomorphisms of G on a topological space X **has a continuous orbit** if there exists a point $x_0 \in X$ such that the restricted action $G \times Gx_0 \to Gx_0$ is continuous.

The topological group G has the **Day fixed point property** if every action of G by continuous affine transformations on a compact convex set, with a continuous orbit, has a fixed point.

The following fact is the particular case for topological groups of Theorem 4 of [Day–64], established there for topological semigroups.

Proposition 2. *A topological group G is B-amenable (Definition 3.8) if and only if it has the Day fixed point property.*

This provides for example an easy proof that the symmetric group $\mathrm{Sym}(\mathbf{N})$ is not B-amenable (Proposition 5.5). Indeed, consider the natural action of this group on the set $\mathrm{Mean}(\mathbf{N})$ of all means on $\mathbf{N}$, that is a compact convex subset of the weak dual of the real Banach space $\ell_{\mathbf{R}}^{\infty}(\mathbf{N})$. Observe that, since $\mathbf{N}$ is infinite, the action of $\mathrm{Sym}(\mathbf{N})$ on $\mathbf{N}$ is paradoxical in the sense of Tarski, so that there is not any $\mathrm{Sym}(\mathbf{N})$-invariant mean on $\mathbf{N}$, i.e., there is not any point in $\mathrm{Mean}(\mathbf{N})$ fixed by $\mathrm{Sym}(\mathbf{N})$. Denote by D the subset of $\mathrm{Mean}(\mathbf{N})$ consisting of the Dirac measures; it is a $\mathrm{Sym}(\mathbf{N})$-orbit. The action of $\mathrm{Sym}(\mathbf{N})$ on D is continuous. It follows that $\mathrm{Sym}(\mathbf{N})$ is not B-amenable.

In [Day–61, § 4], Day considers one more property: topological groups (or semi-groups) G for which every *separately continuous* affine action on a compact convex set has a fixed point. This is a property weaker than B-amenability, and stronger than amenability.

Finally, let us mention that the fixed point property of Definition 3.3 is equivalent for locally compact groups to a formally stronger property introduced in [Simo–72], that is a fixed point property for weakly measurable affine actions on compact convex sets.

References

[Day–61] Mahlon Marsh DAY, *Fixed-point theorems for compact sets*, Illinois J. Math. **5** (1961), 585–590.

[Day–64] Mahlon Marsh DAY, *Correction to my paper "Fixed-point theorems for compact sets"*, Illinois J. Math. **8** (1964), 713.

[Simo–72] Barry SIMON, *A remark on groups with the fixed point property*, Proc. Amer. Math. Soc. **32** (1972), 623–624.

10

SCHREIER GRAPHS OF GRIGORCHUK'S GROUP AND A SUBSHIFT ASSOCIATED TO A NONPRIMITIVE SUBSTITUTION

ROSTISLAV GRIGORCHUK[1], DANIEL LENZ[2], AND TATIANA NAGNIBEDA[3]

[1]Mathematics Department, Texas A&M University, College Station, TX 77843-3368, USA
[2]Mathematisches Institut, Friedrich Schiller Universität Jena, 07743 Jena, Germany
[3]Section de Mathématiques, University of Geneva, 2-4, Rue du Lièvre, Case Postale 64, 1211 Genève 4, Suisse

To Wolfgang Woess on the occasion of his sixtieth birthday

Abstract

There is a recently discovered connection between the spectral theory of Schrödinger operators whose potentials exhibit aperiodic order and that of Laplacians associated with actions of groups on regular rooted trees, as Grigorchuk's group of intermediate growth. In this chapter we give an overview of corresponding results, such as different spectral types in the isotropic and anisotropic cases, including Cantor spectrum of Lebesgue measure zero and absence of eigenvalues. Moreover, we discuss the relevant background as well as the combinatorial and dynamical tools that allow one to establish the aforementioned connection. The main such tool is the subshift associated with a substitution over a finite alphabet that defines the group algebraically via a recursive presentation by generators and relators.

Contents

Key words and phrases. substitutional subshift, self-similar group, Schreier graph, Laplacian, spectrum of Schrödinger operators.

Introduction

The study of spectra of graphs associated with finitely generated groups, such as Cayley graphs or Schreier graphs (natural analogues of Cayley graphs associated to not necessarily free group actions), has a long history and is related to many problems in modern mathematics. Still, very little is known about the dependence of the spectrum of the Laplacian on the choice of generators in the group and on the weight on these generators. In a recent paper [39] we addressed the issue of dependence on the weights on generators in the example of Grigorchuk's group of intermediate growth. More specifically, we determined the spectral type of the Laplacian on the Schreier graphs describing the action of Grigorchuk's group on the boundary of the infinite binary tree and showed that it is different in the isotropic and anisotropic cases. In fact, the spectrum is shown to be a Cantor set of Lebesgue measure zero in the anisotropic case, whereas, as has been known for a long time, it consists of one or two intervals in the isotropic case. Moreover, we showed almost surely (with respect to a natural measure on the boundary of the tree) the absence of eigenvalues for the Laplacians in question.

Our investigation in [39] provides (and relies on) a surprising link between discrete Schrödinger operators with aperiodic order and the substitutional dynamics arising from a presentation of the group by generators and relators.

The purpose of this chapter is twofold. First, we want to survey the spectral theoretic results of [39]. Second, we want to put these results in wider perspective by discussing the background on aperiodic order in dimension one. In this context we present a discussion of subshifts and aperiodic order in Section 1 and of Schrödinger operators arising in models of (dis)ordered solids in Section 2.

We also continue our study of the substitution that is instrumental for the results in [39]. It first appeared in 1985 in the presentation of Grigorchuk's group by generators and relators found by Lysenok [66] (such infinite recursive presentations are now called L-presentations). A remarkable fact is that the same substitution also describes basic dynamical properties of the group. Here we review its combinatorial properties and carry out a detailed study of the factor map to its maximal equicontinuous factor, which is the binary odometer. This factor map is proven to be one-to-one everywhere except on three orbits. This, in turn, can be linked to the phenomenon of pure point diffraction, which is at the core of aperiodic order. All these discussions concerning the substitution are contained in Section 3.

The necessary background from graphs and dynamical systems is discussed in Section 4 and basics on Grigorchuk's group and its Schreier graphs can be found in Section 5.

The connection between Schrödinger operators and Laplacians on Schreier graphs revealed in [39] (and reviewed in Sections 6 and 7) can also be used to show that the Kesten-von-Neumann trace and the integrated density of states agree. This seems to be somewhat folklore. We provide a proof in Section 8.

Our approach can certainly be carried out for various further families of groups generalizing Grigorchuk's group. In particular, our results fully extend to a larger family of self-similar groups acting on the infinite binary tree considered by Sunic in [76]. Moreover, each of these self-similar groups belongs to an uncountable family of groups parametrized by sequences in a certain finite alphabet, in the same way that Grigorchuk's group belongs to the family of groups $(G_\omega)_{\omega \in \{0,1,2\}^{\mathbb{N}}}$ constructed by the first named author in [37]. The Schreier graphs of the groups in the same family look the same, but their labeling by generators, and thus their spectra, depend on the sequence ω. The associated subshifts are also different; in particular, they don't have to come from a substitution in the case when the corresponding group is not self-similar. This more general setup will be considered in a separate chapter.

Acknowledgments

R. G. is supported by the NSF within the grant DMS-1207669. The authors acknowledge support of the Swiss National Science Foundation. Part of this research was carried out while D. L. and R. G. were visiting the Department of Mathematics of the University of Geneva. The hospitality of the department is gratefully acknowledged. The authors also thank Fabien Durand and Ian Putnam for most enlightening discussions concerning the material gathered in Section 3.7, Yaroslav Vorobets for allowing them to use his figures (10.3 and 10.4) and Olga Klimecki for help in preparing Figure 10.1. Finally, the authors would like to thank the anonymous referee for a very careful reading of the manuscript and several helpful suggestions.

1. Subshifts and Aperiodic Order in One Dimension

Long-range aperiodic order (or aperiodic order for short) denotes an intermediate regime of order between periodicity and randomness. It has received a lot of attention over the last thirty years or so; see e.g. the article collections and monographs [5, 8, 53, 69, 73].

This interest in aperiodic order has various reasons. On the one hand, it is due to the many remarkable and previously unknown features and phenomena arising from aperiodic order in various branches of mathematics. On the other hand, it is also due to the relevance of aperiodic order in physics and chemistry. Indeed, aperiodic order provides a mathematical theory for the description of a new type of matter discovered in 1982 by Shechtman via diffraction experiments [74]. These experiments showed sharp peaks in the diffraction pattern indicating long-range order and at the same time a fivefold symmetry indicating absence of periodicity. The discovery of solids combining both long-range order with aperiodicity came as a complete surprise to physicists and chemists, and Shechtman was honored with a Nobel Prize in 2011. By now, the solids in question are known as quasicrystals.

In one dimension aperiodic order is commonly modeled by subshifts of low complexity. In higher dimensions it is modeled by dynamical systems consisting of point sets with suitable regularity features (which are known as Delone dynamical systems). Here, we present the necessary notation in order to deal with the one-dimensional situation. *When we speak about aperiodic order subsequently, this will always mean that we have a subshift with suitable minimality features at our disposal.*

Let a finite set $\mathcal{A}$ be given. We call $\mathcal{A}$ the *alphabet* and refer to its elements as *letters*. We will consider the set $\mathcal{A}^*$ of finite words (including the empty word) over the alphabet $\mathcal{A}$, viewed as a free monoid (with the multiplication given by concatenation of words and the empty word representing the identity). Elements of $\mathcal{A}^*$ will often be written as $w = w_1 \ldots w_n$ with $w_j \in \mathcal{A}$. The length of a word is the number of its letters. It will be denoted by $|\cdot|$. The empty word has length zero. Then, $\mathcal{A}^{\mathbb{Z}}$ denotes the set of functions from $\mathbb{Z}$ to $\mathcal{A}$. We think of the elements of $\mathcal{A}^{\mathbb{Z}}$ as bi-infinite words over the alphabet $\mathcal{A}$. The set $\mathcal{A}^{\mathbb{N}}$ denotes the set of functions from $\mathbb{N}$ to $\mathcal{A}$. We think of its elements of one-sided infinite words over $\mathcal{A}$. They will often be denoted by $\xi = \xi_1 \xi_2 \ldots$

If v, w are finite words and $\omega \in \mathcal{A}^{\mathbb{Z}}$ satisfies

$$\omega_1 \ldots \omega_{|v|} = v \text{ and } \omega_{-|w|+1} \ldots \omega_0 = w$$

we write

$$\omega = \ldots w | v \ldots$$

and say that $|$ *denotes the position of the origin.*

We equip $\mathcal{A}$ with the discrete topology and $\mathcal{A}^{\mathbb{Z}}$ with the product topology. By the Tychonoff theorem, $\mathcal{A}^{\mathbb{Z}}$ is then compact. In fact, it is homeomorphic to the Cantor set. A pair (Ω, T) is called a *subshift* over $\mathcal{A}$ if Ω is a closed subset of $\mathcal{A}^{\mathbb{Z}}$ which is invariant under the *shift transformation*

$$T : \mathcal{A}^{\mathbb{Z}} \longrightarrow \mathcal{A}^{\mathbb{Z}}, \ (T\omega)(n) := \omega(n+1).$$

If there exists a natural number $N \neq 0$ with $T^N \omega = \omega$ for all $\omega \in \Omega$, then (Ω, T) is called *periodic*; otherwise it is called *nonperiodic*.

Whenever ω is a word over $\mathcal{A}$ (finite or infinite, indexed by $\mathbb{N}$ or by $\mathbb{Z}$) we define

$$\mathrm{Sub}(\omega) := \text{Finite subwords of } \omega.$$

By convention, the set of finite subwords includes the empty word. Every subshift (Ω, T) comes naturally with the set $\mathrm{Sub}(\Omega)$ of associated finite words given by

$$\mathrm{Sub}(\Omega) := \bigcup_{\omega \in \Omega} \mathrm{Sub}(\omega).$$

A word $v \in \mathrm{Sub}(\Omega)$ is said to *occur with bounded gaps* if there exists an $L_v > 0$ such that every $w \in \mathrm{Sub}(\Omega)$ with $|w| \geq L_v$ contains a copy of v. As is well known (and not hard to see) (Ω, T) is minimal if and only if every $v \in \mathrm{Sub}(\Omega)$ occurs with bounded gaps. For proofs and further discussion we refer to standard textbooks such as [65, 81]. We will be concerned here with the following strengthening of the bounded gaps condition. It concerns the case that L_v can be chosen as $C|v|$ with fixed C (independent of v).

Definition 1.1 (Linearly repetitive). A subshift (Ω, T) is called *linearly repetitive* (LR), if there exists a constant $C > 0$ such that any word $v \in \mathrm{Sub}(\Omega)$ occurs in any word $w \in \mathrm{Sub}(\Omega)$ of length at least $C|v|$.

Remark 1.2. This notion has been discussed under various names by various people. In particular it was studied by Durand, Host, and Skau [32] in the setting of subshifts (under the name "linearly recurrent"). For Delone dynamical systems it was brought forward at about the same time by Lagarias and Pleasants [56] under the name "linearly repetitive." It has also featured in the work of Boris Solomyak [75] (under the name "uniformly repetitive"). It was also already discussed in an unpublished work of Boshernitzan in the 1990s. That work also contains a characterization in terms of positivity of weights. A corresponding result for Delone systems was recently given in [14].

Durand [30] gives a characterization of such subshifts in terms of primitive S-adic systems and shows the following (which was already known to Boshernitzan).

Theorem 1.3. *Let (Ω, T) be a linearly repetitive subshift. Then, the subshift is uniquely ergodic.*

Remark 1.4. In fact, linear repetitivity implies a strong form of subadditive ergodic theorem [61]. Validity of such a result together with the fundamental work of Kotani [55] allows to prove the Cantor spectrum of Lebesgue measure zero, see [59]. Our proof uses an extension of this approach worked out in [12].

2. Schrödinger Operators with Aperiodic Order

In this section, we present (parts of) the spectral theory of discrete Schrödinger operators associated to minimal dynamical systems.

Schrödinger operators occupy a prominent position in the theory of aperiodic order. Indeed, they arise in the quantum mechanical description of conductance properties of quasicrystals and exhibit quite interesting mathematical properties. In fact, already the first two papers on them written by physicists suggest that the corresponding spectral measures are purely singular continuous and the spectrum is a Cantor set of Lebesgue measure zero [54, 71]. By now these features, as well as other conductance-related properties known as anomalous transport, have been thoroughly studied in a variety of models by various authors; see the survey articles [21, 22, 77]. The phenomenon that the underlying spectrum is a Cantor set of Lebesgue measure zero is usually referred to as *Cantor spectrum of Lebesgue measure zero* and this is how we will refer to it subsequently.

The Laplacians on Schreier graphs discussed later will turn out to be unitarily equivalent to certain such Schrödinger operators.

In this section, we first discuss basic mathematical features of Schrödinger operators associated to dynamical systems in Section 2.1, then turn to absolutely continuous spectrum and the spectrum as a set for subshifts in Section 2.2, and finally give some background from physics in Section 2.3.

2.1. Constancy of the Spectrum and the Integrated Density of States (IDS)

In this section we review some basic theory of discrete Schrödinger operators (or rather Jacobi matrices) associated with minimal topological dynamical systems. This framework is slightly more general than the framework of minimal subshifts. The results we discuss include constancy of the spectrum and uniform convergence of the so-called integrated density of states. All of these results are well known.

Whenever T is a homeomorphism of the compact space Ω we will refer to (Ω, T) as a *topological dynamical system*. Later we will meet an even more general definition of a dynamical system. To continuous functions $f, g : \Omega \longrightarrow \mathbb{R}$ we then associate a family of *discrete operators* $(H_\omega)_{\omega \in \Omega}$. Specifically, for each $\omega \in \Omega$, H_ω is a bounded self-adjoint operator from $\ell^2(\mathbb{Z})$ to $\ell^2(\mathbb{Z})$ acting via

$$(H_\omega u)(n) = f(T^n \omega)u(n-1) + f(T^{n+1}\omega)u(n+1) + g(T^n \omega)u(n)$$

for $u \in \ell^2(\mathbb{Z})$ and $n \in \mathbb{Z}$.

In the case $f \equiv 1$ the above operators are known as *discrete Schrödinger operators*. For general f the name *Jacobi matrices* is often used in the literature. Here, we will deal with the case $f \neq 1$ but we will primarily refer to the arising operators as Schrödinger operators.

As the operator H_ω is self-adjoint, the operator $H_\omega - z$ is bijective with continuous inverse $(H_\omega - z)^{-1}$ for any $z \in \mathbb{C} \setminus \mathbb{R}$. Moreover, for any $\varphi \in \ell^2(\mathbb{Z})$ there exists a unique positive Borel measure μ_ω^φ on $\mathbb{R}$ with

$$\int_\mathbb{R} \frac{1}{t - z} d\mu_\omega^\varphi(t) = \langle \varphi, (H_\omega - z)^{-1} \varphi \rangle$$

for any $z \in \mathbb{C} \setminus \mathbb{R}$. This measure is finite and assigns the value $\|\varphi\|^2$ to the set $\mathbb{R}$.

For fixed $\omega \in \Omega$, the measures μ_ω^φ, $\varphi \in \ell^2(\mathbb{Z})$ are called the *spectral measures* of H_ω. The spectrum of H_ω is defined as

$$\sigma(H_\omega) := \{z \in \mathbb{C} : (H_\omega - zI) \text{ lacks a bounded two-sided inverse.}\}$$

It is the smallest set containing the support of any μ_ω^φ (see e.g. [82]). The spectrum is said to be *purely absolutely continuous* if the spectral measures for all $\varphi \in \ell^2(\mathbb{Z})$ are absolutely continuous with respect to Lebesgue measure. The spectrum is said to be *purely singular continuous* if all spectral measures are both continuous (i.e. do not have discrete parts) and singular with respect to Lebesgue measure.

The following result is well known. It can be found in various places; see e.g. [60]. Recall that (Ω, T) is called *minimal* if $\{T^n \omega : n \in \mathbb{Z}\}$ is dense in Ω for any $\omega \in \Omega$.

Theorem 2.1 (Constancy of the spectrum). *Let (Ω, T) be minimal and $f, g : \Omega \longrightarrow \mathbb{R}$ continuous. Then, there exists a closed subset $\Sigma \subset \mathbb{R}$ such that the spectrum $\sigma(H_\omega)$ of H_ω equals Σ for any $\omega \in \Omega$.*

We will refer to the set Σ in the previous theorem as the *spectrum of the Schrödinger operator associated to (Ω, T) (and (f,g))*. The spectrum Σ is one of the main objects of interest in our study.

Before turning to a finer analysis of the spectrum we will introduce a further quantity of interest to the so-called integrated density of states. In order to do so, we will assume that the underlying dynamical system (Ω, T) is not only minimal but also *uniquely ergodic* i.e. possesses a unique invariant probability measure, which we call λ. Then, we can associate to the family (H_ω) the positive measure k on $\mathbb{R}$ defined via

$$\int_\mathbb{R} F(x) dk(x) := \int_\Omega \langle f(H_\omega)\delta_0, \delta_0 \rangle d\lambda(\omega)$$

(for F any continuous function on $\mathbb{R}$ with compact support). Here, $\delta_0 \in \ell^2(\mathbb{Z})$ is just the characteristic function of $0 \in \mathbb{Z}$. This measure k is called the *integrated density of states*. Let

$$N : \mathbb{R} \longrightarrow [0,1], N(E) := \int_{(-\infty,E]} dk,$$

be the distribution function of k.

There is a direct relation between the measure k and the spectrum of the H_ω.

Theorem 2.2. *Let (Ω, T) be minimal and uniquely ergodic. Then, the set Σ is the support of the measure k. If the function f does not vanish anywhere, then k does not have atoms (i.e. it assigns the value zero to any set containing only one element).*

Remark 2.3. This is rather standard in the theory of random operators. Specific variants of it can be found in many places. In particular, the first statement of the theorem can be found in [60]. In the case of f, which does not vanish anywhere, the statements of the theorem are contained in Section 5 of [78]. Given the constancy of the spectrum, Theorem 2.1, the statements are also very special cases of the results of Section 5 of [63]. The key ingredient for the absence of atoms is amenability of the underlying group $\mathbb{Z}$. The statement on the support does not even need this property.

As is well known, it is possible to "calculate" k via an approximation procedure. This will be discussed next. For $\in \mathbb{Z}$ let

$$j_n : \ell^2(\{1, \ldots, n\}) \longrightarrow \ell^2(\mathbb{Z})$$

be the canonical inclusion and let p_n be the adjoint of j_n. Thus,

$$p_n : \ell^2(\mathbb{Z}) \longrightarrow \ell^2(\{1, \ldots, n\})$$

is the canonical projection. Define for $\omega \in \Omega$ then

$$H_\omega^n := p_n H_\omega j_n.$$

We will be concerned with the spectral theory of these operators. Let the measure k_ω^n on $\mathbb{R}$ be defined as

$$\int_{\mathbb{R}} F(x)\, dk_\omega^n(x) := \frac{1}{n} \sum_{k=1}^{n} \langle F(H_\omega^n)\delta_k, \delta_k \rangle$$

(for any continuous F on $\mathbb{R}$ with compact support) and let

$$N_\omega^n : \mathbb{R} \longrightarrow [0,1], \ N_\omega^n(E) := \int_{(\infty,E]} dk_\omega^n(x),$$

be its distribution function. Let $E_1, \ldots, E_n$ be the eigenvalues of H_ω^n counted with multiplicity. Then, straightforward linear algebra (diagonalization of H_ω^n and independence of the trace of the chosen orthonormal basis) shows that

$$\int_\mathbb{R} F(x)\, dk_\omega^n(x) = \frac{1}{n} \sum_j F(E_j)$$

holds for any continuous F on $\mathbb{R}$ with compact support and that the distribution functions of the measures k_ω^n are given by

$$N_\omega^n(E) = \frac{\#\{\text{Eigenvalues of } H_\omega^n \text{ not exceeding } E\}}{n},$$

where $\sharp$ denotes the cardinality of a set. In this sense, k_ω^n is just an averaged eigenvalue counting.

Theorem 2.4 (Convergence of the integrated density of states).   *Let (Ω, T) be minimal and uniquely ergodic. Then, for any continuous F on $\mathbb{R}$ with compact support and any $\varepsilon > 0$ there exists an $N \in \mathbb{N}$ with*

$$\left| \int_\mathbb{R} F(x)\, dk(x) - \int_\mathbb{R} F(x)\, dk_\omega^n(x) \right| \le \varepsilon$$

for all $\omega \in \Omega$ and all $n \in \mathbb{Z}$ with $n \ge N$.

Proof.   It is well known that the measures $(k_\omega^n)_n$ converge weakly toward k for $n \to \infty$ for almost every $\omega \in \Omega$. This can be found in many places, see e.g. Lemma 5.12 in [78]. (That lemma assumes that f does not vanish anywhere, but its proof does not use this assumption.) A key step in the proof is the use of the Birkhoff ergodic theorem. The desired statement now follows by replacing the Birkhoff ergodic theorem with the uniform ergodic theorem (Oxtoby theorem) valid for uniquely ergodic systems [81].   □

The operators H_ω^n are sometimes thought of as arising out of the H_ω by some form of "Dirichlet boundary condition." The previous result is stable under taking different "boundary conditions." In fact, even a more general statement is true as we will discuss next. (The more general statement will even save us from saying what we mean by boundary condition.) Let for any $n \in \mathbb{Z}$ and $\omega \in \Omega$ be a self-adjoint operator C_ω^n on $\ell^2(\{1, \ldots, n\})$ be given. Then, the statement of the theorem essentially continues to hold if the operators H_ω^n are replaced by the operators

$$\widetilde{H}_\omega^n := H_\omega^n + C_\omega^n$$

provided the rank of the $C's$ is not too big. Here, the rank of an operator C on a finite dimensional space, denoted by $\mathrm{rk}(C)$, is just the dimension of the range of C.

In order to be more specific, we introduce the measure $\widetilde{k}^n_\omega$ on $\mathbb{R}$ defined as

$$\int_{\mathbb{R}} F(x)\,d\widetilde{k}^n_\omega(x) := \frac{1}{n}\sum_{k=1}^{n}\langle F(\widetilde{H}^n_\omega)\delta_k, \delta_k\rangle$$

(for any continuous F on $\mathbb{R}$ with compact support) and its distribution function given by

$$\widetilde{N}^n_\omega : \mathbb{R} \longrightarrow [0,1], \ \widetilde{N}^n_\omega(E) := \frac{1}{n}\sharp\{\text{Eigenvalues of } \widetilde{H}^n_\omega \text{ not exceeding } E\}.$$

Corollary 2.5. *Consider the situation just described. Let $\omega \in \Omega$ be given with*

$$\frac{1}{n}\mathrm{rk}(C^n_\omega) \to 0, n \to \infty.$$

Then, for any continuous F on $\mathbb{R}$ with compact support and any $\varepsilon > 0$ there exists an $N \in \mathbb{N}$ with

$$\left| \int_{\mathbb{R}} F(x)\,dk(x) - \int_{\mathbb{R}} F(x)\,d\widetilde{k}^n_\omega(x) \right| \le \varepsilon$$

for all $n \ge N$.

Proof. A consequence of the min-max principle, see e.g. Theorem 4.3.6 in [51], shows

$$|\widetilde{N}^n_\omega(E) - N^n_\omega(E)| \le \frac{1}{n}\mathrm{rk}(C^n_\omega)$$

independent of $E \in \mathbb{R}$. This directly gives the desired statement. $\qquad\square$

The theorem allows one to obtain an inclusion formula for the spectrum Σ. Denote the spectrum of the operator $\widetilde{H}^n_\omega$ by $\widetilde{\Sigma}^n_\omega$. By construction, $\widetilde{\Sigma}^n_\omega$ is just the support of the measure $\widetilde{k}^n_\omega$.

Corollary 2.6. *Assume the situation of the previous theorem. Then, the inclusion*

$$\Sigma \subset \bigcap_n \overline{\bigcup_{k\ge n} \widetilde{\Sigma}^k_\omega}$$

holds for all $\omega \in \Omega$.

The corollary is somewhat unsatisfactory in that it only gives an inclusion. In certain cases more is known. This is further discussed in Section 8.1. For a general result on how to construct approximations whose spectra converge with respect to the Hausdorff distance we refer the reader to [13].

2.2. The Spectrum as a Set and the Absolute Continuity of Spectral Measures

In this section we consider Schrödinger operators associated to locally constant functions on minimal subshifts. Here, a function h on a subshift Ω over a finite alphabet is *locally constant* if there exists an $N > 0$ such that the value of $h(\omega)$ depends only on the word $\omega(-N)\ldots\omega(N)$. A key distinction in our considerations will then be whether (f, g) is periodic or not.

The overall structure of the spectrum in the periodic case is well-known. This can be found in many references; see e.g. the monograph [78].

Theorem 2.7 (Periodic case). *Let (Ω, T) be a minimal subshift and $f, g : \Omega \longrightarrow \mathbb{R}$ locally constant with $f(\omega) \neq 0$ for all $\omega \in \Omega$. If (f, g) is periodic (with period N), then the spectra Σ of the associated Schrödinger operators consist of finitely many (and not more than N) closed intervals of positive length and all spectral measures are absolutely continuous with respect to Lebesgue measure.*

Remark 2.8.   Note that periodicity of (f, g) may have its origin in both properties of (Ω, T) and properties of (f, g). For example periodicity always occurs if $f = g = 1$ irrespective of the nature of (Ω, T). Also, periodicity occurs for arbitrary f, g if (Ω, T) is periodic (i.e. there exists a natural number $N \neq 0$ with $T^N \omega = \omega$ for all $\omega \in \Omega$).

The previous theorem gives rather complete information on Σ in the periodic case. In order to deal with the nonperiodic case, we will need a further assumption on (Ω, T). This condition is linear repetitivity.

Theorem 2.9 (Aperiodic case [12]).   *Let (Ω, T) be a linearly repetitive subshift and $f, g : \Omega \longrightarrow \mathbb{R}$ locally constant with $f(\omega) \neq 0$ for all $\omega \in \Omega$. If (f, g) is nonperiodic, then there exists a Cantor set Σ of Lebesgue measure zero in $\mathbb{R}$ such that*

$$\sigma(H_\omega) = \Sigma$$

for all $\omega \in \Omega$.

Remark 2.10.   The above theorem was first proven in [59] in the case $f \equiv 1$. This result was then extended in [25] from linearly repetitive subshifts to arbitrary subshifts satisfying a certain condition known as Boshernitzan condition (B) (again for the case $f \equiv 1$). In the form stated above it can be inferred from the recent work [12], Corollary 4. This corollary treats the even more general situation, where condition (B) is satisfied. Condition (B) was introduced by Boshernitzan as a sufficient condition for unique ergodicity [16] (see [25] for an alternative approach as well). In our context, we do not actually need its definition here. It suffices to know that linear repetitivity implies (B) (see e.g. [25]).

The previous result deals with the appearance of Σ as a set. It also gives some information on the spectral type.

Corollary 2.11. *Assume the situation of the previous theorem. Then, no spectral measure can be absolutely continuous with respect to the Lebesgue measure.*

2.3. Aperiodic Order and Discrete Random Schrödinger Operators

Schrödinger operators with aperiodic order can be considered within the context of random Schrödinger operators. Indeed, they arise in the quantum mechanical treatment of solids. As this may be revealing we briefly present this context in this section. Further discussion and references can be found e.g. in the textbooks [18, 19].

Consider a subshift (Ω, T) over the finite alphabet $\mathcal{A}$. Assume without loss of generality that $\mathcal{A}$ is a subset of the real numbers. Let a T-invariant probability measure μ on Ω be given. To these data we can associate the family $(H_\omega)_{\omega \in \Omega}$ of bounded self-adjoint operators on $\ell^2(\mathbb{Z})$ acting via

$$(H_\omega u)(n) = u(n+1) + u(n-1) + \omega(n)u(n).$$

Such operators are (slightly special) cases of the operators considered in the previous section. They arise in the quantum mechanical treatment of disordered solids. The solid in question is modeled by the sequence $\omega \in \Omega$. The operator H_ω then describes the behavior of one electron under the influence of this ω. More specifically, if the state of the electron is $u_0 \in \ell^2(\mathbb{Z})$ at time $t = 0$ then the time evolution is governed by the *Schrödinger equation*

$$(\partial_t u)(t) = -iH_\omega u(t), \; u(0) = u_0.$$

This equation has a unique solution given by

$$u(t) = e^{-itH_\omega} u_0.$$

The behavior of this solution is then linked to the spectral properties of H_ω.

The two basic pieces of "philosophy" underlying the investigations are now the following:

- Increasing regularity of the spectral measures increases the conductance properties of the solid in question.
- The more disordered the subshift is, the more singular the spectral measures are.

Of course, this has to be taken with (more than) a grain of salt. In particular, precise meaning has to be given to what is meant by *regularity and singularity of the spectral measures* and *conductance properties* and *disorder in the subshift*. A large part of the theory is then devoted to giving precise sense to these concepts and then prove (or disprove) specific formulations of the mentioned two pieces of philosophy.

Regarding the first point of the philosophy we mention [48–50, 57] as basic references for proofs of lower bounds on transport via quantitative continuity of the spectral measures.

As for the second point of the philosophy, the two, in some sense, most extreme cases are given by periodic subshifts representing the maximally ordered case on the one hand and the Bernoulli subshift (with uniform measure) on the other hand representing the maximally disordered case.

The periodic situation can be thought of as one with maximal order. As discussed in the previous section the spectral measures are all absolutely continuous (hence not at all singular) and the spectrum consists of nontrivial intervals. These intervals are known in the physics literature as (conductance) bands.

The Bernoulli subshift can be thought of as having maximal disorder. In this case all spectral measures turn out to be pure point measures and the spectrum is pure point spectrum with the eigenvalues densely filling suitable intervals. We refer to the monographs [18, 19] for details and further references.

The subshifts considered in the previous section are characterized by some intermediate form of disorder. They are not periodic. However, they are still minimal and uniquely ergodic and have very low complexity. So they are close to the periodic situation (or rather well approximable by periodic models with bigger and bigger periods). Accordingly, one can expect the following spectral features of the associated Schrödinger operators:

- Absence of absolutely continuous spectral measures (due to the presence of disorder i.e. the lack of periodicity).
- Absence of point spectrum (due to the closeness to the periodic case).
- Cantor spectrum of Lebesgue measure zero (as a consequence of approximation by periodic models with bigger and bigger periods and, hence, more and more gaps).

Indeed, a large part of the theory for Schrödinger operators with aperiodic order is devoted to proving these features (as well as more subtle properties) for specific classes of models. Further details and references can be found in the surveys [21, 22]. Here, we emphasize that also our discussion of the operators associated to a certain substitution generated subshift below will be focused on establishing the above features.

3. The Substitution τ, its Finite Words Sub_τ and its Subshift (Ω_τ, T)

In this section we study the two-sided subshift (Ω_τ, T) induced by a particular substitution τ on $\mathcal{A} = \{a, x, y, z\}$ with

$$\tau(a) = axa, \tau(x) = y, \ \tau(y) = z, \ \tau(z) = x.$$

The one-sided subshift induced by this substitution had already been studied by Vorobets [79]. Some of our results can be seen as providing the two-sided counterparts to his investigations. The key ingredient in the investigations of [79] is that the arising one-sided sequences can be considered as Toeplitz sequences. This is equally true in our case of two-sided sequences. Thus, it seems very likely that one could also base our analysis of the corresponding features on the connection to Toeplitz sequenes. Here, we will present a different approach based on the n-decomposition and n-partition introduced in [39].

The subshift (Ω_τ, T) will be of crucial importance for us as it will turn out that the Schrödinger operators associated to it are unitarily equivalent to the Laplacians on the Schreier graphs of the Grigorchuk's group G.

While we do not use it in the sequel, we would like to highlight that the substitution in question has appeared earlier in the study of Grigorchuk's group G. Indeed, it is (a version of) the substitution used by Lysenok [66] for getting a presentation of Grigorchuk's group G. More specifically, [66] gives that

$$G = \langle a, b, c, d \mid 1 = a^2 = b^2 = c^2 = d^2 = bcd = \kappa^k((ad)^4) \rangle$$
$$= \kappa^k((adacac)^4), k = 0, 1, 2, ... \rangle,$$

where κ is the substitution on $\{a, b, c, d\}$ obtained from τ by replacing x by c, y by b, and z by d.

3.1. The Substitution τ and Its Subshift: Basic Features

Let the alphabet $\mathcal{A} = \{a, x, y, z\}$ be given and let τ be the substitution mentioned above mapping $a \mapsto axa$, $x \mapsto y$, $y \mapsto z$, $z \mapsto x$. Let Sub_τ be the associated set of words given by

$$\mathrm{Sub}_\tau = \bigcup_{w \in \mathcal{A}, n \in \mathbb{N} \cup \{0\}} \mathrm{Sub}(\tau^n(w)).$$

Then, the following three properties obviously hold true:

- The letter a is a prefix of $\tau^n(a)$ for any $n \in \mathbb{N} \cup \{0\}$.
- The lengths $|\tau^n(a)|$ converge to ∞ for $n \to \infty$.
- Any letter of $\mathcal{A}$ occurs in $\tau^n(a)$ for some n.

By the first two properties $\tau^n(a)$ is a prefix of $\tau^{n+1}(a)$ for any $n \in \mathbb{N} \cup \{0\}$. Thus, there exists a unique one-sided infinite word η such that $\tau^n(a)$ is a prefix of η for any $n \in \mathbb{N} \cup \{0\}$. This η is then a fixed point of τ, i.e. $\tau(\eta) = \eta$. We will refer to it as *the fixed point of the substitution τ.* Clearly, η is then a fixed point of τ^n as well for any natural number n.

By the third property we then have

$$\mathrm{Sub}_\tau = \mathrm{Sub}(\eta).$$

We can now associate to τ the subshift

$$\Omega_\tau := \{\omega \in \mathcal{A}^{\mathbb{Z}} : \mathrm{Sub}(\omega) \subset \mathrm{Sub}_\tau\}.$$

Note that every other letter of η is an a (as can easily be seen). Thus, a occurs in η with bounded gaps. This implies that any word of Sub_τ occurs with bounded gaps (as the word is a subword of $\tau^n(a)$ and η is a fixed point of τ^n). For this reason (Ω_τ, T) is minimal and $\mathrm{Sub}(\omega) = \mathrm{Sub}_\tau$ holds for any $\omega \in \Omega_\tau$. We can then apply Theorem 1 of [26] to obtain the following.

Theorem 3.1. *The subshift (Ω_τ, T) is linearly repetitive. In particular, (Ω_τ, T) is uniquely ergodic and minimal.*

Remark 3.2. It is well known that subshifts associated to primitive substitutions are linearly repetitive (see e.g. [27, 32]). Theorem 1 of [26] shows that linear repetitivity in fact holds for subshifts associated to any substitution provided minimality holds. Unique ergodicity is then a direct consequence of linear repetitivity due to Theorem 1.3.

Our further considerations will be based on a more careful study of the $\tau^n(a)$. We set

$$p^{(0)} := a \text{ and } p^{(n)} := \tau^n(a) \text{ for } n \in \mathbb{N}.$$

A direct calculation gives

$$p^{(n+1)} = \tau^{n+1}(a) = \tau^n(axa) = \tau^n(a)\tau^n(x)\tau^n(a) = p^{(n)}\tau^n(x)p^{(n)}.$$

Thus, the following *recursion formula* for the $p^{(n)}$

$$(RF) \qquad p^{(n+1)} = p^{(n)} s_n p^{(n)}$$

with

$$s_n = \tau^n(x) = \begin{cases} x & : & n = 3k, k \in \mathbb{N} \cup \{0\} \\ y & : & n = 3k+1, k \in \mathbb{N} \cup \{0\} \\ z & : & n = 3k+2, k \in \mathbb{N} \cup \{0\} \end{cases}$$

is valid.

This recursion formula is a very powerful tool. This will become clear in the subsequent sections. Here we first note that it implies

$$|p^{(n)}| = 2^{n+1} - 1$$

for all $n \geq 0$. We will now use it to present a formula for the occurrences of the x, y, z in η and to introduce some special elements in Ω_τ.

Proposition 3.3 (Positions of a, x, y, z in η). *Consider the fixed point $\eta = \eta_1\eta_2\ldots$ of τ on $\mathcal{A}^{\mathbb{Z}}$. Then the following holds.*

- *The letter a occurs exactly at the positions $1 + 2k$, $k \in \mathbb{N} \cup \{0\}$ (i.e. at the odd positions).*
- *The letter x occurs exactly at the positions of the form $2^{3n+1} + k \cdot 2^{3n+2}$, $n, k \in \mathbb{N} \cup \{0\}$ (i.e. at the positions of the form $2^{3n+1} \cdot m$ with m an odd integer and $n \in \mathbb{N} \cup \{0\}$ arbitrary).*
- *The letter y occurs exactly at the positions of the form $2^{3n+2} + k \cdot 2^{3n+3}$, $n, k \in \mathbb{N} \cup \{0\}$ (i.e. at the positions of the form $2^{3n+2} \cdot m$ with m an odd integer and $n \in \mathbb{N} \cup \{0\}$ arbitrary).*
- *The letter z occurs exactly at the positions of the form $2^{3n+3} + k \cdot 2^{3n+4}$, $n, k \in \mathbb{N} \cup \{0\}$ (i.e. at the positions of the form $2^{3n+3} \cdot m$ with m an odd integer and $n \in \mathbb{N} \cup \{0\}$ arbitrary).*

Proof. We first note that the given sets of positions are pairwise disjoint and cover $\mathbb{N}$. Thus, it suffices to show that the mentioned letters occur at these positions.

The statement for a is clear. The statements for x, y, z can all be proven similarly. Thus, we only discuss the statement for x. Repeated application of (RF) shows that

$$\eta = p^{(3n+1)} r_1 p^{(3n+1)} r_2 p^{(3n+1)} r_3 \ldots$$

with $r_1, r_2, \ldots \in \{x, y, z\}$. Moreover, (RF) implies

$$p^{(3n+1)} = p^{(3n)} x p^{(3n)}.$$

Combining these formula we see that x must occur at all positions of the form

$$|p^{(3n)}| + 1 + k(|p^{(3n+1)}| + 1) = 2^{3n+1} + k \cdot 2^{3n+2}.$$

This finishes the proof. □

Remark 3.4. The previous proposition shows that η is a *Toeplitz sequence* (with periods of the form 2^l for $l \in \mathbb{N}$). As mentioned already the analysis of the one-sided subshift in [79] is based on this property.

We now head further to use (RF) to introduce some special two-sided sequences. As is not hard to see from (RF), for any $n \in \mathbb{N} \cup \{0\}$ and any single letter $s \in \{x, y, z\}$ the word $p^{(n)} s p^{(n)}$ occurs in η. Thus, for any $s \in \{x, y, z\}$ there exists a unique element $\omega^{(s)} \in \Omega_\tau$ such that

$$\omega^{(s)} = \ldots p^{(n)} s | p^{(n)} \ldots$$

holds for all natural numbers n, where the $|$ denotes the position of the origin. The elements $\omega^{(x)}, \omega^{(y)}, \omega^{(z)} \in \Omega_\tau$ will play an important role in our subsequent analysis. They clearly have the property that they agree on $\mathbb{N}$ with η. Indeed, they can be shown to be exactly those elements of Ω_τ which agree with η on $\mathbb{N}$ (see below). Here, we already note that these three sequences are different. Thus, Ω_τ contains different sequences, which agree on $\mathbb{N}$. Hence, Ω is not periodic.

We finish this section by noting a certain reflection invariance of our system. Recall that a nonempty word $w = w_1 \ldots w_n \in \mathcal{A}^*$ with $w_j \in \mathcal{A}$ is called a *palindrome* if $w = w_n \ldots w_1$. An easy induction using (RF) shows that for any $n \in \mathbb{N} \cup \{0\}$ the word $p^{(n)}$ is a palindrome. It starts and ends with $p^{(k)}$ for any $k \in \mathbb{N} \cup \{0\}$ with $k \leq n$. As each $p^{(n)}$ is a palindrome and any word belonging to Sub_τ is a subword of some $p^{(n)}$ we immediately infer that Sub_τ is closed under reflections in the sense that the following proposition holds.

Proposition 3.5. *Whenever $w = w_1 \ldots w_n \in \mathcal{A}^*$ with $w_j \in \mathcal{A}$ belongs to Sub_τ then so does $\widetilde{w} := w_n \ldots w_1$.*

3.2. The Main Ingredient for Our Further Analysis: n-Partition and n-Decomposition

As a direct consequence of the definitions we obtain that for any $n \in \mathbb{N} \cup \{0\}$ the word η has a (unique) decomposition as

$$\eta = p^{(n)} r_1^{(n)} p^{(n)} r_2^{(n)} \ldots$$

with $r_j^{(n)} \in \{x, y, z\}$. Clearly, this decomposition can be thought of as a way of writing η with "letters" from the alphabet $\mathcal{A}_n = \{\tau^n(a), \tau^n(x), \tau^n(y), \tau^n(z)\} = \{p^{(n)}, x, y, z\}$. Moreover, setting $r_j := r_j^{(0)}$ we have $r_j^{(n)} = \tau^n(r_j)$ for any $j \in \mathbb{N}$. This way of writing η will be called the *n-decomposition of η*. It turns out that an analogous decomposition can actually be given for any element $\omega \in \Omega_\tau$. This will be discussed in this section.

Specifically, we will discuss next that each $\omega \in \Omega_\tau$ admits for each $n \in \mathbb{N} \cup \{0\}$ a unique decomposition of the form

$$\omega = \ldots p^{(n)} s_0 p^{(n)} s_1 p^{(n)} s_2 \ldots$$

with

- $s_k \in \{x, y, z\}$ for all $k \in \mathbb{Z}$,
- the origin ω_0 belongs to $s_0 p^{(n)}$.

Such a decomposition will be referred to as *n-decomposition of ω*. A short moment's thought reveals that if such a decomposition exists at all, then it is uniquely determined by the position of any of the s_j's in ω. Moreover, the positions of the s_j's are given by $p + 2^{n+1}\mathbb{Z}$ with $p \in \{0, \ldots, 2^{n+1} - 1\}$. Thus, the positions are given by an element of $\mathbb{Z}/2^{n+1}\mathbb{Z}$. This suggests the following definition.

Definition 3.6 (*n-partition*). For $n \in \mathbb{N} \cup \{0\}$ we call an element $P \in \mathbb{Z}/2^{n+1}\mathbb{Z}$ an *n-partition* of $\omega \in \Omega_\tau$ if for any $q \in P$ both

- $\omega_q \in \{x, y, z\}$ and
- $\omega_{q+1} \ldots \omega_{q+2^{n+1}-1} = p^{(n)}$

hold.

Clearly, for each $\omega \in \Omega_\tau$, existence (uniqueness) of an n-partition is equivalent to existence (uniqueness) of an n-decomposition. In this sense these two concepts are equivalent. It is not apparent that such an n-partition exists at all. Here is our corresponding result.

Theorem 3.7 (Existence and Uniqueness of n-partitions [39]).   *Let $n \in \mathbb{N} \cup \{0\}$ be given. Then any $\omega \in \Omega_\tau$ admits a unique n-partition $P^{(n)}(\omega)$ and the map*

$$P^{(n)} : \Omega_\tau \longrightarrow \mathbb{Z}/2^{n+1}\mathbb{Z}, \ \omega \mapsto P^{(n)}(\omega),$$

is continuous and equivariant (i.e. $P^{(n)}(T\omega) = P^{(n)}(\omega) + 1$).

Based on n-partitions (and n-decompositions) and the previous theorem one can then study the dynamical system (Ω_τ, T) as well as various questions on the structure of Sub_τ. This is the content of the next sections.

3.3.   The Maximal Equicontinuous Factor of the Dynamical System (Ω_τ, T)

In this section we use n-partitions to study the structure of the dynamical system (Ω_τ, T).

For any $n \in \mathbb{N}$ we can consider the cyclic group $\mathbb{J}^{(n)} := \mathbb{Z}/2^n\mathbb{Z}$ together with the map $A^{(n)}$, called *addition map*, which sends $m + 2^n\mathbb{Z}$ to $m + 1 + 2^n\mathbb{Z}$. Then, $(\mathbb{J}^{(n)}, A^{(n)})$ is a periodic minimal dynamical system. Moreover, there are natural maps

$$\pi_n : \mathbb{J}^{(n+1)} \longrightarrow \mathbb{J}^{(n)}, \ m + 2^{n+1}\mathbb{Z} \mapsto m + 2^n\mathbb{Z},$$

for any $n \in \mathbb{N} \cup \{0\}$. These maps allow one to construct the topological abelian group $\mathbb{Z}_2$ as the inverse limit of the system $(\mathbb{J}^{(n+1)}, \pi_n)$, $n \in \mathbb{N}$. Specifically, the elements of $\mathbb{Z}_2$ are sequences (m_n) with $m_n \in \mathbb{J}^{(n)}$ and $\pi_n(m_{n+1}) = m_n$ for all $n \in \mathbb{N}$. This group is called the *group of dyadic integers*. The addition maps $A^{(n)}$ are compatible with the inverse limit and lift to a map A_2 on $\mathbb{Z}_2$ (which is just addition by 1 on each member of the sequence in question). In this way, we obtain a dynamical system $(\mathbb{Z}_2, A_2)$. It is known as the *binary odometer*. As A_2 is just addition one can think of this system as an "adding machine." It is well known that this dynamical system is minimal.

Now Theorem 3.7 can be rephrased as saying that the dynamical system $(\mathbb{J}^{(n+1)}, A^{(n+1)})$ is a factor of (Ω_τ, T) via the factor map $P^{(n)}$. Clearly, the factor maps $P^{(n)}$ are compatible with the natural canonical projections π_n in the sense that

$$\pi_n \circ P^{(n)} = P^{(n-1)}$$

holds for all $n \geq 1$. Thus, we can "combine" the $P^{(n)}$ for all $n \in \mathbb{N}$ to get a factor map

$$P_2 : (\Omega_\tau, T) \longrightarrow (\mathbb{Z}_2, A_2), \omega \mapsto (n \in \mathbb{N} \mapsto P^{(n-1)}(\omega)).$$

We first use this to study continuous eigenvalues. Let (Y, R) be a dynamical system (i.e. Y is a compact space and R is a homeomorphism). Denote the unit circle in $\mathbb{C}$ by $\mathbb{S}^1$. Then, $k \in \mathbb{C}$ is called a *continuous eigenvalue* of the dynamical system (Y, R) if there exists a continuous not everywhere vanishing function f with values in $\mathbb{C}$ on Y satisfying

$$f(R(y)) = kf(y)$$

for all $y \in Y$. Such a function is called a *continuous eigenfunction*. Then, any continuous eigenvalue belongs to $\mathbb{S}^1$ and the continuous eigenvalues form a group under multiplication whenever the underlying dynamical system is minimal. Indeed, by minimality any continuous eigenfunction has constant (nonvanishing) modulus. Then, the product of two eigenfunctions is an eigenfunction to the product of the corresponding eigenvalues. The complex conjugate of an eigenfunction is an eigenfunction to the inverse of the corresponding eigenvalue and the constant function is an eigenfunction to the eigenvalue 1. Moreover, it is not hard to see that minimality implies that the multiplicity of each continuous eigenvalue is one (i.e. for any two continuous eigenfunctions f, g to the same eigenvalue there exists a complex number c with $f = cg$).

Let now $\mathcal{E}_n$ be the group of continuous eigenvalues of $\mathbb{J}^{(n)}$. This is just the subgroup of $\mathbb{S}^1$ given by $\{e^{2\pi i \frac{k}{2^n}} : 0 \leq k \leq 2^n - 1\}$. Then, clearly the groups $\mathcal{E}_n$ and $\mathbb{J}^{(n)}$ are dual to each other via

$$\mathcal{E}_n \times \mathbb{J}^{(n)} \longrightarrow \mathbb{S}^1, (z, m + 2^n\mathbb{Z}) \mapsto z^m.$$

The dual maps to the canonical projections $\pi_n : \mathbb{J}^{(n+1)} \longrightarrow \mathbb{J}^{(n)}$ are then the canonical embeddings

$$\iota_n : \mathcal{E}_n \longrightarrow \mathcal{E}_{n+1}, z \mapsto z.$$

Thus, $\mathcal{E}_n$ is a subgroup of $\mathcal{E}_{n+1}$. Let $\mathcal{E}$ be the group arising as the union of the $\mathcal{E}_n$, i.e.

$$\mathcal{E} := \bigcup_n \mathcal{E}_n.$$

This group is often denoted as $\mathbb{Z}(2^\infty)$. Equip it with the discrete topology and denote its Pontryagin dual $\widehat{\mathcal{E}}$ by $\mathbb{J}$.

Proposition 3.8. *The group $\mathbb{J}$ is canonically isomorphic to the group $\mathbb{Z}_2$ via*

$$\mathbb{Z}_2 \longrightarrow \mathbb{J}, (m_n + 2^n\mathbb{Z})_n \mapsto (z \mapsto z^{m_n}, \text{ whenever } z \in \mathcal{E} \text{ belongs to } \mathcal{E}_n).$$

Proof. By construction $\mathbb{Z}_2$ comes about as the inverse limit of the $(\mathbb{J}^{(n+1)}, \pi_n)$, $n \geq 1$. Then, the dual group of $\mathbb{Z}_2$ arises as the direct limit of the dual system $(\mathcal{E}_n, \iota_n)$. This limit is just the group $\mathcal{E}$. Dualizing once more we find that $\mathbb{J}$ is indeed canonically isomorphic to the dual of $\mathbb{Z}_2$. To obtain the actual formula we can proceed as follows:

Define $\varepsilon_n := e^{2\pi i \frac{1}{2^n}}$. Then, each ε_n is a complex primitive 2^nth root of 1 with $\varepsilon_{n+1}^2 = \epsilon_n$. Now, consider an arbitrary element $\gamma \in \mathbb{J}$ i.e. a character $\gamma : \mathcal{E} \longrightarrow \mathbb{S}^1$. Then, γ is completely determined by its values on the ε_n, $n = 1, 2 \ldots$. Moreover, for each n we have

$$(\gamma(\varepsilon_{n+1}))^2 = \varepsilon_n^{m_n}$$

for a unique $m_n \in \mathbb{Z}/2^n\mathbb{Z}$ as

$$\left((\gamma(\varepsilon_{n+1})^2)\right)^{2^n} = (\gamma(\varepsilon_{n+1}^2))^{2^n} = (\gamma(\varepsilon_n))^{2^n} = \gamma(1) = 1.$$

It is not hard to see that m_{n+1} goes to m_n under the natural surjection π_n. Thus, to each character $\gamma : \mathcal{E} \longrightarrow \mathbb{S}^1$ there corresponds a sequence $(m_1, m_2, \ldots)$ with $m_n \in \mathbb{Z}/2^n\mathbb{Z}$ and $\pi_n(m_{n+1}) = m_n$ for all $n \in \mathbb{N}$ (and vice versa). The set of such sequences is exactly $\mathbb{Z}_2$. $\square$

As any eigenvalue belongs to $\mathbb{S}^1$, there is a canonical embedding of groups $\mathcal{E} \longrightarrow \mathbb{S}^1$. In fact, as things are set up here this embedding is just inclusion of subsets of $\mathbb{C}$.

By duality, this gives rise to a group homomorphism $j : \mathbb{Z} \longrightarrow \mathbb{J}$ with dense range. This homomorphism induces then an action A of $\mathbb{Z}$ on $\mathbb{J}$ via

$$A : \mathbb{J} \longrightarrow \mathbb{J}, A\gamma := j(1)\gamma.$$

It is not hard to see that this A corresponds to A_2 if $\mathbb{J}$ is identified with $\mathbb{Z}_2$ according to Proposition 3.8.

Disentangling definitions, we also infer that the action is given by

$$(A\gamma)(k) = k \, \gamma(k)$$

(for $\gamma \in \mathbb{J}$ and $k \in \mathcal{E}$). We denote the arising dynamical system as $(\mathbb{J}, A)$. It is isomorphic to the binary odometer. By its very construction it is what is called a *rotation on a compact abelian group* (viz the action A comes about by multiplication with $j(1)$, where $j : \mathbb{Z} \longrightarrow \mathbb{J}$ is a group homomorphism). Thus, by standard theory (see e.g. [4, 17]) its group of continuous eigenvalues is exactly given by the dual of $\mathbb{J}$ i.e. by $\mathcal{E}$.

Now, obviously any eigenfunction of $\mathbb{J}^{(n)}$ gives immediately rise to an eigenfunction of Ω_τ for the same eigenvalue (by composing with the factor map). As the factor map is continuous, we obtain in this way continuous eigenfunctions to each of the elements from $\mathcal{E}$. Minimality easily shows that (up to an overall scaling) each of these eigenfunctions is unique. Thus, we obtain a family of continuous eigenfunctions. At this point it is not clear that all continuous eigenvalues of (Ω_τ, T) belongs to $\mathcal{E}$. However, this (and more) will be shown later.

We can use the preceding considerations to introduce a closed equivalence relation $\approx$ on Ω_τ via

$$\omega \approx \omega' :\Longleftrightarrow f(\omega) = f(\omega')$$

for all eigenfunctions corresponding to eigenvalues from $\mathcal{E}$.

Then, $\Omega_\tau / \approx$ is a compact topological space when equipped with the quotient topology.

Clearly, $\omega \approx \omega'$ if and only if $T\omega \approx T\omega'$. Thus, the relation $\approx$ is compatible with the shift operation. Hence, the quotient $\Omega_\tau / \approx$ becomes a dynamical system with the operation $T^\approx$ induced by the shift.

Fix now an $\omega_0 \in \Omega_\tau$. As discussed above continuous eigenfunctions do not vanish anywhere and the multiplicity of each continuous eigenvalue is one. Thus, for each $k \in \mathcal{E}$ there exists a unique eigenfunction f_k to k on Ω_τ with $f_k(\omega_0) = 1$. Then, the arising system of eigenfunctions will have the property that

$$f_{k_1} f_{k_2} = f_{k_1 + k_2}, \ \ f_{-k} = \overline{f_k}$$

for all $k, k_1, k_2 \in \mathcal{E}$. Thus, any $\omega \in \Omega_\tau$ will give rise to an element of $\mathbb{J} = \widehat{\mathcal{E}}$ via

$$\mathcal{E} \longrightarrow \mathbb{S}^1, k \mapsto f_k(\omega).$$

Even more is true and the following result holds. It is well known and can be found in various places in the literature. Recent discussions are given in [4, 6, 9].

Lemma 3.9. *The dynamical systems* $(\Omega_\tau / \approx, T^\approx)$ *and* $(\mathbb{J}, A)$ *are conjugate via the map*

$$[\omega] \mapsto (k \mapsto f_k(\omega)).$$

In particular, the eigenvalues of $(\Omega_\tau / \approx, T^\approx)$ *are exactly given by the elements of* $\mathcal{E}$.

We now further investigate $\approx$ and provide a characterization of $\omega \approx \omega'$. Here, we will again use the special words $\omega^{(x)}, \omega^{(y)}, \omega^{(z)}$ introduced above.

Proposition 3.10 (Characterization $\approx$). *For* $\omega, \omega' \in \Omega_\tau$ *the relation* $\omega \approx \omega'$ *holds if and only if one of the following two properties hold:*

- $\omega = \omega'$.
- *There exist* $s, s' \in \{x, y, z\}$ *with* $s \neq s'$ *and* $N \in \mathbb{Z}$ *with* $\omega = T^N \omega^{(s)}$ *and* $\omega' = T^N \omega^{(s')}$.

Proof. Let ω, ω' with $\omega \neq \omega'$ and $\omega \approx \omega'$ be given. By definition of $\approx$ and the above construction of the eigenfunctions of (Ω_τ, T), we then have that

$$P^{(n)}(\omega) = P^{(n)}(\omega')$$

for all $n \in \mathbb{N} \cup \{0\}$. Call this quantity $P^{(n)}$. In the remaining part of the proof we will identify such a $P^{(n)}$ with its unique representative in $\{0, \ldots, 2^{n+1} - 1\}$.

As $\omega \neq \omega'$ we infer that one of the sequences $(P^{(n)})_{n \in \mathbb{N} \cup \{0\}}$ or $(2^{n+1} - P^{(n)})_{n \in \mathbb{N} \cup \{0\}}$ must be bounded. (Otherwise, ω and ω' would agree on larger and larger pieces around the origin and then had to be equal.) Assume without loss of generality that $(P^{(n)})$ is bounded. By restricting attention to a subsequence we can then assume without loss of generality that $P^{(n)} = P$ for all n. After shifting the sequences by P to the left we can then assume without loss of generality that $P^{(n)} = 0$ for all n. By definition of P, there then exist letters $s, s' \in \{x, y, z\}$ with

$$\omega = \ldots s | p^{(n)} \ldots \quad \text{and} \quad \omega' = \ldots s' | p^{(n)} \ldots$$

for all $n \geq 0$, where $|$ denotes the position of the origin. This gives, by definition of the n-partition that in fact

$$\omega = \ldots p^{(n)} s | p^{(n)} \quad \text{and} \quad \omega' = \ldots p^{(n)} s' | p^{(n)} \ldots$$

for all $n \geq 0$. Thus, we obtain $\omega = \omega^{(s)}$ and $\omega' = \omega^{(s')}$. As $\omega \neq \omega'$ we infer that $s \neq s'$. This finishes the proof. $\quad\square$

The previous result shows that the factor map from Ω_τ to $\Omega_\tau / \approx$ is one-to-one except on three orbits. This has strong consequences as will be discussed next.

As (Ω_τ, T) is uniquely ergodic, there exists a unique T-invariant probability measure λ on Ω_τ. The operation T then induces a unitary operator U_T on the associated L^2-space via

$$U_T : L^2(\Omega, \lambda) \longrightarrow L^2(\Omega, \lambda), \ U_T f = f \circ T.$$

An element $f \in L^2(\Omega, \lambda)$ (with $f \not\equiv 0$) is called a *measurable eigenfunction* to $k \in \mathbb{S}^1$ if $U_T f = kf$. The subshift is said to have *pure point spectrum* if there exists an orthonormal basis of measurable eigenfunctions. From the two previous results we immediately infer the following.

Theorem 3.11. *The dynamical system (Ω_τ, T) has pure point spectrum and any measurable eigenvalue is a continuous eigenvalue and belongs to $\mathcal{E}$.*

Proof. The dynamical system $(\mathbb{J}, A)$ has pure point spectrum with all eigenvalues being continuous and belonging to $\mathcal{E}$ as it is a shift on the compact abelian group $\mathbb{J}$ which is the Pontryagin dual of $\mathcal{E}$, [81]. As the dynamical system $\Omega_\tau / \approx$ is conjugate to $(\mathbb{J}, A)$ due to Lemma 3.9 it has also pure point spectrum with all eigenvalues being continuous and belonging to $\mathcal{E}$.

Now, Proposition 3.10 shows that factor map from Ω_τ to $\Omega_\tau / \approx$ is one-to-one except on three countable orbits. This implies that in terms of measures the associated L^2-spaces are isomorphic. This easily gives the desired result. $\quad\square$

Remark 3.12. The occurrence of pure point dynamical spectrum is a key feature in the investigation of aperiodic order. In fact, while there is no axiomatic framework for aperiodic order a distinctive feature is (pure) point diffraction. Now, pure point diffraction has been shown to be equivalent to pure point dynamical spectrum. For the case of subshifts at hand this can be inferred (after some work) from [72]. A more general result (dealing with uniquely ergodic Delone systems) was then given in [58]. The result can even further be generalized to measure dynamical systems and even processes [7, 62, 64].

The previous theorem implies that $(\mathbb{J}, A)$ is exactly the *maximal equicontinuous factor* of (Ω_τ, T) (see e.g. [3] for definition). Indeed, one of the many equivalent ways to describe this factor is as the dual group of the group of continuous eigenvalues. A recent discussion of this and various related facts can be found in [4]. Henceforth, we will denote the maximal equicontinuous factor of (Ω_τ, T) by $(\Omega_\tau^{\max}, T^{\max})$ and the corresponding factor map by $\pi^{\max}$. Then, our findings so far provide the following theorem.

Theorem 3.13 (Factor map onto Ω_τ^{max}). *The three dynamical systems* $(\Omega_\tau/\approx, T^\approx)$, $(\mathbb{J}, A)$ *and* $(\Omega_\tau^{\max}, T^{\max})$ *are topologically conjugate. The factor map*

$$\pi^{\max} : \Omega_\tau \longrightarrow \Omega_\tau^{\max}$$

is one-to-one in all points except on the images of the points of the orbits of $\omega^{(x)}, \omega^{(y)}, \omega^{(z)}$. *In these points it is three-to-one.*

Remark 3.14.

- Minimal systems with the property that their factor map to the maximal equicontinuous factor is one-to-one in at least one point are known as *almost automorphic systems* (see e.g. [4] for further details). Their study has attracted a lot of attention. As the previous result shows, (Ω_τ, T) is an almost automorphic system. In fact, as Ω_τ is uncountable, the factor map is one-to-one in almost every point with respect to the unique invariant probability measure λ on Ω_τ. This is sometimes expressed as *the factor map from* (Ω_τ, T) *to its maximal equicontinuous factor is almost-everywhere one-to-one.*

- There is an alternative description of the relation $\approx$ for almost automorphic systems. More specifically, define the *proximality relation* $\sim$ by

$$\omega \sim \omega' \iff \inf_{n \in \mathbb{Z}} d(T^n \omega, T^n \omega') = 0,$$

where d is any metric on Ω_τ inducing the topology. (Due to compactness of Ω_τ the relation is indeed independent of the chosen metric.) Note that the proximality relation can be thought of as describing asymptotic agreement. Then, for almost automorphic systems the proximality relation $\sim$ and the relation $\approx$ agree. This can be found in the book of Auslander [3]. A recent discussion is given in [4]. In fact, this result is even more general in that one does not need almost automorphy but only a weaker condition called *coincidence rank one*. We refrain from further discussion of this concept and refer the reader to e.g. [9] for further investigation. We just note here that in our situation, equality of $\sim$ and $\approx$ and the characterization of $\approx$ in Proposition 3.10 imply that sequences that are proximal (i.e. asymptotically equal) are in fact equal everywhere up to one position.

- In [79] Vorobets shows that the one-dimensional subshift associated to τ has the binary odometer $(\mathbb{J}, A)$ as a factor with the factor map being $1:1$ in all points except three orbits. He uses this to conclude pure point spectrum and unique ergodicity. Our corresponding results above for the two-sided subshift can therefore be seen as analogues to his results. However, our approach is different: It is based on n-partition whereas his approach is based on Toeplitz sequences.

3.4. Powers and the Index (Critical Exponent) of Sub_τ

In this section we have a closer look at the structure of Sub_τ. The main focus will be on occurrences of three blocks and the index of words (also known as critical exponent). We will use n-partitions in our study in a spirit similar to [23, 24].

We start by investigating occurrences of almost four blocks. An easy inspection of η gives the following lemma.

Lemma 3.15. *The word $axaxaxa$ belongs to Sub_τ.*

The previous result deals with occurrence of a cube (and even a bit more) of the special word ax. As Sub_τ is invariant under τ this then yields the occurrence of many more cubes. This can be used to exclude eigenvalues for Schrödinger operators via the so-called Gordon argument. This is discussed in [39] for the case at hand. Such an application of the Gordon arguments for subshifts coming from substitution goes back to [20], see [21] for a survey as well.

Next we show there are no fourth powers occurring in Sub_τ.

Let us recall from the considerations on n-partitions in Section 3.2 that there exists a sequence $r_1^{(n)} r_2^{(n)} \ldots \in \{x, y, z\}^{\mathbb{N}}$ such that the fixed point η of τ can be written as

$$\eta = p^{(n)} r_1^{(n)} p^{(n)} r_2^{(n)} \ldots$$

with $r_j^{(n)} = \tau^n(r_j) \in \{x, y, z\}$ for any $n \in \mathbb{N} \cup \{0\}$. This way of writing η is referred to as the n-decomposition of η. Call the sequence

$$r^{(n)} = r_1^{(n)} r_2^{(n)} \ldots \in \{x, y, z\}^{\mathbb{N}}$$

the nth *derived sequence of* η. Note that for any natural number n the combinatorial properties of the sequence $r^{(n)}$ are exactly the same as the combinatorial properties of the sequence $r^{(1)}$ as τ^n is injective on $\{x, y, z\}$ and $r^{(n)} = \tau^n(r^{(1)})$ holds.

Proposition 3.16. [39] *In the derived sequence* $r = r^{(1)}$ *the letters* y *and* z *always occur isolated preceded and followed by an* x*. The letter* x *always occurs either isolated (i.e. preceded and followed by elements of* $\{y, z\}$*) or in the form* xxx*. In particular, there is no occurrence of* $xxxx$*. The analogue statements hold for any natural number* n *for the sequence* $r^{(n)}$ *(with* x, y, z *replaced by* $\tau^n(x), \tau^n(y)$ *and* $\tau^n(z)$*).*

Remark 3.17. In terms of information the derived sequences $r^{(n)}$ are as useful as the original sequence. We will base our subsequent investigations on the relatively simple properties of the derived sequence $r^{(1)}$ given in the preceding proposition. More information should be obtainable from a more detailed study of the derived sequences.

The n-decomposition of η gives a way of writing η as a concatenation of the words $p^{(n)}$ and elements from $\{x, y, z\}$. For example, η can be written as

$$\eta = (axa)y(axa)z(axa)y(ax\ \underbrace{a)x(a}_{axa}\ xa)\ldots$$

where we have put brackets around the $p^{(1)} = axa$. Still, η can contain further occurrences of $p^{(1)}$ as indicated in the preceding formula. More generally, it is not true that a $p^{(n)}$ occurring somewhere in η is in fact one of the words $p^{(n)}$ appearing in the n-decomposition of η. However, it turns out that whenever $p^{(n)}sp^{(n)}$ occurs in η then both of its $p^{(n)}$ actually stem from the n-partition. In this sense, there is some form of alignment. This is the content of the next proposition [39].

Proposition 3.18 (Alignment of the $p^{(n)}sp^{(n)}$). *Consider a natural number* n *and* $s \in \{x, y, z\}$*. If* $p^{(n)}sp^{(n)}$ *occurs in* η *at the position* l *(i.e.* $\eta_l \eta_{l+1} \cdots \eta_{l+|p^{(n)}sp^{(n)}|-1} = p^{(n)}sp^{(n)}$ *holds), then* l *is of the form* $1 + k2^{n+1}$ *for some* $k \in \mathbb{N} \cup \{0\}$*. This means that if* $p^{(n)}sp^{(n)}$ *occurs somewhere in* η *then both of its words* $p^{(n)}$ *actually agree with blocks* $p^{(n)}$ *appearing in the n-decomposition* $\eta = p^{(n)}r_1^{(n)}p^{(n)}r_2^{(n)}p^{(n)}\ldots$

If w is a finite word in Sub_τ and v is a prefix of w and N is a natural number we define the *index of the word* w *in* $w^N v$ by $N + \frac{|v|}{|w|}$ and denote it by $Ind(w, w^N v)$. We then define the *index of the word* w by

$$Ind(w) := \max\{Ind(w, w^N v) : v \text{ prefix of } w, N \in \mathbb{N}, w^N v \in \text{Sub}_\tau\}.$$

As our subshift is minimal and aperiodic the index of every word can easily be seen to be finite.

Theorem 3.19 (Index of Ω_τ). *(a) For every $w \in Sub_\tau$ the inequality $Ind(w) < 4$ holds. In particular, η does not contain a fourth power.*
(b) We have $4 = \sup\{Ind(w) : w \in Sub_\tau\}$.

Remark. The supremum over all the indices is sometimes known as the critical exponent.

Proof. A proof can be found in [39]. Here, we only sketch the idea. By Lemma 3.15 the word $w = axaxaxa = v^3a$ (with $v = ax$) belongs to Sub_τ. For each $n \in \mathbb{N}$ we then have

$$\tau^n(v^3a) = (\tau^n(v))^3\tau^n(a) = p^{(n)}\tau^n(x)p^{(n)}\tau^n(x)p^{(n)}\tau^n(x)p(n)$$

and we infer that the index must be at least 4. Thus, it suffices to show that η does not contain a fourth power. To do so, it suffices to consider occurrences of powers of words w in η. For short words the statement can easily be checked. Consider now the case $|p^{(n)}| + 1 = 2^{n+1} \leq |w| \leq |p^{(n+1)}|$ for some $n \geq 1$. Assume that www occurs in Sub_τ. Then, the proposition on alignment gives that the length of w is given by $|w| = |p^{(n)}| + 1 = 2^{n+1}$. This then easily implies the desired statement. $\square$

Remark 3.20. The proof of the theorem shows that the length of any word $w \in Sub_\tau$ whose cube www also belongs to Sub_τ is given by 2^n for some $n \in \mathbb{N}$.

3.5. The Word Complexity of Sub_τ

In this section we present a result on the word complexity of Sub_τ. A detailed proof can be found in [40].

We define the *word complexity* of the subshift (Ω_τ, T) as

$$\mathcal{C} : \mathbb{N} \cup \{0\} \longrightarrow \mathbb{N} \quad \mathcal{C}(L) = \text{number of elements of } Sub_\tau \text{ of length } L.$$

Recall that a word $w \in Sub_\tau$ is called *right special* if the set of its extensions

$$\{s \in \{a, x, y, z\} : ws \in Sub_\tau\}$$

has more than one element.

Theorem 3.21 (Complexity theorem). *(a) For any $n \geq 2$ and $L = 2^n + k$ with $0 \leq k < 2^n$ we have*

$$\mathcal{C}(L+1) - \mathcal{C}(L) = \begin{cases} 3 : 0 \leq k < 2^{n-1} \\ 2 : 2^{n-1} \leq k < 2^n. \end{cases}$$

(b) The complexity function $\mathcal{C}$ satisfies

$$\mathcal{C}(1) = 4, \mathcal{C}(2) = 6, \mathcal{C}(3) = 8$$

and then for any $n \geq 2$ and $L = 2^n + k$ with $0 \leq k < 2^n$

$$C(L) = \begin{cases} 2^{n+1} + 2^{n-1} + 3k : 0 \leq k < 2^{n-1} \\ 2^{n+1} + 2^n + 2k : 2^{n-1} \leq k < 2^n. \end{cases}$$

(c) Consider $n \geq 2$ and $L = 2^n + k$ with $0 \leq k < 2^n$.

- *If $0 \leq k < 2^{n-1}$, then there exist exactly two right special words of length L. These are given by the suffix of $p^{(n)}$ of length L (which can be extended by x, y, z) and the suffix of $p^{(n-2)}\tau^{n-2}(x)p^{(n-1)}$ of length L (which can be extended by $\tau^{n-2}(x)$ and by $\tau^{n-1}(x)$).*
- *If $2^{n-1} \leq k < 2^{n-1}$, then there exists exactly one right special words of length L. This is given by the suffix of $p^{(n)}$ of length L (which can be extended by x, y, z).*

Proof. The proof relies on a detailed investigation of the n-partition of η. This allows one to directly determine all words of length $|p^{(n)}| = 2^{n+1} - 1$ in Sub_τ and this gives $C(2^{n+1} - 1)$ for all $n \in \mathbb{N} \cup \{0\}$. At the same time the study of the n-partition allows one to obtain a lower bound on the difference $C(L+1) - C(L)$ for all $L \in \mathbb{N}$. This in turn gives a lower bound on C. Combining the lower bound and the precise values we obtain the statements of the theorem. □

Remark 3.22. As Sub_τ is closed under reflections by Proposition 3.5 the above statements about right special words easily translate on corresponding statements about left special words. (Here, a word $w \in \mathrm{Sub}_\tau$ is called *left special* if the set $\{s \in \{a, x, y, z\} : sw \in \mathrm{Sub}_\tau\}$ has more than one element.) This shows in particular that the words $p^{(n)}$, $n \in \mathbb{N} \cup \{0\}$, are both right special and left special (and are the only words with this property).

3.6. Generating the Fixed Point η by an Automaton

In this section we present an automaton that generates the fixed point η of the substitution τ. This is well in line with general theory on how to exhibit fixed points of substitutions by automata, see e.g. the monograph [2] to which we also refer for background on automata. There are numerous applications of automatic sequences in group theory. For a recent example and a list of further references we refer the reader to [41].

Consider the automaton $\mathcal{A}$ from Figure 10.1. It is an automaton over the alphabet $\{0, 1\}$ with four states q_0, q_1, q_2, q_3, which are labeled by a, x, y, z respectively. Then, the infinite sequence

$$\mathcal{A}_{q_0} : \mathbb{N} \cup \{0\} \longrightarrow \{a, x, y, z\}$$

generated by the automaton with initial state q_0 is defined as follows: Write $n \in \mathbb{N} \cup \{0\}$ in its binary expansion as

$$n = x_0 2^i + x_1 2^{i-1} + \cdots + x_{i-1} 2 + x_i$$

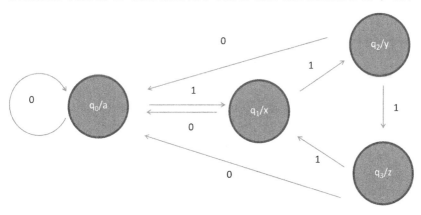

Figure 10.1 The automaton generating η

with $i \in \mathbb{N} \cup \{0\}$ and $x_j \in \{0, 1\}$, $j = 0, \ldots, i$. Consider now the path p_n in the automaton starting in q_0 and following the sequence $x_0 x_1 \ldots x_i$. Then, $\mathcal{A}_{q_0}(n)$ is defined as the label of the state where this path ends.

Theorem 3.23. *The fixed point η of τ agrees with $\mathcal{A}_{q_0}$ (where the fixed point is considered as a map from $\mathbb{N} \cup \{0\}$ to $\{a, x, y, z\}$).*

This theorem is an immediate consequence of the next proposition. To state the proposition we will need some further pieces of notation. For each $n \in \mathbb{N} \cup \{0\}$ and each state q of the automaton we define $f^{(n)}(q)$ to be the word over $\{a, x, y, z\}$ of length 2^n obtained in the following way: Let $v_1, \ldots, v_{2^n}$ be the list of all words of length n over $\{0, 1\}$ in lexicographic order (where $0 < 1$). Consider now for each $k = 1, \ldots, 2^n$ the path in the automaton starting at q and following the word v_k. Then, the kth letter of $f^{(n)}(q)$ is defined as the label of the state where this path ends.

Define the letters

$$s_n = \tau^n(x) = \begin{cases} x & : & n = 3k, k \in \mathbb{N} \cup \{0\} \\ y & : & n = 3k + 1, k \in \mathbb{N} \cup \{0\} \\ z & : & n = 3k + 2, k \in \mathbb{N} \cup \{0\} \end{cases}$$

and recall the recursion formula

$$p^{(n+1)} = p^{(n)} s_n p^{(n)}$$

with $p^{(n)} := \tau^n(a)$.

Proposition 3.24. *For each natural number n and each $i = 0, 1, 2, 3$ we have*

$$f^{(n+1)}(q_i) = p^{(n)} s_{n+i}.$$

Proof. This is proven by induction. The case $n = 1$ follows by inspection. Assume now that the statement is true for some $n \geq 1$ and consider $n + 1$. Then the lexicographic ordering of the words of length $n + 2$ over $\{0, 1\}$ is given by

$$0v_1, \ldots 0v_{2^{n+1}}, 1v_1, \ldots 1, v_{2^{n+1}},$$

where $v_1, \ldots, v_{2^{n+1}}$ is the lexicographic ordering of the words of length $n + 1$ over $\{0, 1\}$. Thus, from the rules of the automaton we obtain for each $i = 0, 1, 2, 3$

$$f^{(n+2)}(q_i) = f^{(n+1)}(q_0)f^{(n+1)}(q_{i+1}),$$

where we set $q_4 = q_1$. From our assumption for n and the recursion we then find

$$f^{(n+2)}(q_i) = p^{(n)}s_n p^{(n)}s_{n+1+i} = p^{(n+1)}s_{n+1+i}$$

for each $i = 0, 1, 2, 3$ and this is the desired statement. $\square$

3.7. Replacing τ by a Primitive Substitution

The substitution τ arises naturally in the study of Grigorchuk groups G and its Schreier graphs (see below). From the point of view of subshifts it has the (slight) disadvantage of not being primitive. It turns out that it is possible to find a primitive substitution ξ with the same fixed point— and hence the same subshift—as τ. This can then be used to obtain alternative proofs for the (proven above) linear repetitivity and pure discreteness of the spectrum. This is discussed at the end of this section. The material presented here was pointed out to us by Fabien Durand [31].

Consider the substitution ζ on the alphabet $\{a, x, y, z\}$ with

$$\zeta(a) = ax, \zeta(y) = ay, \zeta(y) = az, \zeta(z) = ax(= \zeta(a)).$$

As any letter of the alphabet is contained in $\zeta^4(s)$ for any letter s, this is a primitive substitution.

To relate it to τ we use (again) for $n \in \mathbb{N} \cup \{0\}$ the letters $s_n = \tau^n(x)$ as well as the recursion formula $p^{(n+1)} = p^{(n)}s_n p^{(n)}$ with $p^{(n)} := \tau^n(a)$.

Proposition 3.25. *For any natural number n the equality*

$$\zeta^n(a) = \tau^{n-1}(a)s_{n-1}$$

holds. In particular, the fixed point η of τ agrees with the fixed point of ζ.

Proof. This is shown by induction. The cases $n = 1, 2, 3$ are easily checked by direct inspection. Assume now that the statement is true for all integers up to some $n \geq 3$. Then, we can compute

$$\zeta^{n+1}(a) = \zeta^n(ax) = \zeta^n(a)\zeta^n(x) = \zeta^n(a)\zeta^{n-1}(ay)$$
$$= \cdots = \zeta^n(a)\zeta^{n-1}(a)\zeta^{n-2}(a)\zeta^{n-3}(ax).$$

By $\zeta(a) = ax$ we then find

$$\zeta^{n+1}(a) = \zeta^n(a)\zeta^{n-1}(a)\zeta^{n-2}(a)\zeta^{n-2}(a).$$

From our assumption on n we then infer

$$\zeta^{n+1}(a) = \tau^{n-1}(a)s_{n-1}\tau^{n-2}(a)s_{n-2}\tau^{n-3}(a)s_{n-3}\tau^{n-3}(a)s_{n-3}.$$

Successive application of the recursion formula and the fact that $s_{n-3} = s_n$ then give

$$\begin{aligned}
\zeta^{n+1}(a) &= \tau^{n-1}(a)s_{n-1}\tau^{n-2}(a)s_{n-2}(\tau^{n-3}(a)s_{n-3}\tau^{n-3}(a))s_{n-3} \\
&= \tau^{n-1}(a)s_{n-1}\tau^{n-2}(a)s_{n-2}\tau^{n-2}(a)s_{n-3} \\
&= \tau^{n-1}(a)s_{n-1}\tau^{n-1}(a)s_{n-3} \\
&= \tau^n(a)s_{n-3} \\
&= \tau^n(a)s_n.
\end{aligned}$$

This is the desired statement. $\qquad\qquad\qquad\qquad\qquad\qquad\qquad\qquad$ $\square$

Corollary 3.26. *The subshift (Ω_ζ, T) generated by the primitive substitution ζ agrees with the subshift (Ω_τ, T).*

As it is well known see e.g. [27, 30], that subshifts associated to primitive substitutions are linearly repetitive, an immediate consequence of the previous corollary is that (Ω_τ, T) is linearly repetitive. Also, as ζ has constant length (i.e. the length of $\zeta(t)$ is the same for any letter t) and $\zeta(t)$ starts with a for any letter t, we can apply a result of Dekking [29] to obtain purely discrete spectrum.

4. Background on Graphs and Dynamical Systems

In this section we recall some basic notions and concepts from the theory of graphs, dynamical systems, and Schreier graphs. In the next section we will meet all these abstract concepts in the context of a particular group.

Here we first recall some terminology from the theory of graphs and introduce the topological space of (isomorphism classes of) rooted labeled graphs.

Let $\mathcal{B}$ be a finite nonempty set together with an involution $\mathcal{B} \longrightarrow \mathcal{B}$, $b \mapsto \overline{b}$. A *graph with edges labeled by* $\mathcal{B}$ is a pair (V, E) consisting of a set V and a set $E \subset V \times V \times \mathcal{B}$ such that (v, w, b) belongs to E if $(w, v, \overline{b})$ belongs to E. The elements of V are called *vertices* and the elements of E are called *edges*. Whenever $e = (v, w, b)$ is an edge, then b is called the *label*, $v = o(e)$ the *origin* and $w = t(e)$ the *terminal vertex* of the edge. We say that *there is an edge from the vertex v to the vertex w with label b* if (v, w, b) belongs to E. An edge e is said to *emanate* from the vertex v if $v = o(e)$. The number of edges emanating from a vertex is called the *degree* of the vertex. An edge of the form (v, v, b) is called a *loop* at v (with label b).

We will need the *combinatorial distance* on a graph given as follows. Each vertex has distance 0 to itself. The distance between different vertices v and w is 1 if and only if there exists a label b such that (v, w, b) belongs to E. More generally the distance between different vertices v and w is then defined inductively as the smallest natural number n such that there exists a vertex v' with distance $n-1$ to v and distance 1 to w. If no such n exists, the combinatorial distance is defined to be ∞. The graph is called *connected* if the distance between any two if its vertices is finite. Likewise the *connected component* of a vertex is the set of all vertices with finite distance to it.

A *ray* in an infinite graph is an infinite sequence $v_0, v_1, \ldots$ of pairwise different vertices with distance 1 between consecutive vertices. Two rays are equivalent if there exists a third ray containing infinitely many vertices of each of the rays. An equivalence class of rays is called an *end* of the graph. For example, finite graphs have 0 ends, the Cayley graphs of $\mathbb{Z}$, $\mathbb{Z}^2$ and $\mathbb{F}_2$ (the free group on two letters) have $2, 1$ and infinitely many ends, respectively. In general, the Cayley graph of a group may have $0, 1, 2$ or infinitely many ends.

A *rooted graph* is a pair consisting of a graph and a vertex belonging to the vertex set of the graph. This vertex is then called the *root*.

Two rooted graphs (G_1, v_1) and (G_2, v_2) labeled by the same set $\mathcal{B}$ are called isomorphic if there exists a bijective map β from the vertices of G_1 to the vertices of G_2 taking v_1 to v_2 such that the vertices x and y in G_1 are connected by an edge of color b if and only if their images in $V(G_2)$ are connected by an edge of color b. In this case we write $(G_1, v_1) \cong (G_2, v_2)$.

Let us now consider the set $\mathcal{G}_*(\mathcal{B})$ of isomorphism classes of connected rooted graphs labeled with elements from $\mathcal{B}$ that we endow with the following natural metric. The distance between the isomorphism classes of the two rooted graphs (Y_1, v_1) and (Y_2, v_2) is defined as

$$\text{dist}([(Y_1, v_1)], [(Y_2, v_2)]) := \inf\left\{\frac{1}{r+1} : B_{Y_1}(v_1, r) \cong B_{Y_2}(v_2, r)\right\}$$

where $B_Y(v, r)$ is the (labeled) ball of radius r centered in v in the combinatorial metric on Y. If we only consider graphs of uniformly bounded degree (as we will in this chapter), the space $\mathcal{G}_*(\mathcal{B})$ is compact.

We now turn to dynamical systems. Whenever the group H acts on a compact space Y via the continuous map $\alpha : H \times Y \longrightarrow Y$ we call (Y, H, α) a *dynamical system*. We will mostly suppress the α in the notation. In particular, we will write (Y, H) instead of (Y, H, α) and we will write ty for $\alpha_t(y)$ (with $y \in Y$ and $t \in H$). If H is the infinite cyclic group $\mathbb{Z}$, then the action of H on Y is determined by $T := \alpha(1)$

and we then just write (Y, T) instead of $(Y, \mathbb{Z})$ (and this is well in line with the notation used in the previous sections).

Whenever a dynamical system (Y, H) and $y \in Y$ are given we call

$$\{ty : t \in H\}$$

the *orbit of y (under H)*.

The dynamical system (Y, H) is *minimal* if every orbit is dense in Y. The dynamical system (Y, H) is *uniquely ergodic* if there exists exactly one H-invariant probability measure on Y.

The dynamical system (Y, H) is called a *factor* of the dynamical system (Y', H) if there exists a continuous surjective map $\chi : Y' \longrightarrow Y$ with $\chi(ty) = t\chi(y)$ for all $t \in H$ and $y \in Y'$. This map χ is then called a *factor map*. The dynamical system (Y, H) is then also referred to as an *extension* of the dynamical system (Y', H).

We will deal with graphs arising from dynamical systems. More specifically, consider a group H generated by a symmetric finite set S and assume that H acts on a compact space Y. Then any point $y \in Y$ gives rise to the *orbital Schreier graph of y* denoted by Γ_y. This is a rooted graph labeled by S, which is equipped with the involution $S \longrightarrow S, s \mapsto s^{-1}$. The set of vertices of Γ_y is given by the points in the orbit of y. The root is given by y and there is an edge from v to w with label $s \in S$ if $sx = y$. Note that by the required symmetry of S we then also have an edge from w to v with label s^{-1} (as is needed according to our definition of a labeled graph).

As an example we note the Cayley graph of a group G with generating set S. This Cayley graph (with the neutral element of G as the root) is the orbital Schreier graph of G corresponding to the action of G on itself via left multiplication.

If the elements of S happen to all be involutions then there will be an edge of label s from v to w if and only if there is an edge of label s from w to v. This is the situation we will encounter in the next section.

5. Grigorchuk's Group G, its Schreier Graphs and the Associated Laplacians

In this section we introduce the main object of our interest: the first group of intermediate growth G and the Laplacians on the associated Schreier graphs. The group G is the first group with intermediate word growth and was introduced by the first author in [36, 37]. By now it is generally known as Grigorchuk's group G and this is how we will refer to it.[1] It is generated by four involutions a, b, c, d. With notation to be introduced presently, the group G can be viewed as a group acting by automorphisms on the full infinite binary tree $\mathcal{T}$. The action extends

---

[1] This is in spite of the first author's reluctance.

by continuity to an action by homeomorphisms on the boundary $\partial \mathcal{T}$. The action of G gives rise to Schreier graphs.

The Schreier graphs arising from actions of automorphism groups of infinite regular rooted trees have attracted substantial attention in recent years [28, 43, 70]. Of particular interest are so-called self-similar groups (as G) whose action reflects the self-similar structure of the tree. Their Schreier graphs also have self-similarity features. Some are closely related to Julia sets [11, 28, 70]; others are fractal sets close to e.g. the Sierpinski gasket or the Apollonian gasket [10, 44, 45].

The spectra of the Laplacians on such graphs have been described in some cases [10, 42, 44–46]. The spectrum can be a union of intervals [42], a Cantor set [10], or a union of a Cantor set together with an infinite set of isolated points that accumulate to it, as in the case of the so-called Hanoi tower group [45]. These investigations all use the method introduced in [10]. Here we study the spectra of the Laplacians on the Schreier graphs by a different method via the connection to aperiodic order. This allows us to determine the spectral type of the Laplacians for arbitrary choice of weights on the generators of the group. We thus obtain one of the rare examples where we understand how the spectrum of the Laplacian depends on the weights. In general this dependence is very poorly understood, as well as the dependence of the spectrum on the generating set in the group. Let us mention here a recent result of Grabowski and Virag who show in [35] that there is a weighted Laplacian on the lamplighter group that has singular continous spectrum, whereas the unweighted Laplacian on the same generators is known to have pure point spectrum [46].

5.1. Grigorchuk's Group G

Let us denote by $\mathcal{T}_q$, $q \in \mathbb{N}$ with $q \geq 2$, the *rooted regular tree of degree q*. The vertex set of $\mathcal{T}_q$ is given by $\{0, \ldots, q-1\}^*$, i.e. the set of all words over the alphabet $\{0, \ldots, q-1\}$. The root of $\mathcal{T}_q$ is the empty word. There is an edge between v and w whenever $w = vk$ or $v = wk$ holds for some $k \in \{0, \ldots q-1\}$. The words $w \in \{0, \ldots, q-1\}^n$ constitute the *nth level* of the tree. (In the tree, they are at combinatorial distance exactly n from the root.)

The boundary $\partial \mathcal{T}_q$ of $\mathcal{T}_q$ consisting of infinite geodesic rays in $\mathcal{T}_q$ emanating from the root (i.e. infinite paths starting in the root all of whose edges are pairwise different) can then be identified with the set $\{0, 1, \ldots, q-1\}^{\mathbb{N}}$ of one-sided infinite words over $\{0, 1, \ldots, q-1\}$. As mentioned above, the set $\{0, 1, \ldots, q-1\}^{\mathbb{N}}$ is equipped with the product topology and is thus a compact space homeomorphic to the Cantor set.

Any automorphism of $\mathcal{T}_q$ necessarily preserves the root (which is the only vertex with degree q) and maps paths starting in the root to paths starting in the root. This readily implies that any automorphisms group action on $\mathcal{T}_q$ is level preserving, i.e. maps words of length n to words of length n. Any such action then extends to an action of the same group by homeomorphisms on the boundary $\partial \mathcal{T}_q$.

A regular rooted tree is a self-similar object. Indeed, the subtree rooted at an arbitrary vertex of the tree is isomorphic to the whole tree $\mathcal{T}_q$. The full group of automorphisms inherits this self-similarity property in the following sense: any automorphism of $\mathcal{T}_q$ is completely determined by the permutation it induces on the q branches growing from the root (an element of $Sym(q)$) and the collection of q automorphisms $(g_0, \ldots, g_{q-1})$ which coincide with the restrictions of g on the corresponding branches.

However, if one is interested in a subgroup $H < Aug(T_q)$ and wants it to be self-similar, one has to impose the condition that all the restrictions $(g_0, \ldots, g_{q-1})$ are again elements of the same group H, so that every $g \in H$ can be represented as

$$g = \alpha(g_0, \ldots, g_{q-1}),$$

where α belongs to $Sym(q)$ and describes the action of g on the first level of $\mathcal{T}_q$ and $g_i \in G, i = 0, \ldots, q-1$ is the restriction of g on the full subtree of $\mathcal{T}_q$ rooted at the vertex i of the first level of $\mathcal{T}_q$. This leads to the following definition.

Definition 5.1. A group H of automorphisms of $\mathcal{T}_q$ is *self-similar* if, for all $g \in H, x \in \{0, \ldots, q-1\}$, there exist $h \in H, y \in \{0, \ldots, q-1\}$ such that

$$g(xw) = yh(w),$$

for all finite words w over the alphabet $\{0, \ldots, q-1\}$.

We refer the interested reader to [43, 70] for more information about self-similar groups.

We now turn our attention to one particular example of a self-similar group that will be the central object of our study, the Grigorchuk group G. It is generated by four automorphisms a, b, c, d of the rooted binary tree $\mathcal{T} = \mathcal{T}_2$ as follows:

$a(0w) = 1w, a(1w) = 0w;$

$b(0w) = 0a(w), b(1w) = 1c(w);$

$c(0w) = 0a(w), c(1w) = 1d(w);$

$d(0w) = 0w, d(1w) = 1b(w),$

for an arbitrary word w over $\{0, 1\}$. These automorphisms can also be expressed in the self-similar form, as above:

$$a = \epsilon(id, id), \qquad b = e(a, c), \qquad c = e(a, d), \qquad d = e(id, b),$$

where e and ϵ are, respectively, the trivial and the nontrivial permutations in the group $Sym(2)$.

Remark 5.2.   Observe that all the generators are involutions and that $\{1, b, c, d\}$ commute and constitute a group isomorphic to the Klein group $\mathbb{Z}/2\mathbb{Z} \times \mathbb{Z}/2\mathbb{Z}$. Let us also mention that there are many more relations and the group is not finitely presented.

For our subsequent discussion it will be important that G acts *transitively* on each level, i.e. for arbitrary words w, u over $\{0, 1\}$ with the same length there exists a $g \in G$ with $gu = w$.

5.2. The Schreier Graphs of G and the Dynamical System (X, G)

The action of G on the set $V = \{0, 1\}^*$ of vertices of the rooted binary tree and on its boundary $\partial \mathcal{T} = \{0, 1\}^{\mathbb{N}}$ induces on this set the structure of a graph labeled with $\{a, b, c, d\} \subset G$. Specifically, the vertex set of this graph is given by $\{0, 1\}^* \cup \{0, 1\}^{\mathbb{N}}$ and there is an edge with label $s \in \{a, b, c, d\}$ and origin v and terminal vertex w if and only if $sv = w$ holds. Note that the set of arising edges has indeed the symmetry property required in our definition of a labeled graph as any $s \in \{a, b, c, d\}$ is an involution. For the first three levels of the tree the resulting graphs are shown in Figure 10.2. Thus, the rooted graphs consisting of such a connected component together with a root are orbital Schreier graphs (in the sense defined above). The connected components corresponding to the first three levels of the tree are shown in Figure 10.2.

There are two kinds of such connected components viz finite and infinite components. The finite ones correspond to the levels of the tree (recall that the action of G on the levels of the tree is transitive). We will refer to them simply as *Schreier graphs*. The infinite ones correspond to the orbits of the action on the boundary. The corresponding rooted graphs will be referred to as *orbital Schreier graphs*.

We will need the isomorphism classes of Schreier graphs and therefore introduce the map

$$\mathcal{F} : V(\mathcal{T}) \cup \partial \mathcal{T} \longrightarrow \mathcal{G}_*(\{a, b, c, d\}), \quad \mathcal{F}(v) := [(\Gamma_v, v)],$$

where Γ_v is the connected component of v.

The graphs Γ_w and Γ_v coincide (as nonrooted graphs) whenever v and w are in the same orbit of the action of G. In particular, as G acts transitively on each level of the tree, for $n \in \mathbb{N}$ we can therefore define

$$\Gamma_n := \Gamma_{1^n}$$

which coincides with Γ_w for all $w \in V(\mathcal{T})$ with $|w| = n$. In general, the graph Γ_n has a linear shape; it has 2^{n-1} simple edges, all labeled by a, and $2^{n-1} - 1$ cycles of length 2 whose edges are labeled by b, c, d. It is regular of degree 4, with one loop at each edge. The loop contributes 1 to the degree of the vertex because all generators are elements of order 2, and the labeling of the loop is uniquely determined by the labeling of the

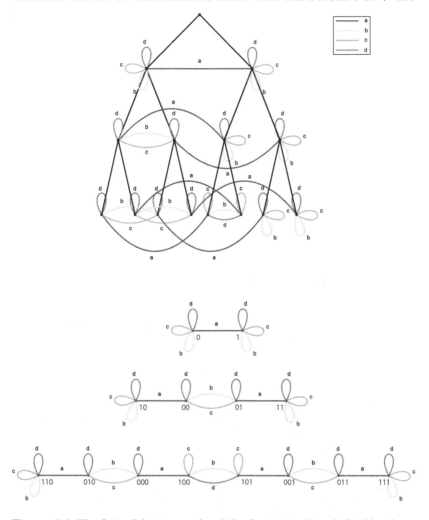

Figure 10.2 The finite Schreier graphs of the first, second, and third level

other edges around the vertex, as edges around one vertex are labeled by $\{a, b, c, d\}$.

The graphs Γ_ξ, $\xi \in \partial \mathcal{T}$ corresponding to the action of G on the boundary are infinite and have either two ends or one end. The graph Γ_{1^∞} corresponding to the orbit of the rightmost infinite ray, is one-ended (see Figure 10.3), and so are then all graphs in the same orbit.

All the other graphs $\Gamma_\xi, \xi \notin G \cdot 1^\infty$, are two-ended. They are all isomorphic as unlabeled graphs.

In [80], Vorobets studied the closure $\overline{\mathcal{F}(\partial \mathcal{T})}$ of the space of Schreier graphs in the space of isomorphism classes of rooted labeled graphs. We recall some of his results next. He showed that the one-ended graphs

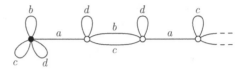

Figure 10.3 The one-ended graph $\Gamma_{1\infty}$

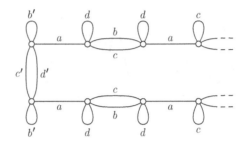

Figure 10.4 Connecting two copies of $\Gamma_{1\infty}$

are exactly the isolated points of this closure $\overline{\mathcal{F}(\partial\mathcal{T})}$, and that the other points in $\overline{\mathcal{F}(\partial\mathcal{T})}$ are two-ended graphs. This suggests we consider the compact set

$$X := \overline{\mathcal{F}(\partial\mathcal{T})} \setminus \{\text{isolated points}\}.$$

The group G acts on X by changing the root of the graph. The arising dynamical system is denoted as (X, G). It is minimal and uniquely ergodic and its invariant probability measure will be denoted as ν.

A precise description of X can be given as follows. The space X consists of all isomorphism classes of two-ended rooted Schreier graphs $\{(\Gamma_\xi, \xi) : \xi \in \partial\mathcal{T} \setminus G \cdot 1^\infty\}$ and of three additional countable families of isomorphism classes of two-ended graphs. These families are obtained by gluing two copies of the one-ended graph $\Gamma_\xi, \xi \in G \cdot 1^\infty$, at the root in three possible ways corresponding to choosing a pair (b, c), (b, d) or (c, d) and then choosing an arbitrary vertex of the arising graph as the root. One of these three possibilities is shown in Figure 10.4. There, the chosen pair is (c, d) and to avoid confusion with other edges with the same labels, labels at the gluing point are denoted with a prime.

The decomposition of X into isomorphism classes of the (Γ_ξ, ξ) and the three families mentioned above immediately gives rise to a factor map

$$\phi : X \to \partial\mathcal{T},$$

which is one-to-one except in a countable set of points, where it is three-to-one. In fact, the inverse map ϕ^{-1} exists on the complement $\partial\mathcal{T} \setminus G \cdot 1^\infty$ of the orbit of the point 1^∞ and agrees there with $\mathcal{F}$.

Under this factor map ϕ the unique G-invariant probability measure ν on X is mapped to the uniform Bernoulli measure μ on $\partial\mathcal{T} = \{0,1\}^{\mathbb{N}}$.

5.3. Laplacian Associated to the Schreier Graphs of G

Whenever a group H acts on a measure space (Y, m) by measure preserving transformations, one obtains the *Koopman representation* ϱ of H on $L^2(Y, m)$ via

$$\varrho(h) : L^2(Y, m) \longrightarrow L^2(Y, m), (\varrho(h)f)(y) = f(h^{-1}y),$$

(for $h \in H$). Any $\varrho(h)$ is then a unitary operator (as the action is measure preserving).

In our situation when $H = G$, we have moreover that for $s \in \{a, b, c, d\}$ the unitary operator $\varrho(s)$ is its own inverse (as s is an involution) and hence must be self-adjoint. In particular, for any set of parameters $t, u, v, w \in \mathbb{R}$ we obtain a self-adjoint operator

$$M_\varrho(t, u, v, w) := t\varrho(a) + u\varrho(b) + v\varrho(c) + w\varrho(d).$$

Consider first an arbitrary $\xi \in \partial\mathcal{T}$. Then, there is an action of G on the (countable) vertex set of $V(\Gamma_\xi)$ of Γ_ξ. Specifically, $s \in \{a, b, c, d\}$ maps the vertex $x \in V(\Gamma_\xi)$ to the vertex sx, which is the unique vertex of $V(\Gamma_\xi)$ connected to v by an edge of label s. Clearly, this action preserves the counting measure on $V(\Gamma_\xi)$. Thus, we obtain a representation ϱ_ξ of G on $\ell^2(V(\Gamma_\xi))$.

Definition 5.3 (Laplacian on the Schreier graph). An operator $M_\xi(t, u, v, w)$ defined by

$$M_\xi(t, u, v, w) := M_{\varrho_\xi}(t, u, v, w) = t\varrho_\xi(a) + u\varrho_\xi(b) + v\varrho_\xi(c) + w\varrho_\xi(d)$$

with $\xi \in \partial\mathcal{T}$ and $t, u, v, w \in \mathbb{R}$ will be called *(weighted) Laplacian on the Schreier graph Γ_ξ*.

Remark 5.4.

- It is possible to understand $V(\Gamma_\xi)$ as G/G_ξ, where G_ξ is the stabilizer of ξ in the action of G on $\partial\mathcal{T}$, and then ϱ_ξ is the quasi-regular representation ϱ_{G/G_ξ} associated to G/G_ξ.
- If t, u, v, w are positive with $1 = t + u + v + w$ then it is possible to interpret the operators M_ξ as the Markov operators of a random walk on the graph Γ_ξ. In the general case, the operator $M_\xi(t, u, v, w)$ can still be seen as the natural weighted "adjacency matrix" or "Laplacian" associated to the the graph Γ_ξ.

We can also equip $\partial\mathcal{T} = \{0,1\}^{\mathbb{N}}$ with the uniform Bernoulli measure μ and consider the Koopman representation π of G on $L^2(\partial\mathcal{T}, \mu)$ given via

$$\pi(g) : L^2(\partial\mathcal{T}, \mu) \longrightarrow L^2(\partial\mathcal{T}, \mu), \pi(g)f(x) = f(g^{-1}x).$$

This is a unitary representation of G and any $\pi(s)$, $s \in \{a, b, c, c\}$, is a unitary self-adjoint involution. For $t, u, v, w \in \mathbb{R}$ we then obtain the operator $M_\pi(t, u, v, w)$ via

$$M_\pi(t, u, v, w) = t\pi(a) + u\pi(b) + v\pi(c) + w\pi(d).$$

The following is a crucial result on the spectral theory of the above operators.

Theorem 5.5 (Independence of spectrum (Bartholdi / Grigorchuk [10])). *For any given set of parameters $t, u, v, w \in \mathbb{R}$ the spectrum of $M_\xi(t, u, v, w)$ does not depend on $\xi \in \partial \mathcal{T}$ and coincides with the spectrum of $M_\pi(t, u, v, w)$.*

Of course, there is the question what the spectrum is in terms of the parameters t, u, v, w. A complete answer was given in [10] in the case $u = v = w$. The spectrum then consists of two points or one or two intervals, and an explicit description of the spectrum can be given in terms of the parameters t and $u = v = w$. In fact, the case $u = v = w$ is the case of periodic Schrödinger-type operators and can easily be treated by classical means (Floquet decomposition). It can also be treated by the method suggested in [10]. Below we will come to a main result of [39], which treats the case, where $u = v = w$ does not hold. We will see that in this case, the spectrum is a Cantor set of Lebesgue measure zero.

6. The Connection between (X, G) and (Ω_τ, T)

We will now provide a map gr from Ω_τ to (isomorphism classes of) labeled rooted graphs with labels belonging to the alphabet $\{a, b, c, d\}$. To get a better understanding of this map it will be useful to think of the letters x, y, z as encoding the pairs

$$\begin{pmatrix} b \\ c \end{pmatrix}; \quad \begin{pmatrix} b \\ d \end{pmatrix}; \quad \begin{pmatrix} c \\ d \end{pmatrix}$$

respectively. Roughly speaking the map will convert $\omega \in \Omega_\tau$ into a graph with vertex set $\mathbb{Z}$ and root 1 and edges between n and $n+1$ with labels from $\{a, b, c, d\}$ according to the value of $\omega(n)$. We now provide the precise **definition of the labeled rooted graph gr(ω)** associated to $\omega \in \Omega_\tau$ as follows:

Vertices: The set of vertices is $\mathbb{Z}$.

Root: The number 1 is chosen as the root.

Edges: There are edges between vertices n, k if and only if $|n - k| \leq 1$. Specifically, edges are assigned between n and $n+1$ and from n to itself and from $n + 1$ to itself in the following ways:

- If $\omega(n) = a$, then there is an edge between n and $n + 1$ and this edge carries the label a.

- If $\omega(n) = x$, then there are two edges between n and $n+1$; one carries the label b and the other carries the label c. Moreover, there is an additional edge from n to itself labeled with d and an additional edge from $n+1$ to itself labeled with d.
- If $\omega(n) = y$, then there are two edges between n and $n+1$; one carries the label b and the other carries the label d. Moreover, there is an additional edge from n to itself labeled with c and an additional edge from $n+1$ to itself labeled with c.
- If $\omega(n) = z$, then there are two edges between n and $n+1$; one carries the label c and the other carries the label d. Moreover, there is an additional edge from n to itself labeled with b and an additional edge from $n+1$ to itself labeled with b.

We note that in this way any vertex has for each label $\{a, b, c, d\}$ exactly one edge of this color emanating from it. We also note that the arising graphs have a "linear structure" in a natural sense.

Let $\mathcal{G}_*(\{a, b, c, d\})$ be the metric space of isomorphism classes of connected rooted graphs labeled with elements from $\{a, b, c, d\}$. Then, gr gives rise to a map Gr via

$$\mathrm{Gr} : \Omega_\tau \longrightarrow \mathcal{G}_*(\{a, b, c, d\}), \omega \mapsto [\mathrm{gr}(\omega)],$$

where $[(\Gamma, v)]$ is the equivalence class of (Γ, v) in the space of isomorphism classes of the rooted connected labeled graphs.

Recall that X denotes the closure of $\mathcal{F}(\mathcal{T})$ in $\mathcal{G}_*(\{a, b, c, d\})$ without its isolated points (see Section 5.2). Then, it turns out that the image of Ω_τ under Gr is exactly X. In fact, much more is true and this is our main result on the connection of the subshift (Ω_τ, T) and the dynamical system (X, G).

Define the maps A, B, C, D from Ω_τ into itself by

- $A(\omega) = ...\omega_0\omega_1|\omega_2...$ if $\omega_1 = a$ and $A(\omega) = ...\omega_{-1}|\omega_0\omega_1...$ if $\omega_0 = a$.
- $B(\omega) = ...\omega_0\omega_1|\omega_2...$ if $\omega_1 \in \{x, y\}$, $B(\omega) = ...\omega_{-1}|\omega_0\omega_1...$ if $\omega_0 \in \{x, y\}$ and $B(\omega) = \omega$ in all other cases.
- $C(\omega) = ...\omega_0\omega_1|\omega_2...$ if $\omega_1 \in \{x, z\}$, $C(\omega) = ...\omega_{-1}|\omega_0\omega_1...$ if $\omega_0 \in \{x, z\}$ and $C(\omega) = \omega$ in all other cases.
- $D(\omega) = ...\omega_0\omega_1|\omega_2...$ if $\omega_1 \in \{y, z\}$, $D(\omega) = ...\omega_{-1}|\omega_0\omega_1...$ if $\omega_0 \in \{y, z\}$ and $D(\omega) = \omega$ in all other cases.

Then, clearly, A, B, C, D are homeomorphisms and involutions. Let H be the group generated by A, B, C, D (within the group of homeomorphisms of Ω_τ).

Theorem 6.1 (Factor theorem [39]). *The following statements hold:*

(a) The group G is isomorphic to the group H via $\varrho : G \longrightarrow H$ with $\varrho(a) = A$, $\varrho(b) = B$, $\varrho(c) = C$ and $\varrho(d) = D$. In particular, there is a well defined action α of G on Ω_τ given by $\alpha_g(\omega) := \varrho(g)(\omega)$ for $g \in G$ and $\omega \in \Omega_\tau$ and via this action we obtain a dynamical system (Ω_τ, G).

(b) The dynamical system (X, G) is a factor of the dynamical system (Ω_τ, G) with factor map

$$\psi : \Omega_\tau \longrightarrow X, \omega \mapsto Gr(\omega),$$

which is two-to-one.

(c) For every $\omega \in \Omega_\tau$ the orbits $\{T^n \omega : n \in \mathbb{Z}\}$ and $\{\alpha_g(\omega) : g \in G\}$ coincide.

(d) The dynamical system (Ω_τ, G) is uniquely ergodic and the unique T-invariant probability measure on Ω_τ coincides with the unique G-invariant probability measure on Ω_τ.

The theorem provides a factor map $\psi : (\Omega_\tau, G) \longrightarrow (X, G)$. In Section 5.2 we have already encountered the factor map $\phi : (X, G) \longrightarrow (\partial\mathcal{T}, G)$. Putting together these factor maps provides a tower of extensions of the dynamical system $(\partial\mathcal{T}, G)$ via

$$(\Omega_\tau, G) \overset{\psi}{\longrightarrow} (X, G) \overset{\phi}{\longrightarrow} (\partial\mathcal{T}, G).$$

The remainder of this section is devoted to providing additional perspective on these two factor maps by discussing their respective merits. We will see that ϕ resolves the nontypical points and that ψ allows one to embed the group G into a topological full group.

We will need a bit of notation first. Let (Y, H) be a dynamical system. Let $g \in H$ be given. Then, an $y \in Y$ is called g-typical if either $gy \neq y$ or g acts trivially in some neighborhood of y. If $y \in Y$ is g-typical for any $g \in H$ it is called typical. Typical points have many claims to be indeed typical. For instance, if H is countable the set of typical points has a meager complement [38]. Moreover, the following is shown in [38, 43].

Proposition 6.2. *Let (Y, H) be a minimal dynamical system and H finitely generated. Then, the Schreier graphs of typical points are locally isomorphic (as labeled graphs). Moreover, if $x \in Y$ is typical and $y \in Y$ is not typical, then any ball around a vertex of Γ_x is isomorphic to a ball around some vertex in Γ_y.*

In order to get a further understanding of the typical points we will next discuss another characterization of typical points. Define the *stabilizer* of y as

$$\mathrm{st}_H(y) := \{h \in H : hy = y\}$$

and the *neighborhood stabilizer* of y as

$$\mathrm{st}_H^0 := \{g \in H : h \text{ acts as identity on a neighborhood of } y\}.$$

Then, st_H^0 is a normal subgroup of st_H and the quotient

$$\mathrm{germ}(y) := \mathrm{st}_H(y)/\mathrm{st}_H^0$$

is called the *group of germs* (at y).

Lemma 6.3. *Let (Y, H) be a dynamical system. Then, the following assertions are equivalent for $y \in Y$.*

(i) The point y is typical.

(ii) The group of germs germ(y) is trivial.

Proof. (i)$\Longrightarrow$ (ii): If germ(y) is not trivial, there exists a $g \in \mathrm{st}_H \setminus \mathrm{st}_H^0$. Hence, y is not typical.

(ii)$\Longrightarrow$ (i): If y is not typical then there exists a $g \in G$ with $gy = y$ but g is not the identity on a neighborhood of y. Thus, germ(y) is not trivial (as it contains the class of g). $\qquad\square$

Remark 6.4. The points y with nontrivial germ(y) are sometimes called *singular.*

We will also need the concept, going back to [34], of the (topological) *full group* $[[T]]$ of a homeomorphism $T : Y \longrightarrow Y$ of the Cantor set Y. This is the group of those homeomorphisms of Y that at any point coincide locally with powers of T. It is a countable group with remarkable properties. For example, it is amenable if T is minimal [52], its commutator subgroup is simple [68] and it is finitely generated if and only if T is a minimal subshift over a finite alphabet [68].

After these preparations we can now come back to the situation at hand.

The dynamical system $(\partial\mathcal{T}, G)$ is minimal and uniquely ergodic. However, it does have nontypical points. In fact, a point in $\partial\mathcal{T}$ is nontypical if and only if it belongs to $G \cdot 1^\infty$ (see [80]).

The dynamical system (X, G) is minimal and uniquely ergodic (as discussed in [79]). Moreover, all its points are typical (as can easily be seen). Thus, from this point of view a key merit of (X, G) is to provide an extension of $(\partial\mathcal{T}, G)$, which resolves the nontypical points.

The dynamical system (Ω_τ, G) is minimal and uniquely ergodic and all its points are typical (as can easily be seen). Moreover, as shown in the previous theorem, the dynamical system (Ω_τ, G) has the additional feature that there exists a homeomorphism T on Ω_τ such that the orbits of T agree with the orbits of G. This in turn can be seen to imply that G is a subgroup of the topological full group $[[T]]$ of T. In fact, G embeds into the topological full group of (Ω_τ, T), as the action of generators a, b, c, d on Ω_τ can be represented locally as the action by $T^{\pm 1}$ and $T^0 = id$, so G embeds into $[[T]]$.

Remark 6.5. In this context we also mention a recent article of Matte Bon [67] showing that the group G (and other groups of intermediate growth introduced by the first author in [36]) embed into the topological full group $[[\phi]]$ of a minimal subshift ϕ over a finite alphabet. While his approach is different from ours it leads to the same subshift for the group G.

We finish this section with the question whether there exists a minimal homeomorphism S on X such that G is a subgroup of $[[S]]$.

7. Spectral Theory of the M_ξ for $\xi \in \partial \mathcal{T}$

In this section we discuss spectral theory of the operators $M_\xi(t, u, v, w)$ (defined in Section 5.3) for arbitrary values of the parameters t, u, v, w. We do so by using the results of the previous section to transfer the problem of spectral theory of the operators $M_\xi(t, u, v, w)$, $\xi \in \partial \mathcal{T}$, to the field of the spectral theory of discrete Schrödinger operators with aperiodic order, $(H_\omega)_{\omega \in \Omega_\tau}$.

We consider the subshift (Ω_τ, T). In order to define the Schrödinger operators we will define specific functions f, g on Ω_τ depending on four real parameters t, u, v, w. Given these parameters we set

$$D := u + v + w$$

and define

$$f : \Omega_\tau \longrightarrow \mathbb{R} \text{ by } f(\omega) := \begin{cases} t & : & \omega_0 = a \\ D - w & : & \omega_0 = x \\ D - v & : & \omega_0 = y \\ D - u & : & \omega_0 = z \end{cases}$$

and

$$g : \Omega_\tau \longrightarrow \mathbb{R} \text{ by } g(\omega) := \begin{cases} w & : \omega_0\omega_1 \in \{ax, xa\} \\ v & : \omega_0\omega_1 \in \{ay, ya\} \\ u & : \omega_0\omega_1 \in \{az, za\} \end{cases}.$$

For a given $(t, u, v, w) \in \mathbb{R}^4$ and these f, g we let H_ω, $\omega \in \Omega_\tau$, be the associated operators.

Proposition 7.1. [39] *Let $(t, u, v, w) \in \mathbb{R}^4$ be given. Let $\xi \in \partial \mathcal{T} \setminus G \cdot 1^\infty$ be arbitrary. Then, there exists a ω in Ω_τ such that H_ω is unitarily equivalent to $M_\xi(t, u, v, w)$.*

As a consequence of the previous proposition we can translate results on the spectral theory of Schrödinger operators associated to (Ω_τ, T) into results on the M_ξ. Recall that the spectrum of the $M_\xi(t, u, v, w)$ does not depend on $\xi \in \partial \mathcal{T}$ (due to Theorem 5.5). Define

$$\mathcal{P} := \{(t, u, v, w) \in \mathbb{R}^4 : t \neq 0, u + v \neq 0, u + w \neq 0, v + w \neq 0\}.$$

Theorem 7.2 (Intervals vs Cantor spectrum for the M_ξ [39]). *Let $(t, u, v, w) \in \mathcal{P}$ be given and let $\Sigma = \Sigma(t, u, v, w)$ be the spectrum of the associated family of Laplacians $M_\xi(t, u, v, w)$, $\xi \in \partial \mathcal{T} \setminus G \cdot 1^\infty$. Then, the following holds:*

(a) If $u = v = w$ holds then Σ consists of one or two closed nontrivial intervals and all spectral measures are absolutely continuous.

(b) If $u = v = w$ does not hold then Σ is a Cantor set of Lebesgue measure zero and no spectral measure is absolutely continuous.

Proof. Note that (Ω_τ, T) is linearly repetitive due to Theorem 3.1. Now, the theorem is a direct consequence of the preceding proposition and Theorem 2.7 and Theorem 2.9. □

Remark 7.3. The case $u = v = w$ has already been treated in [10] and an explicit description of the spectrum (in terms of the value of u) has been given there.

We can also use the above considerations to translate results on absence of eigenvalues from the (H_ω) to the M_ξ. Recall that μ denotes the uniform Bernoulli measure on $\partial\mathcal{T} = \{0,1\}^{\mathbb{N}}$.

Theorem 7.4 (Absence of eigenvalues [39]). *Let $(t, u, v, w) \in \mathcal{P}$ be given and assume that $u = v = w$ does not hold.*

(a) For μ-almost every $\xi \in \partial\mathcal{T}$ the operator M_ξ does not have eigenvalues.

(b) For any $x \in \phi^{-1}(G1^\infty)$ the operator M_x does not have eigenvalues.

Remark 7.5. The considerations of this section are concerned with the case $(t, u, v, w) \in \mathcal{P}$. For $(t, u, v, w) \notin \mathcal{P}$ the operators in question can be decomposed as a sum of finitely many different finite dimensional operators each appearing with infinite multiplicity. Thus, the spectrum is pure point with finitely many eigenvalues each with infinite multiplicity.

8. Integrated Density of States and Kesten-von Neumann-Serre Spectral Measure

In this section we show that the Kesten-von Neumann-Serre spectral measure actually agrees with the integrated density of states. This result is folklore and certainly known to experts. Also, a recent discussion of a somewhat more general setting can also be found in [1]. Here, we give a direct reasoning for the case at hand based on our approach.

There is one more representation studied in the context of spectral approximation. This representation comes from the action of G on the nth level of the tree. This action clearly preserves the counting measure on the finite set of vertices of the nth level. It hence gives rise to a representation ϱ_n of G on the finite dimensional $\ell^2(V(\Gamma_n))$. For $t, u, v, w \in \mathbb{R}$ we then set

$$M_n(t, u, v, w) := M_{\varrho_n}(t, u, v, w) = t\varrho_n(a) + u\varrho_n(b) + v\varrho_n(c) + w\varrho_n(d).$$

To each such $M_n(t, u, v, w)$ we associate the *spectral distribution* which is the measure $\mu_n(t, u, v, w)$ on $\mathbb{R}$ given by

$$\mu_n(t, u, v, w) := \frac{1}{|V(\Gamma_n)|} \sum_E \delta_E,$$

where the sum runs over eigenvalues E of $M_n(t, u, v, w)$ counted with multiplicities. In the case $u = v = w$ it is shown in [47] that the measures μ_n, $n \in \mathbb{N}$, converge weakly and the limiting measure

$$\mu_\infty(t, u, v, w) = \lim_{n \to \infty} \mu_n(t, u, v, w)$$

is called the *Kesten-von Neumann-Serre spectral measure* there. Next we will show that the limiting measure exists for any values of the parameters t, u, v, w and coincides with the integrated density of states of the associated Schrodinger operators.

For given $(t, u, v, w) \in \mathbb{R}^4$ we chose the functions f, g as in the previous sections and let H_ω, $\omega \in \Omega_\tau$, be the associated Schrödinger operators.

Theorem 8.1. *Let $(t, u, v, w) \in \mathbb{R}^4$ be given. For $n \in \mathbb{N}$, let $M_n(t, u, v, w)$ be the corresponding operator on Γ_n and μ_n its spectral distribution. Then, the sequence of measures $(\mu_n(t, u, v, w))_n$ converges weakly towards the integrated density of states k of (H_ω). In particular, the integrated density of states of the (H_ω) agrees with the Kesten-von Neumann-Serre spectral measure.*

Proof. The graph arising from restricting $\mathrm{gr}(\omega^{(x)})$ to the vertex set $[1, |\tau^n(a)|]$ and the graph Γ_n differ at most in six loops at the ends (as is clear from the definitions). Thus, a simple variant of the argument in Proposition 7.1 shows that the restriction of the operator $H_{\omega^{(x)}}$ to $[1, |\tau^n(a)|]$ is a perturbation of $M_n(t, u, v, w)$ of rank at most six. Now, Corollary 2.5 gives the convergence of the $\mu_n(t, u, v, w)$ toward the integrated density of states of (H_ω). $\square$

A few comments on this result are in order. They are gathered in the next remark.

Remark 8.2.

- Denote the ξ-independent spectrum of the operators $M_\xi(t, u, v, w)$ by $\Sigma(t, u, v, w)$ and the spectrum of $M_n(t, u, v, w)$ by $\Sigma_n(t, u, v, w)$. Clearly, the support of μ_n is given by $\Sigma_n(t, u, v, w)$. By Theorem 2.2 the support of the integrated density of states k is given by $\Sigma(t, u, v, w)$. Thus, the previous theorem gives in a certain and very weak sense the convergence of the $\Sigma_n(t, u, v, w)$ toward $\Sigma(t, u, v, w)$. In particular, there is an inclusion of spectra as shown in Corollary 2.6.

  In the case at hand, however, convergence of the $\Sigma_n(t, u, v, w)$ toward $\Sigma(t, u, v, w)$ holds in a much stronger sense. More precisely, the results of [10] give the following:

  (a) $\Sigma_n(t, u, v, w) \subset \overline{\Sigma_{n+1}(t, u, v, w)}$ for all $n \in \mathbb{N}$.
  (b) $\Sigma(t, u, v, w) = \overline{\cup_n \Sigma_n(t, u, v, w)}$.

  These are rather remarkable features and not at all true for approximation of the integrated density of states of Schrödinger operators in other cases.

Let us note, however, that there are recent results on approximation of spectra of minimal subshifts with respect to the Hausdorff distance by suitable periodic approximations [13].

- It is also worth pointing out that the approximation of the spectra is done "from below" i.e. via unions. This is in contrast to other approximation schemes used in the investigations of Schrödinger operators, where the approximation is done "from above" i.e. via intersections.

- The feature presented in the preceding two points also explain that the spectrum of $M_\pi(t, u, v, w)$ agrees with $\Sigma(t, u, v, w)$. The reason is simply that the representation π decomposes as a sum of the representations π_n (as shown in [10]). Accordingly, $M_\pi(t, u, v, w)$ is a sum of the finite dimensional operators $M_n(t, u, v, w)$, $n \in \mathbb{N}$, and its spectrum is then given as the closure of the union of the spectra of the $M_n(t, u, v, w)$.

References

[1] M. Abert, B. Virag, and A. Thom: *Benjamini-Schramm convergence and pointwise convergence of the spectral measure*, Preprint (2013).

[2] J.-P. Allouche and J. Shallit: *Automatic Sequences: Theory, Applications, Generalization*, Cambridge University Press, (2003).

[3] J. Auslander: *Minimal Flows and their extensions*, North-Holland Mathematical Studies 153, Elsevier (1988).

[4] J.-B. Aujogue, M. Barge, J. Kellendonk, and D. Lenz: *Equicontinuous factors, proximality and Ellis semigroup for Delone sets*, to appear in [53].

[5] M. Baake and U. Grimm: *Aperiodic Order: Volume 1, A Mathematical Invitation*, Encyclopedia of Mathematics and its Applications **149**, Cambridge University Press, (2014).

[6] M. Baake, D. Lenz, and R.V. Moody: *A characterization of model sets by dynamical systems*, Ergodic Theory Dynam. Systems **27** (2007), 341–82.

[7] M. Baake and D. Lenz: *Dynamical systems on translation bounded measures: Pure point dynamical and diffraction spectra*, Ergod. Th. & Dynam. Systems **24** (2004) 1867–93.

[8] M. Baake and R. Moody (eds.): *Directions in mathematical quasicrystals*, CRM Monogr. Ser. **13**, Amer. Math. Soc., Providence, RI (2000).

[9] M. Barge and J. Kellendonk: *Proximality and pure point spectrum for tiling dynamical systems*, Michigan Journal of Mathematics **62** (2013), 793–822.

[10] L. Bartholdi and R.I. Grigorchuk: *On the spectrum of Hecke type operators related to some fractal groups*, Tr. Mat. Inst. Steklova **231** (2000), Din. Sist., Avtom. i Beskon. Gruppy, 5–45; translation in *Proc. Steklov Inst. Math.* (2000), no. 4 (**231**), 1–41.

[11] L. Bartholdi, R. I. Grigorchuk, and V. Nekrashevych: *From fractal groups to fractal sets* in: "Fractals in Graz" (P. Grabner and W. Woess, eds.), Trends in Mathematics, Birkäuser Verlag, Basel, (2003), 25–118.

[12] S. Beckus and F. Pogorzelski: *Spectrum of Lebesgue measure zero for Jacobi matrices of quasicrystals*, Math. Phys. Anal. Geom. **16** (2013), 289–308.

[13] S. Beckus and J. Bellissard: *Continuity of the spectrum of a field of self-adjoint operators*, arXiv:1507.04641.

[14] A. Besbes, M. Boshernitzan, and D. Lenz: *Delone sets with finite local complexity: Linear repetitivity versus positivity of weights*, Disc. Comput. Geom. **49** (2013), 335–47.

[15] N.M. Bon: *Topological full groups of minimal subshifts with subgroups of intermediate growth*, Preprint 2014 (arXiv:1408.0762).

[16] M. Boshernitzan: *A condition for minimal interval exchange maps to be uniquely ergodic*, Duke Math. J. **52** (1985), 723–52.

[17] J. Buescu and I. Stewart: *Liapunov stability and adding machines*, Ergodic Theory Dynam. Systems **15** (1995), 271–90.

[18] H.L. Cycon, R.G. Froese, W. Kirsch, and B. Simon: *Schrödinger operators with application to quantum mechanics and global geometry*, Texts and Monographs in Physics, Springer, Berlin, 1987.

[19] R. Carmona and J. Lacroix: *Spectral theory of random Schrödinger operators*, Birkhäuser Boston, Boston, MA, (1990).

[20] D. Damanik: *Singular continuous spectrum for a class of substitution Hamiltonians*, Lett. Math. Phys. **46** (1998), 303–11.

[21] D. Damanik: *Gordon-type arguments in the spectral theory of one-dimensional quasicrystals*, in [8], 277–305.

[22] D. Damanik, M. Embree, and A. Gorodetski: *Spectral properties of Schrödinger operators arising in the study of quasicrystals*, to appear in [53].

[23] D. Damanik and D. Lenz: *The index of Sturmian sequences*, European J. Combin. **23** (2002), 23–9.

[24] D. Damanik and D. Lenz: *Powers in Sturmian sequences*, European J. Combin. **24** (2003), 377–90.

[25] D. Damanik and D. Lenz: *A condition of Boshernitzan and uniform convergence in the multiplicative ergodic theorem*, Duke Math. J. **133** (2006), 95–123.

[26] D. Damanik and D. Lenz: *Substitution dynamical systems: characterization of linear repetitivity and applications*, J. Math. Anal. Appl. **321** (2006), 766–80.

[27] D. Damanik and D. Zare: *Palindrome complexity bounds for primitive substitution sequences*, Disc. Math. **222** (2000), 259–67.

[28] D. D'Angeli, A. Donno, M. Matter, and T. Nagnibeda: *Schreier graphs of the Basilica group*, J. Mod. Dyn. **4** (2010), 167–205.

[29] F.M. Dekking: *The spectrum of dynamical systems arising from substitutions of constant length*, Z. Wahrscheinlichkeitstheorie und Verw. Gebiete **41** (1977/78), 221–39.

[30] F. Durand: *Linearly recurrent subshifts have a finite number of non-periodic subshift factors*, Ergod. Th. & Dynam. Sys. **20** (2000), 1061–78.

[31] F. Durand, private communication, 2015.

[32] F. Durand, B. Host, and C. Skau: *Substitution dynamical systems, Bratteli diagrams and dimension groups*, Ergod. Th. & Dynam. Sys. **19** (1999), 953–93.

[33] D. Francoeur, private communication.
[34] T. Giordano, I. Putnam, and C. Skau: *Full groups of Cantor minimal systems*, Israel J. Math. **111** (1999), 285–320.
[35] L. Grabowski:*Group ring elements with large spectral density*, Math. Ann. **363** (2015), 637–56.
[36] R.I. Grigorchuk: *On Burnside's problem on periodic groups. (Russian)* Funktsional. Anal. i Prilozhen. **14** (1980), 53–4.
[37] R.I. Grigorchuk: *Degrees of growth of finitely generated groups and the theory of invariant means (Russian)*, Izv. Akad. Nauk SSSR Ser. Mat. **48** (1984), 939–85.
[38] R.I. Grigorchuk: *Some problems of the dynamics of group actions on rooted trees (Russian)*, Tr. Mat. Inst. Steklova **273** (2011), Sovremennye Problemy Matematiki, 72–191; translation in Proc. Steklov Inst. Math. **273** (2011), 64–175.
[39] R.I. Grigorchuk, D. Lenz, and T. Nagnibeda: *Spectra of Schreier graphs of Grigorchuk's group and Schrödinger operators with aperiodic order*, arXiv:1412.6822.
[40] R. Grigorchuk, D. Lenz, T. Nagnibeda, *Combinatorics of the subshift associated to Grigorchuk's group*, to appear in: Proc. Steklov Inst. Math.
[41] R.I. Grigorchuk, Y. Leonov, V. Nekrashevych, and V. Sushchansky: *Self-similar groups, automatic sequences, and unitriangular representations*, arXiv:1409.5027.
[42] R.I. Grigorchuk and V. Nekrashevych: *Self-similar groups, operator algebras and Schur complements*, Journal of Modern Dynamics, **1**, (2007) 323–70.
[43] R.I. Grigorchuk, V. Nekrashevich, and V. Sushanskii: *Automata, dynamical systems and infinite groups*, Proc. Steklov Inst. Math. **231** (2000), 134–214.
[44] R.I. Grigorchuk and Z. Sunic: *Asymptotic aspects of Schreier graphs and Hanoi Towers groups*, Comptes Rendus Math. Acad. **342** (2006), 545–50.
[45] R.I. Grigorchuk and Z. Sunic: *Schreier spectrum of the Hanoi towers group on three pegs*, Proceedings of Symposia in Pure Mathematics, **77** (2008), 183–98.
[46] R.I. Grigorchuk and A. Zuk: *The lamplighter group as a group generated by a 2-state automaton, and its spectrum*, Geom. Dedicata **87** (2001), 209–44.
[47] R.I. Grigorchuk and A. Zuk. *The Ihara Zeta function of infinite graphs, the KNS spectral measure and integrable maps*, in: Random walks and geometry, V. Kaimanovich (eds.), (2004), 141–80.
[48] I. Guarneri: *Spectral properties of quantum diffusion on discrete lattices*, Europhys. Lett. **10** (1989), 95–100.
[49] I. Guarneri: *On an estimate concerning quantum diffusion in the presence of a fractal spectrum*, Europhys. Lett. **21** (1993), 725–33.
[50] I. Guarneri and H. Schulz-Baldes: *Lower bounds on wave packet propagation by packing dimensions of spectral measures*, Math. Phys. Electron. J. **5** (1999), Paper 1, 16 pp.
[51] R. Horn and C.R. Johnson, *Matrix Analysis*, Cambridge University Press, (1985).

[52] K. Juschenko and N. Monod: *Cantor systems, piecewise translations and simple amenable groups*, Ann. of Math. (2) **178** (2013), 775–87.

[53] J. Kellendonk, D. Lenz, and J. Savinien (eds.): *Directions in aperiodic order*, to appear in: Progress in Mathematics, Birkhäuser.

[54] M. Kohmoto, L.P. Kadanoff, C. Tang: *Localization problem in one dimension: mapping and escape*, Phys. Rev. Lett. **50** (1983), 1870–2.

[55] S. Kotani: *Jacobi matrices with random potentials taking finitely many values*, Rev. Math. Phys. **1** (1989), 129–33.

[56] J. Lagarians and P.A.B. Pleasants: *Repetitive Delone sets and quasicrystals*, Ergod. Th. & Dynam. Sys. **23** (2003), 831–67.

[57] Y. Last: *Quantum dynamics and decompositions of singular continuous spectra*, J. Funct. Anal. **142** (1996), 406–45.

[58] J.-Y. Lee, R.V. Moody, and B. Solomyak: *Pure point dynamical and diffraction spectra*, Annales Henri Poincaré **3** (2002) 1003–18.

[59] D. Lenz: *Singular spectrum of Lebesgue measure zero for one-dimensional quasicrystals*, Comm. Math. Phys. **227** (2002), 119–30.

[60] D. Lenz: *Random operators and crossed products*, Math. Phys. Anal. Geom. **2** (1999), 197–220.

[61] D. Lenz:*Uniform ergodic theorems on subshifts over a finite alphabet*, Ergod. Th. & Dynam. Sys. **22** (2002), 245–55.

[62] D. Lenz and R. V. Moody: *Stationary processes with pure point diffraction*, preprint 2012.

[63] D. Lenz, N. Peyerimhoff, and I. Veselic: *Groupoids, von Neumann algebras and the integrated density of states*, Math. Phys. Anal. Geom. **10** (2007), 1–41.

[64] D. Lenz, N. Strungaru: *Pure point spectrum for measure dynamical systems on locally compact Abelian groups*, J. Math. Pures Appl. **92** (2009) 323–341.

[65] D. Lind and B. Marcus: *An Introduction to Symbolic Dynamics and Coding*, Cambridge University Press, (1995).

[66] I.G. Lysenok: *A set of defining relations for the Grigorchuk group (Russian)*, Mat. Zametki **38** (1985), 503–16. English translation in: Math. Notes **38** (1985), 784–92.

[67] N. Matte Bon: *Topological full groups of minimal subshifts with subgroups of intermediate growth*, Preprint 2014 (arXiv:1408.0762).

[68] H. Matui: *Some remarks on topological full groups of Cantor minimal systems*, Internat. J. Math. **17** (2006), 231–51.

[69] R.V. Moody (ed): *The Mathematics of Long-Range Aperiodic Order*, NATO-ASI Series C 489, Kluwer, Dordrecht (1997) 239–68.

[70] V. Nekrashevych: *Self-similar groups*, Mathematical Surveys and Monographs, 117. American Mathematical Society, Providence, RI, 2005.

[71] S. Ostlund, R. Pandit, D. Rand, H. Schellnhuber, and E. Siggia: *One-dimensional Schrödinger equation with an almost periodic potential*, Phys. Rev. Lett. **50** (1983), 1873–7.

[72] M. Queffelec: *Substitution dynamical systems–spectral analysis*, Lecture Notes in Mathematics, 1294. Springer-Verlag, Berlin, 1987.

[73] M. Senechal: *Quasicrystals and Geometry*, Cambridge University Press, (1995).

[74] D. Shechtman, I. Blech, D. Gratias, and J. W. Cahn: *Metallic phase with long-range orientational order and no translation symmetry*, Phys. Rev. Lett. **53** (1984) 183–5.

[75] B. Solomyak: *Nonperiodicity implies unique composition for self-similar translationally finite tilings*, Discr. Comput. Geom. **20** (1998), 265–79.

[76] Z. Sunic: *Hausdorff dimension in a family of self-similar groups*, Geometriae Dedicata **124** (2007), 213–36.

[77] A. Sütő: *Schrödinger difference equation with deterministic ergodic potentials*, in Beyond Quasicrystals (Les Houches, 1994), Springer, Berlin (1995), 481–549.

[78] G. Teschl, *Jacobi operators and completely integrable nonlinear lattices*, Mathematical Surveys and Monographs **72**, Amer. Math. Soc. Providence (2000).

[79] Y. Vorobets: *On a substitution subshift related to the Grigorchuk group*, preprint 2009, (ArXiv:0910.4800).

[80] Y. Vorobets: *Notes on the Schreier graphs of the Grigorchuk group*, Dynamical systems and group actions, 221–248, Contemp. Math., 567, Amer. Math. Soc., Providence, RI, (2012).

[81] P. Walters, *An Introduction to Ergodic Theory*, Springer, New York (1982).

[82] J. Weidmann: *Linear operators in Hilbert spaces*, Graduate Texts in Mathematics **68**, Springer-Verlag, New York-Berlin (1980).

11

THOMPSON'S GROUP F IS NOT LIOUVILLE

VADIM A. KAIMANOVICH

Department of Mathematics and Statistics, University of Ottawa,
585 King Edward, Ottawa ON, K1N 6N5, Canada

Dedicated to Wolfgang Woess on the occasion of his sixtieth birthday

Abstract

We prove that random walks on Thompson's group F driven by strictly
non-degenerate finitely supported probability measures μ have a non-
trivial Poisson boundary. The proof consists in an explicit construction of
two non-trivial μ-boundaries. Both of them are described in terms of the
'canonical' Schreier graph Γ on the dyadic-rational orbit of the canonical
action of F on the unit interval (in fact, we consider a natural embedding
of F into the group $PLF(\mathbb{R})$ of piecewise linear homeomorphisms of the
real line, and realize Γ on the dyadic-rational orbit in $\mathbb{R}$). However, the
definitions of these μ-boundaries are quite different (in perfect keeping
with the ambivalence concerning amenability of the group F). The first
μ-boundary is similar to the boundaries of the lamplighter groups: it
consists of $\mathbb{Z}$-valued configurations on Γ arising from the stabilization of
logarithmic increments of slopes along the sample paths of the random
walk. The second μ-boundary is more similar to the boundaries of the
groups with hyperbolic properties as it consists of sections ('end fields')
of the end bundle of the graph Γ, i.e., of the collections of the limit ends
of the induced random walk on Γ parameterized by all possible starting
points. The latter construction is more general than the former one, and
is actually applicable to any group which has a transient Schreier graph
with a non-trivial space of ends.

Contents

The author was partially supported by funding from the Canada Research Chairs
program, from NSERC (Canada) and from the European Research Council within
European Union Seventh Framework Programme (FP7/2007-2013), ERC grant
agreement 257110-RAWG. A part of this work was conducted during the trimester
'Random Walks and Asymptotic Geometry of Groups' in 2014 at the Institut Henri
Poincaré, Paris.

Introduction

General Setup

The *group* F introduced by Richard Thompson in 1965 is the group of
the orientation preserving piecewise linear dyadic self-homeomorphisms
of the closed unit interval $[0, 1]$. Arguably, the most important open
question about Thompson's group F is the one about its *amenability*
(see Cannon–Floyd [CF11] for its history). Indeed, due to the plethora
of rather unusual properties of this group (described, for instance,
by Cannon–Floyd–Parry in [CFP96]) either answer would imply very
interesting consequences. This problem has recently attracted a lot of
attention, with an impressive number of failed attempts to prove either
amenability or non-amenability of the group F.

We are inclined to believe that Thompson's group F is non-amenable,
and several recent papers provide circumstantial evidence for that. There
are various computational experiments due to Burillo–Cleary–Wiest
[BCW07], Haagerup–Haagerup–Ramirez Solano [HHRS15], Elder–
Rechnitzer–van Rensburg [ERJvR15]. Besides that, Moore [Moo13]
proved that if the group F is amenable, then its Følner function must
be growing very fast (note, however, that, as it was shown by Erschler
[Ers06b], even the groups of intermediate growth may have the Følner
function growing faster than any given function).

The solution of a conjecture of Furstenberg [Fur73] by Rosenblatt
[Ros81] and by Kaimanovich–Vershik [KV83] provides the following
characterization of amenability in terms of the *Liouville property*.
A countable group is amenable if and only if it carries a non-degenerate
Liouville random walk (i.e., such that its *Poisson boundary* is trivial).
Therefore, a possible approach to proving non-amenability of a given
group consists in showing that there are no non-degenerate Liouville
random walks on it. Here we make the first step in this direction for
Thompson's group F by showing that *the random walk* (F, μ) *is non-
Liouville for any strictly non-degenerate finitely supported measure* μ. In
particular, *the simple random walk on the Cayley graph of F determined
by any finite generating set is non-Liouville.*

Independently of the amenability issue, Thompson's group F
was among the first examples of finitely generated groups with an
intermediate (between 0 and 1) value of the *Hilbert space compression*

which were exhibited by Arzhantseva–Guba–Sapir [AGS06]. They proved that for the group F this value is equal to $1/2$ and asked about existence of a compression function strictly better than the square root [AGS06, Question 1.4]. However, as it has been pointed out by Gournay [Gou14], in view of the results of Naor–Peres [NP08], if such a function exists, then the group must be Liouville. In combination with our result, it allowed Gournay to answer the above question of Arzhantseva–Guba–Sapir in the negative: the best Hilbertian equivariant compression function for Thompson's group F is (up to constants) the square root function.

Main Results

The overwhelming majority of the currently known examples of an explicit non-trivial behavior at infinity for random walks on discrete groups falls into one of the following two classes. The first class consists of the examples, for which this behaviour is due to some kind of *boundary convergence* usually related to more or less pronounced manifestations of hyperbolicity (the most representative example being, of course, the *word hyperbolic groups*), see the papers by Kaimanovich–Vershik [KV83], Kaimanovich [Kai85, Kai89, Kai94, Kai00], Kaimanovich–Masur [KM96], Karlsson–Margulis [KM99], Karlsson–Ledrappier [KL07], Maher–Tiozzo [MT14]. For the examples from the second class a non-trivial behavior at infinity is provided by *pointwise stabilization of random configurations* in a certain way associated with the sample paths of the random walk (this situation is exemplified by the *lamplighter groups*), see Kaimanovich–Vershik [KV83], Kaimanovich [Kai83, Kai91], Erschler [Ers04a, Ers04b, Ers11], Karlsson–Woess [KW07], Sava [Sav10a], Lyons–Peres [LP15], Juschenko–Matte Bon–Monod–de la Salle [JMBNdlS15].

In the case of Thompson's group F, true to its ambivalent nature, we actually construct *asymptotic behaviours (μ-boundaries) of both types*. Note that although the descriptions of these μ-boundaries are quite different, currently we do not know anything about their mutual position in the lattice of all μ-boundaries determined by a fixed step distribution μ. In particular, *a priori* it is not excluded that both these μ-boundaries might in fact coincide with the full Poisson boundary.

Our main tool is the 'canonical' *Schreier graph* Γ of the group F (endowed with the standard generators) on the dyadic-rational orbit of its canonical action on the unit interval. In principle one could argue just in terms of this action. However, it turns out to be much more convenient to 'change the coordinates' and to realize Thompson's group F as a subgroup $\widehat{F} \cong F$ of the *group $PLF(\mathbb{R})$ of piecewise linear homeomorphisms of the real line*. [Actually, the geometry of the graph Γ completely described by Savchuk [Sav10b] is really begging for this coordinate change.] Then one of the generators of F becomes just the translation $\gamma \mapsto \gamma - 1$ on $\mathbb{R}$. The dyadic-rational orbit in the unit interval becomes the dyadic-rational orbit in $\mathbb{R}$, so that we can identify the vertex set of Γ with the dyadic-rational line $\mathbb{Z}\left[\frac{1}{2}\right] \subset \mathbb{R}$.

As we have already said, we construct two non-trivial μ-boundaries of the group F. The first one is inspired by an analogy between Thompson's group $F \cong \widetilde{F}$ and the *lamplighter groups*. Let $\mathrm{fun}(\Gamma, \mathbb{Z})$ (respectively, $\mathrm{Fun}(\Gamma, \mathbb{Z})$) denote the additive group of finitely supported (respectively, of all) $\mathbb{Z}$-*valued configurations on the graph* Γ. We assign to any element g of the group $\widetilde{F} \subset PLF(\mathbb{R})$ a configuration $\mathcal{C}_g \in \mathrm{fun}(\Gamma, \mathbb{Z})$ on $\Gamma \cong \mathbb{Z}\left[\frac{1}{2}\right]$. The value of $\mathcal{C}_g$ at a point γ is the *logarithmic increment of the slope* of g at γ (so that the support of $\mathcal{C}_g$ is precisely the discontinuity set of the derivative of g). Or, in a somewhat different terminology, the support of the configuration $\mathcal{C}_g$ is the set of the *break points* of g, and its values are the logarithms of the *jumps* of g.

By the chain rule the sequence of the configurations $\mathcal{C}_{g_n}$ along a sample path (g_n) of the random walk $(\widetilde{F}, \mu)$ satisfies a simple recursive relation. This relation is completely analogous to that for the random walks on the lamplighter groups. In precisely the same way as with the lamplighter groups [KV83, Kai83], if the measure μ is finitely supported, then the transience of the induced random walk on Γ implies that the sequence $\mathcal{C}_{g_n}$ almost surely converges to a random limit configuration $\mathcal{C}_\infty \in \mathrm{Fun}(\Gamma, \mathbb{Z})$. Then it is not hard to verify that the limit configuration $\mathcal{C}_\infty$ cannot be the same for all sample paths. Therefore, the space $\mathrm{Fun}(\Gamma, \mathbb{Z})$ endowed with the arising hitting distribution is a non-trivial μ-boundary.

This argument (once again, precisely in the same way as for the lamplighter groups) is hinged on the *transience* of the induced random walk on Γ. We establish it by showing that for the simple random walk on $\Gamma \cong \mathbb{Z}\left[\frac{1}{2}\right]$ there is a 'drift' that forces the 2-adic norm to go to infinity along the sample paths (this is the original argument we referred to in [Kai04a]). The classical comparison technique then leads to the transience of all random walks on Γ driven by strictly non-degenerate probability measures μ on $F \cong \widetilde{F}$.

Alternatively, the transience of the simple random walk on Γ can be directly obtained from the explicit description of the geometry of the graph Γ due to Savchuk [Sav10b]. Indeed, after removing from Γ a countable set of rays isomorphic to $\mathbb{Z}_+$ the remaining *skeleton* $\overline{\Gamma}$ is a tree roughly isometric to the standard binary tree. This fact readily implies that Γ is transient.

In a recent preprint [Mis15] Mishchenko gives yet another proof of the transience of the Schreier graph Γ, which is based on an explicit Dirichlet norm estimate. He then notices that due to a specific geometry of Γ this transience implies existence of a non-trivial behaviour at infinity for the simple random walk on Γ. Because it is induced by the simple random walk on Thompson's group F, the latter also has a non-trivial behaviour at infinity. This argument gives a different proof of the absence of the Liouville property for the group F.

Mishchenko settles just for showing non-triviality of the Poisson boundary by exhibiting a non-trivial finite partition of it. However, his observation can be developed much further. Indeed, a transient finite range random walk on any graph necessarily converges in the end compactification. If the graph is endowed with a transitive group action that commutes with the transition operator of the random walk, then the dependence of this limit end on the starting point is equivariant, so that it is the same boundary behaviour of the random walk on the group that is exhibited independently of the starting point in the graph. However, this is not the case in our situation, and here, in order to obtain a μ-boundary (i.e., an *equivariant* quotient of the Poisson boundary), one has to consider the collections of the limit ends of the induced random walk on Γ parameterized by *all* possible starting points. In other words, the resulting μ-boundary is realized on the *space of sections of the end bundle*

$$\Gamma \times \partial\Gamma \to \Gamma, \qquad (\gamma, \omega) \to \gamma$$

over the graph Γ (here $\partial\Gamma$ is the space of ends of Γ). One can consider these sections as '*end fields*' on Γ, or, in yet another terminology, as elements of the *space $Fun(\Gamma, \partial\Gamma)$* $\cong (\partial\Gamma)^\Gamma$ *of $\partial\Gamma$-valued configurations* on Γ. By using certain self-similar features of the graph Γ (although its group of automorphisms is trivial, it has a lot of *partial isomorphisms* between its subsets), it is easy to check that this μ-boundary is non-trivial.

Our construction of a μ-boundary of the group F on the configuration space $Fun(\Gamma, \mathbb{Z})$ heavily used the specific features of this group. On the contrary, the above construction of a μ-boundary on the space of sections $Fun(\Gamma, \partial\Gamma)$ of the end bundle over the Schreier graph Γ is much more general. It produces a non-trivial μ boundary for a strictly non-degenerate finitely supported step distribution μ on an *arbitrary* finitely generated group G whenever there exists a Schreier graph Γ of this group with the following two properties:

(i) the graph Γ is transient with respect to the simple random walk;
(ii) for the induced random walk (Γ, μ) its harmonic (hitting) distributions on the space of ends $\partial\Gamma$ are not one-point measures.

Returning to Thompson's group $F \cong \widetilde{F}$, let us notice that the space of ends of its canonical Schreier graph Γ splits into a disjoint union

$$\partial\Gamma = \partial\overline{\Gamma} \cup \partial_-\Gamma \cup \partial_+\Gamma.$$

Here $\partial\overline{\Gamma}$ is the space of ends of the skeleton $\overline{\Gamma}$, which can be identified with the space $\mathbb{Z}_2^{(1)} = \mathbb{Z}_2 \backslash 2\mathbb{Z}_2$ of 2-adic integers of norm 1. The sets

$\partial_-\Gamma, \partial_+\Gamma$ are countably infinite and correspond to the rays attached to the respective subsets of the skeleton. The components of the above decomposition, on which the hitting measures are actually concentrated, can be further described in terms of the barycentres of the images of the measure μ under its homomorphisms to $\mathbb{Z}$. In particular, in the centered case the hitting measures are supported by $\mathrm{Fun}(\Gamma, \partial\overline{\Gamma})$.

Structure of the Chapter

The chapter has the following structure. In Section 1 we remind the reader of the background definitions concerning the Poisson boundary of a random walk on a discrete group, the Liouville property and its links with amenability. We also remind the reader of how one obtains a non-trivial behaviour at infinity for the lamplighter groups, as this was our source of inspiration for constructing one of the μ-boundaries of Thompson's group (the one realized on the configuration space $\mathrm{Fun}(\Gamma, \mathbb{Z})$).

Further, in Section 2 we introduce Thompson's group F and describe its realization as a subgroup $\widetilde{F} \cong F$ of the group $PLF(\mathbb{R})$ of piecewise linear homeomorphisms of the real line.

In Section 3 we establish the transience of random walks on the Schreier graph Γ. First we do that for the simple random walk (Theorem 3.2), and then, by using a general comparison criterion (Theorem 3.4), for the random walk determined by any strictly non-degenerate step distribution (Theorem 3.7).

In Section 4 we construct a non-trivial μ-boundary of the random walks on Thompson's group $F \cong \widetilde{F}$, which is determined by stabilizing $\mathbb{Z}$-valued configurations on the Schreier graph Γ. It is realized on the configuration space $\mathrm{Fun}(\Gamma, \mathbb{Z})$ (Theorem 4.4).

In Section 5 we establish several geometric properties of the Schreier graph Γ and its space of ends.

These properties are then used in Section 6 to construct another non-trivial μ-boundary of random walks on Thompson's group $F \cong \widetilde{F}$. It is determined by the convergence of the induced random walk on the Schreier graph Γ to its ends, and it is realized on the space of sections $\mathrm{Fun}(\Gamma, \partial\Gamma)$ of the end bundle over Γ (Theorem 6.1). In Theorem 6.2 we generalize this result to arbitrary groups that admit a transient Schreier graph with a non-trivial space of ends. In Theorem 6.3 we further specify the components of the space $\mathrm{Fun}(\Gamma, \partial\Gamma)$, which actually support the μ-boundary constructed in Theorem 6.1.

Finally, Section 7 contains a discussion of possible future developments and ramifications.

Historical Comments

The main result of this chapter (absence of the Liouville property for Thompson's group F) was presented in the talks 'Boundary Behaviour of the Thompson Group' at the conference 'Combinatorial, Geometric, and Dynamical Aspects of Infinite Groups' (1–6 June 2003, Gaeta, Italy) and at the special session 'Probabilistic and Asymptotic Aspects of Group Theory' of the AMS meeting in Athens, Ohio (26–27 March 2004), and the slides of these talks [Kai04a] were circulated at that time. However, it took quite a while to complete the present version, and in the meantime some of its ideas and tools were independently developed by other authors. The geometry of the Schreier graphs of the group F corresponding to its canonical action on the unit interval was completely described by Savchuk [Sav10b, Sav15]; Mishchenko [Mis15] published a new proof of the absence of the Liouville property for the group F based on the analysis of Savchuk.

Acknowledgements

I would like to thank the editors of the present volume, Tullio Ceccherini-Silberstein, Maura Salvatori and Ecaterina Sava-Huss for their infinite patience, understanding and cooperation during the preparation of this chapter. I am grateful to the anonymous referee for a very interesting and competent report. Anna Erschler and Yair Hartman made a number of valuable comments and suggestions. Finally, last but not least, my thanks go to Wolfgang Woess, who is at the origin of this volume, and with whose friendship and wisdom I have been honoured for many years.

1. Preliminaries

1.1. Random Walks on Groups

The (right) random walk (G, μ) on a countable group G determined by a probability measure (the step distribution) μ on G is the Markov chain on G with the transitions

$$g \overset{\mu(h)}{\rightsquigarrow} gh, \qquad g \in G,$$

i.e., its transition probabilities π_g are the translates

$$\pi_g = g\mu, \qquad g \in G,$$

of the measure μ. Any initial distribution θ on G determines the associated Markov measure $\mathbf{P}_\theta$ on the space $G^{\mathbb{Z}_+}$ of sample paths $\mathbf{g} = (g_0, g_1, \dots)$.

By $\mathbf{P} = \mathbf{P}_{\delta_e}$ we denote the probability measure on the path space whose initial distribution is the point measure δ_e concentrated on the identity e of the group G. For the random walk issued from the identity of the group (so that $g_0 = e$), its position g_n at time n is the product $g_n = h_1 h_2 \dots h_n$ of n independent identically μ-distributed increments h_i, and the distribution of g_n (the nth marginal distribution of the measure $\mathbf{P}$) is the n-fold *convolution* of the measure μ.

1.2. Poisson Boundary

The measure $\mathbf{P}_\#$, whose initial distribution is the counting measure $\#$ on the group G, is invariant with respect to the time shift $\mathcal{T}$ on the path space. The Poisson boundary $\partial_\mu G$ of the random walk (G, μ) is the *space of ergodic components* of the time shift $\mathcal{T}$ on the space $(G^{\mathbb{Z}_+}, \mathbf{P}_\#)$, and any initial probability distribution θ on G determines the corresponding harmonic measure ν_θ on $\partial_\mu G$ as the image of the measure $\mathbf{P}_\theta \prec \mathbf{P}_\#$ on the path space under the quotient map $G^{\mathbb{Z}_+} \to \partial_\mu G$. We emphasize that *the Poisson boundary is defined in the measure category only.* Below, when talking about the Poisson boundary $\partial_\mu G$, we shall always (unless otherwise specified) endow it with the measure $\nu = \nu_{\delta_e}$, which is the harmonic measure of the initial distribution δ_e.

The Poisson boundary is equipped with the natural (left) action of the group G induced by the coordinate-wise left translations on the path space, and for an arbitrary initial distribution θ

$$\nu_\theta = \theta * \nu = \sum_g \theta(g)\, g\nu.$$

Although the harmonic measure ν need not be quasi-invariant (if the semigroup $\operatorname{sgr} \mu$ generated by the support $\operatorname{supp} \mu$ of the measure μ is smaller than G), it is μ-stationary with respect to this action in the sense that

$$\mu * \nu = \sum_g \mu(g)\, g\nu = \nu.$$

1.3. Poisson Formula

A function f on G is called μ-harmonic if it is preserved by the Markov operator of the random walk (G, μ)

$$P_\mu f(g) = \sum_h \mu(h) f(gh),$$

i.e., if $P_\mu f = f$. The Poisson formula

$$f(g) = \langle \widehat{f}, g\nu \rangle, \qquad g \in G, \tag{11.1}$$

establishes an isometric isomorphism between the space $H^\infty(G, \mu)$ of bounded μ-harmonic functions f on the group G and the L^∞ space on the Poisson boundary ∂_μ with respect to the quotient measure class determined by the measure $\mathbf{P}_\#$. It consists in assigning to a boundary function $\widehat{f}$ the function f on G whose value at a point $g \in G$ is the result of the integration of the function $\widehat{f}$ against the translate $g\nu$ of the harmonic measure ν.

If g in the Poisson formula (11.1) is only allowed to take values in the semigroup $\operatorname{sgr} \mu$, then (11.1) becomes an isometric isomorphism between the space of bounded μ-harmonic functions on sgr μ and $L^\infty(\partial_\mu G, \nu)$. This property uniquely characterizes the Poisson boundary. Namely, any G-space (B, λ), for which formula (11.1) is an isometric isomorphism between the spaces $H^\infty(\text{sgr } \mu, \mu)$ and $L^\infty(B, \lambda)$, is isomorphic to the Poisson boundary $(\partial_\mu G, \nu)$ in the category of measure G-spaces. See Kaimanovich–Vershik [KV83], Kaimanovich [Kai92, Kai00], Furman [Fur02], Erschler [Ers10] for more background on the Poisson boundary.

1.4. Stability of the Liouville Property

A random walk (G, μ) is called Liouville if the harmonic measure ν is a point measure. The reason for this terminology is that in view of the Poisson formula (11.1) this property means that there are no non-constant bounded μ-harmonic functions on the semigroup sgr μ. One can easily see that the latter property is actually equivalent to the absence of non-constant bounded μ-harmonic functions on the whole group gr μ generated by supp μ as well.

There are numerous examples showing that one can have both Liouville and non-Liouville random walks on the same group (e.g., see Kaimanovich–Vershik [KV83], Kaimanovich [Kai83], Erschler [Ers04b, Ers04a], Bartholdi–Erschler [BE11]), and it is still not clear how the Liouville property for random walks on the same group depends on the step distribution μ.

The main open problem here is

whether all finitely supported symmetric measures μ with gr $\mu = G$ on the same finitely generated group G are either Liouville or non-Liouville simultaneously,

or, in a somewhat different form,

whether the Liouville property is stable with respect to rough isometries.

In the absence of group invariance, for general graphs or Riemannian manifolds, the corresponding couterexamples to the stability of the Liouville property were first constructed by Terry Lyons [Lyo87]. We believe that such counterexamples should exist in the group setup as well. However, the pool of groups, for which the Liouville property has been studied, still remains quite limited in spite of some significant recent progress, see Bartholdi–Virag [BV05], Brieussel [Bri09], Bartholdi–Kaimanovich–Nekrashevych [BKN10], Amir–Angel–Virág [AAV13], Amir–Angel–Matte Bon–Virág [AAMBV13], Matte Bon [MB14], Kotowski–Virág [KV15], Juschenko–Matte Bon–Monod–de la Salle [JMBNdlS15].

1.5. Liouville Groups

Because of the above there are different ways of calling a group G Liouville depending on which class of step distributions μ on G one considers (e.g., all measures, just finitely supported measures, finitely supported symmetric measures, etc.). According to one popular definition a group G is called Liouville if the random walk (G, μ) is Liouville for any (possibly degenerate) finitely supported symmetric probability measure μ on G, see Matte Bon [MB14]. Sometimes one also talks about the Liouville property of a group G with respect to a finite symmetric generating set K (in which case one takes for μ the measure equidistributed on K), see Gournay [Gou14].

1.6. Amenability and the Liouville Property

Liouville groups (no matter which definition one takes) are always amenable, which goes back to Furstenberg [Fur73]. The reason for that is that a random walk (G, μ) is Liouville if and only if the sequence of Cesaro averages of the convolution powers of μ is asymptotically invariant with respect to the left action of the group $\operatorname{gr} \mu$ on itself, see Kaimanovich–Vershik [KV83]. Therefore, the Liouville property of the random walk (G, μ) implies amenability of the group $\operatorname{gr} \mu$ in a very constructive way. Conversely, as it was first conjectured by Furstenberg in [Fur73], and proved by Rosenblatt [Ros81] and Kaimanovich–Vershik [KV83], any amenable group G carries a symmetric measure μ with $\operatorname{supp} \mu = G$ such that the random walk (G, μ) is Liouville. Note that there are amenable groups G, for which such a measure μ cannot be chosen finitely supported, as it was shown by Kaimanovich [Kai83], or even to have finite entropy, as it was shown by Erschler [Ers04b].

Therefore, a possible approach to proving non-amenability of a given group G consists in showing that it carries no Liouville random walks with non-degenerate step distributions μ (i.e., such that $\operatorname{gr} \mu = G$). Actually, in view of the above result of Rosenblatt and Kaimanovich–Vershik, it is enough to consider just the measures with $\operatorname{sgr} \mu = G$ (we shall call such measures strictly non-degenerate; sometimes they are also called adapted), or even just with $\operatorname{supp} \mu = G$. Here we take the first step in this direction for Thompson's group F by showing that random walks on F with finitely supported strictly non-degenerate step distributions are non-Liouville.

1.7. Lamplighter Groups

Our approach to the Liouville property for the group F consists in using the transience of the induced random walk on an auxiliary homogeneous space Γ of the group to prove that certain configurations on Γ associated with the elements of F stabilize along the sample paths of the random walk. It is precisely this idea that was first used for exhibiting a nontrivial boundary behavior for random walks on *lamplighter groups*, see Kaimanovich–Vershik [KV83] and Kaimanovich [Kai83], and, to begin with, we shall outline it in the lamplighter context.

Definition 1.1. The lamplighter group $\mathcal{L}(G)$ over a base group G is the semi-direct product of the group G and the additive group $\mathrm{fun}(G, \mathbb{Z}/2\mathbb{Z})$ of finitely supported $\mathbb{Z}/2\mathbb{Z}$-valued configurations on G endowed with the action of G by translations.

Remark 1.2. In the context of functional and stochastic analysis these groups were first introduced by Vershik and the author [KV83] under the name of the *groups of dynamical configurations*. However, this term did not stick, and the current generally accepted standard is to call them *lamplighter groups*. In the algebraic language $\mathcal{L}(G)$ is the *wreath product* with the *active group* G and the *passive group* $\mathbb{Z}/2\mathbb{Z}$. Since our purpose is just to outline the general idea, we are not discussing here more general wreath products (for which see, for instance, Kaimanovich [Kai91] and Erschler [Ers04a, Ers06a]).

As a set, $\mathcal{L}(G)$ is the usual product of G and $\mathrm{fun}(G, \mathbb{Z}/2\mathbb{Z})$, so that

$$\mathcal{L}(G) = \{(g, \Phi) : g \in G, \ \Phi \in \mathrm{fun}(G, \mathbb{Z}/2\mathbb{Z})\}.$$

However, the group operation in $\mathcal{L}(G)$ is 'skewed' by using the left action of G on $\mathrm{fun}(G, \mathbb{Z}/2\mathbb{Z})$ by the group automorphisms

$$\mathbf{T}^g \Phi(h) = \Phi(g^{-1}h) \tag{11.2}$$

induced by the left action of the group G on itself by translations, so that the group multiplication in $\mathcal{L}(G)$ is

$$(g_1, \Phi_1)(g_2, \Phi_2) = (g_1 g_2, \Phi_1 + \mathbf{T}^{g_1}\Phi_2). \tag{11.3}$$

The identity of $\mathcal{L}(G)$ is $(e, \varnothing)$, where e is the identity of G, and $\varnothing$ (the empty configuration defined as $\varnothing(h) = 0$ for any $h \in G$) is the identity of $\mathrm{fun}(G, \mathbb{Z}/2\mathbb{Z})$.

1.8. Stabilization of Configurations

Given a probability measure μ on the group $\mathcal{L}(G)$, the positions (g_n, Φ_n) of the corresponding random walk at two consecutive time moments are related as

$$(g_{n+1}, \Phi_{n+1}) = (g_n, \Phi_n)(h_{n+1}, \varphi_{n+1}),$$

where (h_i, φ_i) are the increments of the random walk. Therefore, by formula (11.3),

$$\begin{cases} g_{n+1} = g_n h_{n+1}, \\ \Phi_{n+1} = \Phi_n + \mathbf{T}^{g_n}\varphi_{n+1}. \end{cases} \tag{11.4}$$

In particular, the first component $g_n \in G$, i.e., the image of the random walk $(\mathcal{L}(G), \mu)$ under the group homomorphism

$$\mathcal{L}(G) \to G, \qquad (g, \Phi) \mapsto g,$$

performs the quotient random walk on G driven by the image μ' of the measure μ under the homomorphism $\mathcal{L}(G) \to G$.

If the quotient random walk (G, μ') on G is transient (i.e., its sample paths almost surely go to infinity on G), and the support of the measure μ is finite (which implies that there is a finite set $A \subset G$ which contains the supports of all the increments φ_n), then formula (11.4) implies that the configurations Φ_n almost surely stabilize as $n \to \infty$. It means that there exists a pointwise random limit

$$\Phi_\infty = \lim_{n \to \infty} \Phi_n = \varphi_1 + \mathbf{T}^{g_1} \varphi_2 + \mathbf{T}^{g_2} \varphi_3 + \dots,$$

which belongs to the space $\mathrm{Fun}(G, \mathbb{Z}/2\mathbb{Z})$ of all (not necessarily finitely supported) $\mathbb{Z}/2\mathbb{Z}$-valued configurations on G.

It is easy to see that under natural conditions on the measure μ the limit configuration Φ_∞ cannot be the same for a.e. sample path, and therefore the Poisson boundary of the random walk $(\mathcal{L}(G), \mu)$ is non-trivial. This is the case if the measure μ is non-degenerate in $\mathcal{L}(G)$, but actually this assumption can be siginificantly weakened, see Kaimanovich [Kai91]. The map, which assigns to a sample path (g_n, Φ_n) the associated limit configuration $\Phi_\infty \in \mathrm{Fun}(G, \mathbb{Z}/2\mathbb{Z})$, is obviously $\mathcal{L}(G)$-equivariant (with respect to the action on $\mathrm{Fun}(G, \mathbb{Z}/2\mathbb{Z})$ determined by the 'configuration component' of formula (11.3)). Therefore, the configuration space $\mathrm{Fun}(G, \mathbb{Z}/2\mathbb{Z})$ endowed with the resulting hitting distribution is a μ-boundary (an equivariant quotient of the Poisson boundary).

2. Thompson's Group F as a Subgroup of $PLF(\mathbb{R})$

2.1. The Group F and Its Generators

Definition 2.1. The Thompson group F is the group of the orientation preserving piecewise linear self-homeomorphisms g of the closed unit interval $[0, 1]$ that are differentiable except for finitely many break points t_i, which are dyadic rational numbers, and such that the slopes a_i are integer powers of 2:

$$g(t) = a_i t + b_i \quad \text{on} \quad [t_i, t_{i+1}],$$

$$t_i = \frac{k_i}{2^{m_i}} \in \mathbb{Z}\left[\tfrac{1}{2}\right], \quad a_i = 2^{n_i}, \quad b_i \in \mathbb{Z}\left[\tfrac{1}{2}\right],$$

see Figure 11.1.

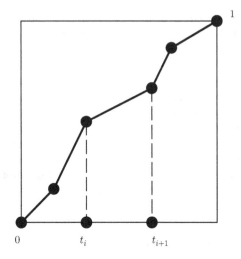

Figure 11.1 The graph of an element of Thompson's group F

We refer to Cannon–Floyd–Parry [CFP96] for the general background on the group F. In particular, it is finitely generated with the **generators**

$$A(t) = \begin{cases} \frac{t}{2}, & 0 \le t \le \frac{1}{2} \\ t - \frac{1}{4}, & \frac{1}{2} \le t \le \frac{3}{4} \\ 2t - 1, & \frac{3}{4} \le t \le 1 \end{cases}$$

and

$$B(t) = \begin{cases} t, & 0 \le t \le \frac{1}{2} \\ \frac{t}{2} + \frac{1}{4}, & \frac{1}{2} \le t \le \frac{3}{4} \\ t - \frac{1}{8}, & \frac{3}{4} \le t \le \frac{7}{8} \\ 2t - 1, & \frac{7}{8} \le t \le 1 \end{cases},$$

see Figure 11.2 (where the arrow indicates that the graph of B restricted to the square $[\frac{1}{2}, 1] \times [\frac{1}{2}, 1]$ is precisely the graph of A rescaled by a factor of 2).

The group F is finitely presented, and it is determined by the relators

$$[AB^{-1}, A^{-1}BA], \qquad [AB^{-1}, A^{-2}BA^2],$$

where $[x, y] = xyx^{-1}y^{-1}$ denotes the usual group theory commutator. The group F is also often described by the infinite presentation

$$\langle x_0, x_1, x_2, \dots \mid x_k^{-1} x_n x_k = x_{n+1} \text{ for all } k < n \rangle,$$

which can be obtained from the previous one by putting $x_0 = A$ and $x_n = A^{1-n}BA^{n-1}$ for $n > 0$.

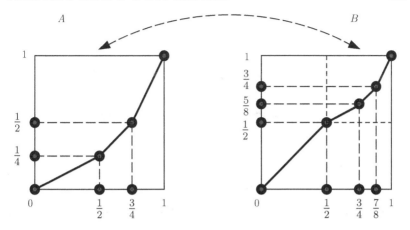

Figure 11.2 The standard generators of Thompson's group F

2.2. Homomorphisms to $\mathbb{Z}$

The abelianization $\mathbb{Z}^2$ of the group F is freely generated by the images of the above generators A and B. By χ_a and χ_b we denote the corresponding homomorphisms of F to $\mathbb{Z}$, i.e., the homomorphisms χ_a and χ_b are defined by putting, respectively,

$$\chi_a(A) = 1 \,, \qquad \chi_a(B) = 0 \,,$$

and

$$\chi_b(A) = 0 \,, \qquad \chi_b(B) = 1 \,.$$

Geometrically, $-\chi_a(g)$ (respectively, $\chi_a(g) + \chi_b(g)$) is the base 2 logarithm of the slope of the graph of g at the endpoint 0 (respectively, at the endpoint 1).

2.3. Change of Variables

The group F can also be realized as a subgroup of the group of piecewise linear homeomorphisms of the real line $PLF(\mathbb{R})$ introduced by Brin–Squier [BS85] (also see Haagerup–Picioroaga [HP11, Remark 2.5]).

More precisely, let $\tau : [0,1] \to \mathbb{R}$ be defined by putting

$$\tau\left(2^n\right) = n + 1 \,, \qquad n \leq -1$$
$$\tau\left(1 - \frac{1}{2^n}\right) = n - 1 \,, \qquad n \geq 1 \tag{11.5}$$

and by linear interpolation in between, see Figure 11.3.

Then one can easily see that τ is a bijection between the set of dyadic-rational numbers in the open unit interval $\mathbb{Z}\left[\frac{1}{2}\right] \cap (0,1)$ and the whole

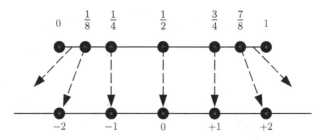

Figure 11.3 The change of variables from $[0,1]$ to $\mathbb{R}$

set of dyadic-rational numbers $\mathbb{Z}\left[\frac{1}{2}\right]$. After this change of coordinates the generators A and B of the group F (see Section **2.1**) take the form

$$\widetilde{A}(\gamma) = \gamma - 1\,, \qquad \gamma \in \mathbb{R}\,,$$

and

$$\widetilde{B}(\gamma) = \begin{cases} \gamma\,, & \gamma \le 0, \\ \frac{\gamma}{2}\,, & 0 \le \gamma \le 2, \\ \gamma - 1\,, & \gamma \ge 2. \end{cases}$$

Their inverses are, respectively,

$$\widetilde{A}^{-1}(\gamma) = \gamma + 1\,, \qquad \gamma \in \mathbb{R}\,,$$

and

$$\widetilde{B}^{-1}(\gamma) = \begin{cases} \gamma\,, & \gamma \le 0, \\ 2\gamma\,, & 0 \le \gamma \le 1, \\ \gamma + 1\,, & \gamma \ge 1\,, \end{cases}$$

see Figure 11.4, where the graphs of the generators $\widetilde{A}$ and $\widetilde{B}$ themselves are drawn with solid lines, and the graphs of their inverses are drawn with dashed lines.

We shall denote the image of the group F under the change of coordinates (11.5) by $\widetilde{F} \cong F$, i.e., $\widetilde{F}$ is the subgroup of $PLF(\mathbb{R})$ generated by the transformations $\widetilde{A}$ and $\widetilde{B}$.

2.4. Intrinsic Description of $\widetilde{F}$

One can easily see that $\widetilde{F}$, as a subgroup of $PLF(\mathbb{R})$, consists precisely of those transformations $g \in PLF(\mathbb{R})$, for which the following three conditions are satisfied:

    (i) The discontinuity points of the derivative g' are all dyadic rational.
    (ii) The slopes of g are integer powers of 2.
    (iii) For a sufficiently large $M > 0$ the transformation g has the form

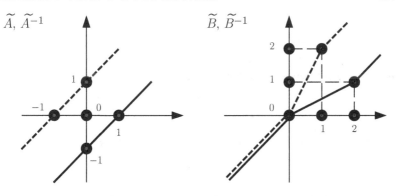

$\widetilde{A}, \widetilde{A}^{-1}$ $\widetilde{B}, \widetilde{B}^{-1}$

Figure 11.4 The graphs of the standard generators of the group $\widetilde{F}$

$$g(\gamma) = \gamma + C_- \quad \text{for} \quad \gamma \in (-\infty, -M) \,,$$

and

$$g(\gamma) = \gamma + C_+ \quad \text{for} \quad \gamma \in (M, \infty) \,,$$

where the constants $C_- = C_-(g)$ and $C_+ = C_+(g)$ are integers.

Clearly, the maps $C_\pm : \widetilde{F} \to \mathbb{Z}$ are group homomorphisms, and for the generators of the group $\widetilde{F}$

$$C_-(\widetilde{A}) = C_+(\widetilde{A}) = -1 \,,$$

whereas

$$C_-(\widetilde{B}) = 0 \,, \qquad C_+(\widetilde{B}) = -1 \,,$$

so that in terms of the homomorphisms $\chi_a, \chi_b : F \cong \widetilde{F} \to \mathbb{Z}$ introduced in Section **2.2**

$$C_- = -\chi_a \quad \text{and} \quad C_+ = -\chi_a - \chi_b \,.$$

3.   Random Walk on the Schreier Graph $\Gamma \cong \mathbb{Z}\left[\frac{1}{2}\right]$

3.1.   The Dyadic-Rational Schreier Graph

For technical reasons (since traditionally one considers *right* random walks on groups, for which the increments are added on the right), from now on we shall use the *postfix notation* for the group $\widetilde{F}$. Thus, the group operation in $\widetilde{F}$ will be defined as

$$(g_1 g_2)(\gamma) = g_2(g_1(\gamma)) \,, \tag{11.6}$$

(rather than the more common $(g_1 g_2)(\gamma) = g_1(g_2(\gamma))$ in the *prefix notation* which we used in Section **2.1**). Then $\widetilde{F}$ will naturally act on $\mathbb{R}$ on the *right* as

$$\gamma.g = g(\gamma)\,, \qquad \gamma \in \mathbb{R},\ g \in \widetilde{F}\,. \tag{11.7}$$

Our notation agrees with the one used when dealing with random walks on *groupoids* (in particular, on the *groupoids associated with group actions*), see Kaimanovich [Kai05].

The set of dyadic rational numbers $\mathbb{Z}\left[\frac{1}{2}\right] \subset \mathbb{R}$ is obviously transitive for this action, so that $\mathbb{Z}\left[\frac{1}{2}\right]$ can be endowed with the structure of the Schreier graph Γ of the group $\widetilde{F}$ with respect to the generating set

$$K = \{\widetilde{A}, \widetilde{B}, \widetilde{A}^{-1}, \widetilde{B}^{-1}\}\,. \tag{11.8}$$

3.2. Simple Random Walk on the Schreier Graph

The simple random walk on the graph Γ has the transitions

$$\gamma \overset{\mu(g)}{\rightsquigarrow} \gamma.g\,, \tag{11.9}$$

where $\mu = \mu_K$ is the probability measure equidistributed on the generating set K, i.e.,

$$\gamma \rightsquigarrow \begin{cases} \overset{\frac{1}{4}}{\rightsquigarrow} \widetilde{A}(\gamma) \\[4pt] \overset{\frac{1}{4}}{\rightsquigarrow} \widetilde{A}^{-1}(\gamma) \\[4pt] \overset{\frac{1}{4}}{\rightsquigarrow} \widetilde{B}(\gamma) \\[4pt] \overset{\frac{1}{4}}{\rightsquigarrow} \widetilde{B}^{-1}(\gamma). \end{cases}$$

More concretely, depending on the position of the point γ these transitions are

$$\gamma \rightsquigarrow \begin{cases} \overset{\frac{1}{2}}{\rightsquigarrow} \gamma + 1 \\[4pt] \overset{\frac{1}{2}}{\rightsquigarrow} \gamma - 1 \end{cases}, \quad \gamma \in (\infty, 0] \cup [2, \infty)\,,$$

$$\gamma \rightsquigarrow \begin{cases} \overset{\frac{1}{4}}{\rightsquigarrow} \gamma + 1 \\[4pt] \overset{\frac{1}{4}}{\rightsquigarrow} \gamma - 1 \\[4pt] \overset{\frac{1}{4}}{\rightsquigarrow} \frac{\gamma}{2} \\[4pt] \overset{\frac{1}{4}}{\rightsquigarrow} 2\gamma \end{cases}, \quad \gamma \in [0, 1]\,,$$

$$\gamma \rightsquigarrow \begin{cases} \overset{\frac{1}{2}}{\rightsquigarrow} \gamma + 1 \\[4pt] \overset{\frac{1}{4}}{\rightsquigarrow} \gamma - 1 \\[4pt] \overset{\frac{1}{4}}{\rightsquigarrow} \frac{\gamma}{2} \end{cases}, \quad \gamma \in [1, 2]\,.$$

We shall denote by $\mathbf{Q}$ the probability measure on the path space of the simple random walk on Γ corresponding to the starting point 0.

Let $|\gamma|_2$ be the usual 2-adic norm of a number $\gamma \in \Gamma \cong \mathbb{Z}\left[\frac{1}{2}\right]$, i.e.,

$$|\gamma|_2 = 2^{-n} \quad \text{for } \gamma = (2m+1)2^n \, ,$$

and $|0|_2 = 0$.

Theorem 3.1. *For $\mathbf{Q}$-a.e. sample path (γ_n) of the simple random walk* (11.9) *on Γ*

$$|\gamma_n|_2 \xrightarrow[n\to\infty]{} \infty \, .$$

As a consequence we immediately obtain

Theorem 3.2. *The simple random walk on Γ is transient.*

Proof of Theorem 3.1. Let us first of all notice that, outside of the interval $(0,2)$, the transition probabilities of the random walk (11.9) are just

$$p(\gamma, \gamma \pm 1) = \frac{1}{2} \, .$$

Therefore, the interval $[0,2)$ is recurrent by the classical properties of the simple random walk on $\mathbb{Z}$. Moreover, starting from any point $\gamma \in [1,2)$ the random walk will eventually hit the interval $[0,1)$ (and the corresponding hitting distribution is supported by the points $\gamma-1$ and $\frac{\gamma}{2}$ with the equal weights $\frac{1}{2}$). Thus, the interval $I = [0,1)$ is also recurrent. The transitions of the induced random walk on I are, for $\gamma \in \left[0, \frac{1}{2}\right)$,

$$\gamma \rightsquigarrow \begin{cases} \xrightsquigarrow{\frac{1}{4}} \gamma \\[4pt] \xrightsquigarrow{\frac{1}{4}} \gamma + 1 \rightsquigarrow \begin{cases} \xrightsquigarrow{\frac{1}{2}} \gamma \\[4pt] \xrightsquigarrow{\frac{1}{2}} \frac{\gamma}{2} + \frac{1}{2} \, , \end{cases} \\[4pt] \xrightsquigarrow{\frac{1}{4}} \frac{\gamma}{2} \\[4pt] \xrightsquigarrow{\frac{1}{4}} 2\gamma \end{cases}$$

i.e.,

$$\gamma \rightsquigarrow \begin{cases} \xrightsquigarrow{\frac{3}{8}} \gamma \\[4pt] \xrightsquigarrow{\frac{1}{4}} \frac{\gamma}{2} \\[4pt] \xrightsquigarrow{\frac{1}{8}} \frac{\gamma}{2} + \frac{1}{2} \\[4pt] \xrightsquigarrow{\frac{1}{4}} 2\gamma \end{cases} ,$$

and, for $\gamma \in \left[\frac{1}{2}, 1\right)$,

$$
\gamma \rightsquigarrow
\begin{cases}
\xrightarrow{\frac{1}{4}} \gamma \\[4pt]
\xrightarrow{\frac{1}{4}} \gamma + 1 \rightsquigarrow
\begin{cases}
\xrightarrow{\frac{1}{2}} \gamma \\[4pt]
\xrightarrow{\frac{1}{2}} \frac{\gamma}{2} + \frac{1}{2}
\end{cases} \\[20pt]
\xrightarrow{\frac{1}{4}} \frac{\gamma}{2} \\[4pt]
\xrightarrow{\frac{1}{4}} 2\gamma \rightsquigarrow
\begin{cases}
\xrightarrow{\frac{1}{2}} 2\gamma - 1 \\[4pt]
\xrightarrow{\frac{1}{2}} \gamma
\end{cases}
\end{cases}
,
$$

i.e.,

$$
\gamma \rightsquigarrow
\begin{cases}
\xrightarrow{\frac{1}{2}} \gamma \\[6pt]
\xrightarrow{\frac{1}{4}} \frac{\gamma}{2} \\[6pt]
\xrightarrow{\frac{1}{8}} \frac{\gamma}{2} + \frac{1}{2} \\[6pt]
\xrightarrow{\frac{1}{8}} 2\gamma - 1
\end{cases}
.
$$

Therefore, for $\gamma \in I \setminus \left\{0, \frac{1}{2}\right\}$ the logarithm $x = \log_2 |\gamma|_2$ of the 2-adic norm changes as

$$
x \rightsquigarrow
\begin{cases}
\xrightarrow{\frac{3}{8}} x \\[6pt]
\xrightarrow{\frac{3}{8}} x + 1 \\[6pt]
\xrightarrow{\frac{1}{4}} x - 1
\end{cases}
, \quad \gamma \in \left(0, \frac{1}{2}\right),
$$

$$
x \rightsquigarrow
\begin{cases}
\xrightarrow{\frac{1}{2}} x \\[6pt]
\xrightarrow{\frac{3}{8}} x + 1 \\[6pt]
\xrightarrow{\frac{1}{8}} x - 1
\end{cases}
, \quad \gamma \in \left(\frac{1}{2}, 1\right),
$$

i.e., these transitions on $\mathbb{Z}_+$ have a uniform positive drift, which implies the claim. $\qquad\square$

Remark 3.3. It might be interesting to study the properties of the induced random walk on I and of its stationary measure (which is most likely unique).

3.3. Comparison Criterion for Transience of Markov Chains

We shall now use a comparison argument in order to deduce the transience of general (not necessarily reversible!) random walks on the Schreier graph Γ from the transience just of the simple random walk on Γ. Its idea goes back to Baldi–Lohoué–Peyrière [BLP77], and it has been quite popular ever since (e.g., see Varopoulos [Var83], Chen [Che91], Woess [Woe94] as well as the exposition in Woess' book [Woe00, Sections 2.C and 3.A]). For the sake of completeness we shall prove it here in the generality sufficient for our purposes by slightly modifying the arguments of Varopoulos from [Var83, Section 4].

Theorem 3.4.   *Let P and P' be two Markov operators on a countable state space X with the respective transition probabilities $p(\cdot,\cdot)$ and $p'(\cdot,\cdot)$, and such that*

(i) *P and P' have a common stationary measure m.*
(ii) *The operator P is reversible with respect to the measure m, i.e., it is self-adjoint as an operator on the space $L^2(X, m)$, or, equivalently,*

$$m(x)p(x, y) = m(y)p(y, x) \qquad \forall\, x, y \in X\,.$$

(iii) *There exists $\varepsilon > 0$ such that $P' \geq \varepsilon P$, i.e.,*

$$p'(x, y) \geq \varepsilon p(x, y) \qquad \forall\, x, y \in X\,.$$

Then the transience of the operator P implies the transience of the operator P'.

Proof.   First of all, let us notice that the operator

$$Q = \frac{P' - \varepsilon P}{1 - \varepsilon}$$

is Markov and preserves the measure m. Therefore, for any $0 < \lambda < 1$ and any $f \in L^2(X, m)$

$$|\lambda\langle f, Qf\rangle| \leq |\langle f, Qf\rangle| \leq \|f\|^2\,.$$

It implies that

$$\langle (I - \lambda P')f, f\rangle \geq \varepsilon\|f\|^2_{P,\lambda} \geq 0\,, \tag{11.10}$$

where $\langle \cdot, \cdot \rangle$ denotes the scalar product on the space $L^2(X, m)$,

$$\langle f, g\rangle_{P,\lambda} = \langle f, (I - \lambda P)g\rangle \tag{11.11}$$

is the positive definite bilinear form on $L^2(X, m)$ determined by the operator $I - \lambda P$, and $\|\cdot\|, \|\cdot\|_{P,\lambda}$ are the respective associated norms.

The Cauchy–Schwarz inequality for the form $\langle \cdot, \cdot \rangle_{P,\lambda}$ then implies that for any $f \in L^2(X, m)$

$$\langle (I - \lambda P')^{-1} f, f \rangle^2 = \langle (I - \lambda P')^{-1} f, (I - \lambda P)(I - \lambda P)^{-1} f \rangle^2$$
$$= \langle (I - \lambda P')^{-1} f, (I - \lambda P)^{-1} f \rangle_{P,\lambda}^2$$
$$\leq \| (I - \lambda P')^{-1} f \|_{P,\lambda}^2 \cdot \| (I - \lambda P)^{-1} f \|_{P,\lambda}^2 .$$

By (11.10),

$$\| (I - \lambda P')^{-1} f \|_{P,\lambda}^2 \leq \frac{1}{\varepsilon} \langle (I - \lambda P')^{-1} f, f \rangle ,$$

whereas

$$\| (I - \lambda P)^{-1} f \|_{P,\lambda}^2 = \langle (I - \lambda P)^{-1} f, f \rangle ,$$

so that

$$\langle (I - \lambda P')^{-1} f, f \rangle^2 \leq \frac{1}{\varepsilon} \langle (I - \lambda P')^{-1} f, f \rangle \cdot \langle (I - \lambda P)^{-1} f, f \rangle . \quad (11.12)$$

Now, if $f = \delta_o$ for a point $o \in X$, then

$$\mathcal{G}_{P,\lambda}(o, o) = \frac{1}{m(o)} \langle (I - \lambda P)^{-1} \delta_o, \delta_o \rangle$$

is the λ-Green kernel of the operator P at the point o, and the same holds for the operator P'. Since $\lambda < 1$,

$$0 < \mathcal{G}_{P',\lambda}(o, o) < \infty ,$$

whence by (11.12)

$$\mathcal{G}_{P',\lambda}(o, o) \leq \frac{1}{\varepsilon} \mathcal{G}_{P,\lambda}(o, o) \leq \frac{1}{\varepsilon} \mathcal{G}_{P,1}(o, o) ,$$

which, by letting $\lambda \to 1$, implies the inequality

$$\mathcal{G}_{P',1}(o, o) \leq \frac{1}{\varepsilon} \mathcal{G}_{P,1}(o, o) . \quad (11.13)$$

Therefore, the finiteness of $\mathcal{G}_{P,1}(o, o)$ ($\equiv$ the transience of P) implies the finiteness of $\mathcal{G}_{P',1}(o, o)$ ($\equiv$ the transience of P'). $\square$

Remark 3.5. We are not aware of any 'elementary' proof of inequality (11.13).

Remark 3.6. The form (11.11) decomposes as

$$\langle \cdot, \cdot \rangle_{P,\lambda} = \mathcal{D}_P + (1 - \lambda) \langle \cdot, \cdot \rangle ,$$

where $\mathcal{D}_P$ is the *Dirichlet form* of the operator P.

3.4. General Random Walks

Given a probability measure μ on G we shall denote by (Γ, μ) the induced random walk on Γ with the transitions

$$\gamma \overset{\mu(g)}{\rightsquigarrow} \gamma.g \ . \tag{11.14}$$

In other words, the measure $\mathbf{Q}_\gamma$ on the space of paths of the induced random walk issued from a point $\gamma \in \Gamma$ is the image of the measure $\mathbf{P}$ on the path space of the random walk (G, μ) (see Section **1.1**) under the map

$$\mathbf{g} = (g_n) \mapsto \gamma.\mathbf{g} = (\gamma.g_n) \ . \tag{11.15}$$

Theorem 3.7. *Let μ be a strictly non-degenerate probability measure on the group $\widetilde{F}$. Then the induced random walk (Γ, μ) is transient.*

Proof. This is a direct application of Theorem 3.4. By the strict non-degeneracy of the measure μ, for any $g \in \widetilde{F}$ there exists $n = n(g)$ such that the n-fold convolution $\mu^{*n}(g)$ is strictly positive. Let

$$n = \max\{n(g) : g \in K\} \ ,$$

and

$$\delta = \min\{\mu^{*n(g)}(g) : g \in K\} > 0 \ ,$$

where K is the generating set (11.8). Then the measure

$$\mu' = \frac{1}{n} \sum_{k=1}^{n} \mu^{*k}$$

has the property that

$$\mu'(g) \geq \varepsilon \mu_K(g) \qquad \forall g \in K \ , \tag{11.16}$$

with $\varepsilon = \delta/n > 0$.

Let us now take for P (respectively, P') the Markov operator of the simple random walk (Γ, μ_K) (respectively, the operator of the random walk (Γ, μ')). If we take for m the counting measure on Γ, then conditions (i) and (ii) of Theorem 3.4 are obviously satisfied, whereas condition (iii) follows from inequality (11.16). Thus, Theorem 3.2 implies transience of the random walk (Γ, μ'), and therefore of the random walk (Γ, μ) as well. $\qquad \square$

Remark 3.8. One should be able to significantly relax the condition sgr $\mu = \widetilde{F}$ imposed on the measure μ in Theorem 3.7. We expect it to hold just under very mild assumptions on the group gr μ.

4. Non-trivial Behaviour at Infinity Determined by Stabilizing Configurations

4.1. Groups of Configurations on Γ

Let fun$(\Gamma, \mathbb{Z})$ (respectively, Fun$(\Gamma, \mathbb{Z})$) denote the additive group of finitely supported (respectively, of all) $\mathbb{Z}$-valued configurations on the graph Γ, cf. Section **1.7**. By using the right action (11.7) of the group $\widetilde{F}$ on $\Gamma \cong \mathbb{Z}\left[\frac{1}{2}\right]$ we shall define the associated *left* action of $\widetilde{F}$ on the configuration space Fun$(\Gamma, \mathbb{Z})$ as

$$\mathbf{S}^g \varphi(\gamma) = \varphi(\gamma.g) = \varphi(g(\gamma)) \,, \qquad g \in \widetilde{F}, \ \gamma \in \Gamma \,. \tag{11.17}$$

4.2. Configurations Associated with the Elements of $\widetilde{F}$

Given a transformation $g \in \widetilde{F}$, we shall denote by $C_g \in$ fun$(\Gamma, \mathbb{Z})$ the associated configuration on Γ defined as

$$C_g(\gamma) = \log_2 g'(\gamma + 0) - \log_2 g'(\gamma - 0) \qquad \forall \gamma \in \Gamma \,.$$

In other words, $C_g(\gamma)$ is the difference between the base 2 logarithms of the left and the right slopes of g at the point γ. The support of C_g is precisely the set of break points of g, i.e., the set of discontinuity points of the derivative g', see Figure 11.5.

Informally we shall say that the value $C_g(\gamma)$ is the logarithmic increment of the slope of g at the point γ. The ratio $\frac{g'(\gamma+0)}{g'(\gamma-0)}$ appears in the literature on Thompson's groups (e.g., see Liousse [Lio08]) under the name of the jump of g at the break point γ, so that in these terms $C_g(\gamma)$ is the logarithm of the jump of g at γ.

Remark 4.1.   Obviously, two different transformations $g_1, g_2 \in \widetilde{F}$ give rise to the same configuration if and only if $g_2 = g_1 + C$ for $C \in \mathbb{Z}$ (for, if the difference between two functions from $\widetilde{F}$ is a constant, then this constant must be an integer, see Section **2.4**). On the other hand,

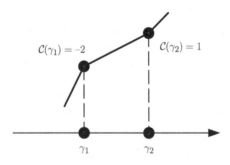

$C(\gamma_1) = -2$ $C(\gamma_2) = 1$

γ_1 γ_2

Figure 11.5 The configuration on $\Gamma \cong \mathbb{Z}\left[\frac{1}{2}\right]$ associated with an element of $\widetilde{F}$

not every configuration $\varphi \in \text{fun}(\Gamma, \mathbb{Z})$ corresponds to an element $g \in \widetilde{F}$. There are two conditions, whose combination is necessary and sufficient for that. The first condition is

$$\sum_{\gamma \in \Gamma} \varphi(\gamma) = 0$$

(for, any $g \in \widetilde{F}$ has slope 1 both at $-\infty$ and at $+\infty$, so that the sum of logarithmic increments of the slope must be 0). It guarantees that there exists a transformation $g \in PLF(\mathbb{R})$ with

$$\begin{cases} g(\tau) = \tau + C_- & \text{at } -\infty, \\ g(\tau) = \tau + C_+ & \text{at } +\infty, \end{cases}$$

where the difference $C_+ - C_-$ is uniquely determined by the configuration φ. Now, the second condition is $C_+ - C_- \in \mathbb{Z}$ (cf. Section **2.4**). For instance, the configuration $\varphi = -\delta_0 + \delta_1$ defined as

$$\varphi(\gamma) = \begin{cases} -1, & \gamma = 0, \\ +1, & \gamma = 1, \\ 0, & \text{othwerwise} \end{cases}$$

satisfies the first condition, but not the second one; the resulting transformation $g \in PLF(\mathbb{R})$ is (up to an additive constant)

$$g(\gamma) = \begin{cases} \gamma, & \gamma \leq 0, \\ \frac{\gamma}{2}, & 0 \leq \gamma \leq 1, \\ \gamma - \frac{1}{2}, & \gamma \geq 1. \end{cases}$$

4.3. Composition of Configurations

By the chain rule

$$(g_1 g_2)'(\gamma) = g_1'(\gamma) \cdot g_2'(g_1(\gamma))$$

(keep in mind that the group multiplication (11.6) in $\widetilde{F}$ is defined by using the 'inverse composition'). Thus,

$$\mathcal{C}_{g_1 g_2}(\gamma) = \mathcal{C}_{g_1}(\gamma) + \mathcal{C}_{g_2}(g_1(\gamma)) \qquad \forall\, g_1, g_2 \in \widetilde{F}, \gamma \in \Gamma, \qquad (11.18)$$

or, in terms of the action (11.17),

$$\mathcal{C}_{g_1 g_2} = \mathcal{C}_{g_1} + \mathbf{S}^{g_1} \mathcal{C}_{g_2} \qquad \forall\, g_1, g_2 \in \widetilde{F}. \qquad (11.19)$$

Remark 4.2. Formula (11.19) looks very similar to the 'configuration component' of formula (11.3) for the multiplication in the lamplighter groups. Note, however, that the actions $\mathbf{T}$ (11.2) and $\mathbf{S}$ (11.17) on the respective configuration spaces, which appear in these formulas, are quite different. The action $\mathbf{T}$ is determined by the action of the group on itself on the *left*, whereas the action **S** is determined by the action on $\Gamma \cong \mathbb{Z}\left[\frac{1}{2}\right]$ on the *right*. In particular, as a result of this

$$\mathbf{T}^g \delta_h = \delta_{gh} \qquad \forall\, g, h \in G$$

in the lamplighter setup of Section **1.7**, whereas for the action (11.17)

$$\mathbf{S}^g \delta_\gamma = \delta_{\gamma.g^{-1}} \qquad \forall\, g \in \tilde{F},\ \gamma \in \Gamma\,.$$

Formula (11.19) allows one to define yet another ('skew') left action of the group $\tilde{F}$ on the configuration space $\mathrm{Fun}(\Gamma, \mathbb{Z})$ as

$$(g, \mathcal{C}) \mapsto \mathcal{C}_g + \mathbf{S}^g \mathcal{C}\,. \tag{11.20}$$

This action is similar to the natural action of the lamplighter group $\mathcal{L}(G)$ on the configuration space $\mathrm{Fun}(G, \mathbb{Z}/2\mathbb{Z})$ defined as

$$(g, \Phi)\,\mathcal{C} = \Phi + \mathbf{T}^g \mathcal{C}\,.$$

It is this action that we mentioned at the end of Section **1.8** saying that it is determined by the 'configuration part' of the multiplication formula in the group $\mathcal{L}(G)$. Note, however, that in the lamplighter case the component g of a group element $(g, \Phi) \in \mathcal{L}(G)$ is responsible just for 'translating' a configuration $\mathcal{C}$, whereas the component Φ is responsible just for adding an 'increment' to $\mathcal{C}$. On the contrary, there is no such 'splitting' in the case of the action (11.20).

Lemma 4.3. *The action* (11.20) *of the group* $\tilde{F}$ *on* $\mathrm{Fun}(\Gamma, \mathbb{Z})$ *has no fixed points.*

Proof. Let $\mathcal{C} \in \mathrm{Fun}(\Gamma, \mathbb{Z})$ be a fixed point of the action, i.e.,

$$\mathcal{C} = \mathcal{C}_g + \mathbf{S}^g \mathcal{C} \qquad \forall\, g \in \tilde{F}\,.$$

Since the configuration $\mathcal{C}_{\tilde{A}}$ associated with the generator $\tilde{A}$ of the group $\tilde{F}$ (see Section **2.3**) is empty, the configuration $\mathcal{C}$ must be then invariant with respect to the transformation $\mathbf{S}^{\tilde{A}}$ determined by the action **S** (11.17). In other words, $\mathcal{C}$ must be periodic on $\Gamma \cong \mathbb{Z}\left[\frac{1}{2}\right]$ with period 1.

In the same way, by taking $g = \tilde{B}$ we arrive at the condition

$$\mathcal{C} = \mathcal{C}_{\tilde{B}} + \mathbf{S}^{\tilde{B}} \mathcal{C}\,,$$

where

$$\mathcal{C}_{\tilde{B}} = -\delta_0 + \delta_2$$

by the definition of the generator $\tilde{B}$ (see Section **2.3**). Since $\tilde{B}$ acts trivially on $\mathbb{Z}\left[\frac{1}{2}\right] \cap (-\infty, 0]$, it implies that $\mathcal{C}$ cannot be 1-periodic, whence a contradiction. $\qquad\square$

4.4. Stabilization of Configurations

Theorem 4.4. *Let μ be a finitely supported strictly non-degenerate probability measure on the group $\widetilde{F}$. Then for a.e. sample path (g_n) of the random walk $(\widetilde{F}, \mu)$ the configurations $\mathcal{C}_n = \mathcal{C}_{g_n}$ pointwise converge to a limit $\mathbb{Z}$-valued configuration $\mathcal{C}_\infty \in \mathrm{Fun}(\Gamma, \mathbb{Z})$, and the space $\mathrm{Fun}(\Gamma, \mathbb{Z})$ endowed with the arising limit distribution λ is a non-trivial μ-boundary. Therefore, the Poisson boundary $(\partial_\mu \widetilde{F}, \nu)$ itself is also non-trivial.*

Proof. By formula (11.18), for any $\gamma \in \Gamma$

$$\mathcal{C}_{n+1}(\gamma) = \mathcal{C}_n(\gamma) + \mathcal{C}_{h_{n+1}}(\gamma \cdot g_n),$$

where, as usual, (h_n) is the sequence of increments of the random walk (g_n). Since $\mathrm{supp}\,\mu$ is finite, the supports of all configurations $\mathcal{C}_h$, $h \in \mathrm{supp}\,\mu$, are contained in a certain finite set $A \subset \Gamma$. Therefore, by Theorem 3.7 the sequence $(\mathcal{C}_n(\gamma))$ almost surely stabilizes for any $\gamma \in \Gamma$, i.e., it converges to the limit configuration

$$\mathcal{C}_\infty = \lim_n \mathcal{C}_n = \mathcal{C}_{h_1} + \mathbf{S}^{g_1}\mathcal{C}_{h_2} + \mathbf{S}^{g_2}\mathcal{C}_{h_3} + \cdots.$$

Then the resulting measure ν on the space $\mathrm{Fun}(\Gamma, \mathbb{Z})$ is μ-stationary with respect to the action (11.20), cf. Section **1.8** for the lamplighter case.

Now it remains to show that the limit configuration $\mathcal{C}_\infty$ cannot be the same for a.e. sample path (g_n), i.e., that there does not exist a configuration $\mathcal{C} \in \mathrm{Fun}(\Gamma, \mathbb{Z})$ such that almost surely $\mathcal{C}_\infty = \mathcal{C}$. Indeed, since $\mathrm{sgr}\,\mu = \widetilde{F}$, if this were the case, then the limit configuration $\mathcal{C}$ would have necessarily been a fixed point of the action (11.20) of the group $\widetilde{F}$ on $\mathrm{Fun}(\Gamma, \mathbb{Z})$. However, this is impossible by Lemma 4.3. $\quad\square$

Remark 4.5. The limit distribution λ constructed in Theorem 4.4 is concentrated on the subset of $\mathrm{Fun}(\Gamma, \mathbb{Z})$ which consists only of infinitely supported configurations. The reason is that its complement $\mathrm{fun}(\Gamma, \mathbb{Z})$ is countable, whereas it is well known that any non-trivial μ-boundary is purely non-atomic, see Kaimanovich [Kai95].

5. Geometry of the Schreier Graph Γ

5.1. The Graph Γ

Our proof of Theorem 3.2 on the transience of the simple random walk on the Schreier graph $\Gamma \cong \mathbb{Z}\left[\frac{1}{2}\right]$ of the group $\widetilde{F}$ was based on purely probabilistic considerations, and this is the argument we had referred to in [Kai04a].

Shortly thereafter Savchuk independently analyzed the geometry of the Schreier graph of the original Thompson group F (endowed with the standard set of generators) on the dyadic-rational orbit in the interval $[0,1]$, and obtained its complete description [Sav10b] (also see [Sav15]

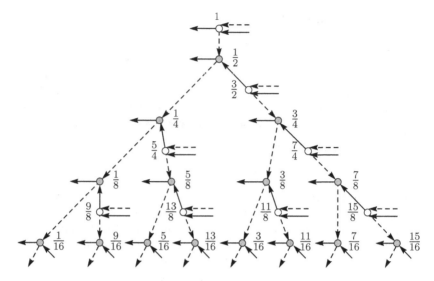

Figure 11.6 The Schreier graph Γ of Thompson's group $F \cong \widetilde{F}$ on the dyadic-rational orbit in $\mathbb{R}$

for the Schreier graphs on the other orbits of the canonical action of the group F). Since the dyadic-rational orbit of our group $\widetilde{F}$ is precisely the image of the dyadic-rational orbit of the original group F under the coordinate change (11.5) (see Section **2.3**), this Schreier graph is isomorphic to our graph Γ.

Figure 11.6 is a somewhat modified version of the picture of the graph Γ which first appeared in [Sav10b] (I am most grateful to Dmytro Savchuk for sharing his graphics source files with me).

The labelling of the vertices of Γ on our Figure 11.6 corresponds to the action of $\widetilde{F}$ on $\Gamma \cong \mathbb{Z}\left[\frac{1}{2}\right]$ rather than to the original action of the group F on $\mathbb{Z}\left[\frac{1}{2}\right] \cap (0,1)$ as in [Sav10b]. Following [Sav10b], we use three different colors for the vertices of the graph Γ in order to better exhibit its structure: black for $\gamma \in \mathbb{Z}\left[\frac{1}{2}\right] \cap (-\infty, 0]$, grey for $\gamma \in \mathbb{Z}\left[\frac{1}{2}\right] \cap (0,1)$, and white for $\gamma \in \mathbb{Z}\left[\frac{1}{2}\right] \cap [1, \infty)$.

The solid (respectively, dashed) arrows represent the oriented edges in Γ corresponding to the generator $\widetilde{A}$ (respectively, $\widetilde{B}$), see Section **2.3**. As is customary for Cayley and Schreier graphs, labels are assigned to *oriented* edges, so that the label of the same edge endowed with the opposite orientation is the letter of the alphabet $\{\widetilde{A}, \widetilde{A}^{-1}, \widetilde{B}, \widetilde{B}^{-1}\}$ inverse to the label of the original edge.

The 'open arrows' $\leftarrow$ (respectively, $\Leftarrow$) on Figure 11.6 represent the infinite **rays of negative** (respectively, **positive**) **type**, see Figure 11.7 and Figure 11.8, respectively. These rays are the connected components of

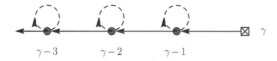

Figure 11.7 Negative type rays in the Schreier graph Γ

Figure 11.8 Positive type rays in the Schreier graph Γ

the subgraph of Γ with the vertex set $(-\infty, 0] \cap \mathbb{Z}\left[\frac{1}{2}\right]$ (respectively, $[2, \infty) \cap \mathbb{Z}\left[\frac{1}{2}\right]$) attached to the starting points from the set $(0, 1] \cap \mathbb{Z}\left[\frac{1}{2}\right]$ (respectively, $[1, 2) \cap \mathbb{Z}\left[\frac{1}{2}\right]$). The only vertex of Γ, to which two rays are attached (both a negative type and a positive type ones), is 1. Note that by the presence of these two families of rays the structure of the graph Γ is really begging for applying the coordinate change (11.5).

5.2. The Skeleton

The skeleton $\overline{\Gamma}$ of the Schreier graph Γ is the tree obtained by removing from Γ all the rays of negative and positive types, see Figure 11.9. The tree $\overline{\Gamma}$ is obviously *roughly isometric* to the usual *rooted binary tree*. Indeed, let B be the binary tree which consists just of the grey vertices of $\overline{\Gamma}$ (i.e., of the points from $(0, 1) \cap \mathbb{Z}\left[\frac{1}{2}\right]$). Then $\overline{\Gamma}$ and B differ only by the presence of an additional edge (the one between the vertices 1 and $\frac{1}{2}$) and of additional white vertices which subdivide some of the edges of B into two halves.

5.3. Transience of Γ

Since the rays removed when passing from Γ to $\overline{\Gamma}$ are clearly recurrent for the simple random walk (we have already referred to this property in the proof of Theorem 3.1), the transience of the simple random walk on the Schreier graph is then a direct consequence of the classically known transience of the rooted binary tree (e.g., see Woess [Woe00]). This argument is self-evident, once the geometry of the Schreier graph Γ has been exhibited. Although the statement about the transience of Γ does not explicitly appear in Savchuk's paper [Sav10b], its author was aware of it and mentioned it in some of his talks given at that time.

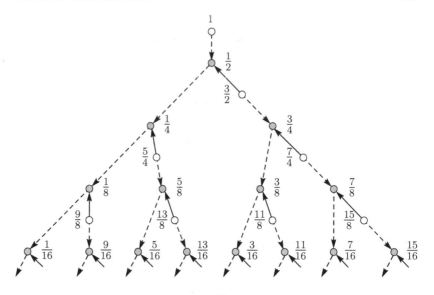

Figure 11.9 The skeleton $\overline{\Gamma}$ of the Schreier graph Γ

5.4. Space of Ends of Γ

Let $\partial\Gamma$ denote the space of ends of the graph Γ. It obviously splits into a disjoint union

$$\partial\Gamma = \partial\overline{\Gamma} \cup \partial_-\Gamma \cup \partial_+\Gamma . \tag{11.21}$$

Here $\partial_-\Gamma$ and $\partial_+\Gamma$ are the infinite sets of ends determined by the rays of the negative and the positive types, respectively, and $\partial\overline{\Gamma}$ is the space of ends of the skeleton $\overline{\Gamma}$.

The sets $\partial_-\Gamma$ and $\partial_+\Gamma$ can be identified with the respective subsets of the skeleton $\overline{\Gamma}$, to which the corresponding rays are attached, i.e., with $\mathbb{Z}\left[\frac{1}{2}\right] \cap (0,1]$ and with $\mathbb{Z}\left[\frac{1}{2}\right] \cap [1,2)$, see Section **5.1**. Note that the topology of the union (11.21) is the same as that of the *end compactification* of the skeleton $\overline{\Gamma}$ (with the only difference that the point $1 \in \overline{\Gamma}$ appears with multiplicity 2).

In what concerns $\partial\overline{\Gamma}$, as a topological space it can be easily identified with the space $\mathbb{Z}_2^{(1)} = \mathbb{Z}_2 \setminus 2\mathbb{Z}_2$ of 2-adic integers of norm 1. For doing that let us first notice that the downward branchings in the 'grey' binary tree B (see Section **5.2**) consist in applying the maps

$$f_0 : \gamma \mapsto \frac{\gamma}{2} \quad \text{and} \quad f_1 : \gamma \mapsto \frac{\gamma}{2} + \frac{1}{2} ,$$

or, equivalently,

$$f_\varepsilon : \gamma \mapsto \frac{\gamma}{2} + \varepsilon\frac{1}{2} \quad \text{with} \quad \varepsilon \in \{0, 1\} .$$

Therefore, any geodesic ray $\gamma = (\gamma_1, \gamma_2, \ldots)$ in the tree B issued from the point $\gamma_1 = \frac{1}{2}$ can be encoded by the corresponding sequence $\varepsilon = (\varepsilon_1, \varepsilon_2, \ldots)$, where ε_n are uniquely determined by the condition

$$\gamma_{n+1} = f_{\varepsilon_n}(\gamma_n)$$
$$= \frac{\varepsilon_n}{2} + \frac{\varepsilon_{n-1}}{4} + \frac{\varepsilon_{n-2}}{8} + \cdots + \frac{\varepsilon_2}{2^{n-1}} + \frac{\varepsilon_1}{2^n} + \frac{1}{2^{n+1}}, \qquad n \geq 1,$$

so that

$$|\gamma_n|_2 = 2^n \qquad \forall\, n \geq 1.$$

Then the sequence

$$|\gamma_n|_2 \cdot \gamma_n = 1 + \varepsilon_1 \cdot 2 + \varepsilon_2 \cdot 4 + \cdots + \varepsilon_{n-1} \cdot 2^{n-1}$$

obviously converges in the 2-adic topology to the 2-adic integer

$$1 + \sum_{n=1}^{\infty} \varepsilon_n \cdot 2^n$$

of norm 1. Conversely, any 2-adic integer of norm 1 gives rise to the corresponding geodesic $\gamma = (\gamma_1, \gamma_2, \ldots)$ in the tree B, i.e., to an end of $\overline{\Gamma}$.

5.5. Sections of the End Bundle

By

$$\mathrm{Fun}(\Gamma, \partial\Gamma) \cong (\partial\Gamma)^{\Gamma}$$

we shall denote the space of $\partial\Gamma$-valued configurations on Γ, or, equivalently, of sections of the end bundle

$$\Gamma \times \partial\Gamma \to \Gamma, \qquad (\gamma, \omega) \to \gamma,$$

over the graph Γ (cf. the definitions of the *hyperbolic boundary bundles* and of the *Poisson bundles* in author's papers [Kai04b] and [Kai05], respectively). One can also consider the configurations from $\mathrm{Fun}(\Gamma, \partial\Gamma)$ as end fields on Γ. In the same way as the space $\mathrm{Fun}(\Gamma, \mathbb{Z})$ of $\mathbb{Z}$-valued configurations on Γ, the space $\mathrm{Fun}(\Gamma, \partial\Gamma)$ is endowed with the left action (11.17) of the group $\widetilde{F}$.

5.6. Partial Isometries of Γ

Generally speaking, groups do not act on their Schreier graphs by graph automorphisms (here and below when talking about graph automorphisms or isomorphisms we mean the maps which preserve the Schreier labelling of edges). In particular, in our case the graph Γ and its skeleton $\overline{\Gamma}$ are both rigid, i.e., their groups of automorphisms are trivial. However, in spite of that the graph Γ still has certain symmetry properties sufficient for our purposes. Namely, it has a rather rich set of partial isomorphisms, i.e., of graph isomorphisms between its subgraphs.

First of all, obviously, all the rays of negative (respectively, of positive) type are pairwise isomorphic. Further, for a tree T and any two vertices $o \neq o' \in T$ let $T_{o \to o'}$ denote the shadow of the vertex o' as seen from the vertex o, i.e., the subtree of T which consists of all vertices $x \in T$ such that o' lies on the geodesic $[o, x]$. Then the skeleton $\overline{\Gamma}$ is self-similar in the sense that the shadows $\overline{\Gamma}_{1 \to \frac{1}{2}}$ and $\overline{\Gamma}_{1 \to \gamma}$ are isomorphic for any $\gamma \in \mathbb{Z}\left[\frac{1}{2}\right] \cap (0, 1)$ (i.e., for any 'grey' vertex $\gamma \in \overline{\Gamma}$). If now $\Gamma_{1 \to \gamma}$ denotes the subgraph of Γ obtained by reattaching all the negative and positive type rays to the vertices of the shadow $\overline{\Gamma}_{1 \to \gamma} \subset \overline{\Gamma}$, then the subgraphs $\Gamma_{1 \to \gamma} \subset \Gamma$ are all pairwise isomorphic for $\gamma \in \mathbb{Z}\left[\frac{1}{2}\right] \cap (0, 1)$.

6. Non-trivial Behaviour at Infinity Determined by the Boundary of the Schreier Graph

6.1. Mishchenko's Work

In a recent preprint [Mis15] Mishchenko gives yet another proof of the transience of the Schreier graph Γ different from the proofs described in Section **3.2** and Section **5.3** (it is based on an explicit estimate of the Dirichlet norm on Γ). He further notices that due to a specific geometry of Γ this transience implies existence of a non-trivial behaviour at infinity ($\equiv$ non-triviality of the Poisson boundary) for the simple random walk on Γ, and therefore for the simple random walk on Thompson's group F as well (cf. the discussion of the relationship between the simple random walks on Γ and on $F \cong \widetilde{F}$ in Section **3.2**). Thus, the Poisson boundary of the simple random walk on Thompson's group F is non-trivial [Mis15, Theorem 2.5], which provides another proof of our result on the absence of the Liouville property for the group F.

The aforementioned geometric argument in [Mis15] is based on the following observation (Theorem 2.3), which we shall quote here in a slightly modified form:

> If a tree T has the property that there exists a vertex $o \in T$ of degree at least 2, and such that for any neighbour o' of o the shadow $T_{o \to o'}$ is transient, then the Poisson boundary of the simple random walk on T is non-trivial.

In fact, it is classically known (e.g., see Woess [Woe00]) that the transience of the simple random walk on a connected graph implies convergence of its sample paths to an *end* of this graph. Moreover, the same is true for any bounded range random walk, and, as we have already established in Theorem 3.7, the random walk on Γ determined by any strictly non-degenerate probability measure μ on $\widetilde{F}$ is transient.

Yet another point is that the boundary behaviour provided by the above argument from [Mis15] depends on the starting point $\gamma \in \Gamma$ of the induced random walk, so that the resulting quotient of the Poisson boundary is not a μ-boundary. Actually, the quotient of the Poisson boundary produced by [Mis15, Theorem 2.3] is finite (it can be identified with the set of neighbours of the vertex o), so that it cannot be a μ-boundary already for this reason (for, as we have already mentioned in Remark 4.5, any non-trivial μ-boundary is purely non-atomic [Kai95]).

6.2. Non-trivial μ-Boundary

Theorem 6.1. *Let μ be a strictly non-degenerate finitely supported probability measure on the group $\widetilde{F}$. Then for a.e. sample path $\mathbf{g} = (g_n)$ of the random walk $(\widetilde{F}, \mu)$ and any point $\gamma \in \Gamma \cong \mathbb{Z}\left[\frac{1}{2}\right]$ the sequence $\gamma.g_n$ converges to an end $\omega_\gamma = \omega_\gamma(\mathbf{g}) \in \partial\Gamma$, which gives rise to a map*

$$\mathbf{g} \mapsto (\omega_\gamma)_{\gamma \in \Gamma} \in \mathrm{Fun}(\Gamma, \partial\Gamma)$$

from the path space $(\widetilde{F}^{\mathbb{Z}_+}, \mathbf{P})$ of the random walk $(\widetilde{F}, \mu)$ to the configuration space $\mathrm{Fun}(\Gamma, \partial\Gamma)$, and the space $\mathrm{Fun}(\Gamma, \partial\Gamma)$ endowed with the arising probability measure λ (the image of the measure $\mathbf{P}$ on the path space under the above map) is a non-trivial μ-boundary.

Proof. As we have already noticed, Theorem 3.7 in combination with the fact that the support of μ is finite implies the convergence

$$\gamma.g_n \to \omega_\gamma \in \partial\Gamma \qquad \text{for all } \gamma \in \Gamma \text{ and a.e. sample path } (g_n) \, .$$

Thus, it only remains to show that the distribution of the configurations $(\omega_\gamma)_{\gamma \in \Gamma}$ is not a point measure.

First of all let us notice that if the configuration (ω_γ) were the same for a.e. sample path (g_n), then it would necessarily have been constant (i.e., taking the same value at all points of Γ), because the group $\widetilde{F}$ acts on $\mathrm{Fun}(\Gamma, \partial\Gamma)$ by 'changing' the $\Gamma \cong \mathbb{Z}\left[\frac{1}{2}\right]$ variable, and $\mathrm{sgr}\,\mu = \widetilde{F}$.

Now let us assume that $\omega_\gamma \equiv \omega \in \partial\Gamma$. It means that for any starting point $\gamma \in \Gamma$ almost every sample path of the random walk (Γ, μ) (11.14) converges to ω. Let us denote by $\mathbf{Q}_{\#}$ the measure on the space of sample paths of this random walk whose initial distribution is the counting measure $\#$ on Γ.

If $\omega \in \partial_-\Gamma \cup \partial_+\Gamma$ is the end corresponding to a ray of the negative or of the positive type, then there is a $\mathbf{Q}_{\#}$-non-negligible set of sample paths which are entirely contained in this ray. However, since all the rays of the same type are isomorphic (see Section 5.6), the same would be true for any other ray of the same type, which means that with positive probability the limit end would be different from ω.

If $\omega \in \partial\overline{\Gamma}$ is a skeleton end, then, in the same way, the set of the sample paths entirely contained in the subgraph $\Gamma_{1 \to \frac{1}{2}}$ is $\mathbf{Q}_{\#}$-non-negligible.

In view of the self-similarity of the graph Γ (see Section **5.6**) it implies that the same would be true for any subgraph $\Gamma_{1\to\gamma}$. However, one can obviously choose γ in such a way that $\omega \notin \partial\Gamma_{1\to\gamma}$. $\square$

The construction of a μ-boundary of the group F on the configuration space $\mathrm{Fun}(\Gamma,\mathbb{Z})$ from Theorem 4.4 heavily used the specifics of this group. On the contrary, the above construction of a μ-boundary on the space of sections $\mathrm{Fun}(\Gamma,\partial\Gamma)$ of the end bundle over the Schreier graph Γ is much more general.

Theorem 6.2.   *If a Schreier graph Γ of a finitely generated group G is transient (with respect to the simple random walk on it), then for any strictly non-degenerate finitely supported probability measure μ on G either*

(i) *there exists an end $\omega \in \partial\Gamma$ such that for any $\gamma \in \Gamma$ a.e. sample path of the induced random walk (Γ,μ) issued from γ converges to ω,*

or

(ii) *the space of sections $\mathrm{Fun}(\Gamma,\partial\Gamma)$ of the end bundle over Γ is a non-trivial μ-boundary, so that the Poisson boundary of the random walk (G,μ) is also non-trivial.*

Proof.   In the same way as in the proof of Theorem 3.7, the transience of the simple random walk on Γ implies the transience of the induced random walk (Γ,μ) for any strictly non-degenerate probability measure μ on G. Further, if μ is finitely supported, then a.e. sample path of the induced random walk on Γ converges to a random end $\omega \in \partial\Gamma$. Let us denote by $\{\varkappa_\gamma\}_{\gamma\in\Gamma}$ the family of the corresponding hitting distributions on the space of ends $\partial\Gamma$. By the strict non-degeneracy of μ, all the measures $\varkappa_\gamma$ are pairwise equivalent. Now, in case (i) all measures $\varkappa_\gamma$ coincide with the delta-measure δ_γ. Otherwise, in case (ii), all measures $\varkappa_\gamma$ are not delta-measures, so that $\mathrm{Fun}(\Gamma,\partial\Gamma)$ endowed with the arising limit distribution is a non-trivial μ-boundary in the same way as in Theorem 6.1. $\square$

6.3. Convergence to Components of the End Space

The key ingredient of our proof of Theorem 6.1 was the observation that for a finitely supported measure μ on $\widetilde{F}$ the transience of the random walk (Γ,μ) (11.14) implies the convergence of a.e. sample path to a random end of the graph Γ (also see Theorem 6.2). By using the homomorphisms

$$\chi_a, \chi_b : F \cong \widetilde{F} \to \mathbb{Z}$$

introduced in Section **2.2** one can easily describe the components of the decomposition (11.21) of the space of ends $\partial\Gamma$, on which the hitting measures $\varkappa_\gamma$ of the random walk (Γ,μ) are actually concentrated.

Since the measure μ is finitely supported, the restriction of the random walk (Γ, μ) to any of the negative (respectively, positive) type rays (see Section **5.1**) coincides, outside of a finite neighbourhood of ray's origin, just with the usual translation invariant random walk on $\mathbb{Z}$. Here and below we identify the negative (respectively, positive) type rays in Γ with the negative (respectively, positive) integer ray $\mathbb{Z}_- = \{\ldots, -2, -1, 0\}$ (respectively, $\mathbb{Z}_+ = \{0, 1, 2, \ldots\}$) of $\mathbb{Z}$. As it follows from the description of the graph Γ (see Figure 11.6), the step distribution of the induced random walk on $\mathbb{Z}_-$ (respectively, on $\mathbb{Z}_+$) is $(-\chi_a)(\mu)$ (respectively, $(-\chi_a - \chi_b)(\mu)$), where by $\chi(\mu)$ we denote the image of the measure μ under a group homomorphism $\chi : \widetilde{F} \to \mathbb{Z}$. Therefore, the drift of the induced random walk on $\mathbb{Z}_-$ ($\equiv$ on the negative type rays) is the barycentre

$$\overline{(-\chi_a)(\mu)} = -\alpha \ ,$$

and the drift of the induced random walk on $\mathbb{Z}_+$ ($\equiv$ on the positive type rays) is the barycentre

$$\overline{(-\chi_a - \chi_b)(\mu)} = -\alpha - \beta \ ,$$

where

$$\alpha = \overline{\chi_a(\mu)} \quad \text{and} \quad \beta = \overline{\chi_b(\mu)}$$

denote the barycentres (the expectations) of the measures $\chi_a(\mu)$ and $\chi_b(\mu)$ on $\mathbb{Z}$, respectively.

It is classically known that the recurrence properties of the random walk on $\mathbb{Z}$ governed by a finitely supported step distribution σ are completely determined by the barycentre $\overline{\sigma}$. In particular, the ray $\mathbb{Z}_-$ (respectively, $\mathbb{Z}_+$) is transient (i.e., the random walk escapes to infinity along this ray) if and only if the drift $\overline{\sigma}$ is negative (respectively, positive). Therefore, we arrive at the following description of the components of the decomposition (11.21), on which the hitting measures $\varkappa_\gamma$ are concentrated, in terms of the parameters $\alpha, \beta \in \mathbb{R}$ (also see Figure 11.10):

Proposition 11.22. *Under the conditions of Theorem 6.1*

(i) *If $\alpha > 0$ and $\alpha + \beta < 0$, then both the negative type and the positive type rays are transient, and the hitting measures $\varkappa_\gamma$ are concentrated on the union $\partial_-\Gamma \cup \partial_+\Gamma$.*

(ii) *If $\alpha > 0$ and $\alpha + \beta \geq 0$, then only the negative type rays are transient, and the hitting measures $\varkappa_\gamma$ are concentrated on $\partial_-\Gamma$.*

(iii) *If $a \leq 0$ and $\alpha + \beta < 0$, then only the positive type rays are transient, and the hitting measures $\varkappa_\gamma$ are concentrated on $\partial_+\Gamma$.*

(iv) *If $a \leq 0$ and $\alpha + \beta \geq 0$, then neither the negative nor the positive type rays are transient, and the hitting measures $\varkappa_\gamma$ are concentrated on $\partial\overline{\Gamma}$.*

Proposition 11.22 allows us to make more precise the description of the μ-boundary of the random walk $(\widetilde{F}, \mu)$ obtained in Theorem 6.1.

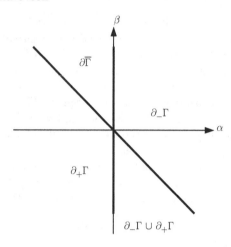

Figure 11.10 The domains in the parameter plane (α, β) corresponding to different supports of the hitting measures $\varkappa_\gamma$

Theorem 6.3. *Under conditions of Theorem 6.1 the arising probability measure λ is concentrated on*

$$\begin{cases} \mathrm{Fun}(\Gamma, \partial_-\Gamma \cup \partial_+\Gamma)\,, & \text{if}\,\alpha > 0,\ \text{and}\,\alpha + \beta < 0\,, \\ \mathrm{Fun}(\Gamma, \partial_-\Gamma)\,, & \text{if}\,\alpha > 0,\ \text{and}\,\alpha + \beta \geq 0\,, \\ \mathrm{Fun}(\Gamma, \partial_+\Gamma)\,, & \text{if}\,a \leq 0,\ \text{and}\,\alpha + \beta < 0\,, \\ \mathrm{Fun}(\Gamma, \partial\overline{\Gamma})\,, & \text{if}\,a \leq 0,\ \text{and}\,\alpha + \beta \geq 0\,. \end{cases}$$

7. Concluding Remarks

7.1. Relaxing Conditions on the Step Distribution

Throughout the chapter we have used the assumption that the measure μ on the group $\widetilde{F} \cong F$ is finitely supported and strictly non-degenerate. It would be interesting to see, to what extent our results can be carried over to more general measures. [Let us remind once again that amenability of the group F is equivalent to existence of a Liouville symmetric probability measure μ with $\mathrm{supp}\,\mu = F$.]

Here one can try to relax both the conditions on *how much the measure μ is allowed to spread out*:

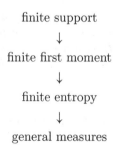

and the conditions on *how non-degenerate the measure μ is allowed to be*:

$$\operatorname{sgr} \mu = \widetilde{F}$$
$$\downarrow$$
$$\operatorname{gr} \mu = \widetilde{F}$$
$$\downarrow$$

$\operatorname{gr} \mu$ is a *non-elementary* subgroup of $\widetilde{F}$.

The meaning of 'non-elementary' of course needs to be specified in our context (for instance, cf. Kaimanovich–Masur [KM96] for the subgroups of the mapping class group, or Kaimanovich [Kai00] for the subgroups of groups with hyperbolic properties).

As for the conditions on how much the measure μ is allowed to spread out, one can expect that an extension to the class of measures with a *finite first moment* should be quite feasible (cf. the case of the lamplighter groups considered by the author [Kai91] and the case of hyperbolic graphs considered by Kaimanovich–Woess [KW92]). Notice, however, that, once the finite first moment condition is dropped, the 'hyperbolic' and the 'lamplighter' models drastically diverge. Boundary convergence in hyperbolic spaces does not require any moment conditions, see Kaimanovich [Kai00], Maher–Tiozzo [MT14], whereas for the lamplighter groups, due to their amenability, the situation is completely different.

For instance, there are examples of measures with infinite first moment on the lamplighter groups, for which the configurations do not stabilize, but the Poisson boundary is still non-trivial, see Kaimanovich [Kai83, Kai91]. In fact, Erschler proved in [Ers04b] that the Poisson boundary on the lamplighter groups is non-trivial for all step distributions with finite entropy provided the quotient random walk is transient. She also described another family of groups (which she calls *groups of Baumslag type*) with this property. These groups are quite similar to the lamplighter groups, but the difference is that for the Baumslag type groups certain infinitely supported configurations are also allowed.

Outside of the class of measures with finite entropy the situation becomes even more complex. There are examples of measures μ on the lamplighter groups, for which the Poisson boundary is trivial in spite of non-triviality of the Poisson boundary for the reflected measure $\check{\mu}(g) = \mu(g^{-1})$, see Kaimanovich [Kai83]. This is impossible for the measures with finite entropy because of the entropy criterion of triviality of the Poisson boundary, as in the finite entropy case the asymptotic entropies of the original and of the reflected random walks are obviously the same.

7.2. Other Subgroups of $PLF(\mathbb{R})$

It would be interesting to understand, to what extent our results can be carried over to other subgroups of $PLF(\mathbb{R})$, for instance, to the group generated by the translation $\gamma \mapsto \gamma + 1$ and the transformation defined as

$$\gamma \mapsto \begin{cases} \gamma \,, & \gamma \le 0 \,, \\ 2\gamma \,, & \gamma \ge 0 \,. \end{cases}$$

Note that by a result of Brin–Squier [BS85] $PLF(\mathbb{R})$ (and therefore all its subgroups as well) does not contain non-abelian free subgroups.

As it has been pointed out by the referee, our 'drifting' technique used in the proof of Theorem 3.1 should be applicable to the 'F-series' of the so-called *Thompson–Stein groups* (they are a generalization of Thompson's groups introduced by Stein [Ste92]). In particular, it almost *verbatim* carries over to the groups $F(p)$ introduced by Burillo–Cleary–Stein [BCS01] (for these groups the break points are allowed to be in $\mathbb{Z}[\frac{1}{p}]$ for an integer $p > 1$). An interesting feature of the family $\{F(p)\}$ is that all these groups embed one into the other, so that they are all either amenable or non-amenable simultaneously.

We are not aware of any explicit description of the Schreier graphs of the canonical action of any of the Thompson–Stein groups other than Savchuk's description of the Schreier graphs for the original Thompson's group F [Sav10b, Sav15].

7.3. The Lattice of μ-Boundaries and the Problem of Identification of the Poisson Boundary

Although the μ-boundaries constructed in Theorem 4.4 and Theorem 6.1 seem to be quite different, currently we do not have any information about their mutual position in the lattice of all μ-boundaries. In particular, *a priori* it is not excluded that both these μ-boundaries might in fact coincide with the full Poisson boundary (which is the maximal element in the lattice of μ-boundaries). If not, what are, for instance, their infimum and supremum? Can one be just a quotient of the other one? And, of course, how are they related to the full Poisson boundary? Will their supremum, one of them, or maybe even both be the Poisson boundary? See author's papers [Kai96, Kai00] for a discussion of the general identification problem for the Poisson boundary.

7.4. Configuration Spaces as μ-Boundaries, and Generating Partitions of the Poisson Boundary.

The μ-boundary from Theorem 6.1 seems to be the first example when a boundary behavior is realized on the space of configurations on a single group orbit in such a way that the group action consists of just permuting the configurations without changing their point values.

There is a well-known construction in ergodic theory that allows one to realize general group actions as actions on configuration spaces by translations (e.g., see Cornfeld–Fomin–Sinai [CFS82]). Namely, let (X, m) be a Lebesgue measure space endowed with a measure class preserving action of a countable group G. Given an at most countable measurable partition ξ, let X_ξ be the corresponding countable quotient space (i.e., the space of the elements of the partition ξ), and let $x \mapsto \xi(x)$ be the corresponding quotient map which assigns to any point $x \in X$ the element of the partition ξ which contains x. The function

$$\Phi_x : G \to X_\xi , \quad g \mapsto \xi(g^{-1}x) ,$$

can be then considered as an X_ξ-valued configuration on G, and the map $x \mapsto \Phi_x$ is obviously G-equivariant if one endows the space of configurations on G with the action (11.2). If the map $x \mapsto \Phi_x$ separates (mod 0) the points of X, i.e., if this map is an isomorphism between the space (X, m) and the space of X_ξ-valued configurations on G, then the partition ξ is called generating.

Generating partitions for $\mathbb{Z}$-actions are a classical tool of ergodic theory. Moving to more general actions, it is, for example, easy to see that if one takes the partition of the boundary of a free group into the cylinder sets determined by the first letter of the infinite words representing boundary points, then this partition is generating even in the topological category. Therefore, this example provides a symbolic realization of the action of the free group on its Poisson boundary whenever the Poisson boundary can be identified with the space of infinite words (for instance, for the step distributions with a finite first moment, see Kaimanovich [Kai00]). However, we are not aware of any work on generating partitions for Poisson boundaries or their quotients in general.

7.5. Poisson Boundary and Sections of Boundary Bundles over Schreier Graphs

The statement of Theorem 6.2 on the construction of a non-trivial μ-boundary for a group G from the space of ends of its transient Schreier graph Γ is, of course, very well known when one takes for Γ just the group G (in which case the action of the group G on itself extends to the space ∂G, and the limit configurations are equivariant as functions from G to ∂G). However, it would be interesting to have other examples in the situation when the stabilizers G_γ, $\gamma \in \Gamma$, are sufficiently far from being normal. It is natural to impose the condition that the Schreier graph is determined by a core-free subgroup $H \subset G$ (i.e., that the action of G on the homogeneous space $\Gamma \cong H \backslash G$ is faithful), as otherwise one can always pass to the quotient of G by the normal core of H. Note that for

Thompson's group $F \cong \tilde{F}$ the latter condition is obviously satisfied as the canonical action of Thompson's group is completely determined by its restriction to the dyadic-rational orbit (cf. Section **2.3**).

Instead of the end bundle one can use the Poisson bundle over the graph Γ, i.e., the map $\Gamma \times \partial_\mu \Gamma \to \Gamma$, where $\partial_\mu \Gamma$ denotes the Poisson boundary of the induced random walk (Γ, μ) on the graph Γ. The space of its sections is the space of configurations $\mathrm{Fun}(\Gamma, \partial_\mu \Gamma)$. Let $\mathrm{bnd} = \mathrm{bnd}_{(\Gamma,\mu)}$ denote the boundary map from the path space of the random walk (Γ, μ) to the Poisson boundary $\partial_\mu \Gamma$. If $\partial_\mu \Gamma$ is non-trivial, then the projection

$$\mathbf{g} = (g_n) \mapsto \{\mathrm{bnd}(\gamma.\mathbf{g})\}_{\gamma \in \Gamma} \ ,$$

where $\mathbf{g} \mapsto \gamma.\mathbf{g}$ is the map (11.15) from the path space of the random walk (G, μ) to the path space of the induced random walk (Γ, μ), makes $\mathrm{Fun}(\Gamma, \partial_\mu \Gamma)$ a non-trivial μ-boundary. It would be interesting to find conditions under which this μ-boundary coincides with the full Poisson boundary of the random walk (G, μ).

References

[AAMBV13] Gideon Amir, Omer Angel, Nicolás Matte Bon and Bálint Virág, *The Liouville property for groups acting on rooted trees*, arXiv:1307.5652, 2013.

[AAV13] Gideon Amir, Omer Angel and Bálint Virág, *Amenability of linear-activity automaton groups*, J. Eur. Math. Soc. (JEMS) **15** (2013), no. 3, 705–30. MR 3085088

[AGS06] G. N. Arzhantseva, V. S. Guba and M. V. Sapir, *Metrics on diagram groups and uniform embeddings in a Hilbert space*, Comment. Math. Helv. **81** (2006), no. 4, 911–29. MR 2271228 (2007k:20084)

[BCS01] J. Burillo, S. Cleary and M. I. Stein, *Metrics and embeddings of generalizations of Thompson's group F*, Trans. Amer. Math. Soc. **353** (2001), no. 4, 1677–89. MR 1806724

[BCW07] José Burillo, Sean Cleary and Bert Wiest, *Computational explorations in Thompson's group F*, Geometric group theory, Trends Math., Birkhäuser, Basel, 2007, pp. 21–35. MR 2395786 (2009c:20071)

[BE11] Laurent Bartholdi and Anna Erschler, *Poisson-Furstenberg boundary and growth of groups*, arXiv:1107.5499, 2011.

[BKN10] Laurent Bartholdi, Vadim A. Kaimanovich and Volodymyr V. Nekrashevych, *On amenability of automata groups*, Duke Math. J. **154** (2010), no. 3, 575–98. MR 2730578

[BLP77] Paolo Baldi, Noël Lohoué and Jacques Peyrière, *Sur la classification des groupes récurrents*, C. R. Acad. Sci. Paris Sér. A-B **285** (1977), no. 16, A1103–A1104. MR 0518008

[Bri09] Jérémie Brieussel, *Amenability and non-uniform growth of some directed automorphism groups of a rooted tree*, Math. Z. **263** (2009), no. 2, 265–93. MR 2534118 (2010g:43002)

[BS85] Matthew G. Brin and Craig C. Squier, *Groups of piecewise linear homeomorphisms of the real line*, Invent. Math. **79** (1985), no. 3, 485–98. MR 782231 (86h:57033)

[BV05] Laurent Bartholdi and Bálint Virág, *Amenability via random walks*, Duke Math. J. **130** (2005), no. 1, 39–56. MR 2176547 (2006h:43001)

[CF11] J. W. Cannon and W. J. Floyd, *What is ... Thompson's group?*, Notices Amer. Math. Soc. **58** (2011), no. 8, 1112–13. MR 2856142

[CFP96] J. W. Cannon, W. J. Floyd and W. R. Parry, *Introductory notes on Richard Thompson's groups*, Enseign. Math. (2) **42** (1996), no. 3-4, 215–56. MR 1426438 (98g:20058)

[CFS82] I. P. Cornfeld, S. V. Fomin and Ya. G. Sinaĭ, *Ergodic theory*, Grundlehren der Mathematischen Wissenschaften [Fundamental Principles of Mathematical Sciences], vol. 245, Springer-Verlag, New York, 1982, Translated from the Russian by A. B. Sosinskiĭ. MR 832433 (87f:28019)

[Che91] Mu Fa Chen, *Comparison theorems for Green functions of Markov chains*, Chinese Ann. Math. Ser. B **12** (1991), no. 3, 237–42, A Chinese summary appears in Chinese Ann. Math. Ser. A **12** (1991), no. 3, 521. MR 1130250

[ERJvR15] Murray Elder, Andrew Rechnitzer and Esaias J. Janse van Rensburg, *Random sampling of trivial words in finitely presented groups*, Exp. Math. **24** (2015), no. 4, 391–409. MR 3383471

[Ers04a] Anna Erschler, *Boundary behavior for groups of subexponential growth*, Ann. of Math. (2) **160** (2004), no. 3, 1183–210. MR 2144977 (2006d:20072)

[Ers04b] _____, *Liouville property for groups and manifolds*, Invent. Math. **155** (2004), no. 1, 55–80. MR 2 025 301

[Ers06a] _____, *Isoperimetry for wreath products of Markov chains and multiplicity of self-intersections of random walks*, Probab. Theory Related Fields **136** (2006), no. 4, 560–86. MR 2257136

[Ers06b] _____, *Piecewise automatic groups*, Duke Math. J. **134** (2006), no. 3, 591–613. MR 2254627

[Ers10] _____, *Poisson-Furstenberg boundaries, large-scale geometry and growth of groups*, Proceedings of the International Congress of Mathematicians. Vol. II, Hindustan Book Agency, New Delhi, 2010, pp. 681–704. MR 2827814 (2012h:60016)

[Ers11] _____, *Poisson-Furstenberg boundary of random walks on wreath products and free metabelian groups*, Comment. Math. Helv. **86** (2011), no. 1, 113–43. MR 2745278 (2011k:60013)

[Fur73] Harry Furstenberg, *Boundary theory and stochastic processes on homogeneous spaces*, Harmonic analysis on homogeneous spaces (Proc. Sympos. Pure Math., vol. XXVI, Williams Coll., Williamstown, MA, 1972), Amer. Math. Soc., Providence, RI, 1973, pp. 193–229. MR 50 no. 815

[Fur02] Alex Furman, *Random walks on groups and random transformations*, Handbook of dynamical systems, vol. 1A, North-Holland, Amsterdam, 2002, pp. 931–1014. MR 1928529 (2003j:60065)

[Gou14] Antoine Gournay, *The Liouville property and Hilbertian compression*, arXiv:1403.1195, 2014.

[HHRS15] Søren Haagerup, Uffe Haagerup and Maria Ramirez-Solano, *A computational approach to the Thompson group F*, Internat. J. Algebra Comput. **25** (2015), no. 3, 381–432. MR 3334642

[HP11] Uffe Haagerup and Gabriel Picioroaga, *New presentations of Thompson's groups and applications*, J. Operator Theory **66** (2011), no. 1, 217–32. MR 2806554 (2012f:46099)

[JMBNdlS15] Kate Juschenko, Nicolás Matte Bon, Monod Nicolas and Mikael de la Salle, *Extensive amenability and an application to interval exchanges*, arXiv:1503.04977, 2015.

[Kai83] Vadim A. Kaimanovich, *Examples of nonabelian discrete groups with nontrivial exit boundary*, Zap. Nauchn. Sem. Leningrad. Otdel. Mat. Inst. Steklov. (LOMI) **123** (1983), 167–84, Differential geometry, Lie groups and mechanics, V. MR 697250 (85b:60008)

[Kai85] _____, *An entropy criterion of maximality for the boundary of random walks on discrete groups*, Soviet Math. Dokl. **31** (1985), 193–7. MR 86m:60025

[Kai89] _____, *Lyapunov exponents, symmetric spaces and a multiplicative ergodic theorem for semisimple Lie groups*, J. Soviet Math. **47** (1989), 2387–98. MR 89m:22006

[Kai91] _____, *Poisson boundaries of random walks on discrete solvable groups*, Probability measures on groups, X (Oberwolfach, 1990), Plenum, New York, 1991, pp. 205–38. MR 1178986 (94m:60014)

[Kai92] _____, *Measure-theoretic boundaries of Markov chains, 0-2 laws and entropy*, Harmonic analysis and discrete potential theory (Frascati, 1991), Plenum, New York, 1992, pp. 145–80. MR 94h:60099

[Kai94] _____, *The Poisson boundary of hyperbolic groups*, C. R. Acad. Sci. Paris Sér. I Math. **318** (1994), no. 1, 59–64. MR 1260536 (94j:60142)

[Kai95] _____, *The Poisson boundary of covering Markov operators*, Israel J. Math. **89** (1995), no. 1-3, 77–134. MR 96k:60194

[Kai96] _____, *Boundaries of invariant Markov operators: the identification problem*, Ergodic theory of $\mathbf{Z}^d$ actions (Warwick, 1993–4), London Math. Soc. Lecture Note Ser., vol. 228, Cambridge University Press, 1996, pp. 127–76. MR 97j:31008

[Kai00] _____, *The Poisson formula for groups with hyperbolic properties*, Ann. of Math. (2) **152** (2000), no. 3, 659–92. MR 1815698 (2002d:60064)

[Kai04a] Vadim Kaimanovich, *Boundary behaviour of the Thompson group*, presentation slides, 2004.

[Kai04b] Vadim A. Kaimanovich, *Boundary amenability of hyperbolic spaces*, Discrete geometric analysis, Contemp. Math., vol. 347, Amer. Math. Soc., Providence, RI, 2004, pp. 83–111. MR 2077032 (2005j:20051)

[Kai05] _____, *Amenability and the Liouville property*, Israel J. Math. **149** (2005), 45–85, Probability in mathematics. MR 2191210 (2007c:43001)

[KL07] Anders Karlsson and François Ledrappier, *Linear drift and Poisson boundary for random walks*, Pure Appl. Math. Q. **3** (2007), no. 4, part 1, 1027–36. MR 2402595

[KM96] Vadim A. Kaimanovich and Howard Masur, *The Poisson boundary of the mapping class group*, Invent. Math. **125** (1996), no. 2, 221–64. MR 1395719 (97m:32033)

[KM99] Anders Karlsson and Gregory A. Margulis, *A multiplicative ergodic theorem and nonpositively curved spaces*, Comm. Math. Phys. **208** (1999), no. 1, 107–23. MR 1729880 (2000m:37031)

[KV83] V. A. Kaimanovich and A. M. Vershik, *Random walks on discrete groups: boundary and entropy*, Ann. Probab. **11** (1983), no. 3, 457–90. MR 85d:60024

[KV15] Mihał Kotowski and Bálint Virág, *Non-Liouville groups with return probability exponent at most 1/2*, Electron. J. Probab. **20** (2015), no. 12, 1–12 (electronic).

[KW92] Vadim A. Kaimanovich and Wolfgang Woess, *The Dirichlet problem at infinity for random walks on graphs with a strong isoperimetric inequality*, Probab. Theory Related Fields **91** (1992), no. 3–4, 445–66. MR 1151805 (93e:60139)

[KW07] Anders Karlsson and Wolfgang Woess, *The Poisson boundary of lamplighter random walks on trees*, Geom. Dedicata **124** (2007), 95–107. MR 2318539 (2009b:60246)

[Lio08] Isabelle Liousse, *Rotation numbers in Thompson-Stein groups and applications*, Geom. Dedicata **131** (2008), 49–71. MR 2369191

[LP15] Russell Lyons and Yuval Peres, *Poisson boundaries of lamplighter groups: proof of the Kaimanovich-Vershik conjecture*, arXiv:1508.01845, 2015.

[Lyo87] Terry Lyons, *Instability of the Liouville property for quasi-isometric Riemannian manifolds and reversible Markov chains*, J. Differential Geom. **26** (1987), no. 1, 33–66. MR 892030 (88k:31012)

[MB14] Nicolás Matte Bon, *Subshifts with slow complexity and simple groups with the Liouville property*, Geom. Funct. Anal. **24** (2014), no. 5, 1637–59. MR 3261637

[Mis15] Pavlo Mishchenko, *Boundary of the action of Thompson's group F on dyadic numbers*, arXiv:1512.03083, 2015.

[Moo13] Justin Tatch Moore, *Fast growth in the Følner function for Thompson's group F*, Groups Geom. Dyn. **7** (2013), no. 3, 633–51. MR 3095713

[MT14] Joseph Maher and Giulio Tiozzo, *Random walks on weakly hyperbolic groups*, arXiv:1410.4173, 2014.

[NP08] Assaf Naor and Yuval Peres, *Embeddings of discrete groups and the speed of random walks*, Int. Math. Res. Not. IMRN (2008), Art. ID rnn 076, 34. MR 2439557

[Ros81] Joseph Rosenblatt, *Ergodic and mixing random walks on locally compact groups*, Math. Ann. **257** (1981), no. 1, 31–42. MR 83f:43002

[Sav10a] Ecaterina Sava, *A note on the Poisson boundary of lamplighter random walks*, Monatsh. Math. **159** (2010), no. 4, 379–96. MR 2600904 (2011b:60308)

[Sav10b] Dmytro Savchuk, *Some graphs related to Thompson's group F*, Combinatorial and geometric group theory, Trends Math., Birkhäuser/Springer Basel AG, Basel, 2010, pp. 279–96. MR 2744025 (2012b:20104)

[Sav15] _____, *Schreier graphs of actions of Thompson's group F on the unit interval and on the Cantor set*, Geom. Dedicata **175** (2015), 355–72. MR 3323646

[Ste92] Melanie Stein, *Groups of piecewise linear homeomorphisms*, Trans. Amer. Math. Soc. **332** (1992), no. 2, 477–514. MR 1094555

[Var83] Nicolas Th. Varopoulos, *Brownian motion and transient groups*, Ann. Inst. Fourier (Grenoble) **33** (1983), no. 2, 241–61. MR 699497

[Woe94] Wolfgang Woess, *Topological groups and recurrence of quasitransitive graphs*, Rend. Sem. Mat. Fis. Milano **64** (1994), 185–213 (1996). MR 1397471 (97i:60092)

[Woe00] _____, *Random walks on infinite graphs and groups*, Cambridge Tracts in Mathematics, vol. 138, Cambridge University Press, (2000). MR 2001k:60006

Added in Proof

According to our Theorem 6.1 the sample paths of the induced random walk on the canonical Schreier graph Γ converge to a limit random end for any non-degenerate *finitely supported* step distribution on Thompson's group F. However, Kate Juschenko and Tianyi Zheng have recently proved ("Infinitely supported Liouville measures of Schreier graphs", arXiv:1608.03554) that this may fail if the finite support condition is dropped, namely, that there are *infinitely supported* measures μ on F, for which the Poisson boundary of the induced random walk on Γ is trivial. Their argument is based on the observation that the strong transitivity of the action of F on Γ implies a certain approximation property which can then be used for constructing the desired measure μ in the same way as Reiter's property was used for constructing a Liouville random walk on any amenable group by Kaimanovich–Vershik [KV83]. We shall return to the relationship between amenability and the approximation property of the action of F on Γ used by Juschenko–Zheng elsewhere.

12

A PROOF OF THE SUBADDITIVE ERGODIC THEOREM

ANDERS KARLSSON

Section de Mathématiques, University of Geneva,
2-4, Rue du Lièvre, Case Postale 64,
1211 Genève 4, Suisse and
Uppsala universitet, Sweden

For Wolfgang Woess on the occasion of his sixtieth birthday

Abstract

This is a presentation of the subadditive ergodic theorem. A proof is given that is an extension of F. Riesz's approach to the Birkhoff ergodic theorem.

Contents

1. Introduction

Between the years 1938 and 1948 F. Riesz published works in ergodic theory [R] and found among other things several extensions of the mean ergodic theorem by elegantly simple and powerful arguments. In October 1931, von Neumann had proved the basic version of the mean ergodic theorem [vN32] inspired by Koopman's observations published in a note in June that same year [Ko31]. In his 1944 lecture notes [R44], Riesz gave an insightful exposition of these developments.[1] In particular, he presented there a new proof of the maximal ergodic lemma needed for Birkhoff's a.e. ergodic theorem, which is a deeper result than the mean ergodic theorem and dates November 1931 [B31]. This became one of the standard proofs of this theorem and was based on a lemma which is harder to formulate than to prove (see Lemma 3.2 below). He also remarked that his work on the mean ergodic theorem was suggested by the method of Carleman [C32]. In fact, Carleman announced in May 1931 results similar to Koopman's as well as a proof of the mean ergodic

---

[1] He was supposed to deliver these lectures in Geneva in the spring of 1944, but he was prevented from coming.

theorem. This announcement was published in June 1931 [C31] and the details can be found in [C32].[2]

The main purpose of this note is to show how Riesz's method extends to give a proof of the subadditive ergodic theorem of Kingman from 1968 [K68]. This was a by-product of the work [KM99] and all the details of the proof are included. Note that this proof at the same time gives a proof of Birkhoff's theorem.

Let throughout this chapter (X, μ) be a measure space with $\mu(X) = 1$ and $T : X \to X$ a measure preserving transformation. Recall the following result:

Theorem (Birkhoff, 1931). *Let $f \in L^1(X)$, then there is an integrable, a.e. T-invariant function $\bar{f}$ such that*

$$\lim_{n \to \infty} \frac{1}{n} \sum_{k=0}^{n-1} f(T^k x) = \bar{f}(x)$$

for a.e. x (the convergence also takes place in L^1). In fact $\|\bar{f}\|_1 \leq \|f\|_1$ and for any T-invariant set A, $\int_A f = \int_A \bar{f}$, in particular $\int_X f = \int_X \bar{f}$.

Note that if

$$c(n, x) = \sum_{k=0}^{n-1} f(T^k x) \qquad (12.1)$$

then

$$c(n + m, x) = c(n, x) + c(m, T^n x),$$

and we then call c an *additive cocycle*. These are all of the form (12.1), for $f(x) = c(1, x)$.

If for a sequence of functions $a(n, .) \in L^1(X)$, with integer $n > 0$ and $a(0, x) = 0$, we instead require

$$a(n + m, x) \leq a(n, x) + a(m, T^n x),$$

then a is here called a *subadditive cocycle*. Assume that

$$\inf \frac{1}{n} \int_X a(n, x) d\mu(x) > -\infty.$$

Then the following generalization of the Birkhoff ergodic theorem holds.

Theorem (Kingman, 1968). *Under the above conditions, there is an integrable, a.e. T-invariant function $\bar{a}$ such that*

$$\lim_{n \to \infty} \frac{1}{n} a(n, x) = \bar{a}(x)$$

---

[2] Several early publications in the subject give due credit to Carleman, for example, E. Hopf's important monograph *Ergodentheorie* from 1938. In most modern works, however, including the standard reference [Kr85], Carleman is not mentioned.

for a.e. x (the convergence also takes place in L^1). Moreover

$$\lim_{n\to\infty} \frac{1}{n} \int_A a(n,x)\,d\mu(x) = \int_A \bar{a}(x)\,d\mu(x)$$

for all T-invariant measurable sets A.

A draft of this chapter was written in 1998 when I was a graduate student at Yale University. Some people have since told me that my text has been useful to them, so I thought that it might perhaps be worthwhile to publish a revised version. In addition, I thank the referee for useful comments leading to an improved text. I dedicate it to Wolfgang Woess, whom I have had the pleasure of knowing for more than a decade. I have at various times benefited from his many insights within, as well as outside of, mathematics.

2. A Few Examples of Subadditive Cocycles

We give a few examples of subadditive cocycles to illustrate that Kingman's theorem is a significant extension of Birkhoff's theorem with many applications. In fact the origin of Kingman's theorem comes from probability theory (theory of percolation) in the works by Hammersley and Welsh.

2.1. Random Products in a Banach Algebra

Let $A : X \to B$ be a measurable map into a Banach algebra. Let

$$u(n,x) = A(T^{n-1}x)A(T^{n-2}x)\dots A(x),$$

then

$$a(n,x) = \log\|u(n,x)\|$$

is a subadditive cocycle, because $\|AB\| \leq \|A\|\,\|B\|$. The corresponding convergence was first proved by Furstenberg and Kesten in 1960 for random products of matrices, of course without the use of the subadditive ergodic theorem. This application is used in some proofs of Oseledets's multiplicative ergodic theorem.

2.2. Random Walks

Let G be a topological group and $h : X \to G$ a Borel measurable map. Let $v(n,x) = h(T^{n-1}x)\dots h(Tx)h(x)$, if $\{h \circ T^k\}$ are independent, then this is usually called a random walk. The range, that is, how many points visited in G,

$$a(n,x) := \mathrm{Card}\{v(i,x) : 1 \leq i \leq n\}$$

is a subadditive cocycle.

Assume that d is a left invariant metric on G, (e.g. a word metric in the case G is finitely generated) then the drift

$$b(n, x) := d(e, v(n, x))$$

is a subadditive cocycle, by the triangle inequality and the invariance of d. A third source of subadditivity is the entropy of the measures of the distribution of random walk after n steps, see [D80].

2.3. Metric Theory of Continued Fractions

See [Ba97]. Let $X = [0, 1)$ and $\mathcal{A}$ be the Borel σ-algebra. For any x write it as

$$x = \frac{1}{\frac{1}{x}} = \frac{1}{\left[\frac{1}{x}\right] + \left\{\frac{1}{x}\right\}}.$$

Continuing this scheme gives the continued fraction expansion of x, $a(x) := \left[\frac{1}{x}\right]$, etc. The relevant transformation of X is

$$Tx = \left\{\frac{1}{x}\right\}$$

unless $x = 0$ in which case $Tx = 0$. So $a_n(x) = a(T^{n-1}x)$.

The corresponding approximants $p_n(x)/q_n(x)$ are defined for all n if x is irrational, they are given by recusion formulas, and by

$$\frac{p_n(x)}{q_n(x)} = \cfrac{1}{a_1(x) + \cfrac{1}{a_2(x) + \cfrac{1}{\ddots + \cfrac{1}{a_n(x)}}}}.$$

There is a unique T-invariant probability measure absolutely continuous with respect to Lebesgue measure, namely

$$\nu(A) = \frac{1}{\log 2} \int_A \frac{1}{1 + x} dx$$

called the Gauss measure.

By use of the subadditive ergodic theorem, Barbolosi proves that for all constants $C > 0$,

$$\lim_{n \to \infty} \frac{1}{n} \mathrm{Card}\left\{1 \le i \le n : |xq_i(x) - p_i(x)| \le \frac{C}{q_i(x)}\right\}$$

exists for a.e. x. The functions are not simply a subadditive cocycle; it is necessary to divide into odd and even index, due to the fact that approximants lie on alternating sides of x.

2.4. Further Examples

There are too many applications of the subadditive ergodic theorem to list here; let us just refer to [H72] and [D80].

3. A Proof of Kingman's Theorem

3.1. Three Elementary Observations

Proposition 3.1. *Let $(v_n)_{n \geq 1}$ be a subadditive sequence of real numbers, that is $v_{n+m} \leq v_n + v_m$. Then the following limit exists*

$$\lim_{n \to \infty} \frac{1}{n} v_n = \inf_{m > 0} \frac{1}{m} v_m \in \mathbb{R} \cup \{-\infty\}.$$

Proof. Given $\varepsilon > 0$, pick M such that $v_M / M \leq \inf v_n / n + \varepsilon$. Decompose $n = k_n M + r_n$, where $0 \leq r_n < M$. Hence $k_n / n \to 1/M$. Using the subadditivity and considering n big enough ($n > N(\varepsilon)$)

$$\inf \frac{1}{m} v_m \leq \frac{1}{n} v_n = \frac{1}{n} v_{k_n M + r_n} \leq \frac{1}{n} (k_n v_M + v_{r_n})$$

$$\leq \frac{1}{M} v_M + \varepsilon \leq \inf \frac{1}{m} v_m + 2\varepsilon.$$

Since ε is at our disposal, the lemma is proved. $\quad\square$

Note that this proposition implies that for any T-invariant set A,

$$\lim_{n \to \infty} \frac{1}{n} \int_A a(n, x) d\mu(x) = \inf \frac{1}{n} \frac{1}{n} \int_A a(n, x) d\mu(x),$$

when $X = A$ denote this value by $\gamma(a)$.

F. Riesz noted in [R44] that the following simple lemma can be used to prove Birkhoff's theorem. The lemma is sometimes called Riesz's combinatorial lemma or the lemma about leaders.

Lemma 3.2. *Call the term c_u a leader in the finite sequence $c_0, c_1, \ldots, c_{n-1}$ if one of the sums*

$$c_u, c_u + c_{u+1}, \ldots, c_u + \cdots + c_{n-1}$$

is negative. Then the sum of the leaders is nonpositive. (An empty sum is by convention 0).

Proof. Proof by induction. If $n = 1$, then either $c_0 \geq 0$, in which case the sum is empty, or $c_0 < 0$, in which case the sum equals $c_0 < 0$.

Assume that the statement is true for integers smaller than n. Consider the two cases, c_0 is or is not a leader. If c_0 is not a leader then all leaders are among $c_1, \ldots, c_{n-1}$ in which case the induction hypothesis applies.

If c_0 is a leader, then pick the smallest integer k such that $c_0 + \cdots + c_k < 0$, then each c_i, $i \leq k$ is a leader. If not then $c_i + \cdots + c_k \geq 0$, but by minimality $c_0 + \cdots + c_{i-1} \geq 0$, which is a contradiction.

Hence $c_0, \ldots, c_k$ are all leaders and $c_0 + \cdots + c_k < 0$, the remaining leaders (if any) are leaders of $c_k, \ldots, c_n$, for which the induction hypothesis applies. □

Proposition 3.3. *Let $a(n,x)$ be a subadditive cocycle as in the introduction. Then the functions*

$$f(x) = \limsup_{n \to \infty} \frac{1}{n} a(n, x)$$

and

$$g(x) = \liminf_{n \to \infty} \frac{1}{n} a(n, x)$$

are a.e. T-invariant.

Proof. Note that $f(Tx) \geq f(x)$ and $g(Tx) \geq g(x)$ because of the subadditivity

$$a(n, Tx) \geq a(n+1, x) - a(1, x)$$

and in the case of limsup (same for liminf)

$$\limsup \left(\frac{1}{n} a(n+1, x) - \frac{1}{n} a(1, x) \right) = f(x).$$

Now integrate

$$\int_X [f(Tx) - f(x)] \, d\mu(x) = 0$$

by the T-invariance, but the integrand is nonnegative, hence $f(Tx) - f(x) = 0$ a.e. □

3.2. The Maximal Ergodic Inequality

The following key lemma will be proved by an extension of the argument of F. Riesz. It thus avoids use of the usual maximal ergodic inequality and it is not more difficult than Derriennic's proof.

Lemma 3.4 (Derriennic, 1975). *Let $a(n,x)$ be a subadditive ergodic cocycle as in the introduction. Let*

$$B = \left\{ x : \liminf_{n \to \infty} \frac{1}{n} a(n, x) < 0 \right\}$$

then

$$\lim_{n \to \infty} \frac{1}{n} \int_B a(n, x) d\mu \leq 0.$$

Proof. For each n, let

$$\Psi_n = \left\{ x : \inf_{1 \leq k \leq n} a(n, x) - a(n - k, T^k x) < 0 \right\}.$$

Note that

$$A_n := \left\{ x : \inf_{1 \le k \le n} a(k, x) < 0 \right\} \subset \Psi_n$$

by subadditivity. Note also that $A_n \subset A_{n+1}$ and $B \subset \bigcup A_n$.
For each n, let

$$b_n(x) = a(n, x) - a(n - 1, Tx).$$

Because of telescoping we have that

$$a(n, x) - a(n - k, T^k x) = b_n(x) + b_{n-1}(Tx) + \cdots + b_{n-k+1}(T^{k-1} x)$$

and in particular, recall that $a(0, x) = 0$,

$$a(n, x) = \sum_{0 \le k \le n-1} b_{n-k}(T^k x).$$

By definition, $T^k x \in \Psi_{n-k}$ means that there is a j, $k \le j \le n - 1$ such
that

$$b_{n-k}(T^k x) + \cdots + b_{n-j}(T^j x) < 0.$$

Hence by the lemma about leaders applied to $c_k = b_{n-k}(T^k x)$ we have

$$\sum_{0 \le k \le n-1, T^k x \in \Psi_{n-k}} b_{n-k}(T^k x) \le 0.$$

Therefore, using the T-invariance of μ and B,

$$0 \ge \int_B \sum_{0 \le k \le n-1, T^k x \in \Psi_{n-k}} b_{n-k}(T^k x) d\mu(x) =$$

$$= \sum_{0 \le k \le n-1} \int_{B \cap T^{-k} \Psi_{n-k}} b_{n-k}(T^k x) d\mu(x) =$$

$$= \sum_{0 \le k \le n-1} \int_{T^k B \cap \Psi_{n-k}} b_{n-k}(x) d\mu(x) =$$

$$= \sum_{i=1}^{n} \int_{B \cap \Psi_i} b_i(x) d\mu(x).$$

On the other hand, again by the T-invariance

$$\frac{1}{n} \int_B a(n, x) d\mu(x) = \frac{1}{n} \sum_{i=1}^{n} \int_B b_i(x) d\mu(x) =$$

$$= \frac{1}{n} \sum_{i=1}^{n} \int_{B \cap \Psi_i} b_i(x) d\mu(x) + \frac{1}{n} \sum_{i=1}^{n} \int_{B - (B \cap \psi_i)} b_i(x) d\mu(x) \le$$

$$\le 0 + \frac{1}{n} \sum_{i=1}^{n} \int_{B - (B \cap A_i)} a^+(1, x) d\mu(x)$$

since $b_j(x) \leq a(1,x) \leq a^+(1,x) := \max\{0, a(1,n)\}$, which is positive and $A_i \subset \Psi_i$. Now since $a^+ \in L^1$, $A_i \subset A_{i+1}$ and $B \subset \bigcup_{i \geq 1} A_i$ it follows when taking limsup, that

$$\limsup \frac{1}{n} \int_B a(n,x)d\mu \leq 0.$$

Since B is invariant, the limsup actually is the limit, by Proposition 3.1. □

Proposition 3.5. *Let $a(n,x)$ be a subadditive ergodic cocycle as in the introduction. Let*

$$B = \left\{ x : \liminf \frac{1}{n} a(n,x) < \lambda \right\}$$

then

$$\lim_{n \to \infty} \frac{1}{n} \int_B a(n,x)d\mu \leq \lambda\mu(B).$$

Proof. Apply the lemma to $a(n,x) - n\lambda$, which is a subadditive cocycle. □

3.3. The Proof of a.e. Convergence

First we establish the result for additive cocycles $c_n = c(n,x)$. The point is that $-c_n$ is again additive, hence in particular the previous proposition applies to $-c_n$ as well.

Let

$$E_{\alpha,\beta} = \left\{ x : \liminf \frac{1}{n} c_n < \alpha < \beta < \limsup \frac{1}{n} c_n \right\}$$

and by Proposition 3.3 this set is T-invariant. Hence we can apply Proposition 3.5 with $X = E_{\alpha,\beta}$. If we let $E := \{x : \liminf \frac{1}{n} c_n < \alpha\}$, then $E \cap E_{\alpha,\beta} = E_{\alpha,\beta}$, and this gives

$$\int_{E_{\alpha,\beta}} c_1 d\mu = \lim_{n \to \infty} \frac{1}{n} \int_{E_{\alpha,\beta}} c_n d\mu \leq \alpha\mu(E_{\alpha,\beta}).$$

And similarly for $-c_n$,

$$-\int_{E_{\alpha,\beta}} c_1 d\mu = \lim_{n \to \infty} \frac{1}{n} \int_{E_{\alpha,\beta}} -c_n d\mu \leq -\beta\mu(E_{\alpha,\beta}).$$

This yields a contradiction unless $\mu(E_{\alpha,\beta}) = 0$, because

$$\beta\mu(E_{\alpha,\beta}) \leq \int_{E_{\alpha,\beta}} c_1 d\mu \leq \alpha\mu(E_{\alpha,\beta})$$

but $\beta > \alpha$.

Now let $a_n(x) = a(n, x)$, be a subadditive cocycle satisfying the integrability assumptions in the introduction. Consider

$$v_n(x) = a_n(x) - \sum_{i=0}^{n-1} a_1(T^i x).$$

Note that v_n is a subadditive cocycle and $v_n \leq 0$. Let $g(x) = \liminf \frac{1}{n} v_n(x)$ and $f(x) = \limsup \frac{1}{n} v_n(x)$. For an arbitrary $\alpha > 0$, we want to show that $B := \{x : f(x) - g(x) > \alpha\}$ has measure zero. By the additive cocycle case above we know that $\frac{1}{n} \sum_{i=0}^{n-1} a_1(T^i x)$ converges a.e., so taken together this would show that $\frac{1}{n} a_n(x)$ converges a.e. as desired.

Fix $\varepsilon > 0$. Pick M large enough so that for all $m > M$,

$$\frac{1}{m} \int_X v_m \leq \gamma(v) + \varepsilon.$$

Let

$$g^M(x) := \liminf_{n \to \infty} \frac{1}{nM} v_{nM}(x)$$

and

$$f^M(x) := \limsup_{n \to \infty} \frac{1}{nM} v_{nM}(x).$$

On the one hand, because nM is a subsequence of n, we have that $g^M(x) \geq g(x)$ and $f^M(x) \leq f(x)$. On the other hand, by the subadditivity and nonpositivity of v_n, for all $0 \leq k < M$,

$$v_{(n+1)M}(x) \leq v_{nM+k}(x) + v_{M-k}(T^{nM+k} x) \leq v_{nM+k}(x)$$

and

$$v_{nM+k}(x) \leq v_{nM}(x) + v_k(T^{nM} x) \leq v_{nM}(x).$$

Therefore in view of that $(n+1)nM/(nM+k) \to 1$ and $nM/(nM+k) \to 1$, we can conclude that $g^M(x) = g(x)$ and $f^M(x) = f(x)$. Let

$$v_n^M(x) := v_{nM}(x) - \sum_{i=0}^{n-1} v_M(T^{iM} x),$$

which again is subadditive and nonpositive. We then have that:

$$f - g = f^M - g^M \leq -\liminf_{n \to \infty} \frac{1}{nM} v_n^M,$$

because $v_n^M \leq 0$ and the established convergence for additive cocycles. This means that

$$B := \{x : f - g > \alpha\} \subset \left\{x : \liminf_{n \to \infty} \frac{1}{n} v_n^M(x) < -M\alpha\right\} =: E.$$

Note that

$$0 \geq \gamma(v^M) = M\gamma(v) - \int_X v_M \geq -M\varepsilon.$$

Now applying Proposition 3.5 we get

$$-M\alpha\mu(E) \geq \lim \frac{1}{n}\int_E v_n^M \geq \lim \frac{1}{n}\int_X v_n^M \geq -M\varepsilon.$$

Hence

$$\mu(E) \leq \frac{\varepsilon}{\alpha}$$

and letting $\varepsilon \to 0$ we conclude that $\mu(B) = 0$ for any $\alpha > 0$ as required.

The limit is almost everywhere T-invariant, by Proposition 3.3 and integrable by Fatou's lemma

$$\int \liminf -\frac{1}{n}v_n \leq \liminf \int -\frac{1}{n}v_n < \infty.$$

In general, $\int |a(n,x)| \leq n\int |a(1,x)|$.

4. Appendix: Garsia's Proof of the Maximal Ergodic Lemma

This appendix, which is not used above, notes that Garsia's celebrated argument for Birkhoff's ergodic theorem [G70] has a minor subadditive extension, although not strong enough to yield Kingman's theorem.

Lemma 4.1. *Let $a(n,x)$ be a subadditive ergodic cocycle as in the introduction. For $1 \leq n \leq \infty$ define*

$$E_n = \left\{ x : \sup_{1 \leq k \leq n} a(k,x) \geq 0 \right\}.$$

Then

$$\int_{E_n} a(1,x)d\mu \geq 0 \text{ and } \int_{E_\infty} a(1,x)d\mu \geq 0.$$

Proof. Let

$$h_n(x) = \sup_{1 \leq k \leq n} a(k,x),$$

so

$$h_n^+(x) - a(k,x) \geq 0$$

for $k \leq n$. By positivity and linearity of T, viewed as (Koopman) operator on functions,

$$Th_n^+ \geq Ta_k$$

for $k \leq n$. ($a_k(x) = a(k,x)$.)

Hence
$$a_1 + Th_n^+ \geq a_1 + Ta_k \geq a_{k+1}$$
by subadditivity. So
$$a_1 \geq a_{k+1} - Th_n^+$$
for all $k, 1 \leq k \leq n$ and trivially for $k = 0$. Therefore
$$a_1 \geq \sup_{0 \leq k \leq n-1} a_{k+1} - Th_n^+ = h_n - Th_n^+.$$

Now integrate
$$\int_{E_n} a_1 \geq \int_{E_n} h_n - \int_{E_n} Th_n^+ = \int_X h_n^+ - \int_{E_n} Th_n^+$$
$$\geq \int_X h_n^+ - \int_X Th_n^+ \geq 0,$$

because T is measure preserving (or contractive as in [G70] would be enough) and $h_n^+ \geq 0$.

The statement about $E_\infty = \bigcup E_n$ follows from passing to the limit.

$\square$

References

[Ba97] Barbolosi, D. Une application du théoréme ergodique sous-additif á la théorie métrique des fractions continues. J. Number Theory 66 (1997), no. 1, 172–82.

[B31] Birkhoff, G.D. Proof of the ergodic theorem. *Proc. Nat. Acad. Sci. USA* 17 (1931) 656–60.

[C31] Carleman, Torsten, Application de la théorie des équations intégrales singuliéres aux équations différentielles de la dynamique, *Ark. Mat. Astr. Fys.* 22, 1931.

[C32] Carleman, Torsten, Application de la théorie des équations intégrales linéaires aux systémes d'équations différentielles non linéaires. *Acta Math.* 59 (1932), no. 1, 63–87.

[D75] Derriennic, Yves, Sur le théoréme ergodique sous-additif. *C. R. Acad. Sci. Paris* Sér. A-B 281 (1975), no. 22, Aii, A985–A988.

[D80] Derriennic, Yves, Quelques applications du théoréme ergodique sous-additif. In: *Conference on Random Walks* (Kleebach, 1979) pp. 183–201, 4, Astérisque, 74, Soc. Math. France, Paris, 1980.

[G70] Garsia, Adriano M. *Topics in almost everywhere convergence.* Lectures in Advanced Mathematics, 4 Markham Publishing Co., Chicago, Ill. 1970 x+154 pp.

[H72] Hammersley, J. M. A few seedlings of research. In: *Proceedings of the Sixth Berkeley Symposium on Mathematical Statistics and Probability* (University of California, Berkeley, Calif., 1970/1971), Vol. I: Theory of statistics, pp. 345–94. University of California Press, Berkeley, Calif., 1972.

[KM99] Karlsson, Anders; Margulis, Gregory A., A multiplicative ergodic theorem and nonpositively curved spaces. *Comm. Math. Phys.* 208 (1999), no. 1, 107–23.

[K68] Kingman, J. F. C. The ergodic theory of subadditive stochastic processes. *J. Roy. Statist. Soc.* Ser. B 30 1968 499–510.

[Ko31] Koopman, B.O. Hamiltionian systems and transformations in Hilbert space. *Proc. Nat. Acad. Sci. USA*, 17 (1931) pp. 315–18.

[Kr85] Krengel, Ulrich, *Ergodic theorems*. With a supplement by Antoine Brunel. de Gruyter Studies in Mathematics, 6. Walter de Gruyter & Co., Berlin, 1985. viii+357 pp.

[vN32] von Neumann, J. Proof of the quasi-ergodic hypothesis. *Proc. Nat. Acad. Sci. USA*, 18 (1932) pp. 70–82.

[R] Riesz, Frigyes, *Oeuvres complétes*. Publiées sur l'ordre de l'Académie des Sciences de Hongrie par kos Csszr. 2 Vols. Akadémiai Kiad, Budapest 1960 1601 pp. (4 plates).

[R44] F. Riesz, Sur la théorie ergodique, *Comment. Math. Helv.* 17 (1944–1945), 221–39.

13

BOUNDARIES OF $\mathbb{Z}^n$-FREE GROUPS

ANDREI MALYUTIN[1], TATIANA NAGNIBEDA[2], AND
DENIS SERBIN[3]

[1]St. Petersburg Department of V.A. Steklov Mathematical Institute,
Fontanka 27, 191023, St. Petersburg, Russia
[2]Section de Mathématiques, University of Geneva, 2-4 Rue du Lièvre,
Case Postale 64, 1211 Genève 4, Suisse
[3]Department of Mathematical Sciences, Stevens Institute of Technology,
1 Castle Point on Hudson, Hoboken, NJ 07030, USA

To Wolfgang Woess on the occasion of his sixtieth birthday

Abstract
In this chapter we study random walks on a finitely generated group G
which has a free action on a $\mathbb{Z}^n$-tree. We show that if G is nonabelian and
acts minimally, freely and without inversions on a locally finite $\mathbb{Z}^n$-tree
Γ with the set of open ends Ends(Γ), then for every nondegenerate
probability measure μ on G there exists a unique μ-stationary
probability measure ν_μ on Ends(Γ), and the space (Ends(Γ), ν_μ) is a
μ-boundary. Moreover, if μ has finite first moment with respect to the
word metric on G (induced by a finite generating set), then the measure
space (Ends(Γ), ν_μ) is isomorphic to the Poisson–Furstenberg boundary
of (G, μ).

Contents

1. Introduction

In this chapter we study random walks on $\mathbb{Z}^n$-free groups and the
boundaries of such groups. This family of groups is a particularly nice and
well-studied subclass of groups acting freely on Λ-trees (Λ-*free groups*),
where Λ is an arbitrary ordered abelian group.

2010 *Mathematics Subject Classification*. 20E08, 20F65, 20F69, 37B05, 60J50.
Key words and phrases. Poisson boundary, Λ-tree, group action, $\mathbb{Z}^n$-free group.
The first author is supported by RNF grant 14-11-00581. The authors acknowledge
support of the Swiss National Science Foundation.

The theory of group actions on Λ-trees goes back to the work [23] of Lyndon who introduced abstract length functions on groups, axiomatizing Nielsen cancellation method; he initiated the study of groups with real valued length functions. Later, in [6], Chiswell related such length functions with group actions on $\mathbb{Z}$- and $\mathbb{R}$-trees, providing a construction of the tree on which the group acts. At about the same time, Tits in [27] gave the first formal definition of an $\mathbb{R}$-tree, which is a geodesic metric space with a tree-like structure. Eventually, in [25], Morgan and Shalen introduced Λ-trees for an arbitrary ordered abelian group Λ and the general form of Chiswell's construction. A Λ-tree is a metric space whose metric takes values in Λ and is subject to certain tree axioms. The theory of group actions on such objects was consistently developed by Alperin and Bass (see [1]), where authors state the fundamental problem: find the group theoretic information carried by a Λ-tree action, in particular, the structure of Λ-free groups. Whereas the case of Archimedean free actions, that is, when $\Lambda = \mathbb{R}$, is basically closed by the Rips' Theorem that describes finitely generated $\mathbb{R}$-free groups (see [3, 11]), the general non-Archimedean case is still open, though a lot of progress was made (see [2, 14, 20, 21]).

The case when $\Lambda = \mathbb{Z}^n$ with the right lexicographic order received a lot of attention due to the natural combinatorial structure of $\mathbb{Z}^n$-trees. This structure was exploited in [19] and [22] to obtain a description of finitely generated $\mathbb{Z}^n$-free groups in terms of free products with amalgamation and HNN extensions of a particular type (see also [2]). The class of $\mathbb{Z}^n$-free groups is a natural generalization of free groups which contains limit groups, $\mathbb{R}$-free groups, etc. and which is closed under taking subgroups, free products, and amalgamated free products along maximal cyclic subgroups (n is not preserved in general). All these groups are hyperbolic relative to noncyclic maximal abelian subgroups (see [9, 14]) (hyperbolic if all maximal abelian subgroups are cyclic), coherent, with nice algorithmic properties.

Let (X, d) be a Λ-tree. One can define an equivalence relation on maximal linear subtrees of X, which are called X-*rays*. The equivalence classes under this relation are called *ends* of X and the subset of *open ends*, that is, ends not ending at a point of X, is denoted by $\text{Ends}(X)$. An *open cut* of X is a pair (X_0, X_1) of open subtrees X_0 and X_1 of X such that $X = X_0 \bigsqcup X_1$. An open cut (X_0, X_1) corresponds to a pair of open ends (e_0, e_1) respectively of X_0 and X_1 and we denote the set of all such pairs (e_0, e_1) arising from open cuts by $\text{Cuts}(X)$ (see Subsection 3.1 for details).

Now that the sets $\text{Ends}(X)$ and $\text{Cuts}(X)$ are defined, we can introduce the set

$$\overline{X} := X \cup \text{Ends}(X) \cup \text{Cuts}(X).$$

In the case when $\Lambda = \mathbb{Z}^n$ the set $\overline{X}$ can be equipped with a metric, so that $\overline{X}$ is compact in the topology induced by this metric (see (13.6), Subsection 4.1).

The main result of the chapter is formulated below.

Theorem 1.1.   *Let G be a countable nonabelian group acting freely and without inversions on a $\mathbb{Z}^n$-tree Γ, for some $n \in \mathbb{N}$. Assume also that the action of G is minimal, that is, Γ has no proper G-invariant subtrees. Then, for every non-degenerate probability measure μ on G:*

(1) *there is a unique μ-stationary measure ν_μ on the space $\mathrm{Ends}(\Gamma)$ of ends of Γ; this measure is continuous (i.e. it takes zero value on each point); the measure space $(\mathrm{Ends}(\Gamma), \nu_\mu)$ is a μ-boundary of (G, μ),*

(2) *for almost every path $\tau = \{\tau_i\}$ of the μ-random walk, all the sequences $\{\tau_i \cdot v\}$, $v \in \Gamma$, converge to a random end $\omega(\tau) \in \mathrm{Ends}(\Gamma)$; the distribution of the limits $\omega(\tau)$ is given by the μ-stationary measure ν_μ,*

(3) *if G is finitely generated and μ has finite first moment with respect to the word metric on G (induced by a finite generating set), then the measure space $(\mathrm{Ends}(\Gamma), \nu_\mu)$ is isomorphic to the Poisson–Furstenberg boundary of (G, μ).*

In the proof of the above theorem we use methods of [10] and [5], and the structure of the proof is as follows. At first, in Section 4, we introduce a metric d_f on $\overline{\Gamma}$ which makes $\overline{\Gamma}$ a compact metric space equipped with a continuous action of G (Theorem 4.8). Next, we prove that for every nondegenerate measure μ on G there exists a unique μ-stationary measure ν_μ on $\overline{\Gamma}$ and $(\overline{\Gamma}, \nu_\mu)$ is a μ-boundary of (G, μ) (see Proposition 5.6), which implies that the space $\overline{\Gamma}$ is μ-proximal. Then, since the measure ν_μ on $\overline{\Gamma}$ is concentrated on $\mathrm{Ends}(\Gamma)$ (see Proposition 5.1), we deduce that $(\mathrm{Ends}(\Gamma), \nu_\mu)$ is a μ-boundary too, and hence μ-proximal (see Corollary 5.7). Finally, we show that if μ has finite first moment with respect to the word metric on G induced by a finite generating set, then $(\mathrm{Ends}(\Gamma), \nu_\mu)$ is a maximal μ-boundary (Proposition 5.11).

Observe that there are other constructions of the Poisson boundaries of $\mathbb{Z}^n$-free groups. First of all, since every $\mathbb{Z}^n$-free group G is hyperbolic relative to its non-cyclic abelian subgroups (follows from results of [14] and [9]), one can study its Floyd boundary (see [12, 13]) which is nontrivial. Hence, from [17] it follows that the Floyd boundary of G is its Poisson boundary. Another approach is to use the fact that G is $CAT(0)$ with isolated flats (see [4]), so one can construct its Poisson boundary using results of [18] (see also [26]).

The advantage of the construction presented here is that it comes naturally from the action of a $\mathbb{Z}^n$-free group G on the underlying $\mathbb{Z}^n$-tree without reference to general results about relatively hyperbolic groups and $CAT(0)$-groups, and provides a description of the boundary in terms of the underlying $\mathbb{Z}^n$-tree.

Acknowledgments

The authors are very grateful to Vadim Kaimanovich and to Anders Karlsson for many helpful comments and suggestions.

2. Preliminaries: Poisson–Furstenberg Boundaries

2.1. Random Walks and Poisson–Furstenberg Boundaries

Let G be a countable group and let μ be a probability measure on G. The measure μ is *non-degenerate* if its support generates G as a semigroup. The *right-hand random walk on G with distribution μ* (or, briefly, μ-*random walk*) is the time-homogeneous Markov chain whose state space is G, the transition probabilities are given by $P(g,h) = \mu(g^{-1}h)$, and the initial distribution is concentrated at the identity 1_G of the group. Realizations of this process are called *paths* of the random walk.

The *path space* of the random walk (G, μ) is the measure space $(G^{\mathbb{N}_0}, P_\mu)$ ($\mathbb{N}_0 := \{0, 1, \dots\}$), where the measure P_μ is determined by the following condition. Let $C_{g_0,\dots,g_p}$, $p \in \mathbb{N}_0$, denote the set of paths

$$C_{g_0,\dots,g_p} := \left\{ \{\tau_i\}_{i \in \mathbb{N}_0} \in G^{\mathbb{N}_0} \mid (\tau_0, \dots, \tau_p) = (g_0, \dots, g_p) \right\}.$$

Then

$$P_\mu(C_{g_0,\dots,g_p}) = \delta_{1_G}(g_0) \times \mu(g_0^{-1}g_1) \times \cdots \times \mu(g_{p-1}^{-1}g_p),$$

where δ_{1_G} is the probability measure on G with $\delta_{1_G}(\{1_G\}) = 1$. The σ-algebra of P_μ-measurable subsets is the P_μ-completion of the σ-algebra generated by the *cylindrical sets* $C_{g_0,\dots,g_p}$. Since G is countable, the path space $(G^{\mathbb{N}_0}, P_\mu)$ is a Lebesgue space.

There are several ways to define the Poisson–Furstenberg boundary for random walks. In particular, the Poisson–Furstenberg boundary of the random walk (G, μ) can be defined as the *quotient measure space* of the path space $(G^{\mathbb{N}_0}, P_\mu)$ with respect to the *measurable hull* (= *measurable envelope*) of the partition ζ whose elements are the classes of the equivalence relation $\sim$ (on the set $G^{\mathbb{N}_0}$ of paths) defined as follows:

$$(\tau_i)_{i \in \mathbb{N}_0} \sim (\tau_i')_{i \in \mathbb{N}_0} \iff \exists k, k': \tau_{k+j} = \tau_{k'+j}' \ \forall j > 0. \tag{13.1}$$

The measurable hull and hence the Poisson–Furstenberg boundary are objects defined *modulo subsets of measure* 0 (mod 0).

The partition ζ is invariant under the action of G on $G^{\mathbb{N}_0}$ defined by the rule $g(\tau_0, \tau_1, \ldots) = (g\tau_0, g\tau_1, \ldots)$. This action induces a canonical action of G on the Poisson–Furstenberg boundary.

2.2. Furstenberg μ-Boundaries

Our exposition of the theory of Furstenberg μ-boundaries is based on [10].

An *action* of a group G on a topological space M is a homomorphism from G to the group Homeo(M) of all homeomorphisms of M. An action is said to be (*topologically*) *minimal* if each orbit of the action is dense (or, equivalently, if the action has no proper closed invariant subsets). A space endowed with an action of the group G is called a *G-space*. A map $f\colon X \to Y$ of G-spaces is said to be *equivariant* if $g(f(x)) = f(g(x))$ for all $x \in X$, $g \in G$.

We denote the space of all probability measures on G by $\mathcal{P}(G)$ and the space of all Borel probability measures on M by $\mathcal{P}(M)$.[1] We endow $\mathcal{P}(M)$ with the *weak* topology* (by duality with the space $C_b(M)$ of all bounded continuous functions on M). A subbase for the weak* topology is the collection of sets of the form

$$\left\{ \nu \in \mathcal{P}(M) \;\middle|\; a < \left(\int_M f \, d\nu \right) < b \right\}, \quad \text{where } f \in C_b(M), \; a, b \in \mathbb{R}.$$

Thus, a sequence $\{\nu_i\}_{i \in \mathbb{N}_0}$ of Borel probability measures converges with respect to the weak* topology (*converges weakly*) to a measure ν if and only if for all $f \in C_b(M)$ the numerical sequence $\left\{ \int_M f \, d\nu_i \right\}_{i \in \mathbb{N}_0}$ converges to $\int_M f \, d\nu$.

Any action of G on M induces an action of G on $\mathcal{P}(M)$:

$$(g \cdot \theta)(E) = \theta(g^{-1} \cdot E).$$

If $x \in M$, then the *Dirac measure* $\delta_x \in \mathcal{P}(M)$ is the measure defined by $\delta_x(E) = 1$ if $x \in E$ and $\delta_x(E) = 0$ otherwise (for a measurable E). We say that a measure is *continuous* if it takes zero value on each point.

Suppose that $\mu \in \mathcal{P}(G)$. A measure $\nu \in \mathcal{P}(M)$ is said to be μ-*stationary* if

$$\mu * \nu = \sum_{g \in G} (g \cdot \nu)\mu(g) = \nu.$$

Assertion 2.1 ([10]). *Let G be a countable group acting on a compact metric space M and let $\mu \in \mathcal{P}(G)$ be an arbitrary measure. Then the set of μ-stationary measures in $\mathcal{P}(M)$ is nonempty.*

---

[1] In [10], the space of *compact-regular* probability measures is considered. We are interested in Polish (separable complete metric) spaces or Borel subsets thereof. All Borel probability measures on these spaces are compact-regular.

Lemma 2.2 ([16]).   *Let G be a countable group acting on a space M, μ a non-degenerate probability measure on G and $\nu \in \mathcal{P}(M)$ be a μ-stationary measure. Let $E \subset M$ be a measurable subset such that for every $g \in G$ we have either $g \cdot E = E$, or $(g \cdot E) \cap E = \varnothing$. Suppose further that there is an infinite family of pairwise-disjoint sets of the form $g \cdot E$, where $g \in G$. Then $\nu(E) = 0$. In particular, if the orbit $G \cdot x$ of every point $x \in M$ is infinite, then ν is continuous.*

Lemma 2.3.   *Let G be a countable group acting minimally on a topological space M, μ be a nondegenerate measure on G, and $\nu \in \mathcal{P}(M)$ be a μ-stationary measure. Let $E \subset M$ be a nonempty open subset. Then $\nu(E) > 0$.*

Proof.   Observe that by definition, the μ-stationary measure ν is μ^{*k}-stationary for every $k \in \mathbb{N}$. Consequently, for every $k \in \mathbb{N}$ we have

$$\nu(E) = \sum_{g \in G} \mu^{*k}(g)(g \cdot \nu)(E) = \sum_{g \in G} \mu^{*k}(g)\nu(g^{-1} \cdot E). \qquad (13.2)$$

Suppose that $\nu(E) = 0$. Then (13.2) implies that $\nu(g^{-1} \cdot E) = 0$ whenever there exists $k \in \mathbb{N}$ such that $\mu^{*k}(g) > 0$. Hence, $\nu(g^{-1} \cdot E) = 0$ for each $g \in G$ because μ is nondegenerate. On the other hand, since G acts on M minimally, and E is open and nonempty, it follows that for each $x \in M$ there exists $g \in G$ such that $g \cdot x \in E$. Hence,

$$\bigcup_{g \in G} g^{-1} \cdot E = \bigcup_{g \in G} g \cdot E = M.$$

Since G is countable, it follows that

$$1 = \nu(M) = \nu \left(\bigcup_{g \in G} g^{-1} \cdot E \right) \leq \sum_{g \in G} \nu(g^{-1} \cdot E) = 0.$$

This contradiction proves that $\nu(E) > 0$.   □

Theorem 2.4 ([10, 16]).   *Let G be a countable group acting on a compact metric space M. Take $\mu \in \mathcal{P}(G)$ and let $\nu \in \mathcal{P}(M)$ be a μ-stationary measure. Then for P_μ-almost every (a. e.) path $\tau = \{\tau_i\}_{i \in \mathbb{N}_0}$ of the μ-random walk , the sequence $\{\tau_i \cdot \nu\}_{i \in \mathbb{N}_0}$ converges to some measure $\lambda(\tau) \in \mathcal{P}(M)$.*

Definition 2.5.   Suppose that M is a compact metric G-space, $\mu \in \mathcal{P}(G)$, and $\nu \in \mathcal{P}(M)$ is a μ-stationary measure. The pair (M, ν) is called a *Furstenberg μ-boundary* for G if for a. e. path $\tau = \{\tau_i\}_{i \in \mathbb{N}_0}$ of the right μ-random walk, the sequence of measures $\{\tau_i \cdot \nu\}_{i \in \mathbb{N}_0}$ converges to some Dirac measure $\delta_{\omega(\tau)}$, where $\omega(\tau) \in M$.

Definition 2.6. Let μ be a probability measure on G and let (X, ν) be the Poisson–Furstenberg boundary of the pair (G, μ). The quotient measure space of (X, ν) with respect to a G-invariant measurable partition will be called a μ-*boundary in the sense of Kaimanovich* for G.

Assertion 2.7. *Let G be a countable group acting on a metric space M. Take a $\mu \in \mathcal{P}(G)$ and let $\nu \in \mathcal{P}(M)$ be a μ-stationary measure. Assume that for P_μ-a. e. path $\{\tau_i\}_{i\in\mathbb{N}_0}$ of the μ-random walk the sequence of measures $\{\tau_i \cdot \nu\}_{i\in\mathbb{N}_0}$ converges weakly to the Dirac measure δ_x for some $x \in M$. Let*

$$\mathrm{bnd}\colon G^{\mathbb{N}_0} \to M$$

be the (P_μ-a. e. defined) map that sends the path $\{\tau_i\}_{i\in\mathbb{N}_0}$ to x whenever $\{\tau_i \cdot \nu\}_{i\in\mathbb{N}_0}$ converges to δ_x. Then bnd is P_μ-measurable and $\mathrm{bnd}(P_\mu) = \nu$.[2]

Proof. Let us first show that for each bounded continuous $f\colon M \to \mathbb{R}$ the composition

$$F\colon = f \circ \mathrm{bnd}\colon G^{\mathbb{N}_0} \to \mathbb{R}$$

is measurable (P_μ-measurable) and

$$\int_{G^{\mathbb{N}_0}} F \, dP_\mu = \int_M f \, d\nu. \tag{13.3}$$

Observe that the sequence $\{F_k\}_{k\in\mathbb{N}_0}$ of measurable functions

$$F_k\colon G^{\mathbb{N}_0} \to \mathbb{R}, \quad F_k(\tau) = \int_M f \, d\tau_k(\nu)$$

is uniformly bounded (because f is bounded) and P_μ-a. e. converges pointwise to F (by the assumption of the assertion). Then by the Lebesgue bounded convergence theorem F is measurable and[3]

$$\int_{G^{\mathbb{N}_0}} F \, dP_\mu = \lim_{k\to\infty} \int_{G^{\mathbb{N}_0}} F_k \, dP_\mu = \lim_{k\to\infty} \int_{G^{\mathbb{N}_0}} \left(\int_M f \, d\tau_k(\nu) \right) dP_\mu(\tau).$$

Observe that the distribution of τ_k is given by μ^{*k} and since ν is μ-stationary we have $\nu = \mu^{*k} * \nu$, whence

$$\int_{G^{\mathbb{N}_0}} \left(\int_M f \, d\tau_k(\nu) \right) dP_\mu(\tau) = \sum_{g\in G} \left(\left(\int_M f \, dg(\nu) \right) \cdot \mu^{*k}(g) \right)$$

$$= \int_M f \, d(\mu^{*k} * \nu) = \int_M f \, d\nu.$$

---

[2] Since bnd is only P_μ-a. e. defined, it makes sense to consider measurability with respect to the P_μ-completion.

[3] Recall that, more generally, the pointwise limit of a sequence of measurable maps from a measurable space into a metric space is measurable.

We have thus proved that F is measurable and

$$\int_{G^{\mathbb{N}_0}} F \, dP_\mu = \lim_{k \to \infty} \int_{G^{\mathbb{N}_0}} F_k \, dP_\mu = \int_M f \, d\nu.$$

Recall that a map $m \colon \Omega \to X$ from a measure space Ω to a metric space X is measurable (with respect to the Borel σ-algebra on X) if and only if the compositions $f \circ m \colon \Omega \to \mathbb{R}$ are measurable for all bounded continuous $f \colon X \to \mathbb{R}$. (This readily follows from the fact that every metric space is perfectly normal.) It then follows from the above that bnd is measurable.

Since bnd is measurable, the Borel measure $\mathrm{bnd}(P_\mu)$ on M is well-defined and for each bounded continuous $f \colon M \to \mathbb{R}$ we have

$$\int_M f \, d\,\mathrm{bnd}(P_\mu) = \int_{G^{\mathbb{N}_0}} f \circ \mathrm{bnd} \, dP_\mu,$$

whence it follows by (13.3) that

$$\int_M f \, d\,\mathrm{bnd}(P_\mu) = \int_M f \, d\nu. \tag{13.4}$$

Recall that, in a metric space, distinct Borel probability measures determine distinct functionals on the space of bounded continuous functions. Therefore, (13.4) implies that $\mathrm{bnd}(P_\mu) = \nu$ as required. $\qquad\square$

Assertion 2.7 means that every Furstenberg μ-boundary is a quotient of the measure space $(G^{\mathbb{N}_0}, P_\mu)$ (as a measure space, disregarding the topology) with the quotient map bnd. Moreover, since bnd sends each class of the equivalence relation $\sim$ (defined in (13.1)) to a point, it follows that a Furstenberg μ-boundary is an equivariant quotient of the Poisson–Furstenberg boundary. This proves the following corollary.

Corollary 2.8. *Every Furstenberg μ-boundary is a μ-boundary in the sense of Kaimanovich.*

Definition 2.9. (See [10].) Suppose that (M, d) is a metric G-space, $\mu \in \mathcal{P}(G)$. Let

$$\mu_{(n)} = (\mu + \mu^{*2} + \cdots + \mu^{*n})/n.$$

M is said to be μ-*proximal* if for all $x,\, y \in M$ and $\varepsilon > 0$ we have

$$\mu_{(n)}\{g \,|\, d(gx, gy) > \varepsilon\} \xrightarrow[n \to \infty]{} 0.$$

M is called *mean-proximal* if it is μ-proximal for all nondegenerate $\mu \in \mathcal{P}(G)$.

Theorem 2.10. *Let G be a countable group acting on a compact metric space M. Then the following conditions are equivalent:*

(a) M is mean-proximal.
(b) For any nondegenerate $\mu \in \mathcal{P}(G)$ and μ-stationary $\nu \in \mathcal{P}(M)$, the pair (M, ν) is a Furstenberg μ-boundary for G.
(c) For each nondegenerate $\mu \in \mathcal{P}(G)$ there exists a unique μ-stationary $\nu \in \mathcal{P}(M)$ and the pair (M, ν) is a Furstenberg μ-boundary for G.

Proof. The equivalence $(a) \Leftrightarrow (b)$ follows directly from [10, Theorem 14.1]. The implication $(a) \Rightarrow (c)$ is proved in [24, Lemma 3.1]. The implication $(c) \Rightarrow (b)$ is trivial. □

2.3. The Strip Criterion of Kaimanovich

We use the following corollary of the 'Strip Criterion', established by Kaimanovich in [15].

Corollary 2.11 ([15]). *Let G be a finitely generated group with the induced word metric $|\cdot|$. Let μ be a probability measure on G with finite first moment $\sum |g| \mu(g)$. Let $\check{\mu}$ be the reflected measure defined by $\check{\mu}(g) = \mu(g^{-1})$. Let (B_+, λ_+) and (B_-, λ_-) be μ- and $\check{\mu}$-boundaries (in the sense of Kaimanovich), respectively. If there exists a measurable G-equivariant map S assigning to pairs of points $(b_-, b_+) \in B_- \times B_+$ nonempty 'strips' $S(b_-, b_+) \subset G$ such that for $(\lambda_- \times \lambda_+)$-almost every $(b_-, b_+) \in B_- \times B_+$ we have*

$$\frac{1}{i} \log \operatorname{card}\{g \in S(b_-, b_+) \mid |g| \le i\} \xrightarrow[i \to \infty]{} 0, \qquad (13.5)$$

then (B_+, λ_+) is isomorphic to the Poisson–Furstenberg boundary of the pair (G, μ).

Remark 2.12. Note that, under the Strip Criterion (Corollary 2.11), the 'strips' $S(b_-, b_+)$ are required to be

   (i) all nonempty,
  (ii') $(\lambda_- \times \lambda_+)$-almost surely 'thin'.

Clearly, since the strips are allowed to meet the 'thinness' requirement $(\lambda_- \times \lambda_+)$-almost surely (not surely), we can handle the 'nonemptiness' property in the same way. In other words, in order to use the Strip Criterion it suffices to construct a (measurable, equivariant) map $S': B_- \times B_+ \to 2^G$ with strips, which are

   (i') $(\lambda_- \times \lambda_+)$-almost surely nonempty,
  (ii') $(\lambda_- \times \lambda_+)$-almost surely thin.

This is clear, because we can pass from S' to a map S with the property (i) by setting $S(b_-, b_+) = G$ if $S'(b_-, b_+) = \varnothing$ and $S(b_-, b_+) = S'(b_-, b_+)$ otherwise.

Note also that, having a map with the properties (i') and (ii'), we can replace all nonthin strips by empty ones and thus obtain a map that has the property (i) and the property

(ii) all strips are 'thin'.

3. Preliminaries: $\mathbb{Z}^n$-Free Groups

Following [1] we give some basic definitions from the theory of Λ-trees and then consider the case when $\Lambda = \mathbb{Z}^n$. All the details can be found in [1] and [7].

3.1. Λ-Trees

A set Λ equipped with addition '+' and a partial order '$\leq$' is called a *partially ordered* abelian group if

(1) $\langle \Lambda, + \rangle$ is an abelian group,
(2) $\langle \Lambda, \leq \rangle$ is a partially ordered set,
(3) for all $\alpha, \beta, \gamma \in \Lambda$, $\alpha \leq \beta$ implies $\alpha + \gamma \leq \beta + \gamma$.

An abelian group Λ is called *orderable* if there exists a linear order '$\leq$' on Λ, satisfying the condition (3) above. In general, the ordering on Λ is not unique.

For elements $\alpha, \beta \in \Lambda$, the *closed segment* $[\alpha, \beta]$ is defined by

$$[\alpha, \beta] = \{\gamma \in \Lambda \mid \alpha \leq \gamma \leq \beta\}.$$

A subset $C \subset \Lambda$ is called *convex* if for every $\alpha, \beta \in C$ the set C contains $[\alpha, \beta]$. In particular, a subgroup C of Λ is convex if $[0, \beta] \subset C$ for every positive $\beta \in C$.

Let Λ be an ordered abelian group. A Λ-*metric space* is a pair (X, d), where X is a nonempty set and d is a Λ-*metric* on X, that is, a function $d \colon X \times X \to \Lambda$ satisfying the usual metric properties. Λ-metric spaces for an arbitrary ordered abelian group Λ were first introduced by Morgan and Shalen in [25].

If Λ_0 is a convex subgroup of Λ, for any point $x_0 \in X$ the subset

$$X_0 = \{y \in X \mid d(x, y) \in \Lambda_0\}$$

of X is a Λ_0-metric space with respect to the metric $d_0 = d_{|X_0}$, called a Λ_0-*metric subspace* of X.

If (X, d) and (X', d') are Λ-metric spaces, an *isometry* from (X, d) to (X', d') is a mapping $f \colon X \to X'$ such that $d(x, y) = d'(f(x), f(y))$ for all $x, y \in X$.

A *(geodesic) segment* in a Λ-metric space is the image of the isometry $\alpha \colon [a, b]_\Lambda \to X$, where $[a, b]_\Lambda$ is a segment in Λ between some $a, b \in \Lambda$. The endpoints of the segment are $\alpha(a), \alpha(b)$.

We call a Λ-metric space (X, d) *geodesic* if for all $x, y \in X$, there is a segment in X with the endpoints x, y. We call (X, d) *geodesically linear* if for all $x, y \in X$, there is a unique segment in X whose set of endpoints is $\{x, y\}$. We denote such a segment by $[x, y]$.

A Λ-*tree* is a nonempty Λ-metric space (X, d) such that

(T1) (X, d) is geodesically linear,
(T2) if $x, y, z \in X$ then $[x, y] \cap [x, z] = [x, w]$ for some $w \in X$; this w is unique and we write $w = Y(y, x, z)$,
(T3) if $x, y, z \in X$ and $[x, y] \cap [y, z] = \{y\}$ then $[x, y] \cup [y, z] = [x, z]$.

A nonempty convex subset X_0 of a Λ-tree X is called a *subtree*. Every subtree of a Λ-tree is obviously a Λ-tree itself (the axioms (T1)–(T3) hold for X_0 since it is convex). X_0 is *closed* if its intersection with any closed segment of X is either empty, or a closed segment, otherwise X_0 is *open*.

A Λ-tree is called *linear* if it is isometric to a subtree of Λ.

Let (X, d) be a Λ-tree. A point $e \in X$ is an *endpoint* of X if, whenever $e \in [x, y] \subset X$, we have either $e = x$, or $e = y$. A *linear subtree from* $x \in X$ *(or, originating from x)* is any linear subtree L of X such that x is an endpoint of L. Observe that L carries a natural linear ordering (for $y, z \in L : \ y \le z \iff d(x, y) \le d(x, z)$), with x as the least element. If there exists a maximal point $y \in L$ with respect to this ordering then L is just a closed segment $[x, y]$ in X. If $v \in L$, then $L_v = \{z \in L \mid v \le z\}$ is a linear subtree from v and $L = [x, v] \cup L_v$, $[x, v] \cap L_v = \{v\}$. A maximal linear subtree from x in X is called an X-*ray* from x. By [1, Proposition 2.22], the relation '$L \cap L' = L_v = L'_v$ for some v' is an equivalence relation on the set of X-rays. The equivalence classes under this relation are called *ends* of X. We distinguish *closed ends* and *open ends* of X: an end e is closed if all X-rays associated with e (ending at e) are closed segments $[x, e]$, otherwise e is open. The set of open ends of X will be denoted by $\text{Ends}(X)$.

For any $x \in X$ and $e \in \text{Ends}(X)$ there exists a unique X-ray from x ending at e which we denote by $[x, e)$. Similarly, if $e_1, e_2 \in \text{Ends}(X)$ are distinct ends of X, then there exists a unique linear subtree of X, which we denote by $\langle e_1, e_2 \rangle$, connecting the open ends e_1 and e_2 (see [1, Proposition 2.24]).

An *open cut* of X is a pair (X_0, X_1) of open subtrees X_0 and X_1 of X such that $X = X_0 \bigsqcup X_1$. By [1, Proposition 2.26], every open cut (X_0, X_1) uniquely corresponds to a pair of open ends (e_0, e_1) respectively of X_0 and X_1. Denote the set of all such pairs (e_0, e_1) arising from open cuts by $\text{Cuts}(X)$.

Suppose $Y \subseteq X$ is a subtree of X. Then all open ends of Y are also open ends of X, from which it follows that $\text{Cuts}(Y) \subseteq \text{Cuts}(X)$. At the same time, not all Y-rays are X-rays, so in general $\text{Ends}(Y) \not\subseteq \text{Ends}(X)$.

Next, if $x, y, z \in X$, then $[x, y] \cap [x, z] = [x, w]$ for some $w \in X$ by definition of Λ-tree. We denote the point w by $Y(x, y, z)$. By [1, Proposition 2.11], $Y(x, y, z)$ does not depend on the order in the triple $\{x, y, z\}$, that is, for example, $Y(y, x, z) = Y(x, y, z)$. It also follows that $Y(x, y, z) = [x, y] \cap [x, z] \cap [y, z]$. We are going to call $Y(x, y, z)$ the *median of the triple* $\{x, y, z\}$.

We say that a group G acts on a Λ-tree X if any element $g \in G$ defines an isometry $g \colon X \to X$. An action on X is *nontrivial* if there is no point in X fixed by all elements of G. Note that every group has a *trivial action* on any Λ-tree, when all group elements act as identity. An action of G on X is *minimal* if X does not contain a proper G-invariant subtree.

Observe that an action of G on X induces actions respectively on $\mathrm{Ends}(X)$ and $\mathrm{Cuts}(X)$.

Next, $g \in G$ is called *elliptic* if it has a fixed point. An isometry $g \in G$ is called an *inversion* if it does not have a fixed point, but g^2 does. If g is not elliptic and not an inversion, then it is called *hyperbolic*. For a hyperbolic element $g \in G$ define the characteristic set

$$Axis(g) = \{p \in X \mid [g^{-1} \cdot p, p] \cap [p, g \cdot p] = \{p\}\},$$

which is called the *axis of g*. $Axis(g)$ meets every $\langle g \rangle$-invariant subtree of X.

A group G acts *freely* and *without inversions* on a Λ-tree X if for all $1_G \neq g \in G$, g acts as a hyperbolic isometry. In this case we also say that G is Λ-*free*. Observe that, if G is Λ-free and f, g are hyperbolic, then $[f, g] = 1_G$ implies $Axis(f) = Axis(g)$, hence we denote $Axis(g)$ by $Axis(C_G(g))$.

3.2. Groups Acting on $\mathbb{Z}^n$-Trees

Now suppose $\Lambda = \mathbb{Z}^n$ is considered with the right lexicographic order and let (X, d) be a $\mathbb{Z}^n$-tree. Observe that $\mathbb{Z}^n$ contains the minimal positive element $(1, 0, \ldots, 0)$ which we are going to denote by 1 (it will be clear from the context whether 1 denotes an element of $\mathbb{Z}^n$, or a natural number).

Let $k \in [1, n]$. We say that $x, y \in X$ are $\mathbb{Z}^k$-*equivalent* ($x \sim_{\mathbb{Z}^k} y$) if $d(x, y) \in \mathbb{Z}^k$, that is, $d(x, y) = (a_1, \ldots, a_k, 0, \ldots, 0)$. From metric axioms it follows that '$\sim_{\mathbb{Z}^k}$' is an equivalence relation and we call the corresponding equivalence classes *maximal $\mathbb{Z}^k$-subtrees* of X. Denote by $\Xi_k(X)$ the set of all maximal $\mathbb{Z}^k$-subtrees of X.

Observe that $X / \sim_{\mathbb{Z}^k}$ is a $\mathbb{Z}^{n-k}$-tree with a metric d' induced from d (see, for example, [20] for details). The set of vertices of $X / \sim_{\mathbb{Z}^k}$ can be naturally identified with $\Xi_k(X)$, so one can lift d' to the metric $d_k \colon \Xi_k(X) \to \mathbb{Z}^{n-k}$ on $\Xi_k(X)$ as follows: if $\pi \colon X \to X / \sim_{\mathbb{Z}^k}$ is the canonical projection, then for $S, T \in \Xi_k(X)$ define $d_k(S, T)$: $= d'(\pi(S), \pi(T))$.

Let G be a group acting freely and without inversions on X (in this case G is called $\mathbb{Z}^n$-free). It is not hard to discover the structure of G using Bass-Serre theory. By contracting elements of $\Xi_{n-1}(X)$ to points we obtain from X a $\mathbb{Z}$-tree Y equipped with the action of G inherited from X. From Bass-Serre Theory it follows that $\Psi_G = Y/G$ is a graph in which vertices and edges correspond to G-orbits of vertices and edges in Y and G is isomorphic to the fundamental group of the graph of groups associated with Ψ_G. More precisely, if v ranges through the set of vertices of Ψ_G, then G can be obtained from groups G_v (each G_v is a $\mathbb{Z}^{n-1}$-free group isomorphic to the G-stabilizer of some vertex of Y) by means of amalgamated free products and HNN extensions along free abelian groups G_e of rank not greater than $n-1$. Here, every G_e is isomorphic to the G-stabilizer of some edge (an ordered pair of adjacent vertices) of Y.

In the case when G is finitely generated, Ψ_G is finite (see, for example, [22]). Now, since G and all G_e are finitely generated, and Ψ_G is finite, it follows that all G_v are also finitely generated (see, [8, Proposition 29, Proposition 35]). Using induction on n one can similarly show that the G-stabilizer of any maximal $\mathbb{Z}^k$-subtree of X, where $k \in [1, n-1]$, is also finitely generated. In particular, it follows that if G is finitely generated, then it is finitely presented.

If G is finitely generated and $H \leq G$ is also finitely generated, then by [20, Theorem 65] it follows that H is quasi-isometrically embedded into G. We are going to use this fact in the proof of Proposition 5.11.

In our further investigations we are going to need the following technical definition (see [21] for details).

For every $v_0, v_1 \in X$ such that $d(v_0, v_1) = 1$ we call the ordered pair (v_0, v_1) the *edge* from v_0 to v_1. Here, if $e = (v_0, v_1)$ then denote $v_0 = o(e)$, $v_1 = t(e)$ which are respectively the *origin* and *terminus* of e. Denote by $\mathrm{Ed}(X)$ the set of edges of X.

If $Y \subseteq X$ is a subtree of X then we can define $\mathrm{Ed}(Y)$ as above by replacing X with Y since the metric on Y is induced from X. Obviously, $\mathrm{Ed}(Y) \subseteq \mathrm{Ed}(X)$.

If Y and Z are subtrees of X such that $Y \subseteq Z$ then $\mathrm{Ed}(Y) \subseteq \mathrm{Ed}(Z)$.

4. From $\mathbb{Z}^n$-Trees to Metric Spaces

Let Γ be a countable $\mathbb{Z}^n$-tree with a designated point ε. In this section we construct a compact metric space $\bar{\Gamma}$ associated with Γ. Next, we study various properties of $\bar{\Gamma}$ and show, in particular, that any isometric action of a group G on Γ extends to a continuous action of G on $\bar{\Gamma}$.

4.1. Compactification of a $\mathbb{Z}^n$-Tree

Let (Γ, d) be a countable $\mathbb{Z}^n$-tree. Define

$$\overline{\Gamma} := \Gamma \cup \mathrm{Ends}(\Gamma) \cup \mathrm{Cuts}(\Gamma),$$

where $\mathrm{Ends}(\Gamma)$ and $\mathrm{Cuts}(\Gamma)$ are the sets of open ends and cuts of Γ (see Section 3.1 for the definitions).

Recall that if $x \in \Gamma$ and $e \in \mathrm{Ends}(\Gamma)$, then $[x, e)$ is the unique Γ-ray from x to e. Similarly, $\langle e_1, e_2 \rangle$ is the unique linear subtree connecting distinct open ends e_1, e_2 of Γ.

Now, let $a, b \in \overline{\Gamma}$. If $a = b$, in the case when $a \notin \Gamma$, define $\lfloor a, b \rfloor = \varnothing$, and when $a \in \Gamma$, define $\lfloor a, b \rfloor = \{a\} = \{b\}$. Otherwise, if $a \neq b$ then we define $\lfloor a, b \rfloor$ to be a linear subtree of Γ as follows.

- If $a, b \in \Gamma$ then $\lfloor a, b \rfloor = [a, b]$.
- If $a, b \in \mathrm{Ends}(\Gamma)$ then $\lfloor a, b \rfloor = \langle a, b \rangle$.
- If $a \in \Gamma$, $b \in \mathrm{Ends}(\Gamma)$ then $\lfloor a, b \rfloor = [a, b)$.
- Let $a \in \Gamma$ and $b = (e_0, e_1) \in \mathrm{Cuts}(\Gamma)$, where $e_i \in \mathrm{Ends}(\Gamma_i)$, $\Gamma = \Gamma_0 \bigsqcup \Gamma_1$ and $a \in \Gamma_0$. Then define $\lfloor a, b \rfloor = [a, e_0)$.
- Let $a \in \mathrm{Ends}(\Gamma)$ and $b = (e_0, e_1) \in \mathrm{Cuts}(\Gamma)$, where $e_i \in \mathrm{Ends}(\Gamma_i)$, $\Gamma = \Gamma_0 \bigsqcup \Gamma_1$. Then a is an end of one of the subtrees Γ_0, Γ_1, say $a \in \mathrm{Ends}(\Gamma_0)$, so, we define $\lfloor a, b \rfloor = \langle a, e_0 \rangle$.
- Let $a = (e_0, e_1)$, where $e_i \in \mathrm{Ends}(\Gamma_i), \Gamma = \Gamma_0 \bigsqcup \Gamma_1$, and $b = (f_0, f_1) \in \mathrm{Cuts}(\Gamma)$. It follows that b is an open cut of one of the subtrees Γ_0, Γ_1, say $b \in \mathrm{Cuts}(\Gamma_0)$. Now, since $e_0 \in \mathrm{Ends}(\Gamma_0)$ and $b \in \mathrm{Cuts}(\Gamma_0)$, replacing Γ with Γ_0 in the definition case above we can assume that $\lfloor e_0, b \rfloor$ is defined for Γ_0 and set $\lfloor a, b \rfloor = \lfloor e_0, b \rfloor$.

It is easy to see from the definition above that $\lfloor a, b \rfloor = \lfloor b, a \rfloor$ for any $a, b \in \overline{\Gamma}$. Also, if $\lfloor a, b \rfloor \neq \varnothing$ then a and b can be considered as ends of $\lfloor a, b \rfloor$ (closed if points in Γ, open otherwise).

Lemma 4.1. *If $a \in \Gamma$, then the map $\psi : b \mapsto \lfloor a, b \rfloor$ establishes a one-to-one correspondence between $\overline{\Gamma}$ and the set of all linear subtrees from a in Γ.*

Proof. From the definition above it follows that $\lfloor a, b_1 \rfloor \neq \lfloor a, b_2 \rfloor$ if $b_1 \neq b_2$. Hence, ψ is injective. In order to show that it is also surjective, for each linear subtree L of Γ from a we show that there exists $b \in \overline{\Gamma}$ such that $L = \lfloor a, b \rfloor$.

If L is closed, then $L = [a, b]$ for some $b \in \Gamma$. That is, $L = \lfloor a, b \rfloor$.

Now, let L be open. If L is maximal, then it is a Γ-ray and $L = [a, b)$ for some $b \in \mathrm{Ends}(\Gamma)$ and we have $L = \lfloor a, b \rfloor$.

If L is not maximal, then there exists a subtree Γ' of Γ such that L is a Γ'-ray from a (for example, L itself is such a subtree). The union of all such subtrees Γ' is a subtree Γ_0 of Γ since every Γ' contains L. In fact, from the construction of Γ_0 it follows that it is the maximal subtree (by inclusion) of Γ such that L is a Γ_0-ray from a.

Observe that Γ_0 is open. Indeed, since L is not maximal in Γ, it can be extended to a Γ-ray L' from a and there exists $x \in L' \smallsetminus L$. Hence, $x \notin \Gamma_0$ (because L is maximal in Γ_0) and $\Gamma_0 \cap [a,x] = L$ which is not a closed segment. That is, Γ_0 cannot be closed, hence, it is open.

Next, consider $\Gamma_1 = \Gamma \smallsetminus \Gamma_0$. We are going to show that Γ_1 is a subtree of Γ. Observe that $L_1 = L' \smallsetminus L$ is a linear tree contained in Γ_1 and x defined above belongs to L_1. If $y \in \Gamma_1$ is such that $[x,y] \cap \Gamma_0 \neq \varnothing$, then $y \notin L_1$ and $\Gamma' = [y,a] \cup L$ is a subtree of Γ such that L is a Γ'-ray from a. But then we get a contradiction with the construction of Γ_0 since $\Gamma' \nsubseteq \Gamma_0$. The contradiction comes from the assumption that $[x,y] \cap \Gamma_0 \neq \varnothing$ and it follows that $[x,y] \in \Gamma_1$ for every $y \in \Gamma_1$. In other words, Γ_1 is convex, hence, it is a subtree of Γ.

Moreover, Γ_1 is open. Assume on the contrary that Γ_1 is closed and consider $L_1 \subseteq \Gamma_1$. Since Γ_1 is closed, L_1 is also closed and there exists $z \in L_1$ such that $[a,x] \cap L_1 = [z,x]$. Now, since the metric on Γ takes values in $\mathbb{Z}^n$, there exists a point $z_0 \in L'$ such that $d(a,z) = d(a,z_0)+1$, where 1 is the minimal positive element of $\mathbb{Z}^n$. Obviously, $z_0 \notin L_1$ because otherwise $[a,x] \cap L_1 = [z_0,x]$ – a contradiction. Hence, $z_0 \in L$ and $L = [a,z_0]$ – a contradiction with the fact that L is open.

Thus, Γ_1 is open and it follows that $b = (\Gamma_0, \Gamma_1)$ is an open cut such that $L = \lfloor a,b \rfloor$.

From the argument above it follows that for any linear subtree L from a there exists $b \in \overline{\Gamma}$ such that $L = \lfloor a,b \rfloor$ and the statement of the lemma follows. □

Below we are going to use the following convention: for an edge $e \in$ Ed(Γ) and $a, b \in \overline{\Gamma}$ we write $e \in \lfloor a,b \rfloor$ meaning that $e \in \mathrm{Ed}\lfloor a,b \rfloor$. Also, if $a = b$, then obviously $\mathrm{Ed}\lfloor a,b \rfloor = \varnothing$ since $\lfloor a,b \rfloor$ in this case is either empty, or consist of a single point of Γ.

Lemma 4.2. *If $a, b, c \in \overline{\Gamma}$ then the following hold.*

(a) *If $\lfloor a,b \rfloor \neq \varnothing$ and $x \in \lfloor a,b \rfloor$, then $\lfloor a,b \rfloor = \lfloor a,x \rfloor \cup \lfloor x,b \rfloor$ and $\lfloor a,x \rfloor \cap \lfloor x,b \rfloor = \{x\}$.*

(b) *If $x \in \Gamma$ and $\lfloor a,x \rfloor \cap \lfloor x,b \rfloor = \{x\}$, then $\lfloor a,x \rfloor \cup \lfloor x,b \rfloor = \lfloor a,b \rfloor$.*

(c) *If a,b,c are pairwise distinct, then $\lfloor a,c \rfloor \cap \lfloor c,b \rfloor \neq \varnothing$ unless $c = (e_0,e_1) \in \mathrm{Cuts}(\Gamma)$, where $e_i \in \mathrm{Ends}(\Gamma_i), \Gamma = \Gamma_0 \bigsqcup \Gamma_1$, is such that $\lfloor a,c \rfloor \subseteq \Gamma_0$ and $\lfloor c,b \rfloor \subseteq \Gamma_1$.*

(d) *$\lfloor a,c \rfloor \cup \lfloor c,b \rfloor$ is a subtree of Γ.*

(e) *$\lfloor a,b \rfloor \subseteq \lfloor a,c \rfloor \cup \lfloor c,b \rfloor$.*

(f) *$\mathrm{Ed}\lfloor a,b \rfloor \subseteq \mathrm{Ed}\lfloor a,c \rfloor \cup \mathrm{Ed}\lfloor c,b \rfloor$.*

(g) *$\mathrm{Ed}\lfloor a,b \rfloor \cap \mathrm{Ed}\lfloor a,c \rfloor \cap \mathrm{Ed}\lfloor b,c \rfloor = \varnothing$.*

(h) *If $c \in \Gamma$ and $\lfloor a, b \rfloor \neq \varnothing$, then*

$$\lfloor c, a \rfloor \cap \lfloor c, b \rfloor \subseteq \lfloor c, z \rfloor$$

for any $z \in \lfloor a, b \rfloor$.

Proof. (a) If $a, b \in \Gamma$, then by definition of '$\lfloor$, $\rfloor$' we have $\lfloor a, b \rfloor = [a, b]$. Since $x \in [a, b]$, we have $[a, x] \cap [x, b] = \{x\}$ and $[a, b] = [a, x] \cup [x, b]$ by definition of Λ-tree. At the same time, $\lfloor a, x \rfloor = [a, x]$, $\lfloor x, b \rfloor = [x, b]$, and the statement follows.

If $a \in \Gamma$ and $b \in \text{Ends}(\Gamma)$, then by definition of "$\lfloor$, $\rfloor$", we have $\lfloor a, b \rfloor = [a, b)$ which is a linear subtree from a. Since $x \in [a, b)$, it follows that $[x, b)$ is a linear subtree from x such that we have $[a, b) = [a, x] \cup [x, b)$ and $[a, x] \cap [x, b) = \{x\}$ (see [1, Definitions 2.21]). Now the statement follows since $[a, x] = \lfloor a, x \rfloor$ and $[x, b) = \lfloor x, b \rfloor$.

If $a \in \Gamma$ and $b = (e_0, e_1) \in \text{Cuts}(\Gamma)$, where $e_i \in \text{Ends}(\Gamma_i), \Gamma = \Gamma_0 \bigsqcup \Gamma_1$ and $a \in \Gamma_0$, then by definition of "$\lfloor$, $\rfloor$", we have $\lfloor a, b \rfloor = [a, e_0)$. This is a linear subtree from a in Γ_0, so the required statement follows from the argument given in the previous paragraph with Γ replaced by Γ_0.

Suppose $a, b \in \text{Ends}(\Gamma)$. Hence, by definition of '$\lfloor$, $\rfloor$', we have $\lfloor a, b \rfloor = \langle a, b \rangle$. Since $x \in \langle a, b \rangle$, it follows that $\langle a, x] \cap [x, b) = \{x\}$. Now from [1, Proposition 2.24], it follows that $\langle a, x] \cup [x, b) = \langle a, b \rangle$. Hence, the statement holds since $\langle a, x] = \lfloor a, x \rfloor$ and $[x, b) = \lfloor x, b \rfloor$ by definition of '$\lfloor$, $\rfloor$'.

Suppose $a \in \text{Ends}(\Gamma)$ and $b = (e_0, e_1) \in \text{Cuts}(\Gamma)$, where $e_i \in \text{Ends}(\Gamma_i), \Gamma = \Gamma_0 \bigsqcup \Gamma_1$. Then, assuming that $a \in \text{Ends}(\Gamma_0)$, by definition of '$\lfloor$, $\rfloor$', we have $\lfloor a, b \rfloor = \langle a, e_0 \rangle$. Since $e_0, a \in \text{Ends}(\Gamma_0)$, the statement follows from the argument above with Γ replaced by Γ_0.

Finally, suppose that $a = (e_0, e_1)$, where $e_i \in \text{Ends}(\Gamma_i), \Gamma = \Gamma_0 \bigsqcup \Gamma_1$, and $b = (f_0, f_1) \in \text{Cuts}(\Gamma)$. Assuming that $b \in \text{Cuts}(\Gamma_0)$, by definition of '$\lfloor$, $\rfloor$', we have $\lfloor a, b \rfloor = \lfloor e_0, b \rfloor$ and the statement follows from the argument above with Γ replaced by Γ_0.

(b) If $a, b \in \Gamma$ then the statement holds by definition of Λ-tree.

If $a \in \Gamma$ and $b \in \text{Ends}(\Gamma)$, then $\lfloor a, x \rfloor = [a, x]$, $\lfloor x, b \rfloor = [x, b)$. In particular, $[x, b)$ is a linear subtree from x and $[a, x] \cap [x, b) = \{x\}$. Hence, $[a, x] \cup [x, b)$ is a linear subtree from a which ends at b (see [1, Definitions 2.21]). That is, $[a, x] \cup [x, b) = [a, b) = \lfloor a, b \rfloor$ and the statement follows.

If $a \in \Gamma$ and $b = (e_0, e_1) \in \text{Cuts}(\Gamma)$, where $e_i \in \text{Ends}(\Gamma_i), \Gamma = \Gamma_0 \bigsqcup \Gamma_1$ and $a \in \Gamma_0$, then by definition we have $\lfloor a, b \rfloor = [a, e_0)$. It follows that $x \in \Gamma_0$ because otherwise we have a contradiction with $[a, x] \cap [x, b) = \{x\}$. Now, the statement follows from the argument above with Γ replaced by Γ_0.

Suppose $a, b \in \text{Ends}(\Gamma)$. By [1, Proposition 2.24], for any $z \in \Gamma$ we have $[z, a) \cup [z, b) = \langle a, b \rangle$ whenever $[z, a) \cap [z, b) = \{z\}$. This implies the statement because we have $\lfloor x, a \rfloor = [x, a)$, $\lfloor x, b \rfloor = [x, b)$, and $\lfloor a, b \rfloor = \langle a, b \rangle$ by definition of '$\lfloor$, $\rfloor$'

Suppose $a \in \text{Ends}(\Gamma)$ and $b = (e_0, e_1) \in \text{Cuts}(\Gamma)$, where $e_i \in \text{Ends}(\Gamma_i), \Gamma = \Gamma_0 \bigsqcup \Gamma_1$. Assume that $a \in \text{Ends}(\Gamma_0)$. Hence, $x \in \Gamma_0$ because otherwise we have a contradiction with $[x, a\rangle \cap [x, b\rangle = \{x\}$. Now, the statement follows from the argument above with Γ replaced by Γ_0.

The case when $a, b \in \text{Cuts}(\Gamma)$ can be similarly reduced to one of the previous cases.

(c) If $c \in \Gamma$, then both $\lfloor a, c \rfloor$ and $\lfloor c, b \rfloor$ are linear subtrees of Γ from c. Hence, $c \in \lfloor a, c \rfloor \cap \lfloor c, b \rfloor$ and the statement follows.

Suppose $c \in \text{Ends}(\Gamma)$ and let $x \in \lfloor a, c \rfloor$, $y \in \lfloor c, b \rfloor$ be arbitrary (such x and y exist since both $\lfloor a, c \rfloor$ and $\lfloor c, b \rfloor$ are nonempty by assumption). Hence, $[x, c\rangle$ and $[y, c\rangle$ represent the same end of Γ and by definition of ends it follows that $[x, c\rangle \cap [y, c\rangle$ is nonempty. Hence, $\lfloor a, c \rfloor \cap \lfloor c, b \rfloor$ is also nonempty.

Now, let $c = (e_0, e_1) \in \text{Cuts}(\Gamma)$, where $e_i \in \text{Ends}(\Gamma_i), \Gamma = \Gamma_0 \bigsqcup \Gamma_1$. Suppose that both $\lfloor a, c \rfloor$ and $\lfloor c, b \rfloor$ belong to one of the subtrees Γ_0, Γ_1, say Γ_0. Hence, repeating the argument above with Γ replaced by Γ_0 we obtain that $\lfloor a, c \rfloor \cap \lfloor c, b \rfloor \neq \varnothing$.

(d) If at least two of the points a, b, c coincide, then (d) trivially holds. So, suppose that a, b, c are pairwise distinct.

From (c) it follows that $\lfloor a, c \rfloor \cup \lfloor c, b \rfloor$ is connected (hence, it is a subtree of Γ) unless $c = (e_0, e_1) \in \text{Cuts}(\Gamma)$, where $e_i \in \text{Ends}(\Gamma_i), \Gamma = \Gamma_0 \bigsqcup \Gamma_1$, is such that $\lfloor a, c \rfloor \subseteq \Gamma_0$ and $\lfloor c, b \rfloor \subseteq \Gamma_1$. Assume the latter and take arbitrary points $x \in \lfloor a, c \rfloor$, $y \in \lfloor c, b \rfloor$. Then, by [1, Proposition 2.26], we have $[x, y] = [x, e_0\rangle \cup [y, e_1\rangle$. Next, by definition of '$\lfloor$, $\rfloor$', we have $\lfloor x, c \rfloor = [x, e_0\rangle$, $\lfloor c, y \rfloor = \langle e_1, y]$, hence,

$$[x, y] = \lfloor x, c \rfloor \cup \lfloor c, y \rfloor.$$

Applying (a) we have

$$\lfloor a, c \rfloor \cup \lfloor c, b \rfloor = (\lfloor a, x \rfloor \cup \lfloor x, c \rfloor) \cup (\lfloor c, y \rfloor \cup \lfloor y, b \rfloor) = \lfloor a, x \rfloor \cup [x, y] \cup \lfloor y, b \rfloor$$

which is connected. Hence, $\lfloor a, c \rfloor \cup \lfloor c, b \rfloor$ is a subtree of Γ.

(e) Again, if at least two of the points a, b, c coincide, then (e) trivially holds. So, suppose that a, b, c are pairwise distinct.

Suppose at first that one of the subtrees $\lfloor a, c \rfloor$, $\lfloor c, b \rfloor$ is contained in the other one, say $\lfloor a, c \rfloor \subset \lfloor c, b \rfloor$. If $a \in \Gamma$ then $a \in \lfloor c, b \rfloor$. Hence, from (a) we have

$$\lfloor c, b \rfloor = \lfloor c, a \rfloor \cup \lfloor a, b \rfloor$$

and the statement follows. Suppose $a \notin \Gamma$. Then it follows that $a \in \text{Cuts}(\Gamma)$. Indeed, a cannot be an end of Γ since for every $z \in \lfloor c, a \rfloor$, the linear subtree $\lfloor z, a \rfloor$ is not a maximal linear subtree from z. Let $a = (e_0, e_1)$, where $e_i \in \text{Ends}(\Gamma_i), \Gamma = \Gamma_0 \bigsqcup \Gamma_1$. Assume, without loss of generality, that c belongs to Γ_0 and b belongs to Γ_1 (as points, ends, or open cuts). From (c) it follows that there exists $x \in \lfloor c, b \rfloor \cap \lfloor a, b \rfloor$. Indeed, since $a \in \text{Cuts}\lfloor c, b \rfloor$, the ends e_0 and e_1 uniquely correspond to each other and they cannot be separated by b because $b \neq a$. By (a) we have

$$\lfloor c, b \rfloor = \lfloor c, x \rfloor \cup \lfloor x, b \rfloor, \quad \lfloor a, b \rfloor = \lfloor a, x \rfloor \cup \lfloor x, b \rfloor$$

and it is enough to show that $\lfloor a, x \rfloor \subset \lfloor c, x \rfloor$. But $\lfloor a, x \rfloor = \langle a, x]$ by definition of '$\lfloor$, $\rfloor$' and for every $y \in \lfloor c, x \rfloor \cap \Gamma_0$, by [1, Proposition 2.26], we have $[y, x] = [y, e_0) \cup \langle e_1, x]$. In other words,

$$\lfloor a, x \rfloor = \langle e_1, x] \subset [y, x] \subseteq \lfloor c, x \rfloor.$$

Now suppose that $\lfloor a, c \rfloor \not\subset \lfloor c, b \rfloor$, $\lfloor c, b \rfloor \not\subset \lfloor a, c \rfloor$. From (c) we have the following two cases.

Case I. $\lfloor a, c \rfloor \cap \lfloor c, b \rfloor \neq \varnothing$.

Let $x \in \lfloor a, c \rfloor \smallsetminus \lfloor c, b \rfloor$ and $y \in \lfloor c, b \rfloor \smallsetminus \lfloor a, c \rfloor$. Observe that $\lfloor x, c \rfloor$ and $\lfloor y, c \rfloor$ are linear subtrees of $\lfloor a, c \rfloor \cup \lfloor c, b \rfloor$ (which is a subtree of Γ by (d)). Hence, by [1, Proposition 2.22], there exists $v \in \lfloor a, c \rfloor \cup \lfloor c, b \rfloor$ such that either $\lfloor a, c \rfloor \cap \lfloor c, b \rfloor = [c, v]$ in the case when $c \in \Gamma$, or $\lfloor a, c \rfloor \cap \lfloor c, b \rfloor = \langle c, v]$ in the case when $c \notin \Gamma$. In both cases, by definition of '$\lfloor$, $\rfloor$', we have

$$\lfloor a, c \rfloor \cap \lfloor c, b \rfloor = \lfloor c, v \rfloor$$

and $\lfloor a, v \rfloor \cap \lfloor b, v \rfloor = \{v\}$. Now, by (b) we have $\lfloor a, v \rfloor \cup \lfloor b, v \rfloor = \lfloor a, b \rfloor$ and

$$\lfloor a, b \rfloor = \lfloor a, v \rfloor \cup \lfloor b, v \rfloor \subset \lfloor a, c \rfloor \cup \lfloor c, b \rfloor,$$

where the last inclusion follows from (a).

Case II. $c = (e_0, e_1) \in \mathrm{Cuts}(\Gamma)$, where $e_i \in \mathrm{Ends}(\Gamma_i)$, $\Gamma = \Gamma_0 \bigsqcup \Gamma_1$, is such that $\lfloor a, c \rfloor \subseteq \Gamma_0$ and $\lfloor c, b \rfloor \subseteq \Gamma_1$.

By (c) there exist $x \in \lfloor a, c \rfloor \cap \lfloor a, b \rfloor$ and $y \in \lfloor c, b \rfloor \cap \lfloor a, b \rfloor$. Hence, by (a) we have

$$\lfloor a, b \rfloor = \lfloor a, x \rfloor \cup \lfloor x, y \rfloor \cup \lfloor y, b \rfloor,$$

$$\lfloor a, c \rfloor = \lfloor a, x \rfloor \cup \lfloor x, c \rfloor, \quad \lfloor c, b \rfloor = \lfloor c, y \rfloor \cup \lfloor y, b \rfloor.$$

Hence, in order to prove $\lfloor a, b \rfloor \subseteq \lfloor a, c \rfloor \cup \lfloor c, b \rfloor$ it is enough to show that $\lfloor x, y \rfloor \subseteq \lfloor x, c \rfloor \cup \lfloor c, y \rfloor$. But we have $\lfloor x, y \rfloor = [x, y]$, $\lfloor x, c \rfloor = [x, c\rangle$, $\lfloor c, y \rfloor = \langle c, y]$, so the required follows from [1, Proposition 2.26].

(f) Suppose on the contrary that there exists an edge (v_0, v_1) in the set $\mathrm{Ed}\lfloor a, b \rfloor \smallsetminus (\mathrm{Ed}\lfloor a, c \rfloor \cup \mathrm{Ed}\lfloor c, b \rfloor)$. This implies that one of the points v_0 and v_1 lies in $\lfloor c, a \rfloor \smallsetminus \lfloor c, b \rfloor$, while the other one is in $\lfloor c, b \rfloor \smallsetminus \lfloor c, a \rfloor$. Then v_0 is not in $\lfloor v_1, c \rfloor$, while v_1 is not in $\lfloor v_0, c \rfloor$.

If $\lfloor v_1, c \rfloor \cap \lfloor v_0, c \rfloor$ is empty, then (c) implies that v_0 and v_1 are in distinct $\mathbb{Z}$-subtrees of Γ so that (v_0, v_1) is not an edge — a contradiction.

If $\lfloor v_1, c \rfloor \cap \lfloor v_0, c \rfloor$ is not empty, then [1, Proposition 2.28] implies that

$$\lfloor v_0, v_1 \rfloor \cap \lfloor v_0, c \rfloor \cap \lfloor v_1, c \rfloor \neq \varnothing.$$

Suppose $v \in \lfloor v_0, v_1 \rfloor \cap \lfloor v_0, c \rfloor \cap \lfloor v_1, c \rfloor \neq \varnothing$. Then $v \notin \{v_0, v_1\}$ because v_0 is not in $\lfloor v_1, c \rfloor$, while v_1 is not in $\lfloor v_0, c \rfloor$. Therefore, $[v_0, v_1] = [v_0, v, v_1]$, which means that (v_0, v_1) is not an edge — a contradiction.

(g) If at least one of the intersections $\lfloor a, b \rfloor \cap \lfloor a, c \rfloor$, $\lfloor a, b \rfloor \cap \lfloor b, c \rfloor$, $\lfloor a, c \rfloor \cap \lfloor b, c \rfloor$ is empty, then $\lfloor a, b \rfloor \cap \lfloor a, c \rfloor \cap \lfloor b, c \rfloor$ is also empty and the statement follows.

Suppose that all three intersections $\lfloor a, b \rfloor \cap \lfloor a, c \rfloor$, $\lfloor a, b \rfloor \cap \lfloor b, c \rfloor$, $\lfloor a, c \rfloor \cap \lfloor b, c \rfloor$ are nonempty. By [1, Proposition 2.28] it follows that

$$\lfloor a, b \rfloor \cap \lfloor a, c \rfloor \cap \lfloor b, c \rfloor \neq \varnothing.$$

Suppose $v \in \lfloor a, b \rfloor \cap \lfloor a, c \rfloor \cap \lfloor b, c \rfloor$. Hence, from (a) we have

$$\lfloor a, b \rfloor = \lfloor a, v \rfloor \cup \lfloor v, b \rfloor, \quad \lfloor a, c \rfloor = \lfloor a, v \rfloor \cup \lfloor v, c \rfloor, \quad \lfloor b, c \rfloor = \lfloor b, v \rfloor \cup \lfloor v, c \rfloor$$

and

$$\lfloor a, b \rfloor \cap \lfloor a, c \rfloor = (\lfloor a, v \rfloor \cup \lfloor v, b \rfloor) \cap (\lfloor a, v \rfloor \cup \lfloor v, c \rfloor)$$
$$= \lfloor a, v \rfloor \cup (\lfloor a, v \rfloor \cap \lfloor v, c \rfloor) \cup (\lfloor b, v \rfloor \cap \lfloor v, a \rfloor) \cup (\lfloor b, v \rfloor \cap \lfloor v, c \rfloor)$$
$$= \lfloor a, v \rfloor \cup \{v\} \cup \{v\} \cup \{v\} = \lfloor a, v \rfloor.$$

Hence, apply (a) again and obtain

$$\lfloor a, b \rfloor \cap \lfloor a, c \rfloor \cap \lfloor b, c \rfloor = \lfloor a, v \rfloor \cap \lfloor b, c \rfloor = \lfloor a, v \rfloor \cap (\lfloor b, v \rfloor \cup \lfloor v, c \rfloor)$$
$$= (\lfloor a, v \rfloor \cap \lfloor b, v \rfloor) \cup (\lfloor a, v \rfloor \cap \lfloor v, c \rfloor) = \{v\}.$$

Since the intersection $\lfloor a, b \rfloor \cap \lfloor a, c \rfloor \cap \lfloor b, c \rfloor$ is just a single point, the required follows.

(h) Suppose on the contrary that there exists $x \in \lfloor c, a \rfloor \cap \lfloor c, b \rfloor$ such that $x \notin \lfloor c, z \rfloor$. By (e), we have $\lfloor c, b \rfloor \subseteq \lfloor c, z \rfloor \cup \lfloor z, b \rfloor$. Since $x \in \lfloor c, b \rfloor$ and $x \notin \lfloor c, z \rfloor$, it follows that $x \in \lfloor z, b \rfloor$. At the same time, (e) implies that we have $\lfloor c, a \rfloor \subseteq \lfloor c, z \rfloor \cup \lfloor z, a \rfloor$. Since $x \in \lfloor c, a \rfloor$ and $x \notin \lfloor c, z \rfloor$, it follows that $x \in \lfloor z, a \rfloor$. Thus, it is shown that $x \in \lfloor z, b \rfloor \cap \lfloor z, a \rfloor$. However, $\lfloor z, b \rfloor \cap \lfloor z, a \rfloor = \{z\}$ by (a). Consequently, we have $x \in \{z\}$ — a contradiction. $\square$

Next, recall that in Section 3.1 we introduced the notion of the median $Y(x, y, z)$ for any triple of points $\{x, y, z\} \subset \Gamma$. We extend this notion to $\overline{\Gamma}$: for $a, b, c \in \overline{\Gamma}$

- set $Y(a, b, c) = a$ if $a = b$ or $a = c$, and set $Y(a, b, c) = b$ if $b = c$,
- if one of the points a, b, c of $\overline{\Gamma}$, say c, is an open cut $c = (e_0, e_1) \in$ Cuts(Γ), where $e_i \in$ Ends(Γ_i), $\Gamma = \Gamma_0 \bigsqcup \Gamma_1$, such that $\lfloor a, c \rfloor \subseteq \Gamma_0$ and $\lfloor c, b \rfloor \subseteq \Gamma_1$, then set $Y(a, b, c) = c$,
- set $Y(a, b, c) = \lfloor a, b \rfloor \cap \lfloor a, c \rfloor \cap \lfloor b, c \rfloor$ otherwise.

Observe that $Y(a, b, c)$ is correctly defined for any triple $a, b, c \in \overline{\Gamma}$. Indeed, suppose $a, b,$ and c are all pairwise distinct and one of the intersections

$$\lfloor a, b \rfloor \cap \lfloor a, c \rfloor, \quad \lfloor a, b \rfloor \cap \lfloor b, c \rfloor, \quad \lfloor a, c \rfloor \cap \lfloor b, c \rfloor,$$

say $\lfloor a, c \rfloor \cap \lfloor b, c \rfloor$, is empty. In this case, by the assertion (c) of Lemma 4.2 we have the case covered in the second part of the definition above. Now, if all three intersections $\lfloor a, b \rfloor \cap \lfloor a, c \rfloor$, $\lfloor a, b \rfloor \cap \lfloor b, c \rfloor$, $\lfloor a, c \rfloor \cap \lfloor b, c \rfloor$ are nonempty, by the proof of assertion (g) of Lemma 4.2, $\lfloor a, b \rfloor \cap \lfloor a, c \rfloor \cap \lfloor b, c \rfloor$ is a single point $v \in \Gamma$ and by the definition above we have $Y(a, b, c) = v$.

Now we define a metric on $\overline{\Gamma}$. Consider the set $\mathrm{Ed}(\Gamma)$ which is countable by assumption. Let $f \colon \mathrm{Ed}(\Gamma) \to \mathbb{R}_+$ be a *summable* positive real-valued function, that is

$$\sum_{e \in \mathrm{Ed}(\Gamma)} f(e) < +\infty.$$

We define the function

$$d_f \colon \overline{\Gamma} \times \overline{\Gamma} \to \mathbb{R} \tag{13.6}$$

by setting

$$d_f(a, b) = \sum_{e \in \mathrm{Ed}\lfloor a, b \rfloor} f(e).$$

Then d_f is clearly a metric on $\overline{\Gamma}$ (the triangle inequality for d_f follows from the assertion (f) of Lemma 4.2).

Lemma 4.3. *Let $\{a_i\}$ and $\{b_i\}$ be two sequences of points in the metric space $(\overline{\Gamma}, d_f)$. Then*

(1) *the sequence $\{a_i\}$ is fundamental (i.e. Cauchy sequence) if and only if for every edge $e \in \mathrm{Ed}(\Gamma)$ there exists $N > 0$ such that $e \notin \mathrm{Ed}\lfloor a_j, a_k \rfloor$ whenever $j > N$ and $k > N$,*

(2) *the sequence $\{d_f(a_i, b_i)\}$ tends to zero if and only if for each $e \in \mathrm{Ed}(\Gamma)$, the set $\{i \in \mathbb{N} \mid e \in \mathrm{Ed}\lfloor a_i, b_i \rfloor\}$ is at most finite,*

(3) *the sequence $\{a_i\}$ converges to a point $t \in \overline{\Gamma}$ if and only if for each $e \in \mathrm{Ed}(\Gamma)$ the set $\{i \in \mathbb{N} \mid e \in \mathrm{Ed}\lfloor a_i, t \rfloor\}$ is at most finite.*

Proof. (1) If $\{a_i\}$ is fundamental and $e \in \mathrm{Ed}(\Gamma)$, then there exists $N > 0$ such that $d_f(a_j, a_k) < f(e)$ whenever $j > N$ and $k > N$. It follows that $e \notin \mathrm{Ed}\lfloor a_j, a_k \rfloor$ whenever $j > N$ and $k > N$.

Conversely, assume that for every $e \in \mathrm{Ed}(\Gamma)$ there exists $N(e) > 0$ such that $e \notin \mathrm{Ed}\lfloor a_j, a_k \rfloor$ whenever $j > N(e)$ and $k > N(e)$. Observe that, since f is summable, for any $\delta > 0$ there exists a finite set $E_f(\delta) \subset \mathrm{Ed}(\Gamma)$ of edges such that

$$\sum_{e \in \mathrm{Ed}(\Gamma) \setminus E_f(\delta)} f(e) < \delta.$$

Let $N = \max\{N(e) \mid e \in E_f(\delta)\}$. Then $d_f(a_j, a_k) < \delta$ whenever $j > N$ and $k > N$. This means that $\{a_i\}$ is fundamental.

(2) If $\{d_f(a_i, b_i)\}$ tends to zero and $e \in \mathrm{Ed}(\Gamma)$, then there exists $N > 0$ such that $d_f(a_j, b_j) < f(e)$ whenever $j > N$. It follows that the set $\{i \in \mathbb{N} \mid e \in \mathrm{Ed}\lfloor a_i, b_i \rfloor\}$ consists of at most N elements.

Conversely, assume that for every $e \in \mathrm{Ed}(\Gamma)$ there exists $N(e) > 0$ such that $e \notin \mathrm{Ed}\lfloor a_i, b_i \rfloor$ whenever $j > N(e)$. Let $E_f(\delta)$ be the finite set described in the proof of (1) and let $N = \max\{N(e) \mid e \in E_f(\delta)\}$. Then $d_f(a_j, b_j) < \delta$ whenever $j > N$. This means that $d_f(a_i, b_i) \to 0$ as $i \to \infty$, as required.

(3) follows from the previous assertion by setting $b_i = t$ for every i. $\quad\square$

Proposition 4.4. *The topology of the metric space $(\overline{\Gamma}, d_f)$ is independent of the choice of the summable function $f \colon \mathrm{Ed}(\Gamma) \to \mathbb{R}_+$.*

Proof. It is well known (and easy to check) that the topology of a metrizable space is determined by its convergent sequences. By Lemma 4.3(3), the set of convergent sequences in $(\overline{\Gamma}, d_f)$ is independent of the choice of f. $\quad\square$

For an edge $e \in \mathrm{Ed}(\Gamma)$ denote by R_e the binary relation on the set of points of $\overline{\Gamma}$ defined by

$$(a, b) \in R_e \quad \Leftrightarrow \quad e \notin \mathrm{Ed}\lfloor a, b \rfloor.$$

It is easy to see that R_e is an equivalence relation (transitivity follows from the assertion (f) of Lemma 4.2), which has precisely two equivalence classes (follows from the assertion (g) of Lemma 4.2).

Proposition 4.5. $\overline{\Gamma}$ *is compact in the topology induced by the metric d_f.*

Proof. It is enough to show that every infinite sequence $\{x_i\}$ of points in $(\overline{\Gamma}, d_f)$ has a convergent subsequence.

Let us first show that $\{x_i\}$ has a fundamental subsequence. Since $\mathrm{Ed}(\Gamma)$ is countable, $\{x_i\}$ has a subsequence $\{x_{i_j}\}$ such that for each $e \in \mathrm{Ed}(\Gamma)$ there exists $N(e) > 0$ such that $(x_{i_j}, x_{i_k}) \in R_e$ (which is equivalent to $e \notin \mathrm{Ed}\lfloor x_{i_j}, x_{i_k} \rfloor$) whenever $j > N(e)$ and $k > N(e)$. Then, $\{x_{i_j}\}$ is fundamental by the assertion (1) of Lemma 4.3.

Now, we show that every fundamental sequence $\{y_i\}$ in $(\overline{\Gamma}, d_f)$ converges. Fix an arbitrary point $y \in \Gamma$ and consider the linear subtree

$$L = \bigcup_{k \in \mathbb{N}} \bigcap_{j > k} \lfloor y, y_j \rfloor.$$

There exists a unique point $t \in \overline{\Gamma}$ such that $L = \lfloor y, t \rfloor$ (see Lemma 4.1). We show that $\{y_i\}$ converges to t. Assume that it does not. Then, by the assertion (3) of Lemma 4.3 there exists an edge e and an infinite subsequence $\{y_{i_j}\}$ in $\{y_i\}$ such that $e \in \lfloor t, y_{i_j} \rfloor$ for all j. Since $\{y_i\}$ is fundamental, it then follows by the assertion (1) of Lemma 4.3 (and by the assertion (f) of Lemma 4.2) that $e \in \lfloor t, y_i \rfloor$ for all i sufficiently large.

In the case where $e \notin \lfloor y, t \rfloor$, we have $e \in \lfloor y, y_i \rfloor$ for all sufficiently large i (we use the assertion (f) of Lemma 4.2), whence $e \in \lfloor y, t \rfloor$ (by definition of L and t), a contradiction.

In the case when $e \in \lfloor y, t \rfloor$ we have $e \notin \lfloor y, y_i \rfloor$ for all sufficiently large i (we use the assertion (g) of Lemma 4.2 and the fact that $e \in \lfloor t, y_i \rfloor$) whence $e \notin \lfloor y, t \rfloor$ (by definition of L and t), a contradiction.

This proves that $\{y_i\}$ converges to t and the proposition is thus proved. □

Remark 4.6.   It is obvious from the construction that Γ is dense in $(\overline{\Gamma}, d_f)$. Therefore, Proposition 4.5 implies that $\overline{\Gamma}$ is the compactification of Γ with respect to d_f.

4.2. Group Action on $\overline{\Gamma}$

Let Γ be a countable $\mathbb{Z}^n$-tree and $\overline{\Gamma}$ be the compactification of Γ with respect to the metric d_f constructed in Subsection 4.1.

Observe that an automorphism $\alpha \colon \Gamma \to \Gamma$ sends Γ-rays to Γ-rays, ends to ends, open cuts to open cuts, and therefore induces an automorphism $\bar{\alpha} \colon \overline{\Gamma} \to \overline{\Gamma}$.

Proposition 4.7.   *Every automorphism of $\overline{\Gamma}$ is a homeomorphism with respect to the metric d_f.*

Proof.   If $\alpha \colon \overline{\Gamma} \to \overline{\Gamma}$ is an automorphism, let $f_\alpha \colon \mathrm{Ed}(\Gamma) \to \mathbb{R}_+$ be the function defined by

$$f_\alpha(e) := f(\alpha^{-1}(e)).$$

Then α is an isometry between $(\overline{\Gamma}, d_f)$ and $(\overline{\Gamma}, d_{f_\alpha})$. Therefore, α is a homeomorphism because d_f and d_{f_α} induce one and the same topology by Proposition 4.4. □

Theorem 4.8.   *If a group G acts on Γ by isometries, then the induced action of G on $\overline{\Gamma}$ is continuous with respect to the topology generated by the metric d_f. Moreover, if G is nonabelian and acts on Γ freely, without inversions, and so that Γ has no proper G-invariant subtrees, then the orbit of every point in $\overline{\Gamma}$ is infinite.*

Proof.   The first part of the statement follows from Proposition 4.7.

Now suppose there exists $v \in \overline{\Gamma}$ such that $\{G \cdot v\}$ is finite. It follows that $|G : Stab_G(v)| < \infty$.

If $v \in \Gamma$, then Γ must be spanned by $\{G \cdot v\}$ because otherwise Γ contains a proper G-invariant subtree. But then, since the action is isometric, it follows that either there is a fixed point, or an inversion—a contradiction in either case. Thus we have $v \in \mathrm{Ends}(\Gamma) \cup \mathrm{Cuts}(\Gamma)$.

Suppose $v \in \mathrm{Ends}(\Gamma)$. By [7, Lemma 3.1.9], if v is fixed by $g \in G$, then v is an end of $Axis(g)$, which is a maximal linear g-invariant subtree of Γ. If v is fixed by another element $h \in G$, then v is an end of $Axis(h)$ too, so

$Axis(g) = Axis(h)$, from which it follows that $[g, h] = 1$. Thus, $Stab_G(v)$ is abelian and G is virtually abelian. But G is $\mathbb{Z}^n$-free, so commutation is transitive and it follows that G is abelian – a contradiction with our assumption.

The case when $v \in \text{Cuts}(\Gamma)$ is considered similarly. $\square$

Theorem 4.9. *Assume that a group G acts on Γ by isometries and Γ has no proper G-invariant subtrees. Then the induced action of G on the set of ends $\text{Ends}(\Gamma)$ either has a global fixed point, or is topologically minimal (with respect to the topology induced by the metric d_f).*

Proof. Assume on the contrary that the action of G on $\text{Ends}(\Gamma)$ neither is topologically minimal, nor has a global fixed point. Then $\text{Ends}(\Gamma)$ has a proper closed (in $\text{Ends}(\Gamma)$) G-invariant subset K containing at least two points. Set

$$\Gamma_K = \bigcup_{a,b \in K} \lfloor a, b \rfloor.$$

We are going to obtain a contradiction by showing that Γ_K is a proper G-invariant subtree of Γ.

First, observe that $\Gamma_K \neq \varnothing$. Indeed, since K contains at least two points $a \neq b$, we have $\Gamma_K \supset \lfloor a, b \rfloor \neq \varnothing$.

Next, observe that Γ_K is a subtree in Γ. Indeed, if $x, y \in \Gamma_K$, then $x \in \lfloor a, b \rfloor$ and $y \in \lfloor c, d \rfloor$ for some $a, b, c, d \in K$. Interchanging a with b if necessary, we can assume that $b \neq c$. Then

$$\lfloor a, b \rfloor \cap \lfloor b, c \rfloor \neq \varnothing \neq \lfloor b, c \rfloor \cap \lfloor c, d \rfloor,$$

so that the union

$$N = \lfloor a, b \rfloor \cup \lfloor b, c \rfloor \cup \lfloor c, d \rfloor$$

is a subtree by [1, Lemma 2.13]. Therefore, we have $[x, y] \subset N \subset \Gamma_K$. It is thus shown that $[x, y] \subset \Gamma_K$ whenever $x, y \in \Gamma_K$, which means that Γ_K is convex and hence (being nonempty) is a subtree of Γ.

Since K is G-invariant, so is Γ_K. It remains to show that $\Gamma_K \neq \Gamma$.

Since K is a proper closed subset in $\text{Ends}(\Gamma)$, we can choose $p \in \text{Ends}(\Gamma)$ and $\varepsilon > 0$ such that $\text{dist}_{d_f}(p, K) > \varepsilon$. We show that there exists an edge e in Γ such that $\lfloor k, p \rfloor$ contains e for each $k \in K$. Since f is summable, there exists a finite set $E_f(\varepsilon) \subset \text{Ed}(\Gamma)$ of edges such that

$$\sum_{e \in \text{Ed}(\Gamma) \smallsetminus E_f(\varepsilon)} f(e) < \varepsilon.$$

We denote by M the set of all the vertices of the edges in $E_f(\varepsilon)$. Observe that the intersection

$$L = \bigcap_{v \in M} [v, p\rangle$$

of Γ-rays is a Γ-ray because M is finite. Let e be an edge in L. If $k \in K$, then $\lfloor k, p \rfloor$ contains an edge of $E_f(\varepsilon)$ because $\operatorname{dist}_{d_f}(p, K) > \varepsilon$ by construction. Therefore, $\lfloor k, p \rfloor$ contains a point $v \in M$, whence we have

$$e \in \operatorname{Ed}(L) \subset \operatorname{Ed}[v, p) \subset \operatorname{Ed}\lfloor k, p \rfloor.$$

Now, we can show that $\lfloor x, p \rfloor$ contains e for each $x \in \Gamma_K$. Indeed, if $x \in \Gamma_K$, then by the definition of Γ_K there exist a and b in K such that $x \in \lfloor a, b \rfloor$. We see that $\lfloor a, b \rfloor$ does not contain e by the assertion (g) of Lemma 4.2 because $\lfloor a, p \rfloor$ and $\lfloor b, p \rfloor$ both contain e as proved above. Consequently, $\lfloor x, a \rfloor$ does not contain e also, being a subset in $\lfloor a, b \rfloor$. Therefore, $\lfloor x, a \rfloor$ does not contain e while $\lfloor a, p \rfloor$ contains e. Then $\lfloor x, p \rfloor$ contains e because we have $\lfloor a, p \rfloor \subset \lfloor x, a \rfloor \cup \lfloor x, p \rfloor$ by the assertion (e) of Lemma 4.2.

Thus, we have shown that $\operatorname{dist}_{d_f}(\Gamma_K, p) \geq f(e) > 0$. Consequently, $\Gamma_K \neq \Gamma$. This completes the proof. $\qquad \square$

4.3. Convergence in $\overline{\Gamma}$

In the space $\overline{\Gamma}$ under investigation, there is a strong relationship between the convergence of points and the weak convergence of measures. For example, if Γ is an ordinary $\mathbb{Z}^1$-tree ($n = 1$), v_0 a vertex in Γ, z is an end in $\operatorname{Ends}(\Gamma)$, ν is a continuous Borel probability measure on $\overline{\Gamma}$, and $\{\alpha_i\}$ is a sequence of automorphisms of Γ, then the following conditions are equivalent:

(i) the sequence $\{\alpha_i \cdot v_0\}$ converges to z,
(ii) the sequence $\{\alpha_i \cdot \nu\}$ weakly converges to the Dirac measure δ_z.

If $n > 1$, then (i) and (ii) are no longer equivalent. However, in the general case, the relationship mentioned above shows up when we impose appropriate restrictions on z and/or $\{\alpha_i\}$. In fact, our proof of the main theorem is based on this relationship (see Corollary 5.4 and Proposition 5.8). In this subsection, we do some preliminary work that will allow us to prove Corollary 5.4 and Proposition 5.8.

Recall from Subsection 3.2 that $\Xi_{n-1}(\Gamma)$ is the set of all maximal $\mathbb{Z}^{n-1}$-subtrees of Γ.

Definition 4.10. Define the function $\hbar_n \colon \overline{\Gamma} \times \overline{\Gamma} \to [0, +\infty]$ as follows. Set $\hbar_n(a, a) = 0$ for each $a \in \overline{\Gamma}$ and

$$\hbar_n(a, b) = \operatorname{card}\{S \in \Xi_{n-1}(\Gamma) \mid S \cap \lfloor a, b \rfloor \neq \varnothing\},$$

for distinct $a, b \in \overline{\Gamma}$.

That is, $\hbar_n(a, b)$ is the number of distinct maximal $\mathbb{Z}^{n-1}$-subtrees of Γ intersecting with $\lfloor a, b \rfloor$. Observe that if a and b are distinct ends of Γ, then $\hbar_n(a, b)$ may be equal to $+\infty$.

Lemma 4.11. *The function*

$$\hbar_n \colon \overline{\Gamma} \times \overline{\Gamma} \to [0, +\infty]$$

is an extended metric *on* $\overline{\Gamma}$, *that is,* $\hbar_n$ *satisfies the standard axioms of metric, but can attain the value* $+\infty$. *This extended metric is G-invariant. The restriction of this extended metric to* Γ *is a metric.*

Proof. Assertion (e) of Lemma 4.2 implies that $\hbar_n$ satisfies the triangle inequality. The other required properties of $\hbar_n$ immediately follow from its definition. □

Recall that ε is a designated point in Γ. By abuse of notation, for $p \in \overline{\Gamma}$, we set $\hbar_n(p) = \hbar_n(\varepsilon, p)$, and for $g \in G$, we set

$$\hbar_n(g) = \hbar_n(g \cdot \varepsilon) = \hbar_n(\varepsilon, g \cdot \varepsilon).$$

Lemma 4.12. *Let* $\{g_i\}$ *be a sequence in* G, *let* v_0 *be a point in* Γ, *and let* z *be a point in* $\overline{\Gamma}$. *Assume that the sequence* $\{g_i \cdot v_0\}$ *converges to* z *and that the sequence* $\{\hbar_n(g_i)\}$ *tends to infinity as* $i \to \infty$. *Then all the sequences* $\{g_i \cdot v\}$ *with* $v \in \Gamma$ *converge to* z.

Proof. Since the action of G on $\overline{\Gamma}$ is isometric with respect to the extended metric $\hbar_n$, the assumption that $\{\hbar_n(g_i)\} = \{\hbar_n(g_i \cdot \varepsilon, \varepsilon)\}$ tends to infinity implies that $\{\hbar_n(g_i \cdot v)\}$ tends to infinity for each $v \in \Gamma$. Indeed, we observe that $\hbar_n(g_i \cdot \varepsilon, g_i \cdot v) = \hbar_n(\varepsilon, v) = \hbar_n(v)$ is finite, while the triangle inequality (see Lemma 4.11) implies that

$$\hbar_n(g_i \cdot v) = \hbar_n(g_i \cdot v, \varepsilon) \geq \hbar_n(g_i \cdot \varepsilon, \varepsilon) - \hbar_n(g_i \cdot \varepsilon, g_i \cdot v) = \hbar_n(g_i) - \hbar_n(v).$$

In particular, $\{\hbar_n(g_i \cdot v_0)\}$ tends to infinity.

Now, we take $v \in \Gamma$, put $k := \hbar_n(v_0, v)$, and observe that for each $i \in \mathbb{N}$ the geodesic segment $[g_i \cdot v_0, g_i \cdot v]$ is contained in the $\hbar_n$-ball of $\hbar_n$-radius k centered at $g_i \cdot v_0$. Since $\{\hbar_n(g_i \cdot v_0)\}$ tends to infinity, this implies that the $\hbar_n$-distance between ε and $[g_i \cdot v_0, g_i \cdot v]$ tends to infinity. Therefore, for each $e \in \mathrm{Ed}(\Gamma)$ the set $\{i \in \mathbb{N} \mid e \in \mathrm{Ed}[g_i \cdot v_0, g_i \cdot v]\}$ is at most finite. The result then follows by the assertion (2) of Lemma 4.3. □

Lemma 4.13. *Let* $\{a_i\}$ *and* $\{b_i\}$ *be two sequences of points in* $\overline{\Gamma}$. *Suppose that* $\hbar_n(Y(\varepsilon, a_i, b_i)) \to \infty$ *as* $i \to \infty$. *Then* $d_{\overline{\Gamma}}(a_i, b_i) \to 0$ *as* $i \to \infty$. *In particular,* $\{a_i\}$ *converges to a point* $\omega \in \overline{\Gamma}$ *if and only if* $\{b_i\}$ *converges to* ω.

Proof. Observe that if $x \in \overline{\Gamma}$ and $y \in \overline{\Gamma}$ belongs to $\lfloor \varepsilon, x \rfloor$ (as a point or an open cut), then $\hbar_n(y) \leq \hbar_n(z)$ for every $z \in \lfloor y, x \rfloor$ (follows from definition of $\hbar_n$).

Next, from the assertion (h) of Lemma 4.2 it follows that for any $a, b \in \overline{\Gamma}$ such that $\lfloor a, b \rfloor \neq \varnothing$ we have

$$\lfloor \varepsilon, a \rfloor \cap \lfloor \varepsilon, b \rfloor \subset \lfloor \varepsilon, z \rfloor$$

for any $z \in \lfloor a, b \rfloor$. Hence, it follows that

$$\min_{z \in \lfloor a,b \rfloor} \hbar_n(z) = \hbar_n(Y(\varepsilon, a, b)).$$

Therefore, we have $\min_{z \in \lfloor a,b \rfloor} \hbar_n(z) = \hbar_n(Y(\varepsilon, a_i, b_i))$ for all $i \in \mathbb{N}$. Since we assume $\hbar_n(Y(\varepsilon, a_i, b_i)) \to \infty$ as $i \to \infty$, it follows that for each $v \in \Gamma$ the set $\{i \in \mathbb{N} \mid v \in [a_i, b_i]\}$ is at most finite (because we have $v \notin [a_i, b_i]$ whenever $\min_{z \in \lfloor a,b \rfloor} \hbar_n(z) > \hbar_n(v)$). Then the result follows from the assertion (2) of Lemma 4.3. □

Lemma 4.14. *For every vertex v of a $\mathbb{Z}^n$-tree Γ, the set*

$$I_v = \{(a, b) \in \mathrm{Ends}(\Gamma) \times \mathrm{Ends}(\Gamma) \mid v \in \lfloor a, b \rfloor\}$$

is open in $\mathrm{Ends}(\Gamma) \times \mathrm{Ends}(\Gamma)$ *with respect to the product topology induced by the metric* d_f.

Proof. Let $(a, b) \in I_v$ and let $p \in \mathrm{Ends}(\Gamma)$ be such that $d_f(b, p) < d_f(v, b)$. Then $v \notin \lfloor p, b \rfloor$. However, we have $\lfloor a, b \rfloor \subseteq \lfloor a, p \rfloor \cup \lfloor p, b \rfloor$ by the assertion (e) of Lemma 4.2. This implies that $v \in \lfloor a, p \rfloor$, that is, $(a, p) \in I_v$.

By repeating the same argument we readily see that $(r, p) \in I_v$ whenever $d_f(b, p) < d_f(v, b)$ and $d_f(a, r) < d_f(a, v)$. This obviously implies the lemma. □

Remark 4.15. It can be shown that I_v from Lemma 4.14 is also closed.

5. Proof of Theorem 1.1

In this section, we prove Theorem 1.1 according to the scheme outlined in the introduction. For the rest of this section, we fix $n \in \mathbb{N}$ and a countable nonabelian $\mathbb{Z}^n$-free group G acting minimally, freely, and without inversion on a $\mathbb{Z}^n$-tree Γ. Also, we let $(\overline{\Gamma}, d_f)$ denote the compact metric space constructed in Subsection 4.1 for Γ. Recall that $\mathrm{Ends}(\Gamma)$ denote the set of open ends of Γ. We also fix a non-degenerate probability measure μ on G.

5.1. μ-proximality of $\overline{\Gamma}$ and $\mathrm{Ends}(\Gamma)$

Proposition 5.1. *If ν is a μ-stationary measure on $\overline{\Gamma}$ then*

(i) ν *is continuous,*
(ii) $\nu(\mathrm{Ends}(\Gamma)) = 1$,
(iii) $\nu(E) > 0$ *for every nonempty open set $E \subset \overline{\Gamma}$ with $E \cap \mathrm{Ends}(\Gamma) \neq \varnothing$.*

Proof. (i) Continuity of ν follows from Lemma 2.2 because the action of G on $\overline{\Gamma}$ has no finite orbits (see Theorem 4.8).

(ii) The set Γ is countable by assumption. Therefore, Cuts(Γ) is countable. Then, since ν is continuous, we have $\nu(\Gamma) = 0$ and $\nu(\text{Cuts}(\Gamma)) = 0$. Hence, $\nu(\text{Ends}(\Gamma)) = 1$.

(iii) From Theorem 4.9 it follows that either G acts on Ends(Γ) minimally, or there is a global fixed point. If all elements of G fix $e \in \text{Ends}(\Gamma)$ then G is abelian and we have a contradiction with the set of assumptions on G. Hence, G acts on Ends(Γ) minimally and now (iii) follows from (ii) by Lemma 2.3. $\qquad\square$

Define $\mathfrak{S}$ to be the subset in $G^{\mathbb{N}_0}$ consisting of all sequences $\{g_i\}$ with the following three properties:

(i) $\{g_i \cdot \varepsilon\}$ converges in $(\overline{\Gamma}, d_f)$,

(ii) $\{g_i^{-1} \cdot \varepsilon\}$ converges in $(\overline{\Gamma}, d_f)$,

(iii) $\{\hbar_n(g_i)\}$ tends to infinity as $i \to \infty$ (recall Definition 4.10).

Proposition 5.2. *A. e. path* $\{\tau_i\}$ *of the* μ-*random walk contains a subsequence* $\{\tau_{i_j}\} \in \mathfrak{S}$.

Proof. First, we prove that a. e. path $\{\tau_i\}$ of the μ-random walk contains a subsequence with the property (iii) from the definition of $\mathfrak{S}$.

Let $d \in \mathbb{N}$. Since G acts on Γ minimally, it follows that there exists an element $g_{2d} \in G$ such that $\hbar_n(g_{2d}) > 2d$. Since μ is a nondegenerate measure, there exists $k \in \mathbb{N}$ such that $\mu^{*k}(g_{2d}) > 0$. It follows that for a. e. path $\{\tau_i\}$ of the μ-random walk there exists $m \in \mathbb{N}$ such that $\tau_{m+k} = \tau_m g_{2d}$ (to see this, use ergodicity of the Bernoully shift in the space of sequences of μ^{*k}-equidistributed random variables $\{\tau_{jk}^{-1}\tau_{jk+k}\}_{j \in \mathbb{N}_0}$, or observe that the measure P_μ of the set of all paths of the μ-random walk with $\tau_{jk+k} \neq \tau_{jk}g_{2d}$ for each $j \in \mathbb{N}$ equals $\lim_{s \to \infty}(1 - \delta)^s = 0$, where $\delta = \mu^{*k}(g_{2d}) > 0$). Then

$$\hbar_n(\tau_m^{-1}\tau_{m+k}) = \hbar_n(g_{2d}) > 2d.$$

Now, we apply the triangle inequality (see Lemma 4.11) and deduce that

$$\max\{\hbar_n(\tau_{m+k}), \hbar_n(\tau_m)\} > d.$$

We have thus shown that for any $d \in \mathbb{N}$, the path $\{\tau_i\}$ a. s. has an element τ_i such that $\hbar_n(\tau_i) > d$. It obviously follows that $\{\tau_i\}$ a. s. has a subsequence $\{\tau_i'\}$ such that $\{\hbar_n(\tau_i')\}$ tends to infinity (the property (iii)).

Since $(\overline{\Gamma}, d_f)$ is compact, each infinite sequence in G (and $\{\tau_i'\}$ in particular) contains a subsequence having both properties (i) and (ii). $\qquad\square$

Lemma 5.3. *Let* $\{g_i\}$ *be a sequence from* $\mathfrak{S}$. *Let* ω *and* $\check{\omega}$ *be points from* $\overline{\Gamma}$ *that the sequences* $\{g_i \cdot \varepsilon\}$ *and* $\{g_i^{-1} \cdot \varepsilon\}$ *converge to. Then, for each* $p \in \overline{\Gamma} \smallsetminus \check{\omega}$, *the sequence* $\{g_i \cdot p\}$ *converges to* ω.

Proof. In order to prove the lemma, we will show that for every $p \in \overline{\Gamma} \smallsetminus \breve{\omega}$ we have

$$\hbar_n(p_i) \xrightarrow[i \to \infty]{} \infty, \tag{13.7}$$

where $p_i = Y(\varepsilon, g_i \cdot \varepsilon, g_i \cdot p)$, and then apply Lemma 4.13.

Observe that for each i we have $g_i^{-1} \cdot p_i = Y(\varepsilon, g_i^{-1} \cdot \varepsilon, p)$ because

$$g_i^{-1} \cdot p_i = g_i^{-1} \cdot Y(\varepsilon, g_i \cdot \varepsilon, g_i \cdot p) = Y(g_i^{-1} \cdot \varepsilon, \varepsilon, p) = Y(\varepsilon, g_i^{-1} \cdot \varepsilon, p).$$

Let us prove that the sequence $\{\hbar_n(Y(\varepsilon, g_i^{-1} \cdot \varepsilon, p))\}$ is bounded. Assume that it is not. Then it has a subsequence $\{\hbar_n(Y(\varepsilon, g_{i_j}^{-1} \cdot \varepsilon, p))\}$ which tends to infinity. Then, by Lemma 4.13, $\{g_{i_j}^{-1} \cdot \varepsilon\}$ converges to p, which contradicts the assumption that $g_i^{-1} \cdot \varepsilon \to \breve{\omega} \neq p$. Thus, $\{\hbar_n(Y(\varepsilon, g_i^{-1} \cdot \varepsilon, p))\}$ is bounded, that is, there exists $N \in \mathbb{N}$ such that

$$\hbar_n(Y(\varepsilon, g_i^{-1} \cdot \varepsilon, p)) < N \quad \text{for each} \quad i \in \mathbb{N}.$$

Therefore, for each $i \in \mathbb{N}$ we have

$$\hbar_n(g_i \cdot \varepsilon, p_i) = \hbar_n(\varepsilon, g_i^{-1} \cdot p_i) \overset{\text{def}}{=} \hbar_n(g_i^{-1} \cdot p_i) = \hbar_n(Y(\varepsilon, g_i^{-1} \cdot \varepsilon, p)) < N.$$

Now, $\{g_i\} \in \mathfrak{S}$, so

$$\hbar_n(\varepsilon, g_i \cdot \varepsilon) \overset{\text{def}}{=} \hbar_n(g_i \cdot \varepsilon) \overset{\text{def}}{=} \hbar_n(g_i)$$

tends to ∞ (the property (iii)), while we have

$$\hbar_n(\varepsilon, p_i) + \hbar_n(p_i, g_i \cdot \varepsilon) \geq \hbar_n(\varepsilon, g_i \cdot \varepsilon)$$

by the triangle inequality (see Lemma 4.11), whence it follows that $\hbar_n(p_i) \to \infty$. Finally, by Lemma 4.13, this implies that $\{g_i \cdot p\}$ converges to ω since $\{g_i \cdot \varepsilon\}$ does. □

Lemma 5.3 obviously implies the following corollary.

Corollary 5.4. *Let $\{g_i\}$ be a sequence from $\mathfrak{S}$ with $\{g_i \cdot \varepsilon\}$ converging to a point $\omega \in \overline{\Gamma}$. Then, for any continuous Borel probability measure λ on $\overline{\Gamma}$, the sequence $\{g_i \cdot \lambda\}$ converges to the Dirac measure δ_ω.*

Corollary 5.5. *If ν is a μ-stationary measure on $\overline{\Gamma}$, then $(\overline{\Gamma}, \nu)$ is a μ-boundary of (G, μ) in the sense of Furstenberg. In other words, the G-space $\overline{\Gamma}$ is mean-proximal.*

Proof. Let $\{\tau_i\}$ be a path of the μ-random walk. Then by Theorem 2.4, the sequence $\{\tau_i \cdot \nu\}$ a.s. converges to some limit. By Proposition 5.2, $\{\tau_i\}$ a.s. contains a subsequence $\{\tau_{i_j}\} \in \mathfrak{S}$. By Corollary 5.4, the subsequence $\{\tau_{i_j} \cdot \nu\}$ converges to a Dirac measure (since ν is continuous by Proposition 5.1). Therefore, the limit of $\{\tau_i \cdot \nu\}$ is a.s. a Dirac measure. This means by definition that $(\overline{\Gamma}, \nu)$ is a μ-boundary of (G, μ) in the sense of Furstenberg. Since μ is an arbitrary nondegenerate measure, it follows by Theorem 2.10 that the G-space $\overline{\Gamma}$ is mean-proximal. $\quad\square$

Now, we can draw the main result of this subsection from the results above.

Proposition 5.6. *There exists a unique μ-stationary measure ν_μ on $\overline{\Gamma}$. This measure is continuous and concentrated on $\mathrm{Ends}(\Gamma)$. The measure space $(\overline{\Gamma}, \nu_\mu)$ is a μ-boundary of G in the sense of Furstenberg.*

Proof. Observe that uniqueness of a μ-stationary measure ν_μ follows from Corollary 5.5 by Theorem 2.10 (the last one is applicable, because $\overline{\Gamma}$ is compact and metrizable). The rest follows from Proposition 5.1 and Corollary 5.5. $\quad\square$

Proposition 5.6 yields the following corollary (see Assertion 2.7 and Corollary 2.8 concerning relationship between μ-boundaries in the sense of Furstenberg and in the sense of Kaimanovich).

Corollary 5.7. *There exists a unique μ-stationary measure ν_μ on $\mathrm{Ends}(\Gamma)$. This measure is continuous. The measure space $(\mathrm{Ends}(\Gamma), \nu_\mu)$ is a μ-boundary of G in the sense of Kaimanovich.*

5.2. Stability of Paths in $\mathbb{Z}^n$-Free Groups

Proposition 5.8. *Let $\{g_i\}$ be a sequence in G, and let z be a point in $\overline{\Gamma}$. Assume that the sequence $\{g_i \cdot \nu_\mu\}$ converges to the Dirac measure δ_z. Then all the sequences $\{g_i \cdot v\}$ with $v \in \Gamma$ converge to z.*

Proof. Let $v \in \Gamma$. Since Γ is countable and admits a minimal action of a group, it follows that there exists a pair of ends $\omega_1, \omega_2 \in \mathrm{Ends}(\Gamma)$ such that $v \in \langle \omega_1, \omega_2 \rangle$ (indeed, otherwise Γ has closed ends and the subset of vertices that are not closed ends forms a proper invariant subtree, which contradicts the minimality assumption).

Lemma 4.14 implies that there exist nonempty open subsets A and B in $\mathrm{Ends}(\Gamma)$ containing respectively ω_1 and ω_2 such that $v \in \langle a, b \rangle$ whenever $a \in A$, $b \in B$. By assertion (iii) of Proposition 5.1 we have $\nu_\mu(A) > 0$ and $\nu_\mu(B) > 0$. Therefore, since $\{g_i \cdot \nu_\mu\}$ converges (weakly) to δ_z, it follows that for every $r > 0$ there exists $N > 0$ such that $d_f(z, g_i \cdot a_i) < r$ and $d_f(z, g_i \cdot b_i) < r$ for some $a_i \in A$ and $b_i \in B$ whenever $i > N$.

Note that the balls of the metric d_f are *geodesically convex*. In particular, the conditions

$$g_i \cdot v \in \langle g_i \cdot a_i, g_i \cdot b_i \rangle, \quad d_f(z, g_i \cdot a_i) < r, \quad \text{and} \quad d_f(z, g_i \cdot b_i) < r$$

imply that $d_f(z, g_i \cdot v) < r$. Indeed, we have

$$g_i \cdot v \in \langle g_i \cdot a_i, g_i \cdot b_i \rangle = \lfloor g_i \cdot a_i, g_i \cdot b_i \rfloor \subset \lfloor z, g_i \cdot a_i \rfloor \cup \lfloor z, g_i \cdot b_i \rfloor$$

by Lemma 4.2, so that

$$d_f(z, g_i \cdot v) \leq d_f(z, g_i \cdot a_i) < r \quad \text{if} \quad g_i \cdot v \in \lfloor z, g_i \cdot a_i \rfloor,$$

$$d_f(z, g_i \cdot v) \leq d_f(z, g_i \cdot b_i) < r \quad \text{if} \quad g_i \cdot v \in \lfloor z, y_i \cdot b_i \rfloor.$$

It is thus shown that for any $r > 0$ there exists N such that $d_f(z, g_i \cdot v)$ $< r$ whenever $i > N$, that is, the sequence $\{g_i \cdot v\}$ converges to z. □

Corollary 5.9. *For almost every path $\tau = \{\tau_i\}$ of the μ-random walk, all the sequences $\{\tau_i \cdot v\}$, where $v \in \Gamma$, converge to a random end $\omega(\tau) \in$ Ends(Γ); the distribution of the limits $\omega(\tau)$ is given by the μ-stationary measure ν_μ.*

Proof. Proposition 5.6 says that the measure space $(\overline{\Gamma}, \nu_\mu)$ is a μ-boundary of G in the sense of Furstenberg. This means by definition that, for almost every path $\tau = \{\tau_i\}$ of the μ-random walk, the sequence $\{\tau_i \cdot \nu_\mu\}$ converges weakly to a random Dirac measure δ_z, where $z = z(\tau) \in \overline{\Gamma}$.

Then Proposition 5.8 implies that all the sequences $\{\tau_i \cdot v\}$, where $v \in \Gamma$, converge to $z(\tau)$. By Assertion 2.7, the distribution of the limits $z(\tau)$ is given by the measure ν_μ. Since $\nu_\mu(\text{Ends}(\Gamma)) = 1$ (see Proposition 5.6), this implies the required assertion. □

5.3. Maximality of the Boundary Ends(Γ)

According to Corollary 5.7, the pair $(\text{Ends}(\Gamma), \nu_\mu)$ is a μ-boundary of G. Now we would like to show that it is a maximal μ-boundary.

Let X be a measurable G-space with a (quasi-invariant) measure ν. Recall that the action of G on X is called *ergodic* if $\nu(Y) \in \{0, 1\}$ for each G-invariant measurable subset $Y \subset X$.

Lemma 5.10 ([15]). *Let G be a countable group, $\mu \in \mathcal{P}(G)$ and $\check{\mu} \in \mathcal{P}(G)$ the reflected measure defined by $\check{\mu}(g) = \mu(g^{-1})$. Let (M_+, ν_+) and (M_-, ν_-) be respectively a μ-boundary and a $\check{\mu}$-boundary of G. Then the action of G on the product $(M_- \times M_+, \nu_- \times \nu_+)$ is ergodic.*

Observe that in [15], Lemma 5.10 is stated and proved in the case when (M_+, ν_+) and (M_-, ν_-) are maximal, respectively, μ- and $\check{\mu}$-boundaries. But the proof works verbatim for the general case.

Proposition 5.11. *Let G be a finitely generated nonabelian $\mathbb{Z}^n$-free group acting on a $\mathbb{Z}^n$-tree Γ for some $n \in \mathbb{N}$ so that Γ has no proper G-invariant subtrees. Let μ be a nondegenerate probability measure on G and ν_μ the unique μ-stationary measure on the space $\text{Ends}(\Gamma)$ of open ends of Γ. If μ has finite first moment with respect to a finite word metric on G then the measure space $(\text{Ends}(\Gamma), \nu_\mu)$ is a maximal μ-boundary of G.*

Proof. In order to prove the theorem, we are going to construct a map

$$S\colon \mathrm{Ends}(\Gamma) \times \mathrm{Ends}(\Gamma) \to 2^G$$

and then show that it satisfies all the requirements of the Strip Criterion (see Corollary 2.11 and Remark 2.12).

Let $a, b \in \mathrm{Ends}(\Gamma)$. There exists a unique bi-infinite geodesic $\langle a, b\rangle$ in Γ from a to b. We define

$$S(a, b) = \{g \in G \mid g \cdot \varepsilon \in \langle a, b\rangle\}.$$

Obviously, the map $S\colon \mathrm{Ends}(\Gamma) \times \mathrm{Ends}(\Gamma) \to 2^G$ is G-equivariant.

Claim 0. *The map S is measurable in the sense that $S^{-1}(M)$ is measurable for any $M \subset G$.*

For a vertex $v \in \Gamma$, define

$$\Omega_v = \{(a, b) \in \mathrm{Ends}(\Gamma) \times \mathrm{Ends}(\Gamma) \mid v \in \langle a, b\rangle\}.$$

Our definitions immediately imply that

$$S(a, b) = \{g \in G \mid (a, b) \in \Omega_{g\cdot\varepsilon} = g \cdot \Omega_\varepsilon\} \tag{13.8}$$

so that for each $g \in G$ we have

$$S^{-1}(g) = \Omega_{g^{-1}\cdot\varepsilon}.$$

By Lemma 4.14, for all $v \in \Gamma$ the sets Ω_v are open and hence Borel measurable. In particular, the set $S^{-1}(g) = \Omega_{g^{-1}\cdot\varepsilon}$ is Borel measurable for every $g \in G$. Hence, **Claim 0** follows by the countability of G.

Claim 1. *Let ν_+ be a μ-stationary measure on $\mathrm{Ends}(\Gamma)$ and ν_- be a $\check{\mu}$-stationary measure on $\mathrm{Ends}(\Gamma)$, where $\check{\mu}$ is the reflected measure of μ defined by $\check{\mu}(g) = \mu(g^{-1})$. Then for $(\nu_- \times \nu_+)$-a. e. pair $(a, b) \in \mathrm{Ends}(\Gamma) \times \mathrm{Ends}(\Gamma)$, the set $S(a, b)$ is non-empty.*

It follows from (13.8) that

$$S(a, b) = \{g \in G \mid g^{-1} \cdot (a, b) \in \Omega_\varepsilon\}. \tag{13.9}$$

Given that Γ is countable and admits a minimal action of a group (while $\mathrm{Ends}(\Gamma) \neq \varnothing$), it follows that Ω_ε is not empty (indeed, otherwise Γ would have closed ends and the subset of vertices that are not closed ends would form a proper invariant subtree, which contradicts the minimality assumption).

Lemma 4.14 says that Ω_ε is open with respect to the product topology. Therefore, since Ω_ε is not empty, there exist nonempty open sets $A, B \subset \mathrm{Ends}(\Gamma)$ with $A \times B \subset \Omega_\varepsilon$. Applying the assertion (iii) of Proposition 5.1 to the measures ν_- and ν_+, we see that $(\nu_- \times \nu_+)(\Omega_\varepsilon) > 0$.

Recall that the action of G on the space $(\mathrm{Ends}(\Gamma) \times \mathrm{Ends}(\Gamma), \nu_- \times \nu_+)$ is ergodic (Lemma 5.10). Consequently, since $(\nu_- \times \nu_+)(\Omega_\varepsilon) > 0$, it follows by (13.9) that for $(\nu_- \times \nu_+)$-a. e. pair $(a, b) \in \mathrm{Ends}(\Gamma) \times \mathrm{Ends}(\Gamma)$, the set $S(a, b)$ is nonempty and **Claim 1** follows.

Now, recall that by assumption G is a finitely generated nonabelian group acting freely and without inversions on Γ, and Γ has no proper G-invariant subtrees. Let Z be a finite generating set of G and let $|\cdot|_G$ be the word metric on G with respect to Z.

To finish the proof of the theorem we are going to prove the following claim.

Claim 2. *For any $m \in \{1, \ldots, n\}$ there exists $C(m) \in \mathbb{N}$ such that for any maximal $\mathbb{Z}^m$-subtree T of Γ, and any distinct $a, b \in \text{Ends}(T)$ we have*

$$card(\{g \in G \mid g \cdot \varepsilon \in \langle a, b \rangle\} \cap B_G(k)) \leq C(m)k^m,$$

where $B_G(k) = \{g \in G \mid |g|_G \leq k\}$.

Fix a maximal $\mathbb{Z}^m$-subtree T and assume that there exists at least one $g \in G$ such that $g \cdot \varepsilon \in \langle a, b \rangle$, otherwise the claim holds trivially for T. Define $H := Stab_G(T)$. Since G is finitely generated, H is also finitely generated, and it follows that H is quasi-isometrically embedded into G (see Subsection 3.2). Let R be a finite generating set of H and let $|\cdot|_H$ be the word metric on H with respect to R.

Denote by $T_{\varepsilon,m}$ the maximal $\mathbb{Z}^m$-subtree of Γ containing ε. In order to prove the claim we use induction on m.

If $m = 1$ then T is a $\mathbb{Z}$-tree and $\langle a, b \rangle$ is isometric to $\mathbb{Z}$. Consider two cases.

(1.1) Assume that $T = T_{\varepsilon,1}$, that is, $\varepsilon \in T$. If $g \in G$ is such that $g \cdot \varepsilon \in \langle a, b \rangle$, then $g \in H$ and we have

$$\{g \in G \mid g \cdot \varepsilon \in \langle a, b \rangle\} = \{g \in H \mid g \cdot \varepsilon \in \langle a, b \rangle\}.$$

Next, observe that H is a finitely generated free group which quasi-isometrically embeds into T by means of the map $h \to h \cdot \varepsilon$ for any $h \in H$.

Now, let $g \in H$ be such that $g \in B_G(k)$, that is, $|g|_G \leq k$. Since H quasi-isometrically embeds into G, it follows that $|g|_H \leq C_1 k + C_2$ for some constants $C_1, C_2 \in \mathbb{N}$ which depend only on H and G. Then, it follows that

$$d_T(\varepsilon, g \cdot \varepsilon) \leq C_3(C_1 k + C_2) + C_4 \leq C'k,$$

where d_T is the distance on T inherited from Γ and C' depends only on H and T. But now observe that since $\langle a, b \rangle$ is a linear subtree of T, the number of points on $\langle a, b \rangle$ at distance not greater than $C'k$ from ε is bounded by $2C'k + 1$. At the same time, the action of H on T is free, so, the number of elements $g \in H$ such that $g \in B_G(k)$ and $g \cdot \varepsilon \in \langle a, b \rangle$ cannot be greater than $2C'k + 1$. It means that the required inequality

$$card(\{g \in G \mid g \cdot \varepsilon \in \langle a, b \rangle\} \cap B_G(k)) \leq C(T_{\varepsilon,1})k$$

holds for $C(T_{\varepsilon,1}) = 2C' + 1$.

(1.2) Assume that $T \neq T_{\varepsilon,1}$. Observe that if

$$g_0 \in \{g \in G \mid g \cdot \varepsilon \in \langle a, b \rangle\} \cap B_G(k)$$

then $g_0^{-1} \cdot a$ and $g_0^{-1} \cdot b$ are distinct ends of $T_{\varepsilon,1}$ and we have

$$\text{card}\left(\{g \in G \mid g \cdot \varepsilon \in \langle a, b \rangle\} \cap B_G(k)\right)$$

$$\leq \text{card}\left(\{g \in G \mid g \cdot \varepsilon \in \langle g_0^{-1} \cdot a, g_0^{-1} \cdot b \rangle\} \cap B_G(2k)\right).$$

Indeed, for every $g_1 \in \{g \in G \mid g \cdot \varepsilon \in \langle a, b \rangle\} \cap B_G(k)$, the element $g_0^{-1} g_1$ belongs to $B_G(2k)$ since

$$|g_0^{-1} g_1|_G \leq |g_0|_G + |g_1|_G \leq 2k,$$

and

$$(g_0^{-1} g_1) \cdot \varepsilon \in \langle g_0^{-1} \cdot a, g_0^{-1} \cdot b \rangle.$$

In other words, $g_0^{-1} g_1 \in \{g \in G \mid g \cdot \varepsilon \in \langle g_0^{-1} \cdot a, g_0^{-1} \cdot b \rangle\} \cap B_G(2k)$ and the required inequality follows. Finally, from (1.1) we have that

$$\text{card}\left(\{g \in G \mid g \cdot \varepsilon \in \langle g_0^{-1} \cdot a, g_0^{-1} \cdot b \rangle\} \cap B_G(2k)\right) \leq C(T_{\varepsilon,1}) \cdot (2k).$$

That is, $C(T) = 2C(T_{\varepsilon,1})$.

Hence, the constant $C(1)$, which works for every maximal $\mathbb{Z}$-subtree of Γ can be taken to be $2C(T_{\varepsilon,1})$. This concludes the proof of the claim in the case when $m = 1$.

Suppose $m > 1$ and assume that **Claim 2** holds for $m - 1$.

Recall from Subsection 3.2 that one can define the metric d_{m-1} on the set $\Xi_{m-1}(T)$ of all maximal $\mathbb{Z}^{m-1}$-subtrees of T and d_{m-1} takes values in $\mathbb{Z}$. Consider two cases.

(m.1) Assume that $T = T_{\varepsilon,m}$, that is, $\varepsilon \in T$. In particular, T contains $T_{\varepsilon,m-1}$.

Observe that $\langle a, b \rangle$ may have nonempty intersection with at most countably many $\mathbb{Z}^{m-1}$-subtrees of T, which we can enumerate by integers

$$\ldots, T_{-r}, \ldots, T_{-1}, T_0, T_1, \ldots, T_r, \ldots,$$

where T_0 is the 'closest' one to $T_{\varepsilon,m-1}$ with respect to the metric d_{m-1}.

There exists a constant $M > 0$ such that for every $h \in \{g \in G \mid g \cdot \varepsilon \in \langle a, b \rangle\} \cap B_G(k)$ we have

$$d_{m-1}(T_{\varepsilon,m-1}, h \cdot T_{\varepsilon,m-1}) \leq Mk.$$

Indeed, because $h \cdot \varepsilon \in \langle a, b \rangle$, it follows that $h \in H$. Next, since H quasi-isometrically embeds into G, from $|h|_G \leq k$ it follows that $|h|_H \leq C_1 k + C_2$ for some constants $C_1, C_2 \in \mathbb{N}$ which depend only on H and G. Now, d_{m-1} satisfies the triangle inequality, so if

$$L = \max_{h_i \in R} d_{m-1}(T_{\varepsilon,m-1}, h_i \cdot T_{\varepsilon,m-1})$$

then

$$d_{m-1}(T_{\varepsilon,m-1}, h \cdot T_{\varepsilon,m-1}) \leq L \cdot (C_1 k + C_2)$$

and existence of the required M follows.

Let

$$\mathcal{T}_k := \{S \in \{T_i\} \mid d_{m-1}(S, T_{\varepsilon,m-1}) \leq Mk\}.$$

Obviously, $\mathrm{card}(\mathcal{T}_k) \leq 2Mk + 1$ and for every $h \in \{g \in G \mid g \cdot \varepsilon \in \langle a, b \rangle\} \cap B_G(k)$ there exists $S \in \mathcal{T}_k$ such that $h \cdot \varepsilon \in \langle a, b \rangle \cap S = \langle a_S, b_S \rangle$, where $a_S, b_S \in \mathrm{Ends}(S)$. Thus, we have

$$\mathrm{card}\left(\{g \in G \mid g \cdot \varepsilon \in \langle a, b \rangle\} \cap B_G(k)\right)$$

$$\leq \sum_{S \in \mathcal{T}_k} \mathrm{card}\left(\{g \in G \mid g \cdot \varepsilon \in \langle a_S, b_S \rangle\} \cap B_G(k)\right)$$

$$\leq (2Mk + 1)(C(m-1)k^{m-1}) \leq C(T_{\varepsilon,m})k^m,$$

where $C(T_{\varepsilon,m})$ is a constant which depends on $T_{\varepsilon,m}$ and Z, but not on a and b.

(m.2) Assume that $T \neq T_{\varepsilon,m}$, that is, $\varepsilon \notin T$.

If

$$g_0 \in \{g \in G \mid g \cdot \varepsilon \in \langle a, b \rangle\} \cap B_G(k)$$

then using the same argument as in (1.2) we can show that

$$\mathrm{card}\left(\{g \in G \mid g \cdot \varepsilon \in \langle a, b \rangle\} \cap B_G(k)\right)$$

$$\leq \mathrm{card}\left(\{g \in G \mid g \cdot \varepsilon \in \langle g_0^{-1} \cdot a, g_0^{-1} \cdot b \rangle\} \cap B_G(2k)\right).$$

Then from (m.1) we obtain

$$\mathrm{card}\left(\{g \in G \mid g \cdot \varepsilon \in \langle g_0^{-1} \cdot a, g_0^{-1} \cdot b \rangle\} \cap B_G(2k)\right)$$
$$\leq C(T_{\varepsilon,m}) \cdot (2k)^m.$$

That is, $C(T) = 2^m C(T_{\varepsilon,m})$.

By setting $C(m) := 2^m C(T_{\varepsilon,m})$ we finish the induction step and **Claim 2** is proved. $\square$

References

[1] R. Alperin and H. Bass. Length functions of group actions on Λ-trees. In *Combinatorial group theory and topology, ed. S. M. Gersten and J. R. Stallings*, vol. 111 of *Annals of Math. Studies*, pp. 265–378. Princeton University Press, 1987.

[2] H. Bass. Group actions on non-Archimedean trees. In *Arboreal group theory*, vol. 19 of *MSRI Publications*, pp. 69–131, New York: Springer-Verlag, 1991.

[3] M. Bestvina and M. Feighn Stable actions of groups on real trees. *Invent. Math.*, 121(2):287–321, 1995.

[4] I. Bumagin and O. Kharlampovich. $\mathbb{Z}^n$-free groups are $CAT(0)$. *J. London Math. Soc.*, 88(3):761–78, 2013.

[5] D. I. Cartwright and P. M. Soardi. Convergence to ends for random walks on the automorphism group of a tree. *Proc. Amer. Math. Soc.*, 107:817–23, 1989.

[6] I. Chiswell. Abstract length functions in groups. *Math. Proc. Cambridge Philos. Soc.*, 80(3):451–63, 1976.

[7] I. Chiswell. *Introduction to Λ-trees.* Singapore: World Scientific, 2001.

[8] D. E. Cohen. *Combinatorial group theory: a topological approach*, vol. 14 of *London Mathematical Society Student Texts.* Cambridge University Press, 1989.

[9] F. Dahmani. Combination of convergence groups. *Geom. Topol.*, 7:933–63, 2003.

[10] H. Furstenberg. Boundary theory and stochastic processes on homogeneous spaces. In *Harmonic analysis on homogeneous spaces*, vol. 26 of *Proc. Sympos. Pure Math.*, pp. 193–229, Providence, RI: American Mathematical Society, 1973.

[11] D. Gaboriau, G. Levitt, and F. Paulin. Pseudogroups of isometries of $\mathbb{R}$ and Rips' Theorem on free actions on $\mathbb{R}$-trees. *Israel. J. Math.*, 87:403–28, 1994.

[12] V. Gerasimov. Floyd maps for relatively hyperbolic groups. *GAFA*, 22(5):1361–99, 2012.

[13] V. Gerasimov and L. Potyagailo. Quasi-isometric maps and Floyd boundaries of relatively hyperbolic groups. *J. Eur. Math. Soc.*, 15(6):2115–37, 2013.

[14] V. Guirardel. Limit groups and groups acting freely on $\mathbb{R}^n$-trees. *Geom. Topol.*, 8:1427–70, 2004.

[15] V. A. Kaimanovich. The Poisson formula for groups with hyperbolic properties. *Ann. of Math.*, 152:659–92, 2000.

[16] V. A. Kaimanovich and H. Masur. The Poisson boundary of the mapping class group. *Invent. Math.*, 125(2):221–64, 1996.

[17] A. Karlsson. Free subgroups of groups with non-trivial Floyd boundary. *Comm. Algebra*, 31:5361–76, 2003.

[18] A. Karlsson and G. Margulis. A Multiplicative Ergodic Theorem and Nonpositively Curved Spaces. *Commun. Math. Phys.*, 208:107–23, 1999.

[19] O. Kharlampovich, A. G. Myasnikov, V. N. Remeslennikov, and D. Serbin. Groups with free regular length functions in $\mathbb{Z}^n$. *Trans. Amer. Math. Soc.*, 364:2847–82, 2012.

[20] O. Kharlampovich, A. G. Myasnikov, and D. Serbin. Actions, length functions, and non-archemedian words. *Internat. J. Algebra Comput.*, 23(2):325–455, 2013.

[21] O. Kharlampovich, A. G. Myasnikov, and D. Serbin. Infinite words and universal free actions. *Groups, Complexity, Cryptology*, 6(1):55–69, 2014.

[22] O. Kharlampovich, A. G. Myasnikov, and D. Serbin. Regular completions of $\mathbb{Z}^n$-free groups. Sumbitted, Available at http://arxiv.org/abs/1208.4640, 2016.

[23] R. Lyndon. Length functions in groups. *Math. Scand.*, 12:209–34, 1963.

[24] A. V. Malyutin and A. M. Vershik. Boundaries of braid groups and the Markov–Ivanovsky normal form. *Izv. RAN. Ser. Mat.*, 72(6):105–32, 2008.

[25] J. Morgan and P. Shalen. Valuations, trees, and degenerations of hyperbolic structures, I. *Ann. of Math.*, 120:401–76, 1984.

[26] A. Nevo and M. Sageev. The Poisson boundary of CAT(0) cube complex groups. *Groups, Geometry, and Dynamics*, 7(3):653–95, 2013.

[27] J. Tits. A 'theorem of Lie-Kolchin' for trees. In *Contributions to algebra: a collection of papers dedicated to Ellis Kolchin*, pp. 377–88. New York: Academic Press: 1977.

14

BUILDINGS, GROUPS OF LIE TYPE AND RANDOM WALKS

JAMES PARKINSON*

School of Mathematics and Statistics, University of Sydney,
Carslaw Building, F07, NSW, 2006, Australia

To Professor Woess on the occasion of his sixtieth birthday

Abstract

In this chapter we survey the theory of random walks on buildings and associated groups of Lie type and Kac–Moody groups. We begin with an introduction to the theory of Coxeter systems and buildings, taking a largely combinatorial perspective. We then survey the theory of random walks on buildings, and show how this theory leads to limit theorems for random walks on the associated groups.

Contents

Introduction

Probability theory on real Lie groups is a classical area, with beautiful results obtained in the 1960s, 1970s, and 1980s (see, for example [5, 22, 23, 29, 48, 57, 58]). It is the purpose of this chapter to survey more recent results dealing with probability theory on groups of Lie type defined over other fields, and extensions of these results into the setting of Kac–Moody groups. A unifying feature of these works is the use of a combinatorial/geometric object called the *building* of the group, which in some ways plays a role analogous to the symmetric space of a real Lie group. In fact, the building becomes the main object of interest, and so

*Research partly supported under the Australian Research Council (ARC) discovery grant DP110103205.

this survey is really about random walks on buildings, with applications to random walks on the associated groups.

Buildings were invented by Jacques Tits in the 1950s in an attempt to give a uniform geometric interpretation of semi-simple Lie groups. He achieved this goal spectacularly by classifying the class of irreducible thick *spherical* buildings of rank at least 3, showing that this class of buildings is essentially equivalent to the class of simple linear algebraic groups of relative rank at least 3, simple classical linear groups, and certain related groups called groups of mixed type (see [49]). Since their invention the scope of building theory has expanded immensely, with *affine* buildings playing an important role in the study of Lie groups over p-adic fields, and *twin* buildings utilised extensively in the theory of Kac–Moody groups.

For the purpose of this introduction a *building* consists of a set Δ (whose elements are called *chambers*) and a way of measuring distance between chambers. This measurement is not simply a numerical distance, instead the distance between two chambers is an element of a *Coxeter group* W associated to the building. Thus, there is a 'W-valued distance function' $\delta : \Delta \times \Delta \to W$, satisfying various axioms making Δ into a kind of 'W-metric space' (see Definition 2.1 and Remark 2.3 for more details).

Buildings arise naturally in connection with groups originating in Lie theory, and the axioms satisfied by δ are in essence capturing the combinatorics of the 'Bruhat decomposition' in these groups. More precisely, if G is a group with a *Tits system* (B, N, W, S) then setting $\Delta = G/B$ and $\delta(gB, hB) = w$ if and only if $g^{-1}h \in BwB$ produces a building (see Section 2.2 for details). While this connection to group theory is the raison d'être for buildings, there are many buildings that are not associated in any nice way to groups (see, for example, [43]). This motivates the philosophy of treating the building, rather than the group, as the primary object of interest.

We will give an introduction to the theory of buildings in the first two sections of the chapter, with Section 1 devoted to the theory of Coxeter groups, a necessary prerequisite to the theory of buildings. Section 2 is devoted to the buildings themselves and to the related group theoretic notion of a Tits system in a group. We will focus on the classes of buildings on which random walks have been studied, including:

(1) The *spherical buildings*, where W is a finite reflection group. By Tits' classification [49] these buildings are closely related to groups of Lie type such as $SL_n(\mathbb{F})$ where $\mathbb{F}$ is a field.

(2) The *affine buildings*, where W is an affine reflection group. By the Tits–Weiss classification [52, 59] these buildings are closely related to groups of Lie type defined over fields with discrete valuation, such as $SL_n(\mathbb{Q}_p)$ or $SL_n(\mathbb{K}((t)))$. The simplest affine buildings are trees with no leaves, for example, homogeneous trees.

(3) The *Fuchsian buildings*, where W is generated by reflections in the hyperbolic disc $\mathbb{H}^2$. These buildings do not admit a classification (see Section 2.5), however some of them are related to certain 'Kac–Moody groups'.

In Section 3 we survey results on random walks on buildings and associated groups. There are various types of random walks that we will consider. A particularly neat class consists of the *isotropic random walks* on the chambers of a building. These are the random walks $(X_n)_{n \geq 0}$ on the set Δ of chambers such that the transition probabilities $p(x, y)$ depend only on the W-distance $\delta(x, y)$. These random walks arise naturally from bi-B-invariant probability measures on groups admitting Tits systems, and any limit theorems established for the random walks on the building imply limit theorems for these measures.

In Section 3.1 we outline the beautiful algebraic theory of isotropic random walks on Δ. Put briefly, the transition operator of an isotropic random walk is an element of an algebra called a *Hecke algebra*. These algebras have been extensively studied, (largely due to their connections with groups of Lie type and p-adic Lie groups), and their representation theory plays a key role in the theory of random walks on buildings.

In Section 3.2 we specialise to the case of finite spherical buildings. In this case the Hecke algebra is finite dimensional, and we give an overview of how the representation theory of this algebra can be applied to investigate isotropic random walks. In particular we provide tractable upper bounds for mixing times for isotropic walks on finite spherical buildings. The analysis follows, in spirit, the work of Diaconis and Ram [18] where the representation theory of finite dimensional Hecke algebras is applied to investigate the systematic scan Metropolis algorithm. It turns out that this theory is related to random walks on spherical buildings, and so Section 3.2 is really a translation of [18] into the language of buildings. Other works related to random walks on spherical buildings can be found in Brown [8, 9] and Brown and Diaconis [7].

Next we consider random walks on affine buildings. In this context it is also natural to consider random walks on the 'vertices' of the building (these walks arise from bi-K-invariant measures on p-adic Lie groups, where K is a maximal compact subgroup). We will survey results on these random walks, and random walks on associated groups, drawing from the works of Cartwright and Woess [14], Lindlbauer and Voit [33], Parkinson [38], Parkinson and Schapira [39], Parkinson and Woess [41], Schapira [46], Tolli [54] and Trojan [55]. These works include precise limit theorems for isotropic random walks on the vertices and chambers of affine buildings, as well as theorems for random walks on groups associated to these buildings. Hecke algebras again play a key role in the analysis. Homogeneous trees are the simplest examples of affine buildings (the 'rank 2 case'), and in this direction we mention the fundamental works of Cartwright, Kaimanovich and Woess [13], Lalley [32] and Sawyer [45]. The literature relating to probability theory and harmonic analysis on homogeneous trees is extensive, and here we will focus on the higher rank cases.

The study of random walks on non-spherical, non-affine buildings is very open territory. In Section 3.4 we survey recent results of Gilch, Müller and Parkinson [21] concerning isotropic random walks on Fuchsian buildings. In this context a law of large numbers and a central limit theorem are available, with interesting formulae for the speed and variance in terms of an underlying automatic structure related to the building.

We conclude our survey by listing some future directions in the theory, and providing some appendices. In the first appendix we carry through a 'by-hand' computation outlining the general theory of isotropic random walks on the vertices of affine buildings in the special case of $\widetilde{C}_2$ buildings. In this basic case we can minimise some of the heavy (although beautiful) machinery used for the general case, thus making the analysis more accessible. In the second appendix we outline the representation theory of rank 2 spherical Hecke algebras, and show how a precise knowledge of the representation theory allows for accurate mixing time estimates for random walks on generalised polygons (that is, rank 2 spherical buildings). As a byproduct we recover a proof of the celebrated Feit-Higman theorem (this approach is due to Kilmoyer and Solomon [31], and is, in turn, an adaptation of Feit and Higman's original proof from 1964 [19]).

On a personal note, it is an absolute pleasure to dedicate this chapter to my friend and collaborator Wolfgang Woess on the occasion of his 60th birthday. Wolfgang's tireless support of young mathematicians has been a true gift to the mathematical community, a gift from which I have greatly benefited.

1. Coxeter Systems

Coxeter systems form the backbone of the higher objects of buildings and groups of Lie type. In this section we recall some basic theory of Coxeter systems, focussing on examples and important classes. Standard references include [1, 4, 26].

1.1. Definitions

Definition 1.1. A *Coxeter system* (W, S) is a group W generated by a finite set S with relations

$$s^2 = 1 \quad \text{and} \quad (st)^{m_{st}} = 1 \quad \text{for all } s, t \in S \text{ with } s \neq t,$$

where $m_{st} = m_{ts} \in \mathbb{Z}_{\geq 2} \cup \{\infty\}$ for all $s \neq t$ (if $m_{st} = \infty$ then it is understood that there is no relation between s and t). We sometimes say that W is a *Coxeter group* when the generating set S is implied.

Let (W, S) be a Coxeter system. The *rank* of (W, S) is $|S|$. The *length* of $w \in W$ is

$$\ell(w) = \min\{n \geq 0 \mid w = s_1 \cdots s_n \text{ with } s_1, \ldots, s_n \in S\},$$

and an expression $w = s_1 \cdots s_n$ with n minimal (that is, $n = \ell(w)$) is called a *reduced expression* for w. It is useful to note that if $w \in W$ and $s \in S$ then $\ell(ws) = \ell(w) \pm 1$. In particular, $\ell(ws) = \ell(w)$ is not possible.

For each $I \subseteq S$ the *standard I-parabolic subgroup* of W is the subgroup $W_I = \langle \{s \mid s \in I\} \rangle$. Then (W_I, I) is a Coxeter system, and hence Coxeter systems 'contain' other Coxeter systems of lower rank. This fact is very important in the theory, facilitating inductive arguments on the rank of the group. This makes the rank 2 systems particularly important as the base case. The rank 2 Coxeter group $W = \langle s, t \mid s^2 = t^2 = (st)^m = 1 \rangle$ is just the dihedral group of order $2m$ (or the infinite dihedral group if $m = \infty$), and is denoted by $I_2(m)$.

The data required to define a Coxeter system is conveniently encoded in a graph $\Gamma(W, S)$ with labelled edges called the *Coxeter graph*. This graph has vertex set S, and vertices $s, t \in S$ are joined by an edge if and only if $m_{st} \geq 3$. If $m_{st} \geq 4$ then the corresponding edge is given the label m_{st} (thus edges with no label have $m_{st} = 3$, and if s and t are not joined by an edge then $m_{st} = 2$). A Coxeter system (W, S) is called *irreducible* if the Coxeter graph $\Gamma(W, S)$ is connected. Note that if $\Gamma(W, S)$ is not connected, and if $S = S_1 \cup \cdots \cup S_k$ is the decomposition into connected components, then W is the direct product of parabolic subgroups $W = W_{S_1} \times \cdots \times W_{S_k}$. Thus, irreducibility is a natural assumption to make in the theory of Coxeter systems.

1.2. The Coxeter Complex and Examples

The Coxeter complex of a Coxeter system is a natural simplicial complex on which the Coxeter group acts, and plays an important role in the general theory.

Recall that a *simplicial complex* with vertex set V is a collection Σ of finite subsets of V (called *simplices*) such that for every $v \in V$, the singleton $\{v\}$ is a simplex (called a *vertex*), and every subset of a simplex σ is a simplex (a *face* of σ). If σ is a simplex which is not a proper subset of any other simplex then σ is a *chamber* of Σ.

Let Σ can be simplicial complex, and let $\leq$ be the face relation (that is, $\sigma' \leq \sigma$ if and only if σ' is a face of σ). Then $(\Sigma, \leq)$ is a partially ordered set satisfying:

(P1) For each pair $\sigma, \sigma' \in \Sigma$ there exists a greatest lower bound $\sigma \cap \sigma'$.

(P2) For each $\sigma \in \Sigma$ the poset $\{\sigma' \mid \sigma' \leq \sigma\}$ is isomorphic to the poset of subsets of $\{1, 2, \ldots, r\}$ for some r.

On the other hand, *any* partially ordered set $(\Sigma, \leq)$ satisfying (P1) and (P2) can be identified with a simplicial complex Σ by taking the vertex to be the set V of all elements $v \in \Sigma$ such that $r = 1$ in (P2), and identifying each element $\sigma \in \Sigma$ with the simplex $\{v \in V \mid v \leq \sigma\}$.

Definition 1.2. Let (W, S) be a Coxeter system. The *Coxeter complex* $\Sigma(W, S)$ is the simplicial complex constructed as above from the poset of all cosets of the form wW_I with $w \in W$ and $I \subseteq S$, ordered by reverse inclusion (we emphasise the reverse inclusion here: $wW_I \leq vW_J$ if and only if $wW_I \supseteq vW_J$).

Explicitly, the vertex set of the simplicial complex $\Sigma(W, S)$ is

$$V = \{wW_{S\setminus\{s\}} \mid w \in W, s \in S\},$$

and $c_0 = \{W_{S\setminus\{s\}} \mid s \in S\}$ is a chamber. The set of all chambers is $\{wc_0 \mid w \in W\}$. We have that $wc_0 = vc_0$ if and only if $w = v$, and so the set of all chambers can be identified with W by $wc_0 \leftrightarrow w$. Each chamber has exactly $|S|$ vertices (namely, the chamber w has vertices $wW_{S\setminus\{s\}}$ for $s \in S$). The Coxeter complex comes equipped with a natural *type function* $\tau : \Sigma(W, S) \to 2^S$ given by $\tau(wW_I) = S\setminus I$. Thus, the vertex $wW_{S\setminus\{s\}}$ has type s (more accurately, type $\{s\}$), and each chamber has exactly one vertex of each type.

Example 1.3. Let (W, S) be the dihedral group of order 6 with $S = \{s, t\}$. Write $W_s = W_{\{s\}} = \{1, s\}$, and similarly for W_t. The Coxeter complex $\Sigma(W, S)$ has six vertices, marked in the diagram below by $\bullet$ (vertices of type s) and $\circ$ (vertices of type t). Each chamber has $|S| = 2$ vertices, and thus chambers are represented as edges in the diagram. Similarly, the Coxeter complex of a dihedral group of order $2m$ is a $2m$-gon, and the Coxeter complex of the infinite dihedral group is a two sided infinite path with alternating vertex types.

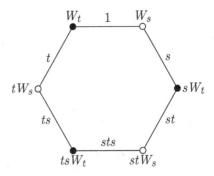

A Coxeter system is called:

(1) *spherical* if $|W| < \infty$,
(2) *affine* if W is infinite and contains a normal abelian subgroup Q such that W/Q is finite,
(3) *Fuchsian* if W is generated by the reflections in the sides of a polygon in $\mathbb{H}^2$.

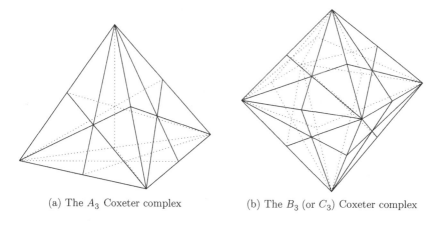

(a) The A_3 Coxeter complex   (b) The B_3 (or C_3) Coxeter complex

Figure 14.1 Examples of rank 3 spherical Coxeter systems

If (W, S) is an irreducible spherical Coxeter system then W can be realised as a group generated by linear reflections in $E = \mathbb{R}^{|S|}$. The action of W decomposes E into $|W|$ geometric cones based at the origin, and by intersecting these cones with the unit sphere we can visualise the Coxeter complex $\Sigma(W, S)$ as a tessellation of the $(|S| - 1)$-sphere (some examples are illustrated in Figure 14.1). There is a well-known classification of the irreducible spherical Coxeter systems due to Coxeter [15]. The nomenclature of this classification has its origins in the Cartan-Killing classification of simple Lie algebras over $\mathbb{C}$. The list of spherical Coxeter systems is as follows (in each case the subscript denotes the rank of the system, see Figure 14.10 in Appendix C for the Coxeter graphs).

(1) *Crystallographic* systems: A_n ($n \geq 1$), $B_n = C_n$ ($n \geq 2$), D_n ($n \geq 4$), E_6, E_7, E_8, F_4, G_2.
(2) *Non-crystallographic* systems: H_3, H_4, $I_2(m)$ (with $m = 5$ or $m \geq 7$).

If (W, S) is an irreducible affine Coxeter system then W can be realised as a group generated by affine reflections in $E = \mathbb{R}^{|S|-1}$ (see Section 1.3). The action of W decomposes E into geometric simplices, and so the Coxeter complex $\Sigma(W, S)$ may be visualied as a tessellation of $\mathbb{R}^{|S|-1}$ (some examples are illustrated in Figure 14.2, the additional information in the figure will be explained in Section 1.3). The classification of irreducible affine Coxeter systems is closely related to the classification of irreducible spherical Coxeter systems. Specifically, to each irreducible crystallographic spherical Coxeter system (of type X_n, say) there is an associated affine Coxeter system of type $\widetilde{X}_n$ obtained by adding one additional generator to the spherical system. In the case of spherical systems of type $B_n = C_n$ there are two associated affine

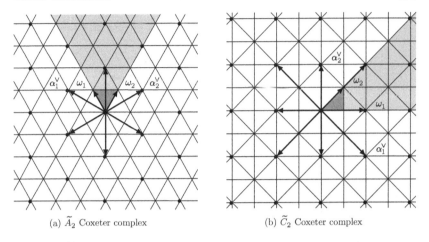

(a) $\widetilde{A}_2$ Coxeter complex　　　　　　　(b) $\widetilde{C}_2$ Coxeter complex

Figure 14.2 Rank 3 affine Coxeter systems, and associated (dual) root systems

systems, called $\widetilde{B}_n$ and $\widetilde{C}_n$, and these are non-isomorphic if $n > 2$. See Figure 14.11 in Appendix C for the Coxeter graphs of the irreducible affine Coxeter systems.

If (W, S) is a Fuchsian Coxeter system then W may be realised as a group generated by reflections in the sides of a polygon in the hyperbolic disc $\mathbb{H}^2$, and thus W is a cocompact discrete subgroup of $PGL_2(\mathbb{R})$ (hence the term 'Fuchsian'). Specifically, let $n \geq 3$ be an integer, and let $k_1, \ldots, k_n \geq 2$ be integers satisfying

$$\sum_{i=1}^{n} \frac{1}{k_i} < n - 2. \tag{14.1}$$

Assign the angles π/k_i to the vertices of a combinatorial n-gon F. There is a convex realisation of F (which we also call F) in the hyperbolic disc $\mathbb{H}^2$. Let W be the subgroup of $PGL_2(\mathbb{R})$ generated by the set S of reflections in the sides of F. Then (W, S) is a Coxeter system (see [16, Example 6.5.3]), and if $s_1, \ldots, s_n$ are the reflections in the sides of F (arranged cyclically), then the order of $s_i s_j$ is

$$m_{ij} = \begin{cases} k_i & \text{if } j = i + 1 \\ \infty & \text{if } |i - j| > 1, \end{cases} \tag{14.2}$$

where the indices are read cyclically. We denote this Coxeter system by $F(k_1, \ldots, k_n)$. The group W acts on $\mathbb{H}^2$ with fundamental domain F, and thus induces a tessellation of $\mathbb{H}^2$ by isometric polygons wF, $w \in W$. Examples are shown in Figure 14.3. We note that this is *not* a depiction of the Coxeter complex of these groups unless $|S| = 3$. For example, each chamber of the Coxeter complex of the group represented in

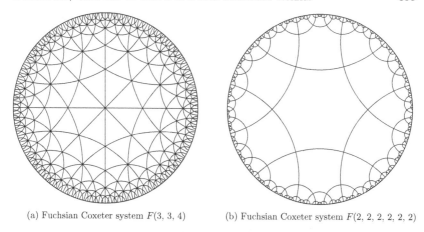

(a) Fuchsian Coxeter system $F(3, 3, 4)$ (b) Fuchsian Coxeter system $F(2, 2, 2, 2, 2, 2)$

Figure 14.3 Fuchsian Coxeter systems

Figure 14.3(b) is a $|S| - 1 = 5$ dimensional simplex. Instead the pictures are (essentially) the *Davis complex* of the group (see [1, Example 12.43]). We will not go into further details, however we simply remark that this is a much more convenient way to visualise Fuchsian Coxeter systems (and their buildings).

1.3. The Coxeter Complex of an Affine Coxeter System

For our later discussions it is necessary to have a concrete description of the Coxeter complex of an affine Coxeter system in terms of *root systems*. The standard reference for this theory is [4]. Let E be a d-dimensional real vector space with inner product $\langle \cdot, \cdot \rangle$. The hyperplane orthogonal to the vector $\alpha \in E \backslash \{0\}$ is $H_\alpha = \{x \in E \mid \langle x, \alpha \rangle = 0\}$, and the *reflection* in H_α is given by $s_\alpha(x) = x - \langle x, \alpha \rangle \alpha^\vee$ where $\alpha^\vee = 2\alpha/\langle \alpha, \alpha \rangle$.

A *root system* in E is a finite set R of non-zero vectors (called *roots*) such that: (1) R spans E, (2) if $\alpha \in R$ and $k\alpha \in R$ then $k = \pm 1$, (3) if $\alpha, \beta \in R$ then $s_\alpha(\beta) \in R$, and (4) if $\alpha, \beta \in R$ then $\langle \alpha, \beta^\vee \rangle \in \mathbb{Z}$. The *rank* of R is $d = \dim(E)$. Figure 14.4 illustrates three rank 2 root systems.

Let R be a rank d root system. There exists a subset $\{\alpha_1, \ldots, \alpha_d\} \subseteq R$ of *simple roots* with the property that every $\alpha \in R$ can be written as a linear combination of $\alpha_1, \ldots, \alpha_d$ with integer coefficients which are either all nonpositive, or all nonnegative. Those roots whose coefficients are all nonnegative are called *positive roots* (with respect to the fixed chosen set of simple roots), and the set of all positive roots is denoted R^+. Then $R = R^+ \cup (-R^+)$.

The *Weyl group* of R is the finite subgroup W_0 of $GL(E)$ generated by the reflections s_α with $\alpha \in R$. For each $i = 1, \ldots, d$ write $s_i = s_{\alpha_i}$, and let $S_0 = \{s_1, \ldots, s_d\}$. Then (W_0, S_0) is a spherical Coxeter system,

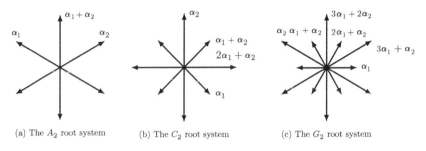

(a) The A_2 root system   (b) The C_2 root system   (c) The G_2 root system

Figure 14.4 The irreducible rank 2 root systems

with the order of $s_i s_j$ being m_{ij}, where $\pi - \pi/m_{ij}$ is the angle between α_i and α_j. The Coxeter system (W_0, S_0) is irreducible if and only if the root system R is irreducible (where the latter means that there is no partition $R = R_1 \cup R_2$ with R_1 and R_2 nonempty such that $\langle \alpha, \beta \rangle = 0$ for all $\alpha \in R_1$ and all $\beta \in R_2$).

The irreducible root systems admit a complete classification, and explicit descriptions of each system can be found in [4, Plates I–IV]. They fall into four infinite families A_d $(d \geq 1)$, B_d $(d \geq 2)$, C_d $(d \geq 2)$ and D_d $(d \geq 4)$, and 5 exceptional types E_6, E_7, E_8, F_4 and G_2. If R is an irreducible root system of type X_d then the Coxeter system (W_0, S_0) is also of type X_d, and hence every irreducible crystallographic spherical Coxeter group can be realised as the Weyl group of an irreducible root system.

For each $\alpha \in R$ and each $k \in \mathbb{Z}$ let $H_{\alpha,k} = \{x \in E \mid \langle x, \alpha \rangle = k\}$. Thus, the affine hyperplane $H_{\alpha,k}$ is a translate of the linear hyperplane $H_\alpha = H_{\alpha,0}$. The orthogonal (affine) reflection in the hyperplane $H_{\alpha,k}$ is given by the formula $s_{\alpha,k}(x) = x - (\langle x, \alpha \rangle - k)\alpha^\vee$ for $x \in E$, and the *affine Weyl group* of R is the subgroup W_{aff} of $\text{Aff}(E)$ generated by all reflections $s_{\alpha,k}$ with $\alpha \in R$ and $k \in \mathbb{Z}$. The root system R has a unique *highest root* φ (the *height* of the root $\alpha = a_1\alpha_1 + \cdots + a_d\alpha_d$ is $a_1 + \cdots + a_d$, and φ is the unique root of greatest height). Let $s_0 = s_{\varphi,1}$. Then W_{aff} is generated by $S_{\text{aff}} = \{s_0, s_1, \ldots, s_d\}$ and the order of $s_0 s_j$ is m_{0j}, where $\pi - \pi/m_{0j}$ is the angle between $-\varphi$ and α_j. Thus, $(W_{\text{aff}}, S_{\text{aff}})$ is a Coxeter system, and (W_0, S_0) is a parabolic subsystem of $(W_{\text{aff}}, S_{\text{aff}})$. For $\lambda \in E$ let $t_\lambda \in \text{Aff}(E)$ be the translation $t_\lambda(x) = x + \lambda$. Since $s_{\alpha,k} = t_{k\alpha^\vee} s_\alpha$ we have

$$W_{\text{aff}} = Q \rtimes W_0, \quad \text{where } Q = \mathbb{Z}\alpha_1^\vee + \cdots + \mathbb{Z}\alpha_d^\vee \text{ is the } coroot\ lattice.$$

Thus, $(W_{\text{aff}}, S_{\text{aff}})$ is an affine Coxeter system. All irreducible affine Coxeter systems arise in this way.

Let $\omega_1, \ldots, \omega_d \in E$ be the dual basis to $\alpha_1, \ldots, \alpha_d$, given by $\langle \omega_i, \alpha_j \rangle = \delta_{i,j}$. The *coweight lattice* P of R is

$$P = \{\lambda \in E \mid \langle \lambda, \alpha \rangle \in \mathbb{Z} \text{ for all } \alpha \in R\} = \mathbb{Z}\omega_1 + \cdots + \mathbb{Z}\omega_d.$$

Note that $Q \subseteq P$, and that Q and P are both W_{aff}-invariant lattices. The set of *dominant coweights* is

$$P^+ = \mathbb{N}\omega_1 + \cdots + \mathbb{N}\omega_d.$$

The family of hyperplanes $H_{\alpha,k}$, $\alpha \in R$, $k \in \mathbb{Z}$, tessellates E into d-dimensional geometric simplices (the *chambers*). The extreme points of the chambers are *vertices*, and each chamber has exactly $d+1$ vertices. The resulting simplicial complex is isomorphic to the Coxeter complex $\Sigma(W_{\mathrm{aff}}, S_{\mathrm{aff}})$. The affine Weyl group W_{aff} acts simply transitively on the set of all chambers. The *fundamental chamber* is

$$\mathfrak{c}_0 = \{x \in E \mid \langle x, \alpha_i \rangle \geq 0 \text{ for } 1 \leq i \leq d, \text{ and } \langle x, \varphi \rangle \leq 1\}$$

and we often identify W_{aff} with the set of chambers by $w \leftrightarrow w\mathfrak{c}_0$. The set P of coweights is a subset of the set of all vertices of $\Sigma(W_{\mathrm{aff}}, S_{\mathrm{aff}})$, called the *special vertices*. The set Q is the set of type 0 vertices, and every chamber has exactly one type 0 vertex.

The action of W_0 decomposes E into $|W_0|$ geometric cones based at the origin, and the translates of these cones by elements of P are called *sectors*. The *fundamental sector* is

$$\mathfrak{s}_0 = \{x \in E \mid \langle x, \alpha_i \rangle \geq 0 \text{ for all } 1 \leq i \leq d\}.$$

We also write $E^+ = \mathfrak{s}_0$ (roughly speaking, we write $\mathfrak{s}_0$ when we are interested in the simplicial structure, and E^+ when we are interested in the metric structure).

Examples of the above construction of affine Coxeter systems are illustrated in Figure 14.2, where the sector $\mathfrak{s}_0$ and the chamber $\mathfrak{c}_0$ are shaded (light and dark, respectively). The coroot lattice Q is indicted with heavy dots. In Figure 14.2(a) all vertices are special, while in Figure 14.2(b) only the vertices of valency 8 are special.

2. Buildings

Buildings were introduced by Jacques Tits in the 1950s. 'The origin of the notions of buildings and BN-pairs lies in an attempt to give a systematic procedure for geometric interpretation of the semi-simple Lie groups and, in particular, the exceptional groups' [49, Introduction]. Over the past 60 years the theory has grown immensely, and has had diverse applications in geometry, group theory, representation theory, and geometric group theory. In this section we give a brief introduction to the theory, with our main references being [1, 42, 49].

2.1. Definitions and Basic Properties

Buildings are defined axiomatically, and historically there have been two main approaches to the theory, both due to Jacques Tits. The initial approach was via simplicial complexes, and later an approach was developed using 'chamber systems' (see [51] for an enlightening historical discussion). Both approaches are relevant and useful, however here we have chosen to adopt Tits' original simplicial complex definition (see Remark 2.3 for the other approach).

Definition 2.1. Let (W, S) be a Coxeter system with Coxeter complex $\Sigma(W, S)$. A *building of type* (W, S) is a nonempty simplicial complex Σ with a family $\mathbf{A}$ of subcomplexes (called *apartments*) such that

(B1) each apartment $A \in \mathbf{A}$ is isomorphic to the Coxeter complex $\Sigma(W, S)$,

(B2) given any two simplices of Σ there is an apartment $A \in \mathbf{A}$ containing both of them, and

(B3) if $A, A' \in \mathbf{A}$ are apartments containing a common chamber then there is a unique simplicial complex isomorphism $\psi : A' \to A$ fixing each simplex of the intersection $A \cap A'$.

Let Σ be a building of type (W, S). The *rank* of Σ is $|S|$. Fix, once and for all, an apartment of Σ and identify it with $\Sigma(W, S)$. Thus, we regard $\Sigma(W, S)$ as an apartment of Σ, the 'standard' (or 'base') apartment. The type function on the Coxeter complex $\Sigma(W, S)$ extends uniquely to a type function $\tau : \Sigma \to 2^S$ on the building making Σ into a labelled simplicial complex. The isomorphism in (B3) is then necessarily type preserving. Let Δ be the set of all chambers of Σ.

Example 2.2. Buildings of type $\widetilde{A}_1 = I_2(\infty)$ are equivalent to trees in which every vertex has valency at least 2. The apartments are two sided infinite geodesics in the tree, and the chambers are the edges of the tree. There are two types of vertices, indicated by • and ○ in the picture. Buildings of higher rank are considerably more sophisticated objects, although the tree example is very instructive.

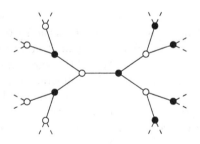

The *dimension* of $\sigma \in \Sigma$ is $|\sigma| - 1$, and the *codimension* of $\sigma \in \Sigma$ is $|S| - |\sigma|$. Each chamber of Σ has dimension $|S| - 1$ (that is, has $|S|$ vertices). A *panel* of Σ is a codimension 1 simplex. Chambers $x, y \in \Delta$ are *s-adjacent* (written $x \sim_s y$) if and only if they share a panel of type $S\backslash\{s\}$ (that is, if either $x = y$ or $x \cap y$ is a panel of type $S\backslash\{s\}$, or equivalently, if either $x = y$ or $x\backslash y$ is a vertex of type s).

For example, in a rank 3 building the chambers are triangles with the three edges (panels) of the triangle corresponding to the three types of adjacency. Chambers are 'glued together' along their s-edges if and only if they are s-adjacent. In a rank d building the chambers are $(d-1)$-simplices, and are glued together along their panels (codimension 1 faces). An alternate way to visualise a higher rank building is to imagine each chamber as a d-gon, with the sides in bijection with S, and these polygons are glued together along their edges according adjacency. This is particularly useful when (W, S) is Fuchsian.

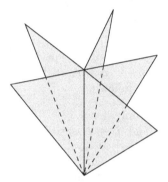

A *gallery of type* $(s_1, \ldots, s_n) \in S^n$ joining $x \in \Delta$ to $y \in \Delta$ is a sequence $x_0, x_1, \ldots, x_n \in \Delta$ of chambers such that

$$x = x_0 \sim_{s_1} x_1 \sim_{s_2} \cdots \sim_{s_n} x_n = y \quad \text{with } x_{j-1} \neq x_j \text{ for all } 1 \leq j \leq n.$$

This gallery has *length* n.

The W-*distance function* $\delta : \Delta \times \Delta \to W$ on Δ is defined as follows: If $x, y \in \Delta$ and if there is a minimal length gallery of type $(s_1, \ldots, s_n)$ from x to y, then let

$$\delta(x, y) = s_1 s_2 \cdots s_n.$$

This does not depend on the particular minimal length gallery chosen.

A building is called *thick* if $|\{y \in \Delta \mid x \sim_s y\}| \geq 2$ for all chambers $x \in \Delta$ and all $s \in S$, and *thin* if $|\{y \in \Delta \mid x \sim_s y\}| = 1$ for all chambers $x \in \Delta$ and all $s \in S$. It is clear that the Coxeter complex $\Sigma(W, S)$ is a thin building of type (W, S), and that all thin buildings are Coxeter complexes. Typically we are interested in thick buildings.

A building is *regular* if for each $s \in S$ the cardinality

$$q_s = |\{y \in \Delta \mid x \sim_s y\}| \quad \text{is finite and does not depend on } x \in \Delta.$$

All locally finite thick buildings whose Coxeter group (W, S) has $m_{st} < \infty$ for all $s, t \in S$ are necessarily regular (see [36, Theorem 2.4]). Here *locally finite* means that $|\{y \in \Delta \mid x \sim_s y\}| < \infty$ for all $x \in \Delta$ and

$s \in S$. The numbers $(q_s)_{s \in S}$ are called the *thickness parameters* (or just the *parameters*) of the (regular) building.

For each $x \in \Delta$ and each $w \in W$ let

$$\Delta_w(x) = \{y \in \Delta \mid \delta(x,y) = w\} \quad \text{be the } \textit{sphere of radius } w \textit{ centred at } x.$$

If (Δ, δ) is regular, then by [36, Proposition 2.1] the cardinality $q_w = |\Delta_w(x)|$ does not depend on $x \in \Delta$, and is given by

$$q_w = q_{s_1} \cdots q_{s_k} \quad \text{whenever } w = s_1 \cdots s_k \text{ is a reduced expression.}$$

The adjectives 'spherical', 'affine' and 'Fuchsian' from Coxeter systems carry over to buildings. Thus, the apartments of spherical, affine, or Fuchsian buildings are tessellations of a sphere, Euclidean space, or hyperbolic disc, respectively.

Remark 2.3. Let Σ be a building of type (W, S) with chamber set Δ and Weyl distance function $\delta : \Delta \times \Delta \to W$. It is not hard to see that the pair (Δ, δ) satisfies the following:

(B1)' $\delta(x,y) = 1$ if and only if $x = y$.
(B2)' If $\delta(x,y) = w$ and $z \in \Delta$ satisfies $\delta(y,z) = s$ with $s \in S$, then $\delta(x,z) \in \{w, ws\}$. If, in addition, $\ell(ws) = \ell(w) + 1$, then $\delta(x,z) = ws$.
(B3)' If $\delta(x,y) = w$ and $s \in S$, then there is a chamber $z \in \Delta$ with $\delta(y,z) = s$ and $\delta(x,z) = ws$.

Conversely, suppose that we are given a set Δ and a function $\delta : \Delta \times \Delta \to W$ satisfying (B1)', (B2)', and (B3)'. For each $I \subseteq S$ and each $x \in \Delta$ let $R_I(x) = \{y \in \Delta \mid \delta(x,y) \in W_I\}$. The poset $(\Sigma, \leq)$ of all sets of the form $R_I(x)$ with $I \subseteq S$ and $x \in \Delta$ (ordered by reverse inclusion) satisfies conditions (P1) and (P2) from Section 1.2, and hence we may regard Σ as a simplicial complex. It turns out that this simplicial complex is a building of type (W, S) (the most challenging thing to check is the existence of apartments). This gives a second approach to buildings: Specifically one can take a building of type (W, S) to be a pair (Δ, δ) where Δ is a set and $\delta : \Delta \times \Delta \to W$ is a function satisfying (B1)', (B2)' and (B3)'. See [1, 51] for further details. A certain fluency in both approaches is useful when working with buildings.

2.2. Buildings and Groups

The group theoretic counterpart to a building is the notion of a *Tits system* in a group. This concept has been very influential in group theory due to the existence of Tits systems in many 'Lie theoretic' groups, facilitating a uniform treatment of these groups. We will see that every Tits system gives rise to a building, however not every building results from a Tits system.

Definition 2.4. A *Tits system* in a group G is a quadruple (B, N, W, S) where B and N are subgroups of G, and (W, S) is a Coxeter system, and the following axioms are satisfied:

(T1) The group G is generated by $B \cup N$.

(T3) The group $H = B \cap N$ is a normal subgroup of N, and $N/H \cong W$.

(T3) If $n_s \in N$ maps to $s \in S$ under the natural homomorphism of N onto W then for all $n \in N$ we have $Bn Bn_s B \subseteq BnB \cup Bnn_s B$.

(T4) With n_s as above, $n_s B n_s^{-1} \neq B$ for all $s \in S$.

Since H is a subgroup of B there is no harm in writing wB in place of nB whenever $n \in N$ maps to $w \in W$ under the homomorphism of N onto W, and we will do so throughout.

The axioms of a Tits system may appear as foreign to the reader as the axioms of a building! Thus, we pause to mention some important classes of groups that admit Tits systems. To begin with, the Chevalley groups and twisted Chevalley groups admit natural Tits systems, with the associated Coxeter systems being of spherical type. For excellent treatments of this theory, see [11, 47]. For readers familiar with Chevalley groups, the Tits system is as follows. Let R be an irreducible root system, and let $G(\mathbb{F})$ be the Chevalley group of type R over the field $\mathbb{F}$. Recall that $G(\mathbb{F})$ is generated by elements $x_\alpha(t)$ with $\alpha \in R$ and $t \in \mathbb{F}$. Let

$$n_\alpha(t) = x_\alpha(t) x_{-\alpha}(-t^{-1}) x_\alpha(t) \quad \text{and} \quad h_{\alpha^\vee}(t) = n_\alpha(t) n_\alpha(-1)$$

for $\alpha \in R$ and $t \in \mathbb{F}^\times$. Let N (respectively H) be the subgroup of $G(\mathbb{F})$ generated by the elements $n_\alpha(t)$ (respectively $h_{\alpha^\vee}(t)$) with $\alpha \in R$ and $t \in \mathbb{F}^\times$. Let U be the subgroup of $G(\mathbb{F})$ generated by the elements $x_\alpha(t)$ with $\alpha \in R^+$ and $t \in \mathbb{F}$, and let $B = \langle U, H \rangle$. Then (B, N, W_0, S_0) is a Tits system in $G(\mathbb{F})$.

When the field $\mathbb{F}$ has a discrete valuation Iwahori and Matsumoto [27] discovered that the Chevalley group $G(\mathbb{F})$ admits another Tits system, this time with Coxeter group being the affine Weyl group W_{aff}. Examples of fields with discrete valuation include the p-adic numbers $\mathbb{Q}_p$, and the field of Laurent series $\mathbb{K}((t))$ with $\mathbb{K}$ any field. For concreteness, suppose that $\mathbb{F} = \mathbb{K}((t))$. Let $K = G(\mathbb{K}[[t]])$ be the Chevalley group defined over the ring of power series with coefficients in $\mathbb{K}$ (the *valuation ring* of $\mathbb{F}$). The evaluation map $\theta : \mathbb{K}[[t]] \to \mathbb{K}$, $t \mapsto 0$, induces a group homomorphism $\theta : K \to G(\mathbb{K})$. Let $N = N(\mathbb{F})$ and $B = B(\mathbb{K})$ be the groups from the previous paragraph (for the groups $G(\mathbb{F})$ and $G(\mathbb{K})$ respectively). The *Iwahori subgroup* of $G(\mathbb{F})$ is the inverse image of B under θ. That is, $I = \theta^{-1}(B)$. Then $(I, N, W_{\text{aff}}, S_{\text{aff}})$ is a Tits system in $G(\mathbb{F})$. See [10, 27] for details.

There is a vast generalisation of the notation of a Chevalley group. Recall that Chevalley groups are constructed as automorphism groups of finite dimensional Lie algebras associated to Cartan matrices. In a similar (although highly non-trivial) way there is a construction of groups using infinite dimensional Lie algebras associated to generalised Cartan matrices (so called *Kac–Moody algebras*, see [28]). The associated *Kac–Moody groups* admit Tits systems with more general Coxeter systems (see [53]). In fact every Coxeter system with $m_{st} \in \{2, 3, 4, 6, \infty\}$ arises as the Coxeter group of a Tits system in a Kac–Moody group.

We will now describe the connection between Tits systems and buildings. Let Σ be a building with system of apartments $\mathbf{A}$. Suppose that G is a group acting on Σ by type preserving simplicial complex automorphisms, and that G preserves the apartment system $\mathbf{A}$. We say that G acts *strongly transitively relative to* $\mathbf{A}$ if it is transitive on pairs (A, x) with A an apartment in $\mathbf{A}$ and x a chamber of A. For the statement of the following theorem it is convenient to adopt the approach to buildings from Remark 2.3.

Theorem 2.5. (1) *Let* (B, N, W, S) *be a Tits system in a group* G. *Let* $\Delta = G/B$, *and define* $\delta : \Delta \times \Delta \to W$ *by*

$$\delta(gB, hB) = w \quad \text{if and only if} \quad g^{-1}h \in BwB.$$

Then (Δ, δ) *is a thick building of type* (W, S). *The set* $A = \{wB \mid w \in W\}$ *is an apartment,* $\mathbf{A} = \{gA \mid g \in G\}$ *is a system of apartments, and* G *acts strongly transitively with respect to* $\mathbf{A}$.

(2) *Let* (Δ, δ) *be a thick building of type* (W, S) *and suppose that a group* G *acts strongly transitively with respect to a* G-*invariant apartment system* $\mathbf{A}$. *Let* $o \in \Delta$ *be a chamber, and let* $A \in \mathbf{A}$ *be an apartment containing* o. *Let*

$$B = \{g \in G \mid go = o\} \quad and \quad N = \{g \in G \mid gx \in A \text{ for all } x \in A\}.$$

Then (B, N, W, S) *is a Tits system in* G.

Thus, we have a wealth of examples of thick buildings. In particular, given that every Coxeter system (W, S) with $m_{st} \in \{2, 3, 4, 6, \infty\}$ can occur as the Coxeter system of a Kac–Moody group, there are thick buildings of type (W, S) for every such (W, S). However, we should emphasise that not all buildings arise from this type of construction (see, for example, Theorem 2.14 below).

2.3. Affine Buildings

In this section we discuss some additional structural theory for affine buildings. Let Σ be an affine building of type $(W_{\mathrm{aff}}, S_{\mathrm{aff}})$. The special vertices of the Coxeter complex of $(W_{\mathrm{aff}}, S_{\mathrm{aff}})$ are the elements of the coweight lattice P, and the *special vertices* of Σ are the vertices that are special vertices in some apartment. Let V be the set of all special vertices of Σ. A *sector* in Σ is a subset $\mathfrak{s}$ which is isomorphic to

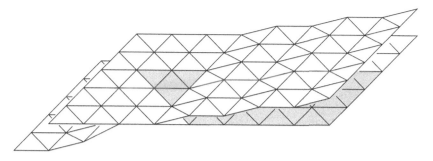

Figure 14.5 A small piece of an $\widetilde{A}_2$ building

a sector in some apartment of the building (where sectors in apartments are as in Section 1.3). Thus, sectors are always based at special vertices. A fundamental fact concerning sectors in affine buildings is (see [1, Chapter 11]):

(S1) If x is a chamber of Σ, and if $\mathfrak{s}$ is a sector of Σ, then there is a subsector $\mathfrak{s}'$ of $\mathfrak{s}$ such that $\mathfrak{s}' \cup x$ is contained in an apartment.

Figure 14.5 shows a simplified picture of an affine building of type $\widetilde{A}_2$; however, note that if the building is thick then the 'branching' actually occurs along *every* wall, and so the picture is rather incomplete. All vertices in this building are special, and a sector is shaded.

The following notion of 'vector distance' gives a refined way of measuring the distance between special vertices in affine buildings.

Definition 2.6. Let $x, y \in V$ be special vertices of Σ. The *vector distance* $\mathbf{d}(x, y) \in P^+$ from x to y is defined as follows. By (B2) there is an apartment A containing x and y, and let $\psi : A \to \Sigma$ be a type preserving isomorphism. Then we define

$$\mathbf{d}(x, y) = \big(\psi(y) - \psi(x)\big)^+,$$

where for $\mu \in P$, we denote by μ^+ the unique element in $W_0\mu \cap P^+$. This value is independent of choice of apartment A and the isomorphism $\psi : A \to \Sigma$ (see [36, Proposition 5.6]).

More intuitively, to compute $\mathbf{d}(x, y)$ one looks at the vector from x to y (in any apartment containing x and y) and takes the dominant representative of this vector under the W_0-action.

If A is an apartment and $\mathfrak{s} \subset A$ is a sector of A, then the *retraction of Σ onto A with centre $\mathfrak{s}$* is the map $\rho_{A,\mathfrak{s}} : \Sigma \to A$ computed as follows: If $x \in \Sigma$, choose (by (S1)) an apartment A' containing x and a subsector of $\mathfrak{s}$. Let $\psi : A' \to A$ be the isomorphism from (B3), and let $\rho_{A,\mathfrak{s}}(x) = \psi(x)$. It is easy to check that this value is independent of the apartment A' chosen. Intuitively speaking, the retraction $\rho_{A,\mathfrak{s}}$ flattens

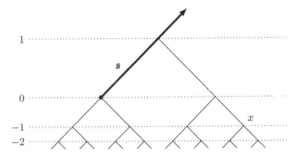

Figure 14.6 Vector Buesmann function for a tree

the building onto the apartment A with the centre of this flattening being 'deep' in the sector $\mathfrak{s}$ (this is illustrated in Figure 14.6 for the rank 2 case of trees).

If $\mathfrak{s}$ is a sector in an apartment A, then there is a unique isomorphism $\psi_{A,\mathfrak{s}} : A \to \Sigma(W_{\text{aff}}, S_{\text{aff}})$ mapping $\mathfrak{s}$ to $\mathfrak{s}_0$ and preserving vector distances (cf. [37, Lemma 3.2]). This gives a canonical way of fixing a Euclidean coordinate system on the apartment A with respect to the sector $\mathfrak{s}$.

Definition 2.7. The *vector Busemann function* associated to the sector $\mathfrak{s}$ is the function

$$h_{\mathfrak{s}} : \Sigma \to P \quad \text{given by} \quad h_{\mathfrak{s}}(x) = \psi_{A,\mathfrak{s}}(\rho_{A,\mathfrak{s}}(x)),$$

where A is any apartment containing $\mathfrak{s}$ (this does not depend on A). We write $h = h_{\mathfrak{s}_0}$.

Example 2.8. Let Σ be a homogeneous tree of degree $q+1$. Thus, Σ is an affine building of type $\tilde{A}_1$ (see Example 2.2). In this case the vector distance $d(x,y)$ is simply the graph distance $d(x,y)$, and the vector Busemann function $h_{\mathfrak{s}} : \Sigma \to P$ is the familiar 'horocycle function' (perhaps with an additional superficial minus sign; see [3, Figure 14.1]). This is illustrated in Figure 14.6 – to compute $h_{\mathfrak{s}}(v)$ for a vertex v one reads off the 'level' of v. For example, $h_{\mathfrak{s}}(x) = -1$. Note that there are infinitely many vertices on each horizontal level. More generally, the vector Busemann functions for an affine building are vector analogues of the usual Busemann functions for a CAT(0) space (see [41, Proposition 2.8]).

Example 2.9. Let R be an irreducible root system, let $G = G(\mathbb{K}((t)))$ be a Chevalley group over the field $\mathbb{K}((t))$, and let $K = G(\mathbb{K}[[t]])$. Following the work of Iwahori and Matsumoto [27] and Bruhat and Tits [10], G/K is the set of type zero vertices of an affine building (thus is a subset of the set of special vertices). The *Cartan* and *Iwasawa* decompositions of G are (respectively):

$$G = \bigsqcup_{\lambda \in Q \cap P^+} K t_\lambda K \qquad \text{and} \qquad G = \bigsqcup_{\mu \in Q} U t_\mu K, \qquad (14.3)$$

(with U being the subgroup of G generated by the elements $x_\alpha(f)$ with $\alpha \in R^+$ and $f \in \mathbb{K}((t))$), and with t_λ given by $t_\lambda = h_{\alpha_1^\vee}(t^{-a_1}) \cdots h_{\alpha_d^\vee}(t^{-a_d})$ if $\lambda = a_1 \alpha_1^\vee + \cdots + a_d \alpha_d^\vee$). The vector distance between vertices gK and hK in Σ is

$$\mathbf{d}(gK, hK) = \lambda \qquad \text{if and only if} \qquad g^{-1} hK \subseteq K t_\lambda K, \qquad (14.4)$$

and since each $u \in U$ stabilises a subsector of the fundamental sector of Σ, the Busemann function $\mathbf{h}$ is given by

$$\mathbf{h}(gK) = \mu \qquad \text{if and only if} \qquad gK \subseteq U t_\mu K. \qquad (14.5)$$

Thus, the vector distance $\mathbf{d}$ and the vector Busemann function $\mathbf{h}$ are natural statistics from the group theoretic context, encoding the Cartan and Iwasawa decompositions (respectively).

2.4. Generalised Polygons

Let Σ be a building of type (W, S) with chamber set Δ and Weyl distance $\delta : \Delta \times \Delta \to W$. Let $I \subseteq S$, and let W_I be the parabolic subgroup of W generated by I. The I-*residue* of a chamber $x \in \Delta$ is

$$R_I(x) = \{y \in \Delta \mid \delta(x, y) \in W_I\}.$$

This residue is a building of type (W_I, I) (it is easiest to check this using the formulation of buildings from Remark 2.3). Thus, general buildings are 'made up of' many other buildings of lower rank. The rank 1 residues have no interesting structure, and so the important case is rank 2.

Hence, buildings of type $I_2(m)$ play a critical role in the theory. We have already discussed buildings of type $I_2(\infty)$ in Example 2.2, and so here we consider the building of type $I_2(m)$ with $2 \leq m < \infty$. It is easy to verify that these buildings are equivalent to bipartite graphs with:

diameter m, and girth $2m$ (where girth is the length of the shortest cycle).

Such graphs are also known in the literature by another name: *generalised* m-*gons*. Thus, buildings of type $I_2(m)$ are equivalent to generalised m-gons. The chambers of the building are the edges of the generalised m-gon, and the apartments of the building are the cycles of length $2m$ in the generalised m-gon. Some examples are shown in Figure 14.7.

It is not hard to see that finite thick generalised m-gons are necesarily biregular (with alternating valencies $q + 1$ and $r + 1$, say), and that if m is odd then necessarily $q = r$. There is a lot to say about generalised m-gons (see, for example, [56]). Let us simply recall some of the key results. Since we are rarely interested in 'ordinary' m-gons, we will usually omit the adjective 'generalised'.

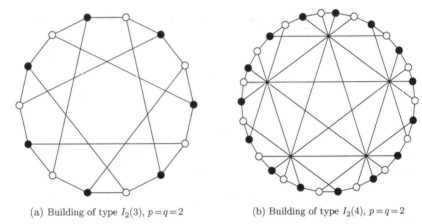

(a) Building of type $I_2(3)$, $p=q=2$ (b) Building of type $I_2(4)$, $p=q=2$

Figure 14.7 Examples of finite thick generalised m-gons

Theorem 2.10. [50] *For each $m \geq 2$ there exist thick m-gons.*

Theorem 2.10 shows that there are many examples of thick m-gons. The proof of the theorem is via a 'free construction' which produces m-gons in which every vertex has infinitely many neighbours. Finite thick m-gons are much more restricted, as shown by the following beautiful and unexpected theorem due to Feit and Higman.

Theorem 2.11. [19] *Finite thick generalsed m-gons exist if and only if $m \in \{2, 3, 4, 6, 8\}$.*

The 'only if' part of the Feit–Higman theorem follows from rather involved character theory (see Appendix B). For the 'if' part of the theorem, the existence of finite thick 2-gons is clear (they are just complete bipartite graphs), and examples of finite thick 3-gons, 4-gons, 6-gons and 8-gons come from the Chevalley groups $A_2(q)$, $B_2(q)$, $G_2(q)$, and the twisted Chevalley group $^2F_4(2^{2m+1})$ via Theorem 2.5.

Thus, in the finite theory the m-gons with $m = 2, 3, 4, 6, 8$ play a special role. We often call 4-gons, 6-gons and 8-gons *quadrangles, hexagons* and *octagons* respectively. Generalised 3-gons are called *projective planes* (this comes from the older language of incidence geometry). There are severe restrictions on the possible thickness parameters of projective planes, quadrangles, hexagons and octagons. Most of these restrictions have character theoretic proofs (we prove some of these in Appendix B). References to the original works can be found in Kantor [30].

Theorem 2.12. *Let Γ be a finite thick m-gon with parameters (q, r).*

(1) If $m = 3$ then $q = r$, and if $q \equiv 1, 2 \mod 4$ then q is a sum of two squares.

(2) If $m = 4$ then $q \leq r^2$, $r \leq q^2$, and $q^2(qr + 1)/(q + r) \in \mathbb{Z}$.

(3) If $m = 6$ then $q \le r^3$, $r \le q^3$, $q^3(q^2r^2 + qr + 1)/(q^2 + qr + r^2) \in \mathbb{Z}$, and $\sqrt{qr} \in \mathbb{Z}$.

(4) If $m = 8$ then $q \le r^2$, $r \le q^2$, $q^4(qr+1)(q^2r^2+1)/(q+r)(q^2+r^2) \in \mathbb{Z}$, and $\sqrt{2qr} \in \mathbb{Z}$.

The number theoretic part of statement (1) is called the *Bruck–Ryser–Chowla theorem*. For example, it prohibits the existence of a projective plane with parameter $q = 6$.

Together, Theorems 2.11 and 2.12 place a lot of conditions on the structure of general locally finite thick buildings. In particular, we immediately have the following corollary by looking at rank 2 residues.

Corollary 2.13. *Let Σ be a locally finite thick building of type (W, S) with parameters $(q_s)_{s \in S}$. For each $s, t \in S$ with $s \ne t$ let m_{st} be the order of st. Then:*

(1) $m_{st} \in \{2, 3, 4, 6, 8, \infty\}$ for all $s, t \in S$ with $s \ne t$, and

(2) if $m_{st} \in \{3, 4, 6, 8\}$ then the pair (q_s, q_t) satisfies the constraints from Theorem 2.12.

It is an open problem to determine the possible parameters of thick projective planes, quadrangles, hexagons and octagons. The *known* examples have the following parameters (where we arrange the parameters (q, r) so that $q \le r$). Projective planes: (q, q) with q a prime power. Quadrangles: (q, q), (q, q^2), (q^2, q^3), $(q-1, q+1)$ with q a prime power. Hexagons: (q, q), (q, q^3) with q a prime power. Octagons: (q, q^2) with $q = 2^{2k+1}$ an odd power of 2. Constructing a thick generalised m-gon with parameters other than these, or proving further restrictions on the possible parameters, would be revolutionary. Finally, we note that for a given value of the parameters there may be multiple non-isomorphic generalised m-gons. For example, there are four distinct projective planes with parameters $(9, 9)$.

2.5. Classification and Free Constructions

Thick irreducible spherical (respectively affine) buildings of rank at least 3 (respectively 4) have been classified by Tits (respectively, Tits and Weiss) (see [49] and [10, 52, 59]). Put very roughly, this classification says that all irreducible thick spherical buildings of rank at least 3 arise from groups of Lie origin via Tits systems, and that all irreducible thick affine buildings of rank at least 4 arise from groups of Lie origin defined over fields (or skew-fields) with discrete valuation via affine Tits systems. The precise statement of these classification theorems is involved (see the above references for details).

On the other hand, the following essentially free construction shows that the situation is very different for Coxeter systems which contain no irreducible rank 3 spherical parabolic subgroups.

Theorem 2.14. [43] *Let (W, S) be a Coxeter system such that every irreducible rank 3 parabolic subgroup of W is infinite. Suppose that $(q_s)_{s\in S}$ is a sequence of integers such that for each pair $s, t \in S$ there exists a generalised m_{st}-gon with parameters (q_s, q_t). Then there exists a locally finite thick regular building of type (W, S) whose rank 2 residues of type $W_{\{s,t\}}$ range through any desired set of generalised m_{st}-gons having parameters (q_s, q_t).*

The above theorem tells us that rank 3 irreducible affine buildings (that is, those of type $\tilde{A}_2$, $\tilde{B}_2$, and $\tilde{G}_2$) cannot be classified (at least not in the spirit of the higher rank classification; for example, one can make a $\tilde{A}_2$ building with thickness parameter $q = 9$ whose rank 2 residues can be chosen freely from the four non-isomorphic projective planes with $q = 9$). The theorem also applies to all Fuchsian Coxeter systems, and so these buildings are also unclassifiable. In other words, many of these buildings are not related in any nice way to groups. Using Corollary 2.13 and Theorem 2.14 we have the following existence result for Fuchsian buildings:

Theorem 2.15. *Let (W, S) be a Fuchsian Coxeter system of type $F(k_1, \ldots, k_n)$. There exists a locally finite thick building of type (W, S) if and only if $k_i \in \{2, 3, 4, 6, 8\}$ for all $i = 1, \ldots, n$ and either $k_i \in \{2, 4\}$ for some $i = 1, \ldots, n$ or $|\{i \mid k_i = 8\}|$ is even.*

Proof. Suppose that there is a locally finite thick building Σ of type $F(k_1, \ldots, k_n)$. By Corollary 2.13 we have $k_i \in \{2, 3, 4, 6, 8\}$ (since $k_i \neq \infty$ by definition for Fuchsian systems). Let $(q_1, \ldots, q_n)$ be the parameters of Σ (arranged cyclically so that the parameters of the rank 2 residue corresponding to k_i are q_i and q_{i+1}). If $k_i \neq 2, 4$ for all i then $k_i \in \{3, 6, 8\}$ for all i. If $k_i \in \{3, 6\}$ then $\sqrt{q_i q_{i+1}} \in \mathbb{Z}$, and if $k_i = 8$ then $\sqrt{2 q_i q_{i+1}} \in \mathbb{Z}$ (see Theorem 2.12). Multiplying these conditions together gives $2^{\alpha/2} q_1 \cdots q_n \in \mathbb{Z}$ where $\alpha = |\{i \mid k_i = 8\}|$, and hence α is even. This proves the 'only if' part of the theorem. We leave the 'if' part of the theorem as an exercise (using Theorem 2.14 and the known examples of generalised m-gons listed after Corollary 2.13). $\square$

3. Random Walks on Buildings

A *random walk* on a finite or countable space X is a sequence $(X_n)_{n\geq 0}$ of X-valued random variables governed by a stochastic *transition matrix* (or *transition operator*) $P = (p(x, y))_{x,y\in X}$. That is,

$$p(x, y) = \mathbb{P}[X_{n+1} = y \mid X_n = x] \quad \text{for all } x, y \in X \text{ and all } n \geq 0,$$

and the transition operator P acts on $\ell^1(X)$ by

$$Pf(x) = \sum_{y\in X} p(x, y)f(y) \quad \text{for all } x \in X.$$

The *n-step transition probabilities* of the walk are

$$p^{(n)}(x, y) = \mathbb{P}[X_n = y \mid X_0 = x],$$

and we have $P^n = (p^{(n)}(x, y))_{x,y \in X}$. The random walk $(X_n)_{n \geq 0}$ is *irreducible* if for each pair $x, y \in X$ there exists $n \geq 0$ such that $p^{(n)}(x, y) > 0$. For the general theory of random walks we refer to [60].

Here we are interested in random walks on buildings Σ (and associated groups). There are a few variations; for example, one might consider random walks on the set Δ of chambers of Σ, or one might consider random walks on the vertices of Σ. This latter case is particularly natural for affine buildings. To begin with we will consider random walks on the chambers of a general (locally finite) building, mainly following the setup from [21, 36].

3.1. Random Walks on Chambers and the Hecke Algebra

Let Σ be a locally finite building of type (W, S) with chamber set Δ and parameters $(q_s)_{s \in S}$. Let $(X_n)_{n \geq 0}$ be a random walk on the chamber set. Without some additional assumptions on the walk there is not so much that one can say. A natural assumption is to assume that the walk is *isotropic*, meaning that the transition probabilities depend only on the Weyl distance:

Definition 3.1. A random walk $(X_n)_{n \geq 0}$ on Δ is *isotropic* if the transition probabilities of the walk satisfy $p(x, y) = p(x', y')$ whenever $\delta(x, y) = \delta(x', y')$.

Isotropic random walks have a beautiful algebraic structure. For each $w \in W$ let $P_w = (p_w(x, y))_{x,y \in \Delta}$ be the transition operator of the isotropic random walk with

$$p_w(x, y) = \begin{cases} q_w^{-1} & \text{if } \delta(x, y) = w \\ 0 & \text{otherwise.} \end{cases}$$

The following elementary proposition shows that every isotropic random walk on Δ is a convex combination of the random walks P_w, $w \in W$.

Proposition 3.2. *A random walk on Δ with transition operator P is isotropic if and only if*

$$P = \sum_{w \in W} a_w P_w \quad \text{where } a_w \geq 0 \text{ and } \sum_{w \in W} a_w = 1,$$

in which case $p(x, y) = a_w q_w^{-1}$ if $\delta(x, y) = w$.

Therefore we are naturally led to consider linear combinations of the (linearly independent) operators P_w, $w \in W$. Let $\mathscr{P}$ be the vector space over $\mathbb{C}$ with basis $\{P_w \mid w \in W\}$. The key facts about $\mathscr{P}$ are summarised below (see [36, § 3]).

Theorem 3.3. [36] *The vector space $\mathscr{P}$ is an associative unital algebra under composition of linear operators, and the multiplication table with respect to the vector space basis $\{P_w \mid w \in W\}$ is given by*

$$P_u P_v = \sum_{w \in W} c_{u,v}^w P_w \quad \text{where} \quad c_{u,v}^w = \frac{q_w}{q_u \, q_v} |\Delta_u(x) \cap \Delta_{v^{-1}}(y)|$$

for any $x, y \in \Delta$ with $\delta(x,y) = w$. In particular, the intersection cardinalities $|\Delta_u(x) \cap \Delta_{v^{-1}}(y)|$ depend only on u, v and $\delta(x,y)$, and for $w \in W$ and $s \in S$ we have

$$P_w P_s = \begin{cases} P_{ws} & \text{if } \ell(ws) = \ell(w) + 1 \\ q_s^{-1} P_{ws} + (1 - q_s^{-1}) P_w & \text{if } \ell(ws) = \ell(w) - 1. \end{cases} \tag{14.6}$$

The algebra $\mathscr{P}$ is called the *Hecke algebra* of the building. If $\delta(x,y) = w$ then the n-step transition probability $p^{(n)}(x,y)$ is given by

$$p^{(n)}(x,y) = q_w^{-1} a_w^{(n)}, \quad \text{where} \quad P^n = \sum_{w \in W} a_w^{(n)} P_w.$$

Thus, finding $p^{(n)}(x,y)$ when $\delta(x,y) = w$ is equivalent to finding the coefficient of P_w in P^n.

Before surveying known results on isotropic random walks, let us briefly indicate how isotropic random walks arise from bi-invariant measures on groups acting on buildings (for example, groups of Lie type, or Kac–Moody groups). Specifically we have the following (see [14, Lemma 8.1] for a proof in a similar context).

Proposition 3.4. *Let G be a locally compact group acting transitively on a regular building Σ, and let B be the stabiliser of a fixed base chamber o. Normalise the Haar measure on G so that B has measure 1. Let φ be the density function of a bi-B-invariant probability measure on G. If the group B acts transitively on each set $\Delta_w(o)$ with $w \in W$, then the assignment*

$$p(go, ho) = \varphi(g^{-1}h)$$

for $g, h \in G$ defines an isotropic random walk on the chambers of Σ.

3.2. Isotropic Random Walks on Spherical Buildings

Let Σ be a locally finite thick spherical building of type (W, S) and let $(X_n)_{n \geq 0}$ be an isotropic random walk on the set Δ of chambers with transition operator $P = (p(x,y))_{x,y \in \Delta}$. Because the set Δ of chambers is finite, natural questions to ask include:

(1) What is the limiting distribution of the walk?
(2) What is the value of $p^{(n)}(x,y)$?
(3) What is the mixing time for the walk?

The first question is very easy to answer:

Proposition 3.5. *Let $(X_n)_{n\geq 0}$ be an irreducible isotropic random walk on the set Δ of chambers of a locally finite thick spherical building. Then the uniform distribution is the unique invariant measure, and*

$$\lim_{n\to\infty} \mu^{(n)}(x) = \frac{1}{|\Delta|} \quad \text{for all } x \in \Delta.$$

Proof. Using the thickness of the building it is not difficult to see that irreducible isotropic random walks on Δ are necessarily aperiodic (see [21, Lemma 4.3]). Thus, an irreducible isotropic walk has a unique stationary measure. To see that this stationary measure is the uniform measure $u : \Delta \to [0,1]$, note that for each $y \in \Delta$,

$$\sum_{x\in\Delta} u(x)p(x,y) = \frac{1}{|\Delta|} \sum_{w\in W} \sum_{x\in\Delta_w(y)} p(x,y)$$

$$= \frac{1}{|\Delta|} \sum_{w\in W} \frac{q_w}{q_{w^{-1}}} a_{w^{-1}} = \frac{1}{|\Delta|} \sum_{w\in W} a_{w^{-1}} = \frac{1}{|\Delta|} = u(y),$$

where we have used the fact that if $\delta(x,y) = w$ then $\delta(y,x) = w^{-1}$, and that $q_{w^{-1}} = q_w$. $\qquad\square$

To address questions (2) and (3) we can apply the techniques from [18], where the representation theory of Hecke algebras is used to analyse convergence of systematic scan Metropolis algorithms. The aims and context of [18] are quite different from our setting here, and so we will give an overview of the basic setup of [18], translating into our building theoretic point of view.

We first recall some basic representation theory of finite dimensional associative unital algebras over $\mathbb{C}$. This theory generalises the more familiar representation theory of finite groups (see, for example, [20]).

Definition 3.6. Let $\mathscr{A}$ be a finite dimensional unital associative algebra over $\mathbb{C}$. A *representation* of $\mathscr{A}$ is a pair (ρ, V) where V is a $\mathbb{C}$-vector space, and $\rho : \mathscr{A} \to \mathrm{End}(V)$ is an algebra homomorphism. The *character* of the representation (ρ, V) is the function $\chi_\rho : \mathscr{A} \to \mathbb{C}$ given by

$$\chi_\rho(A) = \mathrm{tr}(\rho(A)) \quad \text{for all } A \in \mathscr{A}.$$

The *dimension* of (ρ, V) is $\dim(V)$ (assumed to be finite throughout this section). Sometimes it is convenient to simply denote a representation (ρ, V) by ρ, and to write $V = V_\rho$.

Since $\mathrm{End}(V) \cong M_d(\mathbb{C})$ (the algebra of $d \times d$ matrices with entries in $\mathbb{C}$, where $d = \dim(V)$), a representation of $\mathscr{A}$ amounts to 'representing the algebra elements by matrices'. Familiar notions of direct sum, subrepresentations, and irreducibility carry over from the group setting.

An algebra is *semisimple* if every finite dimensional representation decomposes as a direct sum of irreducible representations. In this case the irreducible representations are the 'atomic building blocks' of the representation theory in the sense that every representation can be written as a direct sum of these atoms. Note that the group algebra of a finite group is necessarily semisimple (this is Maschke's theorem), however, general algebras need not be.

3.2.1. Hecke Algebras and the Geometric Representation

Let (W, S) be a spherical Coxeter system, and let $(q_s)_{s \in S}$ be a sequence of numbers with $q_s > 0$ and $q_s = q_t$ if s and t are conjugate in W. Let $\mathscr{H}$ be the algebra over $\mathbb{C}$ generated by symbols T_w (with $w \in W$) with relations

$$T_w T_s = \begin{cases} T_{ws} & \text{if } \ell(ws) > \ell(w) \\ q_s^{-1} T_{ws} + (1 - q_s^{-1}) T_w & \text{if } \ell(ws) < \ell(w). \end{cases} \tag{14.7}$$

The algebra $\mathscr{H}$ is unital (with identity T_e) and associative, and is called an *abstract Hecke algebra* (see [26]).

If $(q_s)_{s \in S}$ are the parameters of a locally finite spherical building of type (W, S) then, comparing (14.6) and (14.7) we see that the Hecke algebra $\mathscr{P}$ of Σ gives a representation of the abstract Hecke algebra $\mathscr{H}$. More specifically, let $V_\Delta = \bigoplus_{x \in \Delta} \mathbb{C}x$ be a vector space with basis indexed by the the chambers of the building, and let $\rho_\Delta : \mathscr{H} \to \text{End}(V_\Delta)$ be the linear map with $\rho_\Delta(T_w) = P_w$ for all $w \in W$. Then

$$(\rho_\Delta, V_\Delta) \quad \text{is a } |\Delta|\text{-dimensional representation of } \mathscr{H}.$$

We call this representation the *geometric representation* of the Hecke algebra $\mathscr{H}$. In fact the map $\rho_\Delta : \mathscr{H} \to \mathscr{P}$ is bijective, and so $\mathscr{P} \cong \mathscr{H}$. We emphasise that the geometric representation of the abstract Hecke algebra $\mathscr{H}$ only exists when there is a building with parameters $(q_s)_{s \in S}$.

It is well known that the algebra $\mathscr{H}$ is semisimple (see [24]). Thus, the geometric representation decomposes into a direct sum of irreducible representations. Writing $\text{Irrep}(\mathscr{H})$ for the set of irreducible representations of $\mathscr{H}$ (more formally, isomorphism classes of irreducible representations) we have

$$V_\Delta = \bigoplus_{\rho \in \text{Irrep}(\mathscr{H})} V_\rho^{\oplus m_\rho}$$

where for each $\rho \in \text{Irrep}(\mathscr{H})$ the integer $m_\rho \geq 0$ is the multiplicity of (ρ, V_ρ) in (ρ_Δ, V_Δ). Thus, the character χ_Δ of the geometric representation is given by

$$\chi_\Delta = \sum_{\rho \in \text{Irrep}(\mathscr{H})} m_\rho \chi_\rho. \tag{14.8}$$

The first fundamental task is to compute the multiplicities m_ρ.

Theorem 3.7.   *The multiplicity m_ρ of the irreducible representation ρ in (ρ_Δ, V_Δ) is*

$$m_\rho = \frac{\dim(\rho)}{\langle \chi_\rho, \chi_\rho \rangle}, \quad where \quad \langle f, g \rangle = \frac{1}{|\Delta|} \sum_{w \in W} q_w f(T_w) g(T_{w^{-1}})$$

for functions $f, g : \mathcal{H} \to \mathbb{C}$.

Proof. It follows from general results on the representation theory of symmetric algebras (see [20, Corollary 7.2.4 and §8.1.8]) that if χ and χ' are irreducible characters of $\mathcal{H}$ then

$$\langle \chi, \chi' \rangle = 0 \quad \text{if and only if } \chi \text{ and } \chi' \text{ are non-isomorphic.} \qquad (14.9)$$

Thus, taking inner products with χ_ρ in (14.8), and using the facts that $\chi_\Delta(T_w) = \operatorname{tr}(P_w) = \delta_{w,e}|\Delta|$ and $\chi_\rho(T_e) = \dim(\rho)$, the result follows.
$\qquad\qquad\qquad\qquad\qquad\qquad\qquad\qquad\qquad\qquad\qquad\qquad\qquad\quad\square$

3.2.2.   The Transition Probabilities $p^{(n)}(x,y)$

We now return to question (2), seeking a formula for the n-step return probabilities $p^{(n)}(x, y)$. In principal one could compute this probability by noting that it is the $(x, y)^{\text{th}}$ entry of the matrix P^n. Of course, this is not a practical method because P is a very large matrix (for example, for the smallest thick F_4 building the matrix P has approximately 2×10^8 rows and columns!). The following theorem gives a more practical solution to the problem.

Theorem 3.8.   *Let $P = \sum a_w P_w \in \mathscr{P}$ be the transition matrix of an isotropic random walk on a regular spherical building, and let $T = \sum a_w T_w \in \mathcal{H}$. Then*

$$p^{(n)}(x, y) = \frac{1}{|\Delta|} \sum_{\rho \in \operatorname{Irrep}(\mathcal{H})} m_\rho \chi_\rho(T^n T_{w^{-1}}) \quad if \, \delta(x,y) = w.$$

Proof. We claim that

$$\chi_\Delta(T_u T_{v^{-1}}) = q_u^{-1}|\Delta|\delta_{u,v} \quad \text{for all } u, v \in W. \qquad (14.10)$$

To see this, note that $\chi_\Delta(T_w) = \operatorname{tr}(\rho_\Delta(T_w)) = \operatorname{tr}(P_w) = |\Delta|\delta_{w,e}$ (because each P_w is a $|\Delta| \times |\Delta|$ matrix, with $(x, y)^{\text{th}}$ entry equal to 1 if $\delta(x, y) = w$ and 0 otherwise). Thus,

$$\chi_\Delta(T_u T_{v^{-1}}) = \operatorname{tr}(P_u P_{v^{-1}}) = \sum_{w \in W} c_{u,v^{-1}}^w \operatorname{tr}(P_w)$$

$$= c_{u,v^{-1}}^e |\Delta| = \frac{1}{q_u q_v}|\Delta_u(o) \cap \Delta_v(o)|,$$

where we have used Theorem 3.3, and (14.10) follows.

Now, if $P^n = \sum a_w^{(n)} P_w$ then $T^n = \sum a_w^{(n)} T_w$, and so

$$\frac{1}{|\Delta|}\chi_\Delta(T^n T_{w^{-1}}) = \frac{1}{|\Delta|}\sum_{v\in W} a_w^{(n)}\chi_\Delta(T_v T_{w^{-1}}) = q_w^{-1} a_w^{(n)}.$$

If $\delta(x,y) = w$ then $p^{(n)}(x,y) = q_w^{-1} a_w^{(n)}$, and thus $p^{(n)}(x,y) = \chi_\Delta(T^n T_{w^{-1}})/|\Delta|$. The result follows from (14.8). $\qquad\square$

3.2.3. Mixing Times

Now we move to the more sophisticated question of mixing times. To begin with we need to define a notion of 'distance' between two measures. In the literature the *total variation distance* has become a standard choice, popularised by Persi Diaconis (see, for example, [17]). This distance is defined as follows: If μ and ν are probability measures on Δ then

$$\|\mu - \nu\|_{\mathrm{tv}} = \max_{A\subseteq\Delta}|\mu(A) - \nu(A)|.$$

To work with the total variation distance it is helpful to observe the following elementary fact.

Lemma 3.9. *Let μ be probability measure on Δ and let $u : \Delta \to [0,1]$ be the uniform distribution. Then*

$$\|\mu - u\|_{\mathrm{tv}} = \frac{1}{2}\|\mu - u\|_1 \le \frac{\sqrt{|\Delta|}}{2}\|\mu - u\|_2$$

where $\|\cdot\|_1$ and $\|\cdot\|_2$ are the ℓ^1 and ℓ^2 norms respectively.

Proof. Clearly $\|\mu - u\|_{\mathrm{tv}}$ equals either $|\mu(A) - u(A)|$ or $|\mu(B) - u(B)|$, where A and B are the sets $A = \{x \in \Delta \mid \mu(x) > u(x)\}$ and $B = \{x \in \Delta \mid \mu(x) < u(x)\}$. In fact

$$|\mu(A) - u(A)| = |1 - \mu(B) - 1 + u(B)| = |\mu(B) - u(B)|$$

and so $\|\mu - u\|_{\mathrm{tv}} = \frac{1}{2}(|\mu(A) - u(A)| + |\mu(B) - u(B)|) = \frac{1}{2}\|\mu - u\|_1$. The final inequality follows from Cauchy–Schwarz. $\qquad\square$

Let $\mu^{(n)} : \Delta \to [0,1]$ be the measure $\mu^{(n)}(x) = p^{(n)}(o,x)$, where $o \in \Delta$ is a fixed chamber of Σ. The following theorem gives a mathematically tractable upper bound estimate for the total variation distance $\|\mu^{(n)} - u\|_{\mathrm{tv}}$, and thus can be used to give upper bounds for mixing times. Define an involution $* : \mathscr{H} \to \mathscr{H}$ by $(\sum a_w T_w)^* = \sum \overline{a}_w T_{w^{-1}}$. Note that $\mathscr{H}$ has a one-dimensional representation ρ_{triv} (the *trivial representation*) given by $\rho_{\mathrm{triv}}(T_s) = 1$ for all $s \in S$ (to check this, simply verify that the defining relations (14.7) are satisfied).

Theorem 3.10. (cf. [18]) *Let* $P = \sum a_w P_w$ *be the transition operator of an isotropic random walk on the chambers of a regular spherical building, and let* $T = \sum a_w T_w \in \mathscr{H}$. *Then*

$$\|\mu^{(n)} - u\|_{\text{tv}}^2 \leq \frac{1}{4} \sum_{\rho \neq \rho_{\text{triv}}} m_\rho \chi_\rho \left(T^n (T^*)^n \right).$$

Proof. We have

$$\|\mu^{(n)} - u\|_{\text{tv}}^2 \leq \frac{|\Delta|}{4} \|\mu^{(n)} - u\|_2^2 = \frac{|\Delta|}{4} \left\langle \mu^{(n)} - u, \mu^{(n)} - u \right\rangle_2$$

$$= \frac{1}{4} \left(|\Delta| \langle \mu^{(n)}, \mu^{(n)} \rangle_2 - 1 \right).$$

Now

$$\langle \mu^{(n)}, \mu^{(n)} \rangle_2 = \sum_{w \in W} q_w^{-1} \left(a_w^{(n)} \right)^2 = \frac{1}{|\Delta|} \chi_\Delta (T^n (T^*)^n),$$

and so

$$\|\mu^{(n)} - u\|_{\text{tv}}^2 \leq \frac{1}{4} \left(\chi_\Delta (T^n (T^*)^n) - 1 \right). \qquad \square$$

Theorems 3.8 and 3.10, together with the multiplicity formula from Theorem 3.7, provide some basic theory for studying isotropic random walks on spherical buildings. To make more practical estimates in given examples one needs to work harder with the representation theory (for example, to give meaningful bounds on the right-hand side of the inequality in Theorem 3.10). Some calculations are made in [18], in a different context, that can be translated to give estimates for certain random walks on buildings Σ of type A_n (see [18, Proposition 7.4 and Theorem 7.5]). However, in the building theoretic context the walks covered in [18] are perhaps not the most natural (for example, the simple random walk is not covered). Thus, there is still a lot to do in this direction, and we hope that the setup provided above might stimulate some future research. In Appendix B we outline the details of the representation theory in the rank 2 case (that is, when the building is a generalised polygon).

3.3. Random Walks on Affine Buildings

Let $(X_n)_{n \geq 0}$ be a random walk on an infinite graph with transition probabilities $p(x, y)$. Natural questions to ask in this setting include:

(1) At what velocity does the random walk move to infinity?
(2) What is the distribution of the fluctuations away from expected distance?
(3) What are the asymptotics of $p^{(n)}(x, y)$?

Appropriate solutions to these problems come in the form of a law of large numbers, a central limit theorem, and a local limit theorem (respectively).

For random walks on affine buildings it is natural to consider both random walks on the chambers of the building, and random walks on the vertices of the building. The latter case now has a rather complete theory. We will consider both cases below.

If the affine building has rank 2 then we are dealing with a random walk on a trees. In this context there is a huge literature which takes us too far afield to discuss here, and so we will focus on the higher rank case.

3.3.1. Random Walks on the Vertices of an Affine Building

Let R be an irreducible root system with coweight lattice P, and let $(W_{\text{aff}}, S_{\text{aff}})$ be the associated affine Coxeter system. Let Σ be a regular affine building of type $(W_{\text{aff}}, S_{\text{aff}})$, and let V be the set of all special vertices of Σ. Recall the definitions of the vector distance function $\boldsymbol{d}(\cdot, \cdot)$ from Definition 2.6. Some of the formulae of this section become more complicated in the case of $\widetilde{C}_n$ buildings with $q_0 \neq q_n$, and so here we will restrict to the case $q_0 = q_n$ for $\widetilde{C}_n$ buildings (see [36–38] for the general case).

We now define isotropic random walks on the set V of all special vertices, and outline the algebraic and analytic theory that is used to analyse them. In Appendix A we will give more details in the specific case of $\widetilde{C}_2$ buildings, where one can carry out the calculations 'by hand'.

Definition 3.11. A random walk $(X_n)_{n \geq 0}$ on V is *isotropic* if its transition probabilities satisfy

$$p(x, y) = p(x', y') \quad \text{whenever } \boldsymbol{d}(x, y) = \boldsymbol{d}(x', y').$$

For each $x \in V$ and $\lambda \in P^+$ let

$$V_\lambda(x) = \{y \in V \mid \boldsymbol{d}(x, y) = \lambda\} \quad \text{be the sphere of 'radius' } \lambda \text{ centred at } x.$$

The cardinality $N_\lambda = |V_\lambda(x)|$ does not depend on $x \in V$ (see [37, Proposition 1.5]). If $(X_n)_{n \geq 0}$ is an isotropic random walk on V then there are numbers $a_\lambda \geq 0$ with $\sum_{\lambda \in P^+} a_\lambda = 1$ such that

$$p(x, y) = \frac{a_\lambda}{N_\lambda} \quad \text{for all } y \in V_\lambda(x). \tag{14.11}$$

In an analogous way to the case of isotropic random walks on chambers (see Proposition 3.2) the transition operator A of an isotropic random walk on the vertices of a regular affine building is of the form

$$A = \sum_{\lambda \in P^+} a_\lambda A_\lambda, \tag{14.12}$$

where the numbers a_λ are as in (14.11) and the operator A_λ acts on functions $f : V \to \mathbb{C}$ by

$$A_\lambda f(x) = \frac{1}{N_\lambda} \sum_{y \in V_\lambda(x)} f(y).$$

Let $\mathscr{A}$ be the vector space over $\mathbb{C}$ with basis $\{A_\lambda \mid \lambda \in P^+\}$. The following is an analogue of Theorem 3.3 (the proof is, however, a little more involved).

Theorem 3.12. [37, Theorem 5.24] *The vector space $\mathscr{A}$ is a commutative associative unital algebra under composition of linear operators.*

The algebra $\mathscr{A}$ plays an important role in understanding isotropic random walks on the vertices of affine buildings. The key feature of Theorem 3.12 is that this algebra is commutative. In fact one can be more precise.

Theorem 3.13. [37, Theorem 6.16] *Let Σ be a regular affine building. Let $\mathscr{P}$ the the algebra of chamber set averaging operators on Σ (c.f. Theorem 3.3) and let $\mathscr{A}$ be the algebra of vertex set averaging operators on Σ (c.f. Theorem 3.12). Then $\mathscr{A}$ is isomorphic to the centre of $\mathscr{P}$.*

For each $\alpha \in R$ we write $q_\alpha = q_i$ if $\alpha \in W_0 \alpha_i$, and let

$$r^\lambda = \prod_{\alpha \in R^+} q_\alpha^{\frac{1}{2}\langle \lambda, \alpha \rangle} \quad \text{for all } \lambda \in P.$$

If $q_i = q$ for all $i = 0, 1, \ldots, d$ then $r^\lambda = q^{\ell(t_\lambda)/2}$ where t_λ is the translation by λ.

If $u \in \mathrm{Hom}(P, \mathbb{C}^\times)$ we write $u^\lambda = u(\lambda)$, and if $w \in W_0$ and $u \in \mathrm{Hom}(P, \mathbb{C}^\times)$ let $wu \in \mathrm{Hom}(P, \mathbb{C}^\times)$ be given by $(wu)^\lambda = u^{w\lambda}$ for all $\lambda \in P$. For each $\lambda \in P^+$ the *Macdonald spherical function* P_λ is the function $P_\lambda : \mathrm{Hom}(P, \mathbb{C}^\times) \to \mathbb{C}$ given by

$$P_\lambda(u) = \frac{r^{-\lambda}}{W_0(q^{-1})} \sum_{w \in W_0} u^{w\lambda} c(wu) \quad \text{where} \quad c(u) = \prod_{\alpha \in R^+} \frac{1 - q_\alpha^{-1} u^{-\alpha^\vee}}{1 - u^{-\alpha^\vee}},$$

where $W_0(q^{-1}) = \sum_{w \in W_0} q_w^{-1}$. This formula requires, of course, that the denominators are nonzero, however it turns out that $P_\lambda(u)$ is a linear combination of terms u^μ with $\mu \in P$ and so the 'singular' cases where a denominator vanishes can be obtained by taking an appropriate limit in the general formula. The Macdonald spherical functions arise in the representation theory of p-adic groups (see [34]).

Theorem 3.13, combined with the *Satake isomorphism*, implies the following result, giving a complete description of the irreducible representations of $\mathscr{A}$.

Theorem 3.14. [37, Proposition 2.1] *For each* $u \in \text{Hom}(P, \mathbb{C}^{\times})$ *there is a one-dimensional representation* π_u *of* $\mathscr{A}$ *given by* $\pi_u(A_\lambda) = P_\lambda(u)$. *Moreover, every one-dimensional representation* π *of* $\mathscr{A}$ *is of the form* $\pi = \pi_u$ *for some* $u \in \text{Hom}(P, \mathbb{C}^{\times})$, *and* $\pi_u = \pi_{u'}$ *if and only if* $u' = wu$ *for some* $w \in W_0$.

Each $A \in \mathscr{A}$ maps $\ell^2(V)$ into itself, and $\|A_\lambda f\|_2 \leq \|f\|_2$. Thus, we may regard $\mathscr{A}$ as a subalgebra of the C^*-algebra $\mathscr{L}(\ell^2(V))$ of bounded linear operators on $\ell^2(V)$. It is not hard to see that $A_\lambda^* = A_{\lambda^*}$ where $\lambda^* = -w_0\lambda$ (with w_0 the longest element of W_0), and thus $\mathscr{A}$ is closed under taking adjoints. Let $\mathscr{A}_2$ be the completion of $\mathscr{A}$ with respect to the ℓ^2-operator norm $\|\cdot\|$. Thus, $\mathscr{A}_2$ is a commutative C^*-algebra. By passing to this completion we ensure that the transition operator A of an isotropic random walk on V is an element of $\mathscr{A}_2$ (it is an element of the 'uncompleted' algebra $\mathscr{A}$ if and only if the walk has bounded range).

The one-dimensional representations of $\mathscr{A}_2$ are precisely the extensions to $\mathscr{A}_2$ of the representations $\pi_u : \mathscr{A} \to \mathbb{C}$ which are continuous with respect to the ℓ^2-operator norm, and in [37, §5] it is shown that these are the representations π_u with $u \in \text{Hom}(P, \mathbb{T})$ where $\mathbb{T} = \{z \in \mathbb{C} \mid |z| = 1\}$. If $u \in \text{Hom}(P, \mathbb{T})$ and $A \in \mathscr{A}_2$ we write $\widehat{A}(u) = \pi_u(A)$ (the *Gelfand transform* of A). In particular, we have $\widehat{A}_\lambda(u) = P_\lambda(u)$, and if A is the transition operator of an isotropic random walk as in (14.12) we have

$$\widehat{A}(u) = \sum_{\lambda \in P^+} a_\lambda P_\lambda(u) \quad \text{for all } u \in \text{Hom}(P, \mathbb{T}).$$

The final ingredient in the analysis of $\mathscr{A}_2$ is the calculation of the *Plancherel measure*.

Theorem 3.15. [34, Theorem 5.1.5], [37, Theorem 5.2] *Let* du *denote normalised Haar measure on* $\mathbb{U} = \text{Hom}(P, \mathbb{T})$, *and let* μ *be the measure on* $\mathbb{U}$ *given by*

$$d\mu(u) = \frac{W_0(q^{-1})}{|W_0|} \frac{1}{|c(u)|^2} \, du.$$

Then

$$\frac{1}{N_\lambda} \int_{\mathbb{U}} \widehat{A}_\lambda(u) \overline{\widehat{A}_{\lambda'}(u)} \, d\mu(u) = \delta_{\lambda, \lambda'} \quad \text{for all } \lambda, \lambda' \in P^+.$$

Theorem 3.15 implies that the n-step transition probabilities of an isotropic random walk with transition operator A are given by

$$p^{(n)}(x, y) = \frac{1}{N_\lambda^2} \int_{\mathbb{U}} \widehat{A}(u)^n \overline{\widehat{A}_\lambda(u)} d\mu(u) \quad \text{if } d(x, y) = \lambda. \qquad (14.13)$$

This is the analogue of Theorem 3.8, and is a key result in studying isotropic random walks on the vertices of affine buildings. Indeed the primary limit theorems (that is, the law of large numbers, the central limit theorem, and the local limit theorem) can all be proven using the above machinery via techniques from classical harmonic analysis. These limit theorems were proved by Lindlebauer and Voit [33] for the case of $\widetilde{A}_2$ buildings, and Cartwright and Woess [14] for the case of $\widetilde{A}_n$ buildings. The general case was settled by Parkinson [38], and the results are summarised below.

Theorem 3.16. [38] *Let* $(X_n)_{n\geq 0}$ *be an isotropic random walk on the vertices of a locally finite thick regular affine building.*

(1) Under the moment assumption $\sum_{\lambda\in P^+} |\lambda| a_\lambda < \infty$ *there exists* $\gamma \in E^+$ *such that*

$$\lim_{n\to\infty} \frac{d(o, X_n)}{n} = \gamma \quad almost\ surely.$$

(2) Under the moment assumption $\sum_{\lambda\in P^+} |\lambda|^2 a_\lambda < \infty$, *the vector*

$$(d(o, X_n) - n\gamma)/\sqrt{n}$$

converges in distribution to the multivariable normal distribution $N(0, \Gamma)$, *where* Γ *is a positive definite matrix.*

(3) Let $y \in V_\lambda(x)$ *and* $n \in \mathbb{N}$. *Suppose that* $(X_n)_{n\geq 0}$ *is irreducible and aperiodic. Then*

$$p^{(n)}(x, y) = CP_\lambda(1)\widehat{A}(1)^n n^{-(|R|+d)/2}\left(1 + O(n^{-1/2})\right),$$

where $C > 0$ *is an explicit constant.*

Remark 3.17.   We remark that the precision of Theorem 3.16 is really quite impressive, with explicit formulae for the speed, variance, radius of convergence, and all asymptotic constants (see [38] for details). In the local limit theorem the assumption of aperiodicity may be removed, see [38].

Heat kernel and Green function estimates for finite range isotropic random walks on affine buildings have been obtained recently by Trojan [55] (with earlier results obtained for $\widetilde{A}_n$ buildings by Anker, Schapira and Trojan [2]). The starting point for this analysis is again formula (14.13). Estimates for the Green function are given within the radius of convergence, and at the radius of convergence. For example, at the radius of convergence Trojan proves:

Theorem 3.18. [55, Theorem 7] *The green function of a finite range isotropic random walk on the special vertices of an affine building of rank* r, *evaluated at the radius of convergence, satisfies*

$$\sum_{n=0}^{\infty} p^{(n)}(x, y)\rho^{-n} \asymp P_\lambda(1)\|\lambda\|^{2-r-2|R^+|}.$$

Finally, convergence results for isotropic random walks on affine buildings to Brownian motion in a Weyl sector have been studied by Schapira, at least in the context of nearest neighbour random walks on $\widetilde{A}_r$ buildings. We now describe this result. Let $(X_n)_{n\geq 0}$ be a symmetric nearest neighbour random walk on the vertices of an $\widetilde{A}_r$ building with transition probabilities $p(x,y)$. Let ρ be the spectral radius of $(X_n)_{n\geq 0}$ and let $(Y_n)_{n\geq 0}$ be the random walk with transition probabilities

$$q(x,y) = p(x,y)\frac{P_{d(o,y)}(1)}{P_{d(o,x)}(1)}\rho^{-1}$$

(it is easily seen that this defines a random walk on the building using [37, Theorem 3.22]). Schapira proves the following (see [46, § 2] for the relevant definitions of Brownian motion):

Theorem 3.19. [46, Theorem 6.1] *With the notation as above, the sequence $(Z_t^n)_{t\geq 0}$ with*

$$Z_t^n = \frac{1}{\sqrt{n}}d\big(o, Y_{\lceil nt\rceil}\big)$$

converges in law to Brownian motion $(I_t)_{t\geq 0}$ in the sector $\mathfrak{s}_0$ as $n\to\infty$.

3.3.2. Random Walks on the Chambers of an Affine Building

The literature on isotropic random walks on the chambers of affine buildings is currently less complete. A key reason for this is that the algebra $\mathscr{P}$ of averaging operators on the chambers of an affine building is noncommutative. Thus, the Plancherel theorem for this infinite dimensional noncommutative algebra is rather sophisticated (see [35] and [40]). The general approach to the primary limit theorems is outlined by Parkinson and Schapira in [39], and the detailed calculations are carried through for $\widetilde{A}_2$ buildings. The general case is in preparation by the author.

Theorem 3.20. [39, Theorem 3.7] *For the simple random walk on the chambers of a thick $\widetilde{A}_2$ building with thickness $q > 1$ we have*

$$p^{(n)}(x,y) = C_w\rho^n n^{-4}\left(1 + O(n^{-1/2})\right) \quad \text{if } \delta(x,y) = w$$

where C_w is an explicitly computable constant (depending on w and q only), and where the spectral radius ρ is given by $\rho = (3(q-1) + \sqrt{q^2 + 34q + 1})/6q$.

Remark 3.21. Assuming a suitably transitive group action, a formula for the spectral radius for an isotropic random walk on the chambers of an affine building can be deduced from results of Saloff-Coste and Woess [44]. In particular, see [44, Example 6].

3.3.3. Regular Sequences in Affine Buildings

Recently Parkinson and Woess [41] proved the 'p-adic analogue' of Kaimanovich's characterisation [29] of *regular sequences* in symmetric spaces. This theory has applications to random walks on buildings and associated groups, and we describe this here. Recall the definition of the vector Busemann functions $h_{\mathfrak{s}}$ from Definition 2.7.

It is convenient to work with a natural 'metric realisation' of the affine building Σ. By the construction in Section 1.3 we may regard the apartments of an affine building as tessellations of a Euclidean space, and thus there is a metric on each apartment. Using axioms (B2) and (B3) it can be shown that these metrics may be 'glued together' to make Σ into a metric space (see [1, §11.2]). By [1, Theorem 11.16] this metric space is a CAT(0) space. In this section, we will regard affine buildings as metric spaces, although we also remember the underlying simplicial complex structure. The vector distance and the Busemann functions (originally only defined for vertices) naturally extend to give a vector distance and Busemann function for any points x, y of the building (see [41] for details).

Let $\lambda \in E^{+}$. A λ-*ray* in Σ is a function $\mathfrak{r} : [0, \infty) \to \Sigma$ such that

$$d(\mathfrak{r}(t_1), \mathfrak{r}(t_2)) = (t_2 - t_1)\lambda \quad \text{for all } t_2 \geq t_1 \geq 0.$$

Because we are specifying both a speed and direction, the notion of a λ-ray is a refinement of the usual notion of a ray in a CAT(0) space.

Theorem 3.22. *[41, Theorem 3.2] Let $(x_n)_{n \geq 0}$ be a sequence in Σ, and let $\lambda \in E^{+}$. Let $\mathfrak{s}$ be a sector of Σ. The following are equivalent:*

(1) There is a λ-ray $\mathfrak{r} : [0, \infty) \to \Sigma$ such that $d(x_n, \mathfrak{r}(n)) = o(n)$.

(2) $d(x_n, x_{n+1}) = o(n)$ and $h_{\mathfrak{s}}(x_n) = n\mu_{\mathfrak{s}} + o(n)$ for some $\mu_{\mathfrak{s}} \in W_0\lambda$ (independent of n).

(3) $d(x_n, x_{n+1}) = o(n)$ and $d(o, x_n) = n\lambda + o(n)$.

A sequence $(x_n)_{n \geq 0}$ satisfying any one of the above equivalent conditions is called a λ-*regular sequence*. This is a direct analogue of Kaimanovich's results on symmetric spaces [29], and a generalisation of results of Cartwright, Kaimanovich and Woess on homogeneous trees [13]. We will discuss applications of Theorem 3.22 to random walks on affine buildings and associated groups in this section and the next.

Since Σ is CAT(0) we define the *visibility boundary* $\partial\Sigma$ in the usual way as the set of equivalence classes of rays (with two rays being *equivalent* if the distance between them is bounded). The standard topology makes $\overline{\Sigma} = \Sigma \cup \partial\Sigma$ into a compact Hausdorff space (see Bridson and Haefliger [6, §II.8.5]). Points of the visibility boundary are called *ideal points* of Σ. Given $\xi \in \partial\Sigma$ and $x \in \Sigma$, there is a unique ray in the class ξ with base point x ([6, Proposition II.8.2] or [1, Lemma 11.72]). We sometimes denote this ray by $[x, \xi)$. Thus, one may think of $\partial\Sigma$ as 'all rays based at x' for any fixed $x \in \Sigma$.

Definition 3.23. A random walk on V is *semi-isotropic* if the transition probabilities of the walk depend only on the vectors $d(x, y)$ and $h(y) - h(x)$.

Clearly isotropic random walks are semi-isotropic, but not vice-versa. For each $\lambda \in P$ let $H_\lambda = \{x \in V \mid h(x) = \lambda\}$. As shown in [41, Proposition 4.6] semi-isotropic random walks are 'factorisable' over P, in the sense that the value of the sum

$$\overline{p}(\lambda, \mu) = \sum_{y \in H_\mu} p(x, y) \quad \text{with } \lambda, \mu \in P \text{ and } x \in H_\lambda$$

does not depend on the particular $x \in H_\lambda$ chosen. Moreover, we have

$$\overline{p}(\lambda + \nu, \mu + \nu) = \overline{p}(\lambda, \mu) \quad \text{for all } \lambda, \mu, \nu \in P.$$

In other words, if $(X_n)_{n \geq 0}$ is semi-isotropic then the sequence $h(X_n) \in P$ is a translation invariant random walk on P with transition probabilities $\overline{p}(\lambda, \mu)$. Since $P \cong \mathbb{Z}^d$ the random walk $(h(X_n))_{n \geq 0}$ is well understood from the classical theory, and using Theorem 3.22 we obtain the following result for the original random walk $(X_n)_{n \geq 0}$ on the building.

Theorem 3.24. [41, Corollary 4.8] *Let $(X_n)_{n \geq 0}$ be a semi-isotropic random walk on V. Under the finite first moment assumption $\sum_{\nu \in P} \overline{p}(0, \nu)|\nu| < \infty$ we have*

$$\lim_{n \to \infty} \frac{1}{n} d(o, X_n) = \lambda \quad \text{almost surely,}$$

where λ is the dominant element in the W_0-orbit of $\mu = \sum_{\nu \in P} \overline{p}(0, \nu)\nu$. Moreover, if $\lambda \neq 0$ then $(X_n)_{n \geq 0}$ converges almost surely to an ideal point X_∞.

The drift-free case (when $\lambda = 0$) is more subtle. A weaker form of convergence of the random walk in this case is established in [41, Theorem 4.15] for nearest neighbour random walks.

3.3.4. Random Walks on Groups Acting on Affine Buildings

Limit theorems for isotropic random walks on the vertices of affine buildings imply limit theorems for bi-K-invariant probability measures on groups acting sufficiently transitively on the building, where K is the stabiliser of a fixed (special) vertex of the building. This is completely analogous to the chamber case of Proposition 3.4 (see [38, Remark 2.19] for some details). For example, If $G = G(\mathbb{Q}_p)$ is a Chevalley group over the p-adic numbers, and if $K = G(\mathbb{Z}_p)$ with $\mathbb{Z}_p$ the ring of p-adic integers, then Theorem 3.16 gives a local limit theorem for the density function of a bi-K-invariant probability measure on G (see Example 2.9).

In the prototypical example of $G = SL_{d+1}(\mathbb{Q}_p)$ one can remove the bi-K-invariance assumption, at the cost of losing explicit formulae for the spectral radius. We expect the following result of Tolli to hold for more general Lie types, however at present it is only available for SL_{d+1}.

Theorem 3.25. [54] *Let $G = SL_{d+1}(\mathbb{F})$ where $\mathbb{F}$ is a local field. Let f be a continuous compactly supported density of a probability measure on G such that*

(1) f is symmetric, that is $f(x) = f(x^{-1})$, and
(2) the support of f is a neighbourhood of the identity that generates G.

Then there exists a number $\rho > 0$ and a positive function $\psi : G \to \mathbb{R}_{\geq 0}$ such that

$$\rho^{-n} n^{d(d+2)/2} f^{(*n)} \to \psi \quad \text{pointwise as } n \to \infty.$$

Let Σ be a regular affine building, and let G be a subgroup of the automorphism group $\mathrm{Aut}(\Sigma)$. Let σ be a Borel probability measure on G, such that the support of σ generates G. We say that σ has *finite first moment* if

$$\int_G d(o, go) \, d\sigma(g) < \infty.$$

Let $(g_n)_{n \geq 0}$ be a stationary sequence of G-valued random variables with joint distribution σ The *right random walk* is the sequence $(X_n)_{n \geq 0}$ with

$$X_0 = o \quad \text{and} \quad X_n = g_1 \cdots g_n o \quad \text{for } n \geq 1.$$

The theory of regular sequences (Theorem 3.22) implies the following result for the right random walk.

Theorem 3.26. [41, Theorem 4.1] *Let G and σ be as above, and suppose that σ has finite first moment. Let $(X_n)_{n \geq 0}$ be the associated right random walk on Σ. There exists $\lambda \in E^+$ such that*

$$\lim_{n \to \infty} \frac{1}{n} d(o, X_n) = \lambda \quad \text{almost surely,}$$

and for each sector $\mathfrak{s}$ of Σ there exists $\mu_{\mathfrak{s}} \in W_0 \lambda$ such that

$$\lim_{n \to \infty} \frac{1}{n} h_{\mathfrak{s}}(X_n) = \mu \quad \text{almost surely.}$$

If $\lambda \neq 0$ then $(X_n)_{n \geq 0}$ converges almost surely to an ideal point X_∞.

3.4. Random Walks on Fuchsian Buildings

Probability theory for buildings and related groups of non-spherical, non-affine type is in its infancy. Recently isotropic random walks on the chambers of Fuchsian buildings have been studied by Gilch, Müller and Parkinson, and a law of large numbers and a central limit theorem have been obtained. In this case the Hecke algebra has less controllable representation theory than in the spherical and affine cases, owing partly to the existence of free group subgroups in the Coxeter systems. Thus, the representation theoretic techniques that have worked so nicely for the spherical and affine cases do not seem to help.

Instead the arguments rely much more heavily on the underlying hyperbolic geometry of the building and the planarity of its apartments, with the general ideas adapted from the work of Haïssinski, Mathieu and Müller [25]. In this work the planarity and hyperbolicity of the Cayley graph of a surface group are exploited to develop a 'renewal theory' related to the automata structure of the group. In the setting of Fuchsian buildings the apartments of the building are planar and hyperbolic, and so similar ideas can be applied to the apartments. To lift this to the entire building requires some more work, and in [21] a theory of cones, cone types and automata for Fuchsian buildings paralleling the more familiar notions in groups is developed to achieve this goal. The idea is to find a decomposition of the trajectory of the walk into aligned pieces in such a way that these pieces are independent and identically distributed. Roughly speaking, one fixes a recurrent cone type $\mathbf{T}$ and sets R_1 to be the first time that the walk visits a cone of type $\mathbf{T}$ and never leaves this cone again. Inductively one defines R_{n+1} to be the first time after R_n that the walk enters a cone of type $\mathbf{T}$ and never leaves it again. The main results of [21] are as follows.

Theorem 3.27. *Let Σ be a regular Fuchsian building and let $(X_n)_{n \geq 0}$ be an isotropic random walk on Δ with bounded range. Then,*

$$\frac{1}{n} d(o, X_n) \xrightarrow{a.s.} v = \frac{\mathbb{E}[d(X_{R_2}, X_{R_1})]}{\mathbb{E}[R_2 - R_1]} > 0 \quad as \ n \to \infty.$$

Theorem 3.28. *Let Σ be a regular Fuchsian building and let $(X_n)_{n \geq 0}$ be an isotropic random walk on Δ with bounded range. Then, with v as in Theorem 3.27,*

$$\frac{d(o, X_n) - nv}{\sqrt{n}} \xrightarrow{\mathcal{D}} \mathcal{N}(0, \sigma^2), \quad where \quad \sigma^2 = \frac{\mathbb{E}[(d(X_{R_2}, X_{R_1}) - (R_2 - R_1)v)^2]}{\mathbb{E}[R_2 - R_1]}.$$

3.5. Future Directions

We conclude the main body of this chapter by listing some future directions and open problems in the theory of random walks on buildings:

(1) Provide sharp mixing time estimates and establish cut-off phenomenon for natural random walks on spherical buildings (in particular, for the simple random walk).

(2) Prove a law of large numbers, a central limit theorem, and a local limit theorem for isotropic random walks on the chambers of affine buildings (generalising [39]).

(3) Establish a local limit theorem for p-adic Lie groups of general type (generalising [54]).

(4) Prove convergence properties for the right random walk on a group acting on an affine building in the drift free case (c.f. [41]).

(5) Give an explicit formula for the spectral radius of a random walk on a Fuchsian building or Coxeter group (or any other non-spherical non-affine building or Coxeter group). There are some trivial 'tree-like' examples, although apart from these no explicit formulae are known. Efficient algorithms, or asymptotic formulae in the thickness parameter, would also be interesting in lieu of an explicit formula.

(6) Prove a precise and explicit local limit theorem for a non-spherical, non-affine building.

(7) Derive heat kernel and Green function estimates for random walks on the chambers of affine buildings (extending [55]).

(8) Generalise the Brownian motion convergence results of [46] to arbitrary type. As a first step one might consider either the rank 2 cases, or remove the nearest neighbour restriction from [46] for walks on $\widetilde{A}_n$ buildings.

A. Isotropic Random Walks on the Vertices of a $\widetilde{C}_2$ Building

In this appendix we carry out the details of the outline given in Section 3.3.1 in the special case of an affine building of type $\widetilde{C}_2$. These calculations were made by the author some years ago, in collaboration with Donald Cartwright, following the calculations made in the $\widetilde{A}_2$ case by Cartwright and Młotkowski [12]. The calculations here are very much 'hands-on', and do not require as much machinery as the general case. See the author's thesis for some further calculations for $\widetilde{G}_2$ buildings and so called $\widetilde{BC}_2$ buildings.

Let Σ be a building of type $\widetilde{C}_2$ with thickness parameters $q_0 = q_2 = q$ and $q_1 = r$. Let V be the set of all special vertices of Σ. The root system and fundamental coweights ω_1 and ω_2 are illustrated in Figure 14.2(b). If $\lambda = k\omega_1 + l\omega_2$ we write $V_\lambda(x) = V_{k,l}(x)$.

Lemma A.1. The cardinalities $N_{k,l} = |V_{k,l}(x)|$ do not depend on $x \in V$, and we have

$$N_{k,l} = (q+1)(r+1)(qr+1)q^2r(q^2r^2)^{k-1}(q^2r)^{l-1}$$
$$N_{k,0} = (r+1)(qr+1)q(q^2r^2)^{k-1}$$
$$N_{0,l} = (q+1)(qr+1)(q^2r)^{l-1}.$$

Let us illustrate Lemma A.1 with an examples (the general argument is an induction). Figure 14.8 shows part of an apartment of Σ. Panels with thickness q are shown as solid lines, and panels with thickness r are shown as dashed lines. The vertex y is in $V_{1,2}(x)$.

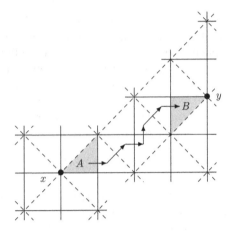

Figure 14.8 Computing the cardinalities $|V_{k,l}(x)|$

Let A be a chamber of the building containing x. There are $q^4 r^2$ galleries in the building starting at A and ending at a chamber in position B, and each of these end chambers contains a vertex in $V_{1,2}(x)$. Moreover, every vertex in $V_{1,2}(x)$ can be reached by such a gallery starting at some chamber containing x, and different starting chambers result in different end vertices in $V_{1,2}(x)$. Thus, $|V_{1,2}(x)| = Kq^4 r^2$, where K is the number of chambers containing x. The set of chambers containing x is a spherical building of type C_2 (that is, a generalised quadrangle) with parameters (q, r). Thus, $K = (q+1)(r+1)(qr+1)$, and hence $N_{1,2} = (q+1)(r+1)(qr+1) q^4 r^2$.

For each pair $k, l \geq 0$ define an operator $A_{k,l}$ acting on functions $f : V \to \mathbb{C}$ by

$$A_{k,l} f(x) = \frac{1}{N_{k,l}} \sum_{y \in V_{k,l}(x)} f(y).$$

Every isotropic random walk $(X_n)_{n \geq 0}$ on V has transition operator A of the form

$$A = \sum_{k,l \geq 0} a_{k,l} A_{k,l}$$

with $a_{k,l} \geq 0$ and $\sum_{k,l \geq 0} a_{k,l} = 1$. Explicitly, $a_{k,l} = \mathbb{P}[X_{n+1} \in V_{k,l}(x) \mid X_n = x] = p(x, y)/N_{k,l}$ for any $y \in V_{k,l}(x)$.

Theorem A.2. *The following formulae hold, where in each case the indices m, n are required to be large enough to ensure that the indices appearing on the right are all at least 0.*

$$A_{1,0}A_{0,1} = A_{0,1}A_{1,0}$$

$$N_{1,0}A_{m,n}A_{1,0} = rA_{m+1,n-2} + (q-1)(r+1)A_{m,n} + q^2r^2A_{m+1,n}$$
$$+ q^2rA_{m-1,n+2} + A_{m-1,n}$$

$$N_{1,0}A_{0,n}A_{1,0} = (r+1)A_{1,n-2} + (q-1)(r+1)A_{0,n} + q^2r(r+1)A_{1,n}$$

$$N_{1,0}A_{m,0}A_{1,0} = qr(q+1)A_{m-1,2} + q^2r^2A_{m+1,0} + (q-1)A_{m,0} + A_{m-1,0}$$

$$N_{1,0}A_{m,1}A_{1,0} = q^2r^2A_{m+1,1} + q^2rA_{m-1,3} + A_{m-1,1} + (qr+q-1)A_{m,1}$$

$$N_{0,1}A_{m,n}A_{0,1} = A_{m,n-1} + qrA_{m+1,n-1} + qA_{m-1,n+1} + q^2rA_{m,n+1}$$

$$N_{0,1}A_{0,n}A_{0,1} = A_{0,n-1} + q^2rA_{0,n+1} + q(r+1)A_{1,n-1}$$

$$N_{0,1}A_{m,0}A_{0,1} = (q+1)A_{m-1,1} + qr(q+1)A_{m,1}.$$

Proof. From the definition of the operators $A_{m,n}$ we have

$$A_{m,n}A_{s,t}f(x) = \sum_{u,v \geq 0} \left(\frac{1}{N_{u,v}} \sum_{y \in V_{u,v}(x)} \frac{N_{u,v}}{N_{m,n}N_{s,t}} |V_{m,n}(x) \cap V_{s,t}(y)| f(y) \right)$$

$$(14.14)$$

(see [36, (3.1)] for some intermediate steps). The formulae in the theorem follow by computing the cardinalities $|V_{m,n}(x) \cap V_{s,t}(y)|$ in the cases $(s,t) = (1,0)$ and $(s,t) = (0,1)$. We will give the calculation for $(s,t) = (0,1)$ and $m, n \geq 1$, leaving the remaining cases as an exercise.

Suppose that $y \in V_{k,l}(x)$, with $k, l \geq 1$, and consider the intersection $V_{i,j}(x) \cap V_{0,1}(y)$. It is clear from Figure 14.9 that if this intersection is nonempty then

$$(i,j) \in \{(k, l-1), (k-1, l+1), (k+1, l-1), (k, l+1)\}$$

(these are the four points marked with o in Figure 14.9). Using some basic building theory we see that $|V_{k,l-1}(x) \cap V_{0,1}(y)| = 1$, $|V_{k-1,l+1}(x) \cap V_{0,1}(y)| = q$, $|V_{k+1,l-1}(x) \cap V_{0,1}(y)| = qr$, and $|V_{k,l+1}(x) \cap V_{0,1}(y)| = q^2r$. Thus, for large enough $m, n \geq 1$ we have

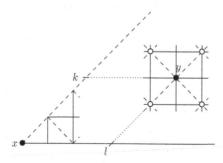

Figure 14.9 Computing the intersection cardinalities

$$|V_{m,n}(x) \cap V_{0,1}(y)| = \begin{cases} 1 & \text{if } y \in V_{m,n+1}(x) \\ q & \text{if } y \in V_{m+1,n-1}(x) \\ qr & \text{if } y \in V_{m-1,n+1}(x) \\ q^2 r & \text{if } y \in V_{m,n-1}(x). \end{cases}$$

It follows from (14.14) and Lemma A.1 that

$$A_{m,n}A_{0,1} = \frac{1}{N_{0,1}}(q^2 r A_{m,n+1} + qr A_{m+1,n-1} + q A_{m-1,n+1} + A_{m,n-1}).$$

The remaining formulae are similar, with some care for small values of m and n.   □

Let $\mathscr{A}$ be the linear span of $\{A_{m,n} \mid m, n \geq 0\}$ over $\mathbb{C}$.

Lemma A.3. *The vector space $\mathscr{A}$ is a commutative unital algebra over $\mathbb{C}$, generated by $A_{1,0}$ and $A_{0,1}$. Moreover, $\mathscr{A}$ is isomorphic to $\mathbb{C}[X, Y]$ (the algebra of polynomials in commuting indeterminates X and Y), with an isomorphism given by $X \mapsto A_{1,0}$ and $Y \mapsto A_{0,1}$.*

Proof. Let $\prec$ be the total order on $\mathbb{N}^2$ given by $(k, l) \prec (m, n)$ if either $k + l < m + n$ or $k + l = m + n$ and $k < m$. An induction using this total order and the formulae in Theorem A.2 shows that for each $(m, n) \in \mathbb{N}^2$ and $(k, l) \in \mathbb{N}^2$ the product $A_{m,n}A_{k,l}$ is a linear combination of terms $A_{i,j}$ with $(i, j) \in \mathbb{N}^2$. Thus, $\mathscr{A}$ is a unital algebra (with unit $A_{0,0} = I$). Moreover, for each $(m, n) \in \mathbb{N}^2$ an induction shows that there is a positive number $c_{m,n} > 0$ such that

$$A_{m,n} = c_{m,n} A_{1,0}^m A_{0,1}^n + \text{a linear combination of } A_{1,0}^k A_{0,1}^l$$

$$\text{with } (k, l) \prec (m, n). \qquad (14.15)$$

Thus, $\mathscr{A}$ is generated by $A_{1,0}$ and $A_{0,1}$, and hence is commutative as $A_{1,0}A_{0,1} = A_{0,1}A_{1,0}$.

It follows that there is a surjective homomorphism $\psi : \mathbb{C}[X, Y] \to \mathscr{A}$ with $\psi(X) = A_{1,0}$ and $\psi(Y) = A_{0,1}$. Suppose that $z = \sum a_{k,l} X^k Y^l \in \ker(\psi)$ is nonzero, and let $(m, n) \in \mathbb{N}^2$ be maximal subject to $a_{m,n} \neq 0$. Then (14.15) implies that

$$0 = \psi(z) = c'_{m,n} A_{m,n} + \text{linear combination of terms } A_{k,l}$$

$$\text{with } (k, l) \prec (m, n)$$

for some $c'_{m,n} \neq 0$, contradicting the linear independence of the operators $A_{k,l}$. Thus, ψ is injective, and so $\mathscr{A} \cong \mathbb{C}[X, Y]$.   □

Let C_2 be the group of signed permutations on two letters, acting on pairs of nonzero complex numbers (z_1, z_2) by permutations and inversions (for example, there is an element $\sigma \in C_2$ with $\sigma(z_1, z_2) = (z_2^{-1}, z_1)$). This group of order 8 is the Weyl group of type C_2.

Theorem A.4. *For each pair (z_1, z_2) of nonzero complex numbers there is a one-dimensional representation π_{z_1,z_2} of $\mathscr{A}$ given by*

$$\pi_{z_1,z_2}(A_{m,n}) = \frac{(qr)^{-m}(q\sqrt{r})^{-n}}{(1+q^{-1})(1+r^{-1})(1+q^{-1}r^{-1})} \sum_{\sigma \in C_2} c(z_{\sigma(1)}, z_{\sigma(2)}) z_{\sigma(1)}^{m+n} z_{\sigma(2)}^{n}$$

where

$$c(z_1, z_2) = \frac{(1 - q^{-1}z_1^{-1}z_2^{-1})(1 - q^{-1}z_1^{-1}z_2)(1 - r^{-1}z_1^{-2})(1 - r^{-1}z_2^{-2})}{(1 - z_1^{-1}z_2^{-1})(1 - z_1^{-1}z_2)(1 - z_1^{-2})(1 - z_2^{-2})}$$

whenever $z_1, z_2, z_1^{-1}, z_2^{-1}$ are pairwise distinct, and if $z_1, z_2, z_1^{-1}, z_2^{-1}$ are not pairwise distinct then the formula for $\pi_{z_1,z_2}(A_{m,n})$ is obtained from the above formula by taking an appropriate limit. Moreover every one-dimensional representation π of $\mathscr{A}$ is of the form $\pi = \pi_{z_1,z_2}$ for some $z_1, z_2 \in \mathbb{C}^\times$, and $\pi_{z_1,z_2} = \pi_{z_1',z_2'}$ if and only if $(z_1', z_2') = \sigma(z_1, z_2)$ for some $\sigma \in C_2$.

Proof. By Lemma A.3 we have $\mathscr{A} \cong \mathbb{C}[X, Y]$. The one-dimensional representations of $\mathbb{C}[X, Y]$ are precisely the evaluation maps $X \mapsto u$ and $Y \mapsto v$, and thus for each $(u, v) \in \mathbb{C}^2$ there is a unique one-dimensional representation $\pi^{(u,v)}$ of $\mathscr{A}$ determined by $\pi^{(u,v)}(A_{1,0}) = u$ and $\pi^{(u,v)}(A_{0,1}) = v$, and all one-dimensional representations are of this form.

Let $(u, v) \in \mathbb{C}^2$ and write $a_{m,n} = (qr)^m (q\sqrt{r})^n \pi^{(u,v)}(A_{m,n})$. Let $u' = (qr)^{-1} N_{1,0} u$ and $v' = (q\sqrt{r})^{-1} N_{0,1} v$. Applying π to the formulae in Theorem A.2 gives:

$$u' a_{m,n} = a_{m+1,n-2} + (1 - q^{-1})(1 + r^{-1})a_{m,n} + a_{m+1,n}$$
$$+ a_{m-1,n+2} + a_{m-1,n} \tag{14.16}$$

$$u' a_{0,n} = (1 + r^{-1})(a_{1,n-2} + (1 - q^{-1})a_{0,n} + a_{1,n}) \tag{14.17}$$

$$u' a_{m,0} = (1 + q^{-1})a_{m-1,2} + a_{m+1,0} + (1 - q^{-1})r^{-1}a_{m,0} + a_{m-1,0} \tag{14.18}$$

$$u' a_{m,1} = a_{m+1,1} + a_{m-1,3} + a_{m-1,1} + (1 + r^{-1} - q^{-1}r^{-1})a_{m,1} \tag{14.19}$$

$$v' a_{m,n} = a_{m,n-1} + a_{m+1,n-1} + a_{m-1,n+1} + a_{m,n+1} \tag{14.20}$$

$$v' a_{0,n} = a_{0,n-1} + a_{0,n+1} + (1 + r^{-1})a_{1,n-1} \tag{14.21}$$

$$v' a_{m,0} = (1 + q^{-1})(a_{m-1,1} + a_{m,1}), \tag{14.22}$$

where in each case the indices m, n are required to be large enough to ensure that the indices appearing on the right are all at least 0. From (14.20) we have

$$a_{m+1,n-1} + a_{m-1,n+1} = v' a_{m,n} - a_{m,n-1} - a_{m,n+1} \quad \text{for all } m, n \geq 1,$$

and using this equation in (14.16) gives

$$
\begin{aligned}
u' a_{m,n} &= (1 - q^{-1})(1 + r^{-1}) a_{m,n} + (a_{m+1,n-2} + a_{m-1,n}) \\
&\quad + (a_{m+1,n} + a_{m-1,n+2}) \\
&= (1 - q^{-1})(1 + r^{-1}) a_{m,n} + (v' a_{m,n-1} - a_{m,n-2} - a_{m,n}) \\
&\quad + (v' a_{m,n+1} - a_{m,n} - a_{m,n+2}),
\end{aligned}
$$

valid for all $m \geq 1$ and $n \geq 2$. A similar calculation using (14.21) and (14.17) shows that the above formula also holds for $m = 0$. By replacing n by $n + 2$ and rearranging we obtain

$$
a_{m,n+4} - v' a_{m,n+3} + \alpha\, a_{m,n+2} - v' a_{m,n+1} + a_{m,n} = 0 \quad \text{for all } m, n \geq 0,
\tag{14.23}
$$

where $\alpha = 2 + u' - (1 - q^{-1})(1 + r^{-1})$. The auxiliary equation of this linear recurrence (in n) factorises as a product of two quadratics: $\lambda^4 - v'\lambda^3 + \alpha\lambda^2 - v'\lambda + 1 = (\lambda^2 - a\lambda + 1)(\lambda^2 - b\lambda + 1)$, and so the roots of the auxiliary equation are of the form $z_1, z_1^{-1}, z_2, z_2^{-1}$ for some numbers $z_1, z_2 \in \mathbb{C}^\times$. By Newton's identities we have

$$
u = \frac{qr}{N_{1,0}} \left((1 - q^{-1})(1 + r^{-1}) + (z_1 + z_1^{-1})(z_2 + z_2^{-1}) \right)
\tag{14.24}
$$

$$
v = \frac{q\sqrt{r}}{N_{0,1}} \left(z_1 + z_1^{-1} + z_2 + z_2^{-1} \right).
\tag{14.25}
$$

Writing $\pi_{z_1,z_2} = \pi^{(u,v)}$ whenever $(z_1, z_2) \in (\mathbb{C}^\times)^2$ and $(u, v) \in \mathbb{C}^2$ are related as above, it follows from (14.24) and (14.25) that $\pi_{z_1,z_2} = \pi_{z_1',z_2'}$ if and only if $(z_1', z_2') = (z_{\sigma(1)}, z_{\sigma(2)})$ for some $\sigma \in C_2$. We now verify that $\pi_{z_1,z_2}(A_{m,n})$ is given by the formula in the statement of the theorem.

Assuming for now that $z_1, z_1^{-1}, z_2, z_2^{-1}$ are pairwise distinct, solving the recurrence (14.23) gives

$$
a_{m,n} = C_{1,m}(z_1, z_2) z_1^n + C_{2,m}(z_1, z_2) z_2^n + C_{3,m}(z_1, z_2) z_1^{-n} + C_{4,m}(z_1, z_2) z_2^{-n}
$$

for suitable functions $C_{i,m}(z_1, z_2)$ (independent of n). Writing $C_{1,m}(z_1, z_2) = C_m(z_1, z_2)$, the invariance under the group C_2 implies that

$$
\begin{aligned}
C_{2,m}(z_1, z_2) &= C_m(z_2, z_1), \quad C_{3,m}(z_1, z_2) = C_m(z_1^{-1}, z_2^{-1}), \\
C_{4,m}(z_1, z_2) &= C_m(z_2^{-1}, z_1^{-1}),
\end{aligned}
$$

and also that $C_m(z_1, z_2^{-1}) = C_m(z_1, z_2)$. Thus, for all $m, n \geq 0$ we have

$$
\begin{aligned}
a_{m,n} = &\; C_m(z_1, z_2) z_1^n + C_m(z_2, z_1) z_2^n + C_m(z_1^{-1}, z_2^{-1}) z_1^{-n} \\
&+ C_m(z_2^{-1}, z_1^{-1}) z_2^{-n}.
\end{aligned}
\tag{14.26}
$$

Writing $C_m = C_m(z_1, z_2)$ it follows from (14.26) and (14.20) that

$$z_1^{-1} C_{m+2} - (z_2 + z_2^{-1}) C_{m+1} + z_1 C_m = 0 \quad \text{for all } m \geq 0 \qquad (14.27)$$

(we have used the fact that $v' = z_1 + z_1^{-1} + z_2 + z_2^{-1}$). The roots of the auxiliary equation of the recurrence (14.27) are $z_1 z_2$ and $z_1 z_2^{-1}$, and these are distinct by hypothesis, and hence

$$C_m = D(z_1, z_2) z_1^m z_2^m + D'(z_1, z_2) z_1^m z_2^{-m} \quad \text{for all } m \geq 0$$

for suitable functions $D(z_1, z_2)$ and $D'(z_1, z_2)$ independent of m. Since $C_m(z_1, z_2) = C_m(z_1, z_2^{-1})$ we have $D'(z_1, z_2) = D(z_1, z_2^{-1})$, and thus by (14.26) we have

$$a_{m,n} = \sum_{\sigma \in C_2} D(z_{\sigma(1)}, z_{\sigma(2)}) z_{\sigma(1)}^{m+n} z_{\sigma(2)}^m \quad \text{for all } m, n \geq 0.$$

To compute $D(z_1, z_2)$ we proceed as follows: Using the recurrence formulae we obtain explicit formulae for $a_{0,0}, a_{0,1}, a_{0,2}$ and $a_{0,3}$ in terms of z_1 and z_2. In particular, $a_{0,0} = 1$, and $a_{1,0} = qru$ and $a_{0,1} = q\sqrt{r}v$ are given by (14.24) and (14.25). Then (14.21) with $n = 1$ gives $a_{0,2} = v'a_{0,1} - 1 - (1 + r^{-1})a_{1,0}$. By (14.21) with $n = 2$ we have $a_{0,3} = v'a_{0,2} - a_{0,1} - (1 + r^{-1})a_{1,1}$, and $a_{1,1}$ is computed using (14.22) with $m = 1$ giving $a_{11} = v'(1 + q^{-1})^{-1}a_{1,0} - a_{0,1}$. This gives the initial conditions of the recurrence (14.23) with $m = 0$, and thus the coefficients in (14.26) (with $m = 0$) can be computed, giving

$$C_0(z_1, z_2) = \frac{(1 - q^{-1} z_1^{-1} z_2^{-1})(1 - q^{-1} z_1^{-1} z_2)(1 - r^{-1} z_1^{-2})}{(1 + q^{-1})(1 + q^{-1} r^{-1})(1 - z_1^{-1} z_2^{-1})(1 - z_1^{-1} z_2)(1 - z_1^{-2})}.$$

From (14.21) and (14.26) we see that $C_1(z_1, z_2) = \frac{z_1(z_2 + z_2^{-1})}{1 + r^{-1}} C_0$, and thus the initial conditions of the recurrence (14.27) are known. Thus, we can solve for $D(z_1, z_2)$, and we find that $D(z_1, z_2) = c(z_1, z_2)/(1 + q^{-1})(1 + r^{-1})(1 + q^{-1}r^{-1})$, with $c(z_1, z_2)$ as in the statement of the theorem (a computer algebra package is recommended for these calculations).

Thus, the formula for π_{z_1, z_2} is verified in the case where $z_1, z_2, z_1^{-1}, z_2^{-1}$ are pairwise distinct. Generally, from (14.24), (14.25), and the fact that $A_{1,0}$ and $A_{0,1}$ generate $\mathscr{A}$ we see that $\pi(A_{m,n})$ is a polynomial in $z_1, z_1^{-1}, z_2, z_2^{-1}$. Thus, in the case where $z_1, z_1^{-1}, z_2, z_2^{-1}$ are not pairwise distinct we can obtain the formula for $\pi_{z_1, z_2}(A_{m,n})$ by taking an appropriate limit. $\qquad \Box$

Let $\mathbb{T} = \{t \in \mathbb{C} \mid |t| = 1\}$, and let dt denote normalised Haar measure on $\mathbb{T}$. Thus, for integrable functions f on $\mathbb{T}$ we have $\int_{\mathbb{T}} f(t)\, dt = \frac{1}{2\pi} \int_0^{2\pi} f(e^{i\theta})\, d\theta$. If $A \in \mathscr{A}$ and $t \in \mathbb{T}^2$ we write $\widehat{A}(t) = \pi_t(A)$. The following theorem establishes the Plancherel formula for the algebra $\mathscr{A}$.

Theorem A.5. *We have*

$$\frac{1}{N_{k,l}} \int_{\mathbb{T}^2} \widehat{A}_{k,l}(t)\widehat{A}_{m,n}(t)\, d\mu(t) = \delta_{(k,l),(m,n)},$$

where $d\mu(t)$ is the measure

$$d\mu(z) = K\frac{1}{|c(z_1,z_2)|^2}\, dt_1\, dt_2, \quad \text{with} \quad K = \frac{1}{8}(1+q^{-1})(1+r^{-1})(1+q^{-1}r^{-1}).$$

Proof. Since $A_{k,l}A_{m,n} = \sum_{i,j\geq 0} c^{(i,j)}_{(k,l),(m,n)} A_{i,j}$ and $c^{(0,0)}_{(k,l),(m,n)} = \delta_{(k,l),(m,n)} N_{k,l}$, it suffices to show that $\int_{\mathbb{T}^2} \widehat{A}_{k,l}(t)\, d\mu(t) = \delta_{(k,l),(0,0)}$. Using the facts that $|c(t_1,t_2)|^2 = c(t_{\sigma(1)}, t_{\sigma(2)})c(t^{-1}_{\sigma(1)}, t^{-1}_{\sigma(2)})$ for $(t_1,t_2) \in \mathbb{T}^2$ and $\sigma \in C_2$, and that $\int_{\mathbb{T}} f(t^{-1})\, dt = \int_{\mathbb{T}} f(t)\, dt$ we have

$$\int_{\mathbb{T}^2} \widehat{A}_{k,l}(t)\, d\mu(t) = \frac{1}{8}(qr)^{-k}(q\sqrt{r})^{-l} \int_{\mathbb{T}^2} \sum_{\sigma \in C_2} \frac{t^{k+l}_{\sigma(1)} t^l_{\sigma(2)}}{c(t^{-1}_{\sigma(1)}, t^{-1}_{\sigma(2)})}\, dt_1\, dt_2$$

$$= (qr)^{-k}(q\sqrt{r})^{-l} \int_{\mathbb{T}} \left(\int_{\mathbb{T}} \frac{t^{k+l}_1 t^l_2}{c(t^{-1}_1, t^{-1}_2)}\, dt_1 \right) dt_2.$$

As a contour integral, the inner integral is

$$\int_{\mathbb{T}} \frac{t^{k+l}_1 t^l_2}{c(t^{-1}_1, t^{-1}_2)}\, dt_1 = \frac{1}{2\pi i} \int_{\Gamma} \frac{z^{k+l}_1 t^l_2}{c(z^{-1}_1, t^{-1}_2)} \frac{dz_1}{z_1}$$

where Γ is the unit circle traversed once counterclockwise. The poles of the function $f(z_1) = 1/c(z^{-1}_1, t^{-1}_2)$ are at $z_1 = qt^{-1}_2, qt_2, \pm\sqrt{r}$ and since $|t_2| = 1$ and $q, r > 1$ we see that $f(z_1)$ has no poles inside the contour Γ, and so by residue calculus we deduce that

$$\int_{\mathbb{T}} \frac{t^{k+l}_1 t^l_2}{c(t^{-1}_1, t^{-1}_2)}\, dt_1 = \delta_{(k,l),(0,0)} \lim_{z_1 \to 0} \frac{1}{c(z^{-1}_1, t^{-1}_2)} = \delta_{(k,l),(0,0)}\frac{1-t^2_2}{1-r^{-1}t^2_2}.$$

Thus,

$$\int_{\mathbb{T}^2} \widehat{A}_{k,l}(t)\, d\mu(t) = \delta_{(k,l),(0,0)} \int_{\mathbb{T}} \frac{1-t^2_2}{1-r^{-1}t^2_2}\, dt_2 = \delta_{(k,l),(0,0)}. \qquad \square$$

Theorem A.6. *Let $(X_n)_{n\geq 0}$ be an isotropic random walk on Σ with transition operator A. Then*

$$p^{(n)}(x,y) = \frac{1}{N^2_{k,l}} \int_{\mathbb{T}^2} \widehat{A}(t)^n \widehat{A}_{k,l}(t)\, d\mu(t) \quad \text{if } y \in V_{k,l}(x).$$

Proof. We have $A^n = \sum_{i,j\geq 0} a^{(n)}_{i,j} A_{i,j}$ where $p^{(n)}(x,y) = a^{(n)}_{i,j}/N_{i,j}$ whenever $y \in V_{i,j}(x)$. Thus, by Theorem A.5 we have, for any pair x, y with $y \in V_{k,l}(x)$,

$$\int_{\mathbb{T}^2} \widehat{A}(t)^n \widehat{A}_{k,l}(t) \, d\mu(t) = \sum_{i,j \geq 0} a_{i,j}^{(n)} \int_{\mathbb{T}^2} \widehat{A}_{i,j}(t) \widehat{A}_{k,l}(t) \, d\mu(t)$$

$$= N_{k,l} a_{k,l}^{(n)} = N_{k,l}^2 p^{(n)}(x,y). \qquad \square$$

The asymptotics for the n-step transition probabilities of the random walk (that is, the local limit theorem) can be extracted in a standard way from Theorem A.6. Let us simply illustrate this in an example. Consider the 'simple random walk' on V with transition operator $A_{0,1}$ (this is the random walk such that if $X_n = y$ in Figure 14.9, then X_{n+1} is one of the $N_{0,1} = (q + 1)(qr + 1)$ vertices marked with o in the figure, each chosen with equal probability $1/N_{0,1}$). We can compute $\widehat{A}_{0,1}(t)$ from the general formula in Theorem A.4, however a shortcut is given by (14.25), giving

$$\widehat{A}_{0,1}(t) = \frac{q\sqrt{r}}{(q+1)(qr+1)} \left(t_1 + t_1^{-1} + t_2 + t_2^{-1} \right).$$

The simple random walk is periodic, with period 2, and so we consider $p^{(2n)}(x,x)$. Writing $e^{i\theta} = (e^{i\theta_1}, e^{i\theta_2})$ we have

$$\widehat{A}_{0,1}(e^{i\theta}) = \frac{2q\sqrt{r}(\cos\theta_1 + \cos\theta_2)}{(q+1)(qr+1)}$$

and

$$\frac{1}{|c(e^{i\theta})|^2} = \frac{4(\theta_1^2 - \theta_2^2)^2 \theta_1^2 \theta_2^2}{(1 - q^{-1})^4(1 - r^{-1})^4} + O(\|\theta\|^4).$$

Let $\rho = 4q\sqrt{r}/((q+1)(qr+1))$ and $K' = K/(\pi^2(1 - q^{-1})^4(1 - r^{-1})^4)$. Some standard tricks from asymptotic analysis now give

$$p^{(2n)}(x,x) = \frac{K}{4\pi^2} \int_{-\pi}^{\pi} \int_{-\pi}^{\pi} \frac{\widehat{A}_{0,1}(e^{i\theta})}{|c(e^{i\theta})|^2} \, d\theta_1 \, d\theta_2$$

$$\sim K' \rho^{2n} \int_{-\pi}^{\pi} \int_{-\pi}^{\pi} \left(\frac{\cos\theta_1 + \cos\theta_2}{2} \right)^{2n} (\theta_1^2 - \theta_2^2)^2 \theta_1^2 \theta_2^2 \, d\theta_1 \, d\theta_2$$

$$\sim 2K' \rho^{2n} \int_{-\epsilon}^{\epsilon} \int_{-\epsilon}^{\epsilon} \left(\frac{\cos\theta_1 + \cos\theta_2}{2} \right)^{2n} (\theta_1^2 - \theta_2^2)^2 \theta_1^2 \theta_2^2 \, d\theta_1 \, d\theta_2$$

$$= \frac{2K'}{n^4} \rho^{2n} \int_{-\sqrt{n}\epsilon}^{\sqrt{n}\epsilon} \int_{-\sqrt{n}\epsilon}^{\sqrt{n}\epsilon} \left(\frac{\cos(\varphi_1/\sqrt{n}) + \cos(\varphi_2/\sqrt{n})}{2} \right)^{2n}$$
$$(\varphi_1^2 - \varphi_2^2)^2 \varphi_1^2 \varphi_2^2 \, d\varphi_1 \, d\varphi_2$$

$$\sim \frac{2K'}{n^4} \rho^{2n} \int_{-\infty}^{\infty} \int_{-\infty}^{\infty} e^{-(\varphi_1^2 + \varphi_2^2)/2} (\varphi_1^2 - \varphi_2^2)^2 \varphi_1^2 \varphi_2^2 \, d\varphi_1 \, d\varphi_2,$$

and thus in conclusion we have

$$p^{(2n)}(x,x) \sim \frac{6(q+1)(r+1)(qr+1)q^2r^2}{\pi(q-1)^4(r-1)^4} \left(\frac{4q\sqrt{r}}{(q+1)(qr+1)}\right)^{2n} n^{-4}.$$

B. Rank 2 Hecke Algebras and the Feit–Higman Theorem

In this appendix we present the representation theory of rank 2 spherical Hecke algebras and apply it to Theorems 3.8 and 3.10. As a byproduct we arrive at a proof of the Feit–Higman Theorem (this proof is due to Kilmoyer and Solomon [31]). Let (W, S) be the Coxeter system of type $I_2(m)$. Write $S = \{s_1, s_2\}$, and write $q_{s_1} = q$ and $q_{s_2} = r$ (with $q = r$ if m is odd). Let $T_i = T_{s_i}$ for $i = 1, 2$ be the generators of the abstract Hecke algebra $\mathcal{H}$. The classification of the irreducible representations of $\mathcal{H}$ is elementary:

Proposition B.1. [20, Theorem 8.3.1] *In the above notation:*

(1) *If m is odd then the complete list of irreducible representations of $\mathcal{H}$ is as follows: There are precisely 2 one-dimensional irreducible representations, given by*

$$\begin{array}{cc} \rho_{\mathrm{triv}}(T_1) = 1 & \rho_{\mathrm{sgn}}(T_1) = -q^{-1} \\ & and \\ \rho_{\mathrm{triv}}(T_2) = 1 & \rho_{\mathrm{sgn}}(T_2) = -q^{-1} \end{array}$$

and precisely $(m-1)/2$ 2-dimensional representations, given by

$$\rho_j(T_1) = \frac{1}{q}\begin{bmatrix} -1 & 0 \\ c_j & q \end{bmatrix} \quad and \quad \rho_j(T_2) = \frac{1}{q}\begin{bmatrix} q & c'_j \\ 0 & -1 \end{bmatrix}$$

for $1 \leq j \leq (m-1)/2$, where c_j and c'_j are any numbers satisfying $c_j c'_j = 4q\cos^2(\pi j/m)$.

(2) *If m is even then the complete list of irreducible representation of $\mathcal{H}$ is as follows: There are precisely 4 1-dimensional representations, given by*

$$\begin{array}{ll} \rho_{\mathrm{triv}}(T_1) = 1 & \rho_{\mathrm{triv}}(T_2) = 1 \\ \rho_{\mathrm{sgn}}(T_1) = -q^{-1} & \rho_{\mathrm{sgn}}(T_2) = -r^{-1} \\ \rho^1(T_1) = 1 & \rho^1(T_2) = -r^{-1} \\ \rho^2(T_1) = -q^{-1} & \rho^2(T_2) = 1. \end{array}$$

There are exactly $(m-2)/2$ two-dimensional representations, given by

$$\rho_j(T_1) = \frac{1}{q}\begin{bmatrix} -1 & 0 \\ c_j & q \end{bmatrix} \quad and \quad \rho_j(T_2) = \frac{1}{r}\begin{bmatrix} r & c'_j \\ 0 & -1 \end{bmatrix}$$

for $1 \leq j \leq (m-2)/2$, *where* c_j *and* c_j' *are any numbers satisfying* $c_j c_j' = q + r + 2\sqrt{qr}\cos(2\pi j/m)$.

Proof. It is a straightforward exercise to show that the claimed formulae produce representations (by checking that the defining relations (14.7) are satisfied). It is also easy to check that the representations are irreducible. To check that they are pairwise non-isomorphic one can compute the inner products and check (14.9), and finally to check we have all irreducible representations one uses the character theoretic fact that $\sum \dim(\rho)^2 = |I_2(m)| = 2m$, where the sum is over all irreducible representations of $\mathscr{H}$. See [20, Theorem 8.3.1] for details. □

The representation theory described in Proposition B.1 allows us to be extremely precise for random walks on generalised polygons. Let us simply illustrate these arguments for the case of generalised quadrangles (that is, $m = 4$):

Corollary B.2. *For the simple random walk on a generalised quadrangle with parameters (q, r) we have*

$$p^{(n)}(o, o) = \frac{1 + k_1\lambda_1^n + k_2\lambda_2^n + k_3\lambda_3^n + k_4(\lambda_+^n + \lambda_-^n)}{(q+1)(r+1)(qr+1)}$$

and

$$\|\mu^{(n)} - u\|_{\mathrm{tv}}^2 \leq \frac{1}{4}\left(k_1\lambda_1^{2n} + k_2\lambda_2^{2n} + k_3\lambda_3^{2n} + k_4(\lambda_+^{2n} + \lambda_-^{2n})\right),$$

where the numbers k_i, λ_i, and $\lambda_\pm$ are given by $k_1 = q^2r^2$, $k_2 = r^2(qr+1)/(q+r)$, $k_3 = q^2(qr+1)/(q+r)$, $k_4 = qr(q+1)(r+1)/(q+r)$, $\lambda_1 = -2/(q+r)$, $\lambda_2 = (q-1)/(q+r)$, $\lambda_3 = (r-1)/(q+r)$, and $\lambda_\pm = (q+r-2 \pm \sqrt{(q-r)^2 + 4(q+r)})/2(q+r)$.

Proof. For the first statement, from Theorem 3.8 we have

$$p^{(n)}(o, o) = \frac{1}{|\Delta|} \sum_{\rho \in \mathrm{Irrep}(\mathscr{H})} m_\rho \chi_\rho(T^n),$$

where $T = \frac{q}{q+r}T_1 + \frac{r}{q+r}T_2$ (since we are considering the simple random walk). We have $|\Delta| = (q+1)(r+1)(qr+1)$, and by Proposition B.1 there are five irreducible representations of $\mathscr{H}$ with respective multiplicities $m_{\mathrm{triv}} = 1$, $m_{\mathrm{sgn}} = q^2r^2$, $m^1 = r^2(qr+1)/(q+r)$, $m^2 = q^2(qr+1)/(q+r)$, and $m_1 = qr(q+1)(r+1)/(q+r)$ (computed using Theorem 3.7). We have $\chi_{\mathrm{triv}}(T^n) = 1$, $\chi_{\mathrm{sgn}}(T^n) = (-2/(q+r))^n$, $\chi^1(T^n) = ((q-1)/(q+r))^n$, $\chi^2(T) = ((r-1)/(q+r))^n$, and a calculation of eigenvalues gives $\chi_1(T^n) = \lambda_+^n + \lambda_-^n$. The result follows, and the second statement follows similarly from Theorem 3.10, noting that $T^* = T$. □

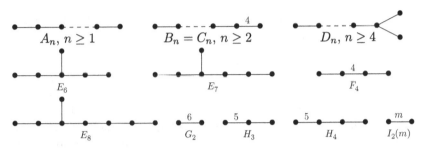

Figure 14.10 Irreducible spherical Coxeter systems

By considering the formulae for the multiplicities of irreducible representations in the geometric representation we obtain a proof of the Feit–Higman theorem and some of the divisibility conditions from Theorem 2.12 (c.f. [31]):

Proof of the Feit–Higman Theorem and Divisibility Conditions. Suppose that a finite thick generalised m-gon exists with parameters (q, r). Let χ_j be the character of the representation ρ_j from Proposition B.1. By Theorem 3.7 we have $\langle \chi_j, \chi_j \rangle \in \mathbb{Q}$. On the other hand we can explicitly compute these inner products. Writing $\theta_j = 2\pi j/m$, a tedious calculation gives

$$
|\Delta| \langle \chi_j, \chi_j \rangle =
\begin{cases}
2m + \dfrac{(q-1)^2 m}{q(1-\cos\theta_j)} & \text{if } m \text{ is odd} \\[2ex]
2m + \dfrac{(r(q-1)^2 + q(r-1)^2)m}{2qr\sin^2\theta_j} + \dfrac{(q-1)(r-1)m\cos\theta_j}{\sqrt{qr}\sin^2\theta_j} & \text{if } m \text{ is even.}
\end{cases}
$$

If m is odd, then the formulae force $\cos\theta_j$ to be rational, and this implies that $m = 3$. If m is even, then the above formulae imply that both $\sin^2(2\pi/m)$ and $\cos(2\pi/m)$ are rational (consider $\langle \chi_1, \chi_1 \rangle + \langle \chi_{(m/2)-j}, \chi_{(m/2)-j} \rangle$ to see that $\sin^2(2\pi/m)$ is rational). Together these facts imply that $m \in \{2, 4, 6, 8\}$.

The divisibility conditions for the cases $m = 4, 6, 8$ follow by computing the multiplicity of ρ^1 using Theorem 3.7. The facts that $\sqrt{qr} \in \mathbb{Z}$ and $\sqrt{2qr} \in \mathbb{Z}$ for hexagons and octagons (respectively) arise from the multiplicity of ρ_1.   □

C. Spherical and Affine Coxeter Systems

In this final appendix we list the Coxeter diagrams of the irreducible spherical and affine Coxeter systems. For the affine systems, the extra generator that is added to the spherical system is indicated by o.

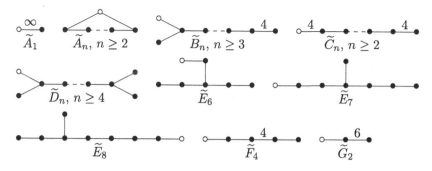

Figure 14.11 Irreducible affine Coxeter systems

References

[1] P. Abramenko, K. Brown *Buildings: theory and applications*, Graduate Texts in Mathematics, **248**, Springer, 2008.

[2] J.-P. Anker, B. Schapira, B. Trojan, *Heat kernel and Green function estimates on affine building of type* $\widetilde{A}_r$, preprint, 2006.

[3] L. Bartholdi, M. Neuhauser, W. Woess, *Horocyclic products of trees*, J. Eur. Math. Soc. **10**, 771–816, 2008.

[4] N. Bourbaki, *Lie groups and lie algebras, Chapters 4–6*, Elements of mathematics, Springer-Verlag, Berlin Heidelberg New York, 2002.

[5] P. Bougerol, *Théorème central limite local sur certains groupes de Lie*, Ann. Sci. École Norm. Sup., **14**, no. 4, 403–32, 1981.

[6] M. Bridson, A. Haefliger, *Metric spaces of non-positive curvature*, Grundlehren **319**, Springer-Verlag, 1999.

[7] K. Brown, P. Diaconis, *Random walks and hyperplane arrangements*, Ann. Probab. **26**, 4, 1813–54, 1998.

[8] K. Brown, *Semigroups, rings, and Markov chains*, J. Theoret. Probab. **13**, 871–938, 2000.

[9] K. Brown, *Semigroup and ring theoretical methods in probability*, Representations of finite dimensional algebras and related topics in Lie theory and geometry, 326, Fields Inst. Commun., **40**, Amer. Math. Soc., Providence, RI, 2004.

[10] F. Bruhat, J. Tits, *Groupes réductifs sur un corps local*, Publ. Math. IHES, **41**, 5–251, 1972.

[11] R. Carter, *Simple Groups of Lie Type*, Wiley, 1989.

[12] D. I. Cartwright, W. Młotkowski, *Harmonic analysis for groups acting on triangle buildings*, Journal of the Australian Mathematical Society A, **56**, 345–83, 1994.

[13] D. I. Cartwright, V. Kaimanovich, W. Woess, *Random walks on the affine group of local fields and of homogeneous trees*, Annales de l'Institut Fourier **44**, no. 4, 1243–88, 1994.

[14] D. I. Cartwright, W. Woess, *Isotropic random walks in a building of type* $\widetilde{A}_d$, Mathematische Zeitschrift, **247**, 101–35, 2004.

[15] H. S. M. Coxeter, *The complete enumeration of finite groups of the form $r_i^2 = (r_i r_j)^{k_{ij}} = 1$*, J. London Math. Soc. **10**, 21–5, 1935.

[16] M. W. Davis, *The geometry and topology of Coxeter groups*, London Mathematical Society Monograph Series, Princeton University Press, **32**, 2008.

[17] P. Diaconis, *Group representations in probability and statistics*, Institute of Mathematical Statistics lecture notes, vol. 11, 1988.

[18] P. Diaconis, A. Ram, *Analysis of systematic scan metropolis algorithms using Iwahori-Hecke algebra techniques*, Michigan Math. J. **48**, 2000.

[19] W. Feit, G. Higman, *The nonexistence of certain generalized polygons*, J. Algebra, **1**, 114–31, 1964.

[20] M. Geck, G. Pfeiffer, *Characters of finite Coxeter groups and Iwahori-Hecke algebras*, Oxford University Press, 2000.

[21] L. Gilch, S. Müller, J. Parkinson, *Limit theorems for random walks on Fuchsian buildings and Kac-Moody groups: Groups, geometry, and dynamics*, to appear 2017.

[22] Y. Guivarc'h, M. Keane, P. Roynette, *Marches aléatoires sur les groupes de Lie*, LNM Vol. **624**, Springer-Verlag, 1977.

[23] Y. Guivarc'h, *Loi des grands nombres et rayon spectral d'une marche aléatoire sur un groupe de Lie*, Astérisque, vol. **74**, 47–98, 1980.

[24] A. Gyoja, K. Uno, *On the semisimplicity of Hecke algebras*, J. Math. Soc. Japan **41**, 75–9, 1989.

[25] P. Haïssinski, P. Mathieu, S. Müller *Renewal theory for random walks on surface groups*, preprint, 2013.

[26] J. E. Humphreys, *Reflection groups and Coxeter groups*, Cambridge Studies in Advanced Mathematics, **29**, 1990.

[27] N. Iwahori, H. Matsumoto, *On some Bruhat decomposition and the structure of the Hecke rings of p-adic Chevalley groups*, Publications Mathématiques de l'IHÉS **25**, 5–48, 1965.

[28] V. Kac, *Infinite dimensional dimensional Lie algebras*, 3rd edn, Cambridge University Press, 1990.

[29] V. Kaimanovich, *Lyapunov exponents, symmetric spaces, and a multiplicative ergodic theorem for semisimple Lie groups*. J. Soviet Math. **47**, 2387–98, 1989.

[30] W. Kantor, *Generalized polygons, SCABs and GABs*, In: Buildings and the Geometry of Diagrams, Proceedings, Como, Lecture Notes in Mathematics, Springer-Verlag, **1181**, 79–158, 1984.

[31] R. Kilmoyer, L. Solomon, *On the theorem of Feit-Higman*, Journal of Combinatorial Theory, Series A, **15**, 3, 310–22, 1973.

[32] S. P. Lalley, *Finite range random walk on free groups and homogeneous trees*, Ann. Probab., **21** (4), 2087–130, 1993.

[33] M. Lindlbauer, M. Voit, *Limit theorems for isotropic random walks on triangle buildings*, J. Aust. Math. Soc, **73**, 301–33, 2002.

[34] I. G. Macdonald, *Spherical functions on a group of p-adic type*, Publications of the Ramanujan Institute, **2**, University of Madras, 1971.

[35] E. Opdam, *On the spectral decomposition of affine Hecke algebras*, J. Inst. Math. Jussieu, **3**, No. 4, 531–648, 2004.

[36] J. Parkinson, *Buildings and Hecke algebras*, J. Algebra, **297**, 1–49, 2006.

[37] J. Parkinson, *Spherical harmonic analysis on affine buildings*, Math. Z. **253**, 571–606, 2006.

[38] J. Parkinson, *Isotropic random walks on affine buildings*, Annales de l'Institut Fourier, **57**, No.2, 379–419, 2007.

[39] J. Parkinson, B. Schapira, *A local limit theorem for random walks on the chambers of $\tilde{A}_2$ buildings*, Progress in Probability, **64**, Birkhäuser, 15–53, 2011.

[40] J. Parkinson, *On calibrated representations and the Plancherel theorem for affine Hecke algebras*, J. Algebraic Combin. **40**, 331–71, 2014.

[41] J. Parkinson, W. Woess, *Random walks and regular sequences in affine buildings*, Ann. Inst. Fourier, **65**, 675–707, 2015.

[42] M. Ronan, *Lectures on buildings*, University of Chicago Press, 2009.

[43] M. Ronan, *A construction of buildings with no rank 3 residues of spherical type*. Buildings and the Geometry of Diagrams, Lecture Notes in Mathematics, **1181**, 242–8, 1986.

[44] L. Saloff-Coste, W. Woess, *Transition operators, groups, norms, and spectral radii*, Pacific Journal of Mathematics, **180**, no. 2, 1997.

[45] S. Sawyer, *Isotropic random walks in a tree*, Z. Wahrsch. Verw. Gebiete **42**, 279–92, 1978.

[46] B. Schapira, *Random walk on a building of type $\tilde{A}_r$ and Brownian motion of the Weyl chamber*, Annales de l'I.H.P. (B) **45**, 289–301, 2009.

[47] R. Steinberg, *Lecture notes on Chevalley groups*, Yale University, 1967.

[48] D. Stroock, S. R. S. Varadhan, *Limit theorems for random walks on Lie groups*, Indian J. of Stat., Sankhyā, Ser. A, **35**, 277–94, 1973.

[49] J. Tits, *Buildings of spherical type and finite BN-pairs. Lecture Notes in Mathematics*, vol. **386** Springer-Verlag, Berlin-New York, 1974.

[50] J. Tits, *Endliche Spiegelungsgruppen, die als Weylgruppen auftreten*, Invent. Math. **43**, 283–95, 1977.

[51] J. Tits, *A local approach to buildings*, The geometric vein: The Coxeter Festschrift, Springer-Verlag, 519–47, 1981.

[52] J. Tits, *Immeubles de type affine*, Buildings and the geometry of diagrams (Como, 1984), 159–90, Lecture Notes in Mathematics, **1181**, Springer, Berlin, 1986.

[53] J. Tits, *Uniqueness and presentation of Kac-Moody groups over fields*, J. Algebra, **105**(2), 542–73, 1987.

[54] F. Tolli, *A local limit theorem on certain p-adic groups and buildings*, Monatsh. Math., **133**, 163–73, 2001.

[55] B. Trojan, *Heat kernel and Green function estimates on affine buildings*, preprint, 2013.

[56] H. Van Maldeghem, *Generalized polygons*, Monographs in Mathematics, **93**, Birkhäuser, Basel, Boston, Berlin, 1998.

[57] A. D. Virtser, *Central limit theorem for semi-simple Lie groups*, Theory Probab. Appl. **15**, 667–87, 1970.

[58] D. Wehn, *Probabilities on Lie groups*, Proc. Nat. Acad. Sci. USA, **48**, 791–5, 1962.

[59] R. Weiss, *The structure of affine buildings*, Annals of Mathematics Studies, **168**, Princeton University Press, 2009.

[60] W. Woess, *Random walks on infinite graphs and groups*. Cambridge Tracts in Mathematics **138**, Cambridge University Press, 2000.

15

ON SOME RANDOM WALKS DRIVEN
BY SPREAD-OUT MEASURES

LAURENT SALOFF-COSTE[1],* AND TIANYI ZHENG[2]

[1]Department of Mathematics, Cornell University, Ithaca, New York
[2]Department of Mathematics, Stanford University, Stanford, California

To Professor Woess on the occasion of his sixtieth birthday

Abstract

Let G be a finitely generated group equipped with a symmetric generating k-tuple S. Let $|\cdot|$ and V be the associated word length and volume growth function. Let ν be a probability measure such that $\nu(g) \simeq [(1+|g|)^2 V(|g|)]^{-1}$. We prove that if G has polynomial volume growth then $\nu^{(n)}(e) \simeq V(\sqrt{n \log n})^{-1}$. We also obtain assorted estimates for other spread-out probability measures.

2010 MSC: 20F65, 60J10, 60J51

Key words: Random walk, polynomial volume growth, local limit theorems

Contents

1. Introduction

This work is concerned with questions related to a number of recent studies where "stable-like" processes and random walks are considered. We focus on random walks on groups, mostly nilpotent groups and groups of polynomial volume growth, associated with various types of spread-out probability measures. Here, "spread out" is used in a nontechnical sense to convey the idea that these measures do not have finite support. We will be interested in two main cases. In the first case, the measure is spread out in a radial way with respect to a fixed word

*Both authors partially supported by NSF grant DMS-1004771 and DMS-1404435.

length on G. In the second case, the measure is spread out along the one parameter groups $\{s_i^m : m \in \mathbb{Z}\}$ corresponding to a given finite list of generators $(s_1, \ldots, s_k)$. In the second case, the measures are "more singular" and the study of their convolution powers is more delicate.

Given a (symmetric) probability measure ν on a (finitely generated) group G, we consider the discrete time random walk $(X_n)_0^\infty$ driven by ν and started at $X_0 = e$. This means that $X_n = \xi_1 \ldots \xi_n$, $n \geq 1$, where $(\xi)_1^\infty$ is an i.i.d. sequence of G-random variables with common law ν. We denote by $\mathbf{P}_g$ the law of the sequence (X_n) with $X_0 = g$. The distribution of X_n itself is the convolution power $\nu^{(n)}$. We also consider the associated continuous-time random walk X_t whose distribution is given by

$$p_t(g) = e^{-t} \sum_0^\infty \frac{t^n}{n!} \nu^{(n)}(g).$$

This continuous-time process will serve as a tool in the study of the discrete random walk driven by ν, a technique that has been used by many authors before.

The question addressed in the present work is the following. Assuming good upper bounds on $\nu^{(n)}(e)$, under which circumstances can one prove matching lower bounds? Further, can one describe (in a certain sense) the region in G where, for a given n, $\nu^{(n)}(g) \simeq \nu^{(n)}(e)$? We provide answers for measures ν that are quite natural and for which well-understood existing techniques are insufficient and/or need to be modified.

1.1. Main Definitions

Definition 1.1. We say that $\| \cdot \| : G \to [0, \infty)$ is a norm on G if $\|g\| = 0$ if and only if $g = e$ and, for all $g, h \in G$, $\|gh\| \leq \|g\| + \|h\|$ and $\|g\| = \|g^{-1}\|$. Given a norm $\| \cdot \|$, we say that $V(r) = \#\{g \in G : \|g\| \leq r\}$ is the associated volume function.

The simplest and most common example is of a norm provided by the word length associated to a given finite symmetric set of generators. We will encounter other norms as well.

The properties studied in this work are the following.

Definition 1.2. Let μ be a symmetric probability measure on a group G. Let $\| \cdot \|$ be a norm with volume function V. Let $r : (0, \infty) \to (0, \infty)$, $t \mapsto r(t)$, be a nondecreasing function. Let $(X_n)_0^\infty$ be the random walk on G driven by μ. We say that μ is $(\| \cdot \|, r)$ controlled if the following properties are satisfied:

1. For all n, $\mu^{(2n)}(e) \simeq V(r(n))^{-1}$.
2. For all $\epsilon > 0$ there exists $\gamma \in (0, \infty)$ such that

$$\mathbf{P}_e \left(\sup_{0 \leq k \leq n} \{\|X_k\|\} \geq \gamma r(n) \right) \leq \epsilon.$$

The first of these two properties is rather straightforward and self-explanatory. It provides a two-sided estimate for the probability of return of the random walk. In more general contexts, this property is also known as a two-sided "on-diagonal" bound. The second property is related to the first insofar as it actually easily implies the lower bound $\mu^{(2n)}(e) \geq V(cr(n))^{-1}$. It also provides a weak control of the behavior of $\mu^{(n)}(g)$ away from the neutral element e.

Definition 1.3. Let μ be a symmetric probability measure on a group G. Let $\| \cdot \|$ be a norm with volume function V. Let $r : (0, \infty) \to (0, \infty)$, $t \mapsto r(t)$, be an increasing continuous function with inverse ρ. Let $(X_n)_0^\infty$ denote the random walk on G driven by μ. We say that μ is strongly $(\| \cdot \|, r)$ controlled if the following properties are satisfied:

1. There exists $C \in (0, \infty)$ and, for any $\kappa > 0$, there exists $c(\kappa) > 0$ such that, for all $n \geq 1$ and g with $\|g\| \leq \kappa r(n)$,

$$c(\kappa) V(r(n)))^{-1} \leq \mu^{(2n)}(g) \leq CV(r(n))^{-1}.$$

2. There exists $\epsilon, \gamma_1, \gamma_2 \in (0, \infty)$, $\gamma_2 \geq 1$, such that, for all n, τ such that $\frac{1}{2}\rho(\tau/\gamma_1) \leq n \leq \rho(\tau/\gamma_1)$

$$\inf_{x:\|x\|\leq\tau} \left\{ \mathbf{P}_x \left(\sup_{0\leq k\leq n} \{\|X_k\|\} \leq \gamma_2\tau; \|X_n\| \leq \tau \right) \right\} \geq \epsilon. \quad (15.1)$$

Strong control implies the following useful estimate. The last section of this chapter gives an application of this estimate to random walks on wreath products.

Proposition 1.4. *Assume that r is continuous, increasing with inverse ρ and that the symmetric probability measure μ is strongly $(\| \cdot \|, r)$ controlled. Then, for any n and τ such that $\gamma_1 r(2n) \geq \tau$, we have*

$$\inf_{x:\|x\|\leq\tau} \left\{ \mathbf{P}_x \left(\sup_{0\leq k\leq n} \{\|X_k\|\} \leq \gamma_2\tau; \|X_n\| \leq \tau \right) \right\} \geq \epsilon^{1+2n/\rho(\tau/\gamma_1)}. \quad (15.2)$$

Proof. By induction on $\ell \geq 1$ such that $1 \leq \ell \leq 2n/\rho(\tau/\gamma_1) < (\ell+1)$, we are going to prove that

$$\inf_{x:\|x\|\leq\tau} \left\{ \mathbf{P}_x \left(\sup_{0\leq k\leq n} \{\|X_k\|\} \leq \gamma_2\tau; \|X_n\| \leq \tau \right) \right\} \geq \epsilon^{1+\ell}.$$

This easily yields the desired result. For $\ell = 1$, the inequality follows from the strong control assumption. Assume the property holds for some $\ell \geq 1$. Let n, τ be such that $(\ell+1) \leq 2n/\rho(\tau/\gamma_1) < (\ell+2)$. Choose n' such that $n - n' = \lceil \rho(\tau/\gamma_1)/2 \rceil$ and note that $2n' \in [1, (\ell+1)\rho(\tau/\gamma_1))$. Write $Z_n = \sup_{k\leq n}\{\|X_k\|\}$ and, for any x such that $\|x\| \leq \tau$,

$$\mathbf{P}_x\left(Z_n \le \gamma_2\tau; \|X_n\| \le \tau\right)$$

$$\ge \mathbf{P}_x\left(Z_n \le \gamma_2\tau; \|X_{n'}\| \le \tau; \|X_n\| \le \tau\right)$$

$$\ge \mathbf{P}_x\left(Z_{n'} \le \gamma_2\tau; \|X_{n'}\| \le \tau; \sup_{n' \le k \le n}\{\|X_k\|\} \le \gamma_2\tau; \|X_n\| \le \tau\right)$$

$$= \mathbf{E}_x\left(1_{\{Z_{n'}\le\gamma_2\tau;\|X_{n'}\|\le\tau\}}\mathbf{P}_{X_{n'}}\left(Z_{n-n'} \le \gamma_2\tau; \|X_{n-n'}\| \le \tau\right)\right)$$

$$\ge \epsilon\mathbf{P}_x\left(Z_{n'} \le \gamma_2\tau; \|X_{n'}\| \le \tau\right) \ge \epsilon^{2+\ell}.$$

This gives the desired property for $\ell + 1$. $\qquad\square$

Example 1.5. Suppose μ is strongly $(\|\cdot\|, r)$ controlled with $r(t) = t^{1/\beta}$ for some fixed $\beta \in (0, 2]$. Then the probability that the random walk driven by μ started at the identity element e exits the ball $\{g : \|g\| \le \tau\}$ before time n is bounded below by $e^{-c(1+n/\tau^\beta)}$.

1.2. Word-Length Radial Measures

Let the group G be equipped with a generating k-tuple

$$S = (s_1, \ldots, s_k)$$

and the associated finite symmetric set of generators $\mathcal{S} = \{s_1^{\pm 1}, \ldots, s_k^{\pm 1}\}$. Let $|g|$ be the associated word length, that is, the minimal k such that $g = u_1 \ldots u_k$ with $u_i \in \mathcal{S}$, $1 \le i \le k$. By definition, the identity element e has length 0. Hence, $|\cdot|$ is a norm and $(x, y) \to |x^{-1}y|$ is a left-invariant distance function on G. Let

$$V_S(r) = \#\{G : |g| \le r\}$$

be the volume of the ball of radius r. We say that G has polynomial volume growth of degree D if $V_S(r) \simeq r^D$ in the sense that the ratio $V_S(r)/r^D$ is bounded away from 0 and ∞ for $r \ge 1$. Finitely generated nilpotent groups have polynomial volume growth and, by Gromov's theorem, any finitely generated group with polynomial volume growth contains a nilpotent subgroup of finite index. More precisely, any finitely generated group G such that there exist constants C, A and a sequence n_k with $V(n_k) \le Cn_k^A$ contains a nilpotent subgroup of finite index and thus has polynomial volume growth of degree D for some integer D. See, e.g., [8].

Example 1.6. Let G be equipped with a word-length function $|\cdot|$ associated with a symmetric finite generating subset. Assume that G has polynomial volume growth. The main results of [10] imply that, for any symmetric probability measure μ with finite generating support, μ is strongly $(|\cdot|, t \mapsto \sqrt{t})$ controlled. The main results of [1, 2] show that, if ν_β is symmetric and satisfies $\nu_\beta(g) \simeq [(1 + |g|)^\beta V(|g|)]^{-1}$ with $\beta \in (0, 2)$, then ν_β is strongly $(|\cdot|, t \mapsto t^{1/\beta})$ controlled. See also [3, 14].

One example that motivates the present work is the case of the measure

$$\nu_2(g) = \frac{c}{(1 + |g|)^2 V(|g|)}.$$

Can one provide good estimates for $\nu_2^{(n)}(e)$ on groups of polynomial volume growth? The following theorem gives a very satisfactory answer to this question and covers not only this particular example but the full range of cases passing through the classical threshold corresponding to the second moment condition.

Theorem 1.7. *Let G be equipped with a word-length function $|\cdot|$ associated with a symmetric finite generating subset. Let V be the associated volume function and assume that G has polynomial volume growth. Let $\phi : [0, \infty) \to [1, \infty)$ be a continuous regularly varying function of positive index. Let r be the inverse function of*

$$t \mapsto t^2 / \int_0^t \frac{sds}{\phi(s)}.$$

Let ν_ϕ be a symmetric probability measure such that

$$\nu_\phi(g) \simeq \frac{1}{\phi(|g|) V(|g|)}. \tag{15.3}$$

Then ν_ϕ is strongly $(|\cdot|, r)$ controlled.

Example 1.8. Assume that $\phi(t) = (1 + t)^\beta \ell(t)$ with ℓ positive continuous and slowly varying (we refer the reader to [4, Chap. I] for the definition and basic properties of slowly and regularly varying functions. The scaling function r of Theorem 1.7 can be described more explicitly as follows.

- If $\beta > 2$, $r(t) \simeq t^{1/2}$.
- If $\beta < 2$, we have $t^2 / \int_0^t \frac{sds}{\phi(s)} \simeq c_\phi \phi(t)$ and r is essentially the inverse of ϕ, namely,

$$r(t) \simeq t^{1/\beta} \left(\ell^{1/\beta} \right)_{\#} \left(t^{1/\beta} \right)$$

where $\ell_\#$ is the de Bruijn conjugate of ℓ. See [4, Prop. 1.5.15]. For example, if ℓ has the property that $\ell(t^a) \simeq \ell(t)$ for all $a > 0$ then $\ell_\# \simeq 1/\ell$.
- The case $\beta = 2$ is more subtle and the proof is more difficult. The function $\psi : t \mapsto \int_0^t \frac{sds}{\phi(s)}$ is slowly varying and satisfies $\psi(t) \geq \frac{c_1}{\ell(t)}$. For example, if $\ell \equiv 1$, we have $\psi(t) \simeq \log t$ and $r(t) \simeq (t \log t)^{1/2}$. When $\ell(t) = (\log t)^\gamma$ with $\gamma \in \mathbb{R}$ then
  - If $\gamma > 1$, $\psi(t) \simeq 1$ and $r(t) \simeq t^{1/2}$;
  - If $\gamma = 1$, $\psi(t) \simeq \log \log t$ and $r(t) \simeq (t \log \log t)^{1/2}$;
  - If $\gamma < 1$, $\psi(t) \simeq (\log t)^{1-\gamma}$ and $r(t) \simeq (t(\log t)^{1-\gamma})^{1/2}$.

1.3. Measures Supported by the Powers of the Generators

In the critical case when ϕ is regularly varying of index 2 and ν_ϕ has infinite second moment (i.e., $\sum |g|^2 \nu_\phi(g) = \infty$), the proof of Theorem 1.7 makes essential use of some of the results from [18], which are related to variations on the following class of examples. Recall that G is equipped with the generating k-tuple $S = (s_1, \ldots, s_k)$. For any k-tuple $a = (\alpha_1, \ldots, \alpha_k) \in (0, \infty)^k$, and consider the probability measure $\mu_{S,a}$ supported on the powers of the generators $s_1, \ldots, s_k$ and defined by

$$\mu_{S,a}(g) = \frac{1}{k} \sum_{1}^{k} \sum_{m \in \mathbb{Z}} \frac{\kappa_i \mathbf{1}_{s_i^m}(g)}{(1 + |m|)^{1+\alpha_i}}, \quad \kappa_i^{-1} = \sum_{m \in \mathbb{Z}} \frac{1}{(1 + |m|)^{1+\alpha_i}}. \quad (15.4)$$

Set

$$\tilde{\alpha}_i = \min\{\alpha_i, 2\} \quad \text{and} \quad \alpha_* = \max\{\tilde{\alpha}_i, 1 \le i \le k\}.$$

Define

$$\|g\|_{S,a} = \min \left\{ r : g = \prod_{j=1}^{m} s_{i_j}^{\epsilon_j} : \epsilon_j = \pm 1, \; \#\{j : i_j = i\} \le r^{\alpha_*/\tilde{\alpha}_i} \right\}. \quad (15.5)$$

Note that $g \mapsto \|g\|_{S,a} : G \to [0, \infty)$ is a norm. Consider also the measure

$$\nu_{S,a,\beta}(g) = \frac{c(G, a, \beta)}{(1 + \|g\|_{S,a})^\beta \, V_{S,a}(\|g\|_{S,a})} \quad (15.6)$$

with $\beta \in (0, 2)$.

Under the assumption that G is nilpotent and $\{s_i : \alpha_i \in (0, 2)\}$ generates a subgroup of finite index in G, it is proved in [18] that there exists a positive real $D_{S,a}$ such that

$$Q_{S,a}(r) = \#\{\|g\|_{S,a} \le r^{1/\alpha_*}\} \simeq r^{D_{S,a}}$$

and

$$\mu_a^{(n)}(e) \le C_{S,a} n^{-D_{S,a}}, \quad \nu_{S,a,\beta}^{(n)}(e) \le C_{S,a,\beta} n^{-\alpha_* D_{S,a}/\beta}.$$

Here we prove the following complementary result.

Theorem 1.9.   *Let G be a finitely generated nilpotent group equipped with a generating k-tuple $S = (s_1, \ldots, s_k)$. Referring to the notation introduced above, fix $a \in (0, \infty)^k$ and assume that $\{s_i : \alpha_i \in (0, 2)\}$ generates a subgroup of finite index in G.*

- *The probability measure $\mu_{S,a}$ is strongly $(\| \cdot \|_{S,a}, t \mapsto t^{1/\alpha_*})$ controlled.*
- *For any $\beta \in (0, 2)$, $\nu_{S,a,\beta}$ is strongly $(\| \cdot \|_{S,a}, t \mapsto t^{1/\beta})$ controlled.*

Remark 1.10. In [18], a detailed analysis of the subadditive function $\|\cdot\|_{S,a}$ and the associated geometry is given. This analysis is key to the above result and to its proper understanding. For instance, it is important to understand that the parameter α_* is not necessarily a significant parameter. It is the quantity $\|\cdot\|_{S,a}^{\alpha_*}$ that is the important expression. Indeed, for any given nilpotent group G, [18] describes conditions on two pairs of tuples (S, a), (S', a'),

$$S = (s_i)_1^k \in G^k, a = (\alpha_i)_1^k \in (0, \infty)^k, S' = (s_i')_1^{k'} \in G^{k'}, a' = (\alpha_i')_1^{k'} \in (0, \infty)^{k'},$$

such that $\|\cdot\|_{S,a}^{\alpha_*} \simeq \|\cdot\|_{S',a'}^{\alpha_*'}$. Because the geometry $\|g\|_{S,a}$ is studied and described rather explicitly in [18], the above results give rather concrete controls of the random walks driven with $\mu_{S,a}$ or $\nu_{S,a,\beta}$.

On the one hand, in the case of the measures $\nu_{S,a,\beta}$ and with much more work, it is possible to improve upon the statement of Theorem 1.9 and obtain a full two-sided pointwise bound on $\nu_{S,a,\beta}$. Indeed, based on the results of [18], it is proved in [14] that, for all $g \in G$ and $n \geq 1$,

$$\nu_{S,a,\beta}^{(n)}(g) \simeq \frac{n}{(n^{1/\beta} + \|g\|_{S,a})^{\alpha_* D_{S,a} + \beta}} \simeq \min\left\{\frac{1}{n^{\alpha_* D_{S,a}/\beta}}, \frac{n}{\|g\|_{S,a}^{\alpha_* D_{S,a} + \beta}}\right\}.$$

On the other hand, in the case of the measures $\mu_{S,a}$, Theorem 1.9 provides the most detailed result available at this time. Indeed, available techniques do not seem to be adequate to provide a sharp two-sided bound for $\mu_{S,a}^{(n)}(g)$ and finding such a two-sided bound is an interesting open problem.

1.4. A Short Guide

Section 2 is based on well-known variations of the celebrated Davies off-diagonal upper-bound technique. Our key observation is that, even in cases where we do not expect to obtain full off-diagonal upper bounds, the Davies technique provides enough information to prove *control* in the sense of Definition 1.2.

Section 3 describes the notion of pointwise pseudo-Poincaré inequality (a variation on the idea introduced in [6]) and shows how, with the help of the underlying group structure, a pseudo-Poincaré inequality allows us to upgrade *control* to *strong control*.

Section 4 applies the earlier results to a family of probability measures and random walks introduced in [18]. These measures are supported on the powers of the given generators. They provide examples for which no good off-diagonal upper bounds are known at this time. Nevertheless, the results developed here apply and capture useful properties of the associated random walks.

Section 5 is concerned with radial-type measures where radial refers to a given norm on the group G. The simplest and most interesting case is when this norm is taken to be the usual word length associated with a finite symmetric set of generators and, in this case, we prove Theorem 1.7.

Section 6 describes the applications to a class of random walks on wreath products. The notion of strong control (on the base group of the wreath product) leads to lower bounds on the probability of return on the wreath product.

2. Davies Method, Tightness, and Control

2.1. Davies Method for the Truncated Process

In this section, we review how Davies's method applies to the continuous-time process associated with truncated jump kernels. We follow [13, Section 5] rather closely even so our setup is somewhat different. The first paper treating jump kernels by Davies method is [5].

Throughout this section G is a discrete group equipped with its counting measure. Fix a norm $g \mapsto \|g\|$ with volume function V and set $d(x, y) = \|x^{-1}y\|$. Note that d is a distance function on G. Consider the left-invariant symmetric jump kernel

$$J(x, y) = \nu(x^{-1}y)$$

associated to a given symmetric probability measure ν on G. For $R > 0$, define

$$\delta_R := \sum_{\|x\| > R} \nu(x) \quad \text{and} \quad \mathcal{G}(R) = \sum_{\|x\| \le R} \|x\|^2 \, \nu(x), \qquad (15.7)$$

$J_R(x, y) = J(x, y)\mathbf{1}_{\{d(x,y) \le R\}}$, $J_R'(x, y) = J(x, y)\mathbf{1}_{\{d(x,y) > R\}}$ and $\nu_R' = J_R'(e, \cdot)$.

Denote by $p(t, x, y)$ and $p_R(t, x, y)$ the transition densities of the continuous-time processes associated to J and J_R, respectively. In particular,

$$p(t, x, y) = p_t(x^{-1}y) = e^{-t} \sum_0^\infty \frac{t^n}{n!} \nu^{(n)}(x^{-1}y).$$

Let

$$\mathcal{E}(f, f) = \mathcal{E}_\nu(f, f) = \frac{1}{2} \sum_{x,y} (f(x) - f(y))^2 J(x, y) \qquad (15.8)$$

be the corresponding Dirichlet form and set also

$$\mathcal{E}_R(f, f) = \frac{1}{2} \sum_{x,y} (f(x) - f(y))^2 J_R(x, y).$$

Note that

$$\mathcal{E}(f,f) - \mathcal{E}_R(f,f) = \frac{1}{2} \sum_{x,y:d(x,y)>R} |f(x)-f(y)|^2 J(x,y)$$

$$\leq \sum_{x,y:d(x,y)>R} (f(x)^2 + f(y)^2)J(x,y) \leq 2\,\|f\|_2^2\,\delta_R.$$

Consider the on-diagonal upper bound given by

$$\forall\, x \in G,\ t > 0,\ \ p_t(e) \leq m(t), \tag{15.9}$$

where $m : [0,\infty) \to [0,\infty)$ is a continuous, regularly varying function of negative index at infinity and $m(0) < \infty$. Since the function $t \mapsto m(t)$ may present a slowly varying factor, we follow [13]. The starting point is the log-Sobolev inequality

$$\sum f^2 \log f \leq \epsilon \mathcal{E}_R(f,f) + (2\epsilon\delta_R + \log m(\epsilon))\,\|f\|_2^2 + \|f\|_2^2 \log \|f\|_2 \tag{15.10}$$

with $\epsilon > 0$ which follows from (15.9) by [7, Theorem 2.2.3]. The following technical proposition is the key to most of the results obtained in later sections. The logarithmic Sobolev inequality (15.10) is one of the main tools used to prove this proposition, which is essentially a corollary of [13, Section 5.1].

Proposition 2.1. *Assume that the on-diagonal upper bound* (15.9) *holds with m regularly varying of negative index. Then there is a constant C such that, for all $R, t > 0$ and $x \in G$ we have*

$$p_R(t,e,x) \leq Ce^{4\delta_R t} m(t) \left(\frac{t}{R^2/\mathcal{G}(R)}\right)^{\|x\|/3R}.$$

Remark 2.2. The bound in this proposition is better than the uniform bound $p_R(t,e,x) \leq Ce^{4\delta_R t} m(t)$ only when $t < R^2/\mathcal{G}(R)$.

Proof. It suffices to consider the case $t < R^2/\mathcal{G}(R)$. Starting with (15.10), we apply Davies method, as described in [13, Section 5.1] to estimate $p_R(t,e,x)$. Let

$$\Lambda_R(\psi)^2 = \max\left\{ \|e^{-2\psi}\Gamma_R(e^\psi, e^\psi)\|_\infty, \|e^{2\psi}\Gamma_R(e^{-\psi}, e^{-\psi})\|_\infty \right\}$$

with

$$\Gamma_R(\psi,\psi)(x) = \sum_y |\psi(x) - \psi(y)|^2 J_R(x,y),$$

then by [13, Corollary 5.3], for any ψ with finite support,

$$p_R(t,x,y) \leq Cm(t) \exp\left(4\delta_R t + 72\Lambda_R(\psi)^2 t - \psi(y) + \psi(x)\right).$$

Consider the case $x = x_0$ and $y = e$. For $\lambda > 0$, set $\psi(z) = \lambda(\|x_0\| - \|z\|)^+$ and write

$$e^{-2\psi(z)} \Gamma_R(e^\psi, e^\psi)(z) = \sum_y \left(e^{\psi(z)-\psi(y)} - 1\right)^2 J_R(z, y)$$

$$\leq e^{2\lambda R} \sum_y (\psi(z) - \psi(y))^2 J_R(z, y)$$

$$\leq \lambda^2 e^{2\lambda R} \sum_{\|y\| \leq R} \|y\|^2 \nu(y) \leq R^{-2} e^{3\lambda R} \mathcal{G}(R).$$

The last inequality uses the elementary inequality $t^2 e^{-t} \leq 1$ with $t = R\lambda$. By inspection, the same estimate holds for $e^{2\psi(z)} \Gamma_R(e^{-\psi}, e^{-\psi})(z)$. As $\psi(e) = \lambda \|x_0\|$, we obtain

$$p_R(t, e, x_0) = p_R(t, x_0, e) \leq Cm(t) \exp\left(4\delta_R t + 72tR^{-2} e^{3\lambda R} \mathcal{G}(R) - \lambda \|x_0\|\right).$$

Since $t < R^2/\mathcal{G}(R)$, we can set

$$\lambda = \frac{1}{3R} \log \frac{R^2}{t\mathcal{G}(R)}$$

so that the second term $72tR^{-2} e^{3\lambda R} \mathcal{G}(R)$ is a constant. This yields the stated upper bound. $\qquad\square$

2.2. Control

Meyer's construction is a useful technique to construct the process X_s by adding big jumps to X_s^R. See, e.g., [12] and [2, Lemma 3.1]. In this section, we combine the off-diagonal upper bound in Proposition 2.1 with Meyer's construction to derive control-type results for the process with jump kernel J. Our goal is to show that, for a certain choice of continuous increasing function $r(t)$, for any $\varepsilon > 0$, there exists constant $\gamma > 1$ such that

$$\mathbf{P}_e\left(\sup_{s \leq t} \|X_s\| \geq \gamma r(t)\right) \leq \varepsilon.$$

Let X_s^R denote the process with truncated kernel J_R. It follows from Meyer's construction that (see [12] and [2, Lemma 3.1])

$$\mathbf{P}_e(X_s \neq X_s^R \text{ for some } s \leq t) \leq t\delta_R.$$

For any $r > 0$, $\gamma > 1$, both to be specified later, we have

$$\mathbf{P}_e \left(\sup_{s \leq t} \|X_s\| \geq \gamma r \right)$$

$$\leq \mathbf{P}_e \left(\sup_{s \leq t} \left\| X_s^R \right\| \geq \gamma r \right) + \mathbf{P}_e \left(X_s \neq X_s^R \text{ for some } s \leq t \right)$$

$$\leq \mathbf{P}_e \left(\sup_{s \leq t} \left\| X_s^R \right\| \geq \gamma r \right) + t\delta_R$$

$$\leq 2 \sup_{s \leq t} \left\{ \mathbf{P}_e \left(\left\| X_s^R \right\| \geq \gamma r/2 \right) \right\} + t\delta_R. \tag{15.11}$$

The inequality

$$\mathbf{P}_e \left(\sup_{s \leq t} \left\| X_s^R \right\| \geq \gamma r \right) \leq 2 \sup_{s \leq t} \left\{ \mathbf{P}_e \left(\left\| X_s^R \right\| \geq \gamma r/2 \right) \right\}$$

is a version of the André reflection principle obtained by considering the stopping time $T_r = \inf\{s : \|X_s^R\| \geq \gamma r)$ and using the (strong) Markov property and the triangle inequality to write

$$\mathbf{P}_e \left(\sup_{s \leq t} \left\| X_s^R \right\| \geq \gamma r \right) \leq \mathbf{P}_e \left(\left\| X_t^R \right\| \geq \gamma r/2 \right)$$

$$+ \mathbf{P}_e \left(\sup_{s \leq t} \left\| X_s^R \right\| \geq \gamma r \text{ and } \left\| X_s^R \right\| \leq \gamma r/2 \right)$$

$$\leq \mathbf{P}_e \left(\left\| X_t^R \right\| \geq \gamma r/2 \right)$$

$$+ \mathbf{E}_e \left(\mathbf{1}_{\{T_r \leq t\}} \mathbf{P}_{X_{T_r}^R} \left(\|X_{t-T_r}\| \geq \gamma r/2 \right) \right)$$

$$\leq 2 \sup_{s \leq t} \left\{ \mathbf{P}_e \left(\left\| X_s^R \right\| \geq \gamma r/2 \right) \right\}.$$

The last inequality uses group invariance. This will be helpful in deriving the following result.

In what follows, we say that a positive increasing function V is doubling if there exists a constant C_{VD} such that, for all $\rho > 0$, $V(2\rho) \leq C_{VD} V(\rho)$.

Proposition 2.3. *Assume that the volume function V is doubling. Assume also that v is such that (15.9) holds where m is regularly varying of negative index. For $\varepsilon > 0$, fix a function $R(t)$ such that*

$$2t\delta_{R(t)} < \varepsilon \quad and \quad \frac{t}{R(t)^2/\mathcal{G}(R(t))} < e^{-1}.$$

Let $r(t) \geq R(t)$ be a positive continuous increasing function such that

$$\sup_{t>0} \left\{ m(t)\, V(r(t))\, e^{-r(t)/6R(t)} \right\} < \infty.$$

Then, for any $\epsilon > 0$ there exists a constant $\gamma \geq 1$ such that

$$\mathbf{P}_e \left(\sup_{s \leq t} \|X_s\| \geq \gamma\, r(t) \right) < \varepsilon.$$

In particular, we have

$$p(t, e, e) \geq \frac{1 - \varepsilon}{V(\gamma\, r(t))}.$$

If, in addition $V(r(t)) \simeq m(t)$, then the measure ν is $(\|\cdot\|, r)$ controlled in continuous time.

Proof. Proposition 2.1 implies that for $s \leq t$,

$$p_R(s, e, x) \leq Cm(s) \left(\frac{s}{R^2/\mathcal{G}(R)} \right)^{\|x\|/3R}$$

$$= Cm(s) \left(\frac{s}{t} \right)^{\|x\|/3R} \left(\frac{t}{R^2/\mathcal{G}(R)} \right)^{\|x\|/3R}.$$

Fix $R = R(t)$, $r = r(t) \geq R$, decompose $\{x : \|x\| \geq \frac{\gamma}{2} r\}$ into dyadic annuli $\{x : \|x\| \simeq 2^i \gamma r\}$ and write

$$\mathbf{P}_e \left(\left\| X_s^R \right\| \geq \frac{\gamma r}{2} \right)$$

$$\leq C \sum_{i=0}^{\infty} m(s) \left(\frac{s}{t} \right)^{2^{i-1}\gamma/3} e^{-2^{i-1}\gamma r/3R} V(2^i \gamma\, r(t))$$

$$= Cm(t)\, V(\gamma r) \sum_{i=0}^{\infty} \frac{m(s)}{m(t)} \left(\frac{s}{t} \right)^{2^{i-1}\gamma/3} e^{-2^{i-1}\gamma r/3R} \left(\frac{V(2^i \gamma r)}{V(\gamma r)} \right).$$

Let C_{VD} denotes the volume-doubling constant of (G, d), then

$$V(\gamma r) \leq C_{VD}^{1+\log\gamma}\, V(r), \quad \frac{V(2^i \gamma r)}{V(\gamma r)} \leq C_{VD}^i.$$

Recall that $m(t)$ is a regularly varying function with negative index. Hence, for γ large enough, we have

$$M = \sup_{0 < s \leq t, i \in \mathbb{N}} \left\{ \frac{m(s)}{m(t)} \left(\frac{s}{t} \right)^{2^{i-1}\gamma/3} \right\} < \infty.$$

Therefore

$$\mathbf{P}_e \left(\left\| X_s^R \right\| \geq \frac{\gamma r}{2} \right) \leq C_1 m(t)\, V(r)\, e^{-r/6R} \sum_{i=0}^{\infty} e^{-2^i \gamma/12} C_{VD}^i.$$

By assumption, $r = r(t)$ and $R = R(t)$ satisfy

$$\sup_{t>0} \left\{ m(t) \, V(r(t)) e^{-r(t)/6R(t)} \right\} < \infty.$$

It follows that for γ sufficiently large, we have

$$\mathbf{P}_e \left(\left\| X_s^R \right\| \geq \frac{\gamma}{2} r(t) \right) < \frac{\varepsilon}{4}.$$

Plugging this estimate into (15.11), we obtain $\mathbf{P}_e \left(\sup_{s \leq t} \{ \| X_s \| \} \geq \gamma r(t) \right)$
$< \varepsilon$. $\qquad \square$

Corollary 2.4. *Under the hypotheses of* Proposition 2.3, *for any* $\epsilon > 0$
there exists $\gamma > 0$ *such that*

$$\mathbf{P}_e \left(\sup_{s \leq t} \| X_s \| \leq 2\gamma r(t), \| X_t \| \leq \gamma r(t) \right) \geq 1 - \varepsilon.$$

Proof. Write

$$\mathbf{P}_e \left(\sup_{s \leq t} \| X_s \| \leq 2\gamma r(t), \| X_t \| \leq \gamma r(t) \right)$$

$$= \mathbf{P}_e \left(\| X_t \| \leq \gamma r(t) \right) - \mathbf{P}_e \left(\sup_{s \leq t} \| X_s \| \geq 2\gamma r(t), \| X_t \| \leq \gamma r(t) \right)$$

$$\geq 1 - 2\mathbf{P}_e \left(\sup_{s \leq t} \| X_s \| \geq \gamma r(t) \right).$$

Note that we have used the fact that, because of space homogeneity (i.e., group invariance), X_t cannot escape to infinity in finite time. $\qquad \square$

Remark 2.5. The conclusions of Proposition 2.3 and Corollary 2.4 apply to the associated discrete-time random walk. To see this, fix a regularly varying function m and note that (up to changing m to cm for some constant c), (15.9) is equivalent to $\nu^{(2n)}(e) \leq m(n)$. Further, it is easy to control the difference between $\mathbf{P}_e(\| X_t \| \geq r)$ and $\mathbf{P}_e(\| X_n \| \geq r)$ with $n = \lfloor t \rfloor$ as long as n is large enough. It follows that the proof above applies the discrete random walk result as well.

3. Pseudo-Poincaré Inequality and Strong Control

3.1. Pseudo-Poincaré Inequality

With some work, the results of the previous section can be extended to the more general context of graphs and discrete spaces. The results presented below make a more significant use of the underlying group structure.

Definition 3.1. Let G be discrete-group equipped with a symmetric probability measure ν, a subadditive function $\|\cdot\|$ and a positive continuous increasing function r with inverse ρ. We say that ν satisfies a pointwise $(\|\cdot\|, r)$ pseudo-Poincaré inequality if, for any f with finite support on G,

$$\forall\, g \in G, \quad \sum_{x \in G} |f(xg) - f(x)|^2 \leq C\rho(\|g\|)\mathcal{E}_\nu(f, f). \tag{15.12}$$

Here $\mathcal{E}_\nu$ is the Dirichlet form of ν defined at (15.8).

Theorem 3.2. *Assume that $(G, \|\cdot\|)$ is such that V is doubling. Let ν be a symmetric probability measure such that $\nu(e) > 0$. Assume that r is a positive doubling continuous increasing function such that*

$$\nu^{(2n)}(e) \simeq V(r(n))^{-1}.$$

Assume further that ν satisfies the $(\|\cdot\|, r)$ pseudo-Poincaré inequality. Then there exists $\eta > 0$ such that for all n and g with $\|g\| \leq \eta r(n)$ we have

$$\nu^{(n)}(g) \simeq V(r(n))^{-1}.$$

Proof. The hypothesis (15.12) and the argument of [10, Theorem 4.2] gives

$$|\nu^{(2n+N)}(x) - \nu^{(2n+N)}(e)| \leq C \left(\frac{\rho(\|x\|)}{N}\right)^{1/2} \nu^{(2n)}(e).$$

Fix x and n such that $\rho(\|x\|) \leq \eta n$ and use the above inequality with $N = 2n$ to obtain

$$\nu^{(4n)}(x) \geq \left(1 - (2C'\eta)^{1/2}\right)\nu^{(4n)}(e).$$

Hence, we can choose $\eta > 0$ such that

$$\nu^{(4n)}(x) \geq c\nu^{(4n)}(e).$$

Because $\nu(e) > 0$, this also holds for $4n + i$, $i = 1, 2, 3$, at the cost of changing the value of the positive constant c. $\square$

3.2. Strong Control

Definition 3.3. We say that $\|\cdot\|$ is well connected if there exists $b \in (0, \infty)$ such that, for any $r > 0$ and $x \in G$ with $\|x\| \leq r$ there exists a finite sequence of points $(x_i)_0^N \in G$ with $\|x_i\| \leq 2r$, $\|x_i^{-1}x_{i+1}\| \leq b$, $x_0 = e$ and $x_N = x$.

Note that in this definition, the number N of points in the sequence is finite but that it depends on x and no upper bound in terms of r is required.

Lemma 3.4. *Assume that $\|\cdot\|$ is well connected and V is doubling. Then for any fixed $\epsilon > 0$ there exists M_ϵ such that for any $r \geq 8b/\epsilon$ and any $\|x\| \leq r$ we can find $(z_i)_0^M$, $z_0 = e$, $z_M = x$, $M \leq M_\epsilon$, such that $\|z_i^{-1} z_{i+1}\| \leq \epsilon r$.*

Proof. Let $\{y_i : 0 \leq i \leq M'\}$ be a maximal $\epsilon r/4$-separated set of points in $B(e, 2r) = \{\|g\| \leq 2r\}$. The ball $B_i = B(y_i, \epsilon r/9)$ are disjoints and have volume $V(\epsilon r/9)$ comparable to $V(2r)$. Hence $M' \leq M_\epsilon'$ for some finite M_ϵ' independent of r. The union of the balls $B_i' = B(y_i, \epsilon r/4)$ covers $B(2r)$ (otherwise, $\{y_i : 0 \leq i \leq M'\}$ would not be maximal). Fix x with $\|x\| \leq r$ and let $(x_i)_0^N$ be a sequence of points witnessing well-connectedness for the point x. By construction, the balls B_i', $1 \leq i \leq M'$, cover the sequence $(x_i)_0^N$ and we can extract a sequence $B_i^* = B_{j_i}'$, $0 \leq i \leq M \leq M_\epsilon'$, such that $B_0^* \ni e$, $B_M^* \ni x$, $\inf\{\|g\| : g \in B_1^*\} \leq b$, $\inf\{\|xg^{-1}\| : g \in B_{M-1}^*\} \leq b$ and

$$\inf\left\{\|h^{-1}g\| : h \in B_i^*, g \in B_{i+1}^*\right\} \leq b, i = 1, \ldots, M-2.$$

Set $z_0 = e$, $z_i = y_{j_i}$, $1 \leq i \leq M-1$, $z_M = x$. Then $\|z_{i-1}^{-1} z_i\| \leq 2\epsilon r/4 + b \leq \epsilon r$ as desired. $\square$

Proposition 3.5. *Assume that the norm $\|\cdot\|$ is such that V is doubling and $\|\cdot\|$ is well connected. Let r be a positive continuous increasing doubling function. Let ν be a symmetric probability measure that is $(\|\cdot\|, r)$ controlled and satisfies $\nu(e) > 0$ and a pointwise $(\|\cdot\|, r)$ pseudo-Poincaré inequality. Then ν is also strongly $(\|\cdot\|, r)$ controlled.*

Proof. First, we show that for any $\kappa > 0$ there exists $c_\kappa > 0$ such that $\|x\| \leq \kappa r(n)$ implies

$$\nu^{(n)}(x) \geq cV(r(n))^{-1}.$$

By Theorem 3.2, there exists η such that $\nu^{(n)}(x) \geq c_1 V(r(n))^{-1}$ for all $\|x\| \leq \eta r(n)$. By Lemma 3.4, for any fixed κ there exists M_κ such that for any $\|x\| \leq \kappa r(n)$ we can find $(z_i)_0^M$, $z_0 = e$, $z_M = x$, $M \leq M_\kappa$, such that $\|z_i^{-1} z_{i+1}\| \leq \eta r(n)/4$. Write $B_i = \{g : \|z_i^{-1} g\| \leq \eta r(n)/4\}$ and

$$\nu^{(Mn)}(x) \geq \sum_{(y_1,\ldots,y_M) \in \otimes_1^M B_i} \nu^{(n)}(y_1) \cdots \nu^{(n)}(y_i^{-1} y_{i+1}) \ldots \nu^{(n)}(y_M^{-1} x)$$

$$\geq c_1^{M+1} V(\eta r(n)/4)^M V(r(n))^{-M-1} \simeq c_1' V(r(n))^{-1}.$$

Given that $\nu(e) > 0$, this shows that $\|x\| \leq \kappa r(n)$ implies $\nu^{(n)}(x) \geq cV(r(n))^{-1}$ as stated. In particular, for any fixed κ, there exists $\epsilon > 0$ such that for any x, n with $\kappa r(n) \leq \tau$ and $\|x\| \leq \tau$,

$$\mathbf{P}_x(\|X_n\| \leq \tau) \geq \epsilon.$$

Now, fix $\gamma_1 \in (1, \infty)$. Let $\epsilon_0 > 0$ be such that, for any x, n, τ with

$$\|x\| \leq \tau \leq \gamma_1 r(2n),$$

we have $\mathbf{P}_x(\|X_n\| \leq \tau) \geq \epsilon_0$. Let $\gamma \geq 1$ be given by Definition 1.2 so that

$$\mathbf{P}_e\left(\sup_{k \leq n}\{\|X_k\|\} \geq \gamma\, r(n)\right) \geq \epsilon_0/2.$$

Set $\gamma_2 = \gamma/\gamma_1 + 1$ and, for any x, n, τ with $\|x\| \leq \tau$ and $\frac{1}{2}\rho(\tau/\gamma_1) \leq n \leq \rho(\tau/\gamma_1)$, write

$$\mathbf{P}_x\left(\sup_{k \leq n}\{\|X_k\|\} \leq \gamma_2\tau, \|X_n\| \leq \tau\right)$$

$$= \mathbf{P}_x\left(\|X_t\| \leq \tau\right) - \mathbf{P}_x\left(\sup_{k \leq n}\|X_k\| \geq \gamma_2\tau, \|X_n\| \leq \tau\right)$$

$$\geq \epsilon_0 - \mathbf{P}_e\left(\sup_{k \leq n}\|X_k\| \geq \gamma\tau/\gamma_1\right) \geq \epsilon_0/2.$$

This proves that ν is strongly $(\|\cdot\|, r)$ controlled. $\qquad\Box$

As a simple illustration of these techniques, consider the case of an arbitrary symmetric measure ν with generating support and finite second moment (with respect to the word-length $|\cdot|$) on a group with polynomial volume growth of degree $D(G)$. It follows from [16] that $\nu^{(n)}(e) \simeq n^{-D(G)/2}$ and satisfies a pointwise classical pseudo-Poincaré inequality (with $\rho(t) = t^2$). Proposition 3.5 yields the following result.

Theorem 3.6.   *Let G be a finitely generated group with polynomial volume growth with word-length $|\cdot|$. Assume that ν is symmetric, satisfies $\nu(e) > 0$, has generating support and satisfies $\sum |g|^2\nu(g) < \infty$. Then ν is strongly $(|\cdot|, t \mapsto t^{1/2})$ controlled.*

4. Measures Supported on Powers of Generators

The goal of this section is to explain the proof of the part of Theorem 1.9 that deals with the measures $\mu_{S,a}$.

4.1. The Measure $\mu_{S,a}$

In this subsection we consider the special case when G is a nilpotent group equipped with a generating k-tuple $S = (s_1, \ldots, s_k)$ and

$$J(x, y) = \mu_{S,a}(x^{-1}y), \quad a = (\alpha_1, \ldots, \alpha_k) \in (0, \infty)^k \qquad (15.13)$$

with $\mu_{S,a}$ given by (15.4). Our aim is to prove the first statement in Theorem 1.9. The study of the random walks driven by this class of measure was initiated by the authors in [18] and we will refer to and use some of the main results of [18].

Following [18, Definition 1.3], let $\mathfrak{w}$ be the power weight system on the formal commutators on the alphabet S associated with setting $w_i = 1/\widetilde{\alpha}_i$, $\widetilde{\alpha}_i = \min\{\alpha_i, 2\}$. Namely, the weight of any commutator c using the sequence of letters $(s_{i_1}, \ldots, s_{i_m})$ from S (or their formal inverse) is $w(c) = \sum_1^m w_{i_j}$. Recall also from Section 1.3 that $\alpha^* = \max\{\widetilde{\alpha}_i\}$ and

$$Q_{S,a}(r) = \#\left\{g : \|g\|_{S,a} \leq r^{1/\alpha^*}\right\}.$$

In [18], the authors proved the following result.

Theorem 4.1 ([18]). *Referring to the above setting and notation, assume that the subgroup of G generated by $\{s_i : \alpha_i < 2\}$ is of finite index. Then there exists a real $D_{S,a} = D(S, \mathfrak{w})$ such that*

$$Q_{S,a}(r) \simeq r^{D_{S,a}}, \quad \mu_{S,a}^{(n)}(e) \simeq n^{-D_{S,a}}.$$

The real $D_{S,a} = D(S, \mathfrak{w})$ is given by [18, Definition 1.7]. Further, there exists a k-tuple $b = (\beta_1, \cdots, \beta_k) \in (0,2)^k$ such that $\beta_i = \alpha_i$ if $\alpha_i < 2$, $D(S, a) = D(S, b)$, and

$$\forall g \in G, \|g\|_{S,a}^{\alpha_*} \simeq \|g\|_{S,\beta}^{\beta_*}.$$

In addition, $\mu_{S,a}$ satisfies a pointwise $(\|\cdot\|_{S,a}, t \mapsto t^{1/\alpha_*})$ pseudo-Poincaré inequality.

The volume estimate $Q_{S,a}(r) \simeq Q_{S,b}(r) \simeq r^{D_{S,a}}$ shows, in particular, that $(G, \|\cdot\|_{S,b})$ has the volume-doubling property. The upper-bound $\mu_{S,a}^{(n)}(e) \leq C n^{-D_{S,a}}$ implies that the continuous time process with jump kernel J defined above satisfies

$$\forall t > 0, x \in G, \quad p(t, x, x) \leq m(t) = C t^{-D_{S,a}}.$$

Note that $\|\cdot\|$ is clearly well connected (Definition 3.3). To apply Propositions 2.3 and 3.5 to the present case and prove Theorem 1.9, it clearly suffices to prove the following lemma, which provides estimates for δ_R and $\mathcal{G}(R)$.

Lemma 4.2. *Referring to the setting and hypotheses of* Theorem 4.1, *for J given by* (15.13), *let $\|\cdot\| = \|\cdot\|_{S,b}$, $D = D_{S,b} = D_{S,a}$, we have*

$$V(r) = \#\{g \in G : \|g\| \leq r\} \simeq r^{D\beta_*},$$
$$\delta_R \simeq R^{-\beta_*},$$
$$\mathcal{G}(R) \simeq R^{2-\beta_*}.$$

Proof. The volume estimate follows immediately from Theorem 4.1. Let $\mathfrak{v}$ be the power weight system associated with b (in particular,

$v_i = 1/\beta_i > 1/2)$. By [18, Proposition 2.17], for each i there exists $0 < \beta_i' \le \beta_i \le \beta_* < 2$ such that

$$\left\| s_i^n \right\| \simeq |n|^{\beta_i'/\beta_*}.$$

In the notation of [18], $\beta_i = \overline{v}_{j_0(s_i)}$. We have

$$\delta_R = \sum_{\|x\| > R} \mu_{S,a}(x) = \sum_{i=1}^{k} \sum_{\|s_i^n\| > R} \frac{\kappa_i}{(1 + |n|)^{1 + \alpha_i}}$$

$$\simeq \sum_{i=1}^{k} \sum_{n > R^{\beta_*/\beta_i'}} \frac{\kappa_i}{(1 + |n|)^{1 + \alpha_i}} \simeq \sum_{i=1}^{k} R^{-\beta_* \alpha_i / \beta_i'} \simeq R^{-\beta_*}.$$

The last estimate uses the fact that there must be some $i \in \{1, \ldots, k\}$ such that $\alpha_i = \beta_i'$ and that, always, $\alpha_i \ge \beta_i'$.

Similarly, since $\alpha_i \ge \beta_i'$ and $\beta_* < 2$, we have $2\beta_i'/\beta_* - \alpha_i > 0$. This yields

$$\mathcal{G}(R) = \sum_{\|x\| \le R} \|x\|^2 \mu_{S,a}(x) = \sum_{i=1}^{k} \sum_{\|s_i^n\| \le R} \frac{\kappa_i \left\| s_i^n \right\|^2}{(1 + |n|)^{1 + \alpha_i}}$$

$$\simeq \sum_{i=1}^{k} \sum_{0 \le n \le R^{\beta_*/\beta_i'}} \frac{\kappa_i |n|^{2\beta_i'/\beta_*}}{(1 + |n|)^{1 + \alpha_i}} \simeq \sum_{i=1}^{k} R^{2 - \beta_* \alpha_i / \beta_i'} \simeq R^{2 - \beta_*}.$$

This proves Lemma 4.2. □

4.2. Some Regular Variation Variants of $\mu_{S,a}$

Consider the class of measure μ of the form

$$\mu(g) = \frac{1}{k} \sum_{1}^{k} \sum_{m \in \mathbb{Z}} \frac{\kappa_i \ell_i(|m|)}{(1 + |m|)^{1 + \alpha_i}} \tag{15.14}$$

where each ℓ_i is a positive slowly varying function satisfying $\ell_i(t^b) \simeq \ell_i(t)$ for all $b > 0$ and $\alpha_i \in (0, 2)$. For each i, let F_i be the inverse function of $r \mapsto r^{\alpha_i}/\ell_i(r)$. Note that F_i is regularly varying of order $1/\alpha_i$ and that $F_i(r) \simeq [r\ell_i(r)]^{1/\alpha_i}$, $r \ge 1$, $i = 1, \ldots, k$. We make the fundamental assumption that the functions F_i have the property that for any $1 \le i, j \le k$, either $F_i(r) \le CF_j(r)$ of $F_j(r) \le CF_i(r)$. For instance, this is clearly the case if all α_i are distinct. Set $a = (\alpha_1, \ldots, \alpha_k) \in (0, 2)^k$ and consider also the power weight system $\mathfrak{v}$ generated by $v_i = 1/\alpha_i$, $1 \le i \le k$, as in [18, Definition 1.3]. Fix $\alpha_0 \in (0, 2)$ such that

$$\alpha_0 > \max\{\alpha_i : 1 \le i \le k\}$$

and $\alpha_0/\alpha_i \notin \mathbb{N}$, $i = 1,\ldots,k$. Observe that there are convex functions $K_i \geq 0$, $i = 0,\ldots,k$, such that $K_i(0) = 0$ and

$$\forall r \geq 1, \quad F_i(r^{\alpha_0}) \simeq K_i(r). \tag{15.15}$$

Indeed, $r \mapsto F_i(r^{\alpha_0})$ is regularly varying of index α_0/α_i with $1 < \alpha_0/\alpha_i \notin \mathbb{N}$. By [4, Theorems 1.8.2–1.8.3] there are smooth positive convex functions $\tilde{K}_i$ such that $\tilde{K}_i(r) \sim F_i(r^{\alpha_0})$. If $\tilde{K}_i(0) > 0$, it is easy to construct a convex function $K_i : [0,\infty) \to [0,\infty)$ such that $K_i \simeq \tilde{K}_i$ on $[1,\infty)$ and $K_i(0) = 0$. Let us use $\mathfrak{K}$ to denote the collection $(K_i)_1^r$.

Now, set

$$\|g\| = \|g\|_{\mathfrak{K}} = \min\left\{ r : g = \prod_{j=1}^m s_{i_j}^{\epsilon_j} : \epsilon_j = \pm 1, \ \#\{j : i_j = i\} \leq K_i(r) \right\}.$$

Because of the convexity property of the K_i, $\|\cdot\|$ is a norm. Note also that it is well connected. The following theorem is proved in [18].

Theorem 4.3. *Referring to the above notation and hypothesis, there exists a real $D = D_{S,a} = D(S,\mathfrak{v})$ and a positive slowly varying function L (explicitly given in [18, Theorem 5.15] and which satisfies $L(t^a) \simeq L(t)$ for all $a > 0$) such that:*

- *For all $r \geq 1$, $V(r) = \#\{g : \|g\| \leq r\} \simeq r^{\alpha_0 D} L(r)$.*
- *For a each $1 \leq i \leq k$, there exists a regularly varying function $\tilde{F}_i$ such that $\|s_i^n\|^{\alpha_0} \leq C\tilde{F}_i^{-1}(n)$ where $\tilde{F}_i \geq F_i$ and with equality for some $1 \leq i \leq r$.*
- *For all $n \geq 1$, $\mu^{(2n)}(e) \leq C(n^D L(n))^{-1}$.*
- *The measure μ satisfies a pointwise $(\|\cdot\|, t \mapsto t^{1/\alpha_0})$ pseudo-Poincaré inequality.*

Here, we prove the following result.

Theorem 4.4. *Let G be a finitely generated nilpotent group equipped with a generating k-tuple $(s_1,\ldots,s_k)$. Assume that μ is a probability measure on G of the form (15.14). Let ℓ_i, F_i, L, $D = D_{S,a}$, $\alpha_0 \in (0,2)$ and $\|\cdot\|$ be as described above. Then μ is strongly $(\|\cdot\|, t \mapsto t^{1/\alpha_0})$ controlled.*

Proof. It suffices to estimate the quantities δ_R and $\mathcal{G}(R)$ in the present context. For δ_R, we have

$$\delta_R \simeq \sum_1^k \sum_{n \geq \tilde{F}_i(R^{\alpha_0})} \frac{1}{nF_i^{-1}(n)} \simeq \sum_1^k \frac{1}{F_i^{-1} \circ \tilde{F}_i(R^{\alpha_0})} \simeq R^{-\alpha_0}.$$

A similar computation gives

$$\log \mathcal{G}(R) \simeq \sum_1^k \frac{R^2}{F_i^{-1} \circ \widetilde{F}_i(R^{\alpha_0})} \simeq R^{2-\alpha_0}.$$

$\square$

4.3. The Critical Case When $\alpha_{i=2}$, $1 \le i \le k$

When $a = \mathbf{2} = (2, \ldots, 2)$, that is, $\alpha_i = 2$ for all $1 \le i \le k$, we work with the usual word-length function $|g|$ associated with the generating set $\mathcal{S} = \{s_1^{\pm 1}, \ldots, s_k^{\pm 1}\}$. In this case, $V(r) = \#\{g : |g| \le r\} \simeq r^{D(G)}$ where $D(G)$ is the classical degree of polynomial growth for the nilpotent group G. It is proved in [18] that $\mu_{S,2}^{(n)}(e) \le C(n \log n)^{-D/2}$ and that $\mu_{S,2}$ satisfies a pointwise $(|\cdot|, t \mapsto (t \log t)^{1/2})$ pseudo-Poincaré inequality. Further, $|s_i^n| \simeq |n|^{1/\beta_i}$ with $\beta_i \ge 1$ and $\beta_i = 1$ for some i. From this it easily follows that

$$\delta_R \simeq R^{-2}, \quad \mathcal{G}(R) \simeq \log R.$$

Applying Proposition 3.5 with $r(t) = (t \log t)^{1/2}$ yields the following theorem.

Theorem 4.5. *Let G be a finitely generated nilpotent group equipped with a generating k-tuple $(s_1, \ldots, s_k)$. Let $D(G)$ be the volume growth degree of G. Then $\mu_{S,2}$ is strongly $(|\cdot|, t \mapsto (t \log t)^{1/2})$ controlled.*

5. Norm-Radial Measures

In this section we assume that G is a finitely generated group with polynomial volume growth of degree $D(G)$ and we consider norm-radial symmetric probability measures.

5.1. Radial Measures with Stable-like Exponent $\alpha \in (0, 2)$

This subsection treats probability measures of the form

$$\nu_\alpha(x) \simeq \frac{1}{(1 + \|x\|)^\alpha V(\|x\|)}, \quad J(x, y) \simeq \nu_\alpha(x^{-1}y), \qquad (15.16)$$

where $\alpha \in (0, 2]$, $\|\cdot\|$ is a norm on G and $V(r) = \#\{g : \|g\| \le r\}$. The case when $\alpha \in (0, 2)$ and $V(r) \simeq r^d$ for some d is treated in [1, 2, 14] where global matching upper and lower bounds are obtained. We note that [1, 2, 14] are set in more general contexts where the group structure plays no role. We start with the following easy observation.

Lemma 5.1. *Referring the situation described above, assume that V satisfies $V(2r) \le C_{VD} V(r)$ for all $r > 0$. Then $\delta_R \simeq R^{-\alpha}$ and*

$$\mathcal{G}(R) \simeq \begin{cases} R^{2-\alpha} & \text{if } \alpha \in (0, 2) \\ \log R & \text{if } \alpha = 2. \end{cases}$$

Proof. This follows by inspection. □

The next lemma follows by application of Proposition 2.3. However, in this lemma, we make a significant hypothesis on $v_\alpha^{(n)}(e)$.

Lemma 5.2. *Set* $r_\alpha(t) = t^{1/\alpha}$ *if* $\alpha \in (0,2)$, $r_2(t) = (t\log t)^{1/2}$. *Referring the situation described above, assume that* V *is regularly varying of positive index and*

$$v_\alpha^{(n)}(e) \le CV(r_\alpha(n))^{-1}. \qquad (15.17)$$

Then v_α *is* $(\|\cdot\|, r_\alpha)$ *controlled.*

The next theorem provides a basic class of examples when the hypothesis (15.17) regarding v_α can indeed be verified. Note that the result is restricted to the case $\alpha \in (0,2)$.

Theorem 5.3. *Referring the situation described above, assume that* V *is regularly varying of positive index and* $\alpha \in (0,2)$. *Then*

$$v_\alpha^{(n)}(e) \simeq V(n^{1/\alpha})^{-1}$$

and v_α *is* $(\|\cdot\|, t \mapsto t^{1/\alpha})$ *controlled.*

Proof. It suffices to prove the upper bound $v_\alpha^{(n)}(e) \le CV(n^{1/\alpha})^{-1}$. Start by checking that

$$v_\alpha(x) \simeq \sum_0^\infty \frac{1}{(1+m)^{1+\alpha}} \frac{\mathbf{1}_{B(m)}(x)}{V(m)}$$

where $B(m) = \{x \in G : \|x\| \le m\}$. Then apply the elementary technique of [3, Section 4.2] to derive the desired upper bound on $v_\alpha^{(n)}(e)$. □

Remark 5.4. In the context of Theorem 5.3, we do not know if $\|\cdot\|$ is well connected and we also do not know if v_α satisfies a *pointwise* $(\|\cdot\|, r_\alpha)$ pseudo-Poincaré inequality. Hence, the techniques used here do not suffice to obtain strong control. However, if $\|\cdot\|$ is well connected and v_α satisfies a pointwise $(\|\cdot\|, r_\alpha)$ pseudo-Poincaré inequality then the strong $(\|\cdot\|, r_\alpha)$ control follows by Proposition 3.5. This proves the second statement in Theorem 1.9 because, referring to the notation of Theorem 1.9, the norm $\|\cdot\|_{S,a}$ is well connected and Theorem 2.10, Lemma 4.4, and Theorem 5.7 in [18] show that the measure $v_{S,a,\beta}$ of Theorem 1.9 with $\beta \in (0,2)$ satisfies the pointwise $(\|\cdot\|_{S,a}, t \mapsto t^{1/\beta})$ pseudo-Poincaré inequality.

5.2. Word-Length Radial Measures

As noticed above, the study of v_α in the case $\alpha = 2$ is significantly more difficult than in the case $\alpha \in (0,2)$. In fact, we do not know how to treat this case in the generality described in the previous subsection.

The following theorem treats the case when $\| \cdot \|$ is the usual word-length function $\| \cdot \| = | \cdot |$ on G.

Theorem 5.5. *Assume that G is a group of polynomial volume growth equipped with generating k-tuple $S = (s_1, \ldots, s_k)$ and the associated word length $| \cdot |$ and volume function V. Let $D(G)$ be the degree of polynomial volume growth of G. Let ν_2 be a symmetric probability measure such that*

$$\nu_2(g) \simeq ((1 + |g|)^2 V(|g|))^{-1}.$$

Then we have

$$\nu_2^{(n)}(e) \simeq (n \log n)^{-D(G)/2}.$$

Further, ν_2 is strongly $(| \cdot |, t \mapsto (t \log t)^{1/2})$ controlled.

Proof. We apply Lemma 5.2 and Proposition 3.5. When G is nilpotent, the upper bound $\nu_2^{(n)}(e) \leq (n \log n)^{-D(G)/2}$ follows from Theorems 4.8 and 5.7 of [18]. Namely, [18, Theorem 5.7] shows that

$$\nu_2^{(n)}(e) \leq C\mu_{S,2}^{(Kn)}(e)$$

and [18, Theorem 4.8] gives $\mu_{S,2}^{(n)}(e) \leq C(n \log n)^{-D(G)/2}$. Further, [18, Lemma 4.4 and Theorem 5.7] shows that ν_2 satisfies a pointwise $(| \cdot |, t \mapsto (t \log t)^{1/2})$ pseudo-Poincaré inequality.

Since any group of polynomial volume growth of degree $D(G)$ contains a nilpotent subgroup of finite index (hence, with the same degree of polynomial volume growth) the upper bound $\nu_2^{(n)}(e) \leq C(n \log n)^{-D(G)/2}$ follows from the comparison theorem [16, Theorem 2.3]. By direct inspection, the desired pseudo-Poincaré inequality also follows. □

Note that Theorem 1.7 includes the result stated in Theorem 5.5 as a special case and provides a very satisfactory result covering the behaviors of word-length radial measures across the second moment threshold.

Proof of Theorem 1.7. The same technique of proof as for Theorem 5.5 gives the much more complete and subtle result stated in the introduction as Theorem 1.7. Namely, let $\phi : [0, \infty) \to [1, \infty)$ be a continuous, regularly increasing function of index 2 and let ν_ϕ be as in (15.3), that is, assume that ν_ϕ is symmetric and satisfies $\nu(g) \simeq [\phi(|g|) V(|g|)]^{-1}$. First, assume that G is nilpotent and let $\mu_{S,\phi}$ be the measure given by

$$\mu_{S,\phi}(g) = \frac{1}{k} \sum_{1}^{k} \frac{\kappa}{(1 + |n|)\phi(n)} \mathbf{1}_{s_i^n}(g).$$

Let r be the inverse function of $t \mapsto t^2 / \int_0^t \frac{s\,ds}{\phi(s)}$. By [18, Lemma 4.4], the measure $\mu_{S,\phi}$ satisfies the pointwise $(| \cdot |, r)$ pseudo-Poincaré inequality.

By [18, Theorem 4.1], it follows that $\mu_{S,\phi}^{(n)}(e) \leq V(r(n))^{-1}$. By [18, Theorem 5.7], we have the Dirichlet form comparison $\mathcal{E}_{\mu_{S,\phi}} \leq C\mathcal{E}_{\nu_\phi}$.

Now, if G has polynomial volume growth then it contains a nilpotent subgroup with finite index, G_0. By inspection, quasi-isometry, and comparison of Dirichlet forms (see [16]), it is easy to transfer both the pointwise $(|\cdot|, r)$ pseudo-Poincaré inequality and the decay $\nu_\phi^{(n)}(e) \leq V(r(n))^{-1}$ from G_0 to G. Further, one checks that the functions δ_R and $\mathcal{G}(R)$ satisfy $\delta_R \simeq 1/\phi(R)$ and $\mathcal{G}(R) \simeq \int_0^R \frac{tdt}{\phi(t)}$. Proposition 2.3 with $r(t) = R(t)$ equals to the inverse function of $s \mapsto s^2/\int_0^s \frac{tdt}{\phi(t)}$ shows that ν_ϕ is $(|\cdot|, r)$ controlled. By Proposition 3.5, ν_ϕ is strongly $(|\cdot|, r)$ controlled. $\qquad\square$

5.3. Assorted Further Applications

The approach presented here is applicable even in cases where we are not able to obtain sharp results, and we illustrate this by an example. Let G be a nilpotent group equipped with a generating k-tuple $S = (s_1, \ldots, s_k)$. Fix $a \in (0,2]^k$ and set $\alpha_* = \max\{\alpha_i, 1 \leq i \leq k\} \leq 2$. Consider the norm $\|\cdot\|_{S,a}$ defined at (15.5). Let ν_* be any symmetric probability measure such that

$$\nu_*(g) \simeq \frac{1}{(1 + \|g\|_{S,a})^2 V(\|g\|_{S,a})}, \quad V(r) = \#\{g : \|g\|_{S,a} \leq r\}.$$

Theorems 3.2, 4.8, and 5.7 of [18] give the following information. There exists two reals $D = D_{S,a}$ and $d = d_{S,a}$ and a constant $C_1 \in (0, \infty)$ such that

$$\nu_*^{(n)}(e) \leq C_1 n^{-\alpha_* D/2} (\log n)^{-d} \qquad (15.18)$$

$$V(r) \simeq r^{\alpha_* D}. \qquad (15.19)$$

Theorem 5.6. *For the probability measure ν_* on a finitely generated nilpotent group as described above, we have*

$$c(\log\log n)^{-\alpha_* D}(n\log n)^{-\alpha_* D/2} \leq \nu_*^{(n)}(e) \leq Cn^{-\alpha_* D/2}(\log n)^{-d}.$$

Proof. The volume estimate (15.19) and Lemma 5.1 gives $\delta_R \simeq R^{-2}$ and $\mathcal{G}(R) \simeq \log R$. In order to apply Proposition 2.3, we set $R(t) \simeq (t\log t)^{1/2}$. Further, we use (15.18)–(15.19) to verify that the choice $r(t) = 6AR(t)\log\log t$ with A large enough satisfies the condition of Proposition 2.3. Indeed, we have $m(t) \simeq t^{-\alpha_* D/2}(\log t)^{-d}$, $V(r) \simeq r^{\alpha_* D}$ so that

$$m(t) V(r(t))e^{-r(t)/6R(t)} \leq C(\log t)^{-d+\alpha_* D/2}(\log\log t)^{\alpha_* D}e^{-A\log\log t}.$$

Clearly, for A large enough, the right-hand side is bounded above by a constant as required by Proposition 2.3, which now gives the stated lower bound on $\nu_*^{(n)}(e)$. $\qquad\square$

5.4. Complementary Off-Diagonal Upper Bounds

In contrast with the case (15.4) of measures supported on powers of generators, for norm-radial kernels of type (15.16), we can use Meyer's construction to derive good off-diagonal bounds for $p(t, e, x)$.

Proposition 5.7. *Let G be a finitely generated group equipped with a norm $\| \cdot \|$. For $\alpha \in (0, 2)$, let ν_α be a symmetric probability measure on G satisfying (15.16). Assume that there exists a positive slowly varying function ℓ and a real $D > 0$ such that:*

1. $\forall\, r > 1, \quad V(r) \simeq r^D \ell(r)$;
2. $\forall\, t > 0, \quad x \in G, \quad p(t, x, x) \leq m(t) \simeq [(1 + t)^{D/\alpha} \ell(t^{1/\alpha})]^{-1}$.

Then there exists C such that, for all $t > 1$ and $x \in G$, we have

$$p(t, e, x) \leq Cm(t) \min \left\{ \left(\frac{t}{\|x\|^\alpha} \right)^{1+D/\alpha} \frac{\ell_1(t^{1/\alpha})}{\ell_1(\|x\|)}, 1 \right\}.$$

Remark 5.8. This proposition is stated here mostly for comparison with the next proposition. In fact, for the measure ν_α with $\alpha \in (0, 2)$, the hypothesis (1) implies automatically that (2) is satisfied as well. See [1, 2, 11, 14]. See [14] for a complete study of this case including two-sided discrete-time estimates.

Proposition 5.9. *Let G be a finitely generated group equipped with a norm $\| \cdot \|$ with volume V. Let ν_2 be a symmetric probability measure on G satisfying (15.16) with $\alpha = 2$. Assume that:*

(1) $\forall\, r > 1, V(r) \simeq r^D$,
(2) $\forall t > 1, x \in G, \quad p(t, x, x) \leq m(t) \simeq (t \log t)^{-D/2}$.

Then there exists C such that, for all $t > 1$ and $x \in G$, we have

$$p(t, e, x) \leq Cm(t) \min \left\{ \left(\frac{t \log \|x\|}{\|x\|^2} \right)^{1+D/2}, 1 \right\}.$$

Further, for any $\gamma \in (0, 2)$, there exists C_γ such that if $1 \leq t \leq \|x\|^\gamma$ then

$$p(t, e, x) \leq \frac{C_\gamma}{t^{D/2}} \left(\frac{t}{\|x\|^2} \right)^{1+D/2}.$$

Proof of Proposition 5.7. Under the stated hypothesis, we have $\delta_R \simeq R^{-\alpha}$ and $\mathcal{G}(R) \simeq R^{2-\alpha}$ and, for $1 \leq t \leq \eta R^\alpha$ (with η to be fixed later, small enough), Proposition 2.1 gives

$$p_R(t, e, x) \leq Cm(t) \left(\frac{t}{\|x\|^\alpha} \right)^{\|x\|/3R}.$$

By Meyer's construction (see the notation introduced after (15.7)), we have

$$p(t, x, y) \leq p_R(t, x, y) + t \left\| v_R' \right\|_\infty$$

$$\leq C_1 \frac{1}{t^{D/\alpha} \ell_1(t^{1/\alpha})} \left(\frac{t}{R^\alpha} \right)^{\|x\|/3R} + \frac{t}{R^{\alpha(1+D/\alpha)} \ell_1(R)}.$$

Choose $R = R(x, t)$ such that the two terms of the sum on the left-hand side are essentially equal, namely, set

$$\left(\log \frac{R^\alpha}{t} \right) \frac{\|x\|}{3R} = \left(\log \frac{R^\alpha}{t} \right) \left(1 + \frac{D}{\alpha} \right) + \log \frac{\ell_1(R)}{\ell_1(t^{1/\alpha})}.$$

As long as η is small enough, this choice of R gives $\|x\| \simeq R$ and

$$p(t, x, y) \leq \frac{2t}{\|x\|^{\alpha(1+D/\alpha)} \ell_1(\|x\|)} \simeq \frac{1}{t^{D/\alpha} \ell_1(t^{1/\alpha})} \left(\frac{t}{\|x\|^\alpha} \right)^{1+D/\alpha} \frac{\ell_1(t^{1/\alpha})}{\ell_1(\|x\|)}.$$

For any t (in particular, $t \geq \eta R^\alpha$) we can also use $m(t)$ for an easy upper bound. This gives

$$p(t, e, x) \leq C m(t) \min \left\{ \left(\frac{t}{\|x\|^\alpha} \right)^{1+D/\alpha} \frac{\ell_1(t^{1/\alpha})}{\ell_1(\|x\|)}, 1 \right\}$$

or, equivalently,

$$p(t, e, x) \leq C \min \left\{ t \nu_\alpha(\|x\|), m(t) \right\}.$$

$\square$

Proof of Proposition 5.9. In the context of Proposition 5.9, we have $\delta_R \simeq R^2$ and $\mathcal{G}(R) \simeq \log R$. For $1 \leq t \leq \eta R^2$, $\eta > 0$ small enough, Proposition 2.1 and Meyer's decomposition give

$$p(t, x, y) \leq p_R(t, x, y) + t \left\| v_R' \right\|_\infty$$

$$\leq C (t \log t)^{-D/2} \left(\frac{t \log R}{R^2} \right)^{\|x\|/3R} + \frac{t}{R^{2+D}}.$$

If $R^2 / \log R \leq t \leq R^2$ then this bound is not better than the easy bound $p(t, x, y) \leq m(t)$. By taking R such that $\|x\| = 3R(1 + D/2)$, we obtain

$$p(t, x, y) \leq C m(t) \min \left\{ \left(\frac{t \log \|x\|}{\|x\|^2} \right)^{1+D/2}, 1 \right\}.$$

However, if $1 \leq t \leq \eta \|x\|^\gamma$ with $\gamma \in (0, 2)$ and η small enough, then we can choose $R \simeq \|x\|$ so that

$$(t \log t)^{-D/2} \left(\frac{t \log R}{R^2} \right)^{\|x\|/3R} = \frac{t}{R^{2+D}},$$

equivalently,

$$\left(\frac{R^2}{t\log R}\right)^{\|x\|/3R} = \left(\frac{R^2}{t\log R}\right)^{1+D/2} \frac{(\log R)^{1+D/2}}{(\log t)^{D/2}}.$$

In the region $t \le \|x\|^\gamma$, this yields

$$p(t,e,x) \le \frac{2t}{\|x\|^{2+D}} \simeq t^{-D/2}\left(\frac{t}{\|x\|^2}\right)^{1+D/2}.$$

$\square$

6. Random Walks on Wreath Products

In this subsection, we illustrate how to use Proposition 1.4 to derive a lower bound for return probability of certain classes of random walks on wreath products.

First, we briefly review the definition of wreath products and a special type of random walks on them. Our notation follows [15]. Let H, K be two finitely generated groups. Denote the identity element of K by e_K and identity element of H by e_H Let K_H denote the direct sum:

$$K_H = \sum_{h \in H} K_h.$$

The elements of K_H are functions $f : H \to K$, $h \mapsto f(h) = k_h$, which have finite support in the sense that $\{h \in H : f(h) = k_h \ne e_K\}$ is finite. Multiplication on K_H is simply coordinate-wise multiplication. The identity element of K_H is the constant function $\mathbf{e}_K : h \mapsto e_K$ which, abusing notation, we denote by e_K. The group H acts on K_H by translation:

$$\tau_h f(h') = f(h^{-1}h'), \quad h, h' \in H.$$

The wreath product $K \wr H$ is defined to be the semidirect product

$$K \wr H = K_H \rtimes_\tau H,$$

$$(f,h)(f',h') = (f\tau_h f', hh').$$

In the lamplighter interpretation of wreath products, H corresponds to the base on which the lamplighter lives and K corresponds to the lamp. We embed K and H naturally in $K \wr H$ via the injective homomorphisms

$$k \longmapsto \underline{k} = (\mathbf{k}_{e_H}, e_H), \quad \mathbf{k}_{e_H}(e_H) = k, \ \mathbf{k}_{e_H}(h) = e_K \text{ if } h \ne e_H$$

$$h \longmapsto \underline{h} = (\mathbf{e}_K, h).$$

Let μ and η be probability measures on H and K respectively. Through the embedding, μ and η can be viewed as probability measures on $K \wr H$. Consider the measure

$$q = \eta * \mu * \eta$$

on $K \wr H$. This is called the switch-walk-switch measure on $K \wr H$ with switch-measure η and walk-measure μ.

Let (X_i) be the random walk on H driven by μ, and let $l(n, h)$ denote the number of visits to h in the first n steps:

$$l(n, h) = \#\{i : 0 \le i \le n, \ X_i = h\}.$$

Set also

$$l_*^g(n, h) = \begin{cases} l(n, h) & \text{if } h \notin \{e_H, g\} \\ l(n, e_H) - 1/2 & \text{if } h = g \\ l(n, e_H) - 1 & \text{if } h = e_H. \end{cases}$$

From [15], probability that the random walk on $K \wr H$ driven by q is at $(h, g) \in K \wr H$ at time n is given by

$$q^{(n)}((f, g)) = \mathbf{E}\left(\prod_{h \in H} \eta^{(2l_*^g(n,h))}(f(h)) \mathbf{1}_{\{X_n = g\}} \right).$$

Note that $\mathbf{E}$ stands for expectation with respect to the random walk $(X_i)_0^\infty$ on H started at e_H.

From now on we assume that η satisfies $\eta(e_K) = \epsilon > 0$ so that

$$\epsilon \eta^{(n-1)}(e_K) \le \eta^{(n)}(e_K) \le \epsilon^{-1} \eta^{(n-1))}(e_K).$$

Write $f \overset{C}{\asymp} g$ if $C^{-1}f \le g \le Cf$. Under these circumstances, we have

$$q^{(n)}((\mathbf{e}_K, g)) \overset{1/\epsilon^3}{\asymp} \mathbf{E}\left(\prod_{h \in H} \eta^{(2l(n,h))}(e_K) \mathbf{1}_{\{X_n = g\}} \right)$$

so that we can essentially ignore the difference between l and l_*.

Set

$$F_K(n) := -\log \eta^{(2n)}(e_K)$$

so that, for any $g \in H$,

$$q^{(n)}((\mathbf{e}_K, g)) \simeq \mathbf{E}\left(e^{-\sum_H F_K(l(n,h))} \mathbf{1}_{\{X_n = g\}} \right). \tag{15.20}$$

Proposition 6.1. *Let H be a finitely generated group equipped with a symmetric measure μ with $\mu(e_H) > 0$. Let K be a finitely generated group equipped with a symmetric measure η with $\eta(e_K) > 0$. Let $\|\cdot\|$ be a norm with volume function V. Let r be a positive continuous increasing function. Assume that:*

(1) The measure μ is strongly $(\|\cdot\|, r)$ controlled and V satisfies $V(t) \simeq t^D$.

(2) The function r satisfies $r(t) = t^{1/\beta} \ell_1(t)$ where ℓ_1 is a positive continuous slowly varying function.

(3) The function $F_K(n) = -\log \eta^{(2n)}(e_K) \simeq n^\gamma \ell_2(n)$ where $\gamma \in [0, 1)$ and ℓ_2 is a positive, continuous, slowly varying function.

Assume also that the slowly varying functions ℓ_i, $i = 1, 2$, are such that $\ell_i(t^a) \simeq \ell_i(t)$ for all $a > 0$. Then the switch-walk-switch measure q on $K \wr H$ associated with the pair η, μ satisfies

$$q^{(n)}(e) \geq \exp\left(-Cn^{\frac{D(1-\gamma)+\gamma\beta}{D(1-\gamma)+\beta}} \ell_1(n)^{\frac{\beta D(1-\gamma)}{D(1-\gamma)+\beta}} \ell_2(n)^{\frac{\beta}{D(1-\gamma)+\beta}} \right).$$

Proof. Let $\mathfrak{m}$ be the spectral measure of $\eta^{(2)}$ in the sense that

$$\int_{[0,1]} t^n d\mathfrak{m}(t) = \nu^{(2n)}(o).$$

For $x \in [0, \infty)$, set

$$F(x) := -\log \int_{[0,1]} t^x d\mathfrak{m}(t).$$

Observe that $F_K(n) = F(n)$ and that F is a concave function. For $\tau > 0$, let $B(\tau) = \{h \in H : \|h\| \leq \tau\}$. Since $q^{(2n)}(e) \geq q^{(2n)}(x)$ for any $x \in K \wr H$, (15.20) yields

$$q^{(n)}(e) \geq \frac{1}{\#B(\tau)} \mathbf{E}\left(e^{-\sum_H F_K(l(n,h))} \mathbf{1}_{\{\|X_n\| \leq \tau\}} \right)$$

$$\geq \frac{1}{\#B(\tau)} \mathbf{E}\left(e^{-\sum_H F_K(l(n,h))} \mathbf{1}_{\{\max_{1 \leq k \leq n}\{\|X_k\|\} \leq \tau\}} \right).$$

Using the concavity of F and the confinement of the walk (X_n) on H in the ball $B(\tau)$ in the last expression, this yields

$$q^{(n)}(e) \geq \frac{1}{V(\tau)} e^{-V(\tau)F(n/V(\tau))} \mathbf{P}_e\left(\max_{1 \leq k \leq n}\{\|X_k\|\} \leq \tau \right).$$

Let τ_n be such that

$$V(\tau_n)F(n/V(\tau_n)) = n/\rho(\tau_n)$$

where ρ is the inverse of r. By our various assumption, this means

$$(\tau_n^D/n)^{1-\gamma} \ell_2(n/\tau_n^D) = \tau_n^{-\beta} \ell_1(\tau_n)^\beta.$$

Hence $n \to \tau_n$ is a regularly varying function of order $(1 - \gamma)/(\beta + D(1 - \gamma)) < 1/\beta$. This shows that $r(n) \gg \tau_n$ and, since μ is strongly $(\|\cdot\|, r)$ controlled, Proposition 1.4 yields

$$q^{(n)}(e) \geq \frac{1}{V(\tau_n)} e^{-V(\tau_n)F(n/V(\tau_n))} e^{-Cn/\rho(\tau_n)} \geq e^{-C_1 n/\rho(\tau_n)}$$

and

$$\frac{n}{\rho(\tau_n)} = n^{\frac{D(1-\gamma)+\gamma\beta}{D(1-\gamma)+\beta}} \ell_1(n)^{\frac{\beta D(1-\gamma)}{D(1-\gamma)+\beta}} \ell_2(n)^{\frac{\beta}{D(1-\gamma)+\beta}}.$$

This gives the stated lower bound on $q^{(n)}(e)$. $\square$

Remark 6.2.   The case $\gamma = 1$ is excluded from these computations. It can be treated by the same method, but τ_n becomes a slowly varying function of n.

Remark 6.3.   In the setting of Proposition 6.1, suppose in addition we have $H = \mathbb{Z}^d$ and μ is in the domain of attraction of an operator-stable law ν on $\mathbb{R}^d$, that is there exists a normalizing sequence $B_n \in GL_d(\mathbb{R})$ such that $B_n^{-1}\mu^{*n} \Rightarrow \nu$. Then the lower bound in Proposition 6.1 is sharp and agrees with [17, Theorem 4.2]. Note that in this case, $\det B_n \simeq V(r(n))$, the scaling relation in [17, Theorem 4.2] reads

$$\frac{a_n \det B_{a_n}}{n} F_K\left(\frac{n}{\det B_{a_n}}\right) \simeq 1$$

and it agrees with

$$V(\tau_n)F(n/V(\tau_n)) = n/\rho(\tau_n),$$

with $a_n \simeq \rho(\tau_n)$.

Example 6.4.   Consider the symmetric probability measure μ on $\mathbb{Z}$ of the form

$$\mu(n) = \sum_{m \in \mathbb{Z}} \frac{\kappa \ell_1(|n|)}{(1+|n|)^{1+\alpha}}$$

where $\alpha \in (0,2)$ and ℓ_1 is a positive, continuous, slowly varying function satisfying $\ell_1(t^b) \simeq \ell_1(t)$ for all $b > 0$. We have

$$\delta_R := \sum_{|n|>R} \mu(n) \sim \frac{\kappa \ell_1(R)}{\alpha R^\alpha} \text{ and } \mathcal{G}(R) = \sum_{|n|\le R} |n|^2 \mu(n) \sim \frac{\kappa}{2-\alpha} R^{2-\alpha} \ell_1(R).$$

Therefore $R^2 \delta_R / \mathcal{G}(R) \to (2-\alpha)/\alpha$. By a classical result (see [9]), μ is in the domain of attraction of an α-stable law on $\mathbb{R}$. The normalizing sequence b_n can be chosen as the solution to the equation $n b_n^{-2} \mathcal{G}(b_n) = 1$, that is $b_n \sim \left(\frac{\kappa}{2-\alpha} n \ell_1(n)\right)^{1/\alpha}$. Let K be a finitely generated group equipped with a symmetric measure η with $\eta(e_K) > 0$. Suppose that the function $F_K(n) = -\log \eta^{(2n)}(e_K) \simeq n^\gamma \ell_2(n)$ where $\gamma \in [0,1)$ and ℓ_2 is a positive continuous slowly varying function. Assume also that $\ell_2(t^a) \simeq \ell_2(t)$ for all $a > 0$. Then [17, Theorem 4.2] (and the remark following that statement in [17]) implies that the switch-walk-switch measure q on $K \wr H$ associated with the pair η, μ satisfies

$$-\log q^{(n)}(e) \simeq n/a_n \simeq n^{\frac{(1-\gamma)+\gamma\beta}{(1-\gamma)+\beta}} \ell_1(n)^{\frac{(1-\gamma)}{(1-\gamma)+\beta}} \ell_2(n)^{\frac{\beta}{(1-\gamma)+\beta}},$$

where a_n is computed from the scaling relation

$$\frac{a_n b_{a_n}}{n} F_K\left(\frac{n}{b_{a_n}}\right) \simeq 1.$$

This agrees with the lower bound in Proposition 6.1. Note that by Proposition 3.5, the measure μ is strongly $(|\cdot|, r)$ controlled where $r(n) = (n\ell_1(n))^{1/\alpha}$.

Acknowledgments

The authors thank Mathav Murugan for his comments and useful remarks.

References

[1] Martin T. Barlow, Richard F. Bass, and Takashi Kumagai, *Parabolic Harnack inequality and heat kernel estimates for random walks with long range jumps*, Math. Z. **261** (2009), no. 2, 297–320. MR 2457301 (2009m:60111)

[2] Martin T. Barlow, Alexander Grigor'yan, and Takashi Kumagai, *Heat kernel upper bounds for jump processes and the first exit time*, J. Reine Angew. Math. **626** (2009), 135–57. MR 2492992 (2009m:58077)

[3] Alexander Bendikov and Laurent Saloff-Coste, *Random walks driven by low moment measures*, Ann. Probab. **40** (2012), no. 6, 2539–88. MR 3050511

[4] N. H. Bingham, C. M. Goldie, and J. L. Teugels, *Regular variation*, Encyclopedia of Mathematics and its Applications, vol. 27, Cambridge University Press, 1987. MR 898871 (88i:26004)

[5] E. A. Carlen, S. Kusuoka, and D. W. Stroock, *Upper bounds for symmetric Markov transition functions*, Ann. Inst. H. Poincaré Probab. Statist. **23** (1987), no. 2, suppl., 245–87. MR 898496 (88i:35066)

[6] Thierry Coulhon and Laurent Saloff-Coste, *Isopérimétrie pour les groupes et les variétés*, Rev. Mat. Iberoamericana **9** (1993), no. 2, 293–314. MR MR1232845 (94g:58263)

[7] E. B. Davies, *Heat kernels and spectral theory*, Cambridge Tracts in Mathematics, vol. 92, Cambridge University Press, 1990. MR 1103113 (92a:35035)

[8] Pierre de la Harpe, *Topics in geometric group theory*, Chicago Lectures in Mathematics, University of Chicago Press, 2000. MR MR1786869 (2001i:20081)

[9] William Feller, *On regular variation and local limit theorems*, Proc. Fifth Berkeley Sympos. Math. Statist. and Probability (Berkeley, Calif., 1965/66), Vol. II: Contributions to Probability Theory, Part 1, University of California Press, Berkeley, Calif., 1967, pp. 373–88. MR MR0219117 (36 #2200)

[10] W. Hebisch and L. Saloff-Coste, *Gaussian estimates for Markov chains and random walks on groups*, Ann. Probab. **21** (1993), no. 2, 673–709. MR MR1217561 (94m:60144)

[11] Jiaxin Hu and Takashi Kumagai, *Nash-type inequalities and heat kernels for non-local Dirichlet forms*, Kyushu J. Math. **60** (2006), no. 2, 245–65. MR 2268236 (2008d:60102)

[12] P. A. Meyer, *Renaissance, recollements, mélanges, ralentissement de processus de Markov*, Ann. Inst. Fourier (Grenoble) **25** (1975), no. 3–4, xxiii, 465–97, Collection of articles dedicated to Marcel Brelot on the occasion of his 70th birthday. MR 0415784 (54 #3862)

[13] Ante Mimica, *Heat kernel upper estimates for symmetric jump processes with small jumps of high intensity*, Potential Anal. **36** (2012), no. 2, 203–22. MR 2886459

[14] M. Murugan and L. Saloff-Coste, *Transition probability estimates for symmetric jump processes*, 2013.

[15] C. Pittet and L. Saloff-Coste, *On random walks on wreath products.*, Ann. Probab. **30, no. 2** (2002), 948–77.

[16] Ch. Pittet and L. Saloff-Coste, *On the stability of the behavior of random walks on groups*, J. Geom. Anal. **10** (2000), no. 4, 713–37. MR MR1817783 (2002m:60012)

[17] L. Saloff-Coste and T. Zheng, *Large deviations for stable like random walks on $\mathbb{Z}^d$ with applications to random walks on wreath products*, 2013.

[18] ———, *Random walks on nilpotent groups driven by measures supported on powers of generators*, 2013.

16
TOPICS ON MATHEMATICAL CRYSTALLOGRAPHY

TOSHIKAZU SUNADA
School of Interdisciplinary Mathematical Sciences, Meiji University,
Nakano 4-21-1, Nakano-ku, Tokyo, 164-8525 Japan

Dedicated to Professor Wolfgang Woess for his sixtieth birthday

Abstract
In July 2012 the General Assembly of the United Nations resolved that 2014 should be the International Year of Crystallography, 100 years since the award of the Nobel Prize for the discovery of X-ray diffraction by crystals. On this special occasion, we address several topics in mathematical crystallography. Especially motivated by the recent development in *systematic design of crystal structures* by both mathematicians and crystallographers, we discuss interesting relationships among seemingly irrelevant subjects; say, *standard crystal models*, *tight frames* in the Euclidean space, *rational points on Grassmannian*, and *quadratic Diophantine equations*. Thus, our view is quite a bit different from the traditional one in mathematical crystallography.

The central object in this article is what we call *crystallographic tight frames*, which are, in a loose sense, considered a generalization of *root systems*. We shall also pass a remark on the connections with *tropical geometry*, a relatively new area in mathematics, specifically with combinatorial analogues of the *Abel-Jacobi map* and *Abel's theorem*.

Contents

2010 *Mathematics Subject Classification.* Primary 74E15, 05C62; Secondary 14T05, 14G05, 17B22.

1. Introduction

It is my pastime to make various models of crystals by juggling a kit that I bought at a downtown stationer's shop. Though it is not always possible to make what I want because of the limited usage of the kit, I can still enjoy playing with it. For instance, my kit allows me to produce the model of the diamond crystal whose beauty, caused by its big symmetry, has intrigued me for some time and motivated me to look for other crystal structures, if any, with the similar symmetric property as the diamond. Actually as shown in [35] there exists the only structure that deserves to be called the *diamond twin*[1] (Fig. 16.1).

In the eyes of mathematicians, a crystal model as a network in space is simply a (piecewise linear) realization of an infinite-fold abelian covering graph over a finite graph. The key in this observation is that the translational action of a lattice group leaving the crystal model invariant yields a finite quotient graph,[2] and that the canonical map onto the quotient graph is a covering map whose covering transformation group is just the lattice group. This simple fact leads to the definition of *topological crystals* of arbitrary dimension, and can be effectively used to enumerate all topological types of crystal structures because an abelian covering graph over a finite graph X_0 corresponds to a subgroup of the first homology group $H_1(X_0, \mathbb{Z})$; thereby the enumeration being reduced to counting finite graphs and subgroups of their first homology groups [37], [38]. Needless to say, however, there are infinitely many ways to

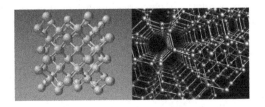

Figure 16.1 Diamond and its twin. From WebElements
http://www.webelements.com.

---

[1] This is what I call the K_4 *crystal* [35], [38] because of the fact that it is, as a graph, the maximal abelian covering graph over the complete graph K_4 consisting of four vertices. The structure was for the first time described by Fritz Laves in 1933. Diamond and its twin are characterized by the "strong isotropic property," the strongest one among all possible meanings of isotropy.

[2] Refer to [9] for this observation made in the community of crystallographers. The *vector method* mentioned in this reference can be interpreted in terms of cohomology of the quotient graph (Sect. 7). Historically A. F. Wells is the crystallographer who initiated a systematic study of crystal structures as 3D networks [42], [43]. See [32] for some recent views on mathematical crystallography.

realize the covering graph. Thus, it is natural to seek a "standard model" having as many symmetries as possible just like diamond and its twin. *Standard realizations* introduced in 2000 by M. Kotani and myself [25], [27] in connection with asymptotic behaviors of random walks may be called standard models. Indeed the standard realization of a topological crystal X has maximal symmetry in the sense that every automorphism of X extends to a congruent transformation leaving the realization invariant. Moreover, *crystal models with "big" symmetry turn out to be the ones obtained by standard realizations* (see Theorem 7.7 in Sect. 7 for the precise formulation). Figure 16.2 illustrates several three-dimensional examples.[3] Classical two-dimensional lattices such as the square lattice, the regular triangular lattice, the honeycomb, and the kagome lattice are also standard realizations. An interesting feature following the tradition of geometry is that standard realizations are characterized by a certain minimal principle, just like the characterization of the round circle by means of the isoperimetric inequality. Furthermore, this notion combined with the enumeration of topological crystals provides a useful method for a systematic design of crystal structures. Actually, there is a simple algorithm for the design with which one can create a computer program to produce the CG images of two- or three-dimensional crystals.[4]

Among all standard models, the simplest one is the *cubic lattice* (the jungle gym-like figure in plain language). As a matter of fact, the cubic lattice is not very interesting as a crystal model,[5] but from some "view," this lends itself to another recreation and gives rise to an interesting mathematical issue that is linked to the standard models of general crystal structures mentioned above.

Let us look at the cubic lattice from a sufficiently remote distance. What we find when we turn it around is that there are some specific directions toward which we may see two-dimensional crystalline patterns (ignoring the effect of perspective). For instance, one can see the square lattice and the regular triangular lattice as such crystalline patterns.

Mathematically, we are looking at the image in the cartesian plane $\mathbb{R}^2$ of the cubic lattice placed in $\mathbb{R}^3$ by the orthogonal projection P : $(x, y, z) \mapsto (x, y)$ (see Figure 16.3). Here the cubic lattice is supposed to be generated by an orthonormal basis $\mathbf{f}_1, \mathbf{f}_2, \mathbf{f}_3$ of $\mathbb{R}^3$. Thus, the set of vertices in it is

---

[3] *Lonsdaleite* (named in honor of Kathleen Lonsdale) in this figure is thought of as a relative of diamond, but is not isotropic.

[4] Due to Hisashi Naito. Crystallographers also sought standard models; see [14], [16], [20] for instance. Some of their models are the same as ours. The algorithm *SYSTRE* created by Delgado–Friedrich in 2004 produces the *barycentric drawing*, which seems to coincide with standard realizations as far as several examples are examined. See also [29].

[5] Sodium chloride (NaCl) crystallizes in a cubic lattice.

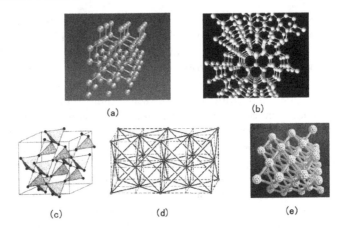

Figure 16.2 (a) Lonsdaleite (from WebElements http://www.webelements.com), (b) ThSi$_2$ structure, (c) three-dimensional kagome lattice, (d) net associated with the face-centered cubic lattice, (e) net associated with the body-centered cubic lattice

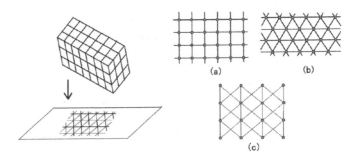

Figure 16.3 Projected images of the cubic lattice

$$\{k_1\mathbf{f}_1 + k_2\mathbf{f}_2 + k_3\mathbf{f}_3 |\ k_1, k_2, k_3 \in \mathbb{Z}\}.$$

We now put $\mathbf{v}_i = (a_i, b_i) = P(\mathbf{f}_i)$. Because

$$\langle \mathbf{x}, \mathbf{f}_1 \rangle \mathbf{f}_1 + \langle \mathbf{x}, \mathbf{f}_2 \rangle \mathbf{f}_2 + \langle \mathbf{x}, \mathbf{f}_3 \rangle \mathbf{e}_3 = \mathbf{x} \quad (\mathbf{x} \in \mathbb{R}^3), \qquad (16.1)$$

and $\langle \mathbf{x}, \mathbf{f}_i \rangle = \langle \mathbf{x}, \mathbf{v}_i \rangle$, $P\mathbf{x} = \mathbf{x}$ for $\mathbf{x} \in \mathbb{R}^2$, we have, by projecting down the equality (16.1) to the x-y plane,

$$\langle \mathbf{x}, \mathbf{v}_1 \rangle \mathbf{v}_1 + \langle \mathbf{x}, \mathbf{v}_2 \rangle \mathbf{v}_2 + \langle \mathbf{x}, \mathbf{v}_3 \rangle \mathbf{v}_3 = \mathbf{x} \quad (\mathbf{x} \in \mathbb{R}^2). \qquad (16.2)$$

The projected image of vertices in the cubic lattice is given by

$$\{k_1\mathbf{v}_1 + k_2\mathbf{v}_2 + k_3\mathbf{v}_3 |\ k_1, k_2, k_3 \in \mathbb{Z}\}.$$

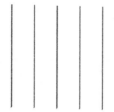

Figure 16.4

What we need to notice here is that the projected image does not always give a crystalline pattern. For instance, Figure 16.4 depicts evenly spaced parallel lines expressing the closure of the projected image of vertices in the case $\mathbf{v}_1 = (1,0)$, $\mathbf{v}_2 = t(0,1)$, $\mathbf{v}_3 = t(1,\sqrt{2})$, where we should note that $\{m + n\sqrt{2}|\ m, n \in \mathbb{Z}\}$ is dense in $\mathbb{R}$ (more generally, given a positive irrational number α, one can find infinitely many positive integers p, q such that $|\alpha - q/p| < 1/p^2$ (Dirichlet's theorem), from which it follows that $\{m + n\alpha|\ m, n \in \mathbb{Z}\}$ is dense).

Actually the image of vertices in question is dense in almost all cases. To have a crystalline pattern, it is necessary (and sufficient) that three vectors $\mathbf{v}_1, \mathbf{v}_2, \mathbf{v}_3$ generate a lattice in $\mathbb{R}^2$, or equivalently there exists a triple of integers $(n_1, n_2, n_3) \neq (0,0,0)$ such that

$$n_1\mathbf{v}_1 + n_2\mathbf{v}_2 + n_3\mathbf{v}_3 = 0, \qquad (16.3)$$

where one may assume without loss of generality that the greatest common divisor of n_1, n_2, n_3 is 1. Then the kernel of the homomorphism $\rho : \mathbb{Z}^3 \longrightarrow \mathbb{R}^2$ defined by $\rho(k_1, k_2, k_3) = k_1\mathbf{v}_1 + k_2\mathbf{v}_2 + k_3\mathbf{v}_3$ coincides with $H := \mathbb{Z}(n_1, n_2, n_3)$. Going back to Figure 16.3, we observe that the square lattice (a), regular triangular lattice (b) and the lattice (c) correspond to $(n_0, n_1, n_2) = (1,0,0)$, $(n_0, n_1, n_2) = (1,1,1)$, $(n_0, n_1, n_2) = (1,1,2)$, respectively.

This expository article, including a few new results, is thought of as a continuation of my book [38] published in 2012. The purpose is, starting from the above elementary observations, to share a link to a few mathematical subjects, say *tight frames* in the Euclidean spaces, *rational points on Grassmannians*, and *quadratic Diophantine equations*. Those subjects are not things of novelty (for instance, tight frames appear in various guises in practical sciences), but turn out to be closely connected with each other in an interesting way.

The protagonist is *crystallographic tight frames* introduced in Sect. 3. This notion generalizes the above-mentioned situation, and is closely related to a systematic design of crystal structures through the notion of standard realizations (Sect. 7). Further, this in a special case is regarded as a generalization of *root systems* whose origin is in the work of

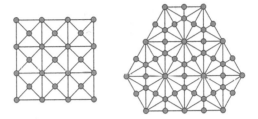

Figure 16.5 Examples of two-dimensional Coxeter complexes

W. Killings, E. Cartan, and H. Weyl on Lie groups. Actually, irreducible root systems yield highly symmetrical crystallographic tight frames. It should be pointed out that root systems pertain to *Euclidean Coxeter complexes* (cf. [7]), very remarkable triangulations of the Euclidean space, named after H. S. M. Coxeter (see Figure 16.5 for two-dimensional examples that correspond to the root systems B_2 and G_2). A remarkable fact is that the one-skeleton of a Coxeter complex is the standard realization of a crystal structure (Sect. 7).

As is well known, root systems are completely classified by means of *Dynkin diagrams*. On the other hand, as will be described in Sect. 4, similarity classes of crystallographic tight frames are parameterized by rational points on Grassmannians (this is by no means surprising if we rephrase the above observation as "the projection of the cubic lattice towards a *rational direction* gives rise to a crystal pattern"). Certain quadratic Diophantine equations show up when we explicitly associate crystallographic tight frames with rational points. A noteworthy situation occurs in the two-dimensional case especially; we may parameterize the (oriented) congruence classes by "rational points" on a certain complex projective quadric. A rational point we mean here is a point in a complex projective space each of whose homogeneous coordinate is represented by a number in an *imaginary quadratic field*. In Sect. 6, we explain a relationship with *tropical geometry*, skeletonized version of algebraic geometry, especially with discrete Abel-Jacobi maps. This unexpected link is brought about via crystallographic tight frames associated with finite graphs. The final section presents a link between discrete Abel-Jacobi maps and standard realizations.

Not surprisingly, the materials here have something to do with, not only the legacy of Eukleides–Archimedes–Kepler's achievements in polyhedral geometry,[6] but also with the *geometry of numbers* and the theory of *quadratic forms* because of the crucial role played by lattice groups in crystallography.

---

[6] Legend has it that Pythagoras derived the notion of regular polyhedra from the shape of a crystal.

Before leaving the introduction, let us fix a few notations used throughout. We express a matrix A by (a_{ij}) for simplicity when the $(i,j)^{\text{th}}$ entry of A is a_{ij}. The transpose of $A = (a_{ij})$ expressed by ${}^t A$ is the matrix whose $(i,j)^{\text{th}}$ entry is a_{ji}. The *trace* of a square matrix A, the sum of the diagonal entries of A, is denoted by $\operatorname{tr} A$. The *determinant* of A is denoted by $\det A$.

Given a field K, we think of K^d as a vector space over K consisting of column vectors $\mathbf{x} = {}^t(x_1, \ldots, x_d)$ with $x_i \in K$. The subspace spanned by vectors $\mathbf{x}_1, \ldots, \mathbf{x}_k \in K^d$ is expressed by $\langle \mathbf{x}_1, \ldots, \mathbf{x}_k \rangle_K$. We denote by $M_{m,n}(K)$ the set of all $m \times n$ matrices whose entries belong to K. We also use the notation $M_n(K)$ for $M_{n,n}(K)$. The identity matrix (δ_{ij}) in $M_n(K)$ is denoted by I_n, or simply I.

A matrix $A \in M_{m,n}(K)$ is identified with the linear operator of K^n into K^m given by $\mathbf{x} \mapsto A\mathbf{x}$. For a linear operator $T : K^n \longrightarrow K^m$, the *kernel* of T is written as $\ker T$. The *image* of T is denoted by $\operatorname{Image} T$. When the matrix $A \in M_{m,n}(K)$ consists of column vectors $\mathbf{a}_i$ $(i = 1, \ldots, n)$, we write $A = (\mathbf{a}_1, \ldots, \mathbf{a}_n)$.

The (standard) inner product $\langle \mathbf{x}, \mathbf{y} \rangle$ of two vectors $\mathbf{x} = {}^t(x_1, \ldots, x_n)$, $\mathbf{y} = {}^t(y_1, \ldots, y_n)$ in $\mathbb{R}^n$ is ${}^t \mathbf{x} \mathbf{y} = x_1 y_1 + \cdots + x_n y_n$. The norm $\|\mathbf{x}\|$ is $\langle \mathbf{x}, \mathbf{x} \rangle^{1/2}$. For a subspace $W \subset \mathbb{R}^n$, the orthogonal complement $\{ \mathbf{x} \in \mathbb{R}^n \mid \langle \mathbf{x}, \mathbf{y} \rangle = 0 \ (\mathbf{y} \in W) \}$ is denoted by $W^\perp$.

Acknowledgement

I would like to thank Peter Kuchment for a valuable hint about the relation between standard realizations and tight frames. I also thank Hisashi Naito and my daughter Kayo for producing several figures.

2. Tight Frames

Property (16.2) tells that $\mathbf{v}_1, \mathbf{v}_2, \mathbf{v}_3$ form a *tight frame* of $\mathbb{R}^2$, a terminology originally used in wavelet analysis. The basic philosophy of tight frames is that representations such as (16.2) are similar to an orthogonal expansion when considering an infinite dimensional Hilbert space such as $L^2(\mathbb{R}^d)$, and that one may have more freedom in choosing the $\mathbf{v}_j$ to have desirable properties such as certain smoothness and small support properties that may be impossible were they to be orthogonal (see [15] for a pioneer work). Tight frames are intimately related to rank-one quantum measurements [19]. In the finite dimensional case they are seen in the study of packet-based communication systems (refer to [22] for instance), and also show up as spherical two-designs in combinatorics [41]. In this chapter, we will give a completely different view to tight frames. Our development is guided by the idea indicated in the Introduction.

Let us start with some fundamental facts on tight frames which are more or less known (cf. [40]). Only some rudiments of linear algebra are required to read this section.

In general, a sequence of N vectors $\mathcal{S} = \{\mathbf{v}_i\}_{i=1}^{N}$ in $\mathbb{R}^d$ is said to be a d-dimensional frame of size N or simply *frame* if it generates $\mathbb{R}^d$. In this definition, some of $\mathbf{v}_i$ allow to be zero or parallel. Given a frame, we may associate a linear operator (called the *frame operator*) $S = S_{\mathcal{S}}$: $\mathbb{R}^d \longrightarrow \mathbb{R}^d$ by setting

$$S(\mathbf{x}) = \sum_{i=1}^{N} \langle \mathbf{x}, \mathbf{v}_i \rangle \mathbf{v}_i,$$

which is symmetric and positive. The matrix for S is given by

$$S = (\mathbf{v}_1, \ldots, \mathbf{v}_N)\, {}^t(\mathbf{v}_1, \ldots, \mathbf{v}_N) = \mathbf{v}_1\,{}^t\mathbf{v}_1 + \cdots + \mathbf{v}_N\,{}^t\mathbf{v}_N,$$

and hence

$$\operatorname{tr} S = \sum_{i=1}^{N} \|\mathbf{v}_i\|^2. \tag{16.4}$$

A frame $\mathcal{S}$ is said to be α-*tight* (or simply tight) if $S = \alpha I_d$ with a positive α, i.e.,

$$\sum_{i=1}^{N} \langle \mathbf{x}, \mathbf{v}_i \rangle \mathbf{v}_i = \alpha \mathbf{x} \quad (\mathbf{x} \in \mathbb{R}^d).$$

In view of (16.4), if $\mathcal{S}$ is 1-tight, then

$$\sum_{i=1}^{N} \|\mathbf{v}_i\|^2 = d.$$

Tightness (resp. 1-tightness) is obviously preserved by similar transformations and permutations of subscripts i in $\mathbf{v}_i$ (resp. by orthogonal transformations). Here two frames $\mathcal{S}_1 = \{\mathbf{u}_i\}_{i=1}^{N}$ and $\mathcal{S}_2 = \{\mathbf{v}_i\}_{i=1}^{N}$ are said to be *similar* if there exists an orthogonal transformation U of $\mathbb{R}^d$ and a positive number λ such that $\mathbf{u}_i = \lambda U(\mathbf{v}_i)$ ($i = 1, \ldots, N$). If $\lambda = 1$ in this relation, $\mathcal{S}_1$ and $\mathcal{S}_2$ are said to be *congruent*.

A tight frame appears in the following situation.

Proposition 2.1. *Suppose that a finite group G acts on $\mathbb{R}^d$ as orthogonal transformations, and let $\mathcal{S} = \{\mathbf{v}_i\}_{i=1}^{N}$ be a frame in $\mathbb{R}^d$ which is invariant under the G-action (precisely speaking, for any $g \in G$, there exists a permutation σ of $\{1, \ldots, N\}$ such that $g\mathbf{v}_i = \mathbf{v}_{\sigma i}$). If the G-action on $\mathbb{R}^d$ is irreducible, then $\mathcal{S}$ is a tight frame satisfying $\sum_{i=1}^{N} \mathbf{v}_i = \mathbf{0}$.*

To check this, we note that the G-action commutes with the frame operator S. Looking at eigenspaces of S, we conclude that $S = \alpha I_d$ for

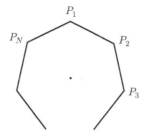

Figure 16.6 A regular polygon

some positive scalar α, as claimed. For the claim $\sum_{i=1}^{N} \mathbf{v}_i = \mathbf{0}$, we only have to notice that the left-hand side is a G-invariant vector.

Using Proposition 2.1, one can prove that for points $P_1, \ldots, P_N$ ($N \geq 3$) in the plane $\mathbb{R}^2$ forming a N-regular polygon with the barycenter O (Fig. 16.6), the vectors $\mathbf{v}_1 = \overrightarrow{OP_1}, \ldots, \mathbf{v}_N = \overrightarrow{OP_N}$ yield a tight frame (of course, one can prove this by a direct computation). By the same reasoning, five Platonic solids (regular convex polyhedra) and thirteen Archimedean solids (semi-regular polyhedra) yield tight frames of $\mathbb{R}^3$ (cf. [13]).

Tight frames have a similar nature to the orthonormal basis seen in the following proposition.

Proposition 2.2. *The following three conditions are equivalent to the 1-tightness of a frame $\mathcal{S} = \{\mathbf{v}_i\}_{i=1}^{N}$, respectively.*

(1) $\displaystyle\sum_{i=1}^{N} \langle \mathbf{v}_i, \mathbf{x} \rangle^2 = \|\mathbf{x}\|^2 \quad (\mathbf{x} \in \mathbb{R}^d),$

(2) $\displaystyle\sum_{i=1}^{N} \langle \mathbf{v}_i, \mathbf{x} \rangle \langle \mathbf{v}_i, \mathbf{y} \rangle = \langle \mathbf{x}, \mathbf{y} \rangle \quad (\mathbf{x}, \mathbf{y} \in \mathbb{R}^d),$

(3) $\displaystyle\sum_{j=1}^{N} \langle T\mathbf{v}_j, \mathbf{v}_j \rangle = \operatorname{tr} T$ *for any linear transformation T of $\mathbb{R}^d$.*

Proof. We only show that (3) implies the 1-tightness because the other claims are easy to check. Consider the operator T defined by $T(\mathbf{x}) = \langle \mathbf{x}, \mathbf{y} \rangle \mathbf{y}$. Evidently $\operatorname{tr} T = \|\mathbf{y}\|^2$ and

$$\sum_{j=1}^{N} \langle T\mathbf{v}_j, \mathbf{v}_j \rangle = \sum_{j=1}^{N} \langle \mathbf{y}, \mathbf{v}_j \rangle^2.$$

Applying the equality (3), we have $\sum_{j=1}^{N} \langle \mathbf{y}, \mathbf{v}_j \rangle^2 = \|\mathbf{y}\|^2$. Thus by (2), $\mathcal{S}$ is 1-tight. $\square$

Recall that the 1-tight frame $\{\mathbf{v}_1, \mathbf{v}_2, \mathbf{v}_3\}$ mentioned in Introduction was obtained as the projected image of an orthonormal basis of $\mathbb{R}^3$. This is true for general 1-tight frames; that is, any 1-tight frame $\mathcal{S} = \{\mathbf{v}_i\}_{i=1}^N$ in $\mathbb{R}^d$ is a projected image of an orthonormal basis of $\mathbb{R}^N$. To see this, we shall introduce an auxiliary operator (matrix). Given a frame $\mathcal{S} = \{\mathbf{v}_i\}_{i=1}^N$, define the *frame projection* $P = P_{\mathcal{S}} : \mathbb{R}^N \longrightarrow \mathbb{R}^d$ by

$$P\big({}^t(x_1, \ldots, x_N)\big) = x_1\mathbf{v}_1 + \cdots + x_N\mathbf{v}_N,$$

i.e., P is the linear operator characterized by $P(\mathbf{f}_i) = \mathbf{v}_i \ (i = 1, \ldots, N)$ where $\{\mathbf{f}_i\}_{i=1}^N$ is the fundamental basis of $\mathbb{R}^N$. The matrix for P is nothing but $(\mathbf{v}_1, \ldots, \mathbf{v}_N)$. Therefore, we have $S = P\,{}^tP$. On the other hand, the matrix for ${}^tPP$ is the *Gramm matrix* $G_{\mathcal{S}} = \big(\langle \mathbf{v}_i, \mathbf{v}_j \rangle\big)$.

The next proposition is considered a special case of *Naimark's dilation theorem* in the theory of quantum measurements.

Proposition 2.3.   *The following four conditions are equivalent:*

(1) $\mathcal{S}$ is 1-tight.

(2) $P = P_{\mathcal{S}}$ satisfies $P\,{}^tP = I$.

(3) P is orthogonal in the sense that the restriction $P\big|(\ker P)^{\perp}$: $\big(\ker P\big)^{\perp} \longrightarrow \mathbb{R}^d$ is an isometry (i.e., it preserves the inner products).

(4) ${}^tPP : \mathbb{R}^N \longrightarrow \mathbb{R}^N$ is a orthogonal projection, or equivalently the Gramm matrix satisfies $G_{\mathcal{S}}^2 = G_{\mathcal{S}}$.

Proof.   (1) $\Leftrightarrow$ (2) is obvious. To show (1) $\Leftrightarrow$ (3), suppose that $\mathcal{S}$ is 1-tight. Since $P\,{}^tP = I$, we find

$$\langle {}^tP(\mathbf{x}), {}^tP(\mathbf{y}) \rangle = \langle P\,{}^tP(\mathbf{x}), \mathbf{y} \rangle = \langle \mathbf{x}, \mathbf{y} \rangle$$

and

$$\langle P\,{}^tP(\mathbf{x}), P\,{}^tP(\mathbf{y}) \rangle = \langle \mathbf{x}, \mathbf{y} \rangle.$$

Hence

$$\langle P\,{}^tP(\mathbf{x}), P\,{}^tP(\mathbf{y}) \rangle = \langle {}^tP(\mathbf{x}), {}^tP(\mathbf{y}) \rangle.$$

Since $\big(\ker P\big)^{\perp} = \text{Image } {}^tP$, we conclude that $P\big|(\ker P)^{\perp} : (\ker P)^{\perp} \longrightarrow \mathbb{R}^d$ is an isometry.

Next suppose that $P\big|(\ker P)^{\perp} : (\ker P)^{\perp} \longrightarrow \mathbb{R}^d$ is an isometry. Again using $\big(\ker P\big)^{\perp} = \text{Image}\,{}^tP$, we have

$$\langle P\,{}^tP(\mathbf{x}), P\,{}^tP(\mathbf{y}) \rangle = \langle {}^tP(\mathbf{x}), {}^tP(\mathbf{y}) \rangle,$$

or equivalently

$$\langle S^2(\mathbf{x}), \mathbf{y} \rangle = \langle S(\mathbf{x}), \mathbf{y} \rangle.$$

This implies $S^2 = S$, and hence $S = I$.

(1) $\Leftrightarrow$ (4): If $\mathcal{S}$ is 1-tight, then

$$({}^tPP)({}^tPP) = {}^tP(P{}^tP)P = {}^tPP.$$

Conversely if $({}^tPP)^2 = {}^tPP$, then $P{}^tPP = P$ since ${}^tP$ is injective. But P is surjective, so $P{}^tP = I$. $\square$

For a frame $\mathcal{S}$ of $\mathbb{R}^d$, the $(N-d)$-dimensional subspace $W(\mathcal{S}) = \ker P_{\mathcal{S}}$ of $\mathbb{R}^N$ is called the *vanishing subspace* for $\mathcal{S}$, which obviously depends only on the similarity class of $\mathcal{S}$.

Proposition 2.4.   (1) *For any subspace $W \subset \mathbb{R}^N$ of dimension $N-d$, there exists a d-dimensional tight frame $\mathcal{S}$ of size N such that $W = W(\mathcal{S})$.*

(2) *Two tight frames $\mathcal{S}_1, \mathcal{S}_2$ are congruent if and only if $W(\mathcal{S}_1) = W(\mathcal{S}_2)$.*

Proof.   (1) Let $p : \mathbb{R}^N \longrightarrow \mathbb{R}^N$ be the orthogonal projection onto $W^\perp$. Choosing an isometry $i : W^\perp \longrightarrow \mathbb{R}^d$, we put $P = ip$, which is obviously a frame projection satisfying (3) in Proposition 2.3 such that $W = \ker P$.

(2) If $\ker P_{\mathcal{S}_1} = \ker P_{\mathcal{S}_2}(= W)$, then since $P_i = P_{\mathcal{S}_i} | W^\perp : W^\perp \longrightarrow \mathbb{R}^d$ is isometry, $U = P_1 P_2{}^{-1}$ is an orthogonal transformation such that $P_{\mathcal{S}_1} = U P_{\mathcal{S}_2}$. $\square$

The symmetric group $\mathfrak{S}_N$ of $\{1, 2, \ldots, N\}$ acts on $\mathbb{R}^N$ as axis permutations[7], i.e.,

$$\sigma(\mathbf{f}_i) = \mathbf{f}_{\sigma(i)} \quad (\sigma \in \mathfrak{S}_N).$$

The *automorphism group* $\mathrm{Aut}(\mathcal{S})$ of a 1-tight frame $\mathcal{S} = \{\mathbf{v}_i\}_{i=1}^N$ is defined to be the subgroup of $\mathfrak{S}_N$ consisting of $\sigma \in \mathfrak{S}_N$ satisfying $\sigma(W(\mathcal{S})) = W(\mathcal{S})$. By virtue of Proposition 2.4 (2), there is an injective homomorphism $U : \mathrm{Aut}(\mathcal{S}) \longrightarrow O(d)$ such that $U(\sigma)\mathbf{v}_i = \mathbf{v}_{\sigma(i)}$.

If $\mathrm{Aut}(\mathcal{S})$ acts transitively on $\{1, 2, \ldots, N\}$, then $\mathcal{S}$ is said to be *isotropic*[8].

Tight frames associated with regular polygons, Platonic solids, and Archimedean solids are isotropic. Among them, the equilateral triangle and the regular tetrahedron are very special in the sense that $\mathrm{Aut}(\mathcal{S})$ agrees with $\mathfrak{S}_N$. We shall say that $\mathcal{S}$ with the property $\mathrm{Aut}(\mathcal{S}) = \mathfrak{S}_N$ is *strongly isotropic*.

The higher dimensional analogue of the equilateral triangle and the regular tetrahedron is the *equilateral simplex*, a simplex Δ in $\mathbb{R}^d$ whose

---

[7] This is called the *standard representation* of $\mathfrak{S}_N$.
[8] An isotropic tight frame $\{\mathbf{v}_i\}_{i=1}^N$ is *uniform* in the sense that $\|\mathbf{v}_1\| = \cdots = \|\mathbf{v}_N\|$. The notion of uniform tight frame appears in various applications. The classification of isotropic frames is obviously related to that of subgroups of $\mathfrak{S}_N$ acting transitively on $\{1, 2, \ldots, N\}$ which has been pursued for over a century since the 1860 Grand Prix of the Académie des Sciences.

edges have equal length. Suppose that its barycenter is the origin O, and let $P_1, \ldots, P_{d+1}$ be its vertices. The symmetric group $\mathfrak{S}_{d+1}$ acts on Δ as orthogonal transformations. Thus, $\{\mathbf{v}_i\}_{i=1}^{d+1}$ defined by $\mathbf{v}_i = \overrightarrow{OP_i}$ is a strongly isotropic tight frame.

Conversely we have the following.

Proposition 2.5. *Let $\mathcal{S}$ be a strongly isotropic 1-tight frame in $\mathbb{R}^d$ $(d \geq 2)$. Then $\mathcal{S}$ is the frame associated with the equilateral simplex.*

Proof. Since $\mathrm{Aut}(\mathcal{S}) = \mathfrak{S}_N$, invariant subspaces for the $\mathrm{Aut}(\mathcal{S})$-action on $\mathbb{R}^N$ are either $W = \{{}^t(x, x, \ldots, x) | x \in \mathbb{R}\}$ or $W^\perp$ (indeed, $\mathbb{R}^N = W \oplus W^\perp$ gives the irreducible decomposition for the $\mathfrak{S}_N$-action). Because $d \geq 2$, the vanishing subspace for $\mathcal{S}$ must be W, and $N = d + 1$. Then the frame projection P is identified with the orthogonal projection of $\mathbb{R}^{d+1}$ onto $W^\perp$. Hence

$$\mathbf{v}_i = P(\mathbf{f}_i) = \mathbf{f}_i - \frac{1}{d+1} \sum_{i=1}^{d+1} \mathbf{f}_i.$$

Since $\langle \mathbf{v}_i, \mathbf{v}_j \rangle = \delta_{ij} - \frac{1}{d+1}$, if $\mathbf{v}_i = \overrightarrow{OP_i}$, then $(P_i P_j)^2 = \|\mathbf{v}_i - \mathbf{v}_j\|^2 = 2$ $(i \neq j)$. We thus conclude that $P_1, \ldots, P_{d+1}$ is the vertices of the equilateral simplex. $\square$

The proof of the following proposition is left as an exercise for the reader.

Proposition 2.6. *If $\{\overrightarrow{OP_i}\}_{i=1}^{d+1}$ is a tight frame of $\mathbb{R}^d$, and $\sum_{i=1}^{d+1} \overrightarrow{OP_i} = \mathbf{0}$, then $P_1, P_2, \ldots, P_{d+1}$ be the vertices of an equilateral simplex.*

We go back to the general case. In terms of matrices, what we have said in Proposition 2.3 is rephrased as

Proposition 2.7. *The row vectors in a matrix*

$$A = \begin{pmatrix} a_{11} & a_{12} & \cdots & a_{1d} \\ a_{21} & a_{22} & \cdots & a_{2d} \\ & & \cdots & \\ & & \cdots & \\ a_{N1} & a_{N2} & \cdots & a_{Nd} \end{pmatrix} \in M_{N,d}(\mathbb{R})$$

give rise to a 1-tight frame if and only if the column vectors of A form an orthonormal system (i.e. ${}^tAA = I_d$).

For if we define $\mathcal{S} = \{\mathbf{v}_i\}_{i=1}^N$ by writing ${}^tA = (\mathbf{v}_1, \ldots, \mathbf{v}_N)$, then ${}^tA$ is the matrix for $P_{\mathcal{S}}$.

Writing $A = (\mathbf{a}_1, \ldots, \mathbf{a}_d)$, we find that $\ker P = \langle \mathbf{a}_1, \ldots, \mathbf{a}_d \rangle_\mathbb{R}^\perp$ because

$$\langle \mathbf{a}_1, \ldots, \mathbf{a}_d \rangle_\mathbb{R}^\perp = (\mathrm{Image}\, A)^\perp = \ker {}^tA.$$

Thus we have the following proposition which rephrases Proposition 2.4.

Proposition 2.8. *Given an $(N - d)$-dimensional subspace W of $\mathbb{R}^N$, there exists a solution $A \in M_{N,d}(\mathbb{R})$ of the equations*

$$^tAA = I_d, \tag{16.5}$$
$$^tA\mathbf{x} = \mathbf{0} \quad (\mathbf{x} \in W). \tag{16.6}$$

If A_1, A_2 are solutions, then there exists $U \in O(d)$ such that $A_1 = A_2 U$.

We now give an explicit parameterization of congruence classes of 1-tight frames. We denote by $\mathrm{T}_N(\mathbb{R}^d)$ the set of congruence classes of 1-tight frames $\{\mathbf{v}_i\}_{i=1}^N$ in $\mathbb{R}^d$. Proposition 2.7 tells us that the set of 1-tight frames is identified with the *Stiefel manifold* $\mathrm{V}_d(\mathbb{R}^N)$ ($= \{A \in M_{N,d}(\mathbb{R}) \mid {}^tAA = I_d\}$). The action of the orthogonal group $O(d)$ on $\mathrm{V}_d(\mathbb{R}^N)$ by $A \in \mathrm{V}_d(\mathbb{R}^N) \mapsto AU^{-1} \in \mathrm{V}_d(\mathbb{R}^N)$ $\left(U \in O(d)\right)$ is compatible with the action of $O(d)$ on the set of 1-tight frames because

$$^t(AU^{-1}) = U\,{}^tA = U(\mathbf{v}_1, \ldots, \mathbf{v}_N) = \left(U(\mathbf{v}_1), \ldots, U(\mathbf{v}_N)\right).$$

Therefore, $\mathrm{T}_N(\mathbb{R}^d)$ is identified with the quotient space $\mathrm{V}_d(\mathbb{R}^N)/O(d)$, where the canonical projection $\varphi : \mathrm{V}_d(\mathbb{R}^N) \longrightarrow \mathrm{V}_d(\mathbb{R}^N)/O(d) = \mathrm{T}_N(\mathbb{R}^d)$ coincides with the map which brings 1-tight frames to their congruence classes.

The quotient space $\mathrm{V}_d(\mathbb{R}^N)/O(d)$ is nothing but the the *Grassmannian* $\mathrm{Gr}_{N-d}(\mathbb{R}^N)$, i.e., the set of $(N-d)$-dimensional subspaces of $\mathbb{R}^N$. Therefore, $\mathrm{T}_N(\mathbb{R}^d)$ is identified with $\mathrm{Gr}_{N-d}(\mathbb{R}^N)$, which is also identified with $\mathrm{Gr}_d(\mathbb{R}^N)$ via the correspondence $W \mapsto W^\perp$. Under these identifications, the canonical projection φ turns out to be nothing but the map $\mathrm{V}_d(\mathbb{R}^N) \longrightarrow \mathrm{Gr}_d(\mathbb{R}^N)$ giving the well-known structure of an $O(d)$-principal bundle.

If we ignore the order of vectors in tight frames, it is natural to take up the quotient space $\mathfrak{S}_N \backslash \mathrm{T}_N(\mathbb{R}^d)$ $\left(= \mathfrak{S}_N \backslash \mathrm{V}_d(\mathbb{R}^N)/O(d)\right)$ where the symmetric group $\mathfrak{S}_N$ acts on $\mathrm{V}_d(\mathbb{R}^N)$ by $(a_{ij}) \mapsto (a_{\sigma^{-1}(i)j})$ $(\sigma \in \mathfrak{S}_N)$. For a 1-tight frame $\mathcal{S} = \{\mathbf{v}_i\}_{i=1}^N$, the isotropy group of the point ${}^t(\mathbf{v}_1, \ldots, \mathbf{v}_N)O(d) \in \mathrm{V}_d(\mathbb{R}^N)/O(d)$ coincides with $\mathrm{Aut}(\mathcal{S})$.

In view of Proposition 2.8, the Stiefel manifold $\mathrm{V}_d(\mathbb{R}^N)$ is regarded as a "quadric" in $M_{N,d}(\mathbb{R})$, and a 1-tight frame with the vanishing group W is obtained as a point in the intersection of the quadric $\mathrm{V}_d(\mathbb{R}^N)$ and the subspace $\left\{A \in M_{N,d}(\mathbb{R}) \mid {}^tA\mathbf{x} = \mathbf{0} \ (\mathbf{x} \in W)\right\}$. Such locution turns out to become more natural when we consider the set of "oriented" congruence classes of two-dimensional tight frames (see Sect. 4). Here the set of orientated congruence classes of 1-tight frames is $\widetilde{\mathrm{T}}_N(\mathbb{R}^d) = \mathrm{V}_d(\mathbb{R}^N)/SO(d)$, which is identified with the *oriented Grassmannian* $\widetilde{\mathrm{Gr}}_d(\mathbb{R}^N)$, the manifold consisting of all oriented d-dimensional subspaces of $\mathbb{R}^N$. It is a double cover over $\mathrm{Gr}_d(\mathbb{R}^N)$.

We close this section by giving a simple remark on Gramm matrices associated with frames.

Proposition 2.9. *Two frames S_1 and S_2 are congruent if and only if* $G_{S_1} = G_{S_2}$.

Proof. It suffices to show that if $\mathbf{b}_1, \ldots, \mathbf{b}_N$ span $\mathbb{R}^d$, and

$$\langle \mathbf{b}_i, \mathbf{b}_j \rangle = \langle \mathbf{c}_i, \mathbf{c}_j \rangle \quad (i, j = 1, \ldots, N)$$

for vectors $\mathbf{c}_1, \ldots, \mathbf{c}_N$ in $\mathbb{R}^d$, then there exists an orthogonal matrix $U \in O(d)$ such that $\mathbf{c}_i = U\mathbf{b}_i$. Certainly this is true when $\mathbf{b}_1, \ldots, \mathbf{b}_N$ ($N = d$) is a basis. In the general case, we take a basis $\mathbf{b}_{i_1}, \ldots, \mathbf{b}_{i_d}$, and $U \in O(d)$ with $\mathbf{c}_{i_k} = U\mathbf{b}_{i_k}$. Then

$$\langle \mathbf{c}_{i_h}, \mathbf{c}_k \rangle = \langle \mathbf{b}_{i_h}, \mathbf{b}_k \rangle = \langle U\mathbf{b}_{i_h}, U\mathbf{b}_k \rangle = \langle \mathbf{c}_{i_h}, U\mathbf{b}_k \rangle,$$

from which it follows that $\mathbf{c}_k = U\mathbf{b}_k$ for every k, as required. $\square$

Proposition 2.3 (4) combined with the above proposition tells us that $\mathrm{T}_N(\mathbb{R}^d)$ is identified with

$$\{ G \in M_N(\mathbb{R}) | \ ^t G = G, \ G^2 = G, \ \mathrm{rnak} \ G = d \}.$$

3.   Crystallographic Tight Frames

We come now to the principal subject of this chapter. Recall that the tight frame $\{\mathbf{v}_1, \mathbf{v}_2, \mathbf{v}_3\}$ mentioned in Introduction forms a crystalline pattern if and only if it generates a lattice in $\mathbb{R}^2$. Having this fact in mind, we shall introduce the notion of *crystallographic tight frame*.

Before going into the subject, we review some items in the theory of lattice groups, which are often used in the rest of this article. In general, for a lattice $\mathcal{L}$ in an n-dimensional vector space M with an inner product $\langle \cdot, \cdot \rangle$, we denote by $\mathrm{vol}(M/\mathcal{L})$ the volume of the flat torus $M/\mathcal{L}$. If $\{\mathbf{atorus}_1, \ldots, \mathbf{a}_n\}$ is a $\mathbb{Z}$-basis of $\mathcal{L}$, then $\mathrm{vol}(M/\mathcal{L})^2 = \det(\langle \mathbf{a}_i, \mathbf{a}_j \rangle)$. Further we have

$$\mathrm{vol}(M/\mathcal{L}_1) = |\mathcal{L}_2/\mathcal{L}_1| \mathrm{vol}(M/\mathcal{L}_2)$$

for two lattices $\mathcal{L}_1, \mathcal{L}_2$ with $\mathcal{L}_1 \subset \mathcal{L}_2$.

We denote by $\mathcal{L}^\#$ the *dual lattice* of $\mathcal{L}$; namely $\mathcal{L}^\# = \{ \mathbf{x} \in M | \ \langle \mathbf{x}, \mathbf{y} \rangle \in \mathbb{Z} \ (\mathbf{y} \in \mathcal{L}) \}$. We observe

$$\mathrm{vol}(M/\mathcal{L}^\#) = \mathrm{vol}(M/\mathcal{L})^{-1}.$$

Therefore, if $\mathcal{L}$ is *integral*, i.e., $\mathcal{L} \subset \mathcal{L}^\#$, then

$$|\mathcal{L}^\#/\mathcal{L}| = \mathrm{vol}(M/\mathcal{L})^2, \tag{16.7}$$

in particular, $\mathrm{vol}(M/\mathcal{L})^2$ is an integer.

A frame $S = \{\mathbf{v}_i\}_{i=1}^N$ of $\mathbb{R}^d$ is said to be *crystallographic* if it generates a lattice in $\mathbb{R}^d$, or what is the same is that $\mathcal{L}_S = \mathrm{Image} \ \rho_S$ is a lattice, where $\rho_{\mathcal{S}} = P_{\mathcal{S}}|\mathbb{Z}^N : \mathbb{Z}^N \longrightarrow \mathbb{R}^d$. We shall designate $\mathcal{L}_{\mathcal{S}}$ as the *periodic lattice* for $\mathcal{S}$.

The *vanishing subgroup* $H(\mathcal{S})$ associated with a frame $\mathcal{S}$ is defined to be the kernel of the homomorphism $\rho_{\mathcal{S}}$, i.e., $^t(k_1,\ldots,k_N) \in H(\mathcal{S})$ if and only if $k_1\mathbf{v}_1 + \cdots + k_N\mathbf{v}_N = \mathbf{0}$.

We have a sufficient condition for a frame being crystallographic (this turns out to be a necessary condition if the frame is 1-tight; see Proposition 4.2).

Lemma 3.1. *A frame* $\mathcal{S} = \{\mathbf{v}_i\}_{i=1}^N$ *is crystallographic provided that the Gramm matrix* $G_{\mathcal{S}}$ *is essentially rational, i.e.,* $\langle \mathbf{v}_i, \mathbf{v}_j \rangle \in \lambda\mathbb{Q}$ *(i, j =* $1,\ldots,N$*) with* $\lambda > 0$.

Proof. Put $v_{ij} = \langle \mathbf{v}_i, \mathbf{v}_j \rangle$. Suppose that $\{\mathbf{v}_i\}_{i=1}^N$ does not generate a lattice. Then $\mathbf{0}$ is an accumulation point in $\mathbb{R}^d$ of the subgroup

$$\left\{ \sum_{i=1}^N k_i\mathbf{v}_i \mid k_i \in \mathbb{Z} \ (i = 1,\ldots,N) \right\},$$

so that one can find a sequence $\{x_{ni}\}_{n=1}^\infty \in \mathbb{Z}$ such that

$$\sum_{i=1}^N x_{ni}\mathbf{v}_i \neq \mathbf{0} \text{ and } \lim_{n\to\infty} \sum_{i=1}^N x_{ni}\mathbf{v}_i = \mathbf{0}.$$

On the other hand,

$$\left\| \sum_{i=1}^N x_{ni}\mathbf{v}_i \right\|^2 = \sum_{i,j=1}^N v_{ij}\, x_{ni}\, x_{nj}.$$

Take a positive integer N such that $\lambda^{-1}Nv_{ij} \in \mathbb{Z}$. Since $\{\lambda^{-1}N\sum_{i,j=1}^N v_{ij}x_ix_j \mid x_i \in \mathbb{Z}\}$ is discrete, we have a contradiction. $\qquad\square$

The period lattice $\mathcal{L}_{\mathcal{S}}$ is just the set of vertices of the network obtained as the projected image by the frame projection of the *hypercubic lattice*. Here hypercubic lattice means the net associated with the N-dimensional standard lattice $\mathbb{Z}^N$, a generalization of the square and cubic lattices.

The hypercubic lattice as a graph is the *Cayley graph* X associated with the free abelian group $\mathbb{Z}^N$ with the set of generators $\{\mathbf{f}_1,\ldots,\mathbf{f}_N\}$ (remember that $\{\mathbf{f}_i\}$ is the fundamental basis of $\mathbb{R}^N$). Thus the quotient graph $X/H(\mathcal{S})$ by the natural $H(\mathcal{S})$-action on X is the Cayley graph associated with the factor group $\mathbb{Z}^N/H(\mathcal{S})$ with the set of generators $\{\mathbf{f}_i + H(\mathcal{S})\}_{i=1}^N$. Furthermore, X is the maximal abelian covering graph over the N-*bouquet graph*, the graph with a single vertex and N loop edges. The projected image of the hypercubic lattice can be thought of as a (periodic) realization of the abstract graph $X/H(\mathcal{S})$ (possibly having degenerate edges, multiple edges, and/or colliding vertices when realized in $\mathbb{R}^d$, like the square lattice in Figure 16.3 (a)). The map (graph morphism) ω of X onto $X/H(\mathcal{S})$ associated with the canonical homomorphism $\mathbb{Z}^N \longrightarrow \mathbb{Z}^N/H(\mathcal{S})$ turns out to be a *covering map*, and is compatible with the frame projection P of the hypercubic lattice onto the projected image.

$$
\begin{array}{ccc}
X & \longrightarrow & \mathbb{R}^N \\
\omega\downarrow & & \downarrow P \\
X/H(\mathcal{S}) & \longrightarrow & \mathbb{R}^d
\end{array}
$$

This story is much generalized in Sect. 7 in terms of topological crystals and their standard realizations.

There are plenty of sources of crystallographic tight frames.

(1) The (isotropic) tight frame associated with the N-regular polygon is crystallographic if and only if $N = 3, 4, 6$. The two-dimensional crystal pattern for $N = 3, 6$ is the regular triangular lattice. For $N = 4$, we have the square lattice.

(2) Platonic solids which yield crystallographic tight frames are the tetrahedron, cube, and octahedron.[9] The tetrahedron and cube yield the net associated with the body-centered lattice (Figure 16.2 (e)), while the crystal net corresponding to the octahedron is the cubic lattice.

Among all Archimedean solids, truncated tetrahedron, cuboctahedron, and truncated octahedron yield crystallographic tight frames; others do not[10] (Figures 16.7 and 16.8).

(3) An advanced example of tight frames is derived from root systems. For the convenience of the reader, let us recall the definition [24].

A *root system* in $\mathbb{R}^d$ is a finite set $\varPhi$ of nonzero vectors (called roots) that satisfy the following conditions:

1. The roots span $\mathbb{R}^d$.
2. The only scalar multiples of a root $\mathbf{x} \in \varPhi$ that belong to $\varPhi$ are $\mathbf{x}$ itself and $-\mathbf{x}$.
3. For every root $\mathbf{x} \in \varPhi$, the set $\varPhi$ is closed under reflection through the hyperplane perpendicular to $\mathbf{x}$.
4. (Integrality) If $\mathbf{x}$ and $\mathbf{y}$ are roots in $\varPhi$, then the projection of $\mathbf{y}$ onto the line through $\mathbf{x}$ is a half-integral multiple of $\mathbf{x}$.

The reflection $\sigma_{\mathbf{x}}$ through the hyperplane perpendicular to $\mathbf{x}$ is explicitly expressed as

---

[9] The reason why restricted polygons and polyhedra appear in (1), (2) is derived from the following general fact: If a finite subgroup G of $GL_d(\mathbb{Z})$ contains an element with order n, then $\varphi(n) \leq d$, where $\varphi(n)$ is the Euler function, the number of positive integers k with $1 \leq k \leq n$ and $\gcd(n, k) = 1$. Thus for $d = 2, 3$, the possible order is $2, 3, 4$, or 6.

[10] We may use Lemma 3.1 to check this (see also Proposition 4.2).

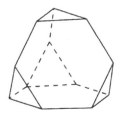
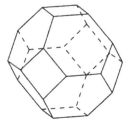

Figure 16.7 Truncated tetrahedron and truncated octahedron

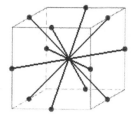
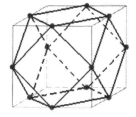

Figure 16.8 A_3 and cuboctahedron

$$\sigma_{\mathbf{x}}(\mathbf{y}) = \mathbf{y} - 2\frac{\langle \mathbf{y}, \mathbf{x} \rangle}{\|\mathbf{x}\|^2}\mathbf{x}.$$

The group of orthogonal transformations of $\mathbb{R}^d$ generated by reflections through hyperplanes associated to the roots of Φ is finite and called the *Weyl group* of Φ.

A root system Φ is called *irreducible* if it cannot be partitioned into the union of proper subsets such that each root in one set is orthogonal to each root on the other. If Φ is irreducible, then the Weyl group acts irreducibly on $\mathbb{R}^d$. Therefore in view of Proposition 2.1, Φ (under any ordering of roots) gives a tight frame. Φ is redundant in the sense that it contains both $\mathbf{x}$ and $-\mathbf{x}$. One can take subset Φ^+ such that $\Phi = \Phi^+ \cup -\Phi^+$ and $\Phi^+ \cap -\Phi^+ = \emptyset$ (the set of *positive roots* in the root system gives such Φ^+). When Φ is irreducible, Φ^+ obviously gives a tight frame.

The tight frames given by Φ and Φ^+ are crystallographic because Φ generates a lattice (called the *root lattice*).[11]

It is known that there are four infinite families of *classical* irreducible root systems designated as A_d $(d \geq 1)$, B_d $(d \geq 2)$, C_d $(d \geq 3)$, and D_d $(d \geq 4)$, and the five *exceptional* root systems E_6, E_7, E_8, F_4, and G_2. Among them, A_d, D_d, E_6, E_7, E_8 give isotropic frames (cf. [17]). The crystal nets for A_2 and A_3 are the regular triangular lattice and the net associated with the face-centered lattice $\big($Figure 16.2 (d)$\big)$, respectively.

---

[11] Φ and Φ^+ yield the same crystal net.

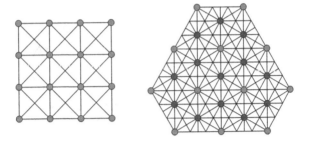

Figure 16.9 The net for B_2 and G_2

The tight frame associated with A_3 coincides with the one coming from *cuboctahedron* (Figure 16.8).

Interestingly A_3 (and the net associated with A_3) comes up in crystallography in various forms; say, in the hexagonal arrangement that gives the densest sphere packing,[12] the diamond crystal, and the diamond twin (see Sect. 6).

Figure 16.9 depicts the nets for B_2 and G_2 (compare with Figure 16.5).

(4) The *Leech lattice* Λ discovered by John Leech in 1967 is a highly symmetrical lattice in $\mathbb{R}^{24}$ characterized by the following properties [10], [30]:

- It is self-dual, i.e., $\Lambda^{\#} = \Lambda$.
- It is even, i.e., the square of the length of any vector in Λ is an even integer.
- The length of any nonzero vector in Λ is at least 2.

The group (called the *Conway group* Co_0) of orthogonal transformations which preserves Λ permutes transitively the 196560 vectors $\mathbf{x} \in \Lambda$ with $\|\mathbf{x}\| = 2$. The action of Co_0 on $\mathbb{R}^{24}$ is irreducible, so that $\Phi = \{\mathbf{x} \in \Lambda | \|\mathbf{x}\| = 2\}$ gives a crystallographic, isotropic tight frames. It is known that Co_0 modulo its center (designated as Co_1) is a simple group of order $4157776806543360000 = 2^{21} \cdot 3^9 \cdot 5^4 \cdot 7^2 \cdot 11 \cdot 13 \cdot 23$. The Leech lattice is not a root lattice, but considered a kin of the exceptional root lattice E_8 in view of the fact that E_8 is the unique self-dual even lattice in eight-dimension.

(5) Let (x, y, z) be a primitive Pythagorean triple; namely x, y, z are coprime positive integers satisfying $x^2 + y^2 = z^2$. Put

$$\mathbf{v_1} = \begin{pmatrix} 1 \\ 0 \end{pmatrix}, \quad \mathbf{v_2} = \begin{pmatrix} x/z \\ y/z \end{pmatrix}, \quad \mathbf{v_3} = \begin{pmatrix} 0 \\ 1 \end{pmatrix}, \quad \mathbf{v_4} = \begin{pmatrix} -y/z \\ x/z \end{pmatrix}.$$

---

[12] This is the Kepler conjecture for which Thomas Hales gave a proof in 1998. Johaness Kepler stated this conjecture in the short pamphlet entitled *New-Year's gift concerning six-cornered snow* ("Strena Seu de Nive Sexangula" in Latin) in 1611.

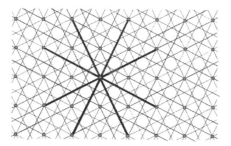

Figure 16.10 A Pythagorean lattice

One can check that $\{\mathbf{v}_i\}_{i=1}^4$ is a crystallographic tight frame whose vanishing subgroup is $\mathbb{Z}^t(z, -x, 0, y) + \mathbb{Z}^t(0, y, -z, x)$. The crystalline pattern associated with this tight frame is what we call a *Pythagorean lattice* [34]. Note that the tight frame $\{\pm\mathbf{v}\}_{i=1}^4$ consisting of eight vectors is isotropic. See Figure 16.10 in the case $(x, y, z) = (3, 4, 5)$.

Related to Pythagorean lattices is the notion of the *coincidence symmetry group*, which originates in the theory of crystalline interfaces and grain boundaries in polycrystalline materials.

In general, the coincidence symmetry group for a lattice $\mathcal{L}$ in $\mathbb{R}^d$ is defined as

$$G(\mathcal{L}) = \{g \in SO(d) \mid g\mathcal{L} \sim \mathcal{L}\},$$

where $\mathcal{L}_1 \sim \mathcal{L}_2$ means that two lattices $\mathcal{L}_1$ and $\mathcal{L}_2$ are *commensurable*, i.e., $\mathcal{L}_1 \cap \mathcal{L}_2$ is a lattice. Because $\sim$ is an equivalence relation, $G(\mathcal{L})$ is actually a subgroup of $SO(d)$. Note that the *symmetry group* $\{g \in SO(d) \mid g\mathcal{L} = \mathcal{L}\}$ is always finite, while $G(\mathcal{L})$ could be infinite in general. In the special case $\mathcal{L} = \mathbb{Z}^d$, we get $G(\mathbb{Z}^d) = SO(d) \cap M_d(\mathbb{Q})$, which is a dense subgroup of $SO(d)$ (see the remark below). In particular,

$$G(\mathbb{Z}^2) = \left\{ \begin{pmatrix} p & -q \\ q & p \end{pmatrix} \;\middle|\; p^2 + q^2 = 1, \; p, q \in \mathbb{Q} \right\}.$$

If $p = x/z, q = y/z$ with a primitive Pythagorean triple (x, y, z), then we have

$$g\mathbb{Z}^2 = \mathbb{Z}\begin{pmatrix} x/z \\ y/z \end{pmatrix} + \mathbb{Z}\begin{pmatrix} -y/z \\ x/z \end{pmatrix}.$$

(6) Let $\mathcal{L}^{\mathrm{hc}}$ be the two-dimensional lattice with the $\mathbf{Z}$-basis $\mathbf{a}_1, \mathbf{a}_2$ such that $\|\mathbf{a}_1\|^2 = \|\mathbf{a}_2\|^2 = 1$, $\mathbf{a}_1 \cdot \mathbf{a}_2 = -1/2$ (note that $\mathcal{L}^{\mathrm{hc}}$ is a lattice whose translational action preserves the honeycomb). We then have

$$G(\mathcal{L}^{\mathrm{hc}}) = \left\{ \begin{pmatrix} p - \frac{1}{2}q & -\frac{\sqrt{3}}{2}q \\ \frac{\sqrt{3}}{2}q & p - \frac{1}{2}q \end{pmatrix} \;\middle|\; p^2 - pq + q^2 = 1, \; p, q \in \mathbb{Q} \right\},$$

which is also a dense subgroup of $SO(2)$.

Put $\mathbf{v}_1 = \mathbf{a}_1$, $\mathbf{v}_2 = \mathbf{a}_2$, $\mathbf{v}_3 = -\mathbf{a}_1 - \mathbf{a}_2$. Obviously $\{\mathbf{v}_1, \mathbf{v}_2, \mathbf{v}_3\}$ is a 1-tight frame associated with the equilateral triangle whose period lattice is $\mathcal{L}^{\mathrm{hc}}$. For any $g \in G(\mathcal{L}^{\mathrm{hc}})$, the frame $\{\mathbf{v}_1, \mathbf{v}_2, \mathbf{v}_3, g\mathbf{v}_1, g\mathbf{v}_2, g\mathbf{v}_3\}$ is crystallographic and tight.

Here are several remarks related to the examples (5), (6).

(1) The coincidence symmetry group $G(\mathcal{L})$ is dense in $SO(d)$ $\Leftrightarrow$ the lattice $\mathcal{L}$ is *essentially rational*, i.e., there exists a positive λ such that $\langle \mathcal{L}, \mathcal{L} \rangle \subset \lambda \mathbb{Q}$. To sketch the proof, we first note that $G(\alpha\mathcal{L}) = G(\mathcal{L})$ for any lattice and $\alpha > 0$. Thus, in the proof of the implication $\Leftarrow$, one may assume that $\mathcal{L}$ is rational. Selecting a $\mathbb{Z}$-basis $\{\mathbf{a}_1, \ldots, \mathbf{a}_d\}$ of $\mathcal{L}$, we consider the symmetric matrix $S = {}^t(\mathbf{a}_1, \ldots, \mathbf{a}_d)(\mathbf{a}_1, \ldots, \mathbf{a}_d)$, and define

$$L_\mathbb{Q}(S) = \{X = (x_{ij}) \in M_d(\mathbb{Q}), \ {}^tXS + SX = O\},$$

which is a vector space over $\mathbb{Q}$ of dimension $\frac{d(d-1)}{2}$ because S is rational. Then putting

$$\varphi(X) = (\mathbf{a}_1, \ldots, \mathbf{a}_d)(I - X)(I + X)^{-1}(\mathbf{a}_1, \ldots, \mathbf{a}_d)^{-1},$$

we have an injective map $\varphi : L_\mathbb{Q}(S) \longrightarrow G(\mathcal{L})$ whose image is dense in $SO(d)$ (φ is what we call *Cayley's parameterization*.)[13] From this argument, it follows that $G(\mathcal{L})$ is dense in $SO(d)$.

Conversely suppose that $G(\mathcal{L})$ is dense in $SO(d)$, which is equivalent to the condition that in the rotation group $SO(S) = \{A \in GL_d(\mathbb{R})| \ {}^tASA = S, \det A > 0\}$ for the symmetric matrix S, the subgroup $SO_\mathbb{Q}(S) = \{A \in GL_d(\mathbb{Q})| \ {}^tASA = S, \ \det A > 0\}$ is dense. We easily find

$$\{S' \in M_d(\mathbb{R})| \ {}^tAS'A = S' \ (A \in SO(S)), \ {}^tS' = S'\} = \mathbb{R}S$$

(this is equivalent to say that any $SO(d)$-invariant symmetric bilinear form on $\mathbb{R}^d$ is a scalar multiple of the standard inner product). On the other hand, the equation for S' given by

$$ {}^tAS'A = S' \ (A \in SO_\mathbb{Q}(S)), \ {}^tS' = S'$$

reduces to a homogeneous linear equation with rational coefficients having a nonzero real solution; say, S, so that one can find a nonzero rational solution S_0. Since $SO_\mathbb{Q}(S)$ is dense in $SO(S)$ by the assumption, we conclude that ${}^tAS_0A = S_0 \ (A \in SO(S))$, ${}^tS_0 = S_0$, thereby $S_0 = \lambda S$ for some $\lambda \neq 0$, and S being essentially rational.

(2) For two frames $\mathcal{S}_1 = \{\mathbf{u}_i\}_{i=1}^M$ and $\mathcal{S}_2 = \{\mathbf{v}_i\}_{i=1}^N$, we define the *join* $\mathcal{S}_1 \vee \mathcal{S}_2 = \{\mathbf{w}_i\}_{i=1}^{M+N}$ by

$$\mathbf{w}_i = \begin{cases} \mathbf{u}_i & (i = 1, \ldots, M) \\ \mathbf{v}_{i-M} & (i = M+1, \ldots, M+N). \end{cases}$$

---

[13] The idea, originally given by A. Cayley, dates back to 1846.

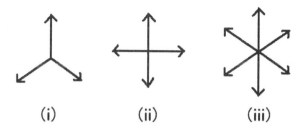

(i)   (ii)   (iii)

Figure 16.11 Two-dimensional crystallographic isotropic tight frames

If both $\mathcal{S}_1$ and $\mathcal{S}_2$ are tight frames, then so is $\mathcal{S}_1 \vee \mathcal{S}_2$. In order that $\mathcal{S}_1 \vee \mathcal{S}_2$ is crystallographic, it is necessary and sufficient that the period lattices $\mathcal{L}_{\mathcal{S}_1}$ and $\mathcal{L}_{\mathcal{S}_2}$ are commensurable. Obviously the period lattice of $\mathcal{S}_1 \vee \mathcal{S}_2$ is $\mathcal{L}_{\mathcal{S}_1} + \mathcal{L}_{\mathcal{S}_2}$.

(3) Because finite subgroups of $O(2)$ are cyclic or dihedral (cf. [33]), two-dimensional crystallographic isotropic tight frames are classified as follows (note that the order of a cyclic subgroup must be less than or equal to 6).

(a) Frames (i), (ii), (iii) depicted in Figure 16.11.

(b) Joins of each frame in (a) and its rotation by an element in the coincidence symmetry group $\big($more precisely $G(\mathbb{Z}^2)$ for (ii) and $G(\mathcal{L}^{\mathrm{hc}})$ for (i) and (iii)$\big)$.

(4) In the terminology of crystallography, the intersection $\mathcal{L}_1 \cap \mathcal{L}_2$ for commensurable lattices $\mathcal{L}_1$ and $\mathcal{L}_2$ is called the *coincidence site lattice* (CSL), while the sum $\mathcal{L}_1 + \mathcal{L}_2$ is called the *displacement shift complete lattice* (DSC lattice). For more about the CSL theory, refer to [23], [1], [44], [45].

4. Parameterizations of Crystallographic Tight Frames

If a frame $\mathcal{S}$ is crystallographic, then the vanishing group $H(\mathcal{S})$ is obviously a direct summand[14] of $\mathbb{Z}^N$, and rank $H(\mathcal{S}) = N - d$. Therefore $H(\mathcal{S})$ is a lattice in the vanishing subspace $W(\mathcal{S}) = \ker(P_{\mathcal{S}} : \mathbb{R}^N \to \mathbb{R}^d)$, and hence the subspace $H(\mathcal{S})_{\mathbb{R}}$ of $\mathbb{R}^N$ spanned by $H(\mathcal{S})$ coincides with $W(\mathcal{S})$.

Conversely suppose that $H(\mathcal{S})$ is a direct summand of $\mathbb{Z}^N$. Take a $\mathbb{Z}$-basis $\{\mathbf{c}_1, \ldots, \mathbf{c}_N\}$ of $\mathbb{Z}^N$ such that $\{\mathbf{c}_{d+1}, \ldots, \mathbf{c}_N\}$ is a $\mathbb{Z}$-basis of $H(\mathcal{S})$. Then $\big\{\rho_{\mathcal{S}}(\mathbf{c}_1), \ldots, \rho_{\mathcal{S}}(\mathbf{c}_d)\big\}$ is a $\mathbb{Z}$-basis of $\mathcal{L}_{\mathcal{S}}$ (remember that $\rho_{\mathcal{S}} = P_{\mathcal{S}}|\mathbb{Z}^N$). Therefore $\mathcal{S}$ is crystallographic.

---

[14] A subgroup B of an additive group A is called a *direct summand* of A if there exists another subgroup B' such that $A = B \oplus B'$, i.e., any element $a \in A$ can be expressed uniquely as $a = b + b'$ with $b \in B, b' \in B'$. In the case $A = \mathbb{Z}^N$, B is a direct summand if and only if the factor group A/B is free abelian.

Proposition 4.1 (Existence and Uniqueness). *Let H be a direct summand of $\mathbb{Z}^N$ of rank $N - d$. Then there exists a unique crystallographic 1-tight frame $\mathcal{S}$ of $\mathbb{R}^d$ up to congruence such that $H(\mathcal{S}) = H$.*

Proof. Let $H_{\mathbb{R}}$ be the subspace of $\mathbb{R}^N$ spanned by H, and let $P : \mathbb{R}^N \longrightarrow \mathbb{R}^d$ be the frame projection giving a 1-tight frame whose kernel is $H_{\mathbb{R}}$. To prove $H(\mathcal{S}) = H$, note $H(\mathcal{S}) = \ker P \cap \mathbb{Z}^N \supset H$ and rank $H(\mathcal{S}) = \operatorname{rank} H$. Hence we have a finite subgroup $H(\mathcal{S})/H$ of $\mathbb{Z}^N/H$. Since H is a direct summand, $H(\mathcal{S})/H = \{0\}$.

The uniqueness follows from Proposition 2.4.   □

There is a one-to-one correspondence between direct summands of rank k in $\mathbb{Z}^N$ and subspaces of dimension k in $\mathbb{Q}^N$. Indeed we have

(1) Let $H \subset \mathbb{Z}^N$ be a direct summand, and let $H_{\mathbb{Q}}$ be the subspace of the vector space $\mathbb{Q}^N$ over $\mathbb{Q}$ spanned by H. Then $H_{\mathbb{Q}} \cap \mathbb{Z}^N = H$.

(2) Let W is a subspace of $\mathbb{Q}^N$ of dimension k. Then $H = W \cap \mathbb{Z}^N$ is a direct summand of $\mathbb{Z}^N$ of rank k.

In fact, (1) is proved in the same way as the proof of the above theorem. For (2), suppose that $\mathbb{Z}^N/(W \cap \mathbb{Z}^N)$ is not free. Then there exists $x \in \mathbb{Z}^N$ not contained in $W \cap \mathbb{Z}^N$ such that $nx \in W \cap \mathbb{Z}^N$. This implies that $x \in W$, and hence $x \in W \cap \mathbb{Z}^N$, thereby a contradiction.

We denote by $\mathrm{T}_N^{\mathrm{cr}}(\mathbb{R}^d)$ the set of congruence classes of crystallographic 1-tight frames $\{\mathbf{v}_i\}_{i=1}^N$ in $\mathbb{R}^d$. Proposition 4.1 together with the above remark tells us that $\mathrm{T}_N^{\mathrm{cr}}(\mathbb{R}^d)$ is parameterized by $\mathrm{Gr}_{N-d}(\mathbb{Q}^N)$.

$\mathrm{Gr}_{N-d}(\mathbb{Q}^N)$ is thought of as the set of "rational points" on the complex Grassmannian $\mathrm{Gr}_{N-d}(\mathbb{C}^N)$ by considering it as a subvariety of the complex projective space $P^M(\mathbb{C})$ ($M = \binom{N}{d} - 1$) by means of the *Plücker embedding*. Here we recall the general definition of rational points. Let K be an algebraic number field, and let V be a projective algebraic variety, defined in some projective space $P^{n-1}(\mathbb{C})$ by homogeneous polynomials $f_1, \ldots, f_m$ with coefficients in K. A K-rational point of V is a point $[z_1, \ldots, z_n]$ in $P^{n-1}(K)\big(\subset P^{n-1}(\mathbb{C})\big)$ that is a common solution of all the equations $f_j = 0$.

Given $W \in \mathrm{Gr}_{N-d}(\mathbb{Q}^N)$, we shall explicitly construct a 1-tight frame whose vanishing group is $H = W \cap \mathbb{Z}^N$ (this is the stage where quadratic Diophantine equations show up). For this sake, we select nonzero vectors $\mathbf{n}_1, \ldots, \mathbf{n}_d$ of W such that $\mathbf{n}_i \in \mathbb{Z}^N$, $\langle \mathbf{n}_i, \mathbf{x} \rangle = 0$ for every $\mathbf{x} \in H$, and $\langle \mathbf{n}_i, \mathbf{n}_j \rangle = 0$ for $i \neq j$ (we may do this because for a subspace V of $\mathbb{Q}^N$, the "rational" orthogonal complement $V^\perp = \big\{ \mathbf{x} \in \mathbb{Q}^N \mid \langle \mathbf{x}, \mathbf{y} \rangle = 0 \ (\mathbf{y} \in V) \big\}$ satisfy $\mathbb{Q}^N = V \oplus V^\perp$; then we argue by induction for the construction of $\mathbf{n}_i$). Write $\|\mathbf{n}_i\|^2 = m_i^2 D_i$ with a square free positive integer D_i and a positive integer m_i. Then putting $\mathbf{a}_i = (m_i \sqrt{D_i})^{-1} \mathbf{n}_i$, we obtain $A = (\mathbf{a}_1, \ldots, \mathbf{a}_d) \in \mathrm{V}_d(\mathbb{R}^N)$. If we define $\mathbf{v}_i$ by the relation $(\mathbf{v}_1, \ldots, \mathbf{v}_N) = {}^t A$, then we obtain a 1-tight frame $\mathcal{S} = \{\mathbf{v}_i\}_{i=1}^N$ in $\mathbb{R}^d$ whose vanishing group is H.

To express the matrix A more explicitly, we write $\mathbf{n}_i = {}^t(n_{1i}, \ldots . n_{Ni})$. Then

$$
A = \begin{pmatrix}
\dfrac{1}{\sqrt{D_1}}\dfrac{n_{11}}{m_1} & \dfrac{1}{\sqrt{D_2}}\dfrac{n_{12}}{m_2} & \cdots & \dfrac{1}{\sqrt{D_d}}\dfrac{n_{1d}}{m_d} \\[2mm]
\dfrac{1}{\sqrt{D_1}}\dfrac{n_{21}}{m_1} & \dfrac{1}{\sqrt{D_2}}\dfrac{n_{22}}{m_2} & \cdots & \dfrac{1}{\sqrt{D_d}}\dfrac{n_{2d}}{m_d} \\[2mm]
& \cdots & \cdots & \\[1mm]
\dfrac{1}{\sqrt{D_1}}\dfrac{n_{N1}}{m_1} & \dfrac{1}{\sqrt{D_2}}\dfrac{n_{N2}}{m_2} & \cdots & \dfrac{1}{\sqrt{D_d}}\dfrac{n_{Nd}}{m_d}
\end{pmatrix}, \tag{16.8}
$$

where n_{ij} and m_i satisfy the following Diophantine equation[15]

$$
n_{1i}{}^2 + \cdots + n_{Ni}{}^2 = D_i m_i{}^2 \quad (i = 1, \ldots, d),
$$
$$
n_{1i}n_{1j} + \cdots + n_{Ni}n_{Nj} = 0 \quad (i \neq j).
$$

Taking a look at (16.8), we have

Proposition 4.2. *The Gramm matrix for a crystallographic 1-tight frame is rational.*

Conversely if we start with a matrix $A \in V_d(\mathbb{R}^N)$ of the form (16.8), then the 1-tight frame associated with A is crystallographic in view of Lemma 3.1. From this observation, it also follows that the set of congruence classes of 1-tight frames is parameterized by

$$
\{G \in M_N(\mathbb{Q}) \mid {}^t G = G, \; G^2 = G, \; \text{rank}\, G = d\}.
$$

We also have

Corollary 4.3. *The period lattice $\mathcal{L}_S$ for a crystallographic 1-tight frame S is rational in the sense that $\langle \mathcal{L}_S, \mathcal{L}_S \rangle \subset \mathbb{Q}$.*

We now restrict ourselves to the case $d = 2$ and $N > 2$, and consider $\widetilde{T}_N(\mathbb{R}^2) = V_2(\mathbb{R}^N)/SO(2)$, the set of oriented congruence classes of 1-tight frames in the plane $\mathbb{R}^2$. As remarked before, $V_2(\mathbb{R}^N)/SO(2)$ is identified with the oriented Grassmannian $\widetilde{\mathrm{Gr}}_2(\mathbb{R}^N)$, which is also identified with the complex quadric

$$
Q_N = \{[z_1, \ldots, z_N] \in P^{N-1}(\mathbb{C}) \mid z_1{}^2 + \cdots + z_N{}^2 = 0\}.
$$

This routine procedure is carried out by the map

$$
\begin{pmatrix}
a_1 & b_1 \\
a_2 & b_2 \\
\cdot & \cdots \\
\cdot & \cdot \\
a_N & b_N
\end{pmatrix} \in V_2(\mathbb{R}^N) \mapsto (a_1 + b_1\sqrt{-1}, \ldots, a_N + b_N\sqrt{-1}) \in \mathbb{C}^N,
$$

---

[15] A trivial remark is that a *ruler-compass construction* of the 1-tight frame is possible once this Diophantine equation would be is solved.

which yields a one-to-one correspondence between $\widetilde{\mathrm{T}}_N(\mathbb{R}^2)$ and Q_N (note that the quotient of Q_N by the conjugation $[z_1,\ldots,z_N] \mapsto [\overline{z_1},\ldots,\overline{z_N}]$ is identified with $\mathrm{T}_N(\mathbb{R}^2)$).

For a direct summand H of rank $N-d$ in $\mathbb{Z}^N$, define the subspace L_H in $P^{N-1}(\mathbb{C})$ by setting

$$L_H = \{[z_1,\ldots,z_N] \in P^{N-1}(\mathbb{C}) \mid k_1 z_1 + \cdots + k_N z_N = 0$$
$$\text{for every } {}^t(k_1,\ldots,k_N) \in H\}.$$

The intersection $Q_N \cap L_H$ consists of two points (one is the conjugate of another), both of which correspond to an oriented similarity class of a crystallographic tight frame $\mathcal{S}$ with the vanishing subgroup $H = H_{\mathcal{S}}$.

Applying the above observation in the general case, one can easily show that there exists a square-free positive integer D such that the two points in $Q_N \cap L_H$ are $\mathbb{Q}(\sqrt{-D})$-rational points on Q_N. We thus have

Theorem 4.4. *The set*

$$Q_N \cap \bigcup_D P^{N-1}(\mathbb{Q}(\sqrt{-D}))$$

is identified with the set of oriented similarity classes of two-dimensional tight frames of size N.

For an illustration, we shall make a brief excursion into $\mathbb{Q}(\sqrt{-D})$-rational points on Q_3. These rational points are related to two-dimensional crystal patterns obtained as projected images of the cubic lattice mentioned in the Introduction.

The crystallographic 1-tight with the vanishing group $H = \mathbb{Z}$ ${}^t(n_1, n_2, n_3)$ corresponds to the point $[z_1, z_2, z_3] \in P^2(\mathbb{C})$ given by

$$z_1 = n_2{}^2 + n_3{}^2,$$
$$z_2 = -n_1 n_2 \pm \sqrt{-(n_1{}^2 + n_2{}^2 + n_3{}^2)},$$
$$z_3 = -n_1 n_3 \mp \sqrt{-(n_1{}^2 + n_2{}^2 + n_3{}^2)}.$$

Therefore, D is the square-free part of $n_1{}^2 + n_2{}^2 + n_3{}^2$.

A question arises: For which square-free D, does the quadric Q_3 have a $\mathbb{Q}(\sqrt{-D})$-rational point? The answer is given in the following.

Proposition 4.5. *The quadric Q_3 has a $\mathbb{Q}(\sqrt{-D})$-rational point if and only if D is not of the form $8k+7$.*

This is, as easily conceived and proved below, a consequence of the theorem of three squares due to Legendre[16] (1798), which says that a positive integer n can be expressed as the sum of three squares if and only if n is not of the form $4^\ell(8k+7)$.

---

[16] Legendre's proof is not complete. It is Gauss who gave a complete proof.

From what we observed above, it follows that the quadric Q_3 has a $\mathbb{Q}(\sqrt{-D})$-rational point if and only if the equation $n_1{}^2 + n_2{}^2 + n_3{}^2 = Dm^2$ has a nontrivial integral solution n_1, n_2, n_3, m.

We first show that if D is not of the form $8k + 7$, then $n_1{}^2 + n_2{}^2 + n_3{}^2 = D$ has an integral solution n_1, n_2, n_3. To this end, suppose that $n_1{}^2 + n_2{}^2 + n_3{}^2 = D$ has no integral solution. Then $D = 4^\ell(8k + 7)$ for some ℓ and k by invoking Legendre's theorem. Since D is square-free, D must be of the form $8k + 7$.

Next suppose that D is of the form $8k+7$. If $n_1{}^2 + n_2{}^2 + n_3{}^2 = Dm^2$ has a nontrivial integral solution n_1, n_2, n_3, m, then writing $m = 2^\ell(2s + 1)$, we have

$$Dm^2 = 4^\ell(8k + 7)(2s + 1)^2 = 4^\ell\big(8(8kt + 7t + k) + 7\big),$$

where $t = s(s + 1)/2$. This contradicts to Legendre's theorem.

5. A Height Function on the Rational Grassmannian

The rational Grassmannian $\mathrm{Gr}_{N-d}(\mathbb{Q}^N)$ is dense in $\mathrm{Gr}_{N-d}(\mathbb{R}^N)$; in particular there are infinitely many congruence classes of d-dimensional crystallographic 1-tight frames of size N. This section is devoted to a brief explanation of how to count congruence classes, with some excursions into a characterization of crystallographic tight frames by means of a certain minimal principle. The tool that we employ is a natural height function on the rational Grassmannian.

We define the *height function* h on $\mathrm{Gr}_k(\mathbb{Q}^N)$ in the following way. For $W \in \mathrm{Gr}_k(\mathbb{Q})$, we let $H = W \cap \mathbb{Z}^N$, which is a direct summand of $\mathbb{Z}^N$ as noticed before, and is an integral lattice in $H_\mathbb{R}(= W_\mathbb{R})$. We then put

$$h(W) = \mathrm{vol}(H_\mathbb{R}/H)\big(= |H^\#/H|^{1/2}\big),$$

where the inner product on $H_\mathbb{R}$ is the one induced from the standard inner product on $\mathbb{R}^N$.

The function h deserves to be called a height function because we have

Proposition 5.1. *For any $c > 0$, there are only finitely many $W \in \mathrm{Gr}_k(\mathbb{Q}^N)$ such that $h(W) \leq c$.*

Proof. It suffices to prove that for every $c > 0$, there are only finitely many direct summands H of rank k such that $\mathrm{vol}(H_\mathbb{R}/H) \leq c$. To this end, we shall introduce the quantity $c(H)$. For a $\mathbb{Z}$-basis $J = \{\mathbf{c}_1, \ldots, \mathbf{c}_k\}$ of H, we put

$$c(J) = \max(\|\mathbf{c}_1\|, \ldots, \|\mathbf{c}_k\|),$$

and define

$$c(H) = \min_J c(J).$$

It is straightforward to check that there are only finitely many direct summands H of rank k such that $c(H) \leq c$. Therefore, it suffices to prove

$$Cc(H) \leq \mathrm{vol}(II_{\mathbb{R}}/H)$$

with a positive constant C not depending on H. For this sake, we invoke the fact that there exists a $\mathbb{Z}$-basis (called a *reduced basis*) $\mathbf{u}_1, \ldots, \mathbf{u}_k$ of H satisfying

(i) $\|\mathbf{u}_1\| \leq \|\mathbf{x}\|$ for all $\mathbf{x} \in H \backslash \{0\}$,
(ii) if $\mathbf{u}_1, \ldots, \mathbf{u}_{i-1}, \mathbf{x}$ is a part of a $\mathbb{Z}$-basis of H, then $\|\mathbf{u}_i\| \leq \|\mathbf{x}\|$.

This property of $\mathbf{u}_1, \ldots, \mathbf{u}_k$ implies $\|\mathbf{u}_1\| \leq \|\mathbf{u}_2\| \leq \cdots \leq \|\mathbf{u}_k\|$. Moreover a theorem in the geometry of numbers due to Minkowski asserts that there exists a positive constant C_k such that

$$\|\mathbf{u}_1\| \cdots \|\mathbf{u}_k\| \leq C_k \mathrm{vol}(H_{\mathbb{R}}/H)$$

(cf. [6]). By the definition of $c(J)$, if $J = \{\mathbf{u}_1, \ldots, \mathbf{u}_k\}$, then $c(J) = \|\mathbf{u}_k\|$, so we find $c(H) \leq \|\mathbf{u}_k\|$. Since $\|\mathbf{u}_i\| \geq 1$ for $i = 1, \ldots, k-1$ (note $H \subset \mathbb{Z}^N$), we have

$$c(H) \leq \|\mathbf{u}_k\| \leq \|\mathbf{u}_1\| \cdots \|\mathbf{u}_k\| \leq C_k \mathrm{vol}(H_{\mathbb{R}}/H).$$

This completes the proof. $\qquad\qquad\qquad\qquad\qquad\qquad\qquad\qquad$ □

Remark 5.2. (1) By Hadamard' inequality, we have

$$\mathrm{vol}(H_{\mathbb{R}}/H) \leq \|\mathbf{c}_1\| \cdots \|\mathbf{c}_k\| \leq c(J)^k$$

for any $\mathbb{Z}$-basis $J = \{\mathbf{c}_1, \ldots, \mathbf{c}_k\}$; therefore

$$\mathrm{vol}(H_{\mathbb{R}}/H) \leq c(H)^k.$$

(2) For a direct summand $H = \mathbb{Z}{}^t(n_1, \ldots, n_N)$ of $\mathbb{Z}^N$,

$$h(H_{\mathbb{Q}}) = \sqrt{n_1{}^2 + \cdots + n_N{}^2}.$$

Using this, we obtain the asymptotic formula

$$\left|\{ W \in P^{N-1}(\mathbb{Q}) = \mathrm{Gr}_1(\mathbb{Q}^N) \,|\, h(W) \leq h \}\right| \sim \frac{1}{2}\zeta(N)^{-1}\omega_N h^N \quad (h \to \infty),$$

where $\omega_N = \pi^{N/2}/\Gamma(1 + N/2)$, the volume of the unit disk in $\mathbb{R}^N$, and $\zeta(s)$ is the Riemann zeta function. It is interesting to ask whether a similar asymptotic formula holds for $\mathrm{Gr}_k(\mathbb{Q}^N)$.

The height function $h(W)$ is closely connected with crystallographic tight frames.

Proposition 5.3. $\mathrm{vol}(\mathbb{R}^d/\mathcal{L}_\mathcal{S}) = h(H_{\mathbb{Q}})^{-1}\big(= \mathrm{vol}(H_{\mathbb{R}}/H)^{-1}\big)$ *for a crystallographic 1-tight frame* $\mathcal{S} = \{\mathbf{v}_i\}_{i=1}^N$ *whose vanishing subgroup is* H.

Proof. Let $\mathcal{S} = \{\mathbf{v}_i\}_{i=1}^N$ be a crystallographic 1-tight frame of $\mathbb{R}^d$, and let $P = P_{\mathcal{S}} : \mathbb{R}^N \longrightarrow \mathbb{R}^n$ be the frame projection. Again take a $\mathbb{Z}$-basis $\{\mathbf{c}_1, \ldots, \mathbf{c}_N\}$ of $\mathbb{Z}^N$ as before such that $\{\mathbf{c}_{d+1}, \ldots, \mathbf{c}_N\}$ is a $\mathbb{Z}$-basis of the vanishing subgroup $H(\mathcal{S})$. Note that $|\det(\mathbf{c}_1, \ldots, \mathbf{c}_N)| = 1$ because $(\mathbf{c}_1, \ldots, \mathbf{c}_N) \in GL_N(\mathbb{Z})$. Thus for the square matrix $C = (\langle \mathbf{c}_i, \mathbf{c}_j \rangle)$ with integral entries, we find that $\det C = (\det(\mathbf{c}_1, \ldots, \mathbf{c}_N))^2 = 1$.

Putting $\mathbf{b}_i = P(\mathbf{c}_i)$ $(i = 1, \ldots, d)$, we have a $\mathbb{Z}$-basis $\{\mathbf{b}_1, \ldots, \mathbf{b}_d\}$ of the period lattice $\mathcal{L}_{\mathcal{S}}$. Using ${}^tPP(\mathbf{c}_i) - \mathbf{c}_i \in \ker P$, we may write

$$
{}^tP(\mathbf{b}_i) - \mathbf{c}_i = \sum_{j=d+1}^N f_{ij}\mathbf{c}_j.
$$

Substituting ${}^tPP(\mathbf{c}_i) = \mathbf{c}_i + \sum_{j=d+1}^N f_{ij}\mathbf{c}_j$ into $\langle {}^tPP(\mathbf{c}_i), \mathbf{c}_k \rangle = 0$ $(k = d + 1, \ldots, b)$, we have

$$
-\langle \mathbf{c}_i, \mathbf{c}_k \rangle = \sum_{j=d+1}^N f_{ij} \langle \mathbf{c}_j, \mathbf{c}_k \rangle.
$$

Now writing

$$
C = \begin{pmatrix} C_{11} & C_{12} \\ C_{21} & C_{22} \end{pmatrix},
$$

where $C_{11} \in M_d(\mathbb{R})$ and using these matrices of small size, we may compute the matrix $F = (f_{ij}) \in M_{d,N-d}(\mathbb{R})$ as

$$
F = -C_{12}C_{22}^{-1}.
$$

Furthermore,

$$
\langle \mathbf{b}_i, \mathbf{b}_j \rangle = \langle {}^tPP(\mathbf{c}_i), \mathbf{c}_j \rangle = \left\langle \mathbf{c}_i + \sum_{k=d+1}^N f_{ik}\mathbf{c}_k, \mathbf{c}_j \right\rangle
$$

$$
= \langle \mathbf{c}_i, \mathbf{c}_j \rangle + \sum_{k=d+1}^N f_{ik} \langle \mathbf{c}_k, \mathbf{c}_j \rangle.
$$

Therefore, the matrix $B = (\langle \mathbf{b}_i, \mathbf{b}_j \rangle) \in M_d(\mathbb{R})$ is computed as

$$
B = C_{11} + FC_{21} = C_{11} - C_{12}C_{22}^{-1}C_{21} \tag{16.9}
$$

(this tells us that B is a rational matrix; thereby giving an alternative proof of Corollary 4.3). It is readily checked that

$$
\begin{pmatrix} C_{11} & C_{12} \\ C_{21} & C_{22} \end{pmatrix} \begin{pmatrix} I & O \\ -C_{22}^{-1}C_{21} & I \end{pmatrix} = \begin{pmatrix} C_{11} - C_{12}C_{22}^{-1}C_{21} & C_{12} \\ O & C_{22} \end{pmatrix},
$$

so $\det C = \det(C_{11} - C_{12}C_{22}^{-1}C_{21}) \cdot \det C_{22} = \det B \cdot \det C_{22}$. Because $\det C = 1$, $\det B = \mathrm{vol}(\mathbb{R}^d/\mathcal{L}_{\mathcal{S}})^2$, and $\det C_{22} = \mathrm{vol}(H_{\mathbb{R}}/H)^2$, our claim follows. $\qquad\square$

The following gives a characterization of crystallographic 1-tight frames by means of a minimal principle.

Proposition 5.4. *For a crystallographic frame* $\mathcal{S} = \{\mathbf{v}_i\}_{i=1}^{N}$ *in* $\mathbb{R}^d$ *with the vanishing group H, we have*

$$\sum_{i=1}^{N} \|\mathbf{v}_i\|^2 \geq d \cdot \text{vol}(\mathbb{R}^d/\mathcal{L}_{\mathcal{S}})^{2/d} h(H_{\mathbb{Q}})^{2/d} \left(= d \cdot \text{vol}(\mathbb{R}^d/\mathcal{L}_{\mathcal{S}})^{2/d} |H^{\#}/H|^{1/d} \right).$$

The equality holds if and only if $\mathcal{S}$ is tight.

It should be noted that the quantity $\text{vol}(\mathbb{R}^d/\mathcal{L}_{\mathcal{S}})^{-2/d} \sum_{i=1}^{N} \|\mathbf{v}_i\|^2$ depends only on the similarity class of the crystallographic frame $\mathcal{S}$.

Proof. For any positive symmetric matrix $B \in M_d(\mathbb{R})$, we have

$$\text{tr } B \geq d \left(\det B \right)^{1/d},$$

where the equality holds if and only if $B = \alpha I_d$ for some $\alpha > 0$ (this is easily deduced from the *inequality of arithmetic and geometric means*). Applying this inequality to $B = {}^{t}AA$, we find that

$$\sum_{i=1}^{N} \|\mathbf{v}_i\|^2 \geq d(\det S_{\mathcal{S}})^{1/d}$$

for any frame $\mathcal{S} = \{\mathbf{v}_i\}_{i=1}^{N}$, where the equality holds if and only if $\mathcal{S}$ is tight.

We now let S be the frame operator associated with $\mathcal{S}$. Recall that S is symmetric and positive, whence one can find a (unique) symmetric positive operator $S^{-1/2}$ with $(S^{-1/2})^2 = S^{-1}$. Then

$$\sum_{i=1}^{N} \langle \mathbf{x}, S^{-1/2}\mathbf{v}_i \rangle S^{-1/2}\mathbf{v}_i = S^{-1/2} \sum_{i=1}^{N} \langle S^{-1/2}\mathbf{x}, \mathbf{v}_i \rangle \mathbf{v}_i$$

$$= S^{-1/2} SS^{-1/2}\mathbf{x} = \mathbf{x}.$$

Therefore, $\mathcal{U} = \{S^{-1/2}\mathbf{u}_i\}_{i=1}^{N}$ is a 1-tight frame whose vanishing subgroup is obviously H. Thus $\rho_{\mathcal{U}} = S^{-1/2}\rho_{\mathcal{S}}$ and

$$\text{vol}(\mathbb{R}^d/\text{Image } \rho_{\mathcal{U}}) = \det S^{-1/2}\text{vol}(\mathbb{R}^d/\text{Image } \rho_{\mathcal{S}}),$$

from which it follows that

$$\det S = \text{vol}(\mathbb{R}^d/\mathcal{L}_{\mathcal{S}})^2 \text{vol}(H_{\mathbb{R}}/H)^2.$$

This completes the proof.   □

In Theorem 4.4, we observed that a square-free positive integer D is associated with each two-dimensional tight frame $\mathcal{S}$. We close this section with establishing a relationship between the integer D and the height of $H(\mathcal{S})_{\mathbb{Q}}$.

Proposition 5.5. Let $[z_1, \dots, z_N] \in Q_N$ be a $\mathbb{Q}(\sqrt{-D})$-rational point corresponding to a two-dimensional tight frame $\mathcal{S} = \{v_i\}_{i=1}^N$ whose vanishing group is H. Then D is the square-free part of $h(H_{\mathbb{Q}})^2 (= \text{vol} (H_{\mathbb{R}}/H)^2)$.

Proof. One may assume $z_i \in \mathbb{Q}(\sqrt{-D})$. Let $w_1, w_2 \in \mathbb{C}$ be a $\mathbb{Z}$-basis of the period lattice $\mathcal{L}_{\mathcal{S}}$ (we are working in $\mathbb{C}$ instead of $\mathbb{R}^2$). Writing $w_i = a_i + b_i \sqrt{-D} \in \mathbb{Q}(\sqrt{-D})$, we find

$$\text{vol}(\mathbb{C}/\mathcal{L}_{\mathcal{S}})^{-1} \sum_{i=1}^N \|v_i\|^2 = \text{vol}(\mathbb{C}/\mathcal{L}_{\mathcal{S}})^{-1}(|z_1|^2 + \cdots + |z_N|^2)$$

$$= \frac{1}{|a_1 b_2 - a_2 b_1|\sqrt{D}}(|z_1|^2 + \cdots + |z_N|^2),$$

which is equal to $2\text{vol}(H_{\mathbb{R}}/H)$ in view of Proposition 5.4. Because $|z_1|^2 + \cdots + |z_N|^2$ is rational, the claim is proved. $\qquad\square$

6. Tight Frames Associated with Finite Graphs and Combinatorial Abel's Theorem

The notions of *Jacobian* and the *Picard group* together with *Abel's theorem* play a significant role in classical algebraic geometry. The aim of this section is to introduce combinatorial analogues of these notions by using certain tight frames associated with finite graphs.

We first fix some basic notations and terminology. A *graph* is represented by an ordered pair $X = (V, E)$ of the set of *vertices* V and the set of all *directed edges* E (note that each edge has just two directions, which are to be expressed by arrows). For an directed edge e, we denote by $o(e)$ the *origin*, and by $t(e)$ the *terminus*. The inversed edge of e is denoted by $\bar{e}$. With these notations, we have $o(\bar{e}) = t(e)$, $t(\bar{e}) = o(e)$. An *orientation* of X is a subset E^o of E such that $E = E^o \cup \overline{E^o}$ and $E^o \cap \overline{E^o} = \emptyset$. We use the notation E_x for the set of directed edges e with $o(e) = x$. Throughout, the degree $\deg x = |E_x|$ is assumed to be greater than or equal to three for every vertex x.

We let $X_0 = (V_0, E_0)$ be a finite connected graph that is regarded as a one-dimensional cell complex. From now on, we shall make use of (co)homology theory of cell complexes. Let K be $\mathbb{Z}, \mathbb{Q}$, or $\mathbb{R}$, and let $\partial : C_1(X_0, K) \longrightarrow C_0(X_0, K)$ be the boundary operators of chain groups; namely the homomorphism defined by $\partial(e) = t(e) - o(e)$ ($e \in E_0$). The 1-homology group $H_1(X_0, \mathbb{Z}) = \ker \partial$ is a direct summand of the 1-chain group $C_1(X_0, \mathbb{Z})$, and is a lattice of $H_1(X_0, \mathbb{R})$. We denote by $b_1 = b_1(X_0)$ the *betti number*, i.e., $b_1 = \dim H_1(X_0, \mathbb{R})$.

Define the natural inner products on $C_0(X_0, \mathbb{R})$ and $C_1(X_0, \mathbb{R})$ by setting

$$\langle x, y \rangle = \begin{cases} 1 & (x = y) \\ 0 & (x \neq y), \end{cases}$$

and

$$\langle e, e' \rangle = \begin{cases} 1 & (e' = e) \\ -1 & (e' = \overline{e}) \\ 0 & \text{(otherwise).} \end{cases}$$

The set of vertices V_0 constitutes an orthonormal basis of $C_0(X_0, \mathbb{R})$, while an orientation E_0^o of X_0 yields an orthonormal basis of $C_1(X_0, \mathbb{R})$. With the inner product on $H_1(X_0, \mathbb{R})$ induced from the one on $C_1(X_0, \mathbb{R})$, the lattice $H_1(X_0, \mathbb{Z})$ is integral, and hence the dual lattice $H_1(X_0, \mathbb{Z})^\#$ contains $H_1(X_0, \mathbb{Z})$.

Let $\partial^* : C_0(X_0, \mathbb{R}) \longrightarrow C_1(X_0, \mathbb{R})$ be the adjoint operator of ∂ with respect to the above inner products. It is straightforward to see

$$\partial^* x = - \sum_{e \in E_{0x}} e \quad (x \in V_0),$$

so $\partial^*\big(C_0(X_0, \mathbb{Z})\big) \subset C_1(X_0, \mathbb{Z})$. We also have $H_1(X_0, \mathbb{R}) = \big(\text{Image } \partial^*\big)^\perp$, and hence $C_1(X_0, \mathbb{R}) = H_1(X_0, \mathbb{R}) \oplus \text{Image } \partial^*$ (orthogonal direct sum).

We denote by $P_0 : C_1(X_0, \mathbb{R}) \longrightarrow H_1(X_0, \mathbb{R})$ the orthogonal projection, and put $\mathbf{v}_0(e) = P_0(e)$ ($e \in E_0$). Since $\ker P_0 = \text{Image } \partial^*$, we get

$$\sum_{e \in E_{0x}} \mathbf{v}_0(e) = \mathbf{0} \quad (x \in V_0). \tag{16.10}$$

Lemma 6.1. $\mathcal{S} = \big\{\mathbf{v}_0(e)\big\}_{e \in E_0^o}$ *is a crystallographic 1-tight frame (and hence* $\big\{\mathbf{v}_0(e)\big\}_{e \in E_0}$ *is 2-tight). Its vanishing group is* $\partial^*\big(C_0(X_0, \mathbb{Z})\big)$*, and its period lattice is* $H_1(X_0, \mathbb{Z})^\#$.

Proof. The first claim is obvious because the orientation E_0^o gives an orthonormal basis of $C_1(X_0, \mathbb{R})$, and $\mathbf{v}_0(e)$'s are the projected images of this orthonormal basis by the orthogonal projection P_0.

For the second claim, it suffices to show that

$$(\text{Image } \partial^*) \cap C_1(X_0, \mathbb{Z}) = \partial^*\big(C_0(X_0, \mathbb{Z})\big).$$

To this end, take any 0-chain $\alpha = \sum_{x \in V_0} a_x x \in C_0(X_0, \mathbb{R})$ such that $\partial^* \alpha \in C_1(X_0, \mathbb{Z})$.

$$\partial^* \alpha = - \sum_{x \in V_0} \sum_{e \in E_{0x}} a_x e = - \sum_{e \in E_0} a_{o(e)} e = \sum_{e \in E_0^o} \big(a_{t(e)} - a_{o(e)}\big) e,$$

so $a_{t(e)} - a_{o(e)} \in \mathbb{Z}$ for every $e \in E_0$. This implies that there exists a real number a with $a_x + a \in \mathbb{Z}$ ($x \in V_0$). Putting

$$\beta = \sum_{x \in V_0} (a_x + a)x \in C_0(X_0, \mathbb{Z}),$$

we obtain $\partial^*\alpha = \partial^*\beta$. This proves the claim.

To prove that $P_0\big(C_1(X_0, \mathbb{Z})\big) = H_1(X_0, \mathbb{Z})^{\#}$, take a spanning tree T of X_0, and let

$$e_1, e_2, \ldots, e_{b_1}, \overline{e_1}, \overline{e_2}, \ldots, \overline{e_{b_1}}$$

be all directed edges not in T. The vectors $\mathbf{v}_0(e_1), \ldots, \mathbf{v}_0(e_{b_1})$ constitute a $\mathbb{Z}$-basis of the lattice group $H_1(X_0, \mathbb{Z})^{\#}$. This is so because we may create a $\mathbb{Z}$-basis of $H_1(X_0, \mathbb{Z})$ consisting of circuits $c_1, \ldots, c_{b_1}$ in X_0 such that c_i contains e_i, and

$$\langle c_i, \mathbf{v}_0(e_j) \rangle = \langle c_i, P_0(e_j) \rangle = \langle P_0(c_i), e_j \rangle = \langle c_i, e_j \rangle = \delta_{ij},$$

namely $\big\{\mathbf{v}_0(e_1), \ldots, \mathbf{v}_0(e_{b_1})\big\}$ is the dual basis of $\{c_1, \ldots, c_{b_1}\}$. From this, our claim immediately follows. □

Let us exhibit two instructive examples.

(1) Let Δ_{d+1} be the graph depicted in Figure 16.12. Then $\{e_1, \ldots, e_{d+1}\}$ is an orthonormal basis of $C_1(\Delta_{d+1}, \mathbb{R})$, and $e_1 - e_2$, $e_2 - e_3, \ldots, e_d - e_{d+1}$ is a $\mathbb{Z}$-basis of $H_1(\Delta_{d+1}, \mathbb{Z})$. One can check

$$\mathbf{v}_0(e_i) = e_i - \frac{1}{d+1} \sum_{j=1}^{d+1} e_j.$$

Thus $\big\{\mathbf{v}_0(e_i)\big\}_{i=1}^{d+1}$ is the tight frame associated with the equilateral simplex (see the proof of Proposition 2.5).

It should be pointed out that $\mathbf{v}_0(e_i) - \mathbf{v}_0(e_j) = e_i - e_j$ $(i \neq j)$ are vectors representing ridges (edges) of the simplex, and that $\Phi = \{e_i - e_j \mid i \neq j\}$ is the irreducible root system of type A_d with *simple roots* $e_1 - e_2$, $e_2 - e_3, \ldots, e_d - e_{d+1}$.

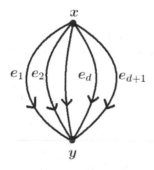

Figure 16.12 Graph Δ_{d+1}

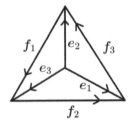

Figure 16.13 Graph K_4

This example has something to do with the diamond crystal (in the case $d = 3$).

(2) Related to the diamond twin mentioned in Introduction is the complete graph K_4 with 4 vertices. See Figure 16.13 for an orientation (the labeling of edges).

Take four closed paths $c_1 = (e_2, f_1, \overline{e_3})$, $c_2 = (e_3, f_2, \overline{e_1})$, $c_3 = (e_1, f_3, \overline{e_2})$, $c_4 = (\overline{f_1}, \overline{f_2}, \overline{f_3})$ in K_4. Then c_1, c_2, c_3, c_4 as 1-chains satisfy $c_1 + c_2 + c_3 + c_4 = 0$, $\|c_1\|^2 = \|c_2\|^2 = \|c_3\|^2 = \|c_4\|^2$ and $\langle c_i, c_j \rangle = -1$ $(i \neq j)$ $\big(c_1, c_2, c_3$ constitute a $\mathbb{Z}$-basis of $H_1(K_4, \mathbb{Z})\big)$. Moreover, c_1, c_2, c_3, c_4 are vectors represented by directed segments joining the origin and vertices of the regular tetrahedron. An easy computation gives

$$\mathbf{v}_0(e_1) = \frac{1}{4}(c_3 - c_2), \quad \mathbf{v}_0(e_2) = \frac{1}{4}(c_1 - c_3), \quad \mathbf{v}_0(e_3) = \frac{1}{4}(c_2 - c_1),$$

$$\mathbf{v}_0(f_1) = \frac{1}{4}(c_1 - c_4), \quad \mathbf{v}_0(f_2) = \frac{1}{4}(c_2 - c_4), \quad \mathbf{v}_0(f_3) = \frac{1}{4}(c_3 - c_4).$$

This implies that $\big\{ \pm \mathbf{v}_0(e_i), \pm\mathbf{v}_0(f_j)\big\}$ is the root system A_3.

Now let us proceed to a combinatorial analogue of Abel's theorem. The classical Abel's theorem in algebraic geometry gives a link between regular (holomorphic) maps from a complex projective algebraic curve into two kinds of complex tori (called the Jacobian and Picard group, respectively). In the graph-theoretic setting, certain finite abelian groups are to act as complex tori, and the counterparts of the regular map into the Jacobian is defined by using the tight frame given in Lemma 6.1. The approach explained from now is somewhat different from the one in [37], [38] (see also [31], [3] for different ways to introduce the concepts).

A key role is played by the direct sum $H_1(X_0, \mathbb{Z}) + \partial^*\big(C_0(X_0, \mathbb{Z})\big)$. Because this is a sublattice of $C_1(X_0, \mathbb{Z})$, the factor group

$$C_1(X_0, \mathbb{Z})/[H_1(X_0, \mathbb{Z}) + \partial^*\big(C_0(X_0, \mathbb{Z})\big)]$$

is a finite abelian group. We shall rewrite this group in two ways; one is $\mathcal{J}(X_0)$, an analogue of Jacobian; another is $\mathrm{Pic}(X_0)$, an analogue of Picard group.

In view of the above lemma, the projection P_0 induces an isomorphism

$$C_1(X_0, \mathbb{Z})/\partial^*\big(C_0(X_0, \mathbb{Z})\big) \longrightarrow H_1(X_0, \mathbb{Z})^\#.$$

Notice that the inclusion

$$H_1(X_0, \mathbb{Z}) \longrightarrow C_1(X_0, \mathbb{Z})$$

induces an injection $H_1(X_0, \mathbb{Z}) \longrightarrow C_1(X_0, \mathbb{Z})/\partial^*\big(C_0(X_0, \mathbb{Z})\big)$, so that $H_1(X_0, \mathbb{Z})$ is identified with a subgroup of $C_1(X_0, \mathbb{Z})/\partial^*\big(C_0(X_0, \mathbb{Z})\big)$. The double factor group

$$[C_1(X_0, \mathbb{Z})/\partial^*\big(C_0(X_0, \mathbb{Z})\big)]/H_1(X_0, \mathbb{Z})$$

is isomorphic to $C_1(X_0, \mathbb{Z})/[H_1(X_0, \mathbb{Z}) + \partial^*\big(C_0(X_0, \mathbb{Z})\big)]$. Thus, we have an isomorphism

$$C_1(X_0, \mathbb{Z})/[H_1(X_0, \mathbb{Z}) + \partial^*\big(C_0(X_0, \mathbb{Z})\big)] \longrightarrow H_1(X_0, \mathbb{Z})^\# / H_1(X_0, \mathbb{Z}),$$

where we should note that $\mathcal{J}(X_0) = H_1(X_0, \mathbb{Z})^\# / H_1(X_0, \mathbb{Z})$ is a finite subgroup of the torus group $J(X_0) = H_1(X_0, \mathbb{R})/H_1(X_0, \mathbb{Z})$. One may think of $\mathcal{J}(X_0)$ as a combinatorial analogue of Jacobian.

On the other hand, the boundary operator induces an isomorphism

$$C_1(X_0, \mathbb{Z})/H_1(X_0, \mathbb{Z}) \longrightarrow \ker \epsilon,$$

where $\epsilon : C_0(X_0, \mathbb{Z}) \longrightarrow \mathbb{Z}$ is the homomorphism (called *argumentation*) defined by

$$\epsilon \left(\sum_{x \in V_0} a_x x \right) = \sum_{x \in V_0} a_x.$$

Thus in a similar way as above, we obtain an isomorphism

$$C_1(X_0, \mathbb{Z})/[H_1(X_0, \mathbb{Z}) + \partial^*\big(C_0(X_0, \mathbb{Z})\big)] \longrightarrow \ker \epsilon/\partial\partial^*\big(C_0(X_0, \mathbb{Z})\big),$$

where $\ker \epsilon$ is regarded as an analogue of the group of divisors with degree 0, so that we denote it by $\mathrm{Div}^0(X_0)$ after the notation in algebraic geometry. Meanwhile $\partial\partial^*\big(C_0(X_0, \mathbb{Z})\big)$ is regarded as an analogue of the group of principal divisors (see [3] for the reason). Therefore, we denote it by $\mathrm{Prin}(X_0)$. Remembering again the terminology in algebraic geometry, it is justified to write $\mathrm{Pic}(X_0)$ for the factor group $\mathrm{Div}^0(X_0)/\mathrm{Prin}(X_0)$ and to call it the Picard group of X_0.

Summarizing the argument, we have

Proposition 6.2. *There is a natural isomorphism φ of the Picard group $\mathrm{Pic}(X_0)$ onto the Jacobian $\mathcal{J}(X_0)$.*

To describe φ more explicitly, pick up a reference vertex $x_0 \in V_0$. For $x \in V_0$, select a path $c = (e_1, \ldots, e_n)$ in X_0 such that $o(c)(= o(e_1)) = x_0$ and $t(c)\big(= t(e_n)\big) = x$. Then regarding c as a 1-chain, we get $\partial c = x - x_0$.

From the way to construct φ, we easily find that φ brings $x - x_0 \in \ker \epsilon$ modulo $\partial\partial^*\big(C_0(X_0,\mathbb{Z})\big)$ to $\mathbf{v}_0(e_1) + \cdots + \mathbf{v}_0(e_n) \in H_1(X_0,\mathbb{Z})^\#$ modulo $H_1(X_0,\mathbb{Z})$.

Now define the *combinatorial Albanese map* $\Phi^{\mathrm{al}} : V \longrightarrow \mathcal{J}(X_0)$ by setting

$$\Phi^{\mathrm{al}}(x) = \mathbf{v}_0(e_1) + \cdots + \mathbf{v}_0(e_n) \in H_1(X_0,\mathbb{Z})^\# \text{ modulo } H_1(X_0,\mathbb{Z}),$$

where the sum on the right-hand side is an analogue of the line integral of a holomorphic 1-form, which appears in the definition of classical Albanese maps (we shall see in the next section that $\mathbf{v}_0$ deserves to be called "harmonic" as a cochain of X_0). We also define the *combinatorial Abel-Jacobi map* $\Phi^{\mathrm{aj}} : V_0 \longrightarrow \mathrm{Pic}(X_0)$ by

$$\Phi^{\mathrm{aj}}(x) = x - x_0 \in \mathrm{Div}^0(X_0) \text{ modulo } \mathrm{Prin}(X_0).$$

By definition, we obtain $\Phi^{\mathrm{al}} = \varphi \circ \Phi^{\mathrm{aj}}$. This is nothing but an analogue of classical Abel's theorem.

To describe the structure of $\mathcal{J}(X_0)$, we consider the integral matrix $A = \big(\langle \alpha_i, \alpha_j \rangle\big) \in M_{b_1}(\mathbb{Z})$ where $\{\alpha_1, \ldots, \alpha_{b_1}\}$ is a $\mathbb{Z}$-basis of $H_1(X_0,\mathbb{Z})$. Applying the theory of elementary divisors to A, we find $P, Q \in GL_{b_1}(\mathbb{Z})$ and positive integers $k_1, \ldots, k_{b_1}$ such that

$$PAQ = \begin{pmatrix} k_1 & 0 & 0 & \cdots & 0 \\ 0 & k_2 & 0 & \cdots & 0 \\ & \cdots & \cdots & & \\ 0 & 0 & 0 & \cdots & k_{b_1} \end{pmatrix}, \tag{16.11}$$

where k_i divides k_{i+1} ($i = 1, \ldots, b_1 - 1$). The array $(k_1, \ldots, k_{b_1})$, which depends only on X_0, determines the structure of $\mathcal{J}(X_0)$; that is, $\mathcal{J}(X_0) = \mathbb{Z}_{k_1} \times \cdots \times \mathbb{Z}_{k_{b_1}}$. For instance, using this fact, we find $\mathcal{J}(\Delta_{d+1}) = \mathbb{Z}_{d+1}$, and $\mathcal{J}(K_n) = (\mathbb{Z}_n)^{n-2}$.

We have more about $\mathcal{J}(X_0)$. Algebraic graph theory [4], [5] allows us to establish the fact that the order of $\mathcal{J}(X_0)$ $\big($and $\mathrm{Pic}(X_0)\big)$ is equal to $\kappa(X_0)$, the *tree number* for X_0, which is defined to be the number of spanning trees in X_0; therefore $k_1 \cdots k_{b_1} = \kappa(X_0)$ (see [38]). Further the canonical inner product on $H_1(X_0,\mathbb{R})$ induces a flat metric on the torus $\mathcal{J}(X_0) = H_1(X_0,\mathbb{R})/H_1(X_0,\mathbb{Z})$ for which, in view of (16.7), we have

$$\mathrm{vol}\big(\mathcal{J}(X_0)\big) = \kappa(X_0)^{1/2}.$$

The Jacobian $\mathcal{J}(X_0)$ has another appendage. A nondegenerate symmetric bilinear form on $\mathcal{J}(X_0)$ with values in $\mathbb{Q}/\mathbb{Z}$ is induced from the inner product on $H_1(X_0,\mathbb{R})$. One may think of this bilinear form as an analogue of "principal polarization"

Before closing this section, we make a minor remark on direct summands of $\mathbb{Z}^N$, which is related to the above discussion. Let H be a direct summand of $\mathbb{Z}^N$ of rank k. We put

$$H^{\perp_\mathbb{Z}} = \{\mathbf{x} \in \mathbb{Z}^N \mid \langle \mathbf{x}, H \rangle = 0\},$$

which is clearly a direct summand of $\mathbb{Z}^N$ of rank $N - k$. We easily observe that $(H^{\perp_\mathbb{Z}})^{\perp_\mathbb{Z}} = H$. Although $(H^{\perp_\mathbb{Z}})_\mathbb{R} = H_\mathbb{R}^\perp$ and $H_\mathbb{R} \oplus H_\mathbb{R}^\perp = \mathbb{R}^N$, the direct sum $H \oplus H^{\perp_\mathbb{Z}}$ does not agree with $\mathbb{Z}^N$ in general. Indeed $\mathbb{Z}^N / (H \oplus H^{\perp_\mathbb{Z}})$ is isomorphic to $H^\#/H$ (this is actually a generalization of what we have seen above). To prove this, let $P : \mathbb{R}^N \longrightarrow \mathbb{R}^N$ be the orthogonal projection whose image is $H_\mathbb{R}$. It suffices to check that $P(\mathbb{Z}^N) = H^\#$; for if this is true, then $P|\mathbb{Z}^N$ induces an isomorphism of $\mathbb{Z}^N / H^{\perp_\mathbb{Z}}$ onto $H^\#$ because $\mathrm{Ker}\, P|\mathbb{Z}^N = H^{\perp_\mathbb{Z}}$. Take a $\mathbb{Z}$-basis $\{\mathbf{a}_1, \dots, \mathbf{a}_N\}$ of $\mathbb{Z}^N$ such that $\{\mathbf{a}_1, \dots, \mathbf{a}_k\}$ is a $\mathbb{Z}$-basis of H. Since $A = (\mathbf{a}_1, \dots, \mathbf{a}_N) \in GL_N(\mathbb{Z})$, there exists $B = (\mathbf{b}_1, \dots, \mathbf{b}_N) \in GL_N(\mathbb{Z})$ such that ${}^t BA = I_N$, or equivalently $\langle \mathbf{b}_i, \mathbf{a}_j \rangle = \delta_{ij}$. Therefore,

$$\langle P(\mathbf{b}_i), \mathbf{a}_j \rangle = \langle \mathbf{b}_i, P(\mathbf{a}_j) \rangle = \langle \mathbf{b}_i, \mathbf{a}_j \rangle = \delta_{ij} \quad (1 \le j \le k),$$

which implies $\{P(\mathbf{b}_1), \dots, P(\mathbf{b}_k)\}$ is a $\mathbb{Z}$-basis of $H^\#$. Therefore, $P(\mathbb{Z}^N) = H^\#$, as required.

Using $\left| \mathbb{Z}^N / (H \oplus H^{\perp_\mathbb{Z}}) \right| = |H^\#/H|$, we find

$$\mathrm{vol}(H_\mathbb{R}/H)\mathrm{vol}(H_\mathbb{R}^\perp/H^{\perp_\mathbb{Z}}) = \mathrm{vol}(\mathbb{R}^N/(H \oplus H^{\perp_\mathbb{Z}}))$$
$$= \left| \mathbb{Z}^N / (H \oplus H^{\perp_\mathbb{Z}}) \right| \mathrm{vol}(\mathbb{R}^N/\mathbb{Z}^N)$$
$$= |H^\#/H|,$$

so

$$\mathrm{vol}(H_\mathbb{R}^\perp/H^{\perp_\mathbb{Z}}) = \mathrm{vol}(H_\mathbb{R}/H) = |H^\#/H|^{1/2}.$$

7. Standard Crystal Models and Tight Frames

The combinatorial Albanese map $\Phi^{\mathrm{al}} : V_0 \longrightarrow J(X_0)$ extends to a piecewise linear map of X_0 into the flat torus $J(X_0) = H_1(X_0, \mathbb{R})/H_1(X_0, \mathbb{Z})$:

$$\Phi^{\mathrm{al}} : X_0 \longrightarrow J(X_0),$$

which, if we think of X_0 as a (singular) Riemannian manifold, turns out to be *harmonic* in the sense of Eells and Sampson [18] (see [25] for the detail).

Let X_0^{ab} be the maximal abelian covering graph over X_0, i.e., the abelian covering graph over X_0 whose covering transformation group is $H_1(X_0, \mathbb{Z})$. Consider a lifting $\Phi_0 : X_0^{\mathrm{ab}} \longrightarrow H_1(X_0, \mathbb{R})$ of Φ^{al}, which obviously satisfies

$$\Phi_0(\sigma x) = \Phi_0(x) + \sigma \quad (\sigma \in H_1(X_0, \mathbb{Z})).$$

The image $\Phi_0(X_0^{\mathrm{ab}})$ is considered a $b(X_0)$-dimensional crystal net. For instance, if $X_0 = \Delta_4$ (resp. $X_0 = K_4$), then $\Phi_0(X_0^{\mathrm{ab}})$ is the diamond crystal (resp. the diamond twin); see the previous section.

Having this observation in mind, we consider general abelian covering graphs over X_0 and their realizations. A d-dimensional *topological graph* is an infinite-fold abelian covering graph $X = (V, E)$ over a finite graph X_0 whose covering transformation group is a free abelian group L of rank d. Theory of covering spaces tells us that there is a subgroup H (called a *vanishing subgroup*) such that $H_1(X_0, \mathbb{Z})/H = L$ (note that H is a direct summand of $H_1(X_0, \mathbb{Z})$). Actually the topological crystal X is the quotient graph of X_0^{ab} over X_0 modulo H. In this view, we call X_0^{ab} the *maximal topological crystal* over X_0.

A (periodic) *realization* is a piecewise linear map $\Phi : X \longrightarrow \mathbb{R}^d$ satisfying

$$\Phi(\sigma x) = \Phi(x) + \rho(\sigma) \qquad (\sigma \in L),$$

where $\rho : L \longrightarrow \mathbb{R}^d$ is an injective homomorphism whose image is a lattice in $\mathbb{R}^d$.[17] We call ρ (resp. $\rho(L)$) the *period homomorphism* (resp. the *period lattice*) for Φ.

By putting $\mathbf{v}(e) = \Phi\big(t(e)\big) - \Phi\big(o(e)\big)$ ($e \in E$), we obtain an L-invariant function $\mathbf{v}$ on E which we may identify with a 1-cochain $\mathbf{v} \in C^1(X_0, \mathbb{R}^d)$ with values in $\mathbb{R}^d$. Because $\mathbf{v}$ determines completely Φ (up to parallel translations), we shall call $\mathbf{v}$ the *building cochain*[18] of Φ. One can check that if we identify the cohomology class $[\mathbf{v}] \in H^1(X_0, \mathbb{R}^d)$ with a homomorphism of $H_1(X_0, \mathbb{Z})$ into $\mathbb{R}^d$ (the *duality of cohomology and homology*), then $[\mathbf{v}] = \rho \circ \mu$, where $\mu : H_1(X_0, \mathbb{Z}) \longrightarrow L$ is the canonical homomorphism. In particular, $\mathrm{Ker}\,[\mathbf{v}] = H$ and $\mathrm{Image}\,[\mathbf{v}] = \rho(L)$.

Lemma 7.1. [37] *Giving a periodic realization of a topological crystal over X_0 is equivalent to giving a 1-cochain $\mathbf{v} \in C^1(X_0, \mathbb{R}^d)$ such that the image of the homomorphism $[\mathbf{v}] : H_1(X_0, \mathbb{Z}) \longrightarrow \mathbb{R}^d$ is a lattice in $\mathbb{R}^d$.*

We are now at the stage to give the definition of standard realizations. Let $H_{\mathbb{R}}$ be the subspace of $H_1(X_0, \mathbb{R})$ spanned by the vanishing group H, and $H_{\mathbb{R}}^{\perp}$ the orthogonal complement of $H_{\mathbb{R}}$ in $H_1(X_0, \mathbb{R})$:

$$H_1(X_0, \mathbb{R}) = H_{\mathbb{R}} \oplus H_{\mathbb{R}}^{\perp}.$$

Then $\dim H_{\mathbb{R}}^{\perp} = \mathrm{rank}\, L = d$. By choosing an orthonormal basis of $H_{\mathbb{R}}^{\perp}$, we identify $H_{\mathbb{R}}^{\perp}$ with the Euclidean space $\mathbb{R}^d$. Using the orthogonal projection $P : H_1(X_0, \mathbb{R}) \longrightarrow H_{\mathbb{R}}^{\perp}$, we put $\mathbf{v}(e) = P\big(\mathbf{v}_0(e)\big)$. Then one can check that $\mathbf{v}$ is the building cochain of a realization $\Phi : X \longrightarrow \mathbb{R}^d$. We call Φ the *normalized standard realization* of X. If we say simply

---

[17] The network $\Phi(X)$ could be "degenerate" in the sense that different vertices of X are realized as one points, or different edges overlap in $\mathbb{R}^d$. But we shall not exclude these possibilities.

[18] In [37], [38], the term "building block" is used. The idea to describe crystal structures by using finite graphs together with vector labeling is due to [9]

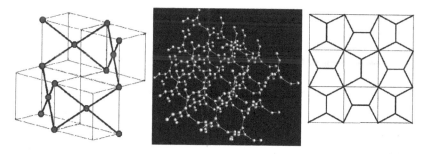

Figure 16.14 Standard realizations

"standard realization", it means a realization obtained by performing a similar transformation to the normalized one.

In view of the properties of $\mathbf{v}_0$ established in the previous section, we find

(1) (**Harmonicity**)

$$\sum_{e \in E_{0x}} \mathbf{v}(e) = \mathbf{0} \quad (x \in V_0), \tag{16.12}$$

(2) (**Tight-frame condition**)

$$\sum_{e \in E_0} \langle \mathbf{x}, \mathbf{v}(e) \rangle \mathbf{v}(e) = 2\mathbf{x} \quad (\mathbf{x} \in \mathbb{R}^d). \tag{16.13}$$

Furthermore $\{\mathbf{v}(e)\}_{e \in E_0}$ is crystallographic; namely it generates a lattice. Since this lattice contains the image of $[\mathbf{v}]$, the period lattice is essentially rational.

Remark 7.2. The direct sum $H \oplus \partial^* \big(C_0(X_0, \mathbb{Z}) \big) \big(\subset C_1(X_0, \mathbb{Z}) \big)$ is contained in the vanishing subgroup for the tight frame $\{\mathbf{v}(e)\}_{e \in E_0^o}$; however they need not coincide with each other.

Figure 16.14 exhibits a few more examples of standard realizations (the picure on the right side is a tiling of pentagons with picturesque properties that has become known as the *Cairo pentagon*).

Remarkable features of standard realizations are embodied in

Proposition 7.3. (1) *The standard realization of X is the unique minimizer, up to similar transformations, of the energy*[19]

$$\mathcal{E}(\Phi) = \text{vol}\big(\mathbb{R}^d / \rho(L)\big)^{-2/d} \sum_{e \in E_0} \|\mathbf{v}(e)\|^2.$$

---

[19] The energy defined here is considered the potential energy per unit cell when we think of the crystal net as a system of *harmonic oscillators*. Clearly this is similarity invariant.

(2) *Let $\Phi : X \longrightarrow \mathbb{R}^d$ be the standard realization. Then there exists a homomorphism κ of the automorphism group $\mathrm{Aut}(X)$ of X into the congruence group $M(d)$ of $\mathbb{R}^d$ such that*

(a) *when we write $\kappa(g) = \big(A(g), b(g)\big) \in O(d) \times \mathbb{R}^d$, we have*

$$\Phi(gx) = A(g)\Phi(x) + b(g) \qquad (x \in V),$$

(b) *the image $\kappa\big(\mathrm{Aut}(X)\big)$ is a crystallographic group, a discrete co-compact subgroup of the motion group of $\mathbb{R}^d$ (see*[8]*).*

The assertion (1) is a direct consequence of Proposition 5.4. The second one tells us that the standard realization has maximal symmetry. See [38] for the proof relying on an asymptotic property of random walks on topological crystals (see also [27],[25],[28]).

Equation (16.12) says that the cochain $\mathbf{v}$ is "harmonic" in the sense that $\delta\mathbf{v} = 0$ where $\delta : C^1(X_0, \mathbb{R}^d) \longrightarrow C^1(X_0, \mathbb{R}^d)$ is the adjoint of the coboundary operator $d : C^0(X_0, \mathbb{R}^d) \longrightarrow C^1(X_0, \mathbb{R}^d)$ with respect to the natural inner products in $C^i(X_0, \mathbb{R}^d)$. A discrete analogue of the Hodge–Kodaira theorem, which is almost trivial, assures us that the correspondence $\mathbf{v} \in \mathrm{Ker}\,\delta \mapsto [\mathbf{v}] \in H^1(X_0, \mathbb{R}^d)$ is a linear isomorphism (hence $\dim \mathrm{Ker}\,\delta = db_1(X_0)$). Thus given ρ, there is a unique harmonic cochain $\mathbf{v}$ with $[\mathbf{v}] = \rho \circ \mu$. A realization satisfying (16.12) is said to be a *harmonic realization* [25] (or an *equilibrium placement* [14]), which is characterized as a minimizer of $\mathcal{E}$ when ρ is fixed.[20]

Let $\mathbf{v} \in C^1(X_0, \mathbb{R}^d)$ be a general building cochain satisfying $[\mathbf{v}] = \rho \circ \mu$. The *distortion* of the realization given by $\mathbf{v}$ from the harmonic one is measured by the 0-cochain $\mathbf{f} \in C^0(X_0, \mathbb{R}^d)$ defined by

$$\mathbf{f}(x) = \sum_{e \in E_{0x}} \mathbf{v}(e)\big(= -(\delta\mathbf{v})(x)\big), \qquad (16.14)$$

which is considered the resultant force acting on the "atom" x when we regard the crystal net as a system of harmonic oscillators. Obviously

$$\sum_{x \in V_0} \mathbf{f}(x) = \mathbf{0}. \qquad (16.15)$$

Conversely if $\mathbf{f} \in C^0(X_0, \mathbb{R}^d)$ satisfies (16.15), then there exists a unique building cochain $\mathbf{v}$ satisfying (16.14) and $[\mathbf{v}] = \rho \circ \mu$. Indeed this is a consequence of $\mathrm{Image}\,\delta = (\mathrm{ker}\,d)^\perp$ and $\mathrm{ker}\,\delta = (\mathrm{Image}\,d)^\perp$.

Eigenvalues of the frame operator $S : \mathbb{R}^d \longrightarrow \mathbb{R}^d$ associated with a building cochain $\mathbf{v}$ gives an information about how much the harmonic realization Φ is distorted from the standard one. More precisely, if $\lambda^{\min}$ (resp. $\lambda^{\max}$) is the minimal (resp. maximal) eigenvalue of S, then the

---

[20] Looking at things through *discrete geometric analysis* [36], one can see lots of conceptual resemblance between crystallography and electric circuits.

ratio $R(\Phi) = \lambda^{\max}/\lambda^{\min}(\geq 1)$ is considered representing the degree of distorsion. Indeed $R(\Phi) = 1$ if and only if Φ is standard.

We associate the flat torus $H_{\mathbb{R}}/H$ with a vanishing subgroup $H \subset H_1(X_0, \mathbb{Z})$. As before, we easily observe that $\mathrm{vol}(H_{\mathbb{R}}/H)^2$ is an integer, and can prove, by modifying slightly the argument used in the proof of Proposition 5.1, that for any $c > 0$, there are only finitely many H of rank $b_1 - d$ such that $\mathrm{vol}(H_{\mathbb{R}}/H) < c$. For this, we just work in $H_1(X_0, \mathbb{Z})$ instead of $\mathbb{Z}^N$.

Let $\mathcal{L}_H$ $(\subset \mathbb{R}^d)$ be the period lattice for the normalized standard realization of the topological crystal corresponding to H, and put $J(X_0, H) = \mathbb{R}^d/\mathcal{L}_H$ $\big($thus $J(X_0, \{0\}) = J(X_0)\big)$. Imitating the proof of Proposition 5.3, one may prove

$$\mathrm{vol}\big(J(X_0, H)\big) = \mathrm{vol}\big(J(X_0)\big)/\mathrm{vol}(H_{\mathbb{R}}/H) = \kappa(X_0)^{1/2}/\mathrm{vol}(H_{\mathbb{R}}/H).$$

As an application of this fact, we take up the issue of "reality" of the standard realization; namely we ask how much part of the family of crystal models is occupied by standard ones which look like genuine crystals. A rough answer is that if we fix the base graph X_0, "most" standard realizations do not look realistic.

To be more precise, we start with the inequality

$$\sum_{e \in E_0} \|\mathbf{v}(e)\|^2 \|\mathbf{x}\|^2 \geq \sum_{e \in E_0} \langle \mathbf{v}(e), \mathbf{x} \rangle^2 = 2\|\mathbf{x}\|^2,$$

from which we get

$$\max_{e \in E_0} \|\mathbf{v}(e)\| \geq (2/|E_0|)^{1/2}. \tag{16.16}$$

This implies that the maximal length of $\mathbf{v}(e)$ is bounded from below by a positive constant depending only on the base graph X_0. On the other hand, there exists a positive constant c_d such that for any lattice group $\mathcal{L}$ in $\mathbb{R}^d$

$$\min_{\mathbf{x} \in \mathcal{L} \setminus \{0\}} \|\mathbf{x}\| \leq c_d \mathrm{vol}(\mathbb{R}^d/\mathcal{L})^{1/d}$$

(this is the celebrated "convex body theorem" due to Minkowski; cf. [30]). Applying this fact to the lattice group $\mathcal{L}_H$, we find a nonzero $\mathbf{x} = \rho(\sigma) \in \mathcal{L}_H$ such that $\|\rho(\sigma)\| \leq c_d \mathrm{vol}\big(J(X_0, H)\big)^{1/d}$, and hence

$$\|\Phi(\sigma x) - \Phi(x)\| \leq c_d \mathrm{vol}\big(J(X_0, H)\big)^{1/d}. \tag{16.17}$$

These facts enable us to establish the following theorem.

Theorem 7.4. *Let c be a positive constant with $c \leq 1$. For a fixed X_0, there are only finitely many d-dimensional topological crystals X over X_0 whose standard realization satisfy*

(i) *$\Phi : X \longrightarrow \mathbb{R}^d$ is injective,*

(ii) $\|\Phi(x) - \Phi(y)\| \leq c \max_{e \in E_x} \|\mathbf{v}(e)\| \implies y$ *is adjacent to* x, *or* $y = x$.

From the nature of genuine crystals, the first condition sounds natural. The second condition roughly means that two "atoms" close enough to each other must be joined by a bond.[21] It is likely that the conclusion in the theorem is true for any reasonable definition of "reality" of crystals.

The proof goes as follows. Suppose that there exist infinitely many X_n satisfying (i) and (ii), and let H_n be the vanishing subgroup corresponding to X_n. Then $\text{vol}((H_n)_{\mathbb{R}}/H_n)$ tends to infinity, so that $\text{vol}(J(X_0, H_n))$ tends to zero as $n \to \infty$. Take an integer k with $k > \max_{x \in X_0} \deg x = \max_{x \in X} \deg x$. For a given $\epsilon > 0$ with $\epsilon < c(2/|E_0|)^{1/2}$, choose n such that $c_d \text{vol}(J(X_0, H_n))^{1/d} < \epsilon/k$. By (16.17), one can find a nonzero $\sigma \in L_n = H_1(X_0, \mathbb{Z})/H_n$ such that $\|\Phi_n(\sigma^i x) - \Phi_n(x)\| < \epsilon$, where Φ_n is the normalized standard realization of X_n. Picking up a vertex x satisfying $\max_{e \in E_x} \|\mathbf{v}(e)\| \geq (2/|E_0|)^{1/2}$ (see (16.16)), and putting $x_i = \sigma^i x$, we obtain k distinct vertices $x_1, \ldots, x_k$ such that

$$\|\Phi(x_i) - \Phi(x)\| < \epsilon < c \max_{e \in E_x} \|\mathbf{v}(e)\|.$$

Therefore by the condition (ii), x_i's are adjacent to x. This implies thet $\deg x \geq k$, thereby a contradiction.

The set of similarity classes of standard realizations of all d-dimensional topological crystals over X_0 is identified with the rational Grassmannian $\text{Gr}_{b_1-d}(H_1(X_0, \mathbb{Q}))$ $(b_1 = b_1(X_0))$. As expected from the discussion in Sect. 4, there is a special feature of the parameterization of two-dimensional standard realizations. To explain this, we introduce the complex vector space

$$\mathbb{H} = \left\{ \mathbf{z} \in C^1(X_0, \mathbb{C}) \mid \sum_{e \in E_{0x}} \mathbf{z}(e) = 0 \ (x \in V_0) \right\}.$$

This is nothing but the space of harmonic cochains (we are identifying $\mathbb{R}^2$ with $\mathbb{C}$), so that we find that $\dim_{\mathbb{C}} \mathbb{H} = b_1(X_0)$ We denote by $P(\mathbb{H})$ the projective space associated with $\mathbb{H}$, and by $Q(X_0)$ the quadric defined by

$$Q(X_0) = \left\{ [\mathbf{z}] \in P(\mathbb{H}) \mid \sum_{e \in E_0} \mathbf{z}(e)^2 = 0 \right\}.$$

---

[21] This condition may not be enough (or may be too strong) to characterize the "reality" of a crytstal model because the physical and chemical aspects of crystals are ignored. In particular, the *electron clouds*, which are responsible for chemical bonding in crystals, are not involved in the simple network models.

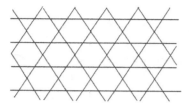

Figure 16.15 The kagome lattice

If we fix an orientation $E_0^o = \{e_1, \ldots, e_N\}$, then by the correspondence $\mathbf{z} \mapsto (\mathbf{z}(e_1), \ldots, \mathbf{z}(e_N))$, we may think of $Q(X_0)$ as a $(b_1(X_0) - 2)$-dimensional subvariety of the complex projective space $P^{N-1}(\mathbb{C})$. Then the intersection

$$Q(X_0) \cap \bigcup_{\substack{D > 0 \\ \text{square-free}}} P^{N-1}\big(\mathbb{Q}(\sqrt{-D})\big)$$

is identified with the family of all oriented similarity classes of standard realizations of two-dimensional topological crystals over X_0 [39]. This is a straightforward generalization of the observation in Sect. 4. We may also prove that D is the square-free part of $\kappa(X_0)\mathrm{vol}(H_{\mathbb{R}}/H)^2$ provided that H is the vanishing subgroup for the standard realization corresponding to a point in $Q(X_0) \cap P^{N-1}\big(\mathbb{Q}(\sqrt{-D})\big)$.

Example 7.5. (1) The *kagome lattice* (Figure 16.15) corresponds to the $\mathbb{Q}(\sqrt{-3})$-rational points

$$\left[\frac{1 \pm \sqrt{-3})}{2}, \frac{1 \mp \sqrt{-3}}{2}, -1, \frac{1 \pm \sqrt{-3}}{2}, \frac{1 \mp \sqrt{-3}}{2}, -1 \right]$$

of the two-dimensional projective quadric

$$\{[z_1, z_2, z_3, z_4, z_5, z_6] \in P^5(\mathbb{C}); \ z_1{}^2 + \cdots + z_6{}^2 = 0,$$

$$z_1 + z_6 = z_3 + z_4, \ z_2 + z_4 = z_1 + z_5, \ z_3 + z_5 = z_2 + z_6 \}.$$

(2) Figure 16.16 is the so-called *dice lattice* (also referred to as the $\mathcal{T}_3$ lattice). This corresponds to $\mathbb{Q}(\sqrt{-3})$-rational points

$$\left[1, \frac{-1 \pm \sqrt{-3}}{2}, \frac{-1 \mp \sqrt{-3}}{2}, -1, \frac{1 \pm \sqrt{-3}}{2}, \frac{1 \mp \sqrt{-3}}{2} \right]$$

of the quadric

$$\{[z_1, \ldots, z_6] \in P^5(\mathbb{C}) | \ z_1{}^2 + \cdots + z_6{}^2 = 0, \ z_1 + z_2 + z_3 = 0, \ z_4 + z_5 + z_6 = 0 \}.$$

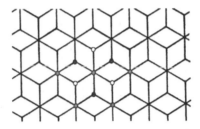

Figure 16.16 Dice lattice

Remark 7.6. Two examples above are realizations of two-dimensional topological crystals which come from periodic tilings (tessellations of tiles in the plane which are periodic with respect to the translational action by lattice groups). Easy topological considerations lead to the fact that there are only finitely many two-dimensional topological crystals over a fixed finite graph X_0 whose standard realizations are 1-skeletons of tilings. In other words, there are only finitely many rational points in $Q(X_0)$ which correspond to tilings. What we need to notice is that a tiling induces a cellular decomposition of the two-dimensional torus, and that there are only finitely many ways (in a topological sense) to attach 2-cells to a finite graph in order to obtain a torus.

Proposition 7.3 somehow claims that the standard realization is a natural concept.[22] To give another justification for the adjective "standard," we shall see that a crystal net $\Phi(X)$ with "big" symmetry is a standard model.

We assume for simplicity that $\Phi : X \longrightarrow \mathbb{R}^d$ is injective. Let Γ be a group of congruence transformations preserving $\Phi(X)$ and containing the period lattice. Clearly Γ is a crystallographic group, so that we have an exact sequence:

$$0 \longrightarrow \mathcal{L} \longrightarrow \Gamma \longrightarrow K \longrightarrow 1,$$

where $\mathcal{L}(= \mathbb{R}^d \cap \Gamma)$ is a lattice containing the period lattice, and $K \subset O(d)$ is what is called the point group. Note that the isotropy group $\Gamma_{\mathbf{x}}$ ($\mathbf{x} \in \mathbb{R}^d$) is identified with a subgroup of K via the (injective) restriction of the homomorphism $\Gamma \longrightarrow K$ to $\Gamma_{\mathbf{x}}$. Under the assumption on Φ, each $g \in \Gamma$ induces an automorphism of X, so Γ is regarded as a subgroup of $\mathrm{Aut}(X)$.

---

[22] The special features of standard realizations might remind the reader of the claim about the *golden ratio* $(1.618033\cdots)$, a root of $x^2 = x + 1$, which overemphasizes its significance in the history of art, architecture, sculpture, and anatomy. I am not going to overrate the significance of standard realizations though quite a few structures in nature and art are explained using standard realizations.

Theorem 7.7. *Suppose that*
(1) *the action of the point group K on $\mathbb{R}^d$ is irreducible, and*
(2) *for any vertex $x \in V$, the fixed point set for $\Gamma_{\Phi(x)}$-action on $\mathbb{R}^d$ is $\{0\}$, i.e., $\{x \in \mathbb{R}^d \mid gx = x \ (g \in \Gamma_{\Phi(x)})\} = \{0\}$.*
Then Φ is a standard realization.

Proof. The group K acts on $\{v(e)\}_{e \in E_0}$ in a natural manner. Indeed, writing $\Phi(\sigma x) = A(\sigma)\Phi(x) + b(\sigma)$ ($\sigma \in \Gamma$), we have

$$v(\sigma e) = \Phi\big(t(\sigma e)\big) - \Phi\big(o(\sigma e)\big) = A(\sigma)[\Phi\big(t(e)\big) - \Phi\big(o(e)\big)] = A(\sigma)v(e).$$

In view of Proposition 2.1, the assumption (1) assures us that $\{v(e)\}_{e \in E_0}$ is tight. On the other hand, the vector $\sum_{e \in E_x} v(e)$ is $\Gamma_{\Phi(x)}$-invariant. Thus, $\sum_{e \in E_x} v(e) = 0$ by the assumption (2). This completes the proof. $\square$

As a corollary, we have

Theorem 7.8. *The 1-skeleton of a Coxeter complex is a standard model.*

In this theorem, a Coxeter complex means a triangulation of $\mathbb{R}^d$ such that
(a) if we denote by $\{\Delta_\alpha^d\}_{\alpha \in A}$ the set of d-dimensional simplices in this triangulation, then the action of the group Γ generated by all reflections fixing facets of Δ_α^d ($\alpha \in A$) preserves the triangulation, and
(b) Γ acts transitively on $\{\Delta_\alpha^d\}_{\alpha \in A}$.
For the proof, we use the fact that Γ is a crystallographic group,[23] which acts on the 1-skeleton as well, and that the point group for Γ is the Weyl group associated with an irreducible root system. Clearly the fixed point set for Γ_x-action is $\{0\}$ for every vertex x, thereby the condition (2) being satisfied. Thus one may apply Theorem 7.7 to complete the proof.

A final remark is in order. The reader might wonder what is the practical use of standard crystal models. Straightforwardly speaking, three-dimensional standard realizations are purely mathematical outgrowths of logical reasoning. Even if a standard realization (or its deformation) looks realistic, it does not necessarily exist in nature; namely it is merely a model of a hypothetical crystal. Once we find a hypothetical crystal, however, a systematic prediction of its physical properties for appropriate atoms can be carried out by *first principles calculations* used in chemistry. The prediction appealing to the computer power encourages (or discourages) material scientists to synthesize the hypothetical crystals.

---

[23] Actually Γ is isomorphic to a semidirect product of $\mathbb{Z}^d$ and the point group.

References

[1] M. Baake, *Solution of the coincidence problem in dimensions d ≤ 4*, in R. V. Moody, ed., The Mathematics of Long-Range Aperiodic Order, Kluwer, Dordrecht, 1997, pp. 199–237.

[2] R. Bacher, P. De La Harpe, and T. Nagnibeda, *The lattice of integral flows and the lattice of integral cuts on a finite graph*, Bull. Soc. Math. France, **125** (1997), 167–98.

[3] M. Baker and S. Norine, *Riemann-Roch and Abel-Jacobi theory on a finite graph*, Adv. in Math., **215** (2007), 766–88.

[4] N. L. Biggs, Algebraic Graph Theory, Cambridge University Press, 1993.

[5] N. L. Biggs, *Algebraic potential theory on graphs*, Bull. London Math. Soc., **29** (1997), 641–82.

[6] A. Borel, Introduction aux groupes arithmetiques, Hermann 1969.

[7] K. S. Brown, Building, Springer, 1989.

[8] L. S. Charlap, Bieberbach Groups and Flat Manifolds, Springer-Verlag, 1986.

[9] S. J. Chung, T. Hahn, and W. E. Klee, *Nomenclature and generation of three-periodic nets: the vector method*, Acta. Cryst., **A40** (1984), 42–50.

[10] J. H. Conway, *A characterisation of Leech's lattice*, Invent. math., **7** (1969), 137–42.

[11] J. H. Conway, H. Burgiel, C. Goodman-Strauss, The Symmetries of Things, A K Peters Ltd, 2008.

[12] H. S. M. Coxeter, Regular Polytopes, Dover 1973.

[13] P. R. Cromwell, Polyhedra, Cambridge University Press, 1997.

[14] O. Delgado-Friedrichs and M. O'Keeffe, *Identification of and symmetry computation for crystal nets*, Acta Cryst., **A59** (2003), 351–60.

[15] R. J. Duffin and A. C. Schaeffer. *A class of nonharmonic Fourier series*, Trans. Amer. Math. Soc., **72** (1952) 341–66.

[16] O. Delgado-Friedrichs, *Barycentric drawings of periodic graphs*, LNCS **2912** (2004), 178–189.

[17] W. Ebeling, Lattices and Codes, Vieweg, 1994.

[18] J. Eells and J. H. Sampson, *Harmonic mappings of Riemannian manifolds*, Amer. J. Math., **86** (1964) 109–160.

[19] Y. C. Eldar, *Optimal tight frames and quantum measurement*, Information Theory, IEEE Transactions on **48** (2002), 599–610.

[20] J-G. Eon, *Archetypes and other embeddings of periodic nets generated by orthogonal projection*, J. Solid State Chem. **147** (1999), 429–37.

[21] J-G. Eon, *Euclidean embeddings of periodic nets: definition of a topologically induced complete set of geometric descriptors for crystal structures*, Acta Cryst. **A67** (2011), 68–86.

[22] V. K. Goyal, J. Kovacevic, and J. A. Kelner, *Quantized frame expansions with erasures*, Applied and Computational Harmonic Analysis **10** (2001), 203–33.

[23] H. Grimmer, *Coincidence rotations for cubic lattices*, Scripta Metallurgica **7** (1973), 1295–1300.

[24] J. E. Humphreys, Introduction to Lie Algebra and Representation theory, Springer, 1970.

[25] M. Kotani and T. Sunada, *Standard realizations of crystal lattices via harmonic maps*, Trans. Amer. Math. Soc., **353** (2000), 1–20.

[26] M. Kotani and T. Sunada, *Jacobian tori associated with a finite graph and its abelian covering graphs*, Advances in Apply. Math., **24** (2000), 89–110.

[27] M. Kotani and T. Sunada, *Albanese maps and off diagonal long time asymptotics for the heat kernel*, Comm. Math. Phys., **209** (2000), 633–70.

[28] M. Kotani and T. Sunada, *Spectral geometry of crystal lattices*, Contemporary Math., **338** (2003), 271–305.

[29] G. McColm, *Generating geometric graphs using automorphisms*, J. of Graph Algorithms and Appl., **16** (2012), 507-41.

[30] J. Milnor and D. Husemoller, Symmetric Bilinear Forms, Springer 1973.

[31] T. Nagnibeda, *The Jacobian of a finite graph*, Contemporary Math., **206** (1997), 149–51.

[32] M. Nespolo, *Does mathematical crystallography still have a role in XXI century?*, Acta Cryst., **A64** (2008), 96–111.

[33] P. M. Neumann, G. A. Stoy, and E. C. Thompson, Groups and Geometry, Oxford University Press, 1994.

[34] T. Sunada, Why do Diamonds Look so Beautiful?, (in Japanese) Springer Japan, 2006.

[35] T. Sunada, *Crystals that nature might miss creating*, Notices Amer. Math. Soc., **55** (2008), 208–15.

[36] T. Sunada, *Discrete geometric analysis*, Proceedings of Symposia in Pure Mathematics, (ed. by P. Exner, J. P. Keating, P. Kuchment, T. Sunada, A. Teplyaev), **77** (2008), 51–86.

[37] T. Sunada, *Lecture on topological crystallography*, Japan. J. Math. **7** (2012), 1–39.

[38] T. Sunada, Topological crystallography –With a View Towards Discrete Geometric Analysis–, Surveys and Tutorials in the Applied Mathematical Sciences, Vol. 6, Springer, 2012.

[39] T. Sunada, *Standard 2D crystalline patterns and rational points in complex quadrics*, arXiv: submit/0620196 [math.CO] 23 Dec 2012.

[40] R. Vale and S. Waldron, *Tight frames and their symmetries*, Constructive Approximation, **21** (2004), 83–112.

[41] S. Waldron, *Generalised Welch bound equality sequences are tight frames*, Information Theory, IEEE Transactions on, **49**, 9 (2003), 2307–9.

[42] A. F. Wells, *The geometrical basis of crystal chemistry*, Acta Cryst. **7** (1954), 535.

[43] A. F. Wells, Three Dimensional Nets and Polyhedra, Wiley (1977).

[44] Y. M. Zou, *Structures of coincidence symmetry groups*, Acta Cryst., **A62** (2006), 109–14.

[45] Y. M. Zou, *Indices of coincidence isometries of the hypercubic lattice $\mathbb{Z}^n$*, Acta Cryst., **A62** (2006), 454–8.

Printed in the United States
By Bookmasters